T0135392

Lecture Notes in Electrical Engineering

Volume 1001

The book series *Lecture Notes in Electrical Engineering* (LNEE) publishes the latest developments in Electrical Engineering—quickly, informally and in high quality. While original research reported in proceedings and monographs has traditionally formed the core of LNEE, we also encourage authors to submit books devoted to supporting student education and professional training in the various fields and applications areas of electrical engineering. The series cover classical and emerging topics concerning:

- Communication Engineering, Information Theory and Networks
- Electronics Engineering and Microelectronics
- Signal, Image and Speech Processing
- Wireless and Mobile Communication
- Circuits and Systems
- Energy Systems, Power Electronics and Electrical Machines
- Electro-optical Engineering
- Instrumentation Engineering
- Avionics Engineering
- Control Systems
- Internet-of-Things and Cybersecurity
- Biomedical Devices, MEMS and NEMS

For general information about this book series, comments or suggestions, please contact leontina.dicecco@springer.com.

To submit a proposal or request further information, please contact the Publishing Editor in your country:

China

Jasmine Dou, Editor (jasmine.dou@springer.com)

India, Japan, Rest of Asia

Swati Meherishi, Editorial Director (Swati.Meherishi@springer.com)

Southeast Asia, Australia, New Zealand

Ramesh Nath Premnath, Editor (ramesh.premnath@springernature.com)

USA, Canada

Michael Luby, Senior Editor (michael.luby@springer.com)

All other Countries

Leontina Di Cecco, Senior Editor (leontina.dicecco@springer.com)

**** This series is indexed by EI Compendex and Scopus databases. ****

Yashwant Singh · Pradeep Kumar Singh ·
Maheshkumar H. Kolekar · Arpan Kumar Kar ·
Paulo J. Sequeira Gonçalves
Editors

Proceedings of International Conference on Recent Innovations in Computing

ICRIC 2022, Volume 1

Editors
Yashwant Singh
Computer Science and IT Department
Central University of Jammu
Jammu, Jammu and Kashmir, India

Pradeep Kumar Singh
Department of Computer Science
KIET Group of Institutions
Ghaziabad, Uttar Pradesh, India

Maheshkumar H. Kolekar
Department of Electrical Engineering
Indian Institute of Technology Patna
Patna, Bihar, India

Arpan Kumar Kar
School of Artificial Intelligence
Indian Institute of Technology Delhi
New Delhi, Delhi, India

Paulo J. Sequeira Gonçalves
IDMEC
Polytechnic Institute of Castelo Branco
Castelo Branco, Portugal

ISSN 1876-1100 ISSN 1876-1119 (electronic)
Lecture Notes in Electrical Engineering
ISBN 978-981-19-9878-2 ISBN 978-981-19-9876-8 (eBook)
https://doi.org/10.1007/978-981-19-9876-8

This Springer imprint is published by the registered company Springer Nature Singapore Pte Ltd.
The registered company address is: 152 Beach Road, #21-01/04 Gateway East, Singapore 189721, Singapore

Preface

The fifth version of international conference was hosted on the theme Recent Innovations in Computing (ICRIC 2022); the conference was hosted by the Central University of Jammu, J&K, India. There were many academic partners for the events including Eötvös Loránd University (ELTE), Hungary Knowledge University, Erbil, WSG University in Bydgoszcz, Poland, and other academic associates, technical societies, from India and abroad. The conference includes the tracks: Cyber Security and Cyber Physical Systems, Internet of Things, Machine Learning, Deep Learning, Big Data Analytics, Robotics Cloud Computing, Computer Networks and Internet Technologies, Artificial Intelligence, Information Security, Database and Distributed Computing, and Digital India. The authors are invited to present their research papers at the Fifth International Conference on Recent Innovations in Computing (ICRIC 2022) in six technical tracks. We are thankful to our valuable authors, along with the Technical Program Committee's for tremendous support for making the 5th ICRIC 2021 a successful event. The conference was opened by presidential address of chief guest Prof. Sanjeev Jain, vice chancellor, Central University of Jammu, J&K, India. Welcome address and conference statistics were shared by conference general chair, Dr. Yashwant Singh, and conference relevance and address to the attendees was given by Co-patron ICRIC 2022, Professor Devanand, Central University of Jammu, J&K, India.

The first keynote address was delivered by Prof. Kusum Deep, Professor, Indian Institute of Technology, Roorkee. Thereafter, the second keynote speech was given by Prof. Srinivas (Sri) Talluri, Professor, Michigan State University, USA. The third keynote was delivered by Prof. Petia Radeva, Professor, University of Barcelona, Spain. All papers of the conference were presented during five parallel technical sessions.

The organizing committee express their sincere thanks to all session chairs—Dr. Brij B Gupta, Dr. Arvind Selwal, Dr. Pradeep Kumar Singh, Dr. Bhavna Arora, Dr. Sudeep Tanwar, Dr. Pradeep Chouksey, Dr. Deepti Malhotra, Dr. Mayank Agarwal, Dr. Neerendra Kumar, Prof. Manu Sood, Prof. Devanand Padha, Dr. Vinod Sharma, Dr. Deepti Malhotra, Dr. Manoj Kumar Gupta, Dr. Aruna Malik, Dr. Samayveer

Singh, Dr. Yugal Kumar, Dr. Kayhan Zrar Ghafoor, Dr. Zdzislaw Polkowski, and Dr. Pljonkin Anton, for extending their help during technical sessions.

The Central University of Jammu, J&K, received the financial support from the funding agency, DST-SERB, for this conference. The university authorities wish to extend their heartfelt thanks for the financial support for the ICRIC 2022 to all officials of DST-SERB team.

We are also grateful to Central University of Jammu, J&K, authorities for their kind approvals from time to time. The committee is thankful to management, board, rectors, vice rectors, deans, and professors from academic associate universities from abroad for extending their support during the conference. Many senior researchers and professors across the world also deserve our gratitude for devoting their valuable time to listen and give feedback and suggestion on the paper presentations. We extend our thanks to Springer, LNEE Series, editorial board for believing in us.

Ghaziabad, India — Pradeep Kumar Singh
Jammu, India — Yashwant Singh
Patna, India — Maheshkumar H. Kolekar
New Delhi, India — Arpan Kumar Kar
Castelo Branco, Portugal — Paulo J. Sequeira Gonçalves

Contents

About the Editors

Dr. Yashwant Singh is a Professor and Head of the Department of Computer Science and Information Technology at the Central University of Jammu where he has been a faculty member since 2017. Prior to this, he was at the Jaypee University of Information Technology for 10 Years.

Yashwant completed his Ph.D. from Himachal Pradesh University Shimla, his Post Graduate study from Punjab Engineering College Chandigarh, and his undergraduate studies from SLIET Longowal. His research interests lie in the area of Internet of Things, Vulnerability Assessment of IoT and Embedded Devices, Wireless Sensor Networks, Secure and Energy Efficient Routing, ICS/SCADA Cyber Security, ranging from theory to design to implementation. He has collaborated actively with researchers in several other disciplines of computer science, particularly Machine Learning, Electrical Engineering.

Yashwant has served on Thirty International Conference and Workshop Program Committees and served as the General Chair for PDGC-2014, ICRIC-2018, ICRIC-2019, ICRIC-2020, and ICRIC-2021. He currently serves as coordinator of Kalam Centre for Science and Technology (KCST), Computational Systems Security Vertical at Central University of Jammu established by DRDO.

Yashwant has published more than 100 Research Papers in International Journals, International Conferences, and Book Chapters of repute that are indexed in SCI and SCOPUS. He has 860 Citations, i10-index 27 and h-index 16. He has Research Projects of worth Rs.1040.9413 Lakhs in his credit from DRDO and Rs. 12.19 Lakhs from NCW. He has guided 4 Ph.D., 24 M.Tech. students and guiding 4 Ph.D.'s and 5 M.Tech.

Dr. Yashwant has visited 8 countries for his academic visits e.g. U.K., Germany, Poland, Chez Republic, Hungary, Slovakia, Austria, Romania. He is Visiting Professor at Jan Wyzykowski University, Polkowice, Poland.

Dr. Pradeep Kumar Singh is currently working as Professor and Head in the Department of CS at KIET Group of Institutions, Delhi-NCR, Ghaziabad, Uttar Pradesh, India. Dr. Singh is Senior Member of Computer Society of India (CSI), IEEE, ACM, and Life Member. He is Associate Editor of the *International Journal of Information*

System Modeling and Design (IJISMD), Indexed by Scopus and Web of Science. He is also Associate Editor of *International Journal of Applied Evolutionary Computation* (IJAEC), IGI Global USA, Security and Privacy, Wiley. He has received three sponsored research projects grant from Government of India and Government of Himachal Pradesh worth Rs. 25 Lakhs. He has edited a total 12 books from Springer and Elsevier. He has Google Scholar citations 1600, H-index 20, and i-10 Index 49. His recently published book titled *Handbook of Wireless Sensor Networks: Issues and Challenges in Current Scenario's* from Springer has reached more than 12,000 downloads in last few months. Recently, Dr. Singh has been nominated as Section Editor for *Discover IoT*, a Springer Journal.

Dr. Maheshkumar H. Kolekar is working as Associate Professor in Department of Electrical Engineering at Indian Institute of Technology Patna, India. He received the Ph.D. degree in Electronics and Electrical Communication Engineering from the Indian Institute of Technology Kharagpur in 2007. During 2008 to 2009, he was Post-doctoral Research Fellow with the Department of Computer Science, University of Missouri, Columbia, USA. During May to July 2017, he worked as DAAD Fellow in Technical University Berlin where he worked in the research area EEG signal analysis using machine learning. His research interests are in the area of digital signal, image and video processing, video surveillance, biomedical signal processing, and deep learning. Recently, he has authored a book titled as *Intelligent Video Surveillance Systems: An Algorithmic Approach*, CRC Press, Taylor and Francis Group, 2018. He has successfully completed R&D Project Sponsored by Principal Scientific Advisor to Government of India on abnormal human activity recognition.

Prof. Arpan Kumar Kar is Amar S Gupta Chair Professor in Data and Decision Science in Indian Institute of Technology Delhi, India, where he shares a joint appointment between Department of Management Studies and School of Artificial Intelligence. His research interests are in the interface of data science, machine learning, digital transformation, internet platforms and public policy. He has authored over 200 peer reviewed articles and edited 9 research monographs. He is the Editor in Chief of *International Journal of Information Management Data Insights*, published by Elsevier. He is also Associate Editor in multiple other established journals. He has undertaken over 40 research, consulting and training projects with funds generated over Rs. 20 Crore. He received the Research Excellence Award by Clarivate Analytics for highest individual Web of Science citations from 2015 to 2020 in India. He received the Basant Kumar Birla Distinguished Researcher Award for the count of ABDC A*/ABS 4 level publications in India between 2014 and 2019. He is also the recipient of the Best Seller Award from Ivey Cases/Harvard Business Publishing in 2020 for his case study on social media analytics. He has received over 20 other national and international awards from reputed organizations like 3 IFIP Best Paper Awards, ACM ICEGOV Best Paper Award, Tata Consultancy Services Gold Medal for Research, Project Management Institute Research Advocacy Award, AIMS JL Batra Research Award, IIT Delhi Teaching Excellence Award, 6 Outstanding Reviewer Awards from Elsevier and many more best paper awards.

Prior to joining IIT Delhi, he has worked in IIM Rohtak, Cognizant Business Consulting and IBM India Research Laboratory. He completed his Graduation in Engineering from Jadavpur University and Doctorate from XLRI Jamshedpur.

Dr. Paulo J. Sequeira Gonçalves is Associate Professor of Electrotechnical and Industrial Engineering with the School of Technology in the Polytechnic Institute of Castelo Branco, Portugal. He is also Senior Researcher at IDMEC, Instituto Superior Técnico, University of Lisbon, Portugal. Dr. Gonçalves was trained in Mechanical Engineering, achieving the M.Sc. and Ph.D. degrees from Instituto Superior Técnico, University of Lisbon, 1998 and 2005, respectively. He presented more than 100 paper in renowned journal and conferences in robotics, vision, and computational intelligence. He serves as Co-Chair of the IEEE-RAS Portuguese chapter. He served as Chair of the IEEE-CIS Portuguese chapter and was Officer of the IEEE RAS working group who developed the first RAS standard and one of the first ontology-based IEEE standards. He received several scientific awards both national and international. He organized and chaired several international conferences and edited its proceedings and organized several workshops in top-tier robotic IEEE conferences, e.g., ICRA, IROS, and ROMAN. He was invited Visiting Professor in European Universities, in the field of robotics. Currently, he is Chair of the IEEE P1872.3—Standard for Ontology Reasoning on Multiple Robots and actively works in other IEEE standards working groups within robotics and automation. He received funding for his research projects, in Principal Investigator and/or Researcher, from the EU, IEEE, private companies, and the Portuguese National Research Funding Agency. He works as Regular Expert for the European Commission and for the National Accreditation of University Study Programs. Dr. Paulo J. Sequeira Gonçalves research keywords include computational intelligence, ontologies, robotics, computer vision, and industrial automation.

Artificial Intelligence

Brain Tumor Segmentation Using Fully Convolution Neural Network

Rupal A. Kapdi, Jigna A. Patel, and Jitali Patel

Abstract Early stage brain tumor diagnosis can lead to proper treatment planning, which improves patient survival chances. A human expert advises an appropriate medical imaging scan based on the symptoms. Diagnosis done by a human expert is time-consuming, non-reproducible, and highly dependent on the expert's expertise. The computerized analysis is preferred to help experts in diagnosis. The paper focuses on implementing a fully convolution neural network to segment brain tumor from MRI images. The proposed network achieves comparable dice similarity with reduced network parameters.

Keywords Convolution neural networks · Brain tumor segmentation · Magnetic resonance imaging · Deep neural networks · Multi-modality images

1 Introduction

Glioma is one of the most life-threatening brain tumors. It occurs in the glial cells of the brain. Glioma grades range from I to IV depending on its severity. Such tumors can further be expressed by its subparts like—necrosis, enhancing tumor, and edema. Medical resonance imaging (MRI) non-invasive imaging technique is preferably used to capture the functioning of soft tissue properly compared to other imaging techniques. A strong magnetic field is created around the brain, which forces protons to align with that field. A radio signal is then passed through the brain, which stimulates protons leading it to move out of equilibrium. When the radio-frequency field is turned off, MRI sensors record the energy released by protons to realign itself

R. A. Kapdi (✉) · J. A. Patel · J. Patel
Computer Science and Engineering Department, Institute of Technology, Nirma University, Ahmedabad, Gujarat, India
e-mail: rupal.kapdi@nirmauni.ac.in

J. A. Patel
e-mail: jignas.patel@nirmauni.ac.in

J. Patel
e-mail: jitali.patel@nirmauni.ac.in

Y. Singh et al. (eds.), *Proceedings of International Conference on Recent Innovations in Computing*, Lecture Notes in Electrical Engineering 1001,
https://doi.org/10.1007/978-981-19-9876-8_1

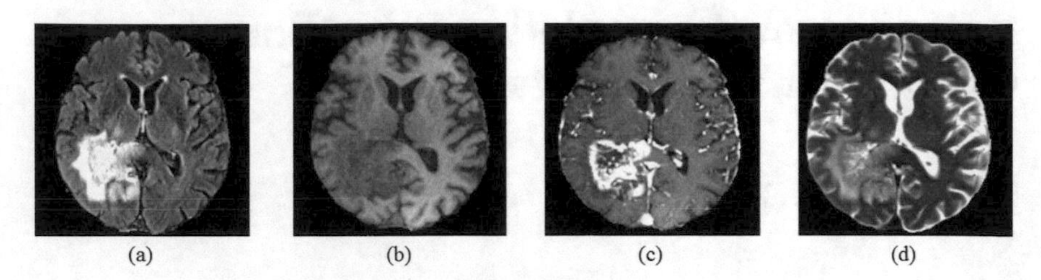

Fig. 1 MRI modalities. **a** FLAIR. **b** T1. **c** T1c. **d** T2

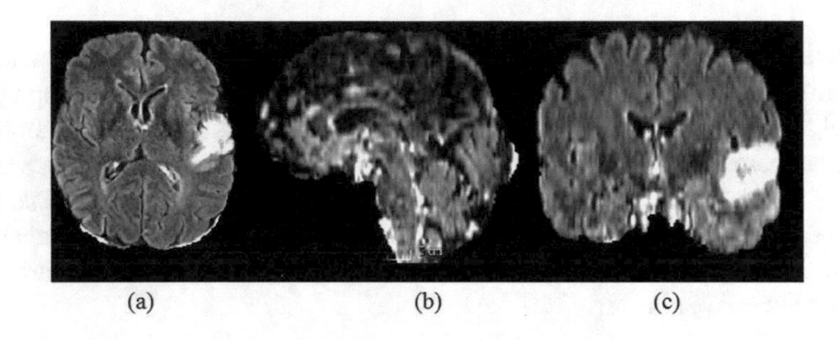

Fig. 2 MRI views. **a** Axial. **b** Sagittal. **c** Coronal

to the magnetic field. Variation in the energy released depends on the environment and chemical nature of the proton. These variations make various modalities of MRI. Figure 1 shows four such modalities. As MRI generates a 3D volumetric image, it allows one to see the brain from three different views: axial, sagittal, and coronal. Figure 2 demonstrates these views.

Human experts use such volumetric MR images to locate the tumor. But such a diagnosis is time-consuming as well as non-reproducible. Accurate diagnosis is highly desirable to plan future treatment and surgery. Computer-aided analysis helps a human expert to locate the tumor. Such analysis consumes less time and helps to get reproducible analysis results. Researchers are working on detection and segmentation methods for a brain tumor.

Literature reports two categories of segmentation methods: semi-automated and automated. Semi-automated methods require some input from the user to segment the tumor. An automated method does not require any external input. The main focus of the paper is automatic brain tumor segmentation. Three categories of automated techniques are: (1) basic, (2) generative, and (3) discriminative [1]. Basic methods include threshold, region-based, texture-based, and edge-based methods. At the same time, generative methods are atlas-based and model-based methods. Discriminative methods are divided into two categories: unsupervised and supervised. Unsupervised discriminative methods include clustering-based methods. The random forest-based methods, support vector machines, and artificial neural network-based methods fall under supervised discriminative methods. Discriminative methods are fully automated soft computing-based methods. Such discriminative methods require an ample amount of data to train the model for an assigned task.

In all the methods specified above, features are extracted from the image and supplied to methods for further processing. Features can be like gray-level intensities histogram, texture features, symmetry-based features, probability distribution, etc. Such feature extraction task heavily depends on the knowledge of the researcher. Improper feature set leads to improper segmentation results.

But the growth of deep learning has removed the burden of extracting features from the data. According to [15], deep learning architectures provide an end-to-end model for solving the tasks like (1) object recognition/classification, (2) classification and localization, (3) object detection, (4) image segmentation, (5) speech recognition, (6) image compression, (7) natural language processing and many more. With the advancement of high configuration computation resources like GPUs, deep learning is evolved in almost all real-life applications. Any deep learning architectures (with a special focus on convolution neural network (CNN)) have multiple layers that perform nonlinear operations to learn features automatically. At shallower layers, it learns low-level features, whereas on deeper layers, it learns complex features. Supervised deep learning algorithms perform well when it is provided with a higher amount of input data with possible variability. CNN keeps the spatial context during feature extraction, which makes it best suitable for interpreting images. Deep neural networks have gained interest in the medical field as well. The paper provides literature survey for automated brain tumor segmentation using CNN architectures and introduces RDUNet network for whole tumor (WT) segmentation. Section 2 shows various CNN architectures for brain tumor segmentation; Sect. 3 demonstrates the implementation of a well-known UNet CNN architecture with minor modifications. Section 4 covers implementation details, followed by conclusion and future work in Sect. 5.

2 CNN Architectures for Brain Tumor Segmentation

The basic architecture of CNN includes three types of layers: convolution, nonlinear activation, and pooling. As per the need, these layers are organized. Fully connected layers may follow such architecture. Various convolution kernels are applied to the input image to extract multiple feature maps. Feature maps as initial layers generate the response for low-level features like lines with different orientations. As convolution operation moves deeper in the network, high-level features are extracted like curves. The nonlinear activation function is applied on these feature maps, which converts the output of the feature map in the range of [0, 1] or [−1,1]. Features with good response are preserved with the maxpooling layer. The maxpool layer extracts the feature with the maximum response from the specified grid. The maxpool layer can also be replaced with the average-pool layer. Here, the values of the features are averaged on the specified grid. The output of the last layer is then fed into a neural network which provides fully connected layers. Various CNN architectures differ by the sequence of layers. Patches are extracted from the image and supplied to the network for further processing. In the case of the brain tumor segmentation, tumorous part in the brain is comparatively smaller than the entire brain, which creates a class imbalance problem. The unbias network training either trains the network on equiprobable patches or uses a weighted loss function. Such network training deals with class imbalance problems. Patches are processed in two ways: patch classification and patch segmentation [10]. Patch classification decides the class of the center voxel of the patch, and patch segmentation densely classifies a number of voxels of the patch.

Authors in [18] used 11 layers deep 3D CNN. This architecture is known as DeepMedic. Here, the concept of local feature maps and global feature maps is used. Multi-scale 3D patches are passed to different CNNs where the feature map size is different. Results of these networks are then combined to pass it to fully connected layers. Training images are normalized with zero mean and unit variance. The segmentation result of CNN is further improved by applying a conditional random field as post-processing.

Authors in [14] proposed various types of 2D CNN architectures. All the proposed architectures use a multi-scale approach. In a multi-scale approach, two path CNN is used. In a local path, smaller receptive fields are used, whereas in a global path, larger receptive fields are used. Another approach which is proposed uses a cascaded network, where an output of one network is passed as input to another network. Images are pre-processed to remove 1% lowest and highest intensity values from the channel, followed by a bias-field correction to T1 and T1c modality images. At last, intensity normalization is applied with channel mean and standard deviation. For post-processing, connected contours are identified to show the connectivity of the voxels. Small blobs which are wrongly classified as tumorous voxels are removed.

Authors in [10] had proposed a multi-stage approach for lesion segmentation; first-stage 3D CNN segmented the entire tumor from the MRI, and the result was concatenated with the second-stage output for tumor substructure segmentation. Both the stages used multi-scale receptive fields to capture local as well as global features. Bias-field correction was applied as a pre-processing step.

Fully convolution neural network (FCNN) is a supervised learning algorithm, which takes as an input arbitrary size input image, and provides an end-to-end model for voxel classification. It takes as an input the entire image and generates dense output. Unlike fully connected layers in CNN, FCNN has convolution layers at the end of the network. FCNN is further improved by introducing a skip connection [22]. Skip connection in FCN is shown in Fig. 3.

Skip connection is the connection between two neural network layers by skipping one or more layers in between, which prevents the degradation of the deeper network at the time of saturation [24]. When the skip connections are introduced at every pooling stage, then such FCNN can be looked as the encoding–decoding architecture, where the encoding side scales down the image by learning various higher-level feature maps at different stages of encoding, decoding side scales up so that output image size is same as an input image. The coarse image at deeper layers is to be

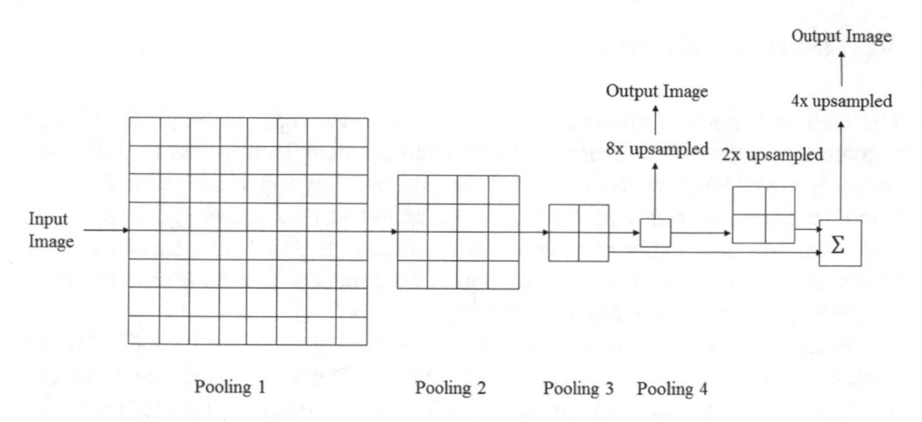

Fig. 3 Skip connection in FCN

reconstructed to match the size of the input image. Reconstruction can be done in two ways: upsampling and deconvolution. SegNet [7] is used by [32] for brain tumor segmentation. They have made two groups of FLAIR 2D images for training SegNet. Group A contains all the 2D FLAIR images which contain the tumor. Group B contains all the 2D FLAIR images with and without tumor. SegNet is trained for group A, group B, and group $A + B$, where in $A + B$ the pretrained network of group A is used to train B, whereas in [17], they have modified their own DeepMedic architecture with residual connections to improve the performance of the network.

RNN is also showing promising results for brain tumor segmentation. Authors of [5] have shown the success of RNN for brain tumor segmentation in [6]. The multi-dimensional signal is processed by considering any one dimension of it, and the output of the gated recurrent unit (GRU) is a summation of the signal processed in the forward and backward direction of the multi-dimensional GRU. Further refinement to the output is done by considering the weights of spatial neighbors. Pre-processing is done using a high-pass filter with a Gaussian kernel.

Generative adversarial network (GAN) is also providing good results in the brain tumor segmentation field. The generator network generates segmentation results, and discriminator networks classify it be correct or incorrect segmentation. Authors in [27] have proposed conditional GANs for brain tumor segmentation. In the generator network, UNet architecture is used, which generates the segmentation result for the specific sub-class of tumor. The discriminator network is based on the Markovian GAN, which discriminates between real or fake segmentation classes. A different approach of GANs is proposed in [21]. The generator and discriminator both use CNN architectures. The segmentation result is post-processed to remove non-tumorous voxels.

2.1 UNet Architecture

The UNet architecture proposed in [28] has gained success in biomedical image segmentation. The FCNN architecture proposed here is five layers deep; five encoding layers have two convolution layers followed by ReLU activation function. Decoding layers use deconvolution/upsampling to learn the upscaling kernel. Every decoding layer has a skip concatenation connection from its peer encoding layer, which allows the network to learn the upscaling properly. Later, the architecture is adopted by [12] for brain tumor segmentation.

Researchers for brain tumor segmentation develop various UNet like FCNN architectures [2]. Authors in [3, 4, 8] are using four 3D UNets to segment the tumor and fine-tune the result of the segmentation. Authors in [13] have applied pre-processing to the images; then, $16 \times 16 \times 16$ patches are extracted from the images. These patches are then supplied to the network for training considering the foreground as tumor and the background as everything else. Once the segmentation is completed, erosion and dilation are applied to remove the false positives of the tumor. In [11], two 3D UNets are used: One is for coarse segmentation of tumor, which gives entire tumor from the image and the second to extract tumor substructure from previous output like tumor core and enhancing tumor. In [19], three different UNet architectures of depth three are proposed with pre-processing on the 2D axial images. Authors in [20] considered multi-modal images taken from the x-axis, y-axis, and z-axis to train UNets. Individual UNet is responsible for dealing with specific axis images for training. On the segmentation result, majority voting is done to label the voxel.

Authors in [25] designed three layers of deep UNet architecture where at each level three convolutional operations are applied. The same architecture is applied to segment the whole tumor and substructures of tumor simply by changing the number of output layers at the last decoding stage. In [9], the authors have used 3D FCN, to segment the tumor. Authors in [30] have implemented FCN where in addition to the input MRI images, a symmetry map for all the modality is also provided as an additional input. In all these methods, bias-field correction is applied as a pre-processing step. Authors in [2–4, 26] used UNet architecture with residual connection for better feature extraction. Modified MobileNetV2 was used in [29] for in-depth feature learning. Authors in [33] used a trans-coder to extract relevant position information of the feature map in UNet architecture. The HNF-Netv2 used various blocks of inter-scale and intra-scale semantic discrimination enhancing for high-resolution features extraction [16]. The following sections provide the details of the proposed network.

3 Proposed Architecture—RDUNet

3.1 *RDUNet Architecture*

Inspired by the UNet architecture, the authors of this chapter have tried to design the UNet architecture which mimics the UNet FCNN except for several layers required. UNet FCNN provides good results for brain tumor segmentation but takes

- longer training time,
- higher number of learnable parameter.

A higher amount of training images are required to train a deeper network, which is a limitation in the medical field. As the dataset for brain tumor segmentation is limited, a reduced number of layers also provides comparable results. In addition to that, less number of learnable parameters reduces the training time drastically. Proposed RDUNet architecture is an extension of [3], where there are four layers in the architecture.

4 Implementation Details

4.1 *Dataset*

The multi-modal brain tumor segmentation (BraTS) 2017 challenge focuses on two tasks: glioma tumor segmentation and prediction of patient overall survival. The challenge provides pre-processed (coregistered, interpolated to $1 \times 1 \times 1$ mm resolution and skull-stripped) images comprised of 210 HGG and 75 LGG samples. With each sample having four 3 T MRI modalities (T1, T2, T1C and FLAIR) along with the ground truth. Each sample has 155 slices with 240×240 pixels per slice. The ground truth for the tumor shows three substructures, namely enhancing tumor (ET), edema (ED), and necrotic/non-enhancing tumor [23]. Also, the overall survival, defined in days, is included in a comma-separated value (.csv) file with 'Patient ID,' and 'Age' for 163 samples. The suggested classes for classification based on the prediction of overall survival were long-survivors (e.g., $>$ 15 months), short-survivors (e.g., $<$ 10 months), and mid-survivors (e.g., between 10 and 15 months).

4.2 *Pre-processing*

Medical image pre-processing is very crucial for computer-assisted analysis. Properly pre-processed images lead to correct results, and as such images show proper voxel relationships. In addition, MR machines are also susceptible to the environment around them. Before processing such images, it is required to pre-process such images to correct MRI artifacts. The following pre-processing techniques are applied to MRI images before supplying them to the network.

- **Image Registration**: Movement of patients in MRI machine results in the images which may not match the exact orientation of the image for analysis. So, MR images should be registered to the template before processing.
- **Skull stripping**: Skull is a hard tissue part surrounding the brain tissues. Skull is not considered to be the brain part when processing. It is desirable to remove the skull portion from the image before further processing.
- **Bias-field correction**: A multiplicative field added to the image due to the magnetic field and radio signal in the MR machine adds noise to the image, such noise is called bias field. Authors in [31] have suggested a bias-field correction technique.
- **Intensity Normalization**: Different modality images have different intensity scales, which are to be transformed on the same scale. Normalization of all the images is basically done by considering zero mean and unit variance.
- **Noise Removal**: Noise which is introduced in the image due to the radio signal or magnetic field should be removed before processing.

BraTS dataset images are registered, skull-stripped, and noiseless. So, according to the architectural requirements, normalization and bias field correction are applied as pre-processing if required.

4.3 Results

We have used NVIDIA Quadro K5200 and Quadro P5000 GPU for the training and testing of CNN. The software used is as follows: Python 2.7 and 3.5, TensorFlow, Keras, Nibabel, N4ITK for UNet and RDUNet. All four modalities (T1, T1c, T2, and FLAIR) are given in input. For UNet as well as RDUNet training, data augmentation is done to increase the size of the dataset. Data augmentation includes: flip horizontal, flip vertical, rotation, shift, shear, zoom, variation in brightness and elastic transformation. ReLU activation follows the convolution layers; the dice loss function is used as the loss function. Stochastic gradient descent optimization is used to reduce the loss with respect to the loss function. Adam optimizer is used to estimate the parameters. The sigmoid function is used for classification. The UNet architecture of [12] is trained for the whole tumor segmentation and edema segmentation, and RDUNet is trained for whole tumor segmentation. Both the networks are trained for 30 epochs. UNet segmentation results for whole tumor as well as edema are shown in Figs. 4 and 5. For RDUNet, whole tumor segmentation is shown in Fig. 6.

The one-vs-all approach is used to segment tumor or its substructures. The whole tumor includes all the sub-classes of the tumor, i.e., necrotic, enhancing tumor, non-enhancing tumor as well as edema. If the whole tumor is segmented, then the entire

Fig. 4 Whole tumor segmentation using UNet, images sequence from left to right: FLAIR, T1, T1c, T2, ground truth, segmentation result

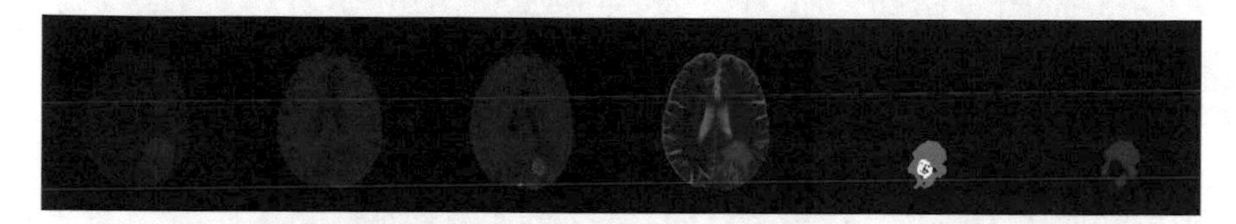

Fig. 5 Edema segmentation using UNet, images sequence from left to right: FLAIR, T1, T1c, T2, ground truth, segmentation result

Fig. 6 Whole tumor segmentation using RDUNet, image sequence from left to right: original FLAIR, segmentation result, ground truth

tumor represents the foreground, and everything else is the background. In Figs. 4 and 6, all the substructures of tumor contribute in the whole tumor. But in Fig. 5, only edema substructure contributes to the foreground, and everything else, including other substructures, is treated as background.

The network is trained for 30 epochs. If the network has achieved the global optimum in less number of epochs, then in that case, training can be stopped because the dice loss will not improve further. But if dice loss sticks to 0 or 1, then in that case network is not improving, leading to wrong segmentation. In that case, the network training is to be restarted. At some times, it has happened that restarting has not improved the network training. In that case, network training should be started with pretrained network weights of other substructures, like whole tumor weights can be used to start the training of edema substructure.

4.4 Evaluation Metrics

Brain tumor segmentation techniques are evaluated on how correctly it labels tumorous voxels as it is. Any similarity measure function which quantifies the similarity between two objects can be used as an evaluation metric. The following evaluation metrics are widely used based on true positives (TP), false positives (FP), true negatives (TN), and false negatives (FN).

- Dice similarity coefficient: It can be defined as the overlap of two segmentations divided by size of the two objects.

$$\mathrm{DSC} = 2\mathrm{TP}/2\mathrm{TP} + \mathrm{FP} + \mathrm{FN} \tag{1}$$

- Sensitivity: measure to correctly identify tumorous voxels.

$$\mathrm{Sensitivity} = \mathrm{TP}/\mathrm{TP} + \mathrm{FN} \tag{2}$$

- Positive predictive value: gives likelihood ration between segmented result and ground truth

$$\mathrm{PPV} = \mathrm{TP}/\mathrm{TP} + \mathrm{FP} \tag{3}$$

We have achieved comparable DSC and sensitivity with other state-of-the-art methods which are shown in Table 1.

Table 1 DSC comparison with other state-of-the-art methods

Reference	DSC—whole tumor	Sensitivity	PPV
[8]	0.88	–	–
[9]	0.91	–	–
[11]	0.82	–	–
[12]	0.86	–	–
[13]	0.89	–	–
[19]	0.84	–	–
[20]	0.84	–	–
[25]	0.85	0.79	0.78
[30]	0.87	0.89	0.85
Proposed	0.88	0.87	–

5 Conclusion and Future Work

The paper covers the working of encoder–decoder-based convolution neural network-based methods that use MRI scans for semantic segmentation of brain tumors. The architecture uses skip connections which allow the network to perform comparably when the deep layers do not contribute to feature learning. The network performance can further be improved by:

- different number of layers,
- variation in skip connection,
- different number of convolution layers at every stage,
- type of images applied for training,
- change in the pre-processing techniques applied,
- different post-processing techniques applied.

The proposed architecture requires 10 MB network parameters and has comparable DSC with other architectures. The network fails for necrosis segmentation due to a high imbalance in the dataset. The segmentation task can be improved with 3D networks which work in ensemble for robust performance.

Acknowledgements Authors would like to thank the support of NVIDIA Corporation with the donation of the Quadro K5200 and Quadro P5000 GPU used for this research, Dr. Krutarth Agravat (Medical Officer, Essar) for his help in discussion on tumors and analysis of the results. Authors are indebted to Mr. Pratik Chaudhary and Mr. Himanshu Budhia for their constant support during GPU installation and implementation.

References

1. Agravat RR, Raval MS (2018) Deep learning for automated brain tumor segmentation in mri images. In: Soft computing based medical image analysis. Elsevier, pp 183–201
2. Agravat RR, Raval MS (2019) Brain tumor segmentation and survival prediction. In: International MICCAI brainlesion workshop. Springer, pp 338–348
3. Agravat RR, Raval MS (2019) Prediction of overall survival of brain tumor patients. In: TENCON 2019–2019 IEEE region 10 conference (TENCON). IEEE, pp 31–35
4. Agravat RR, Raval MS (2020) 3d semantic segmentation of brain tumor for overall survival prediction. In: International MICCAI brainlesion workshop. Springer, pp 215–227
5. Andermatt S, Pezold S, Cattin P (2016) Multi-dimensional gated recurrent units for the segmentation of biomedical 3d-data. In: Deep learning and data labeling for medical applications. Springer, pp 142–151
6. Andermatt S, Pezold S, Cattin P (2017) Multi-dimensional gated recurrent units for brain tumor segmentation. In: MICCAI multimodal brain tumor segmentation challenge (BraTS), pp 15–19
7. Badrinarayanan V, Kendall A, Cipolla R (2017) Segnet: a deep convolutional encoder-decoder architecture for image segmentation. IEEE Trans Pattern Anal Mach Intell 39(12):2481–2495
8. Beers A, Chang K, Brown J, Sartor E, Mammen C, Gerstner E, Rosen B, Kalpathy Cramer J (2017) Sequential 3d u-nets for biologically-informed brain tumor segmentation. arXiv preprint arXiv:1709.02967

9. Casamitjana A, Puch S, Aduriz A, Sayrol E, Vilaplana V (2016) 3d convolutional networks for brain tumor segmentation. In: Proceedings of the MICCAI challenge on multimodal brain tumor image segmentation (BRATS), pp 65–68
10. Chen L, Wu Y, DSouza AM, Abidin AZ, Wismüller A, Xu C (2018) Mri tumor segmentation with densely connected 3d cnn. In: Medical imaging 2018: image processing, vol 10574. International Society for Optics and Photonics, p 105741F
11. Colmeiro RR, Verrastro C, Grosges T (2017) Multimodal brain tumor segmentation using 3d convolutional networks. In: International MICCAI brainlesion workshop. Springer, pp 226–240
12. Dong H, Yang G, Liu F, Mo Y, Guo Y (2017) Automatic brain tumor detection and segmentation using u-net based fully convolutional networks. In: Annual conference on medical image understanding and analysis. Springer, pp 506–517
13. Feng X, Meyer C (2017) Patch-based 3d u-net for brain tumor segmentation. In: 2017 international MICCAI BraTS challenge
14. Havaei M, Davy A, Warde-Farley D, Biard A, Courville A, Bengio Y, Pal C, Jodoin PM, Larochelle H (2017) Brain tumor segmentation with deep neural networks. Med Image Anal 35:18–31
15. Jain K (2015) Analytics Vidhya (2015), https://www.analyticsvidhya.com/blog/2017/08/10-advanced-deep-learningarchitectures-data-scientists/. Accessed 23 May 2018
16. Jia H, Bai C, Cai W, Huang H, Xia Y (2000) Hnf-netv2 for brain tumor segmentation using multi-modal mr imaging. arXiv preprint arXiv:2202.05268
17. Kamnitsas K, Ferrante E, Parisot S, Ledig C, Nori AV, Criminisi A, Rueckert lD, Glocker B, Deepmedic for brain tumor segmentation. In: International
18. Kamnitsas K, Ledig C, Newcombe VF, Simpson JP, Kane AD, Menon DK, Rueckert D, Glocker B (2017) Efficient multi-scale 3d cnn with fully connected crf for accurate brain lesion segmentation. Med Image Anal 36:61–78
19. Kim G (2017) Brain tumor segmentation using deep fully convolutional neural networks. In: International MICCAI brainlesion workshop. Springer, pp 344–357
20. Li Y, Shen L (2017) Deep learning based multimodal brain tumor diagnosis. In: International MICCAI brainlesion workshop. Springer, pp 149–158
21. Li Z, Wang Y, Yu J (2017) Brain tumor segmentation using an adversarial network. In: International MICCAI brainlesion workshop. Springer, pp 123–132
22. Long J, Shelhamer E, Darrell T (2015) Fully convolutional networks for semantic segmentation. In: Proceedings of the IEEE conference on computer vision and pattern recognition, pp 3431–3440
23. Menze BH, Jakab A, Bauer S, Kalpathy-Cramer J, Farahani K, Kirby J, Burren Y, Porz N, Slotboom J, Wiest R, et al (2015) The multimodal brain tumor image segmentation benchmark (brats). IEEE Transactions Med Imaging 34(10):1993–2024
24. Orhan AE, Pitkow X (2017) Skip connections eliminate singularities. arXiv preprint arXiv: 1701.09175
25. Pereira S, Oliveira A, Alves V, Silva CA (2017) On hierarchical brain tumor segmentation in mri using fully convolutional neural networks: a preliminary study. In: Bioengineering (ENBENG), 2017 IEEE 5th Portuguese Meeting on, pp 1–4. IEEE
26. Rajput S, Agravat R, Roy M, Raval MS (2021) Glioblastoma multiforme patient survival prediction. In: International conference on medical imaging and computer-aided diagnosis. Springer, pp 47–58
27. Rezaei M, Harmuth K, Gierke W, Kellermeier T, Fischer M, Yang H, Meinel C (2017) Conditional adversarial network for semantic segmentation of brain tumor. arXiv preprint arXiv: 1708.05227
28. Ronneberger O, Fischer P, Brox T (2015) U-net: Convolutional networks for biomedical image segmentation. In: International conference on medical image computing and computer-assisted intervention. Springer, pp 234–241
29. Saeed MU, Ali G, Bin W̃, Almotiri SH, AlGhamdi MA, Nagra AA, Masood K, Amin RU (2021) Rmu-net: a novel residual mobile u-net model for brain tumor segmentation from mr images. Electronics 10(16):1962

30. Shen H, Jianguo Z, Weishi Z (2017) Efficient symmetry-driven fully convolutional network for multimodal brain tumor segmentation. In: 2017 IEEE International conference on image processing (ICIP). IEEE
31. Tustison NJ, Avants BB, Cook PA, Zheng Y, Egan A, Yushkevich PA, Gee JC (2010) N4itk: improved n3 bias correction. IEEE Trans Med Imaging 29(6):1310–1320
32. Yan T, Ou Y, Huang T (2017) Automatic segmentation of brain tumor from mr images using segnet: selection of training data sets. MICCAI Multimodal Brain Tumor Segmentation Challenge (BraTS) 309–312
33. Zhang T, Xu D, He K, Zhang H, Fu Y (2022) 3d u-net with trans-coder for brain tumor segmentation. In: Thirteenth international conference on graphics and image processing (ICGIP 2021), vol 12083, pp 540–548. SPIE

Artificial Intelligence-Based Lung Nodule Detection: A Survey

Shifa Shah and Anuj Mahajan

Abstract In recent years, the diagnosis and detection of various diseases using image processing systems employing artificial intelligence have become a key interest area for researchers. Numerous studies have empirically shown the relevance and effectiveness of different artificial intelligence frameworks for this cause. The present study presents a comprehensive review of fifteen studies about detecting lung cancer-causing nodules using artificial intelligence. Traditionally, radiologists have widely used diagnostic tools like MRI scans, CT scans, and X-rays to examine lung cancer-causing nodules. The small nodules are often challenging to be detected by radiologists when the results of MRI scans, CT scans, X-rays, pathology, and histopathology are interpreted manually. For more accurate and efficient detection of nodules, various artificial intelligence systems are employed that use digitized glass slides and other medical images as inputs. This study intends to bring about an understanding of different artificial intelligence-based automated approaches used for lung nodule detection. This study focuses mainly on the Deep learning Neural Network-based techniques, the most commonly used state-of-the-art system for object detection and image classification.

Keywords Feature engineering · Lung nodule detection · Deep learning

1 Introduction

Lungs are a vital part of the human body, doing the task of inhaling oxygen and exhaling carbon dioxide. An air bag-like structure called alveoli is present in the lungs, responsible for this air filtration. If there is any hindrance in this process of exchange of gases, then the lungs could be deprived of oxygen which can cause death and several hazardous issues. There are various techniques through which the

S. Shah · A. Mahajan (✉)
School of Computer Science & Engineering, Shri Mata Vaishno Devi University, Katra, India
e-mail: anuj.mahajan@smvdu.ac.in

S. Shah
e-mail: shahshifa909@gmail.com

Y. Singh et al. (eds.), *Proceedings of International Conference on Recent Innovations in Computing*, Lecture Notes in Electrical Engineering 1001,
https://doi.org/10.1007/978-981-19-9876-8_2

functioning and working of lungs can be checked, like scanning lungs through X-rays or CT/MRI scans and then analyzing the reports by a medical professional. The sudden development of some suspicious symptoms can indicate problems in the lungs. Tumors are either benign or malignant. Benign is slow-spreading as the tumor cells grow slowly and are mainly non-cancerous, [1–4] while Malignant is fast-spreading and cancerous. Lung cancer is caused by the abnormal growth of tissues which may vary in shape and size. These abnormal tissues or cells can grow with time and may be cancerous. The manual screening of these scans and detection of cancerous/non-cancerous nodes require excellent expertise and skills on behalf of a medical practitioner. The nodule's shape and size can be identified from the CT scans. Lung nodules are generally opaque and may be round or irregular in shape. The skeptical small nodules-like structures found in the scans are further examined through a biopsy test or pathology test, which is done on the tissues of the lungs before the chemotherapy treatment. If localization of the nodules is achieved in the initial phases of the growth, then the prognosis is good, and the affected lungs can be treated with positive results. Lung cancer is mainly detected at the final stages. Its prognosis is considered low. However, early detection of lung cancer during the initial phases can significantly improve the prognosis of lung cancer patients. Artificial intelligence-based lung cancer detection techniques can help in the early, accurate, and timely diagnosis of lung cancer and can significantly increase the survival time and rate for patients. In this paper, we have surveyed machine learning and deep learning-based approaches for the detection of pulmonary nodules.

At an early stage, it is not easy to identify nodules due to their small size, and it is even more challenging to locate the cancerous ones from non-cancerous ones. The representation of medical images is mostly in grayscale. Cancerous nodules are high in gray values and thus difficult to detect. Radiologists examine the nodules from the CT scans, and it requires lots of knowledge and skills in recognizing the lung nodules. Sometimes, radiologists need a lot of CT scans of a single patient to detect the exact shape and size of nodules [5–10]. As the diagnosis of the nodules is made manually by the radiologists, there is always a possibility of undetected nodules. Although computer-aided diagnosis is used for nodule detection, there are some flaws as well. The detection accuracy is poor. The other drawback is differences in detection proposals and the actual problem area, which may cause the overall detection result to be poor. Radiologists face a challenging, time-consuming, and demanding task in detecting malignant lung nodules at an early stage. The manual examination of many scans proved to be very hectic and took a lot of time with no guarantee of detecting the cancerous nodules. Misdetection is also there in many cases because there are significant similarities between features of benign and malignant nodules. The detection of the nodules is done by finding questionable patches that are likely to contain a lesion and followed by confirmation or denial of the presence of a nodule. This study brings out the understanding of approaches and methodologies, and datasets used in lung cancer detection. Section 2 consists of a survey of the fourteen studies. Section 3 consists of various datasets available for lung cancer detection. Section 4 consists of approaches to lung cancer detection using artificial intelligence. Section 5 consists of performance measures followed by a conclusion in Sect. 6 with references.

2 Literature Survey

Jiang et al. [1] worked on the LIDC/IDR. Each patient data contains metadata with images that give information about a patient. They used two sizes of patches, 32 × 32 and 16 × 16, to scan the lung contour from all directions left, right, down, and up. They used Frangi filter for blurring the background. They used four CNN structures to find the location of nodules. They achieved an accuracy of 80%. Sahu et al. [2] developed a lightweight and small CNN mobile which could be easily used on tablets and phones. Their model examined the lung cancer-causing nodule from different angles. Their model gave a multiview on the suspicious nodule and then combined information via the pooling layer. The authors achieved 93.18% classification accuracy.

Alakwaa et al. [3] used a dataset from Kaggle DSB, which has labeled data of 1300 patients. They used nine hundred seventy images for training and 419 images for testing. They used the label 0 for non-cancerous nodules and 1 for cancer nodules. They performed various techniques like segmentation and sampling on CT scans. They used 3DCNN as a classifier and a U-Net algorithm to find the nodules' location. The authors achieved 89% accuracy. Ying Suet al. [4] proposed a model that used faster RCNN to detect the lung nodules from the CT scans using the LIDC dataset. They have used ZF and VGG16 as their backbone model for feature extraction. Their study found that the VGG 16 model performs better than the ZF model, and they got 68% accuracy for RCNN, 72% for Fast RCNN, and 82% accuracy for Faster RCNN.

Paul et al. [5] conducted their research on the LIDC IDR dataset and compared the results with the ELAP dataset. The authors used a hybrid model for examining features and used SVM for classification. The authors gave a multiple-view CNN model in which suspicious and cancerous nodules present in lungs are examined from different modes of thought, i.e., left to right. They achieved 76% accuracy in their study.

Nasrullah et al. [6] used a hybrid strategy of IoT and deep learning to identify lung cancer. They took some data from patients from hospitals, marked the symptoms of various lung cancer patients, and used and collected patients' family histories. They made an IoT-based system (Wireless body area network to monitor multiple cancer patients). They conducted their research on the LUNA 16 dataset. They performed pre-processing techniques on the dataset and divided it into training and testing. They used CMix Net for nodule detection and Faster RCNN for feature extraction.

Yanfeng et al. [7] conducted their research on the dataset collected from the hospital. They used MRI scans for the identification of cancer-causing nodules. To avoid the problem of overfitting. They used pre-trained models. They used faster RCNN in their study and VGG16 in the feature extraction. They used 97 scans for training and 45 scans for testing.

Gong et al. [8] carried the research on the LUNA16 and ANNODE09 datasets to detect the nodules. They used a 3D filtering approach to detect suspicious nodules and used Random Forest Classifier in the classification process. They achieved an accuracy of 84.62% and a sensitivity of 79.3%. Farahani et al. [9] researched the CT scans, used the MSFCM segmentation clustering algorithm for the lung region's morphological operations, and used a KNN MLP and SVM ensemble to classify the nodules. They achieved a sensitivity of 93.2%. Narayanan et al. [10] proposed a cluster-based classifier architecture for detecting lung nodules. They used the JRST database, a publicly available database with 154 CT scans and one radiologist examining the nodules. They have achieved an accuracy of 75.2%. Zhang et al. [11] proposed their work on the LIDC IDR dataset. They offered a 3D Skeletonization feature named Voxel Remove Rate to differentiate lung nodules from other suspicious things in the CT scans and used an SVM classifier for false-positive reduction. They have received an average sensitivity of 89.7%. Borkowski et al. [12] developed the dataset LC2500 of 25,000 images of lung cancer and colon cancer from the pathological glass slides, 5000 images in each class, by using the augmented library which expanded the images up to 25,000. Mangal et al. [13] conducted their research on the pathological manifestations of lung cancer and trained their neural network with an accuracy of 97.8%. Hatuwal et al. [14] developed their own CNN model on LC25000 and achieved a training accuracy of 96% & validation accuracy of 97%.

3 Datasets

3.1 LIDC/IDR

It is a freely available cancer dataset of lungs consisting of thousands of images of CT scans [1, 2, 4]. Radiologists worked in stages on the dataset to locate the nodules manually. Radiologists examine the CT scans and annotate them for the initial stage. The outcome is further given to a group of radiologists to re-examine and reannotate. As a result, there is an XML file for each patient in the database that records these radiologists' annotations. There are tiny nodules with sizes less than 3 mm in this database and large lesions with diameters greater than 3 mm. CT scans are recorded in the DCM format, widely used in medical image processing. This dataset is 125 GB in size.

3.2 ANODE09

Another publicly accessible database, Automatic Nodule Detection 2009 (ANODE09), contains 55 CT scans of lungs [9]. Five scans are offered to train. The remaining 50 cases are used for testing.

3.3 LC 25000

This dataset was prepared by [12]. This dataset consists of images of the lungs and colon tissue cells, which were collected from pathological glass slides. The 2500 images of lungs and colon tissues were expanded to 25,000 images using the Augmentor library in python. This dataset consists of 5 lung and colon cancer classes with 5000 images in each category. This dataset is 2 GB in size. The summarization of literature survey is given in Table 1 and followed by the various datasets available for the lung nodule detection given in Table 2.

Table 1 Summary of literature survey [1–11, 13, 14]

Methodology	Dataset used	Performance
3D segmentation [8] (Machine learning-based method)	LUNA16 and ANNODE 09	Accuracy = 84.62%
MSMFCM [9] (Machine Learning-based method)	Hospital dataset	Sensitivity = 93.5%
3D skeletonization [10] (Machine learning-based method)	LIDC IDR	Accuracy = 89.7%
Cluster-based algorithm [11] (Machine learning-based method)	JRST	Accuracy = 75.9%
Faster RCNN [4] (Deep learning model)	LIDC IDR	Accuracy = 91%
Small-sized CNN [2] (Deep learning model)	LIDC IDR	Accuracy = 93.18%
3D CNN [3] (Deep learning model)	Kaggle DSB 2017	Accuracy = 86.6%
4 types of CNNs [1] used (Deep learning model)	LIDC IDR	Accuracy = 94%
Multiview CNN [5] (Deep learning model)	LIDC IDR, ELAP	Accuracy = 82%
CMix net and faster RCNN [6] (Deep learning model)	LUNA 16	Accuracy = 91%
CNN [13] (Deep learning model)	LC2500	Accuracy = 97%
Faster RCNN [7] (Deep learning)	Hospital dataset	Accuracy = 85.7%
CNN [14] (Deep learning model)	LC2500	Accuracy = 96%

Table 2 Various datasets [1–5, 7–11, 13, 14]

Dataset	Year	Number of CT scans/pathology images
LIDC IDR [1, 2, 4, 5, 10, 11]	2011	1018 CT scans
LUNA 16 [9]	2013	886 CT scans
ELAP [3]	2003	50 CT scans
Kaggle DSB 2017 [7]	2017	50 CT scans
ANNODE [9]	2009	55 CT scans
JRST [8]	1998	154 CT scans
LC 25000 [13, 14]	2019	250,000 pathology images

4 Approaches

The various approaches in artificial intelligence for the detection of lung nodules are as follows:

Feature engineering approach (traditional machine learning approach). The feature engineering approach consists of the following steps shown in Fig. 1.

Pre-processing step. This step enhances images and removes noise in the images. Feature enhancement using various filters like a median filter and log filter lowers the distortion in the image and increases the classifier's accuracy.

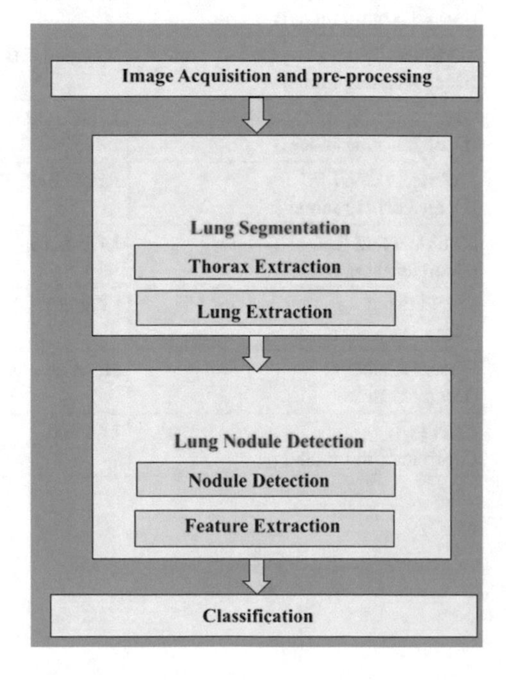

Fig. 1 Feature engineering approach for lung nodule detection [15]

Lung Segmentation. In this step, the main aim is to decrease the searching time by segmenting the lung CT scan.

This step has two sub-steps:

Thorax Extraction. This step involves the extraction of the thorax area and removing associated external things with the patient, e.g., Bed and surroundings.

Lung Extraction. In this step, the region of interest is calculated for the left and right lungs from the CT slices. Various algorithms can be used in the segmentation process. The output extracted from this step was termed the "Biggest and the Next Most Significant Connected Region" [10].

Nodule Detection. The nodules are detected from the images in this stage. The goal is to decrease the false-positive rate and the false-negative rate.

Feature Extraction. In this stage, features are extracted from the output given by the above steps. In this step, the background is blurred, and the region of interest is determined.

Classification. In this step, after detecting the nodules and features, it is classified to which category the nodule belongs (benign or malignant). As the classification of lung nodules is done using a classifier, the performance is dependent on the classifier used. SVM classifiers have performed better than other classifiers [12].

4.1 Deep Learning-Based Approach

Deep learning is a machine learning technique based on algorithms and artificial neural networks that imitate the functioning and structure of a human brain. This subset of artificial intelligence and particularly deep learning CNNs have gained substantial importance in medical imaging due to accuracy on various fronts and capacity for processing massive datasets and detecting objects in images by processing pixel data [16–21]. Using the convolution operations, the convolution layers of the CNN detect the local features in the input images and perform classification as well. The basic architecture Fig. 2 of the neural network approach consists of multiple layers, and this layer-based architecture leads to more accurate classification and detection.

- **Input**: The network takes in data from different sources as input.
- **Input layer**: This layer receives and processes the data.
- **Hidden Layer**: The information is transformed from layer to layer within these layers. Through these layers, the neural network learns the features and performs the computations.
- **Output Layer**: This layer identifies combinations of features and determines the output, such as whether cancer exists or not.

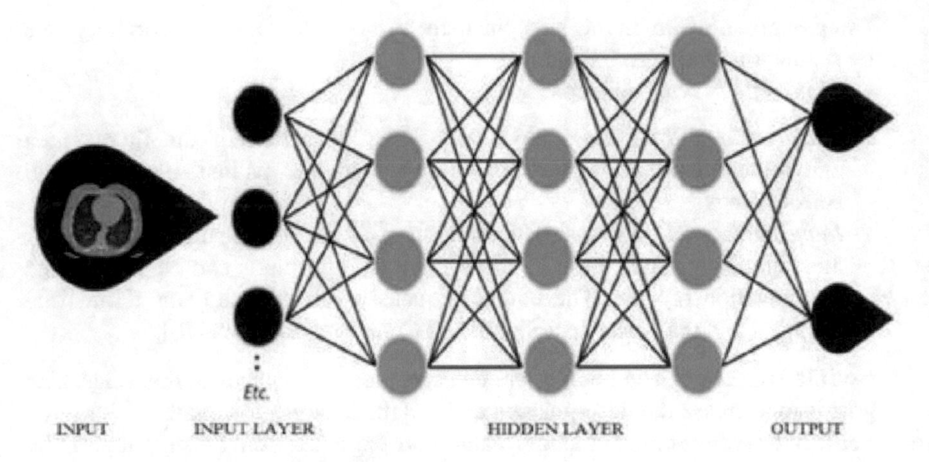

Fig. 2 Basic neural network architecture [8]

- The CNN's basic architecture in Fig. 3 consists of a convolution layer that performs convolution operations, pooling layers, and a fully connected layer that performs classification.

Deep learning models are generally complex and need GPUs or graphics processing units to process the data fast. The various approaches for using deep learning models are as follows:

Building model from Scratch,
Transfer Learning,
Feature Extraction.

- **Building from Scratch**

In this approach, a deep neural network architecture is designed for huge labeled data. This approach requires a high rate of learning, and the models take hours to weeks in the training process.

- **Transfer learning**

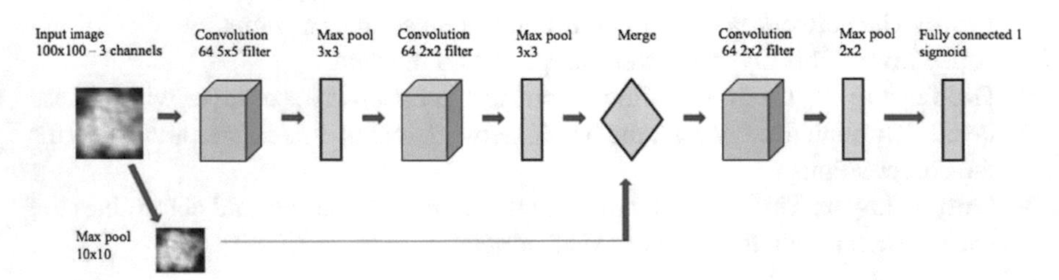

Fig. 3 Basic CNN architecture [5]

This is the most commonly used optimization technique in deep learning which makes the possibility of reusing the existing models for solving different problems. In this approach, the required weights or architecture can be incorporated into another model. The weights of the pre-trained models can be frozen and updated according to the requirements of the new model [22–25]. These pre-trained models can be used in weight initialization, as a classifier as a backbone feature extractor. This approach decreases the training cost and improves accuracy. The most common way to increase the accuracy of the transfer learning model is Fine Tuning. In this approach, architecture of the models is updated as well as retrained. All the transfer learning models are available in Keras [26].

The most common transfer learning algorithm was AlexNet, which first won the ImageNet competition by adding layers in the neural network and making it dense and deep faced the problem of backtracking which was solved by ResNet50.

A. **Resnet50**

Resnet50 won the ILVSR championship. Resnet50 is a deep neural network that is 50 layers deep. It contains 48 layers of convolution and 1 max pool layer and 1 average pool layer. It uses the leave connection or skip connection methods. When any block has a problem, it skips the connection and leaves it while creating shortcuts [26]. The basic architecture of ResNet50 is given in Fig. 4.

B. **Inception V3**

It is the variant of inception architecture, a 48-layered deep convolution neural network introduced by google scholars trained on the ImageNet dataset. This inception network consists of repeated modules that are piled over each other to form a

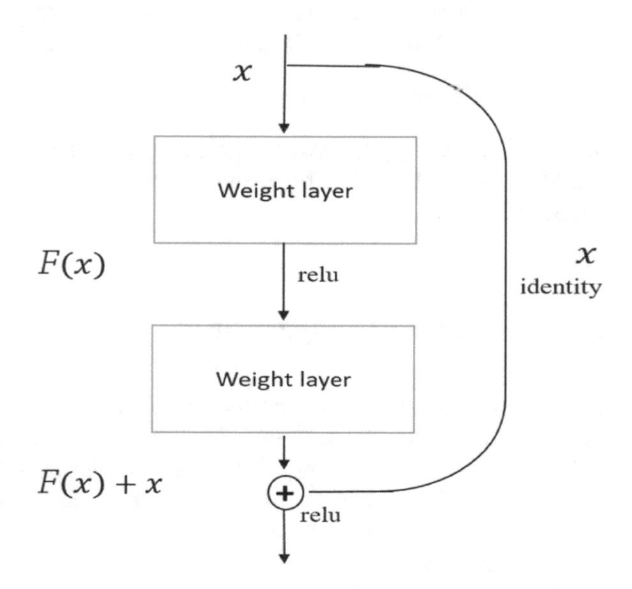

Fig. 4 ResNet 50 architecture [26]

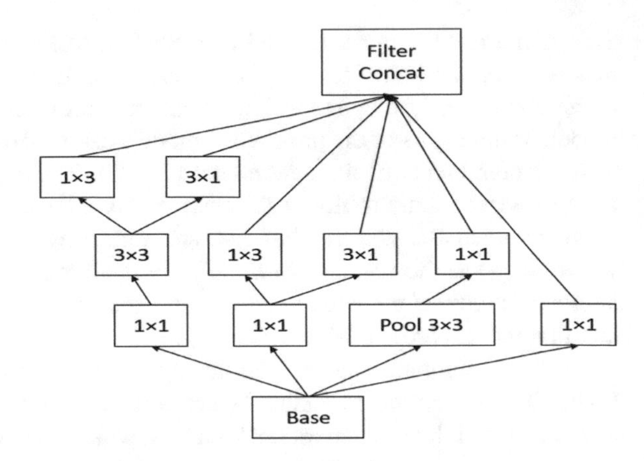

Fig. 5 Inception V3 architecture [26]

deep neural network. It is used in image recognition and object detection. Its architecture consists of convolutions that decrease the number of parameters in the network. It replaces bigger convolutions with smaller ones for fast training. The architecture of Inception V3 is in Fig. 5.

C. **Xception**

This model is inspired by Inception V3 and stands for "Extreme Inception" [14] but it has slightly outperformed the Inception V3 model on the Image Net. Both the models have the same parameters, but the performance measure is dependent upon the use of parameters [14]. It consists of 36 linear stacks of the depth-wise convolution layers with residual connections which are structured into 14 modules that have linear residual connections except for the first and last modules. The entry flows in the architecture of Xception are given in Fig. 6.

D. **Feature Extraction**

This is a less common approach used in deep learning where features can be extracted out of the network and can be given as input to machine learning models like SVM for the classification process.

5 Performance Comparison

In this section, we have given performance (accuracy) measures of surveyed studies in the form of a bar chart. Figure 7 shows the analysis of performance (accuracy) of various techniques existing in the literature. It is clear from the graph, the highest accuracy with deep learning models was seen in the CNN model [13] at 97%. Using the traditional machine learning techniques, the highest performance for detected nodules was seen in MSFCM [9] method with a sensitivity of 93.5%, but these machine learning-based techniques do not scale up well for huge datasets.

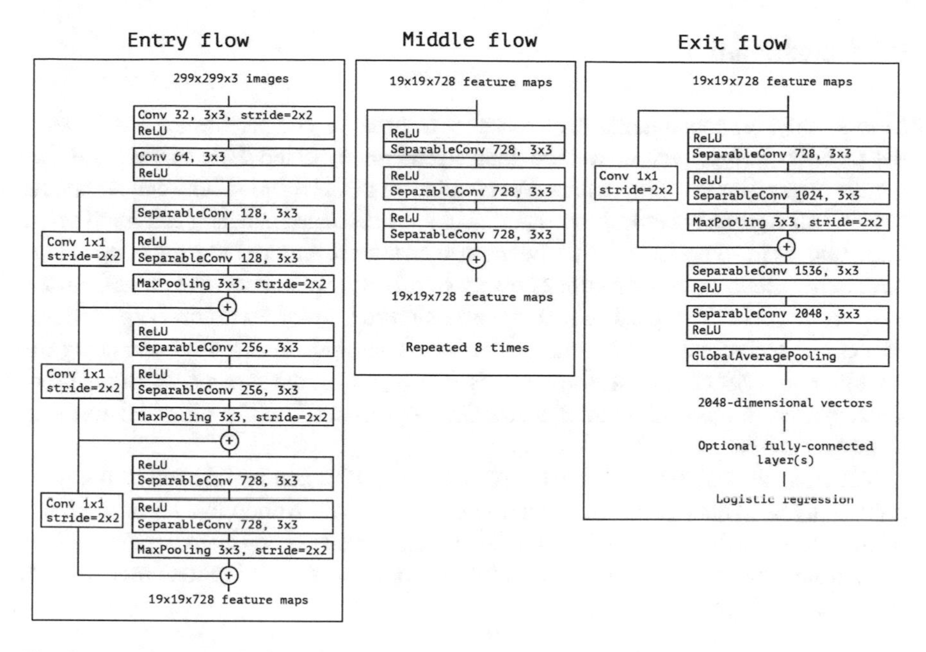

Fig. 6 Architecture of Xception [27]

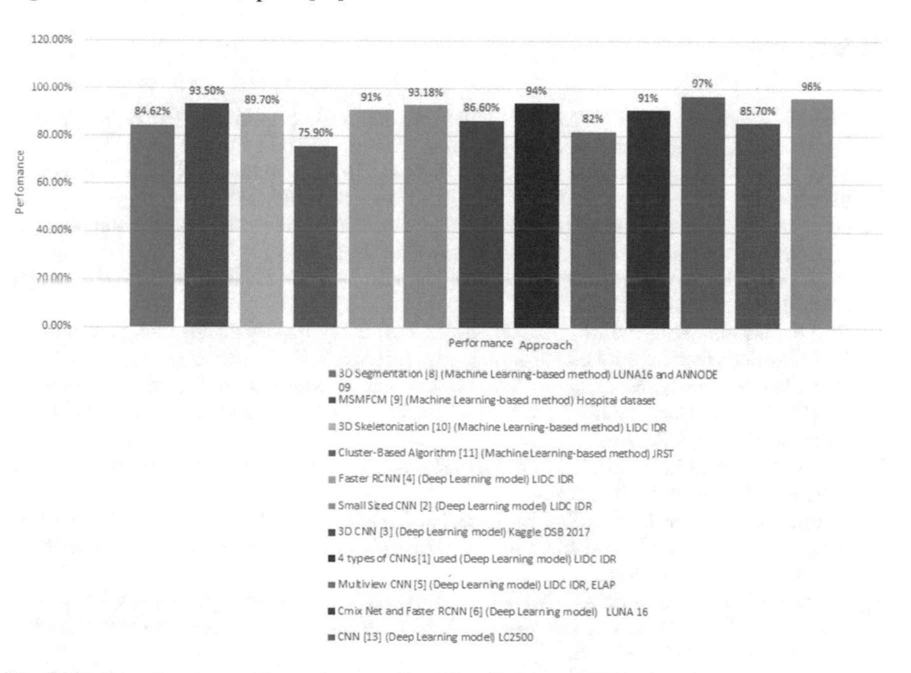

Fig. 7 Performance comparison of surveyed studies [1–11, 13, 14]

6 Conclusion

In this study, we found using deep learning techniques, the performance is better if the model is trained well on a large dataset, and once trained these models can also handle huge volumes of input data. We conclude that CNN-based lung cancer nodule detection has gained popularity in medical image processing and has performed better than humans in some cases [16]. The traditional feature engineering approaches may have better accuracy in some cases; however, the feature extraction and classifications are done as separate steps. The feature extraction step is done with the help of image processing techniques and is time-consuming, Classification step is done using the classifier, and hence, the performance is dependent on the type of classifier used. CNN performs both the feature extraction step as well as the classification step within its layers.

As far as the datasets are concerned, the LIDC IDR can be inferred as the most commonly used dataset for the detection of lung cancer. Although a lot of work has been done using the traditional machine learning as well as deep learning approaches, with promising results, there is still a lot of scope for developing new models with improved performance.

References

1. Jiang H, Ma H, Qian W, Gao M, Li Y (2018) An automatic detection system of lung nodule based on multigroup patch-based deep learning network. IEEE J Biomed Health Inf 22(4)
2. Sahu P, Yu D, Dasari M, Hou F (2019) A lightweight multi-section CNN for lung nodule classification and malignancy estimation. IEEE J Biomed Health Inf
3. Alakwaa W, Badr A, Nassef M (2017) Lung cancer detection and classification with 3D convolutional neural network. Int J Adv Comput Sci Appl 8(8)
4. Su Y, Li D, Chen X (2020) Lung nodule detection based on faster R-CNN framework. Comput Methods Prog Biomed 105866
5. Paul R, Hawkins SH, Schabath MB, Gillies RJ, Hall LO (2018) Predicting malignant nodules by fusing deep features with classical radiomics features. SPIE J Med Imag
6. Nasrullah N, Jun S, Mohammad SA, Mohammad S, Muhammad M, Bin C, Haibo H (2019)Automated lung nodule detection and classification using deep learning combined with mutltiple strategies. Sensors
7. Yanfeng L, Linlin Z, Houjin C, Na Y (2019) Lung nodule detection with deep learning in 3D thoracic MR images. IEEE Access
8. Gong J, Liu JY, Wang LJ, Sun XW, Zheng B, Nie SD (2018) Automatic detection of pulmonary nodules in CT images by incorporating 3D tensor filtering with local image feature analysis. Physica Medica 46:124–133
9. Farahani FV, Ahmadi A, Zarandi MHF (2018) Hybrid intelligent approach for diagnosis of the lung nodule from CT images using spatial kernelized fuzzy c-means and ensemble learning. Math Comput Simul 149:48–68
10. Narayanan BN, Hardie RC, Kebede TM, Sprague MJ (2017) Optimized feature selection-based clustering approach for computer-aided detection of lung nodules in different modalities. Pattern Anal Appl 22(2):559–571
11. Zhang W, Wang X, Li X, Chen J (2017) 3D skeletonization feature based computeraided detection system for pulmonary nodules in CT datasets. Comput Biol Med 92:64–72

12. Borkowski AA, Bui MM, Brannon Thomas L, Wilson CP, DeLand LA, Mastoride SM (2019) Lung and colon cancer histopathological image dataset (LC25000). arXiv:1912.12142
13. Mangal S, Chaurasia A, Khajanchi A (2020) Convolution neural networks for diagnosing colon and lung cancer histopathological images. arXiv:2009.03878
14. Hatuwal BK, Thapa HC (2020) Lung cancer detection using convolutional neural network on histopathological images. Int J Comput Trends Technol 68(10):21–24
15. Halder A, Dey D, Sadhu AK (2020) Lung nodule detection from feature engineering to deep learning in thoracic CT images: a comprehensive review. J Digit Imaging 33(3):655–677
16. Masud M, Sikder N, Al Nahid A, Bairagi AK, Alzain MA (2021) A machine learning approach to diagnosing lung and colon cancer using a deep learning-based classification framework. Sensors (Switzerland) 21(3):1–21
17. Tsai MJ, Tao YH (2021) Deep learning techniques for the classification of colorectal cancer tissue. Electron 10(14)
18. Mandal R, Gupta M (2016) Automated histopathological image analysis: a review on ROI extraction. IOSR J Comput Eng Ver V 17(6):2278–661
19. Dimitriou N, Arandjelović O, Caie PD (2019) Deep learning for whole slide image analysis: an overview. Front Med 6(November)
20. Khvostikov A, Krylov A, Mikhailov I, Malkov P, Danilova N (2020) Tissue type recognition in whole slide histological images. CEUR Workshop Proc 3027:496–507
21. Komura D, Ishikawa S (2018) Machine learning methods for histopathological image analysis. Comput Struct Biotechnol J 16:34–42
22. Bukhari SUK, Syed A, Bokhari SKA, Hussain SS, Armaghan SU, Shah SSH (2020) The histological diagnosis of colonic adenocarcinoma by applying partial self supervised learning. medRxiv, p. 2020.08.15.20175760
23. de Carvalho Filho AO, Silva AC, de Paiva AC, Nunes RA, Gattass M (2018) Classification of patterns of benignity and malignancy based on CT using topology-based phylogenetic diversity index and convolutional neural network. Pattern Recognit 81
24. Selvanambi R, Natarajan J, Karuppiah M, Islam SH, Hassan MM, Fortino G (2020) Lung cancer prediction using higher-order recurrent neural network based on glowworm swarm optimization. Neural Comput Appl 32(9):4373–4386
25. Suresh S, Mohan S (2020) ROI-based feature learning for efficient true positive prediction using convolutional neural network for lung cancer diagnosis. Neural Comput Appl 32(20):15989–16009
26. Online Available: https://keras.io/api/applications
27. Chollet F (2017) Xception: deep learning with depthwise separable convolutions. In: Proceeding—30th IEEE conference computer visual pattern recognition

Correlative Exposition of Various Machine Learning Techniques for Urdu Handwritten Text Recognition

Dhuha Rashid, Naveen Kumar Gondhi, and Chaahat

Abstract One of the most under-researched area in the field of natural language processing (NLP) has been the text classification. Text classification of languages like Urdu is more deprived of research since it is not among the top-most spoken language of the world. Despite that, almost 230 million people around the world speak Urdu. Researchers have worked with various machine learning and deep learning models for the identification of handwritten text of Urdu language. This paper is the comparative analysis of the various algorithms that have been used on the offline as well as online text recognition of Urdu language. From our analysis, CNN is the most widely adopted model for recognition of Urdu handwritten text due to its excellent properties of image classification as well as feature extraction. This paper gives a brief of various machine learning algorithms used for recognition of Urdu text and their comparative analysis.

Keywords Deep learning · Machine learning · Handwritten text recognition · Urdu

1 Introduction

Text recognition is a prime research-oriented field in AI and machine learning. Even in text recognition, the study of offline text recognition holds more significance since its automation is still in a very early phase. Work has been done on various languages using different models of machine learning. Some languages achieved good rate of recognition such as English and Chinese and thus are being practically implemented in multiple fields such as automated document scanning and many more. For some regional or less commonly spoken languages, like Urdu or various Indic languages, the study for their recognition was not tested. In recent years, research is being done

D. Rashid (✉) · N. K. Gondhi
Shri Mata Vaishno Devi University, Katra, J&K, India
e-mail: dhuha.rashid@gmail.com

Chaahat
Model Institute of Engineering and Technology, Jammu, J&K, India

Y. Singh et al. (eds.), *Proceedings of International Conference on Recent Innovations in Computing*, Lecture Notes in Electrical Engineering 1001,
https://doi.org/10.1007/978-981-19-9876-8_3

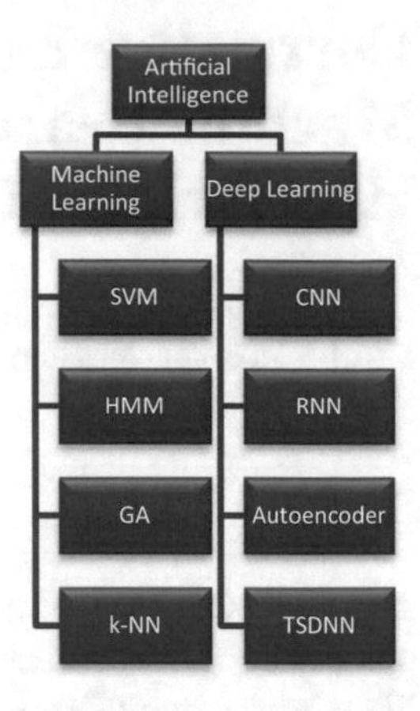

Fig. 1 Various machine learning and deep learning algorithms

on these languages but still have not acquired any remarkable results due to various complexity such as writing styles.

Urdu is also one such language, spoken mostly in the Southeast of Asia. It has a ligature-based and cursive writing style which makes its recognition very complicated. Various researchers have used various models on multiple datasets to improve the recognition of Urdu characters. Machine and deep learning algorithms have been implemented on different datasets of Urdu handwritten as well as printed text to calculate the recognition accuracy of various models. SVM, GA, and *k*-NN are some of the machine learning models used while CNN, RNN, feed-forward neural network (FFNN), and two-stream deep neural network (TSDNN) are some models of deep learning used for identification of Urdu text. Among those, the bidirectional LSTM and CNN had shown some remarkable results. Where one uses the convolutional function in its hidden layer for recognition, the other uses a bidirectional memory in a recurrent neural network for identification of the scripts. Some of the various machine learning and deep learning algorithms used for recognition of Urdu characters are shown in Fig. 1.

1.1 Machine Learning Algorithms

Machine learning algorithms can be supervised as well as unsupervised. Both can be used for character recognition in natural language processing (NLP). SVM is one of the machine learning algorithms. It is classified among the supervised machine learning algorithms. This algorithm finds the hyperplane among the various feature of a dataset. Training of data in Support Vector Machine (SVM) is done from the sample instances randomly selected from the dataset. For text recognition, researchers have suggested pool-based active learning for training of the data [1]. SVM is idealistic for up to three features. Finding of hyperplane for over three features gets quite complicated. Hidden Markov model (HMM) works on assumptions of being a Markov process where working of a hidden state 'A' is observed by another state 'B' which is observable. This model has been applied in the field of NLP for handwritten text recognition also after its success was observed in speech recognition [2]. One of the important assumptions of HMM is that the observation of output is independent of the hidden state depending on certain conditions.

Genetic algorithms (GA) work on fundamental principle of genetics and natural selection. They optimize the search techniques by associating the child with parent based on its features. For text recognition, historical data is used to associate texts with each other. GA when associated with other algorithms such as CNN improve their results dramatically [3]. *K*-Nearest neighbor (*k*-NN) also belongs to supervised machine learning algorithm which is used for both classification and regression. It functions best as a classifier though where it categorizes every new dataset into the group similar to the already available dataset groups present. Euclidean distance is calculated to find the *k*-nearest neighbor between the testing and reference points [4]. A decision tree is a supervised learning technique which is most useful in making decision based on the provided features. Depending on the model, decision tree generates its model with its respective rules [5]. Generally, the decision is in the form of Yes/No. Decision tree has a tree structure with a root and a leaf node.

1.2 Deep Learning Algorithms

CNN or convolutional neural network is a model of deep learning whose algorithm is based on convolution function instead of matrix multiplication in at least one of the layers. Convolutional neural networks (CNNs), originally invented for computer vision, have been shown to achieve strong performance on text classification tasks as well as other traditional natural language processing (NLP) tasks, even when considering relatively simple one-layer models [6]. CNN algorithms work best for image classification than any other deep learning model due to its ability to work efficiently with less pre-processing also.

RNN is cyclic neural network model. RNN has the ability to map input to output with the additional feature of tracing back older computations. BLSTM is the bidirectional long short-term memory which is a recurrent neural network architecture. BLSTM is a concept used on RNN algorithm which can retain computations for longer time in both forward as well as backward direction, which otherwise would have vanished. It usually consists of a cell, an input gate, an output gate, and a forget gate. Memory blocks with additive multiplicative units are used in replacement of memory block used in the hidden layers of RNN model [7]. It can be used quite well for prediction, classification, and processing. Autoencoder, on the other hand, is combination of two sub-models: an encoder and a decoder. The encoder plots the input data to the code, while the decoder plots the code to the remake of input. It has been used in various fields from face recognition to feature detection and thus is also used by researchers for the recognition of Urdu text [8].

2 Literature Review

In 2010, researchers used SVM for the classification of Urdu characters [9]. This method was used on CENPARMI dataset of Urdu characters. Images used were binary as well as grayscale. 64 × 64 as well as 128 × 128 image sizes were used. An accuracy of 97% was achieved. Sobia Tariq et al. [10] used Hidden Markov model and rule-based post-processor to identify Urdu printed characters of font 36. They extracted 1692 ligatures from corpus dictionary. 1596 ligatures were accurately identified. The accuracy of recognition thus was 92.7%. A scale-invariant technique was introduced in one of the papers [11] with segmentation-free approach. They used Hidden Markov model for recognition. Height of each image was normalized to 60 pixels. A total of 2082 components were used for recognition out of which 2017 were high frequency primary ligatures and 11 were secondary ligatures. An accuracy of 97.93% was observed. Ahmed et al. [8] trained 178,573 ligatures. They used Stacked Denoising Autoencoder for feature extraction from raw images. The images were used from UPTI dataset. Their accuracy ranged from 93 to 96%. In another research [12], they suggested the use of Hidden Markov model for primary ligature and then with the diacritics associated with the words. 30 full page documents of Urdu ligatures from CLE database were taken and a total of 10,364 ligatures were obtained out of which 8800 were correctly categorized into 246 clusters. Ligatures were clustered in classes where the errors were manually removed and further were used for training. The obtained ligature recognition was 95% whereas word recognition rate was 89%. Furthermore, in 2017 research for handwritten Urdu text recognition was done using hybrid approach [13]. They used a hybrid model of CNN and RNN on 44 classes of UPTI dataset. CNN extracted low-level translational invariant feature which was then provided to the multi-directional BLSTM which was responsible for classification. Overall accuracy was observed to be of 98.12%. Saeeda et al. [7] gave the idea of using one-directional BLSTM for handwritten text recognition. They used UNHD, which is the largest dataset till date publicly available. 50% of the dataset was used for training, 30% for validation and remaining 20% for testing. 93–94% of accuracy was observed by the research.

Naila Habib et al. [14] suggested multi-level hierarchal clustering for categorizing various ligatures. Corpus of 2430 images was used to implement four different machine learning algorithms, i. e., Decision trees, *k*-NN, Naïve Bayes, and Linear discriminant analysis. Recognition rate of about 62, 90, 73, and 61% was shown by the respective algorithms. In 2019, Hussain et al. [15] published their work on recognition of Urdu text using convolution neural network. They used their own dataset consisting of 38,400 images. 800 images were taken of each 38 characters as well as 10 numeric digits written by almost 250 writers. A flatbed scanner was used, and the images were scanned of size 300 dpi. Manually, they segmented the images into 28 by 28 size. The characters based on their shape were classified into 12 classes, while the numerals were classified into 10 classes. In their experimental setup,75% of the data was used as training data while remaining 25% was used as testing data. They used learning rate as 0.08 and a batch size of 40 for Urdu handwritten characters and

0.0025 learning rate and 132 batch size for Urdu handwritten numerals. Accuracy of 98.03% was obtained for numerals and 96.04% for Urdu characters, respectively. They also conducted experiments using tenfold and eightfold cross-validations on handwritten characters and numerals to invalidate any confusions because of ration in training and testing dataset. Usage of CNN for Urdu text recognition was proposed in one more paper [16]. UPTI and CLE datasets were used for recognition. They proposed their own CNN model with two convolutional layers. In their experimental setup, they have categorized the two datasets into four scenarios: CLE high frequency ligatures (HFL), CLE (Books), UPTI, and UPTI + CLE (Books). Among the four scenarios, CLE HFL showed highest accuracy of about 98.30% in their proposed model. Shahbaz et al. [17] in 2019 proposed a model where CNN was used for feature extraction and bidirectional LSTM was used for classification. This model was used for experimenting over handwritten 1000 lines. Segmentation was used in the lines to segregate the ligatures. Their recognition rate was of 83.69% on 1000 text lines.

A concept of transfer learning using pre-trained CNN networks was also proposed in 2020 [18]. The three pre-trained models used were ResNet18, AlexNet, and GoogleNet. The experiment was conducted on various Urdu as well as Farsi scripts. The dataset used for Urdu handwritten text was UNHD. The three pre-trained CNN models were used as feature extractors. The extracted features were concatenated and forwarded to two fully connected layers where the classification was done. 97.18% of accuracy was observed from this proposed model. OCR-nets were used in one of the researches [19] which are the variants of AlexNet and GoogleNet. An integrated dataset was created using IFHCDB. A total of 54 classes of Urdu characters and 52,380-character images were used. OCR-AlexNet showed the accuracy of 96.3% whereas 94.7% accuracy was shown by OCR-GoogleNet. The recognitions were also tested using different varieties of nib sizes [20]. The dataset for the same was collected from the corpus of NLAIP. They used self-organizing map for character classification after the pre-processing and segmentation were done on the input images. It uses a winner-take-all strategy to work. The result of recognition was 89.03%.

The idea of word spotting was implied for Urdu handwritten text recognition in 2020 [21]. The idea was implemented on UNHD database. HoG feature of ligature images was extracted. Some arbitrary samples were generated using generative adversarial network (GAN) and tested using LSTM model. The results are compared to UNHD dataset. 98.96% accuracy was achieved cycle-GAN models. Naila Habib Khan et al. used genetic algorithm (GA) in one of the proposed papers [22]. Since UPTI dataset was used, the model primarily focused on printed text rather than handwritten. Holistic segmentation was used, and 15 geometric and statistical features were extracted. Intra-feature hierarchal clustering was used to cluster all the datapoints collected during feature extraction. Recognition rate achieved by using GA for classification and recognition was 96.72%. Two-stream deep neural network was also used in one of the researches for recognition of Urdu text in outdoor scenes [23]. A synthetic dataset of 4.2 and 51 k was used. For detection and localization, a custom Faster RCNN was conjoined with pre-trained CNN models like ResNet18,

ResNet50, and SqueezeNet. 1094 real-world images containing 12 k Urdu characters were also tested using TSDNN. 94.90 and 95.20% were the recognition rates acquired for 4.2 and 51 k image datasets. 76.60% was the accuracy for real-world images. A holistic approach with CNN was applied on 5298 samples of 5 words used, viz. tawheed, namaz, hajj, rehmat, and Jannat. Each word was written approximately 1000 times [24]. Three different image sizes of 60 × 60, 50 × 50, and 30 × 30 were used. The highest accuracy achieved was 96% using feed-forward neural network. A dataset called Handwritten Urdu Character dataset (HUCD) was introduced by Mushtaq et al. in their paper [25] which around 106,120 samples of handwritten Urdu text and numerals. These samples were divided into joiners and non-joiners based on their joining properties. 38 basic characters were divided into 27 joiners, 10 non-joiners, and 1 no-joining category. 2 field characters were divided as 1 non-joiner and 1 no-joining. Since the numerals do not show any joining property, they were not placed in any category. A total of 142 forms were created. The dataset was acquired from 750 native writers from Kashmir. The collected data was scanned with 600 dpi. Images were then pre-processed, and the operation for pre-processing used was segmentation, normalization, binarization, and denoising and smoothing. They proposed a CNN architecture, with a 64 × 64 sized input layer, 4 convolutional hidden layers, and 64 arbitrary kernels of size 5 × 5 for the first convolutional layers. The rest of the three layers have 128, 256, and 128 kernels with 3 × 3-unit size. An initial learning rate of 0.001 has been used with batch size of 256 and 30 epochs. They attained accuracy of about 98.82% on their proposed 4-layer CNN model. Multi-layer Perceptron, Support Vector Machine, *k*-NN, CNN, RNN, and random forest algorithms were used on an image dataset of 4668 images [26]. Images of 50 × 50 were used for classification. 80% of the dataset was used for training while 20% was used for testing. 98, 97, 38, 99, 80, and 97% of the accuracies were calculated for the respective algorithms. A new dataset called NUST-UHWD was developed in 2022 [27]. The dataset was created using 1000 different writers, and convolutional recurrent neural network (CRNN) model was used to predict the accuracy on it. The model has three prime parts. The CNN, which is used as feature extractor, LSTM acts as the sequence labeling component, and finally, the CTC is for transcription. Furthermore, an n-gram language model is used to improve the accuracy of the model. It was seen that before n-gram model, the accuracy was about 93% while after the n-gram language model was used, the accuracy improved to 95%. Table 1 gives the tabular representation of the overall literature review of the work done on Urdu text recognition using various machine learning techniques.

3 Comparative Analysis

It is evident from the literature survey that among all the various algorithms used for Urdu text recognition, CNN and RNN are the most approachable algorithms for it. Most of the researchers have opted for CNN and RNN for text recognition in Urdu. After these two, SVM, HMM, and *k*-NN have also been used by many of the

Table 1 Literature survey of Urdu text detection

Ref No.	Year	Dataset	Model	Accuracy	Remarks
[9]	2010	CENPARMI	SVM	97%	Diacritics not included
[10]	2013	1692 ligatures from corpus-based dictionary	HMM and rule-based pre-processor	92.7%	Low accuracy
[11]	2015	2082 components	HMM	97.93%	Focused only on frequently occurring ligatures
[8]	2017	UPTI	Stacked denoising autoencoder	93–96%	Too much focus on feature extraction
[12]	2017	30 pages from CLE	HMM	95% for ligatures 89% for words	Small dataset used
[13]	2017	UPTI	CNN-RNN hybrid	98.12%	Both the models were not used for classification
[7]	2017	UNHD	BLSTM	93–94%	Less accuracy due to line segmentation
[14]	2018	Corpus of 2430 images	Decision Tree Naïve Bayes *k*-NN Linear discriminant analysis	60% 73% 90% 61%	Corpus chosen was very small
15]	2019	38,400 images	CNN	96.04%	Limited dataset
[16]	2019	UPTI and CLE	CNN	98.30%	Focused on printed Urdu text
[17]	2019	1000 text lines	CNN-BLSTM	83.69%	Poor recognition rate
[18]	2020	UNHD	CNN (AlexNet, GoogleNet, ResNet18)	97.13%	Urdu language was not the focus of the model
[19]	2020	Integrated dataset	OCR-AlexNet OCR-GoogleNet	96.3% 94.7%	Poor time constraint for GoogleNet

researchers. The range for CNN has almost been similar to the original text. In most of the researches, CNN has reached accuracy of 98–99% followed by SVM which has accuracy of 97%. *K*-NN has shown least accuracy among all the four algorithms. The state comparative analysis is shown in Table 2.

Out of all the algorithms that have been used for recognition of Urdu text, CNN has shown promising results every time. From the average of accuracies of various

Table 2 Comparative analysis of various ML and DL algorithms

Model	Avg. accuracy (%)	Avg. loss (%)
CNN	98	2
RNN	92	8
K-NN	68	32
SVM	97	3
HMM	94	6

models, CNN has shown the highest accuracy and lowest average loss. This is because of its ability to classify images with good recognition rates. CNN is thus used in various object identification as well as image recognition purposes. Various pre-trained models of CNN have been used in the paper [18, 19]. It can be seen that the recognition rates for pre-trained model are also above 90%. Despite SVM having higher average recognition rate, the next popular choice for character recognition has been RNN due to its exhibiting memory. The memory helps it in recognition of inputs with variable lengths. One of the important derivatives from RNN is bi-directional long short-term memory (BLSTM). It has the ability to predict inputs from backwards (past) as well as forward (future). This helps in much accurate identification of languages where the context is very important. Since it has fairly good recognition rate, it has thus been used for classification and recognition of Urdu handwritten text also.

Apart from using various deep learning models for Urdu handwritten text classification, sometimes the models have also been specifically used for feature extraction. This idea is generally used in hybrid approach which is proposed by various researchers in recent times. A CNN-RNN hybrid has been used in one of the research papers [13] wherein CNN is responsible for feature extraction and RNN is focused on classification. Similarly, CNN-BLSTM has also been used [17] as a hybrid model. In this proposed model, CNN was again used for feature extraction while BLSTM focused on classification. CNN has been preferably used for feature extraction for its excellent property of detecting features without any supervision. This hybrid approach has worked fairly well for handwritten text recognition since it is equally focused on feature extraction and character classification. Despite having good results, hybrid approach has its own limitation. Using one model specifically for feature extraction or classification means losing its property for other purposes. CNN may be idealistic for feature extraction, but CNN is a good image classifier as well. Using it for one purpose limits it from being a good classifier. Also, using any one model for both purposes reduces the chances for a versatile model for classification. Thus, there is a need for a model, which encompasses all the features and yields good recognition rates as well.

4 Conclusion

Text classification is one of the areas of natural language processing (NLP) which is responsible for various task such as sentiment analysis, character recognition, and transfer learning. This paper is about various machine learning algorithms that have been used for recognition of Urdu handwritten text and their comparative analysis with respect each other. Various researchers have used supervised as well as unsupervised learning algorithm for Urdu text recognition. From our study, it is evident that the best-suited algorithm for character recognition in Urdu language is CNN since it has shown the best accuracy on various datasets. CNN has also been alongside of various other models in hybrid approaches. CNN has been used as feature extractor [17, 27] as well as a classifier [15]. Apart from CNN, RNN has also worked well for character recognition purposes. LSTM and BLSTM are the two sub-types of RNN which have also been used for Urdu character recognition [7, 21]. The hybrid model of CNN and RNN has worked considerably well than the individual models since hybrid model collaborates the features of two models in a single model.

References

1. Rachel Ida Buff (2009) The Deported. Am Q 61(2):417–421. https://doi.org/10.1353/aq.0.0077
2. España-Boquera S, Castro-Bleda MJ, Gorbe-Moya J, Zamora-Martinez F (2011) Improving offline handwritten text recognition with hybrid HMM/ANN models. IEEE Trans Pattern Anal Mach Intell 33(4):767–779. https://doi.org/10.1109/TPAMI.2010.141
3. Alsaleh D, Larabi-Marie-Sainte S (2021) Arabic text classification using convolutional neural network and genetic algorithms. IEEE Access 9:91670–91685. https://doi.org/10.1109/ACCESS.2021.3091376
4. Kumar M (2011) k—nearest neighbor based offline handwritten Gurmukhi character recognition (Iciip)
5. Sastry PN, Krishnan R, Venkata B, Ram S (2010) Classıfıcatıon and ıdentıfıcatıon of Telugu handwrıtten characters extracted from palm leaves usıng 5(3):22–32
6. Jacovi A, Shalom OS (2016) Understanding convolutional neural networks for text classification. Published online 2016
7. Bin AS, Naz S, Swati S, Razzak MI (2019) Handwritten Urdu character recognition using one-dimensional BLSTM classifier. Neural Comput Appl 31(4):1143–1151. https://doi.org/10.1007/s00521-017-3146-x
8. Ahmad I, Wang X, Li R, Rasheed S (2017) Offline Urdu Nastaleeq optical character recognition based on stacked denoising autoencoder. China Commun 14(1):146–157. https://doi.org/10.1109/CC.2017.7839765
9. Sagheer MW, He CL, Nobile N, Suen CY (2010) Holistic Urdu handwritten word recognition using support Vector machine. In: Proceeding—International conference pattern recognition. Published online, pp 1900–1903. https://doi.org/10.1109/ICPR.2010.468
10. Javed ST, Hussain S (2013) Segmentation based Urdu Nastalique OCR. Lectures notes computer science (including subscriber lectures notes artificial ıntelligent lectures notes bioinformatics). 8259 LNCS(PART 2):41–49. https://doi.org/10.1007/978-3-642-41827-3_6
11. Khattak IU, Siddiqi I, Khalid S, Djeddi C (2015) Recognition of Urdu ligatures—A holistic approach. In: Proceeding ınternational conference document analysis recognition, ICDAR 2015(Novem), pp 71–75. https://doi.org/10.1109/ICDAR.2015.7333728

12. Chaudhuri A, Mandaviya K, Badelia P, Ghosh SK (2017) Optical character recognition systems. Stud Fuzziness Soft Comput 352(5):9–41. https://doi.org/10.1007/978-3-319-50252-6_2
13. Naz S, Umar AI, Ahmad R et al (2017) Urdu Nastaliq recognition using convolutional–recursive deep learning. Neurocomputing 243:80–87. https://doi.org/10.1016/j.neucom.2017.02.081
14. Khan NH, Adnan A, Basar S (2017) Urdu ligature recognition using multi-level agglomerative hierarchical clustering. Cluster Comput 21(1):503–514. https://doi.org/10.1007/s10586-017-0916-2
15. Husnain M, Missen MMS, Mumtaz S, et al (2019) Recognition of Urdu handwritten characters using convolutional neural network. Appl Sci 9(13). https://doi.org/10.3390/APP9132758
16. Uddin I, Javed N, Siddiqi I, Khalid S, Khurshid K (2019) Recognition of printed Urdu ligatures using convolutional neural networks. J Electron Imag 28(03):1. https://doi.org/10.1117/1.jei.28.3.033004
17. Hassan S, Irfan A, Mirza A, Siddiqi I (2019) Cursive handwritten text recognition using bi-directional LSTMs: a case study on Urdu handwriting. In: Proceeding 2019 ınternational conference deep learning machine learning emerging application deep 2019. Published online 2019, pp 67–72. https://doi.org/10.1109/Deep-ML.2019.00021
18. Kilvisharam Oziuddeen MA, Poruran S, Caffiyar MY (2020) A novel deep convolutional neural network architecture based on transfer learning for handwritten Urdu character recognition. Teh Vjesn 27(4):1160–1165. https://doi.org/10.17559/TV-20190319095323
19. Mohammed Aarif KO, Poruran S (2019) OCR-Nets: variants of pre-trained CNN for Urdu handwritten character recognition via transfer learning. Procedia Comput Sci 2020(171):2294–2301. https://doi.org/10.1016/j.procs.2020.04.248
20. Ahmed Abbasi M, Fareen N, Ahmed AA (2020) Urdu Nastaleeq Nib calligraphy pattern recognition. Am J Neural Networks Appl 6(2):16. https://doi.org/10.11648/j.ajnna.20200602.11
21. Farooqui FF, Hassan M, Siddhu MK (2020) Offline hand written Urdu word spotting using random data generation. Published online 2020, pp 131119–131136. https://doi.org/10.1109/ACCESS.2020.3010166
22. Khan NH, Adnan A, Waheed A, Zareei M, Aldosary A, Mohamed EM (2020) Urdu ligature recognition system: an evolutionary approach. Comput Mater Contin 66(2):1347–1367. https://doi.org/10.32604/cmc.2020.013715
23. Arafat SY, Iqbal MJ (2020) Urdu-text detection and recognition in natural scene images using deep learning. IEEE Access 8:96787–96803. https://doi.org/10.1109/ACCESS.2020.2994214
24. Khan HR (2021) a holistic approach to Urdu language word recognition using deep neural networks 11(3):7140–7145
25. Mushtaq F, Misgar MM, Kumar M, Khurana SS (2021) UrduDeepNet: offline handwritten Urdu character recognition using deep neural network. Neural Comput Appl 33(22):15229–15252. https://doi.org/10.1007/s00521-021-06144-x
26. Chhajro MA (2020) Handwritten Urdu character recognition via images using different machine learning and deep learning techniques. Indian J Sci Technol 13(17):1746–1754. https://doi.org/10.17485/ijst/v13i17.113
27. ul Sehr Zia N, Naeem MF, Raza SMK, Khan MM, Ul-Hasan A, Shafait F (2022) A convolutional recursive deep architecture for unconstrained Urdu handwriting recognition. Neural Comput Appl 34(2):1635–1648. https://doi.org/10.1007/s00521-021-06498-2

Artificial Intelligence and Graph Theory Application for Diagnosis of Neurological Disorder Using fMRI

Bansari Prajapati, Parita Oza, and Smita Agrawal

Abstract In the mid 1990s, Functional Magnetic Resonance Imaging (fMRI) applicable for analyzing brain Functional Magnetic Resonance Imaging (fMRI) was applicable for analyzing brain connectivity and also made it possible for the diagnosis of neurological disorders. Around 10% of the total disease burden is contributed by Neurological disorders in India. Various computational methods like graph theory-based techniques have been used by the research community and have played an important role to understand the connected architecture of the brain. We present applications of graph theory and Artificial Intelligence (AI) to diagnose neurological disorders. We also present various neuroimaging tools to analyze neuronal disabilities with related features. Moreover, this work also presents a summary of publicly available fMRI datasets concerning various neurological disorders such as Autism spectrum disorder (ASD), Alzheimer's disease, and multidisorder. The paper also summarizes various methodologies applied to diagnose the various neurological disorder c.

Keywords Brain connectivity · fMRI · Graph theory · Neuroimaging

1 Introduction

Over the previous few years, neurological disabilities are growing over population like main depression disorder (MCD), obsessive-compulsive disease (OCD), post-traumatic stress disorder (PTSD), Schizophrenia, etc. So there's a need to develop strategies that could early detect and diagnose such disorders. fMRI has proven ability in linking disorders with graph concept. Human brain consists of about 86

B. Prajapati · P. Oza (✉) · S. Agrawal
Department of Computer Science and Engineering, Nirma University, Ahmedabad, India
e-mail: parita.prajapati@nirmauni.ac.in

B. Prajapati
e-mail: 19bce210@nirmauni.ac.in

S. Agrawal
e-mail: smita.agrawal@nirmauni.ac.in

Y. Singh et al. (eds.), *Proceedings of International Conference on Recent Innovations in Computing*, Lecture Notes in Electrical Engineering 1001,
https://doi.org/10.1007/978-981-19-9876-8_4

billion neurons which form positive connections while any task is performed or in resting-state and fMRI can detect such connections and examine it by the use of graph theory [1]. In addition, human connectomes have a significant impact on existential neuroscience with greater understanding of the brain.

1.1 Brain Connectivity

The human brain is a complex network of neurons hence brain connectivity is unintelligible. Brain connectivity is referred to as the evaluation of the relation between different brain areas, these relations are formed when a specific task is related to cognition or resting [2]. Interpreting datasets from fMRI, brain connectivity patterns are classified as a statistical dependency (functional connectivity), a casual interaction (effective connectivity), or an anatomical link (structural connectivity) that defines anatomical connections linking between neuronal units. Structural connectivity consists of undirected links in neuroimaging that are relatively stable on a shorter time-scale and on a longer time-scale, synaptic plasticity experience-dependent changes [3].

Functional connectivity: Functional connectivity is based on statistical dependencies among spatially remote neuronal elements. Functional connectivity is highly dependent on time, fMRI techniques used to derive time-series datasets of such neuronal activity can be determined in many ways such as cross-correlation, mutual information, clustering, statistical parametric mapping, or spectral coherence [3].

Effective connectivity. is a direct connection that influences one neural system over another, either synaptically or statistically. This indicates effective connectivity is possibly shown as a union of functional connectivity and structural connectivity. Techniques that determine effective connectivity datasets include model-based and model-free, such as Granger causality, dynamic causal modeling, Bayesian network, or transfer entropy [4]. Brain connectivity is conventionally represented as graph or matrix format. Many studies are carried out over functional connectivity and effective connectivity, analysis on graph theory and adjacency matrices [4].

Organization of the review paper. Figure 1 shows the paper organization: A brief introduction to the application of graph theory and AI for the diagnosis of neurological disorders using fMRI in presented in Sect. 1. Section 2 discuss various techniques to analyze neurological disorder and brief note on fMRI. Further, Sect. 3 is about a detailed note on graph theory methods for fMRI data analysis and an overview of AI, ML, and DL methods for the classification of neurological disorders. It also consists of the number of publications reviewed in the paper and different fMRI datasets and Tools. There is a short discussion in Sect. 4 about issues and future scope. The last section is about the conclusion of the review paper.

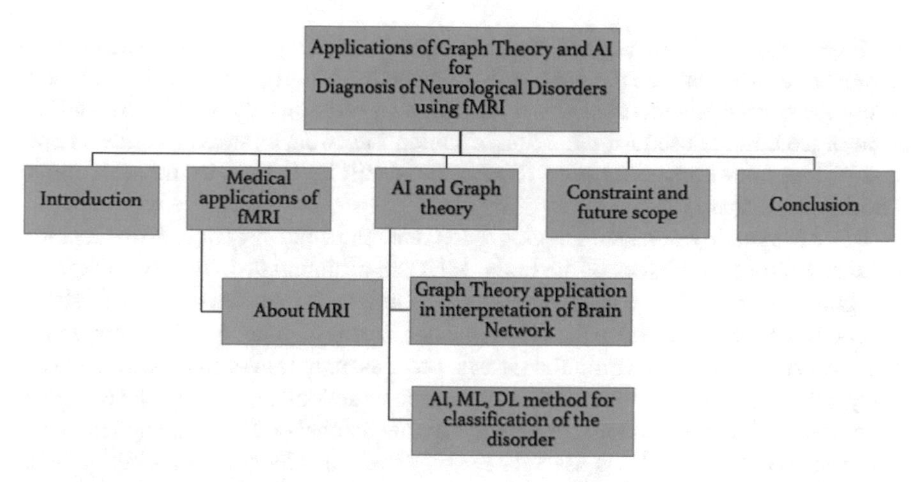

Fig. 1 Organization of paper

2 Medical Applications fMRI

The U.S. National Library of Medicine gives the probability of having more than 600 neuronal diseases. Various neurological disorders are commonly recognized such as epilepsy, Alzheimer's disorder, Autism, major depression disorder and anxiety, schizophrenia, obsessive-compulsive disorder (OCD), multiple sclerosis, and attention deficit hyperactive disorder (ADHD). To analyze neurological disorders there are several types of imaging techniques for brain scans like computed tomography (CT scan), magnetic resonance imaging (MRI), positron emission tomography (PET), and single photon emission computed tomography (SPECT) scans.

CT scan: Computed tomography is the 2-dimensional image of organs, bones, and tissues produced by X-rays. CT scan works as the X-ray tube rotates around a doughnut-shaped circular entrance called a gantry. While scanning, X-rays are emitted and are assimilated in the patient's body lying on the bed. Obtained CT scan data is processed to construct 2-dimensional slice images of the organ bones or specific body part. In general, a CT scan is used for instant detection of brain hemorrhage, irregularities of vascular and bones, brain tumors and cysts, and many more similar disorders [5, 6].

PET: PET positron emission tomography refers to 2-dimensional or 3-dimensional imaging of brain tissue. A radioactive isotope, called a tracer, is injected into the patient's bloodstream. It takes approximately 30–90 min to accumulate the brain tissue and scanning time is about 30–45 min. PET scanner rotates around the body while the computer processes the data to construct a visual image of active areas of the brain. PET scan of the brain is used to diagnose epilepsy, tumor cells, or certain memory disorders [5, 7].

SPECT: SPECT or single photon emission computed tomography is a 2-dimensional image of the 3-dimensional distribution of radionuclide. Similar to a

PET scan, radioactive isotope, a tracer is injected in the patient vein. The Gamma camera rotates around the patient's body lying on the bed. It detects the irregularities while the computer processes the information to construct the image [5]. SPECT scan of the brain is used for detection of tumor, infection, or stress fractures. Especially, dopamine transporter with SPECT (Da-SPECT) is used in the detection of Parkinson's disease [5].

MRI: Magnetic resonance imaging (MRI) technique that produces 3-dimensional detailed anatomical images of the brain. MRI can differentiate between white metal and gray matter. MRI scan is a large cylindrical magnetic tube that generates a strong magnetic field around the patient body. Such strong magnetic fields temporarily realign water molecules within the tissues. Later assimilation of radio waves via the body detects the moving of molecules returns to a random alignment. The computer processes the information and constructs a 2-dimensional or 3-dimensional image of scanned brain tissue. MRI is used for the diagnosis of traumatic brain injury, brain tumors, brain damage caused by epilepsy. It is also used to diagnose neurological disorders such as multiple sclerosis [5]. There are some limitations in MRI that are resolved by fMRI. fMRI is a functional MRI that detects neural activity.

In this review paper, detailed information of fMRI is explained. fMRI is a new emerging field for enthusiasts of data analysis in medical application.

2.1 About Functional Magnetic Resonance Imaging (fMRI)

Early intervention of neurological disorders in patients helps for the rapid treatment by means fMRI comes into the picture. The principle of fMRI is to identify active areas of the brain by Blood Oxygenation Level Dependent (BOLD) in the corresponding region of interest (ROI). fMRI scan is a non-invasive technique based on functional connectivity. Graph theory analyzes the data of fMRI to generate considerable spatial resolution. Functional Magnetic Resonance Imaging (fMRI) is subject to functional connectivity. It means statistical dependencies among physiological events in the brain [8]. Blood Oxygenation Level Dependent (BOLD) identifies the most active areas of the brain that receive a higher level of oxygenated blood. It directly corresponds to higher neural activity in that area and is represented by color-coding [9]. fMRI studies involve two approaches that are either whole brain (WB) or a region of interest (ROI) [10].

fMRI scanner detects signals when protons in nuclei of hydrogen atoms respond by emitting an electromagnetic signal. This tissue signal helps to reconstruct a high-resolution image of the brain. The scanner detects differences in magnetic properties of oxygenated versus deoxygenated blood. Differentiation is denoted by color-coding [9, 11]. fMRI generates an evaluated time-series dataset with the assist of graph theory the usage of Artificial Intelligence (AI), Machine Learning (ML), and Deep Learning (DL) concepts [12].

3 Artificial Intelligence (AI) and Graph Theory

3.1 Graph Theory Application in Interpretation of Brain Network

In a decade, many researchers have been involved themselves with applications of graph theory in brain connectivity having various methodological approaches with different types of datasets [13]. Figure 2 shows a network where a dot is a node and lines connecting nodes are called edges. Formally, there are two kinds of graph presentation: One is an undirected graph that has no specific orientation in any edge and the other is a directed graph that has specific orientation.

Figure 2 graph theory simplifies complex human brain networks and also helps in analysis of neurological disorders. fMRI dataset are divided into a mesh of nodes which describe brain regions, and edges describe functional connectivity between the nodes [14]. Mathematically, a correlation matrix of graph theory represents connectivity between different nodes (brain regions). In graph theory, edges can be weighted or unweighted. Weighted edges show density, size or coherence network while unweighted applies threshold for the weighted edges. Additionally, these links represent effective, anatomical or functional connectivity [13]. Evaluation of fMRI data using graph theory helps in diagnosis of neuronal diseases such as AD, MS, MDD, OCD, ADHD.

Graph theory methods used in data processing: A brain network converts into corresponding graph metrics. Various toolboxes such as GraphVar, a Graph Theoretical Network Analysis Toolbox (GRETNA), Graph Analysis Toolbox (GAT), and many more extracts graph measures. Graph measures are explained below. Nodal degree in an undirected graph is the number of edges incident on the corresponding node. And in a directed graph has two different degrees of a particular node, In-degree represents numbers of the ingoing of edge to the node, Out-degree represents numbers of the outgoing of edge from the node [15]. A characteristic path length is an average number of edges in the shortest path between the given nodes in the graph. It can be calculated both globally and locally. Locally, it is the shortest path

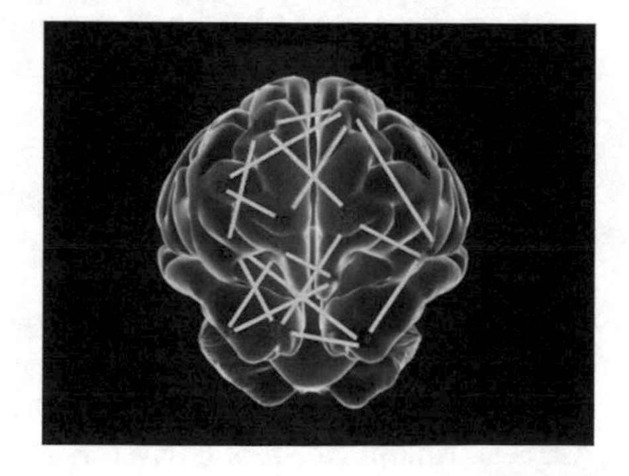

Fig. 2 Brain connectivity

length between a nodes to all other nodes in the network. Efficiency, the strength of the graph, is inversely proportional to the characteristic path length. This implies, shorter the characteristic path length higher the efficiency [16]. Characteristics path length calculated in Eq. (1).

$$l = \frac{1}{n(n-1)} \sum d(v_i, v_j) \tag{1}$$

where n is the number of vertices in the graph, V is the set of vertices and denote the shortest distance between v_i and v_j. The clustering coefficient indicates the degree of nodes that tends to form clusters together in a graph. Research suggests that in a real-world network or social network, nodes tend to create clusters by a relatively high density of bonds. The similarity between clusters in the network is lesser than the similarity within the cluster. There are two different types of clustering coefficients: local and global. The local clustering coefficient is the degree of nodes that tends to form a cluster with its neighborhood. While the global clustering coefficient is a ratio of the number of closed triplets to the number of all triplets including closed and open. The global clustering coefficient is an average of the local clustering coefficient [17]. Global clustering coefficient c is calculated by formula given in Eqs. (2) and (3)

$$c = \frac{\text{number of closed triplets}}{\text{number of all triplets(close and open)}} \tag{2}$$

$$c = \frac{3 * \text{number of triangles}}{\text{Number of all triplets}} \tag{3}$$

3.2 AI, ML, DL Methods for Classification of Disorders

AI in fMRI technology based on Machine Learning (ML), Deep Learning (DL), developing algorithms for achieving high efficiency, accuracy, results, and quality. ML is a pattern recognition technology that is applied in neuroimaging. ML first selects the technique for making predictions or diagnoses. Deep Learning is automatic feature learning with an end-to-end algorithm [1, 18]. AI system helps in managing models and analyzing them. It is the superset of Machine Learning and Deep Learning. AI focuses on success while ML focuses on accuracy and DL attains the highest accuracy for a large amount of dataset. Hence, DL is a subset of ML and ML is subset of AI.

For decades, many ML models have been used for classifying fMRI data. ML algorithms are classified as supervised learning, unsupervised learning, and reinforcement learning. Methods like Support Vector Machine (SVM), Logistic Regression (LR), Artificial Neural Network (ANN), K-nearest neighbor (KNN), Gradient Boosting

Decision Tree (GBDT), and many more are applicable in many papers. Support Vector Machine (SVM) is the most used model by the researcher as it results in more accuracy [19–22]. The research community uses a generalized pipeline for the classification of the disorder. This generalized pipeline is shown below.

Steps for diagnosis of diseases:

1. Selection of dataset
2. Preprocessing of dataset
3. Graph construction
4. Forming graph metrics
5. Statistical analysis
6. Classification

In Table 1, various fMRI datasets are described with features and objectives. ABIDE I and II dataset are most used in research work. ABIDE II is an advanced version of the ABIDE I dataset of 1114 participants with classifying Autism spectrum disorder (ASD). OASIS-3 is another open access fMRI dataset of 1098 participants classifying Alzheimer Disease (AD). SRPBS multi disorder connectivity dataset and 3 T kaggle open access dataset that classifies multi neurological disorders. Since the invention of fMRI, several toolboxes have been developed to analyze neuroimaging using fMRI. Table 2 some primarily toolboxes are briefly described with its objectives and feature. Mainly, neuro-imagining tools are python and MATLAB-based tools. With the help of Datasets and Toolboxes one can diagnose neurological disorder. Table 3 comprises several research papers using various methodologies with the accuracy of a given neurological disorder. Disorder covered in Table 3 are Autism spectrum disorder (ASD), multiple sclerosis (MS), major depression disorder (MDD), obsessive-compulsive disorder (OCD), Alzheimer's disorder, Parkinson's disorder, and Attention deficit/hyperactivity disorder (ADHD). In ASD, most papers have implemented SVM and KNN as classification methods with an average accuracy of 75%. There are few papers on the diagnosis of MS that uses the SVM classification method of accuracy 85%. MDD has less research on the diagnosis that uses T-test (Top-40) and SVM for classification with an accuracy of 92%. In OCD, researchers have used SVM as a classifier with an average accuracy of 75%. Research papers on the diagnosis of Alzheimer's disorder implemented SVM and Gaussian process logistic regression with an accuracy of 68%. Research papers corresponding to Parkinson's disorder use SVM and CNN method with an accuracy of 92.11%. There are several research papers on the diagnosis of ADHD with different methods. We reviewed some of them in our work, out of which, fully connected cascade ANN architecture achieves 90% accuracy.

We present a graph for number of publications per dataset and number of publications per neurological disorder in Figs. 3 and 4, respectively. Most of the research work is done in AS disorder with ABIDE I and II dataset. There are several research work in private dataset for MS, MDD, OCD etc.

Table 1 fMRI public access datasets

Dataset	Feature	Objective	Neurological disorder	References
ABIDE-I and II	Overall donating 1114 datasets from 521 individuals with ASD and 593 controls (age range: 5–64 years)	In regard to measures of core ASD and associated symptoms	Autism spectrum disorder (ASD)	[23]
OASIS-3	1098 participants including 605 cognitively normal adults and 493 individuals at various stages of cognitive decline ranging in age from 42 to 95 years	Hosted by central.xnat.org provide the community with open access to a significant database of neuroimaging and processed imaging data across a broad an easily accessible platform for use in neuroimaging, clinical, and cognitive research on normal aging and cognitive decline	Alzheimer disease	[24]
SRPBS multidisorder connectivity	Resting-state functional connectivity dataset of the patients from the 8 sites, 14 scanners	It is an open access dataset of multidisorder for the researchers	Multidisorder	[25]
3T kaggle	Here is only a subset of the entire dataset is available to minimize the size	Investigate how the brain processes information from faces and the representation of their subjective impression	Multidisorder	[26]

Table 2 Neuroimaging toolboxes

ToolBox	Objective	Feature	References
BCT	It is a brain connectivity toolbox for complex brain network analysis	There different BCT tools for MATLAB, python, and also in C++	[27]
MALINI	It is a MATLAB-based toolbox for analysis of neuroimaging and diseases classification using fMRI	18 different popular classifiers are presented. It works under SPM8	[28]
DPABI	It refers as data processing and analysis for brain image. Objective is to classify the neuronal diseases	Based on BCT, AAL, FSLnet, SPM, and many more that works on MATLAB	[29]
PyMVPA	It stands for multivariate pattern analysis (MVPA) in python. A python package to case statistical learning analyses of large datasets	It runs with many integrated software packages like scikit-learn, shogun, MDP, etc. in python	[30]
PRoNTo	Main objective is to recognize pattern for analysis of brain imaging. The tool is an interaction between machine learning and neuroimaging	Based on MATLAB and SPM compatible, hence it is suitable for both cognitive and clinical neuroscience research	[31]
CONN	CONN is used to analyze rs-fMRI and also cognitive task related. Objective is computation, display, and analysis of functional connectivity magnetic resonance imaging (fcMRI)	It is a MATLAB/SPM-based cross platform open-source software. User-friendly GUI	[32]
BRANT	BRAinNetome Toolkit (BRANT) integrates fMRI data preprocessing, voxel-wise spontaneous activity analysis, functional connectivity analysis, complex network analysis, statistical analysis, data visualization as well as several useful utilities	It is a MATLAB-based toolbox. BRANT are arranged into 7 modules, which are preprocessing, functional connectivity (FC), spontaneous activity (SPON), complex network analysis (NET), statistics (STAT), visualization (View), and utilities	[33]
AAL/AAL2/AAL3	Automated anatomical labeling developed that provided an alternative parcellation of the orbitofrontal cortex	The new atlas is available as a toolbox for SPM, and can be used with MRIcron	[34]

(continued)

Table 2 (continued)

ToolBox	Objective	Feature	References
Nipy	It is a neuroimaging in python (Nipy) Analysis of structural and functional neuroimaging data	It is a python-based tool including many libraries	[35]

4 Constraints and Future Scope

There is difficulty in integrating all reports of brain networks as different factors affect the experiment. For example, patient demographic factors, disease specific characteristics, sample size, and network construction varies frequently. Moreover, the computational approach does not always match the richness of fMRI data. Even though structural pathways have functional connectivity patterns, it cannot be claimed that there is exact match between topological properties in functional and structural organizations like in schizophrenia which may show opposite results over functional and structural organizations. The development of neurodegenerative disorder may not be fully understood and so its treatment can show poor performance.

The areas where researchers need to explore are in development-related issues which are still not answered by existing software. Furthermore in the future, longitudinal studies could be employed for keeping track of brain network topological changes using different therapeutic strategies across longer time durations.

5 Conclusion

Functional connectivity of the brain is the measure of the correlation between the parts of the brain. Graph theory helps in better understanding brain connectivity. Graph theory can be used to form graph metrics of connectivity patterns in the brain and are then classified by AI methods for proper understanding of neural connection. Several AI methods for the different neuronal disorders were reviewed in this paper. Classification models with higher accuracy classify the neurological disorder. As per the doctor's concern, early detection of such neuronal disabilities gives fast rehabilitation to patients. fMRI with AI and graph theory-based concepts provide this characteristic. The main focus of this paper is to serve comprehensive information regarding how AI and graph theory is feasible for detecting neurological disease using fMRI. We have reviewed a total of 26 research papers on different neuronal diseases and classified them based on the dataset used in a publication, the methodology used and its accuracy. We also presented a summary of these papers to provide a quick access to the field.

Table 3 Research papers and their methodologies with accuracy

Research papers	Dataset	Methodology	Accuracy
Autism spectrum disorder (ASD)			
[36]	ABIDE	*t*-test, support vector machine(SVM) and GARCH variance series	71.6% accuracy for male, 93.7% accuracy for female
[37]	ABIDE	Multi-layer CNNs	61.2% of accuracy
[38]	ABIDE	Support vector machine (SVM) and *K*-nearest neighbors (KNN)	85.9% of accuracy of KNN
[39]	ABIDE	Support vector machine (SVM), Logistic Regression (LR) and Ridge	71.4% of accuracy in SVM, 71.79% of accuracy in LR, 71.98% of accuracy in Ridge
[40]	ABIDE I dataset and Kaggle	K-Means clustering technique and classification using SVM.	81% of accuracy
[41]	ABIDE	Multi-site adaption framework via low-rank representation decomposition (maLRR) with SVM and KNN classifiers	71.88% of accuracy with SVM, 73.44% of accuracy with KNN
[42]	ABIDE	Used their own methodology	74.54% of accuracy
[43]	ABIDE	SVM (linear and Gaussian), KNN, LDA, ensemble tree	72.53% of accuracy of SVM linear
Multiple sclerosis (MS)			
[44]	Private	Support vector machine (SVM)	85% of accuracy
[45]	Private	Support vector machine (SVM)	68% of accuracy
Major depression disorder (MDD)			
[46]	Private	T test(Top-40), SVM	92% of accuracy
Obsessive-compulsive disorder (OCD)			
[47]	Private	Pearson's correlation and SVM classifier	80% of accuracy

(continued)

Table 3 (continued)

Research papers	Dataset	Methodology	Accuracy
[48]	Private	Support vector machine (SVM)	66% of accuracy
[49]	Private	Support vector machine (SVM)	74% of accuracy
Alzheimer's disorder			
[50]	ADNI database	Support vector machine (SVM)	62.32% of accuracy in fMRI
[51]	Private	Gaussian process logistic regression (GP-LR)	75% of accuracy
Parkinson's disorder			
[52]	PPMI open data	Support vector machine (SVM)	~95% of accuracy
[53]	InceptionV3, VGG16 and VGG19	Convolutional neural network (CNN)	89.5% in InceptionV3, 88.5% in VGG16 and 91.5% in VGG19
Attention deficit/hyperactivity disorder (ADHD)			
[54]	ADHD-200 global competition database NYU dataset and	Fully connected cascade (FCC) artificial neural network (ANN)	90% of accuracy
[55]	ADHD-200 Global Competitions, NeuroImage dataset	Deep belief networks (DBNs)	63.68% of accuracy in NYU, 69.83% of accuracy in NeuroImage
[56]	ADHD200 consortium dataset	Metaheuristic spatial transformation (MST)	72.10% of accuracy
[57]	ABIDE	Multi-gate mixture-of-experts (MMoE) is applied to the multi-task learning(MTL)	68.70% of accuracy
[58]	ADHD-200 dataset	Gaussian mixture model (GMM) for clustering and multi-network of long short term memory (multi-LSTM)	73.7% of accuracy
[59]	ADHD-200 dataset	Deep belief network(DBN)	44.63% of accuracy
[60]	Peking, KKI, NYU and NI dataset	gcForest method	64.87% of accuracy in Peking, 82.73% of accuracy in KKI, 73.17% of accuracy in NYU, 72% of accuracy in NI dataset
[61]	ADHD-200 dataset	Tensor-based approach	79.6% of accuracy

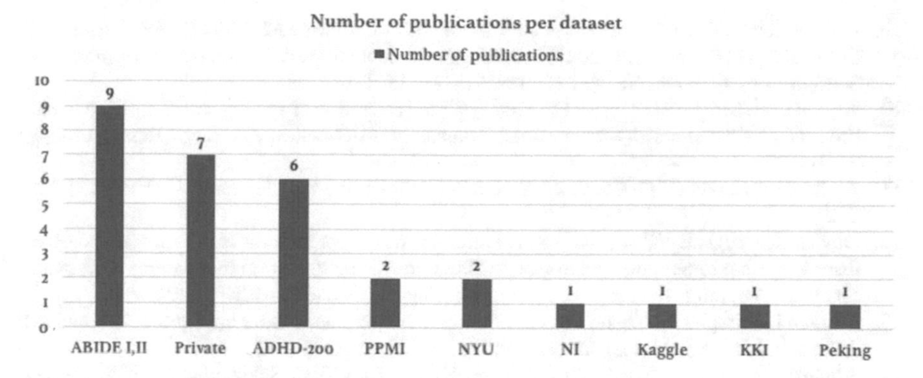

Fig. 3 Number of publications per dataset

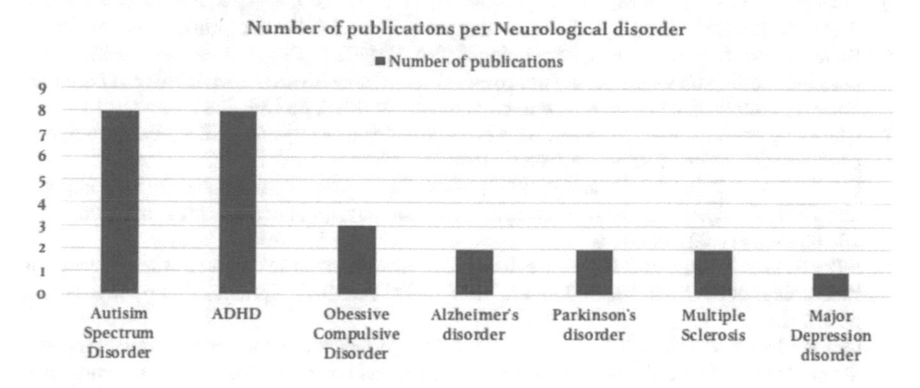

Fig. 4 Number of publications per neurological disorder

References

1. Farahani FV, Karwowski W, Lighthall NR (2019) Application of graph theory for identifying connectivity patterns in human brain networks: a systematic review. Front Neurosci 13:585
2. Sala-Llonch R, Bartrés-Faz D, Junqué C (2015) Reorganization of brain networks in aging: a review of functional connectivity studies. Front Psychol 6:663
3. Sporns O (2007) Brain connectivity. Scholarpedia 2(1):4695
4. Sporns O (2013) Structure and function of complex brain networks. Dialog Clinic Neurosci 15(3):247
5. National Institute of Neurological disorder and Stroke. https://www.nibib.nih.gov/science-education/science-topics/computed-tomography-ct
6. National Institute of Biomedical imaging and bio-engineering. https://www.nibib.nih.gov/science-education/science-topics/computed-tomography-ct
7. Stanford health care. https://stanfordhealthcare.org/medical-tests/p/pet-scan/what-to-expect.html#:~:text=conditions
8. Friston KJ (2011) Functional and effective connectivity: a review. Mary Ann Liebert, Inc. 140 Huguenot Street, 3rd Floor New Rochelle, NY 10801 USA 1(1):13–36

9. Oldrack RA, Baker CI, Durnez J, Gorgolewski KJ, Matthews PM, Munafò MR, Nichols TE, Poline J-B, Vul E, Yarkoni T (2017) Scanning the horizon: towards transparent and reproducible neuroimaging research. Nat Rev Neurosci 18(2):115–126
10. Ashworth E, Brooks SJ, Schiöth HB (2021) Neural activation of anxiety and depression in children and young people: a systematic meta-analysis of fMRI studies. Psychiatry Res: Neuroimag 311:111272
11. Amaro Jr E, Barker GJ (2006) Study design in fMRI: basic principles. Brain Cognit 60(3):220–232
12. Devika K, Oruganti VRM (2021) A machine learning approach for diagnosing neurological disorders using longitudinal resting-state fMRI. In: IEEE, 2021 11th international conference on cloud computing, data science and engineering (confluence), IEEE, pp 494–499
13. Vecchio F, Miraglia F, Rossini PM (2017) Connectome: graph theory application in functional brain network architecture. Clinical Neurophysiol Pract 2:206–213
14. Algunaid RF, Algumaei AH, Rushdi MA, Yassine IA (2018) Schizophrenic patient identification using graph-theoretic features of resting-state fMRI data. Biomed Sig Proc Control 43:289–299
15. Pavlopoulos GA, Secrier M, Moschopoulos CN, Soldatos TG, Kossida S, Aerts J, Schneider R, Bagos PG (2011) Using graph theory to analyze biological networks. BioData Min 4(1):1–27
16. Kazeminejad A, Golbabaei S, Soltanian-Zadeh H (2017) Graph theoretical metrics and machine learning for diagnosis of Parkinson's disease using RS-fMRI. In: IEEE, 2017 artificial intelligence and signal processing conference (AISP). IEEE, pp 134–139
17. Clustering coefficient—Wikipedia, The Free Encyclopedia https://en.wikipedia.org/w/index.php?title=Clustering_coefficient&oldid=1074548258
18. Oza P, Sharma P, Patel S, Kumar P (2022) Deep convolutional neural networks for computer-aided breast cancer diagnostic: a survey. Neural Comput Appl 34:1815–1836. https://doi.org/10.1007/s00521-021-06804-y
19. Pillai R, Oza P, Sharma P (2020) Review of machine learning techniques in health care. In: Proceedings of the ICRIC 2019, Jammu, India, 8–9 March 2019. Springer, Cham, Switzerland, pp 103–111
20. Oza P, Sharma P, Patel S (2021) Machine learning applications for computer-aided medical diagnostics. In: Proceedings of the second international conference on computing, communications, and cyber-security, Ghaziabad, India, 3–4 October. Springer, Singapore, pp 377–392
21. Oza P, Shah Y, Vegda MA (2021) Comprehensive study of mammogram classification techniques. In: Tracking and preventing diseases with artificial intelligence. Springer, Berlin/Heidelberg, Germany, pp 217–238
22. Oza P, Sharma P, Patel S, Bruno A (2021) A bottom-up review of image analysis methods for suspicious region detection in mammograms. J. Imag 7(9)
23. Autism brain imaging data exchange II ABIDE II. ABIDE. (n.d.). http://fcon_1000.projects.nitrc.org/indi/abide/abide_II.html
24. Oasis Brains. OASIS Brains—Open access series of imaging studies (n.d.). https://www.oasis-brains.org/
25. Sage Bionetworks, info@sagebase.org (n.d.). Sage bionetworks. Synapse. https://www.synapse.org/#!Synapse:syn22317078
26. Aché M (2021) 3T fmri dataset. Kaggle. https://www.kaggle.com/mathurinache/3t-fmri-datase
27. Brain Connectivity Toolbox. Google Sites. https://sites.google.com/site/bctnet/
28. https://github.com/pradlanka/malini/blob/master/readme_toolbox.docx
29. Chao-Gan Y (2022) DPABI: a toolbox for data processing and analysis for brain imaging. The R-fMRI Network. http://rfmri.org/dpabi
30. News¶. News—PyMVPA 2.6.5.dev1 documentation. http://www.pymvpa.org/
31. Pattern Recognition for Neuroimaging Toolbox (Pronto). MLNL. http://www.mlnl.cs.ucl.ac.uk/pronto/
32. Conn Toolbox. toolbox. https://web.conn-toolbox.org
33. Welcome to Brant!¶. Welcome to Brant!—BRANT 3.36 documentation. http://brant.brainnetome.org/en/latest/

34. AAL/AAL2/AAL3. Neurofunctional Imaging Group GINIMN. http://www.gin.cnrs.fr/AAL
35. Nipy. "Nipy.org." nipy.org. https://nipy.org/
36. Sartipi S, Shayesteh MG, Kalbkhani H (2018) Diagnosing of autism spectrum disorder based on GARCH variance series for rs-fMRI data. 2018 9th International symposium on telecommunications (IST). IEEE
37. Kiruthigha M, Jaganathan S (2021) Graph convolutional model to diagnose autism spectrum disorder using Rs-Fmri data. 2021 5th International conference on computer, communication and signal processing (ICCCSP). IEEE
38. Al-Hiyali MI et al (2021) Classification of BOLD FMRI signals using wavelet transform and transfer learning for detection of autism spectrum disorder. 2020 IEEE-EMBS conference on biomedical engineering and sciences (IECBES). IEEE
39. Yang X, Islam MS, Khaled AA (2019) Functional connectivity magnetic resonance imaging classification of autism spectrum disorder using the multisite ABIDE dataset. 2019 IEEE EMBS international conference on biomedical & health informatics (BHI). IEEE
40. Ahammed MS et al (2021) Bag-of-features model for ASD fMRI classification using SVM. 2021 Asia-pacific conference on communications technology and computer science (ACCTCS). IEEE
41. Wang M et al (2019) Identifying autism spectrum disorder with multi-site fMRI via low-rank domain adaptation. IEEE Trans Med Imaging 39(3):644–655
42. Byeon K et al (2020) Artificial neural network inspired by neuroimaging connectivity: application in autism spectrum disorder. 2020 IEEE International conference on big data and smart computing (BigComp). IEEE
43. Karampasi A et al (2020) A machine learning fMRI approach in the diagnosis of autism. 2020 IEEE international conference on big data (Big Data). IEEE
44. Ashtiani SN, Behnam H, Daliri MR (2021) Diagnosis of multiple sclerosis using graph-theoretic measures of cognitive-task-based functional connectivity networks. IEEE Trans Cognit Dev Sys 14(3):926–934
45. Van Schependom J et al (2014) SVM aided detection of cognitive impairment in MS. 2014 International workshop on pattern recognition in neuroimaging. IEEE
46. Mousavian M, Chen J, Greening S (2020) Depression detection using atlas from fMRI images. 2020 19th IEEE international conference on machine learning and applications (ICMLA). IEEE
47. Sen Bhaskar et al (2016) Classification of obsessive-compulsive disorder from resting-state fMRI. 2016 38th annual international conference of the ieee engineering in medicine and biology society (EMBC). IEEE
48. Shenas SK, Halici U, Cicek M (2013) Detection of obsessive compulsive disorder using resting-state functional connectivity data. 2013 6th International conference on biomedical engineering and informatics. IEEE
49. Shenas SK, Halici U, Çiçek M (2014) A comparative analysis of functional connectivity data in resting and task-related conditions of the brain for disease signature of OCD. 2014 36th Annual international conference of the ieee engineering in medicine and biology society. IEEE
50. Dachena C et al (2020) Application of MRI, fMRI and cognitive data for Alzheimer's disease detection. 2020 14th European conference on antennas and propagation (EuCAP). IEEE
51. Challis E et al (2015) Gaussian process classification of Alzheimer's disease and mild cognitive impairment from resting-state fMRI. NeuroImage 112:232–243
52. Kazeminejad A, Golbabaei S, Soltanian-Zadeh H (2017) Graph theoretical metrics and machine learning for diagnosis of Parkinson's disease using rs-fMRI. 2017 Artificial intelligence and signal processing conference (AISP). IEEE
53. Sajeeb A et al (2020) Parkinson's disease detection using FMRI images leveraging transfer learning on convolutional neural network. 2020 International conference on machine learning and cybernetics (ICMLC). IEEE
54. Deshpande G et al (2015) Fully connected cascade artificial neural network architecture for attention deficit hyperactivity disorder classification from functional magnetic resonance imaging data. IEEE Trans Cybernet 45(12):2668–2679

55. Farzi S, Kianian S, Rastkhadive I (2017) Diagnosis of attention deficit hyperactivity disorder using deep belief network based on greedy approach. 2017 5th International symposium on computational and business intelligence (ISCBI). IEEE
56. Aradhya AM, Sundaram S, Pratama M (2020) Metaheuristic spatial transformation (MST) for accurate detection of attention deficit hyperactivity disorder (ADHD) using rs-fMRI. 2020 42nd Annual international conference of the ieee engineering in medicine & biology society (EMBC). IEEE
57. Huang ZA, Liu R, Tan KC (2020) Multi-Task learning for efficient diagnosis of ASD and ADHD using Resting-State fMRI data. 2020 International joint conference on neural networks (IJCNN). IEEE
58. Liu R et al (2020) Multi-LSTM networks for accurate classification of attention deficit hyperactivity disorder from resting-state fMRI data. 2020 2nd International conference on industrial artificial intelligence (IAI). IEEE
59. Kuang D, He L (2014) Classification on ADHD with deep learning. 2014 International conference on cloud computing and big data. IEEE
60. Shao L et al (2019) Deep forest in ADHD data classification. IEEE Access 7:137913–137919
61. Li J, Joshi AA, Leahy RM (2020) A network-based approach to study of ADHD using tensor decomposition of resting state fMRI data. 2020 IEEE 17th International symposium on biomedical imaging (ISBI). IEEE

Performance Analysis of Document Similarity-Based DBSCAN and *K*-Means Clustering on Text Datasets

Preeti Kathiria, Vandan Pandya, Harshal Arolkar, and Usha Patel

Abstract The clustering of documents based on their similarity is prolific for an application that wants to extract similar documents and disparate non-similar documents to reduce ambiguity in finding relevant documents. Therefore, we require a robust clustering algorithm that can cluster document efficiently and effectively. In this paper, the performance of two clustering algorithms *K*-Means and DBSCAN with optimal parameters are compared on various textual datasets with distance measures—cosine and hybrid similarity. The challenge of finding the optimal value of epsilon in DBSCAN and value of *K* in *K*-Means is fulfilled by the DMDBSCAN and within-cluster sum of square algorithms, respectively. Hybrid similarity has an impact of single words and phrases, so the shared phrases across the corpus are drawn and the phrase similarity is computed. Then, hybrid similarity is formed using cosine and Phrase. To catch the shared phrases, the document representation model—Document Index Graph—is implemented in the Neo4j graph database. We utilized silhouette score to evaluate the performance of clustering algorithms. Experimental results reflect that DBSCAN performs better than *K*-Means on both the similarity measures.

Keywords Document clustering · Document similarity · DBSCAN · *K*-Means · Phrase similarity · Hybrid similarity

P. Kathiria · V. Pandya · U. Patel (✉)
Institute of Technology, Nirma University, Sarkhej—Gandhinagar Hwy, Gota, Ahmedabad, India
e-mail: ushapatel@nirmauni.ac.in

P. Kathiria
e-mail: preeti.kathiria@nirmauni.ac.in

H. Arolkar
Faculty of Computer Technology, GLS University, Ahmedabad, India
e-mail: harshal.arolkar@glsuniversity.ac.in

Y. Singh et al. (eds.), *Proceedings of International Conference on Recent Innovations in Computing*, Lecture Notes in Electrical Engineering 1001,
https://doi.org/10.1007/978-981-19-9876-8_5

1 Introduction

Data clustering is a method to group or club the data of similar characteristics into a single cluster. Clustering aims to increase intra-cluster similarity and decrease inter-cluster similarity [1]. Many applications of clustering are used in the industry, grouping customers that have similar purchasing behavior into one cluster so that they can be targeted easily with personalized content [2]. In networking, it is used to classify whether traffic is spam or coming from bots. In Information Retrieval, similar documents are grouped so that they can be easily and rapidly searched and filtered [3]. It helps to label unseen documents with categories. Documents can be efficiently organized which helps to extract topics. Clustering is an unsupervised machine learning technique because the model is fed with unlabeled data and automatically divides data into clusters. A lot of problems are faced during the process of clustering documents such as selection of pertinent features of documents, a suitable similarity measure, algorithm selection, finding optimal values for parameters in the algorithm, handling outliers, interpretation of outcomes, evaluation metrics, and many others [4, 5]. Before performing the process of clustering documents, it needs to find similarity measures between different documents which helps to identify how much each document is similar to other documents in the corpus. Similarities of documents can help in finding duplicate documents, matching job descriptions with the CV of employees, to match new patent applications with existing patent applications in the repository. A ton of documents are available on the web. However, sometimes it gets difficult to identify the required documents. This problem is overcome by using clustering techniques that assemble the documents into different clusters based on their similar features and helps in finding relevant documents easily. Many clustering techniques are available, categorized as density-based, partition-based, and Hierarchical Clustering.

This paper mainly focuses on a comparison of the implementation of clustering techniques possessing characteristics such as no predefined determination of some clusters and forms arbitrary shaped clusters (DBSCAN) versus predefined determination of clusters and structures circular-shaped clusters (*K*-Means) on text datasets with different similarity measures such as cosine similarity and proposed hybrid similarity [1, 6, 7]. It also includes a comparative analysis of clusters formed with different similarity measures on these two techniques.

The rest of the paper is structured as follows: The related work, which reviews various clustering techniques related to text mining and its comparative analysis, is discussed in Sect. 2. The experimental workflow, along with the dataset description, is given in Sect. 3. Section 4 discusses the results of the performance of the used clustering techniques on two similarity measures. Section 5 closes the paper with a conclusion and future scope.

2 Related Work

To cluster text documents, there are various proposed clustering algorithms, which use different similarity measuring techniques. Momin et al. has offered a Document Index Graph-based clustering algorithm [8] for clustering based on phrase-based similarity. It allows the inclusion of new documents without affecting existing documents in the cluster. Jin and Bai have introduced a clustering algorithm to cluster text, based on the most common subgraph of text feature. It reduces the high dimension of the text vector and the complexity of the algorithm [9]. Beil et al. have proposed an approach of hierarchical text clustering based on frequent terms [10], which can be useful in searching relevant web pages based on keywords specified by the user. Huan et al. have proposed a *K*-Means clustering algorithm based on KL divergence on text, which improves clustering results on the large dataset and takes minimal time [11]. In 2016, Narayan and Vasumathi have proposed in their paper, a Possibilistic Fuzzy C-Means Algorithm (PFCM) to cluster relevant text features and improve the method by constructing a minimum spanning tree [12]. In 2016, Kathiria and Ahluwalia mainly focused on creating ontology using cosine similarity as an initial step. First, most similar documents were obtained using cosine similarity, then five high-frequency keywords were extracted after applying text summarization to most similar documents, and finally, the ontology was created using those words [13]. Zamir and grouper; Zamir et al. and Zamir and Etzioni have presented a phrase-based document clustering approach based on Suffix tree clustering (STC). The proposed technique uses a compact tree structure to show common suffixes among documents. It identifies the base clusters of documents based on recognized common suffixes. After identifying base clusters, it merges base clusters from a number of similar documents [14–16]. Kang has proposed a new clustering method using the keyword weighting approach [17]. Here, clusters are expanded by keyword relationships. The formation of clusters stops when no more documents are added to the cluster, and irrelevant documents are removed, consisting of no keywords matching with the keywords set of clusters. Kathiria and Arolkar performed a comparative analysis of various document representation models to find phrase-based similarity in 2019 [18]. The comparative analysis of N-gram, Suffix tree, and Document Index Graph (DIG) models is done considering the different factors such as space and time complexity to store and gauge a text and the capability to recognize the dynamic length of phrase and data structure used for storing text. The DIG model generates a graph and matches and identifies phrases in almost linear time. Based on space, time complexity, and dynamic length phrase identification, the DIG model can discover shared phrases effectively. Based on this conclusion, the 2020 DIG model was implemented by the same authors [7] to find shared phrases, and finally, the documents are clustered using the phrase-based and single-term similarity. To implement the DIG, the best suitable graph database is explored by them. For data storage, Neo4j has the support of external memory, main memory, and backend storage. To enhance the performance of the data fetching, it supports the Indexing concept. In Neo4j, nodes and edges are labeled and can have attributes, and edges are directed. ACID properties—Atomicity, Consistency, Isolation, and Durability—are also supported by Neo4j. So, Neo4j is used to implement the DIG model. Based on the shared phrase, the phrase similarity is calculated. From the phrase similarity and single-term similarity, the hybrid similarity is derived. DBSCAN Clustering along with optimal parameters has been applied to measure the performance hybrid similarity measure. From the results of the silhouette score of the clusters, it has been depicted that the hybrid measure has better clustering accuracy than the single-term measure clustering.

In 2015, Mishra et al. proposed a novel clustering method using on inter-passage which groups segments of documents based on similarities. In this, segment score will be calculated for each segment, and based upon segment score, *K*-Means is used to perform inter-document clustering [19]. Patil and Baidari [20] have proposed a novel technique called DeD (depth difference) to approximate the optimal value of clusters by calculating the depth of each point within cluster using the Mahalanobis depth function. As the value of depth difference is maximized, an optimal number of clusters is obtained. In [21], authors have performed a comparative analysis of performances of *K*-Means within Initial Centroid Selection Optimization (ICSO), Genetic, and Chi-square similarity measure on transcribed broadcast news documents. The results reflect that *K*-Means with ICSO and genetic algorithm outperformed the others. Niyaz and Krwan [22] compared the performance of two algorithms *K*-Means and wards with three different scenarios without preprocessing, preprocessing with stemming, and preprocessing without stemming and concluded that wards outperform *K*-Means for all datasets.

Uupyt et al. [23] have proposed a methodology to cluster customers based on similarities of the usage of electricity. They have used *K*-Means for clustering and adequacy measures to select the optimal value of the parameter. Cluster validity indexes are applied to choose the proper value of the number of clusters. Table 1 gives a comparison of different clustering methods existing in the literature based on different parameters. Various clustering techniques are reviewed based on parameters like the capability to handle outliers, applicability for a large dataset, time complexity, advantages, and limitations of the algorithm. The notations used for time complexity are n, k, s, and t which represent the following:

- n stands for a number of data points.
- k stands for a number of clusters.
- s stands for a number of sample data points.
- t stands for a number of iterations.

In 2018, Ogbuabor and Ugwoke performed a comparative analysis of *K*-Means and DBSCAN clustering techniques on healthcare data with evaluation metrics such as silhouette score [29]. No optimal number of clusters or value of epsilon has been determined, so in our paper DMDBSCAN method proposed by Elbatta [30] to determine optimal value of epsilon for DBSCAN and WCSS method proposed by Kodinariya and Makwana [31] to determine optimum value for number of clusters (k) for *K*-Means is opted. Using optimal derived values for both the algorithms, the

Table 1 Comparative Study and Analysis of clustering algorithms

Clustering methods	Clustering algorithm	Outliers handling	Suitable for large dataset	Time complexity	Advantages	Disadvantages
Density-based clustering [24–28]	DBSCAN	Yes	Yes	O(nlogn)	– Outliers easily identified and handled – It can handle clusters of different shapes effectively	– Sensitive to parameters and requires large no. of parameters – Difficult to make clusters if points are scattered
	OPTICS	Yes	Yes	O(nlogn)		
	Incremental DBSCAN	Yes	Yes	O(nlogn)		
Partition-based clustering [24–28]	*K*-Means	No	Yes	O(knt)	– Easy to implement – Produces tighter clusters	– Difficult to identify outliers – Clusters should be predetermined and difficult to choose optimal number for cluster
	PAM	Yes	Yes	$O(k(n-k)^2)$		
	CLARA	Yes	Yes	$O(ks^2 + k(n-k))$		
Hierarchical Clustering [24–28]	Agglomerative	Yes	No	$O(n^2 \text{Log}(n))$	– Dendograms help in understanding data – No need to specify clusters	– It is computationally expensive – Produces clusters of same variance
	BIRCH	Yes	Yes	$O(n)$		

results of DBSCAN and *K*-Means clustering are compared. Based on the comparative analysis, two algorithms, the DBSCAN—density-based algorithm and *K*-Means partition-based algorithm, are identified for the experimental purpose.

3 Experimental Work

The clustering algorithms DBSCAN and *K*-Means with optimal parameter values are used on various text datasets, and text datasets are clustered using two similarity measures—cosine and hybrid similarity. To find out the optimal value of eps-radius in DBSCAN, the DMDBSCAN algorithm is implemented, and to determine the optimum value of clusters (k) in *K*-Means, the WCSS algorithm is implemented. The performance of both the algorithms is equated using the assessment measure silhouette score. The workflow of the experiment is depicted in Fig. 1, followed by a detailed explanation of each stage.

3.1 Dataset Creation and Selection

For the experimental work, six datasets have been used as per Table 2. One public dataset "20 Newsgroup (NG)" available on scikit-learn with five other custom datasets was used on the abstract of research papers using web scraping from the Scopus database. Scopus database consists of categories such as artificial intelligence (AI), image processing (IP), machine learning (ML), wireless sensor networks (WSN), cryptography (CRYPTO), and health (HL). For the experiment purpose, different Scopus dataset categories are combined and used as depicted in Table 2.

3.2 Preprocessing

Preprocessing of data includes removing stop words, common words, and lemmatization of terms, which helps condense the dimensionality of the dataset.

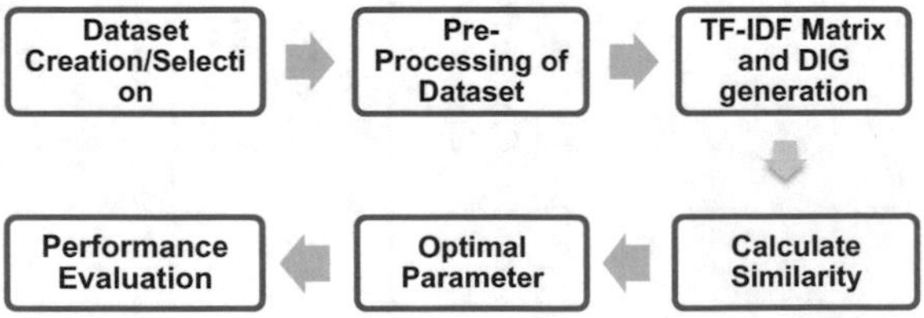

Fig. 1 Workflow of an experiment

Table 2 Dataset description

Dataset name	Categories	No. of documents used for experiment
20 NG	Politics, sports, etc.	18,000
Dataset 1	WSN, ML, IP	547
Dataset 2	WSN, AI, IP	1206
Dataset 3	WSN, CRYPTO, AI	1214
Dataset 4	WSN, HL, ML	405
Dataset 5	WSN, ML, IP	1121

3.3 TF-IDF Matrix Generation and Document Index Graph Generation

After preprocessing, the term frequency-inverse document frequency (TF-IDF) matrix is produced from the dataset. Each document is inserted into a graph called Document Index Graph—DIG [6] for finding shared phrases. DIG model adopts the graph structure to collect words as nodes and adjacency of the words as edges. Each document is tokenized into sentences, and in turn, it is chunked into words. The tokenized words of each sentence are inserted into the graph incrementally, and simultaneously, shared phrases are extracted from the connected edges of the graph's vertices. The edges between nodes store the information regarding the consecutive words belonging to which document, at which sentence, and at which word place. The graph is in a highly connected form. So, for efficient storage and retrieval, the Neo4j graph database is used [32, 33].

3.4 Calculating Similarity Between Documents

Cosine similarity and hybrid similarity are two measures used to calculate similarities between documents [6, 7, 34]. Cosine similarity is a single-term similarity. TF-IDF matrix is used to form cosine similarity matrix as per Eq. 1 for all the datasets, where dot product of documents $d1$ and $d2$ is divided by the product of document length. Phrase represents the proximity of the words, which helps to find syntactic similarity more preciously. The phrase-based similarity is calculated as per Eq. 2 from shared phrases retrieved for all the datasets.

$$\mathrm{Sim}_{\mathrm{df}}(d1, d2) = \frac{(d1 * d2)}{||d1||\cdot||d2||} \tag{1}$$

$$\mathrm{Sim}_{\mathrm{sp}}(d1, d2) = \frac{\sqrt{\sum_{i=1}^{P}\left(\frac{l_i}{\mathrm{avg}(|s_i|)}\right)\cdot(f_{1i} + f_{2i})^2}}{\sum_j |s_{1j}| + \sum_k |s_{2k}|} \tag{2}$$

In Eq. (2), p is the number of shared phrases, l_i is the length of a shared phrase, f_{1i} and f_{2i} are frequency of phrase in each document, and avg(|s$_i$|) is the average length of sentences holding shared phrase i. Hybrid similarity uses a combined form of cosine and phrase-based similarity and is calculated as per Eq. 3 for all the datasets.

$$\mathrm{SIM_{hs}}(d1, d2) = \alpha * \mathrm{sim_{sp}}(d1, d2) + (1 - \alpha) * \mathrm{sim_{df}}(d1, d2) \tag{3}$$

In Eq. (3), α represents similarity blend factor which lies between (0,1), for phrase similarity α is considered as 0.7 because it has more significance and 0.3 for cosine similarity. Here, $\mathrm{sim_{sp}}$ represents phrase-based similarity, and $\mathrm{sim_{df}}$ represents cosine similarity.

3.5 Optimal Parameter Calculation for Clustering

K-Means and DBSCAN are two clustering algorithms used for comparison. For *K*-Means, WCSS (within-cluster sum of squares), and for DBSCAN, DMDBSCAN will be used to calculate the ideal value of parameter [30].

DMDBSCAN: Technique for calculating optimal parameters for DBSCAN

DMDBSCAN proposed and implemented by Elbatta, Rahmah, and Sitanggang [30, 35] is utilized to determine the optimal parameter epsilon of DBSCAN. It is known that density gets distributed from this optimal value of epsilon [36]. Steps to find optimal epsilon value:

- Calculate the distance of each point to its 2nd nearest neighbor for the whole dataset.
- Arrange all points on the x-axis with respect to their distance on the y-axis in ascending order and plot them on the graph.
- Wherever the line gets the closest curve, it is to be decided as the optimal value of epsilon.

Based on this algorithm, the optimal values of eps for each dataset for DBSCAN are calculated. Furthermore, for the selection of optimal minimum number samples, we iterated from 1 to 100 using optimal epsilon value and considered highest silhouette score. Consequently, we obtained optimal epsilon and minimum number of samples.

WCSS (within-cluster sum of squares): Technique for calculating optimal parameters for *K*-Means

The WCSS method proposed by Kodinariya and Makwana [31] is used to obtain the optimal value of *K* (number of clusters). WCSS is calculated using Eq. 4, where *C* is cluster centroids, and *d* is a point in each cluster. Then, summation is performed of squares of all distances from the centroid to the nearest points for each cluster set. Steps to find the optimal value of *K* are as follows:

- WCSS is the sum of squares distances of each point in all clusters to their respective centroids, which is calculated as per Eq. 4.
- Plot graph of WCSS for each value K (number of clusters).
- So, wherever the line gets elbow shape curve, it is decided as optimal no. of a cluster. If the line does not form an elbow, consider the maximum silhouette score of different number clusters at the slope.

$$\text{WCSS} = \sum_{C_k}^{C_n} \left(\sum_{d_i \text{ in } C_i}^{d_m} \text{distance}(d_i, C_k)^2 \right) \tag{4}$$

Based on this algorithm, the optimal value of clusters for each dataset for K-Means is calculated and used for clustering.

3.6 *Clustering*

After obtaining optimal values of parameters for K-Means and DBSCAN algorithm, both the optimal values are used in the algorithms for making clusters, and obtained results are shown as per the silhouette score in Tables 3 and 4. All the datasets are compared for both the algorithms using cosine similarity and hybrid similarity as distance measures.

Table 3 Comparative results of DBSCAN and K-Means as per its optimal parameters with respect to cosine similarity

Cosine similarity	DBSCAN				K-Means	
Name of documents	Optimal epsilon	Minimum no. of samples	No. of clusters	Silhouette score	Optimal No. of clusters	Silhouette score
20 NG	0.61	4	3	0.253	2	0.09
Dataset1	0.64	3	3	0.243	3	0.173
Dataset2	0.58	8	2	0.311	2	0.242
Dataset3	0.61	13	2	0.281	2	0.220
Dataset4	0.62	4	2	0.131	2	0.135
Dataset5	0.57	7	2	0.309	2	0.233

Table 4 Comparative results of DBSCAN and *K*-Means as per its optimal parameters with respect to hybrid similarity

Hybrid similarity	DBSCAN				*K*-Means	
Name of documents	Optimal epsilon	Minimum no. of sample	No. of clusters	Silhouette score	Optimal No. of clusters	Silhouette score
20 NG	0.61	4	3	0.307	2	0.087
Dataset1	0.65	16	3	0.582	3	0.581
Dataset2	0.64	31	3	0.635	3	0.632
Dataset3	0.65	45	3	0.643	3	0.640
Dataset4	0.64	23	2	0.460	2	0.459
Dataset5	0.64	47	2	0.601	2	0.596

4 Results and Discussion

Results obtained using both clustering algorithms have been evaluated using the silhouette score. Silhouette score is an evaluation measure indicating how well a particular document is falling into the right cluster compared to the documents in other clusters. Silhouette score is evaluated in a range of (−1, 1). So, if the silhouette score is 1 then documents of one cluster are far away from another, and documents are well clustered. If the silhouette score is −1, then documents of one cluster are near or mixed with another cluster; the documents are not well clustered. Silhouette score is calculated as per Eq. 5. $a(i)$ is the mean distance of ith document from entirely other documents in the similar cluster, and $b(i)$ is the smallest mean distance of ith document from entirely other documents in the adjacent cluster.

$$s(i) = (b(i) - a(i))/\max\{a(i), b(i)\} \quad (5)$$

Figure 3 shows the performance evaluation between DBSCAN and *K*-Means based on the silhouette score for cosine similarity. As observable, DBSCAN is outperforming *K*-Means most of the time with the only exception in dataset 5. Moreover, Fig. 2 depicts the performance evaluation for hybrid similarity. Here, it is observable that in "20 Newsgroups" DBSCAN is superior with 0.307 silhouette score, which is far better than *K*-Mean score, i.e., 0.087. Also, in custom datasets, it is still on par with *K*-Means.

5 Conclusion and Future Work

In hybrid similarity measure for the standard dataset, DBSCAN shows its supremacy and for custom datasets DBSCAN performs on par with *K*-Means. However, in cosine

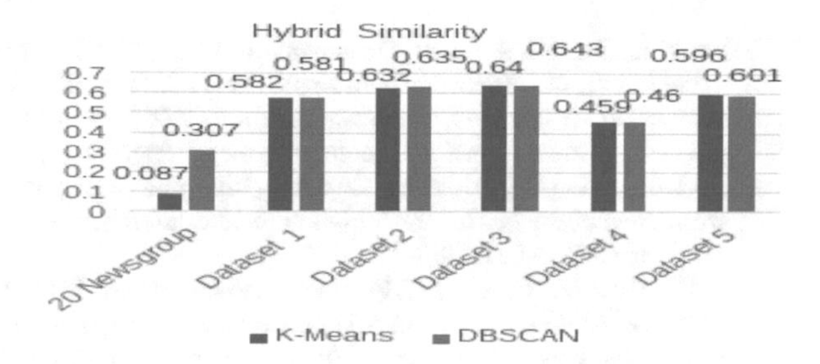

Fig. 2 Comparative analysis of DBSCAN and K-Means for hybrid similarity using silhouette score

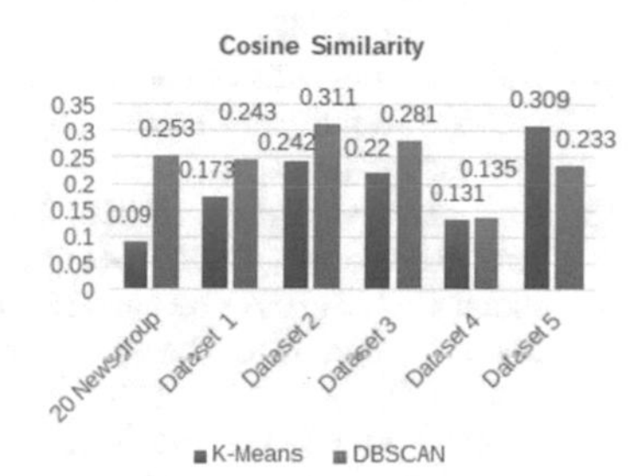

Fig. 3 Comparative analysis of DBSCAN and K-Means for cosine similarity using silhouette score

similarity DBSCAN always outperformed K-Means. So, DBSCAN is more beneficial for the applications that have more importance of single terms for measuring similarity to cluster text data. Additionally, this paper presents a comparative analysis of different clustering algorithms holding different characteristics with their respective advantages and limitations. As a part of future work, the other clustering algorithms can be explored, applied on the same datasets and similarity measures to find their performance. The best-performed clustering technique can be used to generate the clusters of the document and to develop a Recommender system.

References

1. Nagpal A, Jatain A, Gaur D (2013) Review based on data clustering algorithms. In: 2013 IEEE conference on information and communication technologies. IEEE
2. Yu W, Qiang G, Xiao-Li L (2006) A kernel aggregate clustering approach for mixed data set and its application in customer segmentation. In: 2006 international conference on management science and engineering. IEEE
3. Bakr AM, Yousri NA, Ismail MA (2013) Efficient incremental phrase-based document clustering. In: Proceedings of the 21st international conference on pattern recognition (ICPR2012). IEEE
4. Böhm C et al (2009) CoCo: coding cost for parameter-free outlier detection. In: Proceedings of the 15th ACM SIGKDD international conference on knowledge discovery and data mining
5. Cha S-H (2007) Comprehensive survey on distance/similarity measures between probability density functions. City 1(2):1
6. Hammouda KM, Kamel MS (2004) Efficient phrase-based document indexing for web document clustering. IEEE Trans Knowl Data Eng 16(10):1279–1296

7. Preeti K, Harshal A (2020) Document clustering based on phrase and single term similarity using Neo4j. Int J Innov Technol Explor Eng (IJITEE) ISSN 9.3 3188-3192
8. Momin BF, Kulkarni PJ, Chaudhari A (2006) Web document clustering using document index graph. In: 2006 international conference on advanced computing and communications. IEEE
9. Jin C-X, Bai Q-C (2016) Text clustering algorithm based on the graph structures of semantic word co-occurrence. In: 2016 international conference on information system and artificial intelligence (ISAI). IEEE
10. Beil F, Ester M, Xu X (2002) Frequent term-based text clustering. In: Proceedings of the eighth ACM SIGKDD international conference on knowledge discovery and data mining
11. Huan Z, Pengzhou Z, Zeyang G (2018) *K*-means text dynamic clustering algorithm based on KL divergence. In: 2018 IEEE/ACIS 17th international conference on computer and information science (ICIS). IEEE
12. Narayana GS, Vasumathi D (2016) Clustering for high dimensional categorical data based on text similarity. In: Proceedings of the 2nd international conference on communication and information processing
13. Kathiria P, Ahluwalia S (2016) A Naive method for ontology construction. Int J Soft Comput Artif Intell Appl (IJSCAI) 5(1):53–62
14. Zamir O et al (1997) Fast and intuitive clustering of Web documents. KDD 97
15. Zamir O, Etzioni O (1999) Grouper: a dynamic clustering interface to Web search results. Comput Netw 31(11–16):1361–1374
16. Zamir O, Etzioni O (1998) Web document clustering: a feasibility demonstration. In: Proceedings of the 21st annual international ACM SIGIR conference on research and development in information retrieval
17. Kang S-S (2003) Keyword-based document clustering. In: Proceedings of the sixth international workshop on information retrieval with Asian languages
18. Kathiria P, Arolkar H (2019) Study of different document representation models for finding phrase-based similarity. In: Information and communication technology for intelligent systems. Springer, Singapore, pp 455–464
19. Mishra RK, Saini K, Bagri S (2015) Text document clustering on the basis of inter passage approach by using *K*-means. In: International conference on computing, communication and automation. IEEE
20. Patil C, Baidari I (2019) Estimating the optimal number of clusters k in a dataset using data depth. Data Sci Eng 4(2):132–140
21. Maghawry A, Omar YMK, Badr A (2020) Self-organizing map vs initial centroid selection optimization to enhance k-means with genetic algorithm to cluster transcribed broadcast news documents. Int Arab J Inf Technol 17(3):316–324
22. Salih NM, Jacksi K (2020) Semantic document clustering using k-means algorithm and ward's method. In: 2020 international conference on advanced science and engineering (ICOASE). IEEE
23. Užupytė R, Babarskis T, Krilavičius T (2018) The generation of electricity load profiles using k–means clustering algorithm. J Univer Comput Sci. Graz: Graz Univer Technol 24(9)
24. Baser P, Saini JR (2013) A comparative analysis of various clustering techniques used for very large datasets. Int J Comput Sci Commun Netw 3(5):271
25. Gupta MK, Chandra P (2019) A comparative study of clustering algorithms. In: 2019 6th international conference on computing for sustainable global development (INDIACom). IEEE
26. Popat SK, Emmanuel M (2014) Review and comparative study of clustering techniques. Int J Comput Sci Inf Technol 5(1):805–812
27. Rama B, Jayashree P, Jiwani S (2010) A survey on clustering current status and challenging issues. Int J Comput Sci Eng 2(9):2976–2980
28. Xu D, Tian Y (2015) A comprehensive survey of clustering algorithms. Annals Data Sci 2(2):165–193
29. Ogbuabor G, Ugwoke FN (2018) Clustering algorithm for a healthcare dataset using silhouette score value. AIRCC's Int J Comput Sci Inf Technol 10(2):27–37
30. Elbatta MNT (2012) An improvement for DBSCAN algorithm for best results in varied densities

31. Kodinariya TM, Makwana PR (2013) Review on determining number of cluster in *K*-means clustering. Int J 1(6):90–95
32. George S, Sudheep Elayidom M, Santhanakrishnan T (2017) A novel sequence graph representation for searching and retrieving sequences of long text in the domain of information retrieval. Int J Sci Res Comput Sci, Eng Inf Technol 2(5)
33. Chandrababu S, Bastola DR (2018) Comparative analysis of graph and relational databases using herbmicrobeDB. In: 2018 IEEE international conference on healthcare informatics workshop (ICHI-W). IEEE
34. Hoang N, Anh T, Hoang K (2009) Efficient approach for incremental Vietnamese document clustering. In: Proceedings of the eleventh international workshop on Web information and data management
35. Rahmah N, Sitanggang IS (2016) Determination of optimal epsilon (eps) value on dbscan algorithm to clustering data on peatland hotspots in sumatra. In: IOP conference series: earth and environmental science, vol 31, no 1. IOP Publishing
36. Gaonkar MN, Sawant K (2013) AutoEpsDBSCAN: DBSCAN with Eps automatic for large dataset. Int J Adv Comput Theory Eng 2(2):11–16

Evolution of Autonomous Vehicle: An Artificial Intelligence Perspective

Kritika Rana, Gaurav Gupta, Pankaj Vaidya, Abhishek Tomar, and Nagesh Kumar

Abstract Autonomous vehicle is a major application of Artificial Intelligence. Artificial Intelligence systems, that use machine learning techniques to gather, analyze, and transfer data, are used in autonomous vehicles to make judgments that would normally be made by humans. Over the last decade, researchers and the car industry became very active in the deployment of driverless cars as they can greatly increase safety of the vehicle, rate of traffic accidents, and the effect of climate on automobiles. In implementing automated driving, software architecture plays a crucial function. The ability of driverless vehicles to work in real-world situations in a healthy and stable manner is currently being studied in public roads. In this paper, we have explained the background of Artificial Intelligence and how its applications are helping the world getting smarter. The role of Artificial Intelligence in autonomous vehicle industry has been discussed in this paper. Also, we tried to describe the relation of Artificial Intelligence and autonomous vehicles as well as presented the evolution of Artificial Intelligence from 1936 to 2020 with the help of a systematic review and explained how Artificial Intelligence is involved in the success of autonomous vehicles.

Keywords Autonomous vehicle · Artificial Intelligence · Internet of Thing · Security · Threats

---

K. Rana · G. Gupta (✉) · P. Vaidya · A. Tomar
Yogananda School of Artificial Intelligence, Computer, and Data Sciences, Shoolini University Solan, Bajhol, India
e-mail: solan.gaurav@gmail.com

N. Kumar
Chitkara University School of Engineering and Technology, Chitkara University, Baddi, Himachal Pradesh, India

Y. Singh et al. (eds.), *Proceedings of International Conference on Recent Innovations in Computing*, Lecture Notes in Electrical Engineering 1001,
https://doi.org/10.1007/978-981-19-9876-8_6

1 Introduction

Autonomous vehicles (AV) are robotic systems that can regulate their motion based on data obtained from sensors and can behave smartly in their environment [1]. There are various sensors autonomous vehicles rely on for measuring road conditions and for making quick decisions while driving, and most importantly safety depends on the consistency of these sensors. Complicated algorithms, actuators, powerful processors, and machine learning systems are needed to operate software system in autonomous vehicles, in addition to sensors. Using a range of sensors located throughout the vehicle, autonomous cars develop and maintain a map of their surrounding environment. Radar keeps a sharp eye on the movements of vehicles around. Cameras detect traffic signals, read road signs, monitor other vehicles, and look for pedestrians. Lidar estimates distances, detects boundaries of road, and recognizes lane by bouncing light from the vehicle to its surrounding environment. While parking, ultrasonic sensors located in the wheels detect edges and other vehicles in the surroundings.

2 Artificial Intelligence Analysis in Autonomous Vehicles

Artificial Intelligence (AI) lets systems to recognize and attain certain objectives. In today's world, we are surrounded by AI like Alexa by Amazon, Internet predicting what we may buy next. AI is divided into two categories: Narrow AI and General AI. Flagging content online, detecting faces in pictures, and simple customer care inquiries are examples of Narrow AI. General AI till date remains just a concept. The idea behind General AI is to make it as adaptable and flexible as human intelligence.

AI has applications in E-commerce such as personalized shopping and fraud prevention. In field of education, AI has several applications like administrative tasks automated to aid educators, creating smart content, voice assistants, and personalized learning. If we explore more, AI is being used in many other fields like spam filters, facial recognition, recommendation system, navigation systems, robotics, human resources, health care, agriculture, gaming, social media, marketing, chatbots, finance, and automobile industry.

Artificial Intelligence (AI) systems of autonomous vehicles always attempt to recognize traffic signs and road markings, detect vehicles, evaluate speed, and plan the path. Apart from inadvertent hazards such as abrupt malfunctions, these systems are very much prone to intentional attacks intended to interfere with Artificial Intelligence systems and disrupt safety critical functions. These systems are like other IT systems, such that they are vulnerable to attacks that could endanger operation of vehicle. Autonomous vehicles take large amount of data from machine learning, neural networks, and image recognition systems to build a technique which can operate autonomously. AI applications in vehicles are sensor data processing, path planning, path execution, monitoring vehicle's condition, insurance data collection, and many more. Such attacks include splashing paint on the road to confuse navigation or placing stickers on a stop sign to prevent it from being recognized. These alterations could cause the AI system to misidentify objects, causing the autonomous vehicle to behave in potentially dangerous ways. Security assessments of AI components must be undertaken on a frequent basis throughout their lifecycle to improve the security of autonomous vehicles. Artificial Intelligence is used for various very important functions in autonomous vehicle industry, out of which one of the most important functions is route planning. Another important function for AI is the connection of vehicle with the various sensors and processing of data gathered from the sensors [2]. Benefits of Artificial Intelligence are to automate learning using data, expand the smart abilities of present products, adapt smart algorithms to perform programming by data and rational analysis of the data, and refine data correctness [3]. Artificial Intelligence is very much capable of handling autonomous vehicle's massive data including additional conditions such as pedestrians. Traffic data will need to be gathered through different IoT-enabled networks, including LANs (Local Area Networks), WANs (Wide Area Networks), WSNs (Wireless Sensor Networks), and PANs (Personal Area Networks). The massive amount of data/information requires certain components and devices like vehicles, various sensors, and connectivity of networks.

3 History and Approaches of Artificial Intelligence

Artificial Intelligence is focused on collecting huge data. Iterative processing through smart algorithms allows a program to acquire from the features or trends in data, allowing it to process data quickly. AI has been increasingly popular in recent years since it has significantly reduced vision error rate (< 5%) when compared to human vision error rate [4] see Fig. 1.

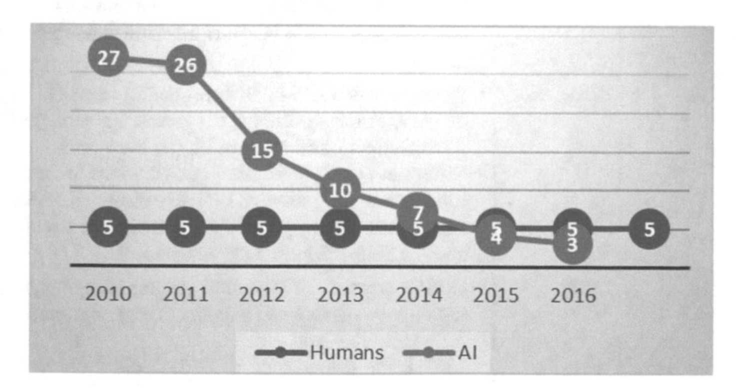

Fig. 1 The error rate of HAI algorithms [4]

In past years, AI has made major progress in almost all its sub-areas like vision, speech recognition and generation, image and video generation, planning, natural language processing, integration of vision, and motor control for robotics. Table 1 demonstrates the inventions that have been made since the year 1936 to 2020.

Artificial Intelligence has undoubtedly made tremendous advances in the human realm, with new theories and practices [6]. Artificial Intelligence is gaining popularity because of the convenience it brings to the entire human race by automating processes that are sometimes cognitively and physically draining for people [7]. Advancements in AI are one of the key enablers of the autonomous vehicle's development. Artificial

Table 1 Advancements in AI from 1936–2020 [5]

S. No.	Year	Inventions	Key points
1	1936	Turing machine	• Turing machines are simple abstract computational devices first defined by Alan Turing in Turing • They are meant to assist examine the scope and constraints of what may be calculated
2	1943	Cybernetics	• In Cybernetics, a monitor compares what is happening to a system at various sample times with some standard of what should be happening, and a controller modifies the system's behavior accordingly
3	1950	Turing imitation game	• The Turing test has a wonderful history in Artificial Intelligence • Despite being nearly 70 years old, it is still discussed and used today • It assesses a machine's intelligence and is named after its developer, Alan Turing
4	1956	Dartmouth conference	• Many specialists met for the first time in Dartmouth • In fact, for a whole consequent period, the main achievements in the field of AI have been obtained by these same scientists or by their students • Because of the Dartmouth Conference, AI turned into an intellectual study field, although one that was controversial, and it began to advance at a rapid pace from that point on
5	1958	Perceptron	• Perceptron is a sort of artificial neural network. Beginning in 1957, Frank Rosenblatt at Cornell University's Cornell Aeronautical Laboratory in Ithaca, New York • Rosenblatt made significant contributions to Artificial Intelligence (AI) through both experimental research of neural network properties (using computer simulations) and rigorous mathematical analysis
6	1960	LISP	• LISP is a family of computer programming languages with a long history and a distinctive prefix notation that is fully parenthesized • LISP is the second oldest still in-use high-level programming language • LISP became a popular programming language for AI

(continued)

Table 1 (continued)

S. No.	Year	Inventions	Key points
7	1970	Expert system	• An expert system is a computer program that uses Artificial Intelligence to resolve problems in a specific domain that would require human expertise • A **knowledge base** and an **inference engine** are the two components of an expert system • A **knowledge base** is a collection of facts concerning the domain of the system • An **inference engine** interprets and assesses the information in the knowledge base
8	1985	Fifth generation project	• The Japanese Ministry of International Trade and Industry (MITI) launched the **Fifth Generation Computer Systems (FGCS)** initiative to develop computers that use massively parallel computing and logic programming • Its goal was to develop an "epoch-making computer" with supercomputer-like performance that would serve as a foundation for future Artificial Intelligence developments
9	1990	Neural networks	• It develops an adaptive system that allows computers to learn from their mistakes and continuously improve • As a result, neural networks seek to solve complex tasks with higher precision, such as summarizing documents or recognizing faces • It creates a way for computers to learn from their errors and improve over time • As a result, artificial neural networks aim to solve complex issues with increased precision, such as summarizing documents or recognizing faces
10	2000–2020	Intelligent systems	• Intelligent systems are high-tech machines that can perceive and respond to their surroundings • From autonomous vacuums like the Roomba to facial recognition programs to Amazon's personalized shopping suggestions, intelligent systems can take many forms • Intelligent systems are also concerned with how technologies interact with human users in changing and dynamic physical and social situations

Intelligence is used by autonomous vehicles to interpret the surroundings, recognize its state, and make driving judgments. It basically imitates the movements of a human driver while operating the car [8].

A model (AI) for autonomous vehicles comprises 3 steps are as follows: collection of data, route planning, and act.

Collection of Data—Autonomous vehicles consists of various sensors like Radar, Cameras, LiDAR, and Ultrasonic sensors which gather massive amount of data from the vehicle as well as from its surroundings. Autonomous vehicle's data includes road conditions, objects/obstacles on the road, parking, traffic, transport, weather conditions, etc. This data is fed to autonomous vehicle for sensor fusion as shown in Fig. 2.

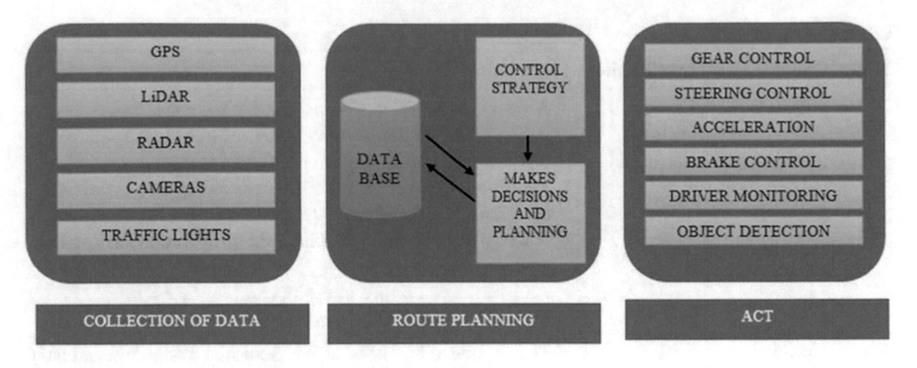

Fig. 2 A model for autonomous vehicles (AI)

Route Planning—Once the data collected by the sensors is fed to the autonomous vehicle, AI agent processes the data and plans the path accordingly.

Act—On the basis of decisions by an AI agent, AV detects objects on the road. Autonomous vehicles are equipped with AI-based control/functional systems such as steering and brake control, gesture controls, speed control, voice recognition, eye tracking, fuel economy, and different driver guidance systems. Autonomous vehicles process the collection of data, route planning, and act in a loop frequently. When the loops are more in number, more intelligent AI agents are needed, which results in more precision of decision-making, specifically in difficult driving circumstances.

In the last decade, different challenges like the famous DARPA Urban Challenge and Intelligent Vehicle Future Challenge (IVFC) have proven that autonomous driving can be a reality in coming years [9].

It is important to understand that autonomous vehicles will take a while to develop and function perfectly. Although this is a heated issue, few individuals oppose it because of the benefits, which include increased free time and improved transportation access. As well as few people stand against this innovation as these include few issues like hacking, manufacturing cost, and many more. We have reviewed papers on the Artificial Intelligence, its advancements, and analysis of Artificial Intelligence for autonomous vehicles in Table 2.

AI advancements are one of the most important facilitators of autonomous vehicle development. Artificial Intelligence is used by autonomous vehicles to interpret the surroundings, recognize its state, and make driving judgments. Since the time AI was used for the first time in autonomous vehicles, it kept adding advancements in autonomous vehicles. According to (SAEs) Society of Automotive Engineers, autonomous vehicles were categorized into 5 levels of automation.

- At level 0, there is no automation or very few features which are automated are provided like warnings signs are provided to support driver and the vehicle is controlled manually.
- At level 1, function-oriented automation is provided such as brake control and steering control. This is the lowest level of automation.

Table 2 Literature review

S. No.	Name of authors	Publication year	Key findings
1	[10]	2019	Examined the technical trend toward autonomous vehicles, discussion of significant difficulties and obstacles confronting the industry. Describes AI and the IoT for AVs and the Artificial Intelligence sector thoroughly
2	[11]	2016	Authors outlined five barriers for developing Verified AI. The authors examined the difficulty of developing and applying formal techniques to systems that are heavily reliant on Artificial Intelligence or machine learning. They have established design and verification principles for each of the five challenges that show promise in addressing that challenge
3	[12]	2019	Authors developed a technique resulting in an online strategy for predicting multiagent interactions based on their SVOs (Social Value Orientation) According to authors, this approach enables autonomous vehicles to track and find approximations of the SVOs of human drivers in real time
4	[13]	2017	Authors have given a general approach to reverse engineering EPS (Electric Power Steering) external controlling. Controlling system of vehicle on which the experiment is being done also involved an expert system which has made that extra adaptable and efficient for the existing application
5	[14]	2018	Authors presented and reviewed developing trends and problems in autonomous cars Authors have focused on recent approaches for perception and planning They explored the current state of autonomous vehicle fleet management and the obstacles
6	[15]	2019	Authors have presented AI algorithms for V2X (Vehicle-to-everything) applications which have shown improved performance over the traditional algorithms. Authors have concluded that different branches of AI can help each other to bring a perfect solution that would not cause problems in the areas they are not intended for
7	[16]	2018	Authors have discussed the need of AI in vehicles and how AI techniques can solve road safety and environment-related issues of traffic congestion. Paper has provided an AIV (Artificial Intelligence for Vehicles) framework which will be a boost factor in the vehicle industry according to the authors

(continued)

Table 2 (continued)

S. No.	Name of authors	Publication year	Key findings
8	[17]	2020	Authors have analyzed categories of CAV communication-based cyber security attacks. Based on UK cyber security attacks, a UML-based CAV framework has been developed
9	[18]	2019	Authors have demonstrated some reflections on Artificial Intelligence (AI). Difference between strong AI and weak AI has been distinguished here Further, they have described the main current AI models and discussed the need to provide knowledge to machines in brief to advance toward the goal of a general AI
10	[19]	2019	This paper proposed an AV system model which will help in representation of much realistic system facing the complication introduced by the human in vehicle system Authors have given conclusion that further research is needed for a stronger impact of safety engineering will promote the agenda of research for Artificial Intelligence-based autonomous vehicles
11	[20]	2020	The aim of this study is to reduce the gap by delivering a thorough survey of important findings in this study This review paper has delivered a brief review of existing practices in which AI is benefitting AV's development and few concerns related to AI for meeting the functionality requirements of AVs are considered by the authors
12	[21]	2020	This paper discusses various crucial discoveries associated to autonomous vehicles and discusses the opinions to convey one of the present concerns in area of AI
13	[22]	2016	This report scratches the surface of the ways AI is leaving its marks in revolution, creating considerable social and economic impacts, and renovating lives According to the authors, policy creators must react to legitimate concerns, also they must not allow pessimists to slow down progress. Rather, they must stay concentrated on speeding up the progress and implementation of AI

- At level 2, vehicle can control steering as well as speed. In level 2, ADAS (Advanced driving assistance system) is introduced.

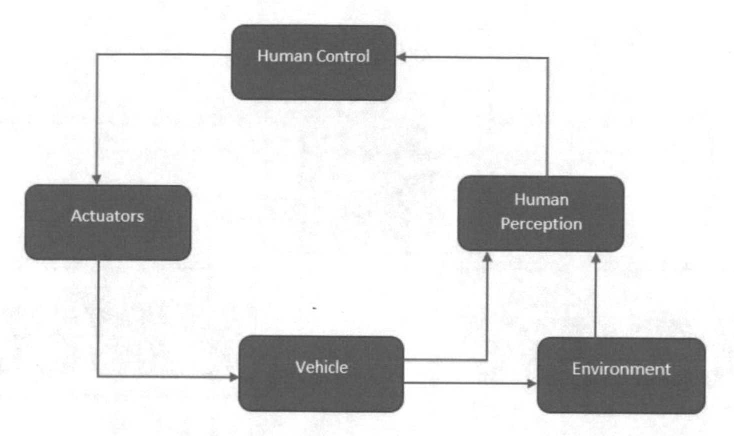

Fig. 3 No automation in autonomous vehicles

- Level 3 includes conditional driving automation; vehicles have capabilities to detect environmental situations which enable it to make decisions such as accelerating and slowing down the speed. In this level, vehicle requires human assistance. Driver must be prepared to take control if the system is unable to execute the intended task.
- Level 4 includes high driving automation; in this level, vehicles can operate on self-driving mode. The difference between level 3 and level 4 is that level 4 vehicles can interfere if things go wrong or in case of system failure.
- Level 5 vehicles are fully automated; there is no need of human attention and can operate without human intervention [23].

Figures 3 and 4 are explaining how the machine has decreased the participation of human beings. As Artificial Intelligence is helping vehicles getting smarter day by day, AI techniques are grouped into different categories: (i) Vehicle control, (ii) Traffic control and prediction, and (iii) Road safety and accident prediction [24]. In the past few years, because of the increasing population and the struggle of their mobility needs [25], autonomous vehicles have integrated different Artificial Intelligence [26] techniques to provide new services to make vehicles much smarter. These services include dealing with a specific amount of data generated by the vehicles and drivers [27, 28]. Autonomous vehicles pursue to overall enhancing traffic safety and sustainability, additionally introducing a positive impression on consumers [29, 30].

4 Conclusion

The literature on Artificial Intelligence for autonomous vehicles is surveyed in this paper, as well as AI history and approaches are reviewed. We have presented the analysis of Artificial Intelligence in AVs. Also, the Artificial Intelligence model for

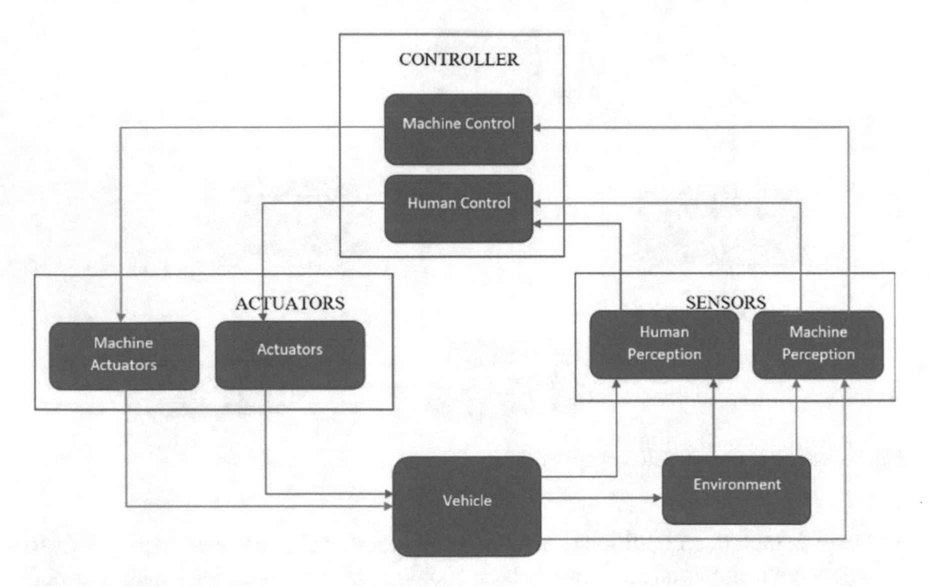

Fig. 4 Semi-automation in autonomous vehicles

autonomous vehicles is explained in this paper. AI has been gradually popular in recent times, and how it has significantly reduced vision error rate as compared to human vision error has been shown with help of a line chart.

References

1. Rana K, Kaur P (2021) Comparative study of automotive sensor technologies used for unmanned driving. In: 2021 2nd international conference on computation, automation and knowledge management (ICCAKM). IEEE, pp 346–350
2. Hristozov A (2020) The role of artificial intelligence in autonomous vehicles. [Online]. Available: https://www.embedded.com/the-role-of-artificial-intelligence-in-autonomous-vehicles/
3. Lin P-H, Wooders A, Wang JT-Y, Yuan WM (2018) Artificial intelligence, the missing piece of online education? IEEE Eng Manage Rev 46(3):25–28
4. Elsayed GF et al (2018) Adversarial examples that fool both computer vision and time-limited humans. arXiv preprint arXiv:1802.08195
5. Cantú-Ortiz FJ, Galeano Sánchez N, Garrido L, Terashima-Marin H, Brena RF (2020) An artificial intelligence educational strategy for the digital transformation. Int J Interact Des Manuf (IJIDeM) 14(4):1195–1209
6. Saini F, Sharma T, Madan S (2022) A comparative analysis of expert opinions on artificial intelligence: evolution, applications, and its future. Adv J Graduate Res 11(1):10–22
7. Borges AFS, Laurindoa FJB, Spínolaa MM, Gonçalvesb RF, Mattos CA (2020) The strategic use of artificial intelligence in the digital era: systematic literature review and future research directions. Int J Inf Manage 57:1–16. https://doi.org/10.1016/j.ijinfomgt.2020.102225
8. Nascimento AM, Vismari LF, Molina CBST, Cugnasca PS, Camargo JB, de Almeida JR, Hata AY et al (2019) A systematic literature review about the impact of artificial intelligence on autonomous vehicle safety. IEEE Trans Intell Transp Syst 21(12):4928–4946

9. Pérez-Gil O, Barea R, López-Guillén E, Bergasa LM, Gómez-Huelamo C, Gutiérrez R, Díaz-Díaz A (2022) Deep reinforcement learning based control for autonomous vehicles in CARLA. Multimedia Tools Appl 1–24
10. Khayyam H, Javadi B, Jalili M, Jazar RN (2020) Artificial intelligence and internet of things for autonomous vehicles. In: Nonlinear approaches in engineering applications, Springer, pp 39–68
11. Seshia SA, Sadigh D, Sastry SS (2016) Towards verified artificial intelligence. arXiv preprint arXiv:1606.08514
12. Schwarting W, Pierson A, Alonso-Mora J, Karaman S, Rus D (2019) Social behaviour for autonomous vehicles. Proc Natl Acad Sci 116(50):24972–24978
13. Shadrin SS, Varlamov OO, Ivanov AM (2017) Experimental autonomous road vehicle with logical artificial intelligence. J Adv Transp 2017
14. Schwarting W, Alonso-Mora J, Rus D (2018) Planning and decision-making for autonomous vehicles. Annual Rev Control, Robot Auton Syst 1:187–210
15. Tong W, Hussain A, Bo WX, Maharjan S (2019) Artificial intelligence for vehicle-to-everything: a survey. IEEE Access 7:10823–10843
16. Li J, Cheng H, Guo H, Qiu S (2018) Survey on artificial intelligence for vehicles. Autom Innov 1(1):2–14
17. He Q, Meng X, Qu R, Xi R (2020) Machine learning-based detection for cyber security attacks on connected and autonomous vehicles. Mathematics 8(8):1311
18. Meyer-Waarden L, Cloarec J (2021) Baby, you can drive my car: psychological antecedents that drive consumers' adoption of AI-powered autonomous vehicles. Technovation 102348
19. Nascimento AM et al (2019) A systematic literature review about the impact of artificial intelligence on autonomous vehicle safety. IEEE Trans Intell Transp Syst 21(12):4928–4946
20. Ma Y, Wang Z, Yang H, Yang L (2020) Artificial intelligence applications in the development of autonomous vehicles: a survey. IEEE/CAA J Automatica Sinica 7(2):315–329
21. Banerjee S (2020) Autonomous vehicles: a review of the ethical, social and economic implications of the AI revolution. Int J Intell Unmanned Syst
22. Castro D, New J (2016) The promise of artificial intelligence. Center Data Innov 115(10):32–35
23. Machin M, Sanguesa JA, Garrido P, Martinez FJ (2018) On the use of artificial intelligence techniques in intelligent transportation systems. In: 2018 IEEE wireless communications and networking conference workshops (WCNCW). IEEE, pp 332–337
24. Taxonomy and Definitions for Terms Related to Driving Automation Systems for On-Road Motor Vehicles, ed: SAE International, 2021
25. Pei Y, Li X, Yu L, Li G, Ng HH, Hoe JKE, Ang CW, Ng WS, Takao K, Shibata H, Okada K (2017) A cloud-based stream processing platform for traffic monitoring using large-scale probe vehicle data. In: IEEE wireless communications and networking conference (WCNC), pp 1–6
26. Russell S, Norvig P (2016) Artificial intelligence: a modern approach, 3rd ed. Pearson
27. Barrachina J, Garrido P, Fogue M, Martinez FJ, Cano JC, Calafate CT, Manzoni P (2012) CAOVA: a car accident ontology for VANETs. In: IEEE wireless communications and networking conference (WCNC), pp 1864–1869
28. Fogue M, Sanguesa JA, Naranjo F, Gallardo J, Garrido P, Martinez FJ (2016) Non-emergency patient transport services planning through genetic algorithms. Expert Syst Appl 61:262–271
29. Qureshi KN, Abdullah AH (2013) A survey on intelligent transportation systems. Middle-East J Sci Res 15(5):629–642
30. Martinez FJ, Toh CK, Cano J-C, Calafate CT, Manzoni P (2012) Determining the representative factors affecting warning message dissemination in VANETs. Wireless Pers Commun 67(2):295–314

Systematic Review of Learning Models for Suicidal Ideation on Social Media

Akshita Sharma and Baijnath Kaushik

Abstract In present-day society, the major critical issues are mental health problems which eventually turn out to be suicidal ideation. Premature detection of suicidal thoughts among people is the solution to avoid suicide in the latter times. With each passing year, the growth rate of suicides is abruptly increasing. The social media platform is the key from which people around the world come across and share their feelings, emotions, and reaction to what they are going through in their lives. The data which is available on social media can be used for the identification and detection of suicidal ideation among people, on social media forums like Twitter, Reddit, Tumblr, Facebook, and Instagram. Natural Language Processing has come a long way in the research of finding the sentiment of individuals and checking the linguistic patterns of text shared by the people. Effective to carry out the research by using machine learning, deep learning, and transfer learning algorithms on the online forum data of people. Algorithms such as SVM, Logistic Regression, Naive Bayes, Convolution Neural Network, Recurrent Neural Network, Bidirectional Long Short-Term Memory, BERT, and RoBERTa, respectively, were enforced to classify or detect the suicides in the social media data.

Keywords Suicidal ideation · Machine learning algorithms · Deep learning algorithms · Transfer learning algorithms · Social media · Suicides

1 Introduction

It is realized that suicide has become the forthcoming happening around the globe. As it is observed that over the last few years the growth of the problem has been apparent. Some risk factors which automatically lead to suicidal ideation are depression, helplessness, anxiety, stress, mental illness, isolation from social life, and the

---

A. Sharma (✉) · B. Kaushik
Shri Mata Vaishno Devi University Katra, Katra, J&K, India
e-mail: akshita.horizon@gmail.com

B. Kaushik
e-mail: baijnath.kaushik@smvdu.ac.in

Y. Singh et al. (eds.), *Proceedings of International Conference on Recent Innovations in Computing*, Lecture Notes in Electrical Engineering 1001,
https://doi.org/10.1007/978-981-19-9876-8_7

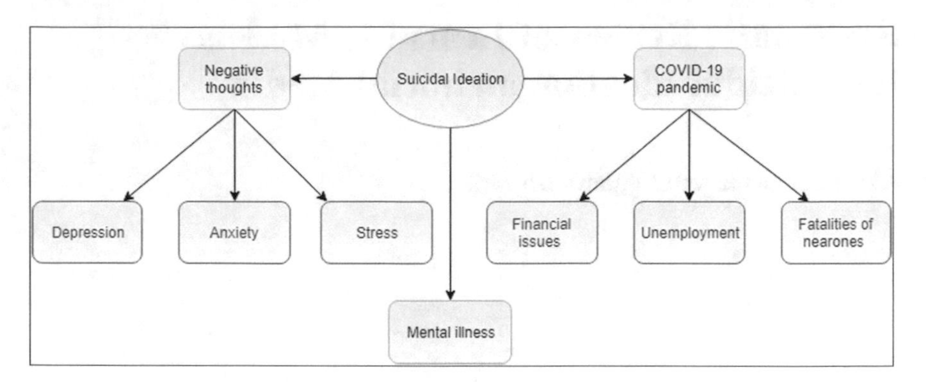

Fig. 1 Some suicidal ideation risk factors

tendency to aggressive behavior. Individuals who are suffering from some kind of stress, anxiety, or depression will isolate themselves from their friends and family. In considering this scenario social media forums are an influential window, where the sentiment and emotions of people are being shared.

According to the World Health Organization (WHO) states the demise of around 700,000 individuals, every year is due to suicide. Among teens (14–19 years) suicide and suicidal thoughts are the fourth prime concern. In adults, it is the second leading concern. Nowadays suicide is the most sensitive subject all over the globe which has an increasing curve. Depressed individuals prefer to express their emotions on social media, rather than in person. There can be numerous causes of suicidal thoughts such as conflicts with family, friends, violence, and abuse. The discriminated groups such as native groups, transgender, LGBTI persons, and guilty people. As per the WHO the previous suicide attempt is a major risk factor [1].

Suicides can be prevented by taking help from family, friends, and psychiatrists but they must be detected first. Most of the depressed users were not detected as depressed even in their regular check-ups with the therapists [2]. It is stated that fatalities due to suicide disturbed the equilibrium of family and public community as it impacts negatively. Social media has a great influence on the lives of people around the world as do online forums like Twitter, Reddit, Tumblr, and Snapchat, suicide notes are being shared on social media like Snapchat nowadays. Figure 1 shows the risk factors of suicidal ideation which eventually leads to suicide.

1.1 Statistical Analysis

The suicide in the year 2020 in India was 153,052, with an increase in the rate of suicide from previous years to 11.3. As to the Indian government, suicide is a major concern in the western state of India, like Maharashtra which attains the highest number of fatalities due to suicide in 2020. As in the rise of the COVID-19 pandemic depression, stress, and anxiety were increased among people and new concerns are

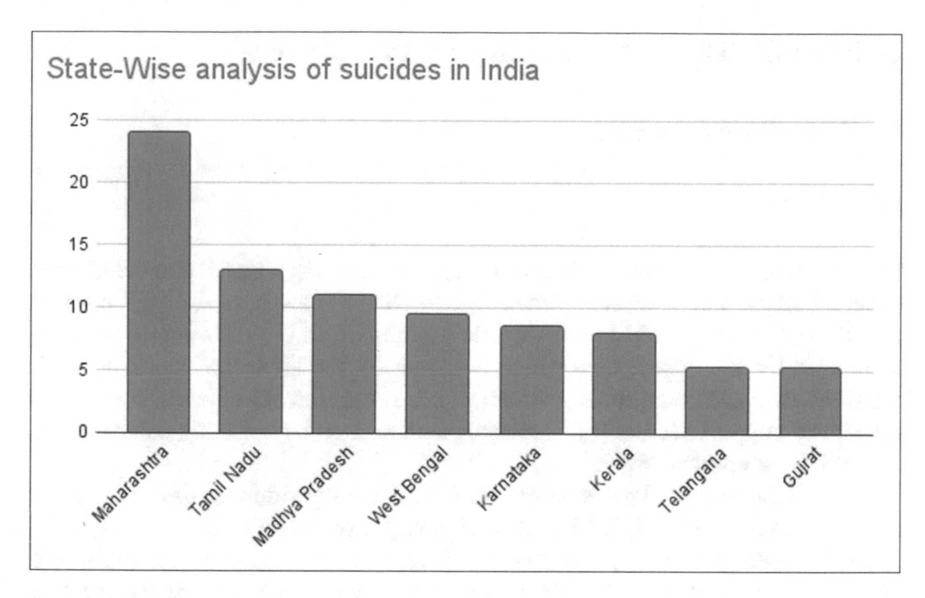

Fig. 2 The suicidal rate in India state-wise

also related to suicide behavior such as financial loss, unemployment due to the pandemic, and the health conditions like mental illness. The mental health crisis is increasing in the pandemic in low-middle income countries. Struggles faced by the people during the pandemic such as loss of income, mental pressure due to unemployment, and fatalities due to COVID-19 are collectively leading to a toll on the mental health of individuals. In comparing the suicides worldwide India attains the annual rate of 10.5 per 100,000 and 11.6 all around the world. Some statistics on suicide in India state-wise are shown in Fig. 2 [3].

For preventing and identifying suicide on social media forums like Twitter, Reddit, Facebook, Instagram, and Tumblr, machine learning and deep learning classifier are applied to the social media data for detecting suicidal behavior among the people on social media.

The residuum of the paper is gathered subsequently. Section 2 describes the distinct machine learning, deep learning, and transfer learning models which were applied on separate datasets, and the associated work that is carried forward in this experiment. Section 3 describes the future work and conclusion which is drawn out after the review of numerous works. And in the future the aspects which can be considered for the respective experiment to proceed with great results.

2 State-Of-The-Art Works in Learning Models

2.1 Machine Learning

2.1.1 SVM

Support vector machine is an approach of machine learning which is extensively used in classification and regression pieces of work. SVM is a supervised ML approach. The SVM is widely used in classification issues. Here the data points are being segregated by the hyper-plane which is drawn by the algorithm to set apart the classified groups. The data points which are close by and are strenuous to be classified are termed support vectors [4]. Hyper-plane is created by selecting some extreme data points (support vectors).

Some authors applied the support vector machine algorithms to the online forum dataset for the purpose to classify suicidal thoughts on social media.

De Choudhury et al. [5] collected the dataset in the form of a questionnaire of Center Epidemiologic Studies Depressive Scale (CES-D) to decide the levels of depressive behavior of individuals who participated in the questionnaire. The author developed the Human Intelligence Tasks with the help of amazon services termed as Mechanical Turk interface. Researchers record that the important trait of depression is a lack of motivation for their daily work and a significant reduction of engagement of individuals in social media online forums. The study employs the SVM classifier for predicting the depression on the dataset of individuals which yields an accuracy of 70%.

Burnap et al. [6] show the study on the lexicons that are generated manually. The data has been gathered from Twitter by the application programming interface (API) of Twitter, making use of 62 keywords. The researcher collected the second dataset via name and the last name of the deceased person. Approximately 1000 tweets were collected, 80% suicidal and other remaining of the deceased dataset. They put in some machine learning algorithms and ensemble techniques. Out of which SVM and Random Forest outperform.

O'Dea et al. [7] collected the data in a tool that was developed by Commonwealth Science and Industrial Research Organization (CSIRO) from Twitter using its API an advantage of using this tool (CSIRO) is that if the tool found some word that is correlated with the suicidal thoughts, it monitored and store automatically with the profile handler name and picture. The study categorized the data into three categorical classes for the classification purpose and to identify the concerned level of tweets for suicidal thoughts and employing the SVM and Logistic Regression (LR) algorithm, SVM attains an accuracy of 76% on the whole dataset.

2.1.2 Logistic Regression

Just like the support vector machine algorithm, Logistic Regression is also used for both Regressions as well as for classification purposes. It comes with a supervised machine learning classification algorithm. It is similar to Linear Regression, but the dissimilarity is the resultant outcome is patterned in binary. Discrete values are the output or the target values of Logistic Regression. In LR, the data is classified according to the threshold which will be set. The threshold classifies the data appropriately, but the threshold is defined based on precision and recall. The respective algorithm is classified depending upon the categories they have such as binomial (only two categories), multinomial (more than two categorics that too are not ordered), ordinal (variables that are targeted, and the categories are in order).

Chadha et al. [8] statuses the classes in which the data is classified among suicidal and non-suicidal. The data has been extracted from Twitter with the use of an application programming interface, i.e., the Twitter API. The basis on which the author collected the data, accordingly, uses the keywords for instance "Want to die", "To kill myself", "To commit suicide", etc. The study shows that 1897 tweets were found relevant for the experiment purpose and employed some machine learning and ensemble learning techniques, among all the machine learning (ML) algorithms LR acquires the best accuracy of 79.65%, which is similar to the Voting Ensemble.

2.1.3 Naive Bayes

The Naive Bayes classifier will work on the Bayes theorem. In Naive Bayes, each feature is independent of further features. For text classification, machine learning (ML) provides two variants of Naive Bayes such as Multimodal Naive Bayes and Bernoulli Naive Bayes. The classifier, i.e., Bernoulli Naive Bayes takes a binary input which is in the form of 0 and 1, probabilistic input, which describes the availability of features. The multimodal Naive Bayes calculates the frequency of data or the features.

Bayes theorem is dependent upon the Bayes formula which is stated as:

$$P(I|J) = \frac{P(J|I)P(I)}{P(J)} \tag{1}$$

where the I and J are the two events.

2.2 Deep Learning

2.2.1 CNN

Convolution neural networks were traditionally outlined for image recognition as having a great performance capability. But CNN is now extensively used in tasks of text classification, as the CNN model comes across hidden patterns that are recognized by the normal feed-forward network. CNN is a type of neural network in which the sliding of the vector over the input vector, i.e., filter, because of which extraction of features is done and is an important aspect. The filter (sliding filter) correlated with some parameters. Output is in the terms of numerous feature maps after the filter is applied over the input and activation is applied for the final output.

CNN adds the padding to avoid the loss of information that resides within the boundaries. During text classification the text data must be embedded, if it's not, embedding can be done in many ways such as Word2vec, Glove.

Tadesse et al. [10] used the data from online forums like Reddit for examining the behavior for the suicidal purposes of individuals. Apply some machine learning and deep learning algorithms to the collected Reddit dataset and proposed the hybrid model of CNN-LSTM, using the word embedding as a feature type, Word2vec on the deep learning models like CNN, LSTM, and a hybrid model of these two. In which CNN attains an accuracy of 90.6% and a combination of CNN-LSTM acquires the accuracy of 93.8%. The combinational approach yields satisfying accuracy in comparison with others.

2.2.2 RNN

In a recurrent neural network the nodes which are connected, are in a form of directed graphs. The working of RNN is quite simple, it takes every previous word, computes it, and propagates the computed information from the initial to the end of the sentence. Computed information consists of the hidden values and the initial word of the sequence (sentence) it carries forward the information, adds the second word of the sequence, and generates the computed information. So, it calculates the latest value by considering the previous values and a hidden value. RNN makes the final predictions or classification by remembering all the previous values; it runs in a loop that affects the model's state by recalling the previous state of affairs of sentences. RNN working is shown in Fig. 3.

The information is multiplied with some weights, let's say $W_{(m)}$.

Chadha et al. [11] collected the data from the forums like Twitter using the application programming interface (API) of Twitter according to the selected features. One hundred twelve features by the author, data is being gathered and applies some machine learning and deep learning algorithm and carry out the comparative analysis, amongst all the models RNN provides the best accuracy on the suicidal dataset which were collected from the Twitter.

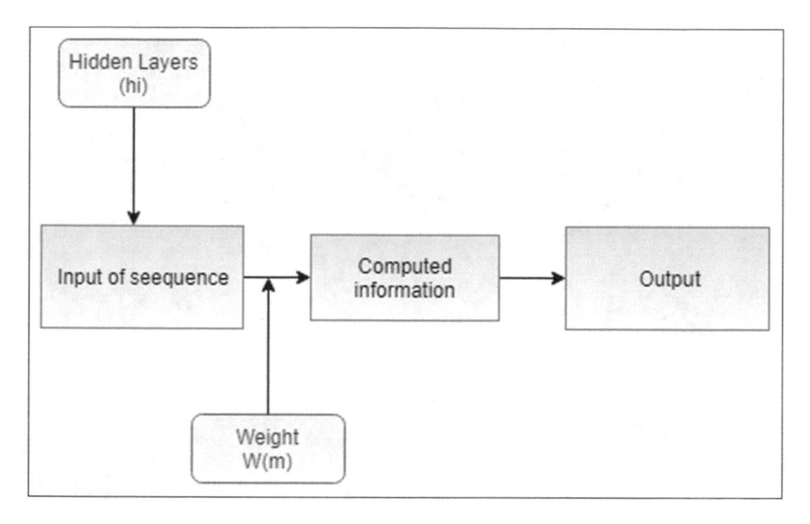

Fig. 3 RNN working

2.2.3 BiLSTM

The combination of two LSTM is termed bidirectional LSTM. It is a two-way sequence algorithm, which flows in both directions, i.e., forward and backward direction. BiLSTMs are the better alternative to simple RNN as the great amount of information sustains in the network. It reduces the difficulty caused by RNN such as vanishing gradient. Figure 4 shows the BiLSTM [12].

2.3 Transfer Learning

2.3.1 BERT

Bidirectional Encoder Representations from Transformer. BERT is developed by Google as a framework for natural language processing (NLP). The main idea behind the BERT is to enhance the search contextually. It is also termed a pre-trained model, which is trained on millions and billions of data (text) that are available on the web in an unannotated form. Unlike other models, BERT is contextual, i.e., it takes both the context, previous, and next context. As it is pre-trained on the tasks which are unsupervised, i.e., Masked LM and Next Sentence Prediction (NSP) [13]. Comparing the BERT with other state-of-the-art methods that are pre-trained is shown in Fig. 5 [14].

Here T_1, T_2, and T_3 is the input word in a contextualized form. BERT is a Transformer pre-trained model having multiple self-attention heads which are assembled

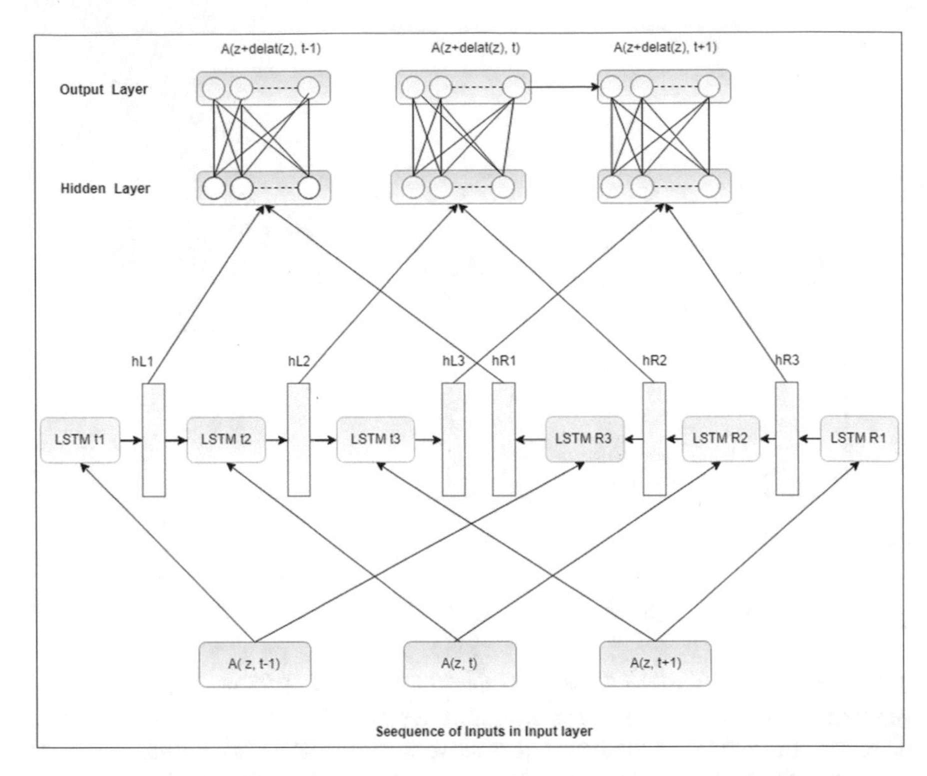

Fig. 4 Shows the BiLSTM architecture which is having a flow of inputs in bi-direction [12]

and several encoders that are stacked on each other. It will give the best results on text analysis as each input data is contextualized in the model.

2.3.2 RoBERTa

Robustly Optimized BERT Pre-training Approach is an improvised version of BERT, which is being improved by Facebook. Extensively used in Generally Language Understanding Evaluation (GLUE) a benchmark of NLP. Roberta's improvised key points are as it detaches the Next Sentence Prediction of BERT, hyperparameters of BERT are altered, learning rates and the batches on which it trains are larger. The masking strategy of BERT is the foundation of RoBERTa, where the hidden context of a text is being predicted [15].

Zogan et al. [16] extracted the data from Twitter for detecting depression among individuals on social media sites. The author tags the data as depressed and non-depressed and proposes a framework for detecting the depressive behavior of an individual based on some history of the individual's posts on the online forum called DepressionNet. Considering the features, the researcher found that behavioral features are the most useful in detecting depression. The proposed framework

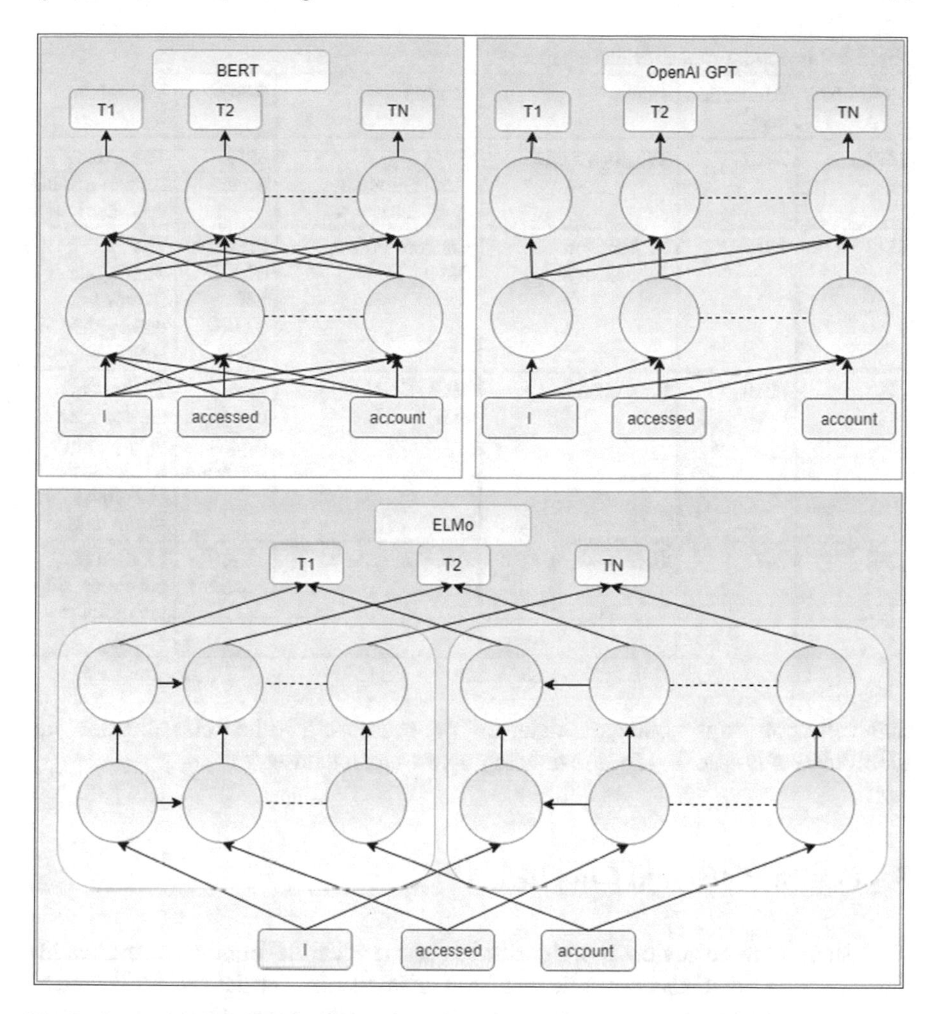

Fig. 5 Comparing BERT with other state-of-the-art pre-trained techniques [14]

is compared with the algorithms such as RoBERTa, BERT, XLNet, and BiGRU, out of which RoBERTa attains great precision, and the proposed framework yields the best accuracy and recall. Table 1 shows the limitations of some reviewed papers.

Table 1 depicts the models (algorithms) applied by various researchers to classify and identify suicidal thoughts by considering the social media publicly available data. The biggest limitation reviewed, is that the majority of experiments have compact data on which the work was executed. The disappointing results may also be due to the poor pre-processing of text which is an important part of the text classification and identification. The unwanted noise in the posts will end by pre-processing.

Figure 6 illustrates the recommended methodology by the author for the experiment like the one and all steps are important for the successful results of algorithms

Table 1 Limitations of the reviewed paper

References	Publication year	Author	Models	Result	Limitations
[17]	2015	Tsugawa S et al	SVM (Topic modeling approach)	69% accuracy	The span of collection data was small
[18]	2018	M. Johnson Vioules et al	Random Forest, SMO	Precision 67.6% RF 82.9% SMO	The Martingale framework was applied to only two users
[19]	2020	Rao. Guozheng et al	MGL-CNN SGL-CNN	56% precision 63% precision	The models are complex in applying on a large scale of online forum data
[20]	2021	Ji. Shaoxiong et al	Relation network	83.85%-accuracy	Method of pre-processing yields some errors

that will apply to the collected data, the pre-processing in text classification and identification is essential for the developing the clean dataset.

3 Future Work and Conclusion

This study aims to review the algorithms for the identification and classification purpose of data that has been collected from social media. The data was collected by the online forums which are publicly available and can be accessed by the Application Programming Interface (API). Datasets are mostly collected by online forums such as Twitter, and Reddit posts, also the interactivity follows through the online groups. The author has concluded after reviewing distant papers that out of all machine learning algorithms Logistic Regression provides great accuracy on the dataset which is having both suicidal ideated posts and everyday user posts. RNN in deep learning algorithm provides great outcomes but the Bidirectional LSTM will also provide considerable results in terms of this respective research as the inputs of text flow in both directions (forward and backward). Since in the transfer learning algorithm BERT and RoBERTa, RoBERTa attains outstanding results, as these models find the hidden context in the text, which can be defined clearly. In the future collected datasets should be large enough to carry out the experiments successfully and the models to apply to the respective dataset will attain the results substantially. Data can be accumulated through surveys in psychiatric hospitals using the questionnaire

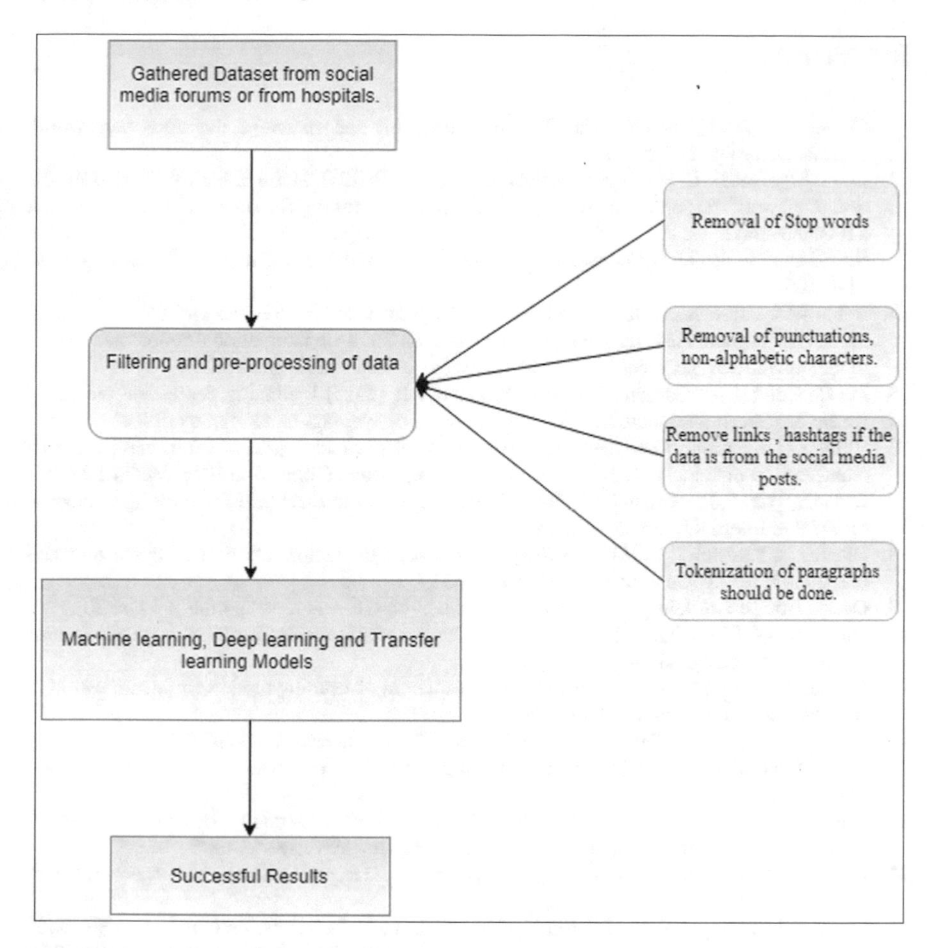

Fig. 6 Suggested methodology to experiment

or by interacting with the patients who are dealing with depression, anxiety, suicidal thoughts, and mental illness. A vital role in the classification of text is pre-processing which removes undesirable features such as hyperlinks, hashtags, punctuation, and stop words. The removal of ASCII numbers and bringing out tokenization on gathered data are also requisite for achieving great results. The consultation with the psychiatrist for a better review of the collected dataset provides clear insights about the data, i.e., congregate from the online forums. According to the dataset distinct models and hybrid models can be applied for gaining great results.

References

1. Chang Q, Yip PSF, Chen Y-Y (2019) Gender inequality and suicide gender ratios in the world. J Affect Disorders 243:297–304
2. Lewis SP, Heath NL, Sornberger MJ, Arbuthnott AE (2012) Helpful or harmful? An examination of viewers' responses to nonsuicidal self-injury videos on YouTube. J Adolescent Health 51(4):380–385
3. Suicides in India. https://ncrb.gov.in/sites/default/files/adsi2020_Chapter-2-Suicides.pdf. 20 Mar 2022
4. Rahat AM, Kahir A, Masum AKM (2019) Comparison of Naive Bayes and SVM algorithm based on sentiment analysis using review dataset. In: 2019 8th international conference system modeling and advancement in research trends (SMART). IEEE, pp 266–270
5. De Choudhury M, Gamon M, Counts S, Horvitz E (2013) Predicting depression via social media. In: Seventh international AAAI conference on weblogs and social media
6. Burnap P, Colombo G, Amery R, Hodorog A, Scourfield J (2017) Multi-class machine classification of suicide-related communication on Twitter. Online Soc Netw Media 2:32–44
7. O'dea B, Wan S, Batterham PJ, Calear AL, Paris C, Christensen H (2015) Detecting suicidality on Twitter. Internet Intervent 2(2):183–188
8. Chadha A, Kaushik B (2021) A Survey on prediction of suicidal ideation using machine and ensemble learning. Comput J 64(11):1617–1632
9. Description of CNN. https://iq.opengenus.org/text-classification-using-cnn/. 19 Mar 2022
10. Tadesse MM, Lin H, Xu B, Yang L (2019) Detection of suicide ideation in social media forums using deep learning. Algorithms 13(1):7
11. Chadha A, Kaushik B (2022) Performance evaluation of learning models for identification of suicidal thoughts. Comput J 65(1):139–154
12. Wang D, Song Y, Li J, Qin J, Yang T, Zhang M, Chen X, Boucouvalas AC (2020) Data-driven optical fiber channel modeling: a deep learning approach. J Lightwave Technol 38(17):4730–4743
13. Devlin J, Chang M-W, Lee K, Toutanova K (2018) Bert: Pre-training of deep bidirectional transformers for language understanding. arXiv preprint arXiv:1810.04805
14. Description of BERT. https://ai.googleblog.com/2018/11/open-sourcing-bert-state-of-art-pre.html. 21 Mar 2022
15. Liu Y, Ott M, Goyal N, Du J, Joshi M, Chen D, Levy O, Lewis M, Zettlemoyer L, Stoyanov V (2019) Roberta: a robustly optimized Bert pretraining approach. arXiv preprint arXiv:1907.11692
16. Zogan H, Razzak I, Jameel S, Xu G (2021) DepressionNet: a novel summarization boosted deep framework for depression detection on social media. arXiv preprint arXiv:2105.10878
17. Tsugawa S, Kikuchi Y, Kishino F, Nakajima K, Itoh Y, Ohsaki H (2015) Recognizing depression from Twitter activity. In: Proceedings of the 33rd annual ACM conference on human factors in computing systems, pp 3187–3196
18. Vioules MJ, Moulahi B, Azé J, Bringay S (2018) Detection of suicide-related posts in Twitter data streams. IBM J Res Dev 62(1):7–1
19. Rao G, Zhang Y, Zhang L, Cong Q, Feng Z (2020) MGL-CNN: a hierarchical posts representations model for identifying depressed individuals in online forums. IEEE Access 8:32395–32403
20. Ji S, Li X, Huang Z, Cambria E (2021) Suicidal ideation and mental disorder detection with attentive relation networks. Neural Comput Appl 1–11

A Novel Recommendation System for Mutual Management on Both OTT Video Streaming and Advertising Using Incremental Learning Approach

Jasmine Samraj and N. Menaka

Abstract In today's trend OTT video service provides a comfortable and luxurious lifestyle and users are easily adapting to it. Advertisements played in middle of the video are the main way of promoting the products or video services. The paper represents a novel approach "Mutual Management over OTT video streaming and Advertisements" (MMOA) based on the incremental learning of user perspective and OTT providers views. The proposed system uses a recommendation system and deep learning to analyze the user and provide service according to their requirements. Here the advertisement influencing is the main target for OTT providers, where the advertisements are chosen to be promoted in exact and minimal time validating the user choice of video streaming. Further, the paper uses machine learning (ML) producing recommendation engines to filter the most viewed video streaming and produces recommendation of ads/videos to user accordingly. Hereby, by the process implementation the OTT platform services and advertisements promotions are improved. An efficient Quality of Experience (QOE) for subscribers in OTT platform is ensured guaranteeing a full-fledged entertainment. An experimental analysis is made from collecting the Kaggle dataset using Github Repository to evaluate the user preference and the effectiveness of the proposed system is produced.

Keywords OTT video streaming · Advertisements · Deep learning · Incremental learning (IL) · Recommendation system · ML and QoE

1 Introduction

The word "Over-The-Top(OTT)" defines the distribution of content or services over an infrastructure that's not under the management of the content provider. Originally, it applied to the distribution of audio and video material. The web connectivity has provided devices with a progressively growing support to digital media and users have a privilege of access media at anytime and anywhere. However, In recent times during

J. Samraj (✉) · N. Menaka
Department of Computer Science, Quaid-E-Millath Government College for Women (Autonomous), Chennai, Tamil Nadu, India
e-mail: dr.jasminesamraj@qmgcw.edu.in

Y. Singh et al. (eds.), *Proceedings of International Conference on Recent Innovations in Computing*, Lecture Notes in Electrical Engineering 1001,
https://doi.org/10.1007/978-981-19-9876-8_8

the lockdown times the usage of OTT has reached an extent that the user consider OTT platform (Netflix, amazon prime, hotstar, etc..) as their primary consideration of entertainment [1]. Hence, the system has been expanded to incorporate data accessible on OTT platforms accounting the sudden growth of users and OTT platforms around the world. And returning to the increase of Over-The-Top (OTT) platforms has modified the wider advertising and entertainment market [2]. And this growth has modified the approach of content creation consumption and delivery on the OTT platforms and therefore the emergence of OTT technology marked a big change in India's trends in media consumption from some years ago. Recommendation has become a part of everyday life, which acts key element to reach the user and full fill their requirement knowing their perspectives. In day-to-day life recommendation system acts as an external knowledge to make decision about an interested item (Example: Doctor Consultation, Subscriptions, movies, etc. …).

- **Recommendation**: Recommendation provides ideas and deliberates information about an object according to the known user preferences. There are several parameters that can influence a person in making above decisions and recommendation plays a major role in it. The recommender systems for OTT platforms are vital to respond with quality and flawless content to the changing needs of an enormous viewer. The system provides an exact recommendation; the system should have a detailed knowledge about the user and their perspectives. For gaining immense amount of knowledge the system uses an incremental learning theory.
- **Advertisements**: Advertising in OTT platforms plays a main role in promoting an object, video or a product. Hence, advertising and entertainment market together forms a complete OTT platform approach. The system saw an acceleration of OTT platform firms delivering a good variety of merchandise and the user expertise is immersive and interactive, the subscriber conduct has changed drastically.

The implementation of the system works are as follows:

(1) Initially the paper proposes an account reference model to analyze the users' interest over OTT platform services. Here the users behavioral classification (age group, watch time, streaming interest, likable shows, etc.)-based on incremental learning is made.
(2) Further, the system proposes a machine learning technique for recommendation. The advertisements are recommended according to the user preference for minimal time.
(3) A classification is made for advertisement likable users and non-likable users. Also, the user preferences are matched and suggestions are deliberated to the subscribers accordingly. This scheme produces benefit for both the OTT services (shows and advertisement) and the users.
(4) In the worst-case scenario, inserting advertisements in the middle of a video may irritates the viewer depends on the user mood. This can be handled by identifying the user's search history using sentimental analysis and recommendation can be done.

The remaining section of the paper is as follows. Section 2 provides a detailed literature review on user behavior classification and OTT ads development. Then, in Sect. 3, the proposed method implementation is focused and the workflow produced. The following results analysis is made in Sect. 4. In Sect. 5 the paper is concluded and the future work is sketched.

2 Literature Survey

This section provides the insight view of previous theories of OTT up gradation related to the present work. Ahmad and Atzoriin [3] stated a QoE-based approach for managing resource and quality in MNO-OTT Collaborative Streaming in 5G. Most relevant studies and data-driven approaches for QoE management and monitoring are designed based on the requirements [4]. The users experience, preferences and content are monitored for providing enhancement in OTT video streaming [5]. Quality of Experience (QoE) to multimedia OTT services and brief discussions on network administration over the streaming services are made [6]. For validating the user perspective and usage of OTT platform in India are calculated in [7] by Veer P Gangwar and others.

Gupta and Komal Singharia [8] analyze OTT media streaming by implementing PLS analysis in COVID-19 lockdown. Rojas et al., in [9] introduce an incremental learning approach for validating the user consumption profile and to overcome the OTT service degradation. Table 1 explains the existing algorithms in recommendation system.

Table 1 Existing recommendation algorithms for OTT platform [8–10]

Existing methods	Description
PLS-SEM	The current study uses partial least squares structural equation modeling (PLS-SEM) to look at the impact of two main antecedents, customer engagement (CE) and Quality of service experience (QoSE), on customers propensity to continue and subscribe to streaming services in the future [8]
Machine learning	Machine learning (MI) in AI may sift through a variety of data points to make appropriate recommendations based on a person's viewing history, preferences and content kind. ML also tracks customer information (demographics, gender, age, viewing history, seasonal viewer-ship, total movies watched) [9]
Personalized video ranking (PVR) algorithm	This is a general—purpose algorithm that usually narrows down the catalog based on a set of criteria and other factors such as user preferences and popularity [10]

3 Method Implementation

3.1 *Workflow of Incremental Learning (IL)*

IL model selection and training are performed in the proposed implementation. IL is made on the premise of continuous learning and adaptation, activating the autonomous progressive development of advanced skills and knowledge. In a machine learning context, IL aims to calmly update the prediction model to account for various tasks and information distributions while still having the ability to re-use and retain knowledge over time. The given Fig. 1, describes the working of Incremental Learning Model [9].

Feature Extraction and Instance Labeling: Here a continuous iterative process is followed, where feature extraction and instance labeling are applied to raw data. The main target is to gain a consolidated dataset.
Revising labels: The dataset created to use to construct IL training and testing. The process results are reviewed by revising the labels and an accurate feedback provided to the IL algorithm for maintaining the settings.
IL algorithm: Then the IL algorithm is activated where the models performs the work of collecting the user preference and send them for recommendation.
Label Prediction: The IL model performs user classification based on their behavior. The labels are predicted and the information about the user perspective and sent for ads settings.

The below algorithm describes an input data uploaded for online streaming (Ds). Then the user's preferences are noted, and data is recommended based on those preferences. (OTTub) is a service that monitors OTT user behavior and automates the video streaming process. There is advertising available. The most viewed video (Vmv) is identified based on incremental learning (AD il), and the maximum advertisement (ADmax) is displayed in-between the video, which is favorable to the advertiser.

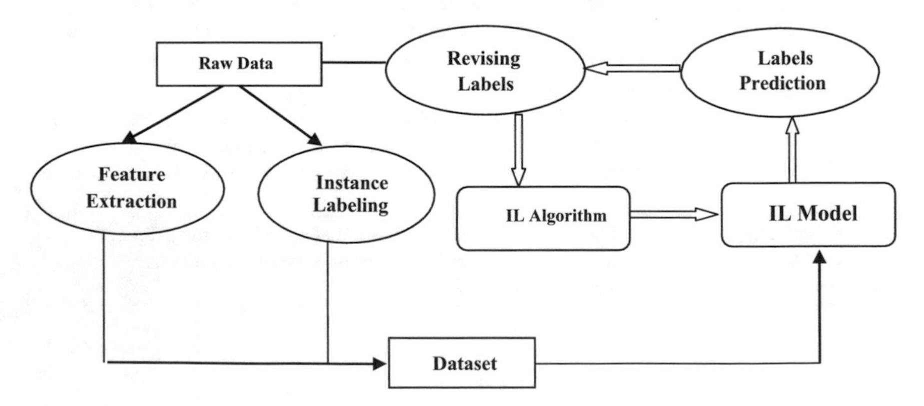

Fig. 1 Illustrate typical IL workflow [9]

Also, the videos with the least amount of views (Vlv) are chosen, and the video quality is improved (DQi). Only a few advertisements (ADmin) are promoted during the video streaming. If the video streaming is larger than the number of advertisements produced, the Mutual Management and automatic recommendation (MAr) is initiated. If video streaming is less popular, fewer advertising is created, which benefits both OTT users and ad promoters. Mutual Management and protected video streaming (MAr) is obtained as a result. The frequency of the user's search history (freqchk) can be used to do sentiment analysis, and advertisement recommendations can be updated based on the viewer's mood (ADrc).

Algorithm 1. Algorithm for Secured Video streaming and Advertisement management

Input: D_s // data upload for online streaming
Output: MA_r // Mutual Management and secured video streaming
Initialization: data uploaded by server
While (D_s) //providing data on user preferences
OTT_{ub} ← tracking OTT user behavior
OTT_{vs} ← OTT video streaming
AD_{il} ← advertisement based on incremental learning
If (V_{mv}) **then** ← // video streaming most viewed
AD_{max} ← maximum no of advertisement insertion
D_s ← security for video and ad
M_{ut} ← multi user preference tracking
OTT_{up} ← OTT video recommendation to user preference
A_m ← advertisement management
End if
If (V_{lv}) **then** // video streaming less viewed
D_{Qi} ← data quality increase
AD_{min} ← minimal advertisement inserted
OTT_{ms} ← more video recommendation made
A_m ← advertisement management
End if
If (MA_r) **then** // Mutual Management and automatic recommendation
If ($V_s$max) **then** // video stream viewed maximum
AD_{ii} ← advertisement level increased. Else
If ($V_s$min) **then** // video stream less viewed
AD_{id} ← advertisement level decreased
End if
If (M_r) **then** // mutual recommendation
V_{ru} ← video recommendation to user preference
AD_{rv} ← advertisement inserted to most viewed video
If (U_{behav}) then ← user behavior
freq(chk) ← frequency of the browsing history is checked
$SA_{(ub)}$ ← sentimental analysis on user behavior
$AD_{(RC)}$ ← advertisement recommendation is controlled
End if
End while

3.2 Mutual Management Over OTT Video Streaming and Advertisements (MMOA) Based on IL

The System model aims to extend the usage of OTT platform by providing an exact recommendation by gaining data relevant to IL-based users' OTT consumption behavior classification. Further the system plans to provide benefit to the ad producers by promoting the advertisement calculating the most viewed content of the subscriber. The system provides a benefit to both the OTT consumers and the advertisement providers.

The system architecture shown in Fig. 2 provides an extension to the OTT services and user preferences. The system focus on providing a beneficial support to the network Administrator by validating the user needs and advertisement utilization mutually at one time.

Application Phase: The application plane contains the details of user subscription over OTT services that are produced to the processing network.
Processing Network: The network validates, collects and stores all the user interest and ads requirements to the server.
MMOA implementation: Initially, in this phase a server is assigned for collecting the OTT consumption behavior and user subscription details along with network traffic are produced. The user consumption dataset containing user subscription profiles and OTT consumption behaviors information are maintained.
Incremental Learning Model: The incremental learning model used for classifying the user's interest and providing advertisement according to the video views.

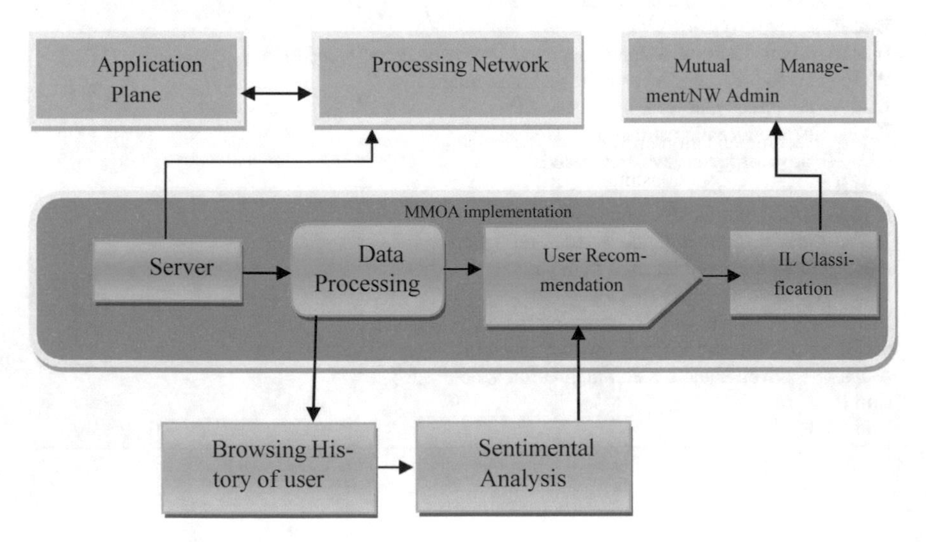

Fig. 2 System architecture

Mutual Management Plane/NW Administrator: The Mutual Management plane consists of the elements of user consumption over OTT services. The most interested OTT video streaming of the users by subscription count are calculated using the IL classification model. Then, the ads are provided by users' interest. Also the advertisement timings are verified where some users may skip the whole content for ads defecting both advertiser and the OTT services. To stop this, the system implies a classifier to classify the most liked and disliked video streaming of user by analyzing the views. Then, the advertisements are assigned in middle-of-the-most viewed videos. So that the advertisement re promoted in an efficient way focusing the user interest too.
Browsing History and Sentimental Analysis: Sentimental analysis can be performed based on the user's search frequency and a threshold value can be set to regulate advertisements and provide recommendation which makes user satisfied.

4 Result Analysis

A large-scale video dataset for Recognizing YouTube videos are taken from Kaggle. The dataset used GitHub Repository for this dataset and trained models. The dataset scraped about 3600 videos from YouTube to evaluate the user preference. The feedbacks, interests, likings/non-liking of the users are validated from the dataset. Table 2 and Fig. 3, describes the recommendation system on various OTT Platforms [1]. Table 3, describes the analysis of existing method with the proposed system.

In Fig. 4, the user and OTT service analysis, time consumption and overall performance between the existing approaches Partial Less Squares (PLS) and ML with the proposed system MMOA-based IL are made. The proposed MMOA-based IL is proved to be efficient in overall performance.

Table 2 Recommendation on various OTT platforms [1]

Source	User consumption (ratings)	Ads recommendation (ratings)	Streaming services (ratings)	QoE (ratings)
Youtube	5	3.5	4	4.5
Amazon videos	3.5	2	3	4
Twitter	5	1	3	3.8
Spotify	4.5	2.8	5	3

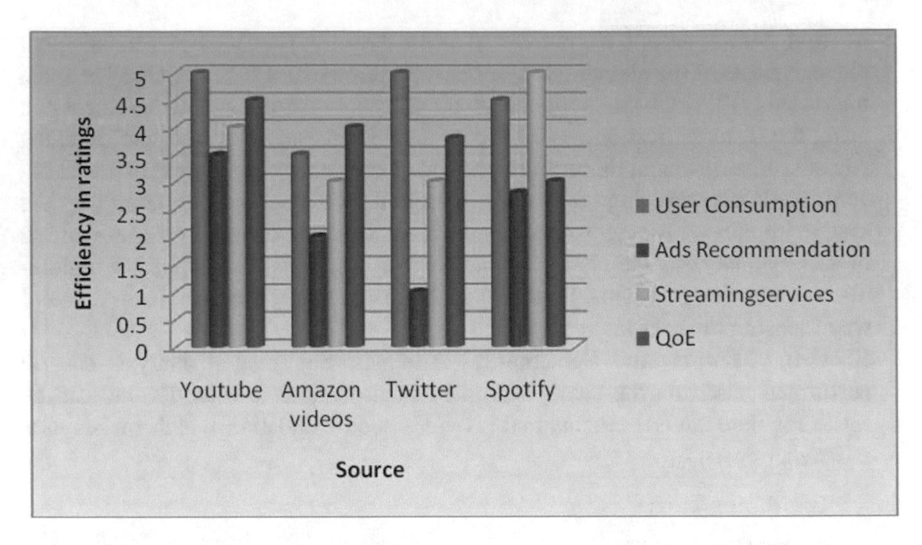

Fig. 3 Recommendation on various OTT platforms [1]

Table 3 Comparison of existing method and proposed method [8, 9]

Parameters/methods	PLS analysis (ratings)	ML approach (ratings)	MMOA-based IL (ratings)
User/OTT Service analysis	2.5	2.5	4
Time consumption	2.5	3.5	4
Overall performance	2	3	5

5 Conclusion and Future Work

The paper provides a deep analysis and Mutual Management on user preference and advertisements. Through a deep incremental learning and classification model the user consumption behavior, OTT services and ADs are well-analyzed and classified accordingly. By implementing the method both the user and the advertisement providers are mutually benefitted. This model increases the efficiency and usage of OTT platform and user are provided with exact recommendation with minimal Ads. Further in future, the implementation of demands on network and OTT service degradation are concentrated. OTT misuses are identified and a rectification model is updated to enhance the security.

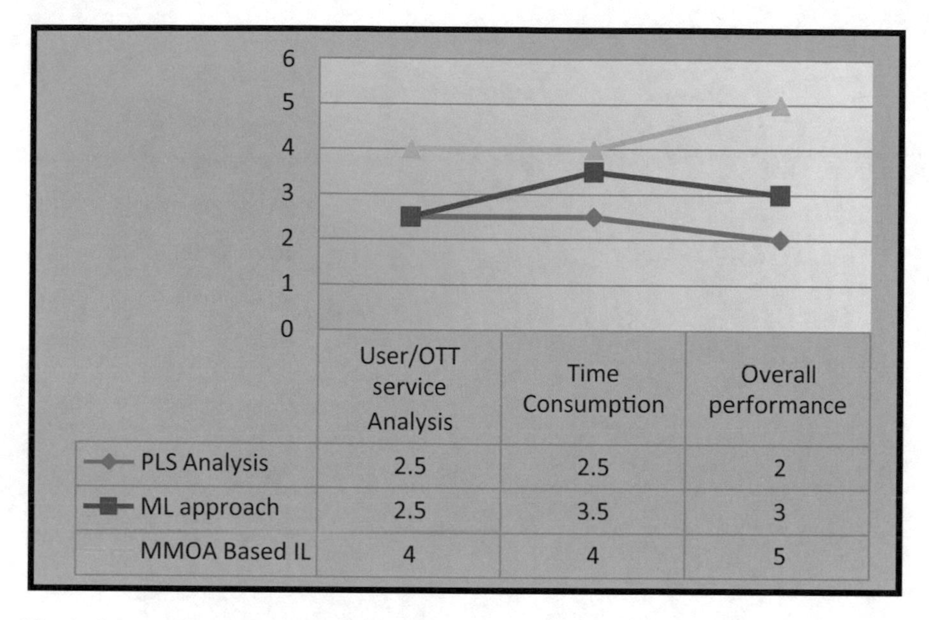

Fig. 4 Comparative analysis of existing system versus proposed systems

References

1. Sundravel E, Elangivan N (2020) Emergence and future of over-the-top (OTT) video services in India: an analytical research, at journal BiNET. Int J Bus Manag Soc Res 08(02):489–499
2. Hutchins B, Li B, Rowe D (2019) Over-the-top sport: live streaming services, changing coverage rights markets and the growth of media sport portals. Media Cult Soc 41(7):975–994. https://doi.org/10.1177/0163443719857623
3. Ahmad, Atzori L (2020) MNO-OTT collaborative video streaming in 5G: the zero-rated QoE approach for quality and resource management. IEEE Trans Netw Service Manag 17(1):361–374. https://doi.org/10.1109/TNSM.2019.2942716
4. Agiwal M, Roy A, Saxena N (2016) Next generation 5G wireless networks: a comprehensive survey. IEEE Commun Surv Tutorials 8(3):1617–1655
5. Taleb T, Samdanis K, Mada B, Flinck H, Dutta S, Sabella D (2017) On multi-access edge computing: a survey of the emerging 5G network edge cloud architecture and orchestration. IEEE Commun Surveys Tutorials 19(3):1657–1681
6. Barakabitze AA (2020) QoE management of multimedia streaming services in future networks: a tutorial and survey. IEEE Commun Surveys Tutorials 22(1):526–565. https://doi.org/10.1109/COMST.2019.2958784
7. Gangwar VP, Sudhagoni VS, Konda STB (2020) Profiles and preferences of OTT users in Indian perspective. Euro J Molecul Clinic Med 7(8). ISSN 2515-8260
8. Gupta Komal Singharia G, Consumption of OTT media streaming in Covid—19 lockdown: insights from PLS analysis, reprints and permissions, in Sagehub.com/Journalspermissions-India https://doi.org/10.1177/0972262921989118
9. Rojas JS, Pekar A, Rendon A, Corrales JC (2020) Smart user consumption profiling: incremental learning—based OTT service degradation. IEEE Access 8:207426–207442. https://doi.org/10.1109/Access.2020.3037971
10. Singh S (2020) Why am I seeing this? How video and E-commerce platforms use recommendation systems to shape user experiences

Early Diagnosis of Cervical Cancer Using AI: A Review

Nahida Nazir, Baljit Singh Saini, and Abid Sarwar

Abstract Cervical cancer is the most frequent cancer in women and is a leading cause of death, especially in developing countries, although it can be efficiently treated if found early. Cervical cancer screening is done through Pap smear, which is highly vulnerable to human errors and is time consuming. Thus, in this review, we present the various automated AI techniques for cervical cancer diagnosis which prevent the human loss, further spread of cancer and are far better than manual analysis approaches. The work presents a comprehensive review of AI techniques in diagnosing of cervical cancer in last 5 years. The datasets, total images, and accuracy achieved by these state-of-the-art techniques have been highlighted properly.

Keywords Cervical · Deep learning · Malignant · Pre-cancerous · Mutation

1 Introduction

Mutation in the DNA of the cervix cell gets developed into cervical cancer which is found in women aged above 14 to 45 years. In this type of cancer, the cells gradually get converted into pre-cancerous cells that have undergone certain abnormal changes; such mutations include cervical intraepithelial neoplasia (CIN) and so on. WHO has recorded 14,480 cervical cancer cases in 2021, and women above 35 are more prone to this cancer [1, 2]. As per the records of cancer statistics, the cervical cancer ranking is the fourth most worldwide and second most cancer among females [3]. There are almost 200 variants of HPV; among them, 13 types are supposed to be dreadful, whereas 70% of the cancer is caused by 16 and 18 type HPV [4, 5]. Severe cervical cancer takes approximately 8 to 10 years to develop from pre-cancerous without having any symptoms [6]. The Pap smear, one of the most common screening procedures for cervical cancer prevention and early diagnosis, is extensively used in

N. Nazir (✉) · B. S. Saini
Lovely Professional University Punjab, Punjab, India
e-mail: nahida.25827@lpu.co.in

A. Sarwar
University of Jammu, Jammu, India

Y. Singh et al. (eds.), *Proceedings of International Conference on Recent Innovations in Computing*, Lecture Notes in Electrical Engineering 1001,
https://doi.org/10.1007/978-981-19-9876-8_9

developed countries and is credited with significantly lowering cervical cancer death rates [6, 7]. However, due to a lack of qualified and skilled health professionals, as well as unavailability of medical facilities to continue screening programs, population-wide screening is still not commonly available in underdeveloped nations. Cervical cancer can be avoided if efficient screening procedures are in place, resulting in lower morbidity and mortality [8]. The paper presents a compressive summary of work done by different researchers in last five years in the domain of diagnosis of cervical cancer using machine learning techniques. The techniques used by researchers have been discussed in detail along with the datasets used and the results obtained. This work will help other researchers and academicians to have a comprehensive insight into the domain of cervical cancer using machine learning and deep learning techniques The first section is based on introduction, section second consists of AI techniques in cervical cancer, section third is based on detailed literature review, and the fourth section contains the conclusion of the entire survey.

2 Techniques for Cervical Cancer Screening

2.1 Convolutional Neural Network

Convolutional neural network (CNN) belongs to deep learning models. CNN comprises three basic layers: convolution, pooling, and fully connected layer. The first two layers are responsible for feature extraction, whereas the third layer performs classification. The convolution layer consists of convolution operations and activation functions. The pooling layer reduces the dimensionality. CNN is designed to function automatically. Learn the spatial hierarchy of features through backpropagation with multiple building blocks.

2.2 Support Vector Machine

Support vector machine is a supervised learning approach. The support vector machine algorithm's problem is to discover a hyperplane in an N-dimensional space that distinguishes between data points. The hyperplane's position and orientation are influenced by support vectors, which are data points that are closer to the hyperplane.

2.3 Multi-layer Perceptron

The multi-layer perceptron (MLP) is a feed forward neural network. The architecture consists of three distinct layers. The input layer receives the signal that will be processed to pass for the next layer. Hidden layer is responsible for computations and pass the information to the output layer. The output layer performs the computational tasks and transmits the data to users.

3 Literature Review

Ashok et al. [9] investigated various feature selection methods for cervical cancer detection using support vector classifier. The architecture design is a multistep process, starting from preprocessing of images (resize and grayscale conversion), segmentation (multi-thresholding), feature extraction (shape and textural), feature selection phase (mutual information, sequential forward selection, sequential floating forward search, and random subset feature selection), and the last step is the classification with the help of SVM classifier. During the segmentation phase, two organelles are segmented nucleus and cytoplasm. Gray-Level Co-occurrence Matrix has been used to mine 14 textural features such as contrast, entropy, and entropy difference. Shape features are extracted from the cytoplasm and nucleus, around 30 in number, like roundness, brightness, etc. Thus, the total features, including texture and shape, were 44. For optimization and to reduce the overfitting problem, feature selection has been introduced. The Pap smear images were collected from Rajah Muthiah Medical College, Annamalainagar; the total images for training and testing purposes were 150. Feature selection techniques are evaluated with accuracy, specificity, and sensitivity metrics. The sequential floating forward selection method has outperformed in comparison with other two techniques by achieving 98.5% accuracy, 98% sensitivity, and specificity 97.5%.

Plissiti et al. [10] presented an annotated image dataset (4049 cells) for cervical cancer diagnosis. These cells are categorized into five distinct classes based on the cytomorphological features. The five cell classes are superficial-intermediate (126 images), parabasal (108 images), koilocytotic (238 images), dyskeratotic (223), and metaplastic (271). The complex nature of Pap smear images makes proper classification of cervical cancer analysis very difficult; thus, authors have investigated the SIPaKMeD database. The region of interest is based on the texture and shape of the cytoplasm and nucleus. The total cells present are 4049, which have been annotated with the help of pathologists, which highlights 26 features for each classifier. Superficial intermediate cells have clear-cut cytoplasm; parabasal cells contain excessive nuclear to cytoplasmic ratio. The abnormal cells contain the HPV, which changes the two cells, namely koilocytosis that have an irregular nuclear membrane, binucleated, or multinucleated cells, and the nucleus degenerates depending on the extent of infection. Dyskeratotic cells are usually present in clusters that create hindrance in the segregation of cytoplasm and nucleus; such cells undergo abnormal keratinization. Metaplastic cells' appearance is also an indication of precancerous lesions. The first classification has been carried out with a support vector machine followed by a multi-layer perceptron, and CNN architecture consists of 3*3 filters, max-pooling of 2*2. CNN has achieved the best performance in comparison with MLP and SVM.

Finally, three distinct features are highlighted such as hand-crafted cell features, image features, and in-depth features.

Sompawong et al. [11] implemented the mask regional convolutional neural network (Mask-RCNN) to screen cervical cancer to detect normal and abnormal nuclear features. The architecture is a complete two-stage-based process; during the first phase, various features are extracted from the area of interest with the help of a feature pyramid network (backbone network) followed by ResNet-50. The features that have been generated will undergo box regression and classification to locate the target object. The second phase achieves three objectives: classifying various objects from the area of interest, reconfiguring the bounding box, and segmentation to produce a mask for each pixel. This approach cannot be carried out on the Herlev dataset as that consists of single-cell images; thus, authors have collected the data from Thammasat University; the total images prepared were 178, since each cell consists of certain artifacts such as the presence of WBC thus cell segregation was of the primary concern that has been done with the help of a doctor into four distinct classes as normal, atypical, low grade, and high-grade (total cells 3460). The model performance has been evaluated with metrics like mean average precision (57.8%), accuracy, sensitivity, and specificity (91.7% per image). The hit nucleus has achieved an accuracy of 89.8%, sensitivity of 72.5%, and 94.3% specificity. This is the first designed instance segmentation method for automatic cervical cancer screening that decreases the manual intervention during the analysis process.

William et al. [12] presented an enhanced fuzzy c means algorithm for automatic cervical detection. This tool can analyze the entire slide within 3 min rather than taking 5–10 min. The authors discovered an intelligent tool (trainable WEKA segmentation tool) for nucleus segmentation. Three datasets have been studied; first one is the Herlev dataset which consists of 917 cells; these cells have been segregated with the help of CHAMP software. Moreover, further classification has been made into seven distinct classes based on the specific features of the cells. The second dataset consists of 497 Pap smear images. The third database consists of 60 Pap smear images from Mbarara Regional Referral Hospital using Olympus BX51 bright-field microscope, followed by segregation into three major classes: normal cells, abnormal cells, and artifacts using trainable WEKA segmentation tool. For classification purposes, 18 features were considered out of 25 features. The three different classifiers single cell achieved 98.88, 99.28, and 97.47% (accuracy, sensitivity, and specificity), multiple cells achieved '97.64% accuracy, 98.08% sensitivity and 97.16%' and specificity, Pap smear slide images achieved '95.00% accuracy, 100% sensitivity, and 90.00% specificity were obtained for each dataset, respectively. The main strength of the developed tool is that it reduces the pathologist's analysis time for normal cells so that more focus should be done on suspicious cells.

Adam et al. [13] investigated the cervical cancer diagnosis with stacked autoencoders and softmax. A stacked autoencoder has been implemented on the dataset to reduce the data dimensions collected from the UCI repository. The proposed automatic system smooths the classification process. The dataset is composed of 668 samples with four target variables. The stacked autoencoder with softmax has been compared with other machine learning algorithms like a decision tree, which has performed better than other techniques. The accuracy achieved by the algorithm is 97.8%. One of the significant strengths of the autoencoder is to discard the unnecessary attributes for reducing the data dimensions. This technique has certain pitfalls like training time is high in comparison with other techniques.

Lin et al. [14] introduced a robust approach for fine-grained classification of cervical cancer cells by combining cell image appearance and cell morphological features. Earlier CNN implementation for cervical cancer classification did not consider morphological features of cervical cells classification; in this work, authors have explored the importance of morphological features for proper cancer classification of Pap smear images. The input to the network is both raw data and segmented mask of cytoplasm and nucleus. GoogleNet has achieved the highest accuracy both in 2 class classification and 7 class classification compared to other networks. The authors have worked on the Herlev public dataset, consisting of 917 publicly available images. The cells are categorized into seven distinct classes, which are further categorized into normal and abnormal cells.

Gorantla et al. [15] introduced the CervixNet methodology for cervical cancer screening with cervigrams, fully automated methods. Authors developed the image enhancement technique for enhancing cervigrams and to evaluate the suggested Hierarchical Convolutional Mixture of Experts (HCME) architecture in the cervix type classification. The classification is carried out with Hierarchical Convolutional Mixture of Experts (HCME). The CervixNet technique enables a flat field correction technique, picture intensity adjustment technique, and local image enhancement to enhance the input image automatically. In addition to that, cross-entropy creates a unique loss function that improves classification performance. Due to the short dataset, the HCME architecture avoids overfitting. The model has been evaluated with three distinct metrics: Dice coefficient, Jacquard index, and pixel accuracy. The technique surpassed existing state-of-the-art methods when validated on an open challenge database of Intel and Mobile-ODT cervical cancer screening. The model has achieved an accuracy of 96.77% and a kappa score of 0.951.

Promworn et al. [16] have investigated five deep learning models to identify abnormal cells for cervical cancer diagnosis. This approach is designed to get the best deep learning technique for custom-made whole slide imager. The Herlev dataset has been used for model training. The dataset has been categorized into a binary classification and a multi-classification. Normal cells constitute 242 images in binary classification, followed by 675 images. In multi-class classification, 98 are columnar cells, 70 are intermediate squamous cells, 74 are superficial squamous, 150 are carcinomas in situ, 182 are mild dysplasia, 146 are moderate dysplasia, and 197 are severe dysplasia. The performance metrics used were specificity, sensitivity, and accuracy. DenseNet 161 has achieved the best performance compared to other algorithms.

Yilmaz et al. [17] presented a comparative analysis of traditional AI and deep learning approaches for cervical cancer classification. Machine learning algorithms such as KNN, SVM, decision tree, random forest, and extreme gradient boosting are explored. From the deep learning domain, authors have considered convolutional neural networks. Model comparison is done on certain performance metrics like accuracy, recall, precision, specificity, and F1 score. The hyperparameters of a model play a vital role in performance since they should be optimized to get good results. The experimental results relieved that CNN has outperformed other existing traditional machine learning techniques for cervical cancer classification. All these models have been trained on a publicly available Herlev dataset. The model has achieved 93% of accuracy, whereas traditional approaches have achieved the highest accuracy, up to 85% only. Features extracted with the help of CNN is much better than hand-crafted features. Apart from this, the training time for the CNN model has been lowered to 22 s for each epoch.

Jia et al. [18] designed a robust strong-feature CNN-SVM model for cervical cancer cell classification. Since several conventional screening methods suffer due to their bad quality, thus combining image processing and deep learning could enhance the performance and proper cell classification. The proposed method combines Gray-Level Co-occurrence Matrix and Gabor to extract the features from cervical cells; the hidden layer of LeNet-5 removed unnecessary features, which prevented the model overfitting. The two distinct feature classes are combined to provide input to the SVM classifier for classifying cells. The model works in three phases: extraction of features, feature fusion followed by classification of cells. The features are extracted with the help of LeNet-5 architecture; epithelial cells are fed as an input to the first layer, which is composed of five convolution codes. Two distinct datasets have been considered Herlev and private datasets. The private data consists of 2000 images collected from 200 patients (Guangdong Province People's Hospital). Different steps were carried out for preprocessing, such as random crop and resize. The model has been evaluated on three different parameters: sensitivity, specificity, and accuracy.

Ghoneim et al. [19] presented a convolutional neural network (CNN-ELM) for detecting and classifying cervical cancer cells. Three distinct architectures of CNN have been explored, one is shallow, which consists of two convolutional layers and two max-pooling layers, the other two are deep CNN (VGG-16 Net and CaffeNet, each of which contains two fully connected layers and a softmax layer, respectively); various classifiers like MLP, autoencoder, and extreme learning method have been examined. Cervical cells are given as an input to the CNN for extracting the deep features, followed by an extreme learning-based machine (ELM) classifier to classify the images. The first layer of CNN consists of 64 filters, whereas two convolutional layers consist of 128 filters. Herlev dataset has been used for model training. The VGG-16 Net and the CaffeNet have achieved the same performance, whereas the external network has achieved less accuracy in comparison with CaffeNet and VGG-16. The accuracy achieved by this model (CNN with ELM) is 99.5 and 91.2% for (2 class and 7 class), respectively.

Alyafeai et al. [20] presented an automatic deep learning approach to diagnose and classify cervical cancer. The model is based on different modules which carry out the particular function. Two separate datasets for training purposes have been collected from a publicly available repository. Intel and Mobile-ODT dataset consists of a total of 1500 annotated images, followed by NCI Guanacaste Project Dataset, which consists of 44,000 cervigram images. Cancer has been categorized into class 0 (negative case) and class 1 (positive case). Class 1 is further categorized into four distinct classes. This technique is 1000 times more efficient than other preexisting techniques.

Wang et al. [21] introduced convolutional neural network (PSINet) for deep Pap smear analysis. Authors used ten convolution layers, followed by three fully connected layers. A convolution kernel is a filter that extracts edge information comparable to a filter. The network model is trained on 389 Pap smear images (primary dataset). This approach has achieved more than 98% accuracy and could prove an effective tool for cervical cancer categorization in clinical settings.

Tripati et al. [22] presented the deep learning techniques for cervical cancer classification. The three distinct models considered for the work are RESNET-50, VGG 16, RESNET-152, and VGG 19. The concept of identity mapping was used to prevent model overfitting. These networks have achieved an accuracy of 94.84%., 95.87%, 94.91%, and 95.36%, respectively. Thus, RESNET-50 has achieved the highest accuracy in comparison with other models. Apart from this precision, recall, F1 score has also been considered for model comparison. The model has been validated on the SIPaKMeD Database (publicly available), which consists of 966 images.

Chandran et al. [23] presented VGG19 and CYNET as two deep learning approaches for cervical cancer diagnosis using colposcopy images called CYNET. The first architecture is incorporated with transfer learning; the other network performs cancer classification. The CYNET architecture is based on fifteen convolution layers, twelve activation layers, five max-pooling layers, and four cross-channel normalization layers. The primary dataset is made up of 5697 colposcopy images which are further preprocessed into three different types. Both the techniques were evaluated on the accuracy, specificity, Cohen's Kappa score, and F1 measure. CYNET outperformed VGG19 by achieving 19% higher accuracy.

Mohammed et al. [24] introduced the classification of single-cell Pap smear images with a pre-trained deep convolutional neural network. Top 10 pre-trained deep CNN models based on normalized performance such as ResNet 101, ResNet 152, DenseNet 169, and so on are used to classify cancer cells from Pap smear images. The approach is based on several steps like acquiring the image, preprocessing, extraction of features, and classification of cells into five distinct classes. The first class (normal cells) consists of only superficial intermediate cells, the second class (normal cells) is based on parabasal cells, the third class (abnormal cells) is composed of koilocytotic cells, the fourth classification (abnormal cells) is made of dyskeratotic cells, and the fifth class (benign) is metaplastic cells. The model has been trained on the SIPaKMeD database, and evaluation is done with accuracy, precision, recall, and F1 score. Model training required 88% of the data. The preprocessing steps are image resize (to reduce the computational overhead), normalization (to balance the pixel intensity), and affine transformation (to enhance the intraclass variation). DenseNet169 has outperformed than other nine mentioned deep CNN techniques. This technique has a minimum size (57 MB) among all other competitors that makes further deployment easier on specific portable devices.

Kano et al. [25] presented the automatic contour segmentation for cervical cancer analysis with high accuracy. This approach would lessen the burden on oncologists, which has been rarely used in clinical practices. The architecture is a combination of

2D UNet and 3D UNet. The main reason for combining these two different networks is that the 3D UNet cannot handle several cases with just a single learning model, so 3D UNet is combined with 2D UNet to handle multiple cases. The 2D UNet consists of a convolutional layer, reverse convolutional layer, rectified linear unit, leaky rectified linear unit, batch normalization followed by a dropout layer. 3D UNet carried automatic contour segmentation based on voxels. Diffusion-weighted images were fed as an input to the architecture, and the final image produced was of $64 \times 64 \times 64$ sizes. Diffusion-weighted images that were considered for investigation which was collected from 98 patients with positive tests. *K* fold cross-validation has been used that divides the training into 65 cases and 33 for testing purposes. The model was evaluated on the Dice similarity coefficient (DSC) and Hausdorff distance (HD).

William et al. [26] have presented a detailed survey of state-of-the-art AI techniques in cervical cancer diagnosis, for which the authors have worked on 30 major publications. The approaches used have achieved good classification in comparison to the traditional methods. The KNN techniques have achieved outstanding classification of cervical cancer cells. However, combining KNN with other approaches such as SVM enhances the accuracy. KNN has achieved 99.27% for a two-classification problem. Peng et al. (2022) presented a survey of breast cancer and cervical cancer detection with the help of deep learning (DL) approaches. The deep learning approaches have achieved almost the similar performance to radiologists for cancer detection. However, the major challenge faced by deep learning approaches is the poor design of the system for proper implementation of disease diagnosis. A summary of the observation made from the paper presented above is given in Table 1. The table presents the summary in the shape of techniques, dataset used, type of dataset public and private, and performance achieved in cervical cancer diagnosis.

Table 1 Dataset, total images, and performance achieved by different techniques

Year	Technique	Dataset	Total images	Performance achieved
[27]	SVM	Real dataset (150 images)	150	Accuracy 98.5%, sensitivity 98%, and specificity 97.5%
[9]	SVM and MLP, CNN	SIPaKMeD	4049	CNN has outperformed than other techniques
[10]	Mask regional convolutional neural network (Mask-RCNN)	Herlev dataset	917 images	Accuracy of 89.8%, sensitivity of 72.5%, and 94.3% specificity
[11]	Enhanced fuzzy c means algorithm	Herlev dataset and two primary datasets	917	95.00% accuracy, 100% sensitivity, and 90.00% specificity

(continued)

Table 1 (continued)

Year	Technique	Dataset	Total images	Performance achieved
[12]	Stacked autoencoders and softmax	Real dataset	Not mentioned	Accuracy of 97.8%
[13]	GoogleNet	Herlev dataset	917	Accuracy of 94.5%
[14]	CNN	Herlev dataset	917	68.0%
[16]	CNN	Challenge database of Intel and Mobile-ODT	Not mentioned	Accuracy 96.77%
[15]	KNN, SVM, decision tree, random forest, and extreme gradient boosting, CNN	Herlev dataset	917	Accuracy 93%
[17]	CNN-SVM	Real dataset	2000	Accuracy of 99.3, sensitivity of 98.9, and specificity of 99.4
[18]	VGG-16 Net and CaffeNet	Herlev dataset	917	Accuracy of 99.5% and 91.2% for (2 class and 7 class) achieved, respectively
[19]	Convolutional neural network (PSINet)	389 pap smear images (primary dataset)	389	Accuracy of 98%
[21]	CNN	Intel and Mobile-ODT dataset consists of a total of 1500 annotated images, followed by NCI Guanacaste project dataset, which consists of 44,000 cervigram images	1500 from first dataset and 44,000 from another dataset	AUC score achieved 0.82
[20]	2D UNet and 3D UNet	Real dataset	98 cases	Not mentioned
[25]	CNN	SIPaKMeD	4049	Achieved 0.990 accuracy
[24]	VGG-19	5697 colposcopies	5697	Achieved 19% higher accuracy
[23]	RESNET-50, VGG 16, RESNET-152, VGG 19	SIPaKMeD	4049	Accuracy of 95.87%,

(continued)

Table 1 (continued)

Year	Technique	Dataset	Total images	Performance achieved
[22]	KNN, SVM, decision tree and other basic machine learning techniques	Herlev dataset	917 Public dataset	Accuracy 99.27%
[26]	Deep learning approaches	Public and private	NA	NA

4 Conclusion

Cervical cancer is among the most diagnosed cancers in women. However, a timely diagnosis could prevent the further spread and mortality rate. The Pap test (Papanicolaou test) has been shown to be the most cost-effective and time-efficient way of diagnosis. Even though it is a basic process, significant research has shown that it is quite comprehensive and contains enough essential information. The review paper presented the basic AI and deep learning-based approaches to segment and classify cervical pathology images. According to the review of cervical cytopathology, image analysis in deep learning is a growing area of interest. Most of the state-of-the-art approaches have been applied to the Herlev dataset, SIPaKMeD dataset, and a few real datasets. The presented article illustrates how a pathologist can use Pap smear image analysis for automatic and speedy diagnostic prediction using deep learning. To our best knowledge, CNN has been frequently used for the cervical cancer diagnosis in contrast to other AI-based approaches.

References

1. Taha B, Dias J, Werghi N (2017) Classification of cervical-cancer using pap-smear images: a convolutional neural network approach. In: Annual conference on medical image understanding and analysis. Springer, Cham, pp 261–272
2. Gadducci A, Barsotti C, Cosio S, Domenici L, Riccardo Genazzani A (2011) Smoking habit, immune suppression, oral contraceptive use, and hormone replacement therapy use and cervical carcinogenesis: a review of the literature. Gynecol Endocrinol 27(8):597–604
3. Bernard HU, Burk RD, Chen Z, Van Doorslaer K, Zur Hausen H, de Villiers EM (2010) Classification of papillomaviruses (PVs) based on 189 PV types and proposal of taxonomic amendments. Virology 401(1):70–79
4. Bosch FX, Burchell AN, Schiffman M, Giuliano AR, de Sanjose S, Bruni L, Munoz N (2008) Epidemiology and natural history of human papillomavirus infections and type-specific implications in cervical neoplasia. Vaccine 26:K1–K16
5. Šarenac T, Mikov M (2019) Cervical cancer, different treatments and importance of bile acids as therapeutic agents in this disease. Front Pharmacol 484

6. Saslow D, Solomon D, Lawson HW, Killackey M, Kulasingam SL, Cain J, Myers ER (2012) American cancer society, American society for colposcopy and cervical pathology, and American society for clinical pathology screening guidelines for the prevention and early detection of cervical cancer. Am J Clin Pathol 137(4):516–542
7. Crum CP, Meserve EE, Peters WA (2018) Cervical squamous neoplasia in diagnostic gynecologic and obstetric pathology. Elsevier, Amsterdam, The Netherlands, pp 298–374
8. Wiley DJ, Monk BJ, Masongsong E, Morgan K (2004) Cervical cancer screening. Curr Oncol Rep 6(6):497–506
9. Plissiti ME, Dimitrakopoulos P, Sfikas G, Nikou C, Krikoni O, Charchanti A (2018) SIPAKMED: a new dataset for feature and image-based classification of normal and pathological cervical cells in Pap smear images. In: 2018 25th IEEE international conference on image processing (ICIP). IEEE, pp 3144–3148
10. Sompawong N, Mopan J, Pooprasert P, Himakhun W, Suwannarurk K, Ngamvirojcharoen J, Tantibundhit C (2019) Automated pap smear cervical cancer screening using deep learning. In: 2019 41st annual international conference of the IEEE engineering in medicine and biology society (EMBC). IEEE, pp 7044–7048
11. William W, Ware A, Basaza-Ejiri AH, Obungoloch J (2019) A pap-smear analysis tool (PAT) for detection of cervical cancer from pap-smear images. Biomed Eng Online 18(1):1–22
12. Adem K, Kilicarslan S, Comert O (2019) Classification and diagnosis of cervical cancer with softmax classification with stacked autoencoder
13. Lin H, Hu Y, Chen S, Yao J, Zhang L (2019) Fine-grained classification of cervical cells using morphological and appearance based convolutional neural networks. IEEE Access 7:71541–71549
14. Promworn Y, Pattanasak S, Pintavirooj C, Piyawattanametha W (2019) Comparisons of pap smear classification with deep learning models. In: 2019 IEEE 14th international conference on nano/micro engineered and molecular systems (NEMS). IEEE, pp 282–285
15. Yilmaz A, Demircali AA, Kocaman S, Uvet H (2020) Comparison of deep learning and traditional machine learning techniques for classification of pap smear images. arXiv preprint arXiv: 2009.06366
16. Gorantla R, Singh RK, Pandey R, Jain M (2019) Cervical cancer diagnosis using cervixnet-a deep learning approach. In: 2019 IEEE 19th international conference on bioinformatics and bioengineering (BIBE). IEEE, pp 397–404
17. Jia AD, Li BZ, Zhang CC (2020) Detection of cervical cancer cells based on strong feature CNN-SVM network. Neurocomputing 411:112–127
18. Ghoneim A, Muhammad G, Hossain MS (2020) Cervical cancer classification using convolutional neural networks and extreme learning machines. Futur Gener Comput Syst 102:643–649
19. Wang P, Wang J, Li Y, Li L, Zhang H (2020) Adaptive pruning of transfer learned deep convolutional neural network for classification of cervical pap smear images. IEEE Access 8:50674–50683
20. Kano Y, Ikushima H, Sasaki M, Haga A (2021) Automatic contour segmentation of cervical cancer using artificial intelligence. J Radiat Res 62(5):934–944
21. Alyafeai Z, Ghouti L (2020) A fully-automated deep learning pipeline for cervical cancer classification. Expert Syst Appl 141:112951
22. William W, Ware A, Basaza-Ejiri AH, Obungoloch J (2018) A review of image analysis and machine learning techniques for automated cervical cancer screening from pap-smear images. Comput Methods Prog Biomed 164:15–22
23. Tripathi A, Arora A, Bhan A (2021) Classification of cervical cancer using deep learning algorithm. In 2021 5th international conference on intelligent computing and control systems (ICICCS). IEEE, pp 1210–1218
24. Chandran V, Sumithra MG, Karthick A, George T, Deivakani M, Elakkiya B, Manoharan S (2021) Diagnosis of cervical cancer based on ensemble deep learning network using colposcopy images. BioMed Res Int
25. Mohammed MA, Abdurahman F, Ayalew YA (2021) Single-cell conventional pap smear image classification using pre-trained deep neural network architectures. BMC Biomed Eng 3(1):1–8

26. Xue P, Wang J, Qin D, Yan H, Qu Y, Seery S, Qiao Y (2022) Deep learning in image-based breast and cervical cancer detection: a systematic review and meta-analysis. NPJ Digit Med 5(1):1–15
27. Ashok B, Aruna P (2016) Comparison of Feature selection methods for diagnosis of cervical cancer using SVM classifier. Int J Eng Res Appl 6:94–99
28. Basu P, Meheus F, Chami Y, Hariprasad R, Zhao F, Sankaranarayanan R (2017) Management algorithms for cervical cancer screening and precancer treatment for resource-limited settings. Int J Gynecol Obstet 138:26–32

Exploring Data-Driven Approaches for Apple Disease Prediction Systems

Nahida Bashir, Himanshu Sharma, Devanand Padha, and Arvind Selwal

Abstract Apple is one of the most popular plants in the Kashmir valley of Indian origin, where it is cultivated in almost half of the horticulture area. The Kashmiri apple is well known for its deliciousness and is exported to different parts every year. However, apple plants are also susceptible to diseases such as apple scab, Alternaria leaf bloch, and apple rot. The timely detection or prediction of these diseases in the apple plants may help the farmers take appropriate measures to control the overall yield. With the emergence of artificial intelligence, machine learning-based techniques can be deployed for more accurate disease prediction. This study presents the data-driven approaches for the automated apple disease prediction system (ADPS). The study aims to explore both traditional and modern deep learning paradigms that have been used to develop ADPS for efficient and accurate disease prediction. A comparative analysis of these techniques is carried out to understand the concept deployed and datasets. Our analysis indicates that most ADPS has been designed using machine learning-based algorithms that show promising performance (i.e., 90–95%). It has been observed that Kashmiri ADPS is a least explored area where only a few studies in the literature are available. Besides, some open research issues have been identified and summarized that will serve as a reference document for the new researchers who want to explore this active field of research.

Keywords Apple disease · Automatic disease prediction system · Machine learning · Deep learning

N. Bashir (✉) · H. Sharma · D. Padha · A. Selwal
Department of Computer Science and Information Technology, Central University of Jammu, Samba, 181143, India
e-mail: nahidabashir2019@gmail.com

D. Padha
e-mail: devanand.csit@cujammu.ac.in

A. Selwal
e-mail: arvind.csit@cujammu.ac.in

Y. Singh et al. (eds.), *Proceedings of International Conference on Recent Innovations in Computing*, Lecture Notes in Electrical Engineering 1001,
https://doi.org/10.1007/978-981-19-9876-8_10

1 Introduction

Plant disease detection is a critical issue that needs to be focused on for productive agriculture and the economy. Diseases in plants cause significant production and economic losses and a reduction in both the quality and quantity of agricultural products. Improved varieties and high-quality seeds are essential requirements for productive agriculture. The effort of both the public and private sectors has made an enormous contribution to global agriculture. Kashmir is famous for its specialty of apples all over India. The plant nursery business has become a significant source of livelihood in various parts of Kashmir. There are a variety of Kashmiri apples, such as golden, delicious, and ambri. Apple cultivation in Kashmir contributes to about 55% of the horticulture area, with an estimated turnover of about 250 million. Apple production is predominantly confined to districts of Kashmir such as Srinagar, Ganderbal, Budgam, Baramulla, Kupwara, Anantnag, and Shopian. Apples are very much susceptible to diseases. The most common diseases are apple scab, apple rot, and Alternaria leaf blotch, as shown in Fig. 1. Apple fruit diseases can cause significant losses in yield and quality. Often, these diseases go unnoticed until harvest and cause significant damage to the quality of the apples. Although there are no curative treatments for infected apples, many diseases can be prevented if detected earlier.

Detecting apple disease using traditional methods is a tedious job as it requires a tremendous amount of work, time, and expertise. Machine learning and deep learning techniques can help identify the apple disease at the initial stage as soon as it appears on plant leaves. A computer vision-based apple disease prediction framework is the need of the hour. Such models can save apple production from major diseases as it is practically impossible for producers to keep track of every apple tree. Given an input image of the plant leaves, an apple disease prediction system extracts the necessary information from the input left. It identifies whether an apple tree is disease-free or not. With more advanced techniques of deep learning such as Convolutional Neural Networks (CNN) and robust datasets, the apple disease prediction systems nowadays identify whether a tree is infected or not and help predict the apple diseases with reli-

Fig. 1 Apple diseases. **a** Apple scab **b** Apple rot **c** Alternaria leaf blotch

able accuracy. Keeping apple-producing disease-free becomes convenient and easy with such apple disease prediction systems. An apple disease prediction framework consists of two phases, namely feature extraction and disease identification. The model first extracts visual features from the input image and then applies some prediction algorithms based on machine learning algorithms (support vector machines, decision trees) to identify whether the apple tree is disease-free or not. Initially, machine learning-based handcrafted features were extracted from the leaf images; however, with the introduction of deep learning-based techniques such as CNN, the traditional feature extraction was replaced by implicit feature extraction. With exponential growth in deep learning-based disease prediction systems, apple prediction systems remain an active research area, with many novel models and datasets being published every day. In addition, massive amounts of research articles are published every year about the apple disease prediction system, with each model having better efficiency and reliability than earlier. Being one of the active research areas, it is essential to survey apple disease prediction systems and datasets comprehensively. Some existing surveys of apple disease prediction systems [1, 2]. Although these studies review apple disease prediction systems comprehensively, they could cover only a specific timeframe. Most of the novel models have been published recently, and hence, there is a need to review these novels models. Therefore, we propose to alleviate the above-discussed research gap by comprehensively reviewing apple disease prediction systems consisting of machine learning and deep learning-based techniques. The contributions of our study are as follows:

i. We analyze apple disease prediction systems along with their characteristics and limitations.
ii. We propose an alternate taxonomy of apple disease prediction systems.
iii. The open research challenges are identified as a result of this study.

The rest of the contents of this article are arranged as follows. Section 2 defines an alternate taxonomy of apple diseases and briefly discusses their attributes and characteristics. A comprehensive literature review of apple disease prediction systems has been described in Sect. 3. In Sect. 4, we identify significant research gaps present in the state-of-the-art apple disease prediction systems.

2 Apple Disease: An Overview

A generic framework for the apple disease prediction system consists of a feature extraction model to identify and extract the features of a given image and a discriminating model that identifies the state of the apple leaves, as shown in Fig. 2. The extracted features are fed to the pre-trained model that discriminates the input leaf images based on whether they are healthy. Based on the type of techniques being used for extracting features and disease prediction, the apple disease prediction models are generally classified into two categories, namely machine learning-based systems

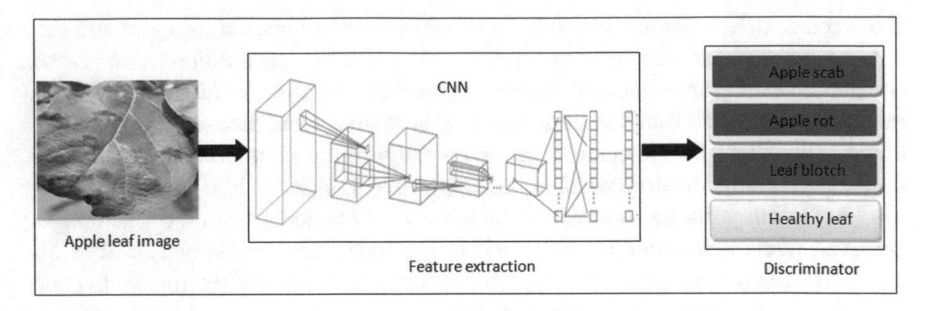

Fig. 2 Generic framework of an apple disease prediction system

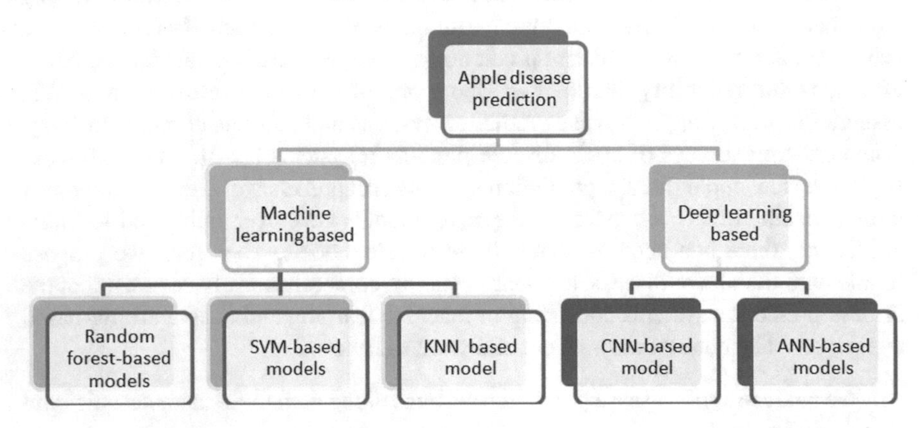

Fig. 3 Proposed taxonomy of apple disease prediction systems

and deep learning-based systems. A complete taxonomy of apple disease prediction systems is shown in Fig. 3. In machine learning-based disease prediction frameworks, handcrafted image features are extracted from the given leaf image, and then, these features are fed to a discriminator unit. Various handcrafted features can be used while extracting features from the given input image. Due to this, the performance of machine learning-based algorithms varies significantly. The extracted image features are analyzed and classified to predict whether the input apple leaf is healthy or not. Based on the type of feature extractor and discriminators used, the machine learning-based frameworks are further classified into the following three essential categories:

i. Random forest-based model: Random forest-based apple disease prediction models define a decision tree using the training samples to identify the state of the apple leaf [3, 4]. The extracted features from the input image are compared against the decision tree nodes based on which a decision about the state of the incoming leaf is taken.
ii. Support vector machine-based models: A support vector machine-based (SVM) apple disease prediction system defines a linear boundary between the healthy and

affected leaves [5, 6]. The SVM-based disease prediction systems are one of the most effective approaches as they produce results with high accuracy. In SVM-based systems, the input features are arranged in some discriminating space. SVM draws inferences about whether the input features belong to the feature space's healthy or affected side and outputs the decision accordingly.

iii. K-nearest neighbor-based models: The K-nearest neighbor-based (KNN) models are much simplified as compared with the above-stated models [5]. These models extract a series of input features from the input leaf image and compare it against our model's training samples. The decision is made based on the class of leaf images that are most similar in terms of features. KNN techniques are suitable only for such instances where the input samples are significantly less.

With the introduction of deep learning-based models such as Artificial Neural Network (ANN) and Convolutional Neural Network (CNN), there is a paradigm shift in feature extraction from traditional handcrafted to modern deep learning-based algorithms. One of the significant advantages of using the deep learning-based feature extraction model is that the model's image features are extracted implicitly. Hence, the chaos of choosing features to be extracted is eliminated. The deep learning-based feature extraction also decreases any chance of bias while extracting features, and hence, direct comparison between the disease prediction frameworks is possible. Based on the type of deep learning model being used for feature extraction, the deep learning-based disease prediction frameworks are classified into two categories:

i. ANN-based models: A neural network model that uses ANN for extracting features and making a decision about the state of the leaf is termed an ANN-based disease prediction model. ANN models produce better results than the classical machine learning-based models.

ii. CNN-based models: CNN is the most efficient deep learning model for recognizing visual information. The CNN-based disease prediction models use CNN to extract semantic information from the input image. The extracted information is then encoded and transformed to the discriminator to identify whether the leaf is healthy or not.

3 Literature Survey

There has been a significant amount of work in apple disease prediction systems by combining machine learning and deep learning techniques, changing the type of features being extracted or varying the mechanism of the descriptor to be used. Kambale et al. [1] provide a solution based on image processing for the classification and grading of apple fruit images. Initially, two databases are prepared, one with standard images and one with defected images. The next steps are image pre-processing and segmentation, followed by feature extraction, classification, and grading. The proposed model generates efficient results to support the automatic classification and grading of apple fruit diseases. Bhavini and Samajpati [5] used a random forest classifier for classifying apple diseases such as apple scab, apple rot, and apple botch. The model extracts low-level features like color, size, and shape and combines them using feature-level fusion. Based on the fused visual details, the discriminator decides the state of the leaf. Kleinen and Leemans [7] suggested a hierarchical technique for evaluating defects in Jonagold apples. The model uses the Bayes theorem to segment the data. The probability distributions of healthy and faulty tissue are classified as non-Gaussian statistics. Dubey and Jalal [4] developed a traditional method for detecting and identifying fruit illness based on the global adaptive threshold method using machine learning-based feature extraction methods. Solanki et al. [8] extracted several low-level image features such as color and texture details from the input image. Using these details, the discriminator model makes its decision.

Singh and Mishra [6] use image segmentation combined with soft computing techniques to predict apple leaf diseases. The model uses a Gabor filter to extract features and an ANN to predict the state of the input leaf. Camargo and Smith [3] identify plant disease's visual context from colored photographs. Kadir et al. [9] proposed a disease prediction system using shape, color, and texture features extracted from the input image. The shape features include geometric and Fourier features. Rathod et al. [10] proposed a leaf disease detection system using a neural network. Initially, edge detection and segmentation algorithms are used to detect salient visual information. The extracted information is then input to the neural network-based discriminator that identifies whether the input leaf is healthy or not. Raut and Fulsunge [11] devised a model to identify apple fruit illnesses such as apple scab, apple rot, and blotch. The model extracts the global and local low-level binary patterns from the input image. The extracted features are compared with the training datasets using an SVM. Based on the feature space where the input image lies, the state of the leaf is predicted using an ANN. Khirade and Patil [5] introduced a plant disease detection model where the RELIEF-F algorithm is used for feature extraction, deep learning, and a combination of neural networks. They extended their work by detecting the disease on grape leaves by suppressing the unwanted noise in the leaf images using digital image processing techniques. The model extracts features from the input image and applies dimensionality reduction techniques to sort the visual details better. The visual information is passed to the discriminator later to predict the outcome. Khan and Banday [12] used a CNN to extract the features from the input leaf image. The CNN is trained on their private leaf datasets. The model analyzes the visual extracted information to predict whether the leaf is healthy or not. Jan and Ahmad [13] also used a CNN-based deep learning approach to classify the incoming leaf in a healthy or infected state. Table 1 illustrates a comprehensive summary of the most prominent leaf disease prediction models.

Table 1 Comprehensive analysis of plant disease detection and prediction techniques

Year	Author(s)	Feature extraction	Deep learning	Dataset	Accuracy (%)
2004	Leemans and Destain [7]	CNN	Yes	Private	89.0
2009	Camargo and Smith [3]	Global and local low-level features	No	Private	87.0
2013	Kadir et al. [9]	Probabilistic neural network (PNN)	Yes	Flavia	93.75
2014	Dubai and Jalal [4]	Local low-level features	No	Private	97.75
2014	Rathod et al. [10]	ANN	Yes	Private	96.0
2015	Kambale and Chougule [1]	Global and local low-level features	No	Private	98.0
2016	Bhavini and Samajpati [5]	Global and local low-level features	No	Private	78.0
2016	Singh and Misra [6]	Global and local low-level features	No	Private	91.0
2017	Khirade and Patil [5]	Global and local low-level features	No	Private	91.6
2020	Khanand Banday [12]	CNN	Yes	Private	97.18
2021	Jan and Ahmad [13]	CNN	Yes	Statistical	99.0

4 Open Research Challenges

Based on the available plant disease detection and prediction literature, we conclude several research challenges presently being active in the state-of-the-art disease prediction systems. The most important of these challenges are described below:

i. **Limited datasets**: As indicated in Table 1, most of the studies and results have been produced over the private datasets that are not shared and not available for public use. As a result, the research community cannot justify their novel results as these models cannot be directly compared with the earlier ones. Also, the unavailability of datasets restricts us from making a robust and efficient model that can work under any circumstances and a variety of input leaf images. In addition, since deep learning-based image recognition techniques need a considerable amount of data for training, having most of the dataset as private restricts us from using such advanced and efficient models in developing plant leaf disease prediction systems.

ii. **Low performance in real-time environments**: The accuracy of apple disease prediction systems is exceptionally good in the training instances [12, 13]; however, the accuracy drops significantly when it comes to natural real-time leaf samples. The primary possible reasons for this are the lack of robust data within the training samples. There are many species of apple trees throughout the world, each having different leaf shapes, sizes, and colors. While the deep learning-based models generate high-efficiency prediction models, the dependence on the training sample is enormous. If the input data is more robust and variable, the performance of such models in real-time situations will be improved. Thus, one of the major research issues nowadays is to alleviate the low performance of such models in real-time situations.

iii. **Multi-disease prediction systems**: A lot of apple disease prediction systems have been developed until now, as seen in Table 1. Out of these many models, only a few [13] can detect multiple diseases within the same model. The segmentation and recognition complexities of multiple diseases incorporated within a single framework are challenging tasks. Some attempts have already been made in this subject using deep learning techniques. However, accuracy remains a challenge. Thus, the research community must focus on developing such multi-disease prediction systems as these may save a lot of human effort and time.

iv. **Infeasibility of acquiring new datasets**: Apple trees grew only in specific regions of the world and India. The species also vary along with this globe dramatically. Obtaining a massive amount of apple leaves for a single researcher is not feasible. In addition, the leaf dataset needs annotated details for training and learning. The annotation of such apple leaves into predefined diseases needs human expertise. Therefore, the infeasibility of producing novel apple leaf datasets is another major research challenge in developing such apple disease prediction systems.

v. **The high dimensionality of input feature space**: The features extracted from the input leaf images are highly complex. The machine learning-based algorithms extract the global and local features from the input image that are heavy for computation and segmentation. Similarly, the deep learning-based methods using ANN or CNN for feature extraction [5, 6, 12, 13] need a lot of resources and time for both training and validation. Thus, the dimensionality of such systems is a significant hurdle in making these systems available for public use. The research must be focused on developing such disease prediction systems using lightweight resources.

5 Conclusions

In this study, a systematic review of ADPS was presented along with the developments in apple disease prediction via intelligent methods. The main contributions of ADPS have been noticed in machine learning, and few techniques were based on deep CNN models. It was also observed that the majority of the ADPS systems had been trained and developed for non-Kashmiri apples. Hence, future research can be oriented toward developing more accurate ADPS through deep learning models. Apart from this, the development of appropriate-sized benchmark datasets covering samples from different classes is another area that requires futuristic attention. Exploring deep learning-based recent methods is an additional area that needs to be thought out by researchers. To countermeasure, the scarcity of datasets, the notion of data augmentation may be further explored.

References

1. Manik KA, Chougule SR, Int J Eng Sci Res Technol Classification Grading Apple Fruit Disease
2. Dharm P, Parth P, Bhavik P, Vatsal S, Yuvraj S, Darji M, A review of apple diseases detection and classification. Int J Eng Res Technol (IJERT) 8(4):2278–0181
3. Camargo A, Smith JS (2009) An image-processing based algorithm to automatically identify plant disease visual symptoms. Biosyst Eng 102(1):9–21
4. Dubey SR, Jalal AS (2013) Detection and classification of apple fruit diseases using complete local binary patterns. In: 2012 Third international conference on computer and communication technology. IEEE, pp 346–351
5. Khirade SD, Patil AB (2015) Plant disease detection using image processing. In: 2015 international conference on computing communication control and automation. IEEE, pp 768–771
6. Singh V, Misra AK (2017) Detection of plant leaf diseases using image segmentation and soft computing techniques. Inf Proc Agric 4(1):41–49
7. Kleynen O, Leemans V, Destain M-F (2005) Development of a multi-spectral vision system for the detection of defects on apples. J Food Eng 69(1):41–49
8. Solanki. Disease prediction descriptor using machine learing. IEEE
9. Kadir A, Nugroho LE, Susanto A, Santosa PI (2013) Leaf classification using shape, color, and texture features. arXiv preprint arXiv:1401.4447
10. Hasan M, Tanawala B, Patel KJ (2019) Deep learning precision farming: tomato leaf disease detection by transfer learning. In: Proceedings of 2nd international conference on advanced computing and software engineering (ICACSE)
11. Raut S, Fulsunge AG (2017) Plant disease detection in image processing using matlab
12. Khan AI, Quadri SMK, Banday S (2021) Deep learning for apple diseases: classification and identification. Int J Comput Intell Stud 10(1):1–12
13. Jan Mahvish, Ahmad Hazik (2020) Image features based intelligent apple disease prediction system: machine learning based apple disease prediction system. Int J Agric Environ Inf Syst (IJAEIS) 11(3):31–47

Deep Learning Techniques Used for Fake News Detection: A Review and Analysis

Yasmeena Akter and Bhavna Arora

Abstract People can quickly obtain and publish the news through many platforms, i.e. social media, blogs, and websites, among others. Everything that is available on these platforms in not credible and it became imperative to check the credibility of articles before it proves to be detrimental for the society. Multiple initiatives have been taken up by platforms like Twitter and Facebook to check the credibility of news on their platforms. Several researches have been undertaken utilizing machine learning (ML) and deep learning (DL) methodologies to address the problem of determining the reliability of news. Traditional media solely employed textual content to spread information. However, with the introduction of Web 2.0, fake images have become more readily circulated. The news piece, along with the graphic statistics, lends credibility to the material. The picture data is occasionally supplemented with the news pieces. For this research, the prime focus is DL-based solutions for text-based fake news detection. This research discusses about various techniques to automated detection of fake news. The paper gives a comparative analysis of various techniques that have been successful in this domain. Various datasets that have been used frequently are also highlighted. Despite various researches have been conducted for tackling fake news, these approaches still lack in some areas like multilingual fake news, early detection and so on.

Keywords Fake news detection · Social nets · Deep learning · Long short-term memory (LSTM) · Text classification · Words embedding technique

1 Introduction

People obtain and share information through social media, which has become an essential information platform. Its growing popularity has also allowed for the widespread distribution of false information, which has enormous detrimental consequences for society. As a result, it is vital to identify and control fake news on these

---

Y. Akter (✉) · B. Arora
Department of Computer Science and IT, Central University of Jammu, Jammu 181143, India
e-mail: yasmeenafateh900@gmail.com

Y. Singh et al. (eds.), *Proceedings of International Conference on Recent Innovations in Computing*, Lecture Notes in Electrical Engineering 1001,
https://doi.org/10.1007/978-981-19-9876-8_11

platforms in order to ensure that customers obtain correct information and social harmony is maintained. The traditional media used only the textual content to spread the data. But with the advent of Web 2.0, the fake images are also widely circulated. The news article along with the visual data makes the content appear credible. The news articles are sometimes complemented with the image data as well. So, the fake content that is spread online can be detected using *Visual-based approaches* [1], *Linguistic-based approaches* [1–5], *multimodal-based approaches* [5–7]. The majority of social media users are naive, and they are influenced by misleading information spread on these sites. They may unwittingly disseminate the misleading content and encourage others to do so by commenting on it. Some political analysts feel that misinformation and rumours had a role in Donald Trump's win in the 2016 US presidential election [2, 3].

Earlier techniques employed a variety of ML algorithms to detect fake news. However, with the widespread usage of internet platforms, the amount of material available online is expanding quickly. The ML techniques are not much efficient to handle this amount of data. DL techniques on the other hand do automatic feature selection are therefore have proved effective for FND problem.

The contribution of the paper is three-folds. The paper discusses about various types of fake news items that are shared on online platforms and provides taxonomy of recognition techniques. The main focus of this paper exists the DL approaches for text-based detection of fake news. A complete review of literature is provided apart from a comparative analysis of various datasets. The paper provides an outline of a general framework of a detection of fake news model. The paper also identifies and addresses various gaps or issues that still exist in detection fake news.

2 Deep Learning Models for Detecting Fake News

Due to the over growing use of social broadcasting over the past decades, and the efficiency of creation of the tempered content, it becomes very difficult to detect the fake content. Existing approaches [8–10] machine learning methods were utilized. But with the amount of data available, the trend has shifted to deep learning-based detection approaches [4, 11, 12], as these are very efficient in extracting the features from huge data. The next sections cover different deep learning approaches that are often used for detecting false news.

Convolutional Neural Networks (CNN): CNNs are capable of detecting patterns and have been effectively used to extract features from both pictures and text. CNN's two basic components are the convolutional layer and the pooling layer. The convolutional layer is made up of a collection of learnable filters that slide over the matrix's rows [13]. In existing papers [9, 14, 15], the authors suggest a Text-Image CNN (TI-CNN) model for detecting false news that takes into account both text and visual information.

A. *Recurrent Neural Network (RNN)*: RNNs are artificial neural networks that build a graph of nodes, either directed or undirected, over time. This enables it to behave in a temporally dynamic manner. RNNs, which are descended from feed forward neural networks, can manage variable length input sequences by utilizing their internal state (memory) [13, 15, 16].
B. *Long short-term memory (LSTM)*: The LSTM is a recurrent neural network. The previous step's output is used as an input in the current RNN stage. It addressed the issue of RNN long-term dependence, which happens when an RNN is unable to predict words stored in long-term memory but can make more accurate predictions based on current data.
C. *Bidirectional LSTM/RNN*: Bidirectional LSTM (BiLSTM) is a two-LSTM sequence processing paradigm with one for forward processing and the other for backward processing. For natural language processing tasks, bidirectional LSTM is a common solution.
D. *Gated recurrent units (GRUs)*: Cho et al. proposed the Gated Recurrent Unit (GRU), a well-known recurrent network version that governs and manages information flow between neural network cells using gating algorithms. The GRU, like an LSTM, has a reset gate, an update gate, but no output gate.

2.1 Fake News Detection Taxonomy

In order to identify fake news using algorithms, we establish a taxonomy described in Fig. 1 that classifies the detection methods as follows: feature-based, ML-based, platform-based, languages-based, and detection level-based.

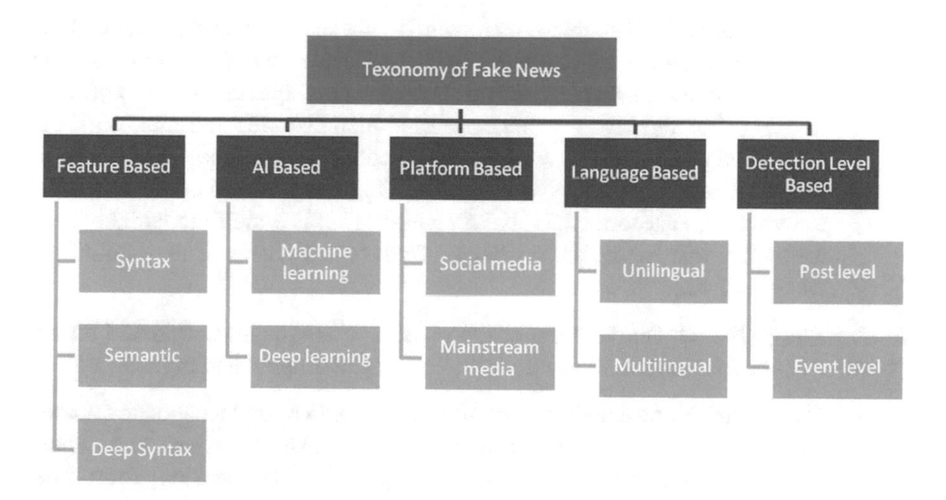

Fig. 1 Taxonomy of paradigms for fake news detection

A. **Feature-based**: This method relies on a person or software algorithm using linguistics to detect fake news [3]. The linguistic-based methods can be further classified on the basis of

   i. *Semantics*: Semantic features are those that produce a collection of shallow meaning representations, such as sentiment, named entities. The researchers propose a unique semantic fake news detection method based on relational features extracted directly from text, i.e. sentiment, entities, and facts.
   ii. *Syntax*: The word analysis is insufficient for predicting false news, other linguistic methodologies, such as syntax and grammar analysis, must be used taken into account.
   iii. *Deep syntax*: Probability Context Grammars [2] are used to implement the deep syntax technique. Probability Context Free Grammars, which perform deep syntactic tasks via parse trees, enable Context Free Grammar analysis.

B. **Artificial Intelligence (AI)-Based**: *V*arious approaches for FND have been studied in the literature. ML and DL methods are two widely used techniques for FND.

   i. *Machine learning* is a sort of synthetic intelligence where computers can learn from previous data and accomplish tasks on their own.
   ii. *Deep learning*: Deep learning is a very sophisticated subset of machine learning. Deep learning is based on a layered structure of algorithms known as an artificial neural network [17].

C. **Platform-Based**: This is about the digital environment in which the dataset's news is shared and distributed to the public. To detect fakes news, the researchers used two distinct sorts of media sites.

   i. *Social Media*: It's a double-edged sword to use social media for news update.
   ii. *Mainstream Media*: Mainstream Media (MSM) is a media of communication may be defined as those that deliver consistent messages in a one-way process to a big, homogeneous audience with similar features and interests. It is concerned with various news websites for collection and analysis of fake data like Washington post, CNN, PolitiFact, etc. Many fake sites are also used to circulate misleading data. FakeNewsNet [18] is a multimodal fake news repository that has its data collected from two websites namely PolitiFact and gossip cop.

D. **Language-Based**: On the basis of the number of languages considered by the models, the FND model is usually bifurcated as unilingual and multilingual.

   i. *Unilingual*: Many existing models have used only a single language for text-based FND. To recognize fake news in Persian texts and tweets, researchers created an LSTM hybrid model using a 14-layer bidirectional long short-term memory (BLSTM) neural network [19].
   ii. *Multilingual*: Complexity, stylometric, and psychological text aspects were investigated using corpora of news articles in American English, Brazilian

Portuguese, and Spanish. Researches in [20, 21] have retrieved traits to distinguish between fraudulent, authentic, and satirical news in multilingual environment.

E. **Detection Level-Based**: On the basis of the type of detection level, a FND model can be either a post-level detection model or event-level detection model.

   i. *Post level*: It only incudes the tweet or post at hand, other auxiliary information is not considered.
   ii. *Event Level*: Apart from the tweet or post, it also considers other auxiliary information like other post that is relevant to the post that is being considered.

Existing deep learning models have made significant progress in tackling the challenge of detecting false news [22, 23].

3 Literature Review

Previous studies and experiments on false news identification using machine learning and deep learning are highlighted in the review of literature. Researchers present in paper [15] about LSTM RNN infrastructures for large scale aural modelling in speech recognition. Various articles have discussed a detection of fake news model created on bidirectional LSTM-intermittent neural network. The result network in [13], two intimately available unshaped newspapers datasets are used to estimate the presentation of the model. Fake news detection using unidirectional LSTM, in terms of [19], strives to illuminate on false update problem and the procedure of relating accuracy, the LSTM-RNN model surpasses the original CNN model [14]. In this learning, a model LSTM and 14-subcaste are utilized. To detect fake news in Persian texts and tweets [16], researchers used a BiLSTM neural network. Accordingly, the current exploration false news using deep literacy approaches. In article, the Text and Image Information Grounded Convolutional Neural Network (TI-CNN) model is examined. TI-CNN is trained with both text and image data at the same time by projecting the unambiguous and idle qualities into a single point space. The usefulness of TI-CNN in solving the fake news detecting challenge has been proved by extensive examination of real-world fake news datasets. The dataset in these paper sweats on the news about [16] American presidential election. In this study [17], researchers learn the problem of fake news discovery. They concentrate on fake news finding styles grounded on text features.

4 Datasets for Detecting Fake News

This section gives an overview of the various datasets available for FND. Table 1 provides a comparative analysis of various frequently used datasets. These datasets are collected from various online platforms and main-stream media websites.

A few patterns emerge when comparing fake news databases. Because the bulk of the datasets are tiny, traditional deep learning models that require massive amounts of training data may become inefficient. Only few datasets have more than half a million samples, the largest of which being CREDBANK and FakeNewsCorpus, which each have millions of samples. Furthermore, many database categorize their data in a restricted number of ways, such as false vs true.

More fine-grained labelling may be found in datasets like, LIAR, and FakeNewsCorpus. While several datasets incorporate data from a number of groups, others focus on specialized topics like politics and GossipCop. Because of the small number of categories, these data samples may have restricted context and writing styles. FakeNewsNet [24], a multimodal dataset that includes text as well as images with it, is not available in their whole but can be retrieved as sample data. This is mostly because the majority of these datasets uses Twitter data for social settings and so is not publicly available under licence laws. The LIAR dataset is generally well-balanced in terms of labels: with the exception of 1050 pants-fire incidents, the occurrences for all other labels vary from 2063 to 2638. The given datasets are outdated and out of date.

Such datasets are insufficient for addressing the issue of fake news data for recent news data since the techniques of fake news producers vary with time.

5 Comparative Analysis and Discussion

A comparison of several fake news detecting techniques: Table 2 summarizes the findings of a study of several false news detection algorithms based on machine learning and deep learning offered by various researchers.

Platform content available on social media is less as compare to the mainstream media, therefore the textual methods for FND performs better is such cases.

When the text data on online platform is complemented with the propagation data, user information for detection, such methods perform better. When huge amount of data is used, DL methods perform better than the ML because feature engineering is done automatically.

The comparison analysis on various basis above table shows that while performing FND for LIAR dataset using LSTM RNN approach has accuracy of 79% means it does not achieve higher accuracy, while researchers developed a hybrid model of long short term memory (LSTM) and a 14-layer bidirectional long short term memory (BiLSTM) neural network [19]. Based on the findings, the suggested model has a 91.08% accuracy rate in detecting fake news and rumours in deep learning.

Table 1 Comparison of various available datasets

Dataset	Year of release	Platform	Content type	Data category	Total No of claims	Label	No. of items
CREDBANK	2015	Twitter	News items posted on twitter	Variety	60,000,000	5-point credibility scale	1049
BuzzFace	2016	Facebook	News items posted to the Facebook by nine new outlets	Political	2263	Mostly true, mostly false, mixture of true and false, no factual content	1656, 104, 244, 259
PHEME	2016	Twitter	Nine different newsworthy events	Variety	330	False, True	250, 4812
LIAR	2017	PolitiFact	Political statements	Political	12,836	Pants-on-fire, false, slightly true, half true, largely true, and true	1050 others 2063–2638
FakeNewsNet	2020	Twitter	US Politics, Entertainment	Political/celebrity	602,659	Fake and true	420, 528 (Politifact), 4947,16,694 (GossipCop)
FakeNewsCorpus	2020	Opensources.co	American politics	Variety	9,400,000	True and fake	5383, 3844

Table 2 Analysis of various fake news detection techniques

Ref	Year	Contributions	Technique used	Dataset	Result (Acc)	Challenges
[1]	2018	LSTM RNN Infrastructures for large scale aural modelling in speech recognition	LSTM, RNN	LIAR	79%	Does not achieve higher accuracy
[2]	2018	The feature extraction approach is long short term memory (LSTM)	LSTM	Kaggle	92%	Only text based characteristics are used to identify fake news
[3]	2018	New features for training classifiers have been added	Transformer-based approach	There are 2482 news stories on the US election	85%	Only a little dataset is available. Authors should be able to employ a large number of datasets in the future, as well as deep learning algorithms with superior fake news forecasting
[4]	2019	By verifying tweets using specific taught characteristics, BiLSTM-CNN can determine whether they are fake news or not	BiLSTM-CNN	5400 twitter tweets	86.11%	Only text-based characteristics may be used to classify if something is fake or not
[5]	2019	BiLSTM is a machine learning algorithm that detects bogus news in Persian texts and tweets	BI-LSTM	Liar dataset	91.07%	To classify whether something is false or real, they can only use text-based characteristics

(continued)

Table 2 (continued)

Ref	Year	Contributions	Technique used	Dataset	Result (Acc)	Challenges
[6]	2019	CNN-LSTM is used to identify fake news based on the relation between article headline and article body	CNN-LSTM	There are 3482 news stories on the US election	71.2%	Does not achieve higher accuracy
[7]	2020	TI-CNN is trained with both text and visual data at the same time	TI-CNN	The dataset efforts on the news about American presidential election	60% to 94%	Does not attain a greater level of accuracy, and is limited to a small data set
[8]	2021	Using an ensemble model that incorporates three popular machine learning models: decision tree, random forest, and extra tree classifier	Classifiers: decision tree, random forest and extra tree	ISOT and LIAR	99.8%-ISOT 99.9%-Liar	Only text-based characteristics may be used to classify if something is fake or not
[9]	2021	Three sets of characteristics linked to linguistic, user-oriented, and temporal propagation were presented	Bi-directional LSTM-recurrent neural network	Twitter and Weibo	90%	The sample was limited to 1111 Twitter posts and 818 Weibo posts
[10]	2021	A classifier was trained using all semantic data, user-based features, structural features, sentiment-based features, and anticipated features	Deep diffusive neural network	400 K official media articles and 60 K erroneous information in an Italian Facebook data collection	90%	Failed to identify the factor that has a detrimental influence on the information

The study examines many DL approaches for FND, among which LSTM may learn long-term dependencies and hence accomplish multiple problems that prior learning algorithms for recurrent neural networks were unable to perform (RNNs). Furthermore, Bidirectional long short-term memory (Bidirectional LSTM) outperforms LSTM since input flows in both directions and data may be used from both sides. It may also be used to simulate the sequential interactions between words and sentences in both directions.

6 General Framework

The next part gives a generic framework for FND (Fig. 2) to define if a given news article is real or fake based on the text data. The process starts from selecting and preprocessing the dataset. The comparison of various datasets is available in above Sect. 4 Table 1.

Apart from the datasets that are available various webs APIs can also be used for collection of raw data. The data collecting technique for developing a fake news detection system is determined by the task specification.

In earlier research, examples of fake news were gathered from a list of suspect websites. Fake detection, fact-checking, truthfulness categorization, and rumour detection are just some of the uses for datasets. The first is a forecast that a certain piece of information (news item, review, remark, etc.) will be purposely false.

The practice of examining and validating true assertions included in a piece of information is known as fact-checking; unlike false detection, fact-checking operates at the statement or claim level. Veracity classification is similar to fake detection in that it seeks to predict whether or not a piece of information is true. Finally,

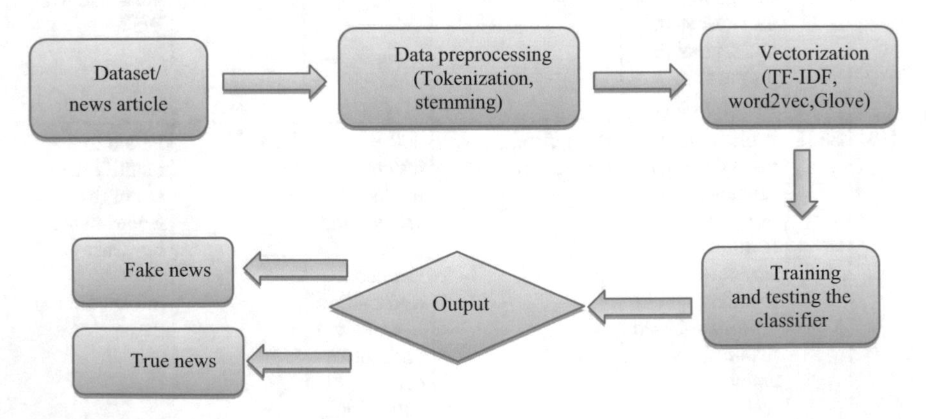

Fig. 2 General framework of the fake news detection system

rumour detection attempts to distinguish between confirmed and unverified information (rather than true or false), with unverified information having the potential to be true or false or remain unresolved [24].

The next process involves data preprocessing to convert raw data into a clean dataset for making analysis easier. Once the dataset is cleaned and ready, the text is converted to numerical sequence or feature vector. Then, word embedding is done which is a learnt text representation in which words with related meanings are represented similarly. Word2Vec, one-hot encoding, GloVe, term frequency-inverse document frequency (TF-IDF), etc., are some of the methods that are frequently used for learning a word embedding from text data.

GloVe embeddings perform better on some data sets, whereas word2vec embeddings perform better on others. They both do an excellent job of capturing the semantics of analogy, which leads us a long way towards lexical semantics in general.

Other pre-trained embedding like Bert and Robustly Optimized BERT (RoBERTa) are also used. BERT and RoBERTa are transformer based models for NLP. RoBERTa architecture reduces the pre training time.

When the feature vector is complete, it is passed into the trained DL model. The CNN (Convolutional Neural Network) is the most basic model in machine learning, although it has the issue of categorization of pictures with varied Positions, Adversarial instances, Coordinate Frame, and other minor limitations such as performance. The other frequently used model is the RNN model, but it suffers with gradient exploding and disappearing problem. Also, it is quite difficult to train an RNN.

Tanh or Relu cannot handle extremely lengthy sequences when employed as an activation function. To overcome these limitations Researchers use LSTM, LSTMs were developed to address the issue of vanishing gradients that might arise while training a regular RNN. Once the model is ready, it will then be evaluated on various performance matrices. There are a set of performance evaluation metrics that are particularly used for performance evaluation in credibility detection task, i.e. confusion matrix, precision, recall or sensitivity, and F1-Score.

7 Open Gaps and Challenges

Although there are several techniques and methods that have been established in the last decade to counter fake news but there are still several open research issues and challenges. The section below provides various challenges and open areas that need to be catered.

A. Multilingual: On the basis of the number of languages considered by the models, the FND model is usually bifurcated as unilingual and multilingual. Multilingual news as news can be in any language nowadays. Earlier the news was unilingual and detection for fake news was a bit easier.
B. Early detection: No early detection of news, it takes almost a month to detect fake news which is very challenging.

C. Quick and real time: The finding of the source in real time is important for controlling the feast of ambiguous information and reducing the negative effect on society.
D. Real-time data collection: It is challenging to collect real-time data, automate rumour detection, and track down the original source. Information pollution, fake news, rumours, disinformation, and insinuation have emerged as a by-product of the digital communication environment, proving to be extremely destructive.
E. Other Challenges: Fake news spreading across several languages and platforms, complex and dynamic network topologies, massive volumes of unlabelled real-time data, and early rumour detection are just a few of the intriguing issues that have yet to be handled and need more investigation. Improving the legitimacy and future of the online information ecosystem is a social community responsibility.

8 Conclusion

Social media have pressed the capability to change information at a much bigger pace, to a far broader audience than ever before. This information is not always credible, because anyone can broadcast anything on the Internet. The existing approaches used ML techniques, but with the amount of data available the trend has shifted to DL-based detection approaches, as these are very efficient in extracting the features from huge data. Various deep learning techniques are used for this purpose which includes CNN, RNN, and LSTM. The CNN and RNN networks particularly are used significantly in commuter vision and NLP tasks, respectively. Variants of RNN, i.e. LSTM, GRU, and other gated networks are also widely used and are proven to be must efficient. For the embedding of data, many techniques like word2vec, GloVe, TF-IDF, one hot encoding are used. GloVe embeddings perform better on some data sets, whereas word2vec embeddings perform better on others. They both do an excellent job of capturing the semantics of analogy. The study explores different datasets that are accessible for detecting false news. Although there are various techniques and methods that have been developed in the last decade to counter fake news but there are still several open research issues and challenges like multilingual content, no early detection of news, tempered images, low accuracy, lack of quick and real time discovery, etc.

References

1. Singh VK, Dasgupta R, Raman K, Ghosh I (2017) Automated fake news detection using linguistic analysis and machine learning automated fake news detection using linguistic analysis and machine learning. https://doi.org/10.13140/RG.2.2.16825.67687
2. De Beer D, Matthee M (2021) Approaches to identify fake news : a systematic literature review, no. Macaulay 2018. Springer International Publishing
3. Raza S, Ding C (2021) Fake news detection based on news content and social contexts: a transformer-based approach. Int J Data Sci Anal. https://doi.org/10.1007/s41060-021-00302-z

4. Caucheteux C, Gramfort A, King J-R (2021) Disentangling syntax and semantics in the brain with deep networks. [Online]. Available: http://arxiv.org/abs/2103.01620
5. Nakamura K, Levy S, Wang WY (2020) r/Fakeddit: a new multimodal benchmark dataset for fine-grained fake news detection. In: Lr. 2020—12th international conference languages resources evaluation conference proceeding, pp 6149–6157
6. Wu XZBJ, Zafarani R (2020) SAFE : similarity-aware multi-modal fake. Springer International Publishing
7. Duong CT, Lebret R, Aberer K (2017) Multimodal classification for analysing social media. [Online]. Available: http://arxiv.org/abs/1708.02099
8. Abedalla A, Al-Sadi A, Abdullah M (2019) A closer look at fake news detection: a deep learning perspective. Pervasive Health Pervasive Comput Technol Healthc 24–28. https://doi.org/10.1145/3369114.3369149
9. Bahad P, Saxena P, Kamal R (2019) Fake news detection using bi-directional LSTM-recurrent neural network. Procedia Comput Sci 165(2019):74–82. https://doi.org/10.1016/j.procs.2020.01.072
10. Zhang J, Dong B, Yu PS (2018) FAKEDETECTOR: effective fake news detection with deep diffusive neural network. [Online]. Available: http://arxiv.org/abs/1805.08751
11. Bangyal WH et al (2021) Detection of fake news text classification on COVID-19 using deep learning approaches. Comput Math Methods Med 2021:1–14. https://doi.org/10.1155/2021/5514220
12. Abedalla A, Al-Sadi A, Abdullah M (2019) A closer look at fake news detection: a deep learning perspective. In: ACM international conference proceeding series, pp 24–28. https://doi.org/10.1145/3369114.3369149
13. Sak HH, Senior A, Google B, Long short-term memory recurrent neural network architectures for large scalc acoustic modeling
14. Yang Y, Zheng L, Zhang J, Cui Q, Li Z, Yu PS (2018) TI-CNN: convolutional neural networks for fake news detection. [Online]. Available: http://arxiv.org/abs/1806.00749
15. Bahad P, Saxena P, Kamal R (2019) Fake news detection using bi-directional LSTM-recurrent neural network. Procedia Comput Sci 165:74–82. https://doi.org/10.1016/j.procs.2020.01.072
16. Education M (2021) The use of LSTM neural network to detect fake news on Persian 12(11):6658–6668
17. Drif A, Giordano S (2019) Fake news detection method based on text-features, no c, pp 26–31
18. Learning to detect misleading Boididou C, Papadopoulos S, Apostolidis L, Kompatsiaris YD, content on Twitter (2017). In: ICMR 2017—Proceeding 2017 ACM international conference multimedia retrieval, pp 278–286. https://doi.org/10.1145/3078971.3078979 "No Title"
19. Goldani MH, Momtazi S, Safabakhsh R (2021) Detecting fake news with capsule neural networks. Appl Soft Comput 101. https://doi.org/10.1016/j.asoc.2020.106991
20. Yesugade T, Kokate S, Patil S, Varma R, Pawar S (2021) Fake news detection using LSTM. Int Res J Eng Technol. [Online]. Available: www.irjet.net
21. Shrestha M (2018) Detecting fake news with sentiment analysis and network metadata
22. FANG: Leveraging Social Context for Fake, Nguyen VH, Sugiyama K, Nakov P, Kan MY (2020) News detection using graph representation. In: International conference information knowledge management proceeding, pp 1165–1174. https://doi.org/10.1145/3340531.3412046. "N"
23. Ma J et al (2016) Detecting rumors from microblogs with recurrent neural networks. In: IJCAI international Jt. conference artificial intelligent, vol 2016-Janua, pp 3818–3824
24. Zhou X, Jain A, Phoha VV, Zafarani R (2020) Fake news early detection. Digit Threat Res Pract 1(2):1–25. https://doi.org/10.1145/33774
25. Li H, Wang H, Liu G, An event correlation filtering method for fake news detection
26. Potthast M (2017) A stylometric inquiry into hyperpartisan and fake news, no. February, 2017
27. AAAA et al (2021) Detecting fake news using machine learning: a systematic literature review. Psychol Educ J 58(1):1932–1939. https://doi.org/10.17762/pae.v58i1.1046
28. Boididou C et al (2018) Verifying information with multimedia content on twitter: a comparative study of automated approaches. Multimed Tools Appl 77(12):15545–15571. https://doi.org/10.1007/s11042-017-5132-9

29. Declare: Debunking fake news and false claims using, Popat K, Mukherjee S, Yates A, Weikum GP, evidence-aware deep learning. In: Proceeding 2018 Conference Empired Methods Natural Language Processing EMNLP 2018, and https://doi.org/10.18653/v1/d1.-1003 22–32, "N"
30. Multimodal fusion with recurrent neural networks for rumor, Jin Z, Cao J, Guo H, Zhang Y, Luo JD, detection on microblogs (2017) MM 2017—Proceeding 2017 ACM Multimedia Conference, pp 795–816. https://doi.org/10.1145/3123266.3123454. "No Title"
31. Mohsen Sadr M, Chalak AM, Ziaei S, Tanha J (2021) The use of LSTM neural network to detect fake news on persian Twitter

Darkness Behind the Darknet: A Comparative Analysis of Darknet Traffic and the Crimes Behind the Darknet Using Machine Learning

Bilal Ahmad Mantoo and Harsh Bansal

Abstract Current era is the age of digital technology with LTE running over it. Ranging from smaller to bigger tasks, every bit of information is being stored and channelized through Internet which becomes a huge reservoir of data. This information became a point of interest for digital users present in the backdoors of Internet, which we call as black hat hackers. These hackers have come up with their own network, where they can perform illegal activities without even being noticed in their network, usually called as darknet. Every unofficial work like drug dealings, ammunition supply, human trafficking, credit card detail sharing, etc., is done here. Traces of these activities are not present because of the inaccessible nature of darknet. The architecture of these Websites is totally different from the normal network we work on, and it is therefore not possible to access such Websites like Tor and Onion directly. Therefore, there is a prior need for the detection of traffic coming from the darknet. In this paper, different algorithms have been used which classifies the darknet traffic based on varied features. The algorithms that are being used are *K*-nearest neighbor (KNN), support vector machine (SVM), Naïve Bayes classifier (NB), and random forest (RF). Out of these algorithms, random forest performs the best with the accuracy of 98.7.

Keywords Darknet · Human trafficking · Tor · Onion

B. A. Mantoo (✉)
Department of Computer Science and Engineering, Presidency University, Itgalpur, Yelahanka, Bangaluru 560064, India
e-mail: bilalbashir136@gmail.com

H. Bansal
Department of Computer Science and Engineering, Sir Padampat Singhania University, Udaipur, Rajasthan, India

Y. Singh et al. (eds.), *Proceedings of International Conference on Recent Innovations in Computing*, Lecture Notes in Electrical Engineering 1001,
https://doi.org/10.1007/978-981-19-9876-8_12

1 Introduction

Darknet is a network inside the normal Internet that must be identified with explicit programming designs, or authorization, [1] and frequently utilizes a one of a kind redid correspondence convention. Two average darknet types are social networks [2] (typically utilized for document facilitating with a shared connection), [3] and obscurity intermediary organizations, for example, Tor through an anonymized series of connections [4].

The "darknet" was promoted by significant media sources to connect with Tor Onion administrations, when the notorious silk road utilized it, [5] in spite of the phrasing being informal. Innovation, for example, Tor, I2P, and Freenet, was planned to guard advanced rights by giving security, namelessness, or oversight obstruction and is utilized for both illicit and real reasons. Unknown correspondence between informants, activists, columnists, and news associations is likewise worked with darknets through utilization of information. The initially detailed online exchange of facts occurred in 1972 and included a modest quantity of cannabis sent between studies utilizing their separate ARPANET accounts [6]. From that point forward, the converging of complex advancements has made the ideal stage for the improvement of unlawful online bootleg trades. Consolidating mysterious Internet perusing, digital currencies, and worldwide conveyance frameworks, darknet markets, or unlawful crypto markets, have arisen and changed the retailing of illegal items and stash. There are genuine utilizations for clients of Tor. In particular, it is being utilized by columnists, activists, and campaigners in the USA to keep up with the security of their correspondences and stay away from retaliations from (their respective) government.

Darknet has been utilized to purchase, sell drugs, weapons, fake archives, for example, visas, driver's licenses, government managed retirement cards, paid killers and service bills, and fake cash, just as to give a medium to contract executioners to request customers, pornography, pirated software. It has likewise been used to purchase and sell MasterCard data (complete with a client's name, address, telephone number, card check worth, and termination date), youngster erotic entertainment, pilfered programming and other protected materials, pernicious programming (or malware), and PC hacking administrations and instruments (to acquire unapproved admittance to accounts and frameworks). There are special markets in the Web which provides commission ranging from three to ten percent.

The blend of online secrecy, pseudo-mysterious exchanges, and modern covertness bundling has established an original climate that restrains law enforcement agencies (LEAs) capacity to explore the exercises of darknet markets. In the previous decade, LEAs have endeavored to control improvement and development of these darknet markets. Because of the overall network of the Internet, these activities have ordinarily elaborate cross-jurisdictional participation among LEAs and coordinated effort as knowledge sharing and joint tasks.

This paper overviews about the process of darknet traffic and analytical ways to find how to distinguish the darknet traffic by describing the data collected. The structure of this paper is as follows: The data is being collected or downloaded form an open source platform, preprocessing the data, applying the dimensionality reduction algorithm to reduce large feature set and then applying the machine learning algorithms.

2 Related Work

The recognition of criminal operations on the Web is some of the time got from a more broad theme characterization. For instance, [7] ordered the substance of Tor covered up administrations into 18 effective classes, just some of which correspond with criminal behavior. Reference [8] consolidated unaided component determination also, a SVM classifier for the characterization of drug deals in a mysterious commercial center [9]. They assessed the ubiquity of onion spaces in the darknet network by analyzing approaching to those areas, whose shortcoming is that the examination can be impeded if this weakness is solved. They drew closer continued in this paper which is completely different. Tor browser is the special browser that is used for darknet, and it passes the information through the victim computer, in order that the traffic should not be traced back to originating user. Tor is actually a layered architecture which uses different layers on different countries worldwide [10]. Tor works on OSI layer 4, and in response to this, onion proxy software works on SOCKS Socket Secure which is layer 5 in networks. This network has lot more redirections and is encrypted so that no one at any point will decrypt the information at any side [11]. [12] caught 30,373,399 parcels in correspondence between subnets at source organization and destination organization. They made 27 classifications of traffic examination profile in a 27-dimensional component vector comprising of bundle count, source IP and port, and objective IP and port. Vindictive parcels were recognized by performing progressive grouping and coordinating malware marks with distinguished bundles. Nonetheless, they could not distinguish new malware tests due to catching at the nearby tap. Distributed reflection denial of service (DRDoS) assault was distinguished by separating extra data like power; rate what is more, geo-area [13]. In another endeavor, the absolute recurrence of bundle, number of source has, and designated port for TCP and UDP conventions were utilized to identify darknet traffic [14].

Tor network not only provides best encryption but it is also being utilizing in mimicking conventional transport of hypertext transport protocol secure (HTTPS), putting together the spotting of Tor passage complex, for all network engineers [15]. A novel method for perceiving Tor traffic by analyzing quantitative gaps in their secure socket layer (SSL) protocol [16].

Table 1 Darknet network data

Traffic category	Application used
Audio-stream	Vimeo and YouTube
Browsing	Firefox
Chat	ICQ, AIM, Skype, Facebook
Email	SMTPS, POP3 and IMAPS
P2P	utTorrent and BitTorrent
Video-streaming	Vimeo and YouTube
VOIP	Facebook, Skype, and Hangout

3 Proposed Model

This section describes three main parts of our research such as (1) feature extraction to select the important features from the data, (2) preprocessing over the selected features, and (3) applying different machine learning models to find the best model.

3.1 Dataset Details

The dataset used in our work is CICDarknet2020 dataset, which contains benign and darknet traffic. The darknet traffic contains audio-streaming, browsing, chat, email, P2P, video-stream, and VOIP Table 1.

The dataset is actually divide into 2 layers labeled as benign to represent normal traffic, and the other one represents malicious traffic from Tor or VPN. Figure 1 shows different traffic protocols used in the dataset. The number of entries in the dataset is approximately 158,659 in total, with 134,348 normal packets and 24,311 darknet packets; out of these, audio-streaming is having the highest packets Table 2.

3.2 Feature Selection

Feature selection is considered to be the most important part in any machine learning application. When you have the best features from the data this not only increases the accuracy but makes the system overall efficient in terms of time and memory. Technically, we can say that reducing the number of features when you are building a predictive model. These features can be selected automatically or chosen manually. If you selected irrelevant information, then your model could learn something wrong. The method that we used in this study for feature selection is feature importance. Feature importance is a inbuilt class that comes with tree-based classifier, and in this study, we implement the extra tree classifier for extracting top 20 features of the data as shown in Fig. 2.

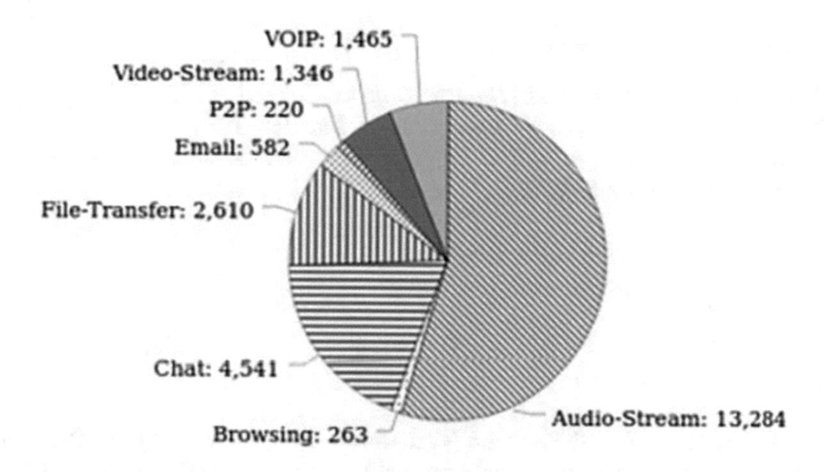

Fig. 1 Distribution of data in CICDarknet dataset 2020 [17]

Table 2 Overall traffic details in darknet dataset

Traffic category	Number of packets
Normal traffic	134,348
Darknet traffic	24,311
Total	158,659

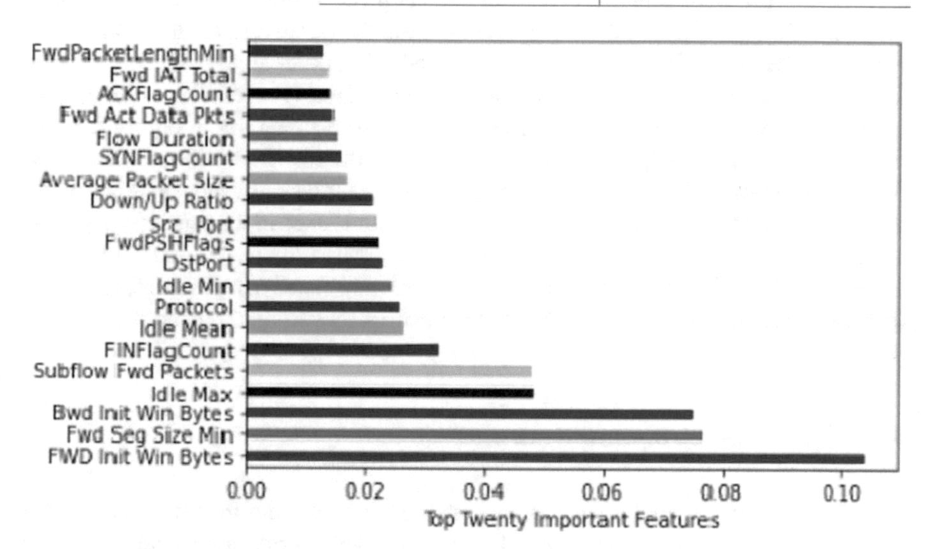

Fig. 2 Top twenty features of the dataset using tree-based classifier

Algorithm 1 Feature selection algorithm

1: Procedure: Feature_Rank (Feature_Colums, Labels)
2: Feature_tree with input (Feature_Colums, Labels)
3: for each node in tree do
4: calculate feature importance score
5: sort node.
6: if feature_importance < 0.01 then
7: Order the features with best values
8: end if
9: end for
10: end procedure

The frequencies of all these features are shown in Fig. 3, which gives the correct information about important features present in darknet area. The impact of backward window size is found high in all four darknet classes except in non-Tor data. Another important feature considered is destination port (DstPort), and network ports provide essential information regarding about open services. Port number present in TCP header is 2 byte field. The destination port is usually targeted by all malwares. The feature analysis gives best information that port 80 (hypertext transfer protocol) is most probed port in darknet network with occurrences of 3000, then port 443 (hypertext transfer protocol over TLS/SSL HTTPS) is probed. Finish flag (Finount) used for connection termination when there are no more data from sender; after this, all the resources are freed. This feature is mostly used by darknet area, where connection termination occurs more frequently and hence contributes more information Fig. 3.

It is clearly visible that only top ten features can be taken out of eighty features rest features vary less as per their scores using Algorithm 1 that is why we considered only top ten features in our study as shown in Fig. 2. Our result improves by removing less importance features from our dataset. Algorithm 1 takes input as two parameters; one is features from the dataset, and other field is target field; a tree is then built using these features, and each node is iterated for calculating feature importance, and these features are then ranked accordingly as their scores by taking a threshold of 0.01.

3.3 Principal Component Analysis

Principal component analysis (PCA) is a method for reducing the dimensions from a large dataset to a lower dimensional data with high information by decomposing the data into perpendicular vector where the spread of information is more. As more variance means, more information in data.

The data is fed to PCA algorithms in the form of matrix of features that are in the dataset. The main intention is to find the main variance in the dataset by calculating the eigen vector of the covariance matrix [18]

The mathematical concept of PCA is as under.

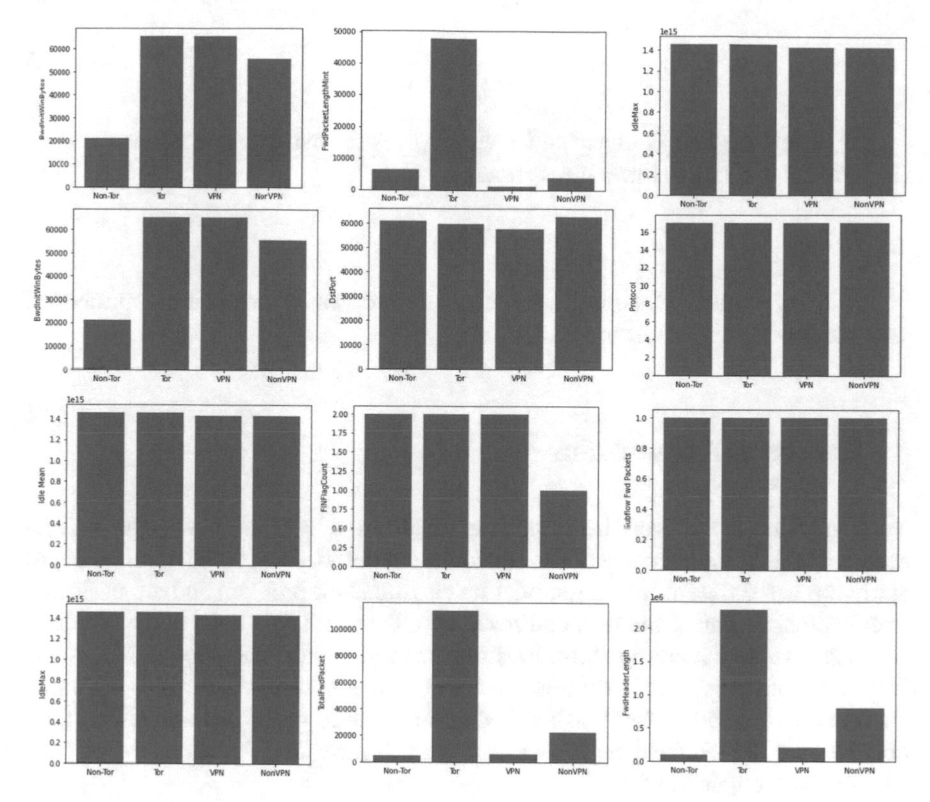

Fig. 3 Frequency of features present in different darknet traffic

(1) For data matrix X, calculate the covariance of X

$$\mu = \mathrm{E}\{\mathrm{X}\} \tag{1}$$

Then calculate the covariance.

$$R = \mathrm{Cov}\{X\} = E\{(X - \mu)(X - \mu)^t\} \tag{2}$$

(2) Find eigen value λ_1, λ_2, λ_3, λ_4... λ_n and eigen vector e_1, e_2, e_3, ..., e_n of the covariance matrix in descending order.

For the covariance R, solve the equation

$$|\lambda i - R| = 0 \tag{3}$$

In order to get the principal component, we have to count the proportion of data covered by the first M eigen values.

$$\frac{\sum_{i}^{m} x}{\sum_{i=0}^{n} y} \tag{4}$$

Select the first M eigen values which gives the maximum information.

(3) Project the data to lower dimensions

$$\mathrm{P} = \mathrm{WtX} \tag{5}$$

By using the first M, respectively, eigen vector, we can reduce the number of variables or dimensions from n to M.

4 Results and Discussion

Main intention in this work is to evaluate our different learning model over darknet data and to increase accuracy of classification and detection of darknet traffic. In our study, we use different learning model on classification problem and are evaluated over confusion matrix and then compared with the results obtained in the literature for other machine learning algorithms like support vector machine (SVM), Naïve Bayes, logistic regression, and random forest. The algorithms are compared based on true positive ratio (TPR) also called as recall, true negative rate (TNR), false positive rate (FPR), false negative rate (FNR), F measure FM calculated on both classes and weighted FM.

$$\mathrm{FM} = (2 * \mathrm{recall} * \mathrm{precision})/(\mathrm{recall} + \mathrm{precision}) \tag{6}$$

Elaborative research was performed to find the impact of variables, and how they added in prediction using Shapley values Fig. 4. Features of each variable are calculated based on Shapley values in whole sample in the dataset. The best features are shown from top to bottom, and their colors show the best fit to which they belong. In Fig. 4, it is clear that "*average packet size*" feature is the most important for prediction the darknet traffic. The Shapley value is also high for this, which means data packet is belonging to darknet traffic. In comparison with this, if the value is less, which indicates that packet comes from non-darknet sites. Different algorithms that are being applied to the dataset give different accuracy for finding the darknet packets. Out of this, random forest is performing best, with KNN in second last shown in Fig. 5 (Table 3).

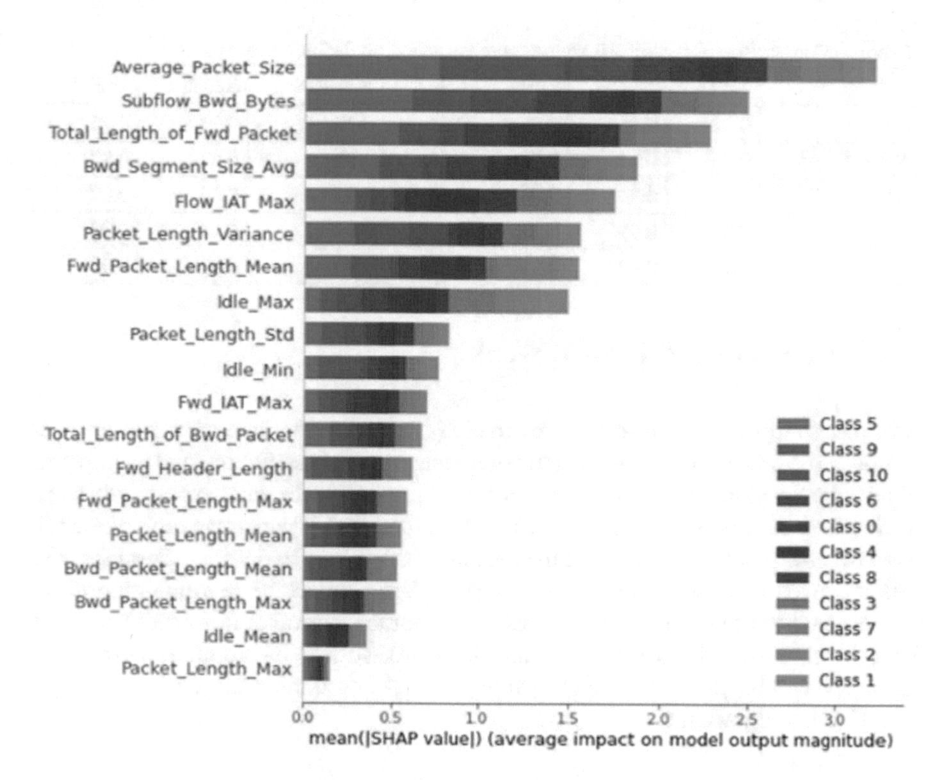

Fig. 4 Shapley values for top features chosen in the dataset

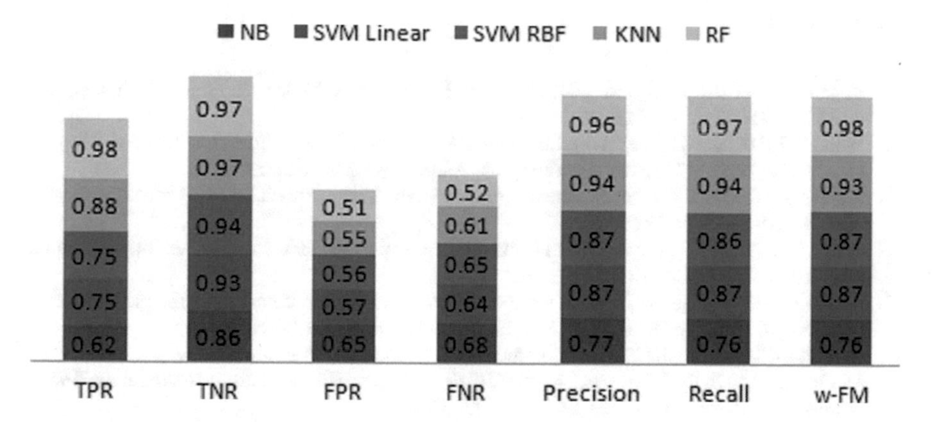

Fig. 5 Comparison of different algorithms

Table 3 Comparison of machine learning models applied on darknet traffic dataset

	TPR	TNR	FPR	FNR	Precision	Recall	w-FM
NB	0.62	0.86	0.65	0.68	0.77	0.76	0.76
SVM linear	0.75	0.93	0.57	0.64	0.87	0.87	0.87
SVM RBF	0.75	0.94	0.56	0.65	0.87	0.86	0.87
KNN	0.88	0.97	0.55	0.61	0.94	0.94	0.93
RF	0.98	0.97	0.51	0.52	0.96	0.97	0.98

5 Conclusion and Future Work

Darknet within Internet is a vast area to study, with different and unknown areas of crimes. The crimes over here vary from betting to organ trafficking and even worse than this. To stop such crimes, we not only require best cyber experts but latest technology as well, to handle the traffic of darknet. In this paper, we took one small step to show that how we can use the available technology to detect such traffic with the approach of using artificial intelligence. Various machine learning algorithms are used within this work which gives best accuracy, but random forest overcomes among all. This field is very vast, and lot more work is unexplored; algorithmic approach can be changed here to in future to make prediction more accurate using deep leering algorithms.

References

1. Gayard L (2018) Darknet: geopolitics and uses. John Wiley & Sons, Hoboken, NJ, pp 158. ISBN 9781786302021
2. Wood J (2010) The darknet: a digital copyright revolution. Richmond J Law Technol 16(4):14; Overhead J, Miller C (2012) Dissecting the Android bouncer. Summerton
3. Mansfield-Devine S (2009) Darknet. Comput Fraud Secur 2009(12):4–6. https://doi.org/10.1016/S1361-3723(09)70150-2
4. Pradhan S (2020) Anonymous. In: The darkest web: the dark side of the internet. India. pp 9. ISBN 9798561755668
5. Martin J (2014) Drugs on the dark net: how cryptomarkets are transforming the global trade in illicit drugs. Palgrave Macmillan, New York, pp 2. ISBN 9781349485666
6. Bartlett J (2014) The dark net: inside the digital underworld. Random House
7. McCann B, Bradbury J, Xiong C, Socher R (2017) Learned in translation: contextualized word vectors. In: Proceeding of NeurIPS, pp 6297–6308
8. Graczyk M, Kinningham K (2015) Automatic product categorization for anonymous market places. Tech Rep Stanford Univer
9. Elahi T, Bauer K, AlSabah M, Dingledine R, Goldberg I (2012) Changing of the guards: a framework for understanding and improving entry guard selection in tor. In: Proceedings of the 2012 ACM workshop on privacy in the electronic society. ACM, pp 43–54
10. Yang Y, Yu H, Yang L, Yang M, Chen L, Zhu G, Wen L (2019) Hadoop-based dark web threat intelligence analysis framework. In: Proceedings of the 2019 IEEE 3rd advanced information management, communicates, electronic and automation control conference (IMCEC). Chongqing, China, pp 1088–1091

11. Sun X, Gui G, Li Y, Liu R, An Y (2018) ResInNet: a novel deep neural network with feature re-use for Internet of Things. IEEE Internet Things J 6
12. Rezaei S, Liu X (2019) Deep learning for encrypted traffic classification: an overview. Data Sci Artif Intell Commun 2019:76–81
13. Fachkha C, Bou-Harb E, Debbabi M (2015) Inferring distributed reflection denial of service attacks from darknet. Comput Commun 62(2015):59–71
14. Gadhia F, Choi J, Cho B, Song J (2015) Comparative analysis of darknet traffic characteristics between darknet sensors. In: International conference on advanced communication technology, pp 59–64
15. Pustokhina I, Pustokhin D, Gupta D, Khanna A, Shankar D, Nhu N (2020) An effective training scheme for deep neural network in edge computing enabled internet of medical things (IoMT) systems. IEEE Access 8
16. Sellappan D, Srinivasan R (2014) Performance comparison for intrusion detection system using neural network with KDD dataset. ICTACT J Soft Comput 4:743–961
17. Lashkari AH, Kaur G, Rahali (2020) DIDrknet: a contemporary approach to detect and characterize the darknet traffic using deep image learning. In: 10th international conference on communication and network security. Tokyo, Japan
18. Xiaio (2010) Principal component analysis for feature extraction of image sequence. In: Proceedings of international conference on computer and communication technologies in agriculture engineering

Development of Chatbot Retrieving Fact-Based Information Using Knowledge Graph

Raghav Dayal, Parv Nangia, Surbhi Vijh, Sumit Kumar, Saurabh Agarwal, and Shivank Saxena

Abstract Chatbot assists users by providing useful responses and not just a conversational system functionalities. The advanced Chatbots such as Siri and Alexa are the results of evolution of different response generation and NLU techniques that have arrived since 1960s. Usually, chatbots are designed to address domain-specific queries; for instance, a medical chatbot requires the user to provide his/her symptoms; in the corporate world, the chatbots designed are mainly for addressing the FAQs asked by their clients/customers. However, state-of-the-art technologies are emerging, and knowledge graph is one of them. The idea of using knowledge graphs is that the data stored in them is linked. The proposed-chatbot addresses the problem of answering factoid questions by retrieving information from knowledge graphs. Initially, the neural machine translation approach was used; however, due to its limitations, keyword extraction approach was adopted for the proposed chatbot. In order to compare the proposed chatbot system with DBpedia metrics, the F-measure quality parameter were used for determining the overall performance of chatbots.

Keywords Knowledge graph · Semantic web · RASA · Keyword extraction · Neural machine translation · SPARQL · DBpedia chatbot

---

R. Dayal · P. Nangia · S. Kumar (✉) · S. Saxena
Amity School of Engineering and Technology, Amity University, Noida, India
e-mail: sumitkumarbsr19@gmail.com

S. Vijh
JSS Academy of Technical Education, Noida, India

S. Agarwal
Kyungil University, Gyeongsangbuk-do, Gyeongsan-si, South Korea

Y. Singh et al. (eds.), *Proceedings of International Conference on Recent Innovations in Computing*, Lecture Notes in Electrical Engineering 1001,
https://doi.org/10.1007/978-981-19-9876-8_13

1 Introduction

Chatbots are the software that mimics human behavior through smart conservations and gives instant responses to users; moreover, they stimulates conversations with the user as if they are talking to another individual. The significant research and development in AI and NLP, presence of huge amounts of unstructured data, and computation powers have led to development of better chatbots. Eliza is the first rule-based chatbot that matches patterns and applies corresponding pre-programmed transformation on input to generate responses. Parry is a modified version of Eliza and was developed in 1971. It has a mental state with affect variables that can manipulate the fear and anger levels of the Chatbot. In 1995, ALICE was developed, utilizing the collection of AIML file as a knowledge base [1]. In 2001, smarter child was created by active buddy containing all the information about sports, news, movie timings, etc., so that it can answer maximum user queries. It is said to be a precursor for SIRI [2] that is virtual assistant which was developed in 2010 by Apple. Advancements in NLP and large data helped in the development of SIRI. Siri can hold realistic general conversations and even understand jokes [3]. Alexa was developed and launched by Amazon in 2016 and is popular in home automation. Further, Google Assistant is a modern virtual assistant that learns from every conversation and query, and it has a friendlier interface than Google Now.

1.1 Semantic Web and Knowledge Graphs

Semantic Web is interpreted as the most comprehensive Knowledge Graph, or—conversely—a knowledge graph that crawls the entire web could be interpreted as self-contained Semantic Web. Berners-Lee et al. [4] introduced the technology referred to as Semantic Web.

A graph-structured data model to combine and use data is called a knowledge graph [5]. Sowa et al. have described semantic nets as a graph-structured representation of knowledge. KGs are a bit different from semantic nets in the sense that knowledge graphs also contain ontologies with their content [6]. KGs have been defined by many authors in diverse ways: Paulheim et al. [7] defined KG as a graph that organizes, describes, and covers world entities, their relationships, their classes, and schema from various domains. According to the Journal of Web Semantics: A network of entities, their properties, relationships, and semantics types is called a knowledge graph [8]. Färber et al. [9] defined KG as a RDF graph which contains set of ordered triples of the form (s, p, o), where 's' stands for subject, 'p' stands for predicate, and 'o' stands for object. Few publicly available knowledge graphs for linking the Web data are: DBpedia, Wikidata, YAGO, Freebase, etc. Following are few graph databases: Neo4J, Apache Tinker pop, Titan, Graph Story, and many more. The RDF graph databases have a standard query language called SPARQL.

Recently, Gremlin, SPARQL, Cypher, and GQL are the most popular graph query languages.

2 Literature Review

2.1 Chatbot Development Platform

A chatbot development platform is software that enables a programmer to utilize software such as Google Dialog Flow, IBM Watson, and Rasa.

2.2 General Architecture of Chatbot

Components of general chatbots are chat interface, NLP engine (consists of NLU and NLG), and dialogue manager. Chat interface is responsible for user-chatbot interaction interface. NLP is a superset of NLU and NLG. Here, NLU enables the virtual agent to understand the semantics of the language spoken by users along with its context. NLG is a process that converts the system-generated results into natural language representations that can be understood by any human [10]. The dialog management module stores information about all conversations. The representations generated by NLU are provided as input to dialogue manager, further dialog manager determines the context and returns a closest semantic response.

2.3 Question Answering Over Linked Data

QA over knowledge graphs involves semantic parsing to convert a natural query language into a formal query language like SPARQL. This parsing can be done using grammars and regular expressions as done by Ochieng et al. [11] and using neural networks as explained by Chakraborty et al. [12]. Neural Network based KGQA Systems are classification based, ranking based and translation models. Moreover, QA over linked data referred to as Knowledge graph question answering (KGQA). The most popular open datasets available for QA over linked data are LC-Quad [13], QALD-9 [14] and monumental data. There are four subtasks involved in a KGQA system. These subtasks are: entity linking, relations identification, logical/numerical operators identification, and generating the correct form/syntactically correct query.

In classification-based KGQA system, the entities and the relations in the user query are identified, and accordingly, the formal query is generated. This technique is used when formal query structure follows a pattern or is fixed and simple. It was used by Mohammed et al. [15] on the SIMPLE-QUESTIONS dataset to generate a state-of-the-art result. Ranking based KGQA is used when the formal queries don't have a fixed structure. All the formal queries related to the NLQ are generated and the candidate whose rank is better than others is selected.

In translation models based KGQA sequence-to-sequence models consisting of encoders and decoders are used to directly translate the given query to the target formal language. The encoder and decoder could be RNNs, CNNs, or any transformers like BERT, etc. SPBERT, a transformer-based language model pre-trained on massive SPARQL query logs was proposed recently by Tran et al. [16]. The authors have custom trained the BERT using masked language model (MLM) and word structural organization (WSO) to understand both English and SPARQL. This custom-BERT they call SPBERT. They have also presented their NMT model for translating English to SPBERT and vice versa. The architecture used in NMT is BERT2SPBERT and SPBERT2BERT. The first one is used to translate natural language queries to their formal SPARQL for retrieving information from DBpedia or Wikidata. The knowledge graphs targeted in the NMT model using SPBERT were DBpedia and Wikidata. In large datasets like SPARQL, QALD-9, question and query pairs from real-world questions and query logs. LC-Quad contains 5000 questions and corresponding queries from the DBpedia, Monument, and VQuAnDa.

2.4 Chabot's Using Knowledge Graphs

2.4.1 Kbot

This Kbot uses the open knowledge graphs, specifically DBpedia, Wikidata, and my personality knowledge graphs for the chatbot. It is a multilingual chatbot which can query the above-mentioned knowledge graphs and retrieve relevant information as per user's queries [17]. The basic components of Kbot were input module, NLU component, information retrieval component, and response selection module. The bot is bilingual and uses speech-to-text API. The NLU component is used for intent classification, NER, and keyword extraction. In information retrieval component, NL queries were converted to SPARQL language using keyword extraction, NER, and intent classification. Text summarization using TF-IDF technique takes place in the response selection module for the responses gathered from multiple open KGs. Information is displayed as a Knowledge Card. It was able to answer about 19 questions correctly out of the total 56 and had an *F*-measure of 0.44 and recall 0.3 outperforming DBpedia, ODC, and ODA chatbot.

2.4.2 DBpedia Chatbot

This chatbot uses its own knowledge graph and modular approach for handling different types of queries. It responds to users through elaborate messages by retrieving results from knowledge graph and responds in short text messages. The core of this chatbot is designed using spring framework. The NLU engine of this chatbot has been developed using Rive script, and it helps in getting the intent from the user query and classifies into one of three categories as listed as factual, related to DBpedia, and Banter [18].

3 Methodology

The objective of this research is to develop a chatbot retrieving fact-based information using "Knowledge Graph". A web application for demonstration and working of the chatbot.

3.1 Prerequisites and Requirements for Proposed Chatbot

NMT approach and keyword extraction approach were experimented. For training and testing NMT approach, Google Collab's GPU instance was used, and for building the chatbot using keyword extraction approach a Linux machine having 16 GB RAM and 500 GB SSD with an 8 GB Nvidia MX-150 GPU was used. The frameworks used were Rasa, Flask, and PyTorch. Various libraries and APIs were used for developing our chatbot. The most important libraries/packages, and API's being used are SPARQL Wrapper, Wikipedia, DBpedia Lookup API, and MySQL package. The databases and endpoints used were DBpedia Open Knowledge Graph, Virtuoso's SPARQL Endpoint, and MySQL database.

3.2 Data Collection

The datasets used to train the proposed chatbot on factoid questions and to experiment with NMT are QALD-9, LCQUAD [13], MON, MON80, MON50, and TREC. QALD-9 dataset is the latest dataset containing approximately 558 questions along with corresponding SPARQL query and the answer. LCQUAD dataset consisted of long and complex questions with corresponding SPARQL query. The MON, MON50, and MON80 contained factual questions regarding monuments and architectures. QALD and LCQUAD datasets were combined to train one instance of NMT model and the MON, MON50, and MON80 datasets were combined to train another instance of the NMT model.

TREC-5500: This dataset was used for training and testing the RASA NLU and Core used for hierarchal-intent classification and NER. This dataset has also been used by KBOT for their intent classification and NER tasks. It originally had around six base classes and approximately 54 total subclasses. The six base classes were: Human, Loc, Enty, Num, Abbr, and Desc. Each question has been categorized into their respected base and sub class. Certain modifications were made to the original dataset, and due to its complexity and ambiguity, two base classes (Enty and Num) along with their subclasses were removed from the dataset. In their place, two new base classes monument and music were added along with their subclasses.

3.3 Data Annotation and Preprocessing

Data preprocessing is a process of making the data noise free and fit for training and testing. The datasets collected for NMT training and testing were annotated beforehand. However, certain modifications were made in the datasets mentioned in previous section. TREC 5500 dataset, on the other hand, was preprocessed and annotated according to Rasa training data format. While preprocessing the data, all the special characters, the punctual symbols, blank lines, other ill-formed sentences, and alpha-numeric characters were removed from the data. Various contractions/slangs like ain't, y'all, can't, don't, etc. were replaced by 'are not', 'you all', 'cannot', and 'do not', respectively. Stop words were not removed from the data as they play an important role in the complete list of contractions. After preprocessing the data, it was put into the nlu.yml file of rasa directory following the Yaml syntax. Afterward, it was annotated for hierarchal intent classification using DIET classifier and response selector and for the named entity recognition task using the spacy transformer language model.

3.4 Rasa NLU

Rasa NLU was used for hierarchal intent classification and named entity recognition task. The hierarchical intent classification was divided into two phases, i.e., base intent classification and sub intent classification. DIET Classifier was used for base intent classification and Response Selector was used for full intent classification, i.e., base intent + sub intent. The named entity recognition was performed using DIET classifier and the SPACY entity extractor. Spacy pre-trained "en_core_web_trf" model was used for the Spacy entity extraction task. A customized NLU pipeline was created in config.yml file for achieving the above tasks. The pipeline consists of SpacyNLP, SpacyTokenizer, SpacyFeaturizer, LexicalSyntactic Featurizer, Count Vector Featurizer, DIET classifier (Base intent classification), Spacy Entity Extractor (en_core_web_trf), and Response selector (Full-Intent-Classification). Spacy NLP is used to initialize Spacy components.

3.5 *Rasa Core*

A component of Rasa open-source framework, consisting of Rasa's dialog manager and action server, which enables the proposed chatbot to predict next action to be taken. The files such as stories.yml, rules.yml, and domain.yml were modified for training Rasa core. In stories.yml, sample flow of some conversations was added covering each base and sub-intent. Adding stories make the defined RASA core more accurate while handling conversations. However, developing a general chatbot representing all the permutations of the conversations that can occur is not possible, and hence, rules.yml file of Rasa along with stories.yml is used. Moreover, inside the rules.yml file the mapping of intent or sub-intent tends to be performed with their respective actions. Further, File domain.yml allows proposed-chatbot to choose the actions from those listed in it, recognize the intents and entities mentioned and influence their behaviour over the course of conversation and choose from the mentioned responses mapped with their respective intents. The base intents along-with retrieval intents (is_retrieval_intent flag was set to true for retrieval intents to separate them from other intents) were mentioned, under the section of intents.

Rasa core also provides the functionality of calling custom actions. For developing the chatbot, six action files were created for each base intent. These action files are action_human.py, action_monument.py, action_music.py, action_loc.py, action_desc.py, and action_abbr.py. Main purpose of these action files is to fetch intents, sub-intents, and entities set from Rasa NLU component for a question.

3.6 *Information Retrieval*

The information-retrieval (IR) component is the core component of the proposed chatbot. It is responsible for fetching the relevant information from the DBpedia. Here, two approaches have been experimented with and compared to choose the best one. The first approach is the NMT approach, and the other one is the keyword extraction approach.

3.6.1 Neural Machine Translation Approach

As the name suggests, NMT model is used to directly translate a question/query into a SPARQL formal query which fetches information/query DBpedia KG. This sequence-to-sequence model was developed using a pre-trained transformer called SPBERT as the decoder. The most probable SPARQL query is selected using beam search technique. For experimentation purposes, two instances of NMT model were trained and tested. Training and testing for one instance were done by combining LcQUAD and the QALD-9 dataset, and the Bleu score for training was 52%, and for testing was 50.04%. The second instance was trained and tested on MON80 dataset, and the BLEU scores were 96.44% and 96.04% for training and testing respectively. Even though the BLUE score for the NMT model trained on Monument data is high, the queries generated by the model were unable to fetch results from the Dbpedia KG. Incorrect query generation is because of incorrect and outdated/old queries present in dataset due to which this approach fails to fetch data from the knowledge graph. Another limitation of this approach is that it requires a lot of hardware resources to produce results in time. Due to all these limitations and drawbacks that were faced in using this approach, the keyword extraction approach for information retrieval was used for developing the final working model of the chatbot proposed.

3.6.2 Keyword-Extraction Approach

This approach depends on intent classification, entity recognition, and keyword extraction. Many state-of-the-art chatbot like KBOT, DBpedia chatbot, etc. use this approach to achieve information retrieval task. In keywords Extraction approach, instead of generating full query for the question, various processes take place to fetch the correct and most relevant information. The processes coded in this approach were URI fetching and caching, local knowledge graph generation, keyword extraction, and related Uris fetching. URI fetching and caching is the process in which the DBpedia URIs for the named entities recognized in the queries are fetched using DBpedia Spotlight and stored in MySQL database if new. The Uri fetched is used to fetch the KG for the resource from DBpedia using SPARQL, and this is called local knowledge graph generation. Further keywords are extracted from the query using the below algorithm and is used to select the specific properties from the local knowledge graph, and this is called Keyword Extraction process. Further the related entities to the named entity recognized are fetched using the Wikipedia API, and their URIs are cached in MySQL database referred to as related URI fetching.

Keywords Extraction Algorithm: Keyword extraction was used to select specific relations from the local knowledge graph of the asked entity. Keywords were also used for searching information for complex questions in which there are no entities or have more than one entity. The algorithm for extracting keywords from user query is as follows:

1. First, the whole query is tokenized using the nltk tokenizer and spacy for the post tagging of each token. If the token is noun and the query was simple, then the token is put into temp_ner variable. All the tokens in the temp_ner list are then added to the ner set which was collected from the Rasa servers.
2. After that, the punctuation, the stop words, and the questions phrases are removed from tokenized user query.
3. Remaining tokens are all added to a key set, and this key set along with ner set is returned.

3.7 Proposed Chatbot Architecture

Architecture is based on the keyword extraction approach for information retrieval for the reasons explained. The main components of the proposed chatbot are the Rasa servers and the Flask server. DBpedia OKG and MySQL databases are being used for the IR task. Flask server consists of the IR module, banter conversations module, the intents-sub-intents and entity fetching module, and the UI component which is a webpage or a chat form. The chatbot is deployed on local machine because of the constraints explained earlier in section A. Rasa framework was used to develop the chatbot. Both the components of Rasa, i.e., NLU and Core are used. The server on which both NLU and Core are running is called the rasa interpreter server. Both Rasa interpreter and the Rasa Action servers used are explained in detail further. Rasa Interpreter Server comprises of Rasa NLU and Core. The Rasa NLU is trained for the tasks of hierarchal classification and NER. When a request is sent from the flask server's 'intent sub-intent and entity fetch module' to the interpreter server. Request in form of JSON object contains the user question and the sender's id. On receiving the request, from the flask server, query is passed to the NLU component for performing Hierarchal intent classification and NER task on the query. In NLU component, tokenization, and factorization are completed, and then classification of the base intents of the question and identification of the named entities in the query are done using the DIET Classifier. Spacy's trf model is also used for named entity recognition in deployed Rasa NLU. After base intent classification, the full retrieval intent classification is done by the response selector model used. This base intent, full retrieval and the entities are sent to the corresponding action process running in the action server. If the question /query is banter then the sub intent classification part is not required, and thus, the base intent with the response is returned to the Flask server. The Rasa action server is used to run the custom action processes. There are six custom action files that were created, and they are action_human, action monument, action_music, action_desc, action_loc, and action_abbr.

These custom actions are responsible for fetching and organizing the intents, full retrieval intents, sub-intents, and named entities from the NLU component into a dictionary and then convert it into a JSON object. This JSON object is returned to the Flask server. Flask Server/Framework was used to create a web application for the proposed chatbot. There were two routes defined in the flask server, and those are /chat and /. The/or the root route points to the index.html file and loads the webpage. This webpage is designed as a chat form, and when users enter their query and submit the form, the js script runs and sends a request containing the user question to /chat route of the flask server. The /chat route contains the information retrieval, 'intents sub-intents and entity fetching' and the banter conversations module. Route /chat is responsible for Information Retrieval tasks using keyword extraction approach as explained and the response/data is then sent to the '/' route or the chat form in a JSON object, which is parsed and the information is displayed as a knowledge card, if the question was factual and as a string if the question was banter question.

Table 1 Results for both proposed chatbot and Dbpedia chatbot

Parameters	Proposed-chatbot	*Dbpedia chatbot*
Correctly answered	37	26
Incorrectly answered	13	24
Pertinent	26	23
Relevant	38	25
Precision	0.52	0.46
Recall	0.684	0.92
F-measure	0.59	0.608

4 Results

RASA NLU and core test results and the comparative study with DBpedia chatbot are discussed based on manually created dataset.

4.1 Chatbots, Rasa NLU, and Core Testing

Rasa NLU and core testing was performed on custom-made test stories. Table shows the macro and micro avg precision, recall, and f1 scores for all the components as shown in Table 1. The macro-avg precision, recall, and f1 scores of the DIET classifier model were all 1.0, and weighted avg precision, recall and f1 scores for this component were also 1.0. The accuracy for this model was also 1.0. Individual precision, recall and f1 scores for all the base intents were also 1.0. These results show that the DIET classifier is predicting the base intents with 100% accuracy.

The micro-avg precision, recall, and f1 score of the DIET (CRF) model for NER task were 1.0, 0.997, and 0.998, respectively. The macro-avg precision, recall, and f1 score were 1.0, 0.999, and 0.999. The weighted-avg precision, recall, and f1 score were 1.0, 0.9973, and 0.998.The macro-avg precision, recall and f1 score for response selector component were 0.9908, 0.9933, and 0.99168, respectively. The weighted-avg precision, recall, and f1 score were 0.9953, 0.9949, and 0.9947, respectively, and accuracy of this component was 0.9949.

The micro-avg precision, recall, and f1 score were 0.9706, 0.63404, and 0.76. The macro-avg precision, recall, and f1 score were 0.86, 0.82, and 0.827. The weighted-avg precision, recall, and f1 score were 0.75, 0.634, and 0.651.

4.2 Comparative Study with DBpedia Chatbot

The test dataset contained factual questions related to architects, monuments, artists, etc., recall, precision, and *F*-measure for the proposed chabot and DBpedia Chatbot were computed. However, certain metrics were required to calculate the $F1$ score, precision, and recall of the chatbot. Therefore, answers were differentiated based on their pertinence and relevance, where in, pertinence means if the answer is exactly to the point and on the other hand relevance means the answer is somewhat related to question asked but not exact. The correctly answered questions are those which can be either relevant or both relevant and pertinent; on the other hand, incorrectly answered questions are those which are neither pertinent nor relevant. Recall is measured as the ratio of number of pertinent answers over relevant answers. Precision is measured as the ratio of number of pertinent answers.

Table 1 shows the results for both proposed chatbot and Dbpedia chatbot. Table predicts that proposed chatbot answered 37 questions correctly out of 50 total questions and answered 13 questions incorrectly, whereas DBpedia chatbot answered only 26 questions out of 50 total questions and answered 24 questions incorrectly. The number of pertinent and relevant answers given by proposed chatbot is 26 and 38, respectively, whereas the pertinent answers and relevant answers given by DBpedia chatbot are 23 and 25, respectively. The precision and recall for the proposed chatbot are 0.52 and 0.684 respectively, whereas the precision and recall for the Dbpedia chatbot are 0.46 and 0.92, respectively. From the precision and recall the F-measure for proposed chatbot was calculated as 0.59, whereas that of DBpedia was calculated as 0.608.

The f-measure for the Dbpedia chatbot is slightly better than the our chatbot; however, the accuracy and precision of our chatbot are both greater than that of the Dbpedia chatbot, which means that our chatbot can answer more factual questions that DBpedia and provide relevant answers if not pertinent.

5 Conclusion

The chatbot developed for retrieving fact-based information using knowledge graphs was successful in achieving the results. A web application was designed to demonstrate the working of this chatbot. The information retrieval component, i.e., neural machine translation (NMT) approach and keyword extraction approach are used to determine the results. However, it is observed that NMT was not found cost-effective for small projects due to huge hardware requirements for the model training, testing, and deployment, and moreover, dataset required for NMT has to be precise to get relevant results. Hence, keyword extraction approach found to be efficient for developing the IR component of proposed chatbot. The task of hierarchal classification and named entity recognition (NER) was successfully completed using the Rasa NLU and Rasa Core. The observation depicts that results based on F measure were as good as DBpedia chatbot. The analysis of the comparison study concludes that the answers given by DBpedia chatbot were for lesser number of questions. It was shows that that the accuracy and precision of our chatbot were higher than that of Dbpedia chatbot.

Proposed chatbot explicitly brings out the reasons as to why keyword approach is better than neural machine translation. Rasa is a strong framework that is used for making industrial-grade chatbot as it offers several functionalities for modeling the various conversation patterns, define responses in the domain, or even write code for custom actions, which are triggered for specific intents according to the predicted intents during conversation with user. Thus, it testifies the results of Rasa NLU and Rasa Core trained for this chatbot. In the future, commercial smaller chatbot using knowledge graph can be developed, also NMT model can be improved.

References

1. Shawar BA, Atwell E (2002) A comparison between Alice and Elizabeth chatbot systems. Univ Leeds, School Comput Res Rep 2002:19
2. Molnár G, Szüts Z (2018) The role of chatbots in formal education. In: 2018 IEEE 16th international symposium on intelligent systems and informatics (SISY). IEEE, pp 000197–000202
3. Pal SN, Singh D (2019) Chatbots and virtual assistant in Indian banks. Industrija 47(4):75–101
4. Berners-Lee T, Fischetti M (1999) Weaving the web. Harpersanfrancisco. Chapter 12. 1999
5. Chen X, Jia S, Xiang Y (2020) A review: knowledge reasoning over knowledge graph. Expert Syst Appl 141:112948
6. Sowa JF (2012) Semantic networks. Encycl Cogn Sci
7. Paulheim H (2017) Knowledge graph refinement: a survey of approaches and evaluation methods. Semantic Web 8(3):489–508
8. Ehrlinger L, Wöß W (2016) Towards a definition of knowledge graphs. SEMANTiCS (Posters, Demos, Success) 48(1–4):2
9. Färber M, Bartscherer F, Menne C, Rettinger A (2018) Linked data quality of dbpedia, freebase, opencyc, wikidata, and yago. Semantic Web 9(1):77–129
10. Bhirud N, Tataale S, Randive S, Nahar S (2019) A literature review on chatbots in healthcare domain. Int J Sci Technol Res 8(7):225–231
11. Ochieng P (2020) Parot: translating natural language to sparql. Exp Syst Appl: X 5:100024
12. Chakraborty N, Lukovnikov D, Maheshwari G, Trivedi P, Lehmann J, Fischer A (2019) Introduction to neural network based approaches for question answering over knowledge graphs. arXiv preprint arXiv:1907.09361
13. Trivedi P, Maheshwari G, Dubey M, Lehmann J (2017) Lc-quad: a corpus for complex question answering over knowledge graphs. In: International semantic web conference. Springer, Cham, pp 210–218
14. Ngomo N (2018) 9th challenge on question answering over linked data (QALD-9). Language 7(1):58–64
15. Mohammed S, Shi P, Lin J (2017) Strong baselines for simple question answering over knowledge graphs with and without neural networks. arXiv preprint arXiv:1712.01969
16. Tran H, Phan L, Anibal J, Nguyen BT, Nguyen TS (2021) SPBERT: an efficient pre-training BERT on SPARQL queries for question answering over knowledge graphs. In: International conference on neural information processing. Springer, Cham, pp 512–523
17. Ait-Mlouk A, Jiang L (2020) KBot: a knowledge graph based chatBot for natural language understanding over linked data. IEEE Access 8:149220–149230
18. Athreya RG, Ngonga Ngomo AC, Usbeck R (2018) Enhancing community interactions with data-driven chatbots—the DBpedia chatbot. In: Companion proceedings of the the web conference, pp 143–146

Short-Term Load Demand Forecasting Using Artificial Neural Network

Temitope M. Adeyemi-Kayode, Hope E. Orovwode, Anthony U. Adoghe, Sanjay Misra, and Akshat Agrawal

Abstract This work proposes a short-term electrical load demand forecaster for the Nigerian power distribution firms in Abuja, Benin, and Enugu. Using artificial neural network, the forecaster is created. Hour of the day, calendar day, day of the week (Sunday-Saturday), load demand of the previous day, load demand of the previous week, and average load demand of the preceding 24 h are the inputs to the neural network. The historical load demand for 2017–2020 includes hourly resolved dates and load demand for Abuja, Benin, and Enugu distribution firms for training purposes, while data for 2020 was used for testing the algorithm. The results generated a mean average percentage error ranging from 0.16 to 0.35. This forecaster is essential to Nigeria's efforts to expand access to power in accordance with Sustainable Development Goal 7.

Keywords Short-term forecasting · Artificial neural network · ANN · Day-ahead forecasting

---

T. M. Adeyemi-Kayode · H. E. Orovwode · A. U. Adoghe
Covenant University, Ota, Nigeria
e-mail: mercy.john@covenantuniversity.edu.ng

H. E. Orovwode
e-mail: hope.orovwode@covenantuniversity.edu.ng

A. U. Adoghe
e-mail: anthony.adoghe@covenantuniversity.edu.ng

S. Misra
Department of Computer Science and Communication, Ostfold University College, Halden, Norway

A. Agrawal (✉)
Amity University Haryana, Haryana, India
e-mail: akshatag20@gmail.com

Y. Singh et al. (eds.), *Proceedings of International Conference on Recent Innovations in Computing*, Lecture Notes in Electrical Engineering 1001,
https://doi.org/10.1007/978-981-19-9876-8_14

1 Introduction

Load forecasting is one of the primary focal points of the conversation around energy management. Telemetry processing, Generation Control, Load forecast, and Network analysis are the four major functional areas summarised in [1], which provide a summary of the Nodal Energy Management System (NEMS) used by the Electric Reliability Council of Texas. [1] also provides a brief explanation of each of these areas. According to the research conducted, load forecasting can be broken down into two primary categories: those that focus on the duration of the prediction and those that concentrate on the evaluation method. The projection of the load is broken down into three distinct time intervals: the short term, the mid term, and the long term. The short-term load forecasting analyses predictions ranging from a few seconds, minutes, and even a few hours; the mid-term load forecast focuses on load forecasting from weeks to months; and the long-term forecast is utilized for load projections of at least one year [2–4].

Estimates of the short-term load are necessary for the day-to-day operations of the power plant, the evaluation of net interchange, the commitment of units, scheduling, and other analyses pertaining to system security. Mid-term load forecasts are utilized in the planning of fuel scheduling as well as maintenance planning. In order to effectively manage the grid and devise a repair schedule, a long-term load estimate is utilised. On the basis of the methodology, there are two primary approaches: the linear approach and the nonlinear approach. Systems are said to be linear if their outputs depend in a linear fashion on the linear combinations of their inputs. Linear models are used to describe linear systems. The existence of linearity is predicated on the fulfilment of two main conditions: homogeneity and superposition. The goal of homogeneity is to assume that the output is a scaled form of the inputs, whereas the goal of superposition is to add the outputs that result from using a variety of distinct inputs. [5] Superposition can be achieved when the outputs of several inputs are added together. A comprehensive classification of the load forecasting approaches is presented in Fig. 1.

This paper seeks to develop a short-term day-ahead load demand forecaster for Abuja, Benin, and Enugu power distribution companies in Nigeria using artificial neural networks. Recent work review presents in Sects. 2 and 3 discussing methodology, Sect. 4 describes results and discussion and Conclusion in Sect. 5.

2 Literature Review

Short-term load forecasting takes into account projections spanning from seconds to hours. In [6], the major methodologies utilized in artificial neural networks (ANN) for short-term load forecasting were extracted from 40 journal articles through a literature review. It was the first work to address the skepticism of many scholars regarding the efficacy of ANN for load forecasting at the time of its publication.

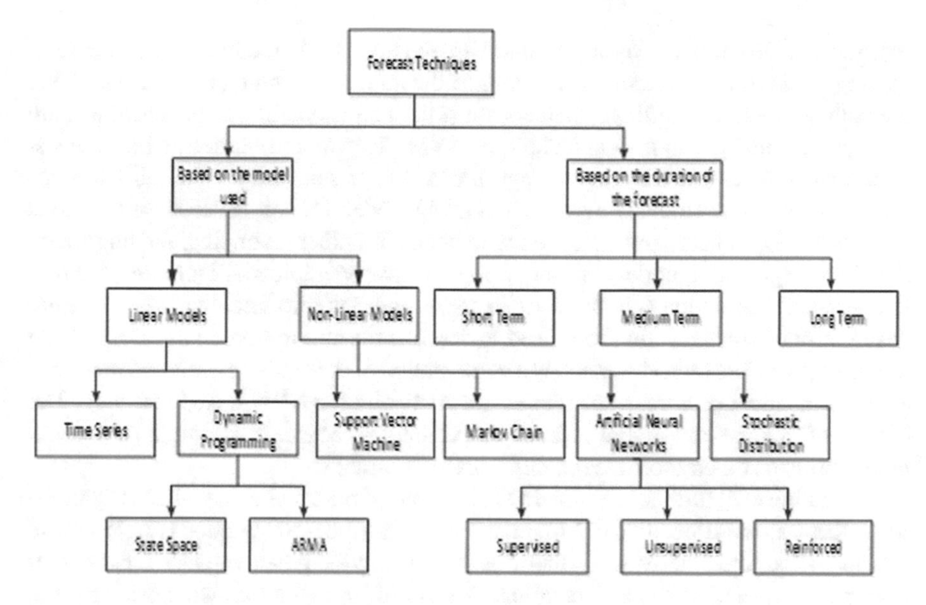

Fig. 1 Description of classification model for load prediction techniques

Based on the papers reviewed in this article, the following can be concluded: (1) the papers presented the variables used for classification based on the position on the calendar (weekday/Weekend/holiday/Month/Season). It should be noted here that many papers reveal a significant difference in load demand based on daily and/or seasonal constraints; (2) The input variables used in the forecasting system were: Load, temperature, humidity, weather variables, nonlinear functions of time, and load parameters; (3) The number of sets of classes were also presented, the values ranged from 1 to 105 classes. However, the majority of the classes were limited to 7—based on the number of days in a week; (4) based on the variable to forecast, most of the reviewed articles forecasted a complete load profile; however, some others forecasted hourly loads, next minute load and peak loads; (5) The number of neurons used in these studies was also presented; these were grouped into input, hidden, and output layers. The highest and least amount of input, hidden, and output layers are (107/5), (60/3), and (48/1), respectively; (6) the activation functions used were Sigmoidal, Linear, and Sinusoidal; (7) The stopping rule employed by the papers preset tolerance level to be reached during training, cross validation, or iterations; (8) The number of parameters in the multilayer perceptron (MLP) ranged from 22 to 6768; (9) The number of MLPs used in forecasting the system also ranged from 1 to 105; (10) Similarly, the total number of parameters which includes weights + threshold ranged from 57 to 25,944; (11) The size of the training sample used in papers reviewed in this study ranged from 7 to 1825; and (12) The size of the test sample ranged from 4 to 365.

In [7], the authors examined the current techniques utilized for day-ahead forecasting of photovoltaic systems. Statistical or time-series-based method, physical method, and ensemble or hybrid method are the three basic methodologies for forecasting photovoltaic systems, according to the authors of the paper. Various ANN algorithms, such as multilayer perceptron (MLP) and support vector machines, are incorporated into the statistical technique (SVM). Regression Models, which consist of the auto-regressive moving average (ARMA) for stationary data and the auto-regressive integrated moving average (ARIMA) models for non-stationary time series [8], the Markov chain, and persistence methods are other examples. Among them, ANN and regression models are the most extensively employed, including Numerical Weather Prediction (NPW), Sky Imagery, and Satellite Imaging [9]. These are based on mathematical equations used to describe the atmosphere in its physical and dynamic state. The hybrid method is a combination of the earlier mentioned methods. Some of its most prominent examples are ANN and the NPW based methods. The benefit of this method is that it leads to excellent forecasting, although they could underperform if meteorological conditions are unstable.

The authors [7] further investigated solar forecast based on data gathered from the SolarTechLab in Politecnico di Milano in 2017 using the ANN method and the hybrid method. Also, the data were further clustered using the mean values of daily solar radiation measured on the PV modules. The authors used the following error indices to assess the validity of their models: mean absolute error (MAE), normalized mean absolute error (NMAE), mean absolute percentage error (MAPE), weighted mean absolute error (WMAE), normalized root mean square error (nRMSE), enveloped-weighted absolute error (EMAE), and objective mean absolute error (OMAE).

The authors endeavour to predict the PV power output for the following day. This paper examined the outcomes of two (2) methods: the ANN method and the hybrid method. Three (3) distinct multi-layer perceptrons (MLPs) were stacked in the ANN approach. Using solar irradiance Gm(d), air temperature Tm(d), and the number of days d, the first MLP, which receives normalized data between -1 and 1, was used to determine the solar irradiance for the following day. This value is then used to determine which model will be utilised to anticipate the electricity that will be generated. If the average predicted irradiance is greater than 150 W/m^2, model 1 is employed; otherwise, model 2 is utilised. The output can be expressed as Eqs. 1 and 2:

$$\{K1(a+1), K2(a+1), \ldots, K24(a+1)\} = f_G(K_m(a), T_m(a), a), \quad (1)$$

$$\{K1(a+1), K2(a+1), \ldots, K24(a+1)\} = f_G(K_m(a), T_m(a), a), \quad (2)$$

The approximation function is denoted by fg, and the hourly values of the predicted solar irradiance are denoted by $K1(a+1)$, $K2(a+1)$, etc., $K24(a+1)$. The sun irradiance K_m (a), the air temperature Tm (a), and the mean power produced Fm are the parameters that go into the MLP in both Model 1 and Model 2. (d). The output layer is comprised of twenty-four output nodes. The output of the neural network can be represented as an equation, as seen here:

$$\{F1(a+1), F2(a+1), \ldots, F24(a+1)\} = f_G(K_m(a), T_m(a), F_m(a)), \quad (3)$$

In the hybrid method, the authors fed the neural network with clear-sky irradiation data and weather forecast data. The method for including the clear sky radiation model (CSRM) is described in [10], while the method for hybridising the ANN is given in [11]. The constructed network is comprised of two layers of 12 and 5 neurons.

The examination of the first example reveals a MAPE of the bright days MLP of 0.236 and a MAPE of 0.54 for overcast days. The second technique provides a MAPE of 0.10 for sunny days and 0.68 for gloomy days. The two techniques predicted sunny days more accurately than cloudy days. The authors of [12, 13] highlighted that forecast error increases significantly when meteorological variables (temperature, humidity, etc.) and exogenous variables (cultural, social event, etc.) are incorporated into the input. Therefore, the authors [2] chose to build an algorithm that improves forecast precision without increasing execution time for meteorological and external variables. Load-data is a matrix including the hour of the day ($h_1, h_2, h_3, \ldots h_m$), the day of the year ($y_1, y_2, y_3, \ldots, y_m$), the dew point temperature data matrix (TDP), the dry bulb temperature data (TDB), and the input time of the day (working day or holiday) (DT). In this investigation, a multi-model forecasting ANN with supervised architecture, sigmoid activation function, and a multi-variable autoregressive algorithm (MARA) for training were utilized. The hidden layer consists of one layer with five neurons and an output layer with twenty-four neurons. In contrast to many other forms of study in day-ahead forecasting, the authors' objective was to reduce forecast inaccuracy rather than execution time. The suggested methodology consists of three modules: pre-preprocessing, forecasting, and optimization.

The authors extracted features from correlated lagged load data, meteorological, and other external variables for the pre-processing module. In addition, irrelevant or redundant input samples were eliminated. In day-ahead load forecasting, the two most common methods are direct and iterative forecasting. It is common for direct and iterative forecasting to result in considerable forecasting errors. This procedure carries out the cascading strategy. The authors implemented an ANN with 24 cascaded forecasters, each of which was responsible for predicting the accuracy of one hour of the following day. Each forecaster in the forecast module is activated by a sigmoid function. MARA, which is faster than both the Levenberg–Marquardt algorithm and Gradient Descent Back Propagation, is the training algorithm employed here. Note also that the Levenberg–Marquardt algorithm can train ANNs 1 to 100 times quicker than the gradient descent back propagation algorithm. Additional information regarding the weight-updating training procedure can be found in [14]. In the optimization module, the mean absolute percent error (MAPE) is minimised. There are numerous techniques for minimising the MAPE, including linear programming, nonlinear programming, quadratic programming, convex optimization, and heuristic optimization. Due to the nonlinear nature of the problem, the linear programming optimization technique would be inappropriate. Nonlinear programming yields precise solutions but requires a lengthy execution time. Heuristics is the optimization method employed in this study. For error forecasting, the heuristics-based optimization technique is utilised. In error minimization, they also employ a modified version of the enhanced differential evolution algorithm (EDE). The authors implemented an altered version of the EDE method. Testing was conducted on Dayton (Ohio, USA) and EKPC datasets (Kentucky, USA). In addition, they compared the outcomes of two current prediction models (bi-level forecast and MI + ANN forecast). This method's precision was 98.76%. The MAPE of the created technique MI + ANN + mEDE was 0.0125 better than the MAPE of the MI + ANN method, which was 0.0227, while the MAPE of the Bi-level forecast was 0.0386. In [15], the authors concentrated on short-term load forecasting for residential and small-to-medium-sized businesses at various levels of aggregation. This study utilized a total of 1700 residential load data and 250 SME load data. These were acquired from smart meters in the past. They used linear regression, gradient-boosted regression tree (GBRT), support vector regression (SVR), multi-layer perceptron (MLP), long short-term memory (LSTM), and deep learning. Authors determined that GBRT was inferior to linear regression, MLP, and SVR. Root Mean Square Error (RMSE), Mean Absolute Error (MAE), and Mean Absolute Percentage Error are prominent evaluation metrics (MAPE). Nevertheless, the authors employed normalized mean absolute error (NMAE). On the Irish utility data, the approach was evaluated. They described some of the challenges associated with working with residential and SME-type data. Numerous variables have a substantial impact on the prediction accuracy of user patterns in residential structures.

3 Methodology

3.1 Data Collection and Description for Load Demand

The information regarding the load demand was collected from the National Control Center of the Transmission Company of Nigeria (TCN), which is located in Oshogbo, Nigeria. The data cover a period of four years, beginning on the first day of 2017 and ending on the 31st day of 2020. The information includes hourly load demand data for three (3) of the eleven (11) Nigerian Distribution Firms (DISCOs), which are the distribution companies located in Abuja, Benin, and Enugu, respectively.

3.2 Load Demand Forecasting Using Artificial Neural Networks

For the short-term load demand forecaster, the author modified ANN codes developed by Ref. [16]. In summary, the short-term load demand forecaster is designed to predict the future power demand of consumers. In order to perform this analysis, the following historical data presented in Table 1 are to be provided. All these parameters directly influence hourly load demand and would constitute the input data to the

Table 1 Input parameters to load demand forecaster

Data parameters	Dataset
Seasonality	Hour
	Day
	Day of week
	Month
Historical load	Prior day load demand
	Prior week load demand
	Average load demand of the last 24 h

forecaster, while the actual load demand would be considered output. Hourly resolved data from 2017 to 2019 was used to train the neural network, while hourly resolved data for 2020 was used to test the neural network.

4 Discussion

4.1 Nigerian Load Demand Data

The average and minimum hourly load demand data for distribution companies in Nigeria are summarized in Table 2. This summary is taken from 2017–2020. From the table, Abuja Distribution Company consumes more electricity than the three distribution companies being evaluated.

In addition, Figs. 2, 3, and 4 show a daily load profile for the Abuja, Benin, and Enugu distribution companies, respectively. It can be observed that the load profile for the Abuja DISCO peaks in the afternoons (between 9 am and 2 pm) and also in the evening time (between 6 and 10 pm), while the Benin and Enugu DISCO peaks from 5 to 11 pm. It should be noted that this load profile is cumulative of the 24 h for the three-year training horizon, and the load profile may vary based on seasonal, weekend, and weekday differences.

Table 2 Average and maximum hourly load demand values from 2017–2020 (MW)

DISCO	Average	Maximum
Abuja	410.37	737.1
Benin	262.53	570.9
Enugu	249.29	575.21

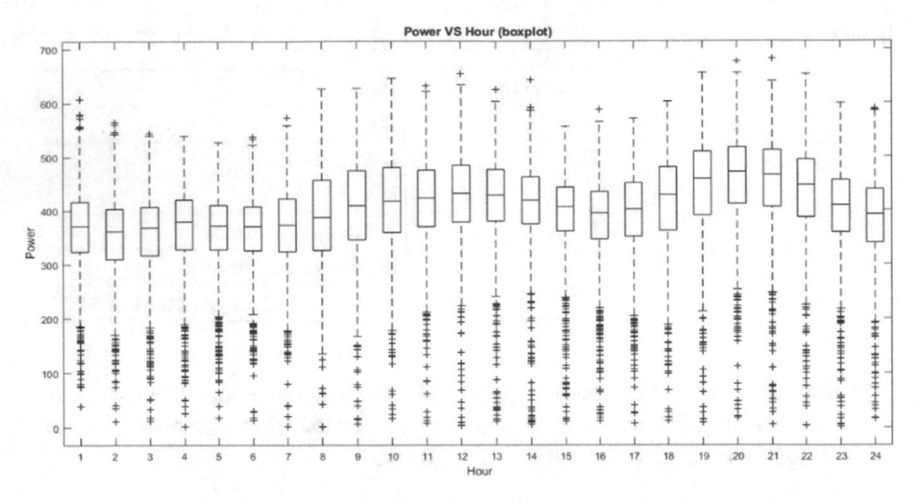

Fig. 2 Abuja load profile for cumulative 24 h period from 2017–2020

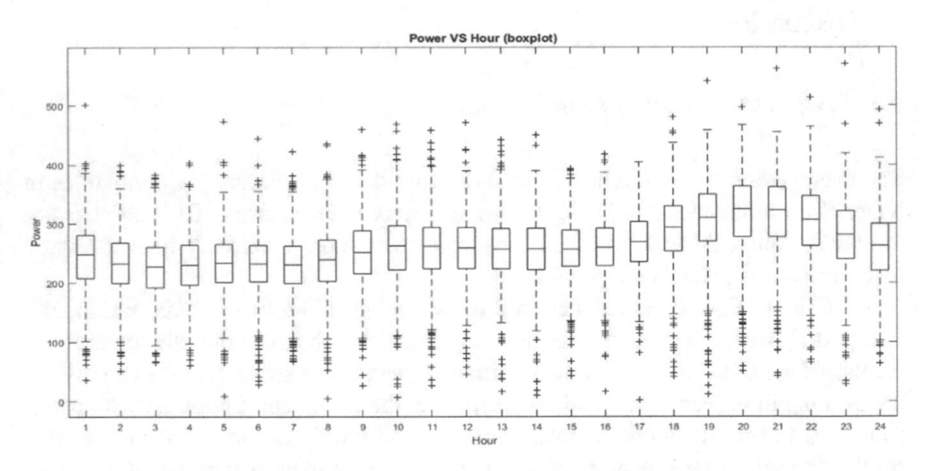

Fig. 3 Benin load profile for cumulative 24 h period from 2017–2020

4.2 Forecasting Accuracy

The result of the day-ahead load demand and historical load forecast for Abuja, Benin, and Enugu are detailed in Table 3. The metric used in evaluating this forecast is the mean absolute percentage error (MAPE), evaluated using Eq. 4. According to Table 3, the MAPE values ranged from 0.16 to 0.35, closely related to some highlighted literature. Figures 5, 6, 7, 8, 9, and 10 show graphs revealing the difference between actual and predicted load demand.

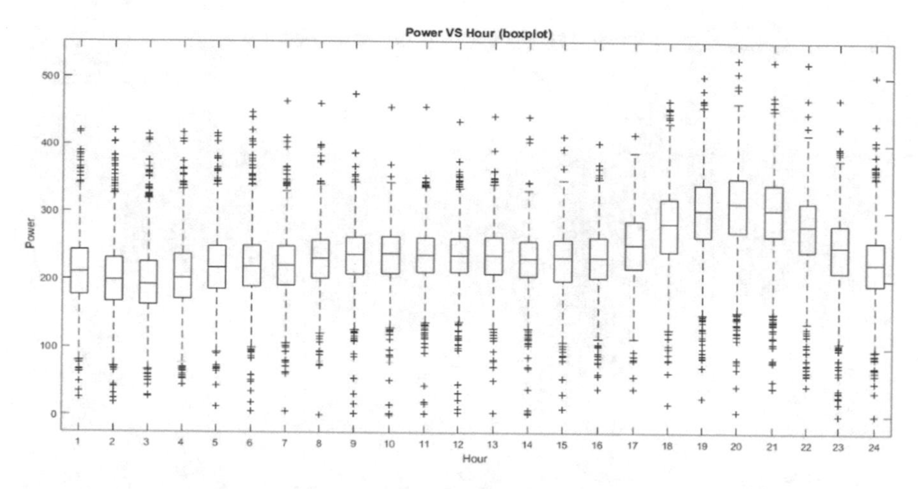

Fig. 4 Enugu load profile for cumulative 24 h period from 2017–2020

$$\mathrm{MAPE} = \frac{1}{n}\sum_{t=1}^{n}\left|\frac{A_t - F_t}{A_t}\right| \tag{4}$$

where:

n = number of points, A_t = actual value, F_t = forecast value.

Table 3 Comparison of results forecasting results

DISCO area	MAPE for forecasted day ahead load	MAPE for historical load forecast
Abuja (this study)	0.22	0.35
Benin (this study)	0.16	0.19
Enugu (this study)	0.20	0.29
[5]	In this study, four scenarios were developed. The scenarios had similar inputs given as: 'day', 'day-1', 'day-2', 'day-7', 'day-8', 'day-14', and 'day-15' but different output given as 'day Ahead Forecast', 'Seven days ahead forecast', 'Day Ahead Forecast with disconnected time', and 'Seven Days Ahead Forecast with disconnected time'. The error criterion AME is 2.3, 3.7, 0.4, and 1.2, respectively	
[7]	The following input parameters are used for the prediction of cloudy and clear days: Ambient temperature in degrees Celsius; global horizontal irradiation in watts per square metre; global irradiation on the plane of the array in watts per square metre; wind speed in metres per second; wind direction in degrees; pressure in millibars; precipitation in millimetres; cloud cover in percentage; and cloud type in low, medium, or high. Error criterion MAPE values range from 0.236 to 0.54 to 0.10 to 0.689 respectively	

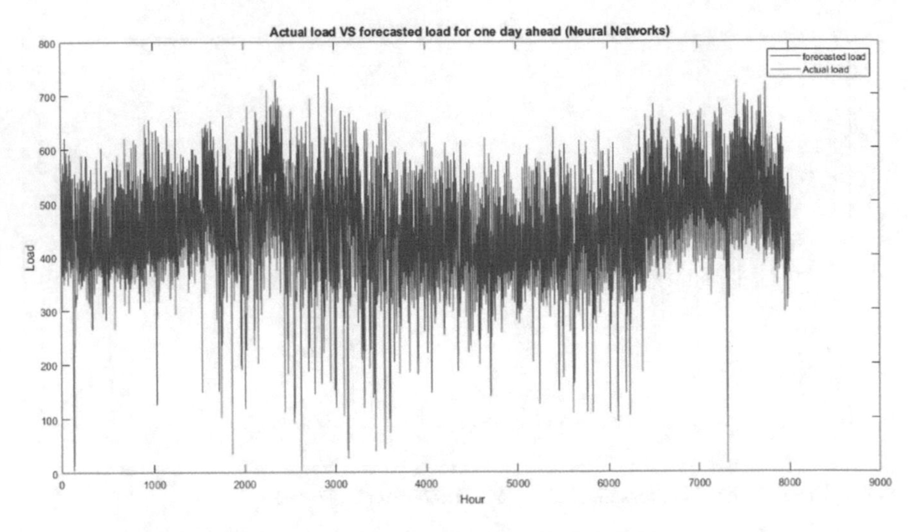

Fig. 5 Actual load versus day ahead forecasted load for Abuja DISCO

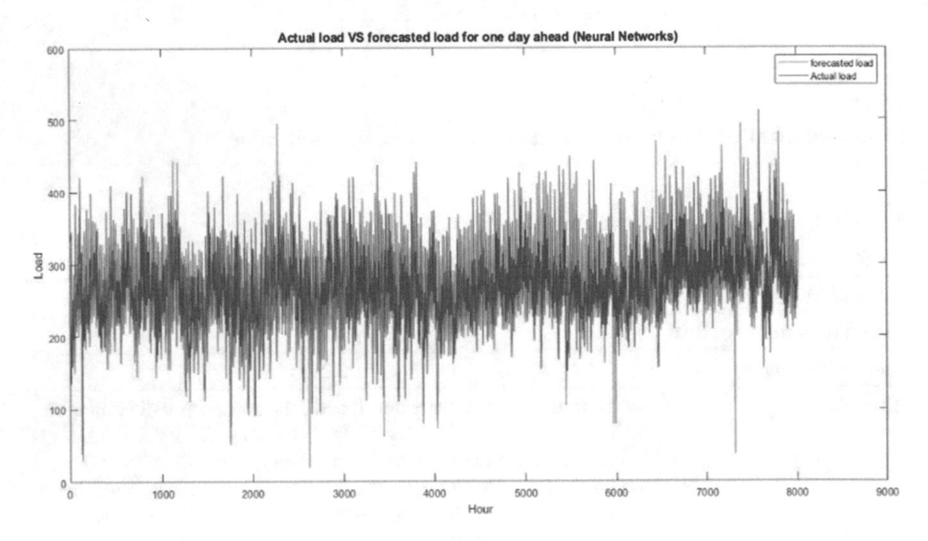

Fig. 6 Actual load versus day ahead forecasted load for Benin DISCO

5 Conclusion

This study was carried out to forecast the day load demand for Abuja, Benin and Enugu distribution companies. The input parameters used in this study are hours, days, day of the week, month, previous day load demand, previous week load demand, and average load demand. The estimation was done using ANN with the MAPE as the error criterion. The MAPE values range from 0.16 to 0.35. This model is essential

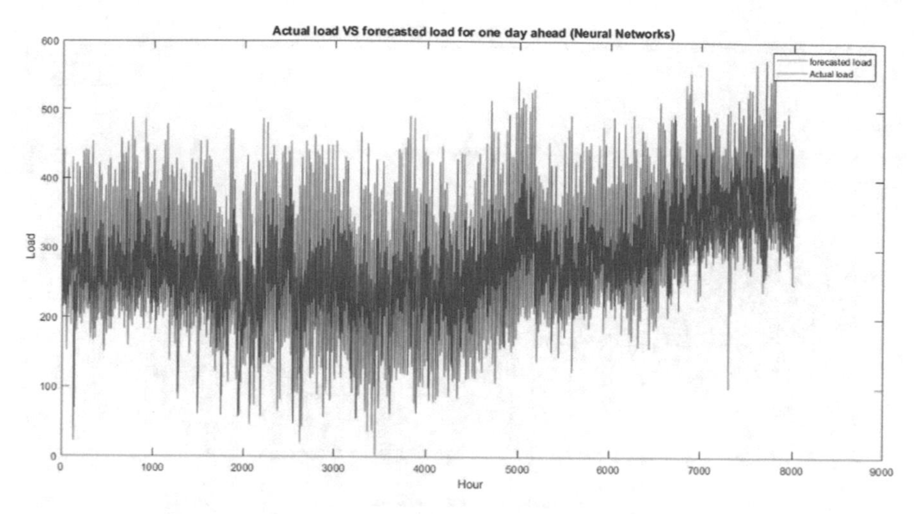

Fig. 7 Actual load versus day ahead forecasted load for Enugu DISCO

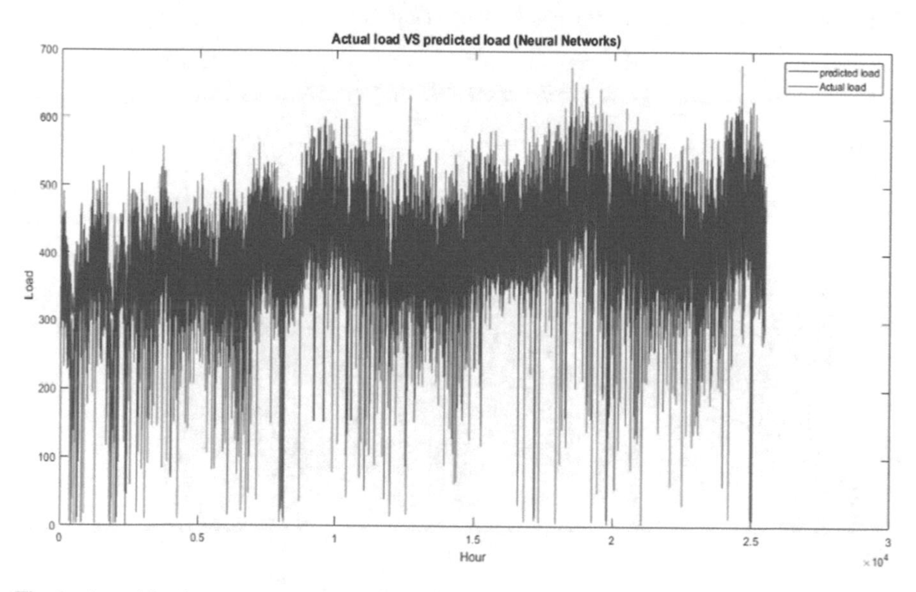

Fig. 8 Actual load versus predicted load for Abuja DISCO

to evaluate the load demand for Nigerians residing in the Abuja, Benin, and Enugu distribution company area to fast-track the attainment of sustainable development goal (SDG) 7.

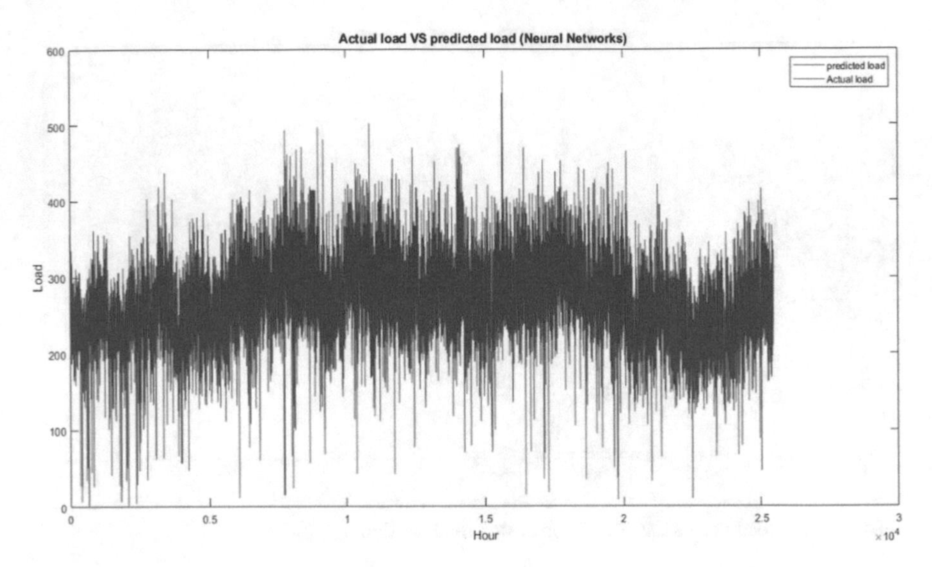

Fig. 9 Actual load versus predicted load for Benin DISCO

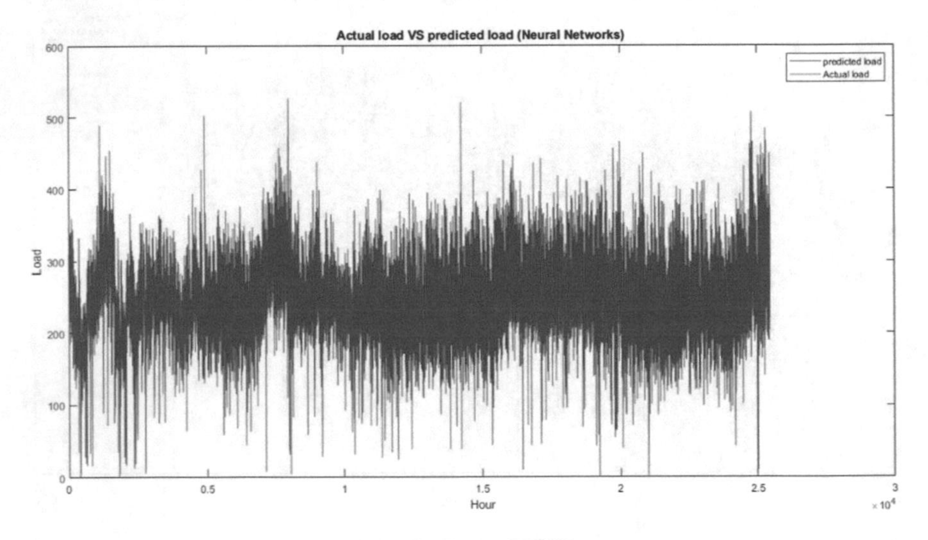

Fig. 10 Actual load versus predicted load for Enugu DISCO

References

1. Xu L, DO, Boddeti M (2009) The roles of energy management system in Texas nodal power market. IEEE Access
2. Ahmad A, Javaid N, Mateen A, Awais M, Khan ZA (2019) Short-Term load forecasting in smart grids: An intelligent modular approach. Energies, 12(1). https://doi.org/10.3390/en12010164
3. Raza MQ, Nadarajah M, Ekanayake C (2016) On recent advances in PV output power forecast.

Sol Energy 136:125–144
4. Olagoke MD, Ayeni A, Hambali MA (2016) Short term electric load forecasting using neural network and genetic algorithm. Int J Appl Inf Syst (IJAIS)
5. Ogunfunmi T (2007) Adaptive nonlinear system identification: the volterra and wiener model approaches: Springer Science & Business Media
6. Hippert HS, Pedreira CE, Souza RC (2001) Neural networks for short-term load forecasting: a review and evaluation. IEEE Trans Power Syst 16(1):44–55
7. Nespoli A et al (2019) Day-ahead photovoltaic forecasting: a comparison of the most effective techniques. Energies 12(9):1621
8. Reikard G (2009) Predicting solar radiation at high resolutions: a comparison of time series forecasts. Sol Energy 83(3):342–349
9. Das UK et al (2018) Forecasting of photovoltaic power generation and model optimization: a review. Renew Sustain Energy Rev 81:912–928
10. Bird RE, RC (1986) Simple solar spectral model for direct and diffuse irradiance on horizontal and tilted planes at the Earth surface for cloudless atmospheres
11. Dolara AG, Leva S, Mussetta M, Ogliaro E (2015) A physucal hybrid artificial neural network for short term forecasting of PV plant power output. Energies 8:1138–1153
12. Amjady NK, Zareipour H (2014) Short-term load forecast of microgrids by a new bilevel prediction strategy. IEEE Trans Smart Grid 1:286–294
13. Liu N, Tang Q, Zhang J, Fan W, Liu J (2014) A hybrid forecasting model with parameter optimization for short-term load forecasting of micro-grids. Appl Energy 129:336–345. https://doi.org/10.1016/j.apenergy.2014.05.023
14. Anderson CWS, EA, Shamsunder S (1998) Multivariate autoregressive models for classification of spontaneous electroencephalographic signals during mental tasks. IEEE Trans Biomed 45:277–286
15. Peng Y, Wang Y, Lu X, Li H, Shi D, Wang Z, Li J (2019) Short-term load forecasting at different aggregation levels with predictability analysis. arXiv preprint arXiv:1903.10679
16. Deoras A (2021) Electricity load and price forecasting webinar case study retrieved from https://www.mathworks.com/matlabcentral/fileexchange/28684-electricity-load-and-price-forecasting-webinar-case-study

Comparative Analysis of Performances of Convolutional Neural Networks for Image Classification Tasks

Abraham Ayegba Alfa, Sanjay Misra, Abubakar Yusuf, and Akshat Agrawal

Abstract The main success of CNNs is in their capability to learn features automatically from domain-specific images unlike the other machine learning schemes. During training phase, CNN architecture utilizes transfer strategy of learned knowledge from a pre-trained network in one previous task to a new task. Traditional classification schemes have numerous weaknesses such as time consumption and subjectivity. Recently, a CNN built on deep learning holds prospects in the estimation and extraction features for improved image classification. This paper experiments the performances of CNNs classification using binary and RGB color images, which were collected from MINST and CIFAR-10. Two CNN-based models were proposed, trained and validated on Google Colaboratory simulator with 50,000 and 10,000 samples of binary and RGB color images, respectively. The results indicated that the CNN model with RGB color images outperformed CNN model with binary images in terms of average simulation time (14.20–29.00 ms). Conversely, the CNN with binary images was superior to the CNN with RGB images for loss function (3.20–88.30%) and accuracy (99.10% to 71.06%), respectively. These outcomes were better than comparable studies using the average duration and accuracy.

Keywords CNN · Binary · Image · RGB color · Accuracy · Simulation time

A. A. Alfa
Confluence University of Science and Technology, Osara, Nigeria
e-mail: alfaaa@custech.edu.ng

S. Misra
Østfold University College, Halden, Norway
e-mail: sanjay.misra@hiof.no

A. Yusuf
Niger State Polytechnic, Zungeru, Nigeria

A. Agrawal (✉)
Amity University, Gurgaon, Hariyana, India
e-mail: akshatag20@gmail.com

Y. Singh et al. (eds.), *Proceedings of International Conference on Recent Innovations in Computing*, Lecture Notes in Electrical Engineering 1001,
https://doi.org/10.1007/978-981-19-9876-8_15

1 Introduction

The availability of large annotated images has demonstrated the superiority of deep learning methods over the traditional machine learning methods. CNN structure is one of the most renowned deep learning methods having great performance in imaging domain. Reason being that CNN is highly capable of autonomously learning features from domain-associated images which is not the case of traditional machine learning methods. Again, CNN architecture during training relies on the transfer learned knowledge from a pre-trained network tactic that accomplished one task into fresh task(s) [1].

Deep learning is a field in machine learning that employs illustration learning by expressing input data in multiple levels of simpler illustrations. In other words, complex concepts such as *person* or *car* are constructed through layers of simpler concepts such as contour or edges. A CNN is a unique kind of deep learning neural networks having unmatched efficiency in image-related tasks. In fact, CNNs have been applied in diverse terrains including image semantic segmentation, image classification, image classification, and object detection. The CNN's input data are processed in a grid-style topology [2].

The effectiveness of CNNs in computer vision-related tasks propeled their adoption in object detection whose results outclass the classical detection methods with manually designed features such as scale invariant feature transform (SIFT) and histogram of oriented gradients (HOG) [3]. CNN is a top-choice method for classifying objects in distinctive specialties because it has an extraordinary capability of unrevealing intricate patterns [4].

In particular, precise detection of farmland obstacles is a crucial environmental perception task for agricultural vehicles. The orchard environment is a complex and unstructured environment, which makes it difficult to detect obstacles accurately and effectively [3]. Therefore, there is the urgency and necessity of detecting objects accurately. Traditional image classification methods are limited by time ineffectiveness and subjectivity [5]. Recently, a CNN founded on deep learning hold prospects in its capability to estimate and extract features in order to improve precision of image classification tasks [6]. Diverse images are available for different applications which can be described by amounts of bands as well as strong correlations in the spatial and spectral domains such as hyperspectral image [7], RGB color images, and binary images [8].

The subsequent sections of this paper are arranged as follows: Section 2 is the literature review; Sect. 3 is the methodology; Sect. 4 is the discussion of results; and Sect. 5 is the conclusion.

2 Literature Review

2.1 *The Concept of Deep Learning*

Knowledge is a key approach for refining performance through participation. Deep learning (DL) is a subfield of artificial intelligence (AI) that uses algorithms inspired by the structural and functionality of the brain neurons. DL methods are intended to reinforce various ML applications [4]. Deep architecture with trainable parameters is utilized in constructing a higher degree of data abstraction through linear and nonlinear transformation functions [9]. In 1965, Alexey Grigorevich Lvakhnenko started DL, which has continued to garner momentum in the GPU flourished computing horsepower and nonlinearity. These provided more profound networks for healthier knowledge utilization [4].

Deep architecture has multiple deep layers, which extend the complexity of artificial neural network (ANN). Following the discovery of backpropagation in 1980s, it has been deployed on ANN in order solve numerous real-time tasks. The conventional backpropagation methods are frequently utilized in image analysis tasks, but findings have revealed that DL methods are most successful in the field of character recognition [10]. Specifically, DL has been widely applied for the Bangla and Manipuri character recognition [11–13]. Also, DL methods are being applied in the computer vision including object detection, localization, classification, and abnormally detection in medical imagery [14].

2.2 *Convolutional Neural Networks*

The idea of CNN was nursed in early 1960s during Hubel and Wiesel investigation of the visual cortex structure in the cat brain. They realized that the biological visual information is transferred across multi-layer receptive domain [15]. This motivated the similar algorithm development for the image recognition. Thereafter, several more CNN architectures evolved steadily. Basically, Simonyan and Zisserman [16] developed VGG16 (13 convolutional layers and three fully connection layers) and VGG19 (16 convolutional layers and three fully connection layers). Szegedy et al. [17] developed GoogLeNet model with small convolution kernel to minimize the computational complexity by adjusting convolution modules.

He et al. [18] built ResNet model capable of accelerating the network convergence speed and raising the network depth for the purpose of advancing the image classification effectiveness. Though, new CNN models adopted several features in the existing models including: Inception-ResNet-v2, DenseNet, and MobileNet. These models leverage on the depth-wise separable convolutions to wrap parameters and increase speed of computation [19].

Tan et al. [20] developed EfficientDet composed of eight model structures (that is, EfficientDet D0–EfficientDet D7) for optimizing the accuracy and speed of image classification tasks. Recently, researchers experiment traditional CNN models with semantic segmentation and object detection. The goal of object detection is to ascertain presence of objects in predefined categories [21]. This is applicable for determining the presence of region of tumors in tissues or organs of medical images known as spatial location [22].

Several tasks can be accomplished with object detection such as lesion tracking, lesion location, and image discernment. The object detection has been extremely applied in medicine. On the other hand, semantic segmentation algorithms enable computer to segment the pixels represented in images. The term semantic means the content of the image, while the segmentation connotes various objects in the image, which are segmented into pixels. The entire process labels each pixel in the image accurately [23].

2.3 Related Studies

Abbas et al. [1] developed a deep CNN model referred to as Decompose, Transfer, and Compose (DeTraC) for COVID-19 chest X-ray images classification. DeTraC uses a class decomposition mechanism for overcoming class boundary problem. The results revealed that DeTraC detected COVID-19 cases from a comprehensive image dataset collected at accuracy and sensitivity of 93.1% and 100%, respectively. However, the effectiveness of the proposed model was affected by dataset irregularities.

A different layout for CNN architecture to overcome the weaknesses in traditional machine learning (ML) techniques by Hazra et al. [4]. It combined feature extraction and classification through nonlinearity in the deep learning models. The architecture of the proposed CNN comprised four layers including convolutional layer, nonlinear activation layer, pooling layer, and fully-connected layer. The model validation was performed using Manipuri Character dataset (or Mayek27) for understanding the functionality of two regional handwritten character identifications.

A deep CNN (DCNN) for apparent and occult scaphoid fractures detection from radiographic images was proposed by Yoon et al. [24]. The radiographic dataset with various scaphoid fractures was collected and randomly split into training, validation, and test datasets. The images were used to train the DCNN model built with EfficientNetB3 architecture for classifying apparent and occult scaphoid fractures. Two-stage examination accurately classified 20 of 22 cases (that is, 90.9%) of occult fracture, which were invisible to human observers. However, there is need to better accuracy of DCNN for medical diagnosis problems.

A CNN-based detection of cellular breakdown in lungs using channels and division methods was proposed by Zhou et al. [25]. The authors computed CT image from cellular fragmentation of patients using digital image-making strategy. The settlement counting approach built with CNN is capable of detecting cellular breakdown at high accuracy.

A study to detect cracks on images from masonry walls was conducted by Dais et al. [26]. A dataset containing photos of masonry structures was generated for multipart backgrounds and numerous sizes and types of cracks. Deep learning networks built on CNNs (that is, FPN, FCN, and U-net) were experimented. It was found that the effect of transfer learning on crack detection of masonry surfaces at patch level produced accuracy of 95.3%, while at pixel level produced F1-score of 79.6%. Future works consider other types of surfaces and objects (such as doors and ornaments).

Yang and Zhang [27] utilized high generalization and adjustment to construct four (4) CNN models for acquiring and analyzing information of damaged building. A sample dataset of damaged buildings was constructed by using multiple disaster images retrieved from the xBD dataset. The CNN model pre-trained with multiple disaster remote sensing images was effective and accurate in extracting collapsed building information from satellite remote sensing data, while the adjusted CNN models tested with adjusted DenseNet121 was the most robust by accuracy of 88.9% from 64.3%.

The study by Fu et al. [28] considered a camera-based basketball scoring detection (BSD) with CNN for object and frame difference motion determination. Author utilized videos of the basketball court as inputs. Subsequently, the position of the basketball hoop was determined using you only look once (YOLO) model in real-time mode for the condition of basketball scoring. The accuracy of models, Faster RCNN, SSD and YOLO, were 89.26, 91.30, and 92.59%. However, larger datasets can be considered for BSD in the future works.

The facial modality was leveraged by Benkaddour et al. [6] in order to develop a gender and age determination system with CNN based on face image or a real-time video. The authors created three CNN models with dissimilar structures (that is, filters, convolution layers, etc.) and validated on standard datasets such WIKI and IMDB. It was found that the CNN models largely improved the determination system in terms of accuracy (age: 86.20–83.97%; gender; and 94.49–93.56% for IMDB and WIKI datasets).

3 Methodology

3.1 Description of the Proposed Model

This paper proposed image-based object classification models with CNN, which is used to classify features available in the RGB color and binary images. The proposed models are comprised three phases: input, process, and output, as shown in Fig. 1.

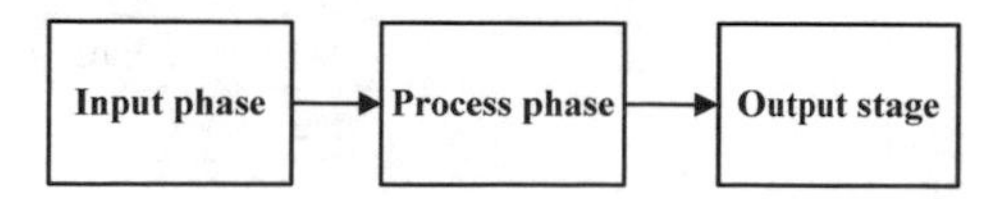

Fig. 1 CNN-based image classification model

From Fig. 1, the explanation of the main phases in the proposed CNN-based image classification models and corresponding activities at each phase is given as follows:

Input phase. It is concerned with the acquisition of images about objects and their categories from the standard databases. Then, the images sampled underwent pre-processing, stemming, and tokenizing (or denoising). This paper chose RGB color and binary images by adopting of the dataset acquired as input for further processing at the next phase.

Process phase. It entails models training and testing through extraction and selection of features available in the distinct images. The model is applied on high dimensional and unstructured data with appropriate hyperparameters for the purpose learning the interrelationships within datasets. The attributes are extracted and compiled accordingly by means of associations and dissociations rules.

Output phase. It tests and evaluates the proposed models, which are typical supervised deep learning algorithms. Standard metrics were used to achieve this including: accuracy, simulation time, training loss, training accuracy, validation loss, and validation loss.

3.2 Experimental Setup

The simulations are performed with Google Colaboratory using minimum system parameters as follows:

Hardware requirements: × 64-based processor, AMD E1-1200 APU with Radeon™, HD Graphics, 1.40 GHz, 4 GB RAM, and 160 HDD.

Software requirements: Windows 8 and 64-bit Operating System.

Hyperparameter selection. The minimal hyperparameters for CNN models are given in Table 1.

In Table 1, the selection of appropriate hyperparameters for the CNN models was achieved manually in order to reduce cost and memory consumption. Also,

Table 1 Minimal hyperparameter setup

Hyperparameter	Value
Number of convolutional layers	2
Number of max pooling layers	2
Number of dense layers	4
Dropout rate	0.2
Optimizer	Adam
Activation function	ReLU
Loss function	Binary-crossentropy
Number of epochs	10
Batch size	64

hyperparameters were set on the basis of the context-specific dataset (that is, binary and RGB color images).

Data Collection. The datasets used for the training and validation of proposed CNN models were collected from standard image databases including MINST and CIFAR-10 for binary images and RGB color images, respectively. A total of 60,000 samples of both image types were divided into 50,000 and 10,000 for training and validation purposes.

Performance evaluation. The effectiveness of the proposed CNN models was computed using metrics represented in Eq. 1 [29, 30]:

$$\text{Accuracy} = \frac{\text{TN} + \text{TP}}{\text{FN} + \text{FP} + \text{TN} + \text{TP}} \tag{1}$$

From Eq. 1, the accuracy measures the rate of correctly classified images to all the samples. True Positive (TP): If the classified image is actually object class, the classification is TP., False Positive (FP): If the classified image is actually object class, the classification is FP, True Negative (TN): If the classified image is actually object class, the classification is TN, False Negative (FN): If the classified image is actually the object class, the classification is FN.

Other metrics [4] utilized in this paper include test loss, test accuracy, validation loss, validation accuracy, image size, and simulation time.

4 Discussion of Results

The outcomes of performing training and validation with selected RGB color image for the proposed CNN model are presented in Table 2.

Table 2 Outcomes of CNN model with RGB color images

Epoch	Simulation time (s)	Training loss	Training accuracy	Validation loss	Validation accuracy	Overall loss	Overall accuracy
1	25	1.5592	0.4297	1.2685	0.5460		
2	13	1.1814	0.5811	1.1101	0.6065		
3	13	1.0207	0.6421	1.0193	0.6423		
4	13	0.9185	0.6768	0.9548	0.6624		
5	13	0.8396	0.7046	0.8999	0.6879		
6	13	0.7812	0.7248	0.8999	0.6904		
7	13	0.7276	0.7467	0.8618	0.7034		
8	13	0.6823	0.7618	0.8607	0.7067		
9	13	0.6418	0.7738	0.9396	0.6769		
10	13	0.6058	0.7874	0.8830	0.7106	0.8830	0.7106

From Table 2, the CNN model with RGB color images achieved loss of 88.30 and 71.06% accuracy after 1 s in epoch 1. These are depicted for training and validation as presented in Fig. 2.

From Fig. 3, the accuracy of the CNN model showed similar trend during training and validation procedures before epoch 2. Both curves increased steadily after epoch 2, but the validation reduced to peak at 0.7-point accuracy at epoch 10, while the training procedure increased steadily through epoch 10 but better than validation curve at 0.8-point accuracy.

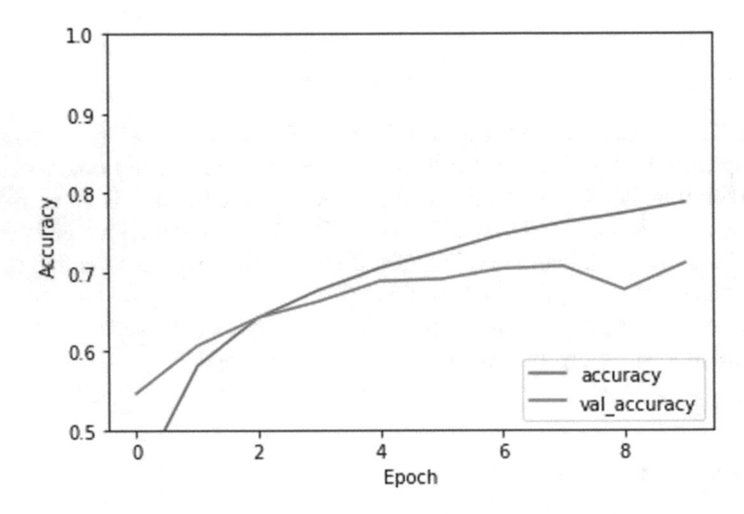

Fig. 2 Performance of CNN model with RGB color images

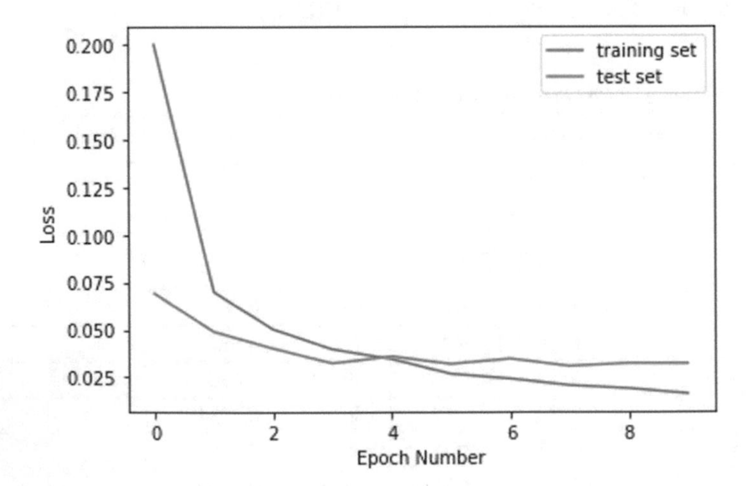

Fig. 3 Loss of CNN model during training and testing procedures

Similarly, the results of the training and validation procedures for the sampled binary images in the objects datasets are given in Table 3.

From Table 3, the CNN model approach for identifying binary images of objects achieved loss of 3.20 and 99.10% accuracy after 1 s in epoch 1. The performance of CNN model used for detecting binary objects during training and validation are presented in Figs. 3 and 4.

From Fig. 3, the loss function was changing during the training, which is expect to get smaller and smaller on every next epoch.

Similarly, in Fig. 4, the accuracy function was changing during the training, which is expect to get larger after epoch 3 better than the test procedure throughout the epoch

Table 3 Performance of the CNN model with binary images

Epoch	Simulation time (s)	Training loss	Training accuracy	Validation loss	Validation accuracy	Overall loss	Overall accuracy
1	29	0.1998	0.9394	0.0690	0.9789		
2	29	0.0696	0.9790	0.0488	0.9841		
3	29	0.0502	0.9845	0.0402	0.9865		
4	29	0.0398	0.9891	0.0323	0.9887		
5	29	0.0345	0.9891	0.0359	0.9886		
6	29	0.0267	0.9915	0.0319	0.9897		
7	29	0.0240	0.9922	0.0346	0.9893		
8	29	0.0206	0.9934	0.0307	0.9916		
9	29	0.0189	0.9939	0.0322	0.9899		
10	29	0.0162	0.9950	0.0320	0.9909	0.0320	0.9910

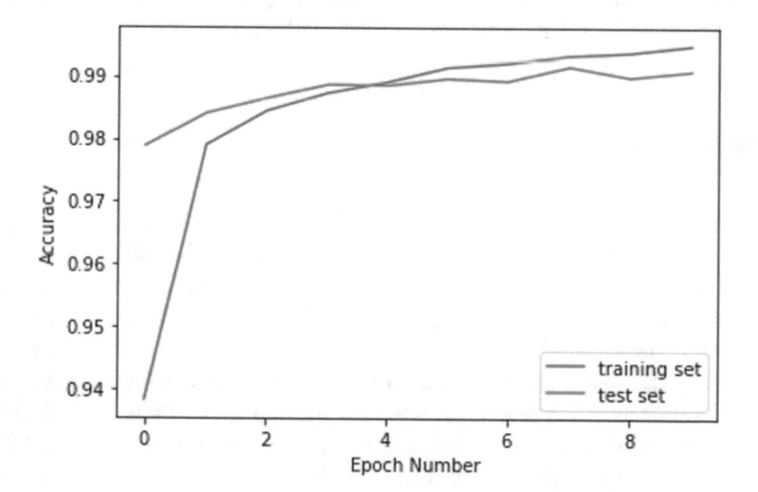

Fig. 4 Accuracy of CNN model during training and testing procedures

Table 4 Performances of image classification tasks compared

Metric	CNN (RGB color image)	CNN (binary images)	Binary images [2]	RGB color images [3]
Simulation time (ms)	14.20	29.00	22.00	13.00
Accuracy (%)	71.06	99.10	97.00	91.76
Image size	60,000	60,000	47,434	4000
Database	CIFAR-10	MINST	AHCD	YOLOvs

10. However, both accuracy curves improved considerably throughout training and testing phases.

4.1 Performance Benchmarking

The performances of the CNN models against comparable studies are presented in Table 4.

From Table 4, the outcomes revealed that the average duration for CNN model with RGB color images was better than CNN model with binary images (that is, 14.20–29.00 ms), which is agreement with previous works in RGB color images [3] and binary images [2]. The reverse was the case of accuracy, and the outcomes revealed that CNN modeled with RGB color images was outperformed by the CNN modeled with binary images by 71.06–99.10%. The same trend was observed in the benchmark studies in RGB color images [3] and binary images [2]. The performance of the two CNN models adopted in this study outperformed existing benchmark studies in terms of accuracy, average duration, and number of images considered.

5 Conclusion

Consequent upon the advancement of artificial intelligence, the CNN has progressed rapidly with high-degree of reliability and performance. CNN and its extension algorithms play important roles on imaging classification, object detection, and semantic segmentation in different fields including medicine, agriculture, surveillance, security, and road transportation. This paper investigated the effectiveness of binary images and RGB color images for classification tasks using two CNN models.

The outcomes revealed that the average duration for CNN model with RGB color images was better than CNN model with binary images (that is, 14.20–29.00 ms), which is agreement with previous works in RGB color images [3] and binary images [2], respectively. In the case of accuracy, the outcomes revealed that CNN modeled with RGB color images was outperformed by the CNN modeled with binary images by 71.06–99.10%. The same trend was observed in the benchmark studies in RGB color images [3] and binary images [2], respectively.

However, the timings and performances of the proposed CNN models can be improved on for real-time geographical matching applications such as Urban 3D, complex facades in museums, and shopping malls. The CNN model accuracy for RGB color images requires improvements against binary images depiction of objects.

References

1. Abbas A, Abdelsamea MM, Gaber MM (2021) Classification of COVID-19 in chest X-ray images using DeTraC deep convolutional neural network. Appl Intell 51(2):854–864
2. Altwaijry N, Al-Turaiki I (2021) Arabic handwriting recognition system using convolutional neural network. Neural Comput Appl 33(7):2249–2261
3. Li Y, Li M, Qi J, Zhou D, Zou Z, Liu K (2021) Detection of typical obstacles in orchards based on deep convolutional neural network. Comput Electron Agric 181:105932. https://doi.org/10.1016/j.compag.2020.105932
4. Hazra A, Choudhary P, Inunganbi S, Adhikari M (2020) Bangla-Meitei Mayek scripts handwritten character recognition using convolutional neural network. Appl Intell 51(4):2291–2311
5. Lu J, Tan L, Jiang H (2021) Review on convolutional neural network (CNN) applied to plant leaf disease classification. Agriculture 11(707):1–18
6. Benkaddour MK, Lahlali S, Trabelsi M (2021) Human age and gender classification using convolutional neural network. In: 2020 2nd international workshop on human-centric smart environments for health and well-being (IHSH), pp 215–220
7. Bera S, Shrivastava VK (2020) Analysis of various optimizers on deep convolutional neural network model in the application of hyperspectral remote sensing image classification. Int J Remote Sens 41(7):2664–2683. https://doi.org/10.1080/01431161.2019.1694725
8. Alfa AA, Ahmed KB, Misra S, Adewumi A, Ahuja R, Ayeni F, Damasevicius R (2019) A comparative study of methods for hiding large size audio file in smaller image carriers. In: ICETCE 2019, CCIS, vol 985, pp 179–191. https://doi.org/10.1007/978-981-13-8300-7
9. Pramanik R, Bag S (2020) Segmentation-based recognition system for handwritten Bangla and Devanagari words using conventional classification and transfer learning. IET Image Process 14(5):959–972
10. Alom MZ, Sidike P, Hasan M, Taha TM, Asari VK (2017) Handwritten Bangla character recognition using the state-of-art deep convolutional neural networks. arXiv Preprint: arXiv: 1712.09872
11. Akhand M, Ahmed M, Rahman MH, Islam MM (2018) Convolutional neural network training incorporating rotation-based generated patterns and handwritten numeral recognition of major Indian scripts. IETE J Res 64(2):176–194
12. Malakar S, Paul S, Kundu S, Bhowmik S, Sarkar R, Nasipuri M (2020) Handwritten word recognition using lottery ticket hypothesis based pruned CNN model: a new benchmark on CMATERdb2. 1.2. Neural Comput Appl 32(18):15209–15220
13. Ghosh R, Vamshi C, Kumar P (2019) RNN based online handwritten word recognition in Devanagari and Bengali scripts using horizontal zoning. Pattern Recogn 92:203–218
14. Chaudhary A, Hazra A, Chaudhary P (2019) Diagnosis of chest diseases in x-ray images using deep convolutional neural network. In: 2019 10th international conference on computing, communication and networking technologies (ICCCNT). IEEE, pp 1–6
15. Hubel DH, Wiesel TN (1962) Receptive field, binocular interaction and functional architecture in the cat's visual cortex. J Physiol 160(1):106–154
16. Simonyan K, Zisserman A (2014) Very deep convolutional networks for large-scale image recognition. arXiv preprint arXiv: 1409.1556

17. Szegedy C, Liu W, Jia Y, Sermanet P, Reed S, Anguelov D, et al (2015) Going Deeper with convolutions. In: Proceedings of the IEEE conference on computer vision and pattern recog, pp 1–9
18. He K, Zhang X, Ren S, Sun J (2016) Deep residual learning for image recognition. In: 2016 IEEE conference on computer vision and pattern recognition (CVPR). IEEE, pp 770–778
19. Yu D, Xu Q, Guo H, Zhao C, Lin Y, Li D (2020) An efficient and lightweight convolutional neural network for remote sensing image scene classification. Sensors 20:1999
20. Tan M, Pang R, Le QV (2020) EfficientDet: scalable and efficient object detection. In: 2020 In IEEE/CVF conference on computer vision and pattern recognition, pp 10778–10787
21. Yang R, Yu Y (2021) Artificial convolutional neural network in object detection and semantic segmentation for medical imaging analysis. Front Oncol 11:573
22. Venhuizen FG, Bram VG, Bart L, Freekje VA, Vivian S, Sascha F et al (2018) Deep learning approach for the detection and quantification of intraretinal cystoid fluid in multivendor optical coherence tomography. Biomed Opt Exp 9:1545
23. Girshick R, Donahue J, Darrell T, Malik J (2014) Rich feature hierarchies for accurate object detection and semantic segmentation. In: 2014 IEEE conference on computer vision and pattern recognition, pp 580–587
24. Yoon AP, Lee Y, Kane RL, Kuo C, Lin C, Chung KC (2021) Development and validation of a deep learning model using convolutional neural networks to identify scaphoid fractures in radiographs. JAMA Netw Open 4(5):e216096–e216096
25. Zhou Y, Lu Y, Pei Z (2021) Microprocessors and microsystems accurate diagnosis of early lung cancer based on the convolutional neural network model of the embedded medical system. Microprocess Microsyst 81:103754
26. Dais D, Bal E, Smyrou E, Sarhosis V (2021) Automatic crack classification and segmentation on masonry surfaces using convolutional neural networks and transfer learning. Autom Constr 125(103606):1–18. https://doi.org/10.1016/j.autcon.2021.103606
27. Yang W, Zhang X (2021) Transferability of convolutional neural network models for identifying damaged buildings due to earthquake. Remote Sensing 13(3):504
28. Fu X-B, Yue S-L, Pan D (2021) Camera-based basketball scoring detection using convolutional neural network. Int J of Auto and Comput 18(2):266–276
29. Raj RJS, Shobana SJ, Pustokhina IV, Pustokhin DA, Gupta D, Shankar K (2020) Optimal feature selection-based medical image classification using deep learning model in internet of medical things. IEEE Access 8:58006–58017
30. Prasetya R, Ridwan A (2019) Data mining application on weather prediction using classification tree, Naïve Bayes and K-nearest neighbor algorithm with model testing of supervised learning probabilistic brier score, confusion matrix and ROC. J Appl Comm Inform Technol 4(2):25–33

IoT and Networking

Detection of Android Malwares on IOT Platform Using PCA and Machine Learning

Bilal Ahmad Mantoo, Sumit Malhotra, and Kiran Gupta

Abstract The Android operating system is considered to be the most advanced and popular smart phone operating system and has a dramatic increase in the market with approximately 1.5 billion Android-based devices shipped by 2021. This spike increased in the market leads to high number of malwares in the Android platform. The reason is simple, because the mutating nature of Android malwares reduces the efficiency of malware detection tools in that are already in the market. Furthermore, the vast number of features in the Android platform provides a big challenge to the current tools to detect the Android malware on IOT-based platform. In this paper, we used a real physical device, i.e., Android smart phone, smart tablets, Android watches instead of protected environment like genymotion for analysis and extract the feature from 10,650 applications of malware and benign. Large spaces of feature set are reduced with the use of principal component analysis approach which extracts the best features from the dataset. The data is then fed to the deep learning model with different hidden layers; the algorithms are applied for dynamic features as well as the combination with permissions. The results reveals that the deep learning model with three hidden layers gives overall best accuracy as compared to the previous works done on the same field using protected environment.

Keywords Android malware · Dynamic features · Static features · PCA · Deep learning

1 Introduction

Android Smartphone has become spectacularly useful tool for everyone's daily life, such as shopping, payment transaction, and instant messaging, with this mounting

B. A. Mantoo (✉)
Department of Computer Science and Engineering, Presidency University, Itgalpur, Yelahanka, Bangaluru 560064, India
e-mail: bilalbashir136@gmail.com

S. Malhotra · K. Gupta
Department of Computer Science and Engineering, Chandigarh University, Gharaun, Mohali, Punjab 193101, India

Y. Singh et al. (eds.), *Proceedings of International Conference on Recent Innovations in Computing*, Lecture Notes in Electrical Engineering 1001,
https://doi.org/10.1007/978-981-19-9876-8_16

market, there is also a matter for its security that whether it is susceptible or there is a loop hole. Open source nature of Android OS, makes intruders write the malware using security loopholes of Android operating system, and this becomes the main reason for rapid increase in malwares in this market. The critical behaviors like stealing private information, sending SMS, bypassing the security protocols, etc., threatening the security and property of user by intruders within this market.

In today's technology, widely Android OS is facing myriad of issues particularly with malicious applications. These malwares are used for number of activities like creating botnets [1], carrying malicious activities, performing crimes, drug dealing, etc. Over the last couple of years, malwares in the market are hiked significantly. According to McAfee, over 3 million Android malwares applications were found in the last year 2020, thus increasing the number of malwares from the previous year by half million [2].

To reduce the spreading of Android malware, search engines are introduced with detection mechanism to its market like Bouncer-Bouncer checks the application submitted in sandbox to detect the harmful behavior, but these can be bypassed by simple methods [3]. Moreover Google has launched a Google Play Protect which is used for detection of malwares in this engine, it scans the applications automatically 24/7.The report of play protect says that there are over 50 billion apps that are verified every day.

Various approaches have been proposed which are used for Android malware detection. These are categorized in static analysis, dynamic analysis, or sometimes both these approaches are combined with intrinsic analysis called hybrid approach. The static analysis approach requires the reverse engineering of malicious application [4]. Similarly dynamic analysis is used for detection of Android malwares in real time by installing the application in protected environment such as sandbox or gennymotion. Various dynamic approaches have been proposed like [5]. Intrinsic feature-based approach with dynamic analysis is also being used for malware detection along with static analysis [6]. As the number of features increases, these algorithms or approaches become somewhat slow and their efficiency degrades, so there is a need that the best features should be used and extracted from the huge dataset using different feature extraction algorithms like Principle Component Analysis and Linear Discriminant Analysis.

In contrary to basic algorithms being used, deep learning (DL) has gained many chances as an important method of AI [7]. These classifiers have many applications in image classification, natural language processing, and speech recognition. Contrary to previous deep learning-based methods, this work experiences new approach coupled with dynamic input, with purpose of getting higher accuracy. In conclusion, this paper contributes to the following:

1. We present a deep learning-based dynamic method for malware detection in Android smart phones, consuming the state-based input.
2. The features selection procedure used in this method are Principle Component Analysis and Liner Discriminant Analysis and best features are selected after comparing the result of the two methods.

3. Extensive and comparative study has been done by using different classifiers in addition to the deep learning in controlled environment. Results obtained show better accuracy compared to the traditional classifiers.

Moreover, the paper is structured as, Sect. 2 overviews about related work, and then followed by methodology and experiments. Sections 3 and 4 give detailed results and discussions.

2 Related Work

Researchers are working hard to make the detection easier and efficient; this section discusses the important work that has been done in the area of Android malware detection. Basically, there are two approaches of malware detection using static analysis and dynamic analysis.

2.1 *Android Malware Analysis Technologies*

The static analysis method refers to extracting the features of Android application by simply decompiling the application using APK tool [8, 9]. The permission field in the APK file provides the necessary information about the increasing risks of malware [10]. Bilal et al. [6] provide the permission and intrinsic feature-based Android malware detection that classifies the malware on permissions and intrinsic features. WHY-PER, a framework based on natural language processing (NLP) presented by Rahul et al. [11] identifies the comments regarding the permissions in the file. Stowaway, a tool to detect the over privileged application, and Felt et al. [12] provide this method. Arora et al. [13] provide another approach for malware detection using permission fields.

API-based Android malware detection is also used as a key instance for detection [14]. Dynamic analysis approach consists of running the applications in controlled environment or emulator like genymotion where the application frequency is measured by system calls with its kernel [15]. In order to have the real picture of communication with kernel, it is better to install the application in real device rather than emulator [16]. Enck et al. [17] provided a tool for analyzing system calls with low CPU overhead.

Fu et al. [18] projected ntLeakSemaic, a structure that can automatically set unusual susceptible network transmissions from mobile apps. Compared to on hand taint analysis approaches, it can attain improved correctness with less false positives. Android malware always try to copy the sensitive behaviors of benign applications, in order to reduce their chance of being detected [19].

Antivirus softwares detect malwares effectively using signature code which is updated on client side; this method is going to be highly efficient for known signatures, however if unknown malware that is not seen before cannot be detected effectively. Researchers have used machine learning algorithms to detect such cases [20]. To improve the performance of machine learning model, Chen et al. [21] presented a novel approach of to make the model protective to be evaded.

Supervised and unsupervised learning introduced by Arora et al. [22, 23] provided a hybrid approach using permissions and traffic analysis over network. Awad et al. [24] proposed modeling malware as a language and assessed the probability of ruling semantics in instances of that language, and they classified malware documents by applying the KNN.

3 Methodology and Experiments

To find the best features in the dataset, we have to install all the application in the device and which is of course a tedious task. Therefore, an automated tool is required to install these applications to extract the features. These features are then used as input to the machine learning algorithms/deep learning model. We used DynaLog for installing all these applications, which is a dynamic analysis framework [25].

DynaLog is an automated tool that is used to install massive number of applications for dynamic analysis, after that this tool helps in finding the malicious behavior of apps. This tool is designed to accept the large number of applications, installs them one by one in sequence in emulator, and logs them for further processing. DynaLog actually uses Droid Box [26] which actually extracts the features from the logs which are present in *logcat*. The logcat contains several dynamic features like API calls, intents, actions/events, and permissions. As of now, most of the Android malware detection tool uses stateless approach means features are extracted using tools like Monkey tool. One of the researches in previous year uses [27] stateless, stateful, and hybrid-based and finds that stateful approach is very robust and enables the greater code coverage than stateless and hybrid-based input model.

In order to perform dynamic analysis, real devices are much better than emulator as shown in Alzylae et al. [25]. Therefore, our approach in this study is to install apps on real devices rather than emulator. Devices that are used in analysis are Android with lollipop version of operating system, 4 GB RAM, 64 GB external SD card, and 2.6 GHz CPU. Each Android device installs approximately 110 applications in a single day. Furthermore, device is also connected to the 4G Internet and can send text and receive calls when requested.

The executed run time of each application is approximately of 3 min and within these minutes each application hits 3–4 K events for a particular app, most of the applications did not provide any other event. The overall process of detection using real devices is shown in Fig. 1.

In Fig. 1, it is clearly mentioned that all the applications are installed in the Android devices, and then their features are extracted in log files. Once the log files

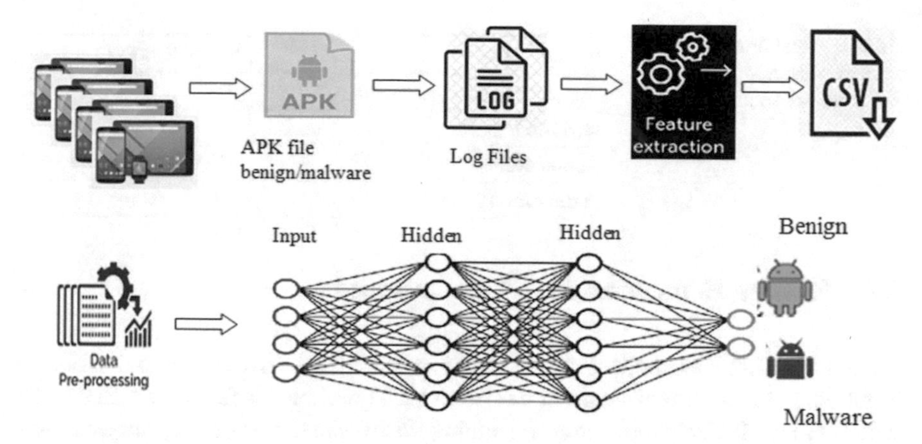

Fig. 1 Experimental setup for feature extraction over real devices with deep learning

are obtained, these log files are combined to form a single file, then features are extracted using Python script, and csv file is obtained which is given as input to our machine learning algorithms.

3.1 Dataset

For evaluating our system performance, we have to collect the data of Android malwares and genuine apps and compare our results with the previous machine learning algorithms. We used a dataset consisting of 10,650 applications of both malware and benign from different sources like Google play Store, Androgaurd, and McAfee Labs. Out of these applications, 5430 applications are malware, and rest applications are benign. Different categories are present like payment apps, gaming apps, chatting, media apps, etc. The dataset is divided into benign and malware sample datasets. The quantity and source of these samples are given in Table 1.

Table 1 Application quantity and source used in this research work

Application source	No. of applications
Google Play Store	5650
Androgaurd	3000
MacAfee Labs	2000

Table 2 Total number of extracted features form different data sources

Feature set	No. of features
Application attribute features	90
Intrinsic features	50
Action events	40
Permission features	200

3.2 Feature Extraction and Preprocessing

Once applications are collected and stored in local hard drive, each application is installed and run on real Android device using DynaLog for feature extraction as shown in Fig. 1. Each application is running for two instances one for stateless and another for state full; the features are logged in .csv file. The .csv file is used as input to the Jupyter notebook for further processing. The features that are present in the applications are set as 1 in .csv file and if the feature is not present is kept as 0. Total number of features that are extracted using DynaLog are 400, out of which only top 10 features are extracted using Principle Component Analysis as a feature extraction algorithm. Out of 380 features, 200 permissions, 50 intrinsic features, 90 application attribute features, 40 event features are used as given in Table 2.

Once the features are extracted from using DynamicLog, all these features are fed to preprocessing algorithm to find the top ranked feature from the whole dataset.

3.3 Principal Component Analysis

Principal component analysis (PCA) is a method for reducing the dimensions from a large dataset to a lower dimensional data with high information by decomposing the data into perpendicular vector where the spread of information is more. As more variance means more information in data.

The data is fed to PCA algorithms in the form matrix of features that are in the dataset. The main intention is to find the main variance in the dataset by calculating the eigen vector of the covariance matrix [28].

The mathematical steps used in of PCA is as under.

(1) For data matrix X, calculate the covariance of X.

$$\mu = E\{X\} \tag{1}$$

Then calculate the covariance.

$$R = \mathrm{Cov}\{X\} = E\{(X - \mu)(X - \mu)t\} \tag{2}$$

(2) Find eigen value $\lambda_1, \lambda_2, \lambda_3, \lambda_4, \ldots, \lambda_n$ and eigen vector $e_1, e_2, e_3, \ldots, e_n$ of the covariance matrix in descending order.

For the covariance R, solve the equation

$$|\lambda i - R| = 0 \tag{3}$$

In order to get the principal component, we have to count the proportion of data covered by the first M eigen values.

$$\frac{\sum_{i}^{m} x}{\sum_{i=0}^{n} y} \tag{4}$$

Select the first M eigen values which gives the maximum information.

(3) Project the data to lower dimensions

$$P = WtX \tag{5}$$

By using the first M respectively eigen vector, we can reduce the number of variables or dimensions from n to M.

Based on the results the features that have been ranked on top are shown in Figs. 2 and 3.

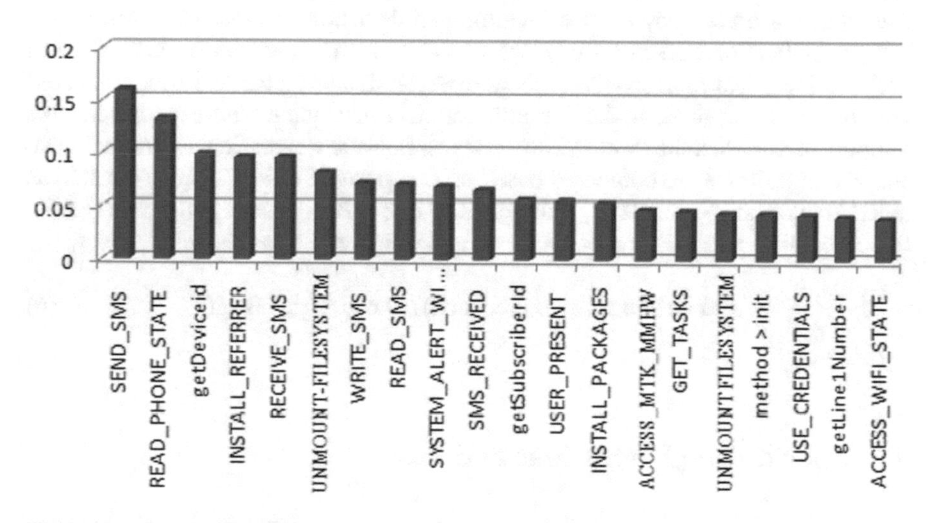

Fig. 2 Top 20 ranked features based on principal component analysis with permissions

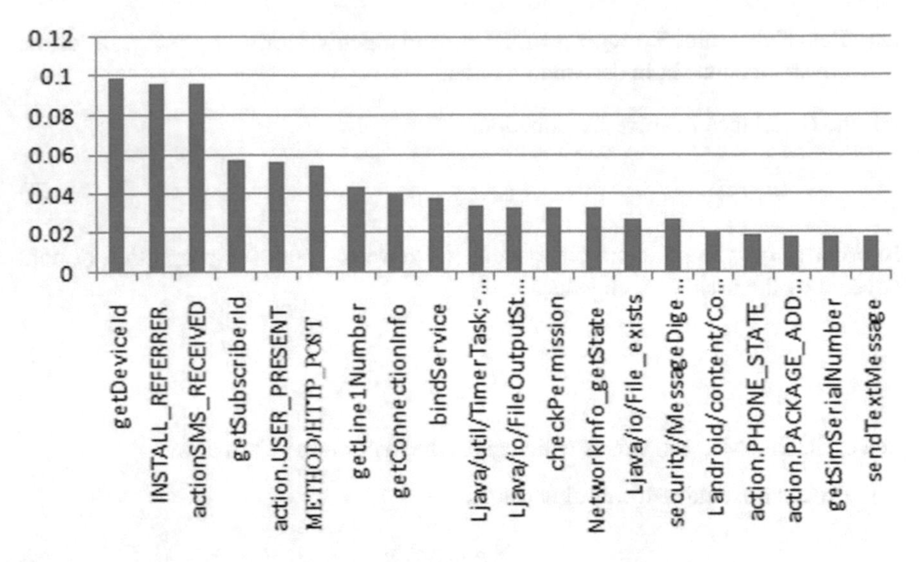

Fig. 3 Top 20 ranked features based on principal component analysis without permissions

3.4 Investigating Deep Learning Classifier

Main intention in this work is to evaluate our deep learning model over real devices and to increase accuracy of classification and detection of Android malware and benign applications. In our study, we use deep learning model on classification problem. Deep learning model is evaluated over confusion matrix and is the compared with the results obtained in the literature for other machine learning algorithms like Support Vector Machine (SVM), Naïve Bayes, Logistic Regression, Random Forest, etc. The algorithms are compared based on true positive ratio (TPR) also called as recall, true negative rate (TNR), false positive rate (FPR), false negative rate (FNR), F measure FM calculated on both classes and weighted FM as shown below.

$$\text{FM} = (2 * \text{recall} * \text{precision})/(\text{recall} + \text{precision}) \tag{6}$$

4 Experimental Discussions and Results

As per the data analysis done, permission-related features of both sample types like malware and benign are frequently used in the application as these features are mostly used for data stealing purposes. However, some interesting API calls are also used by most of the malware categories. It is worth mentioning that event *getdeviceId* is mostly used by apps, for finding the device id of a subscription, for example, the IMEI or GSM and the MEID for CDMA phones. The *Referrer* API is also used for reliable conversation via Google Play. And this feature become key point for hackers to lookout for new ways to game the system, they found loop hole in the referrer method by installing a form of install hijacking. Another problematic feature that is being highlighted *readphonestate*, this feature encompasses everything from something as innocuous as needing to have when a phone call is coming in, it will access your crucial data such as your device's IMEI number, and this permission is mostly observed in malicious apps.

The experimental results obtained using the deep learning model with different combinations of hidden layers are given in Table 3. Deep learning models are significantly better than the previous machine learning algorithms such as SVM, Naïve Bayes, and KNN. The results shown here are for static, dynamic features with 300, 300, 300 combinations perform the best as compared to other. We acquired w-FM as 0.0988 without the use of permissions Fig. 4.

The same scenario is repeated with the addition of state full feature, i.e., permissions and the results are discussed in Table 4.We can see that the same 300 neurons and three hidden layers were used, and the w-FM score to be obtained is approximately 0.99. Table 5 compares the results of this work with the previous works done in this area. Droiddetector [29] which uses both static and dynamic features gives 98.5% as accuracy. Our model outperforms the Deep4MalDroid [30] which uses three hidden layers Fig. 5.

Table 3 Deep learning model results with different combinations of hidden layers with dynamic features only

No. of layers	TPR	TNR	FPR	FNR	Precision	Recall	Accuracy	w-FM	AUC
2	0.95	0.88	0.11	0.05	0.93	0.95	0.92	0.94	0.97
2	0.93	0.88	0.11	0.06	0.93	0.93	0.92	0.93	0.96
2	0.97	0.90	0.09	0.03	0.94	0.96	0.95	0.95	0.98
2	0.96	0.90	0.09	0.03	0.94	0.96	0.94	0.95	0.98
3	0.90	0.90	0.09	0.09	0.95	0.90	0.90	0.92	0.92
3	0.97	0.89	0.10	0.02	0.94	0.97	0.94	0.95	0.98
3	0.97	0.88	0.11	0.02	0.94	0.97	0.94	0.95	0.98
3	0.74	0.95	0.04	0.25	0.97	0.74	0.82	0.84	0.87
3	0.53	0.98	0.01	0.46	0.98	0.53	0.70	0.69	0.76
3	0.45	0.99	0.04	0.54	0.99	0.45	0.65	0.62	0.73
3	0.97	0.90	0.09	0.02	0.94	0.98	0.95	0.96	0.99
3	0.53	0.99	0.04	0.46	0.99	0.54	0.70	0.70	0.77
4	0.95	0.89	0.10	0.04	0.93	0.96	0.93	0.94	0.98
4	0.56	0.99	0.04	0.43	0.99	0.56	0.72	0.72	0.78
4	0.97	0.91	0.08	0.02	0.94	0.97	0.95	0.96	0.95
4	0.98	0.85	0.14	0.01	0.92	0.98	0.93	0.96	0.93
4	0.96	0.89	0.11	0.03	0.93	0.96	0.93	0.95	0.96

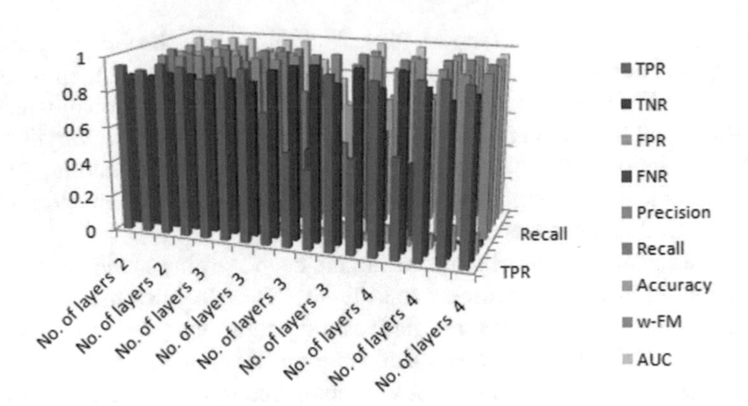

Fig. 4 Graphical representation of deep learning model with multiple layers

Table 4 Deep learning model results with different combinations of hidden layers with dynamic features and permissions

No. of layers	TPR	TNR	FPR	FNR	Precision	Recall	Accuracy	w-FM	AUC
2	0.87	0.93	0.07	0.12	0.93	0.95	0.92	0.90	0.93
2	0.96	0.96	0.08	0.03	0.93	0.93	0.92	0.95	0.98
2	0.95	0.96	0.12	0.03	0.94	0.96	0.95	0.94	0.97
2	0.96	0.95	0.08	0.03	0.94	0.96	0.94	0.95	0.98
3	0.90	0.90	0.09	0.09	0.95	0.90	0.90	0.96	0.92
3	0.97	0.93	0.08	0.02	0.94	0.97	0.94	0.95	0.98
3	0.97	0.88	0.11	0.02	0.94	0.97	0.94	0.95	0.98
3	0.95	0.93	0.04	0.25	0.97	0.97	0.95	0.96	0.98
3	0.94	0.92	0.03	0.46	0.98	0.97	0.95	0.96	0.98
3	0.95	0.93	0.04	0.54	0.96	0.97	0.95	0.96	0.98
3	0.98	0.97	0.03	0.02	0.98	0.99	0.98	0.99	0.99
3	0.93	0.90	0.04	0.46	0.99	0.97	0.95	0.96	0.98
4	0.95	0.91	0.10	0.04	0.93	0.96	0.93	0.95	0.98
4	0.96	0.92	0.07	0.33	0.95	0.96	0.95	0.95	0.98
4	0.97	0.91	0.07	0.02	0.94	0.97	0.95	0.96	0.95
4	0.98	0.91	0.87	0.01	0.92	0.98	0.93	0.96	0.98
4	0.96	0.90	0.09	0.03	0.93	0.96	0.93	0.95	0.97

5 Conclusion

In this study, a deep learning model for malware detection in Android market is presented. This model works on the real Android devices unlike previous work which uses emulators. We evaluated the Android framework over 10,500 Android

Table 5 Comparison of deep learning classifier with previous work done in the same field

	TPR	TNR	FPR	FNR	Precision	Recall	w-FM
NB	0.62	0.86	0.15	0.38	0.77	0.76	0.76
SVM linear	0.75	0.93	0.07	0.24	0.87	0.87	0.87
SVM RBF	0.75	0.94	0.06	0.25	0.87	0.86	0.87
RF	0.88	0.97	0.03	0.12	0.94	0.94	0.93
DL (300, 300, 300)	0.98	0.97	0.03	0.02	0.96	0.97	0.98

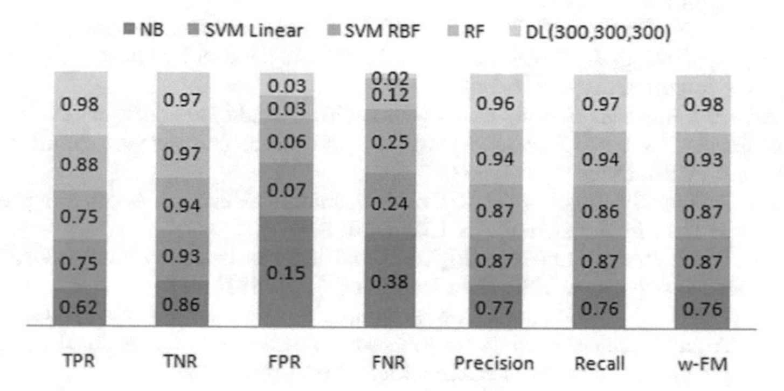

Fig. 5 Comparison of deep learning classifier with previous work done in the same field

applications with 390 features of static and dynamic. Using PCA as feature extraction tool, we reduced our large space of feature to only top 20 features, which helps in increasing the accuracy of machine learning model. The results shown clearly show that deep learning classifier performs better with three hidden layers. Perhaps this work is considered to be the most recent work using real devices and feature extraction algorithms to enhance the accuracy.

References

1. Anagnostopoulos M, Kambourakis G, Gritzalis S (2016) New facets of mobile botnet: architecture and evaluation. Int J Inf Secur 15(5):455–473
2. McAfee Labs Threats Predictions Report. McAfee Labs
3. Oberheide J, Miller C (2012) Dissecting the android bouncer. Summercon 2012
4. Arp D, Spreitzenbarth M, Hubner M, Gascon H, Rieck K, Siemens C (2014) Drebin: effective and explainable detection of android malware in your pocket. In: NDSS'14, pp 23–26
5. Enck W, Gilbert P, Chun B-G, Cox LP, Jung J, McDaniel P, Sheth AN (2010) Taintdroid: an information-flow tracking system for realtime privacy monitoring on smartphones. In: OSDI'10, vol 49, pp 1–6. https://doi.org/10.1145/2494522
6. Mantoo B, Khurana S (2020) Static, dynamic and intrinsic features based android malware detection using machine learning. In: Proceedings of the third international conference in computing. Central University of Jammu, March 2019, pp 31–46

7. Hao S, Liu B, Nath S, Halfond WG, Govindan R (2014) Puma: programmable ui-automation for large-scale dynamic analysis of mobile apps. In: Proceedings of the 12th annual international conference on mobile systems, applications, and services. ACM, pp 204–217
8. Bakour K, Never HM, Ghanem R (2018) Android malware static analysis: techniques, limitations, and open challenges. In: Proceedings of the 3rd international conference on computer science and engineering (UBMK'18), Sarajevo, Bosnia-Herzegovina, September 2018, pp 586–593
9. Riad K, Ke L (2018) RoughDroid: operative scheme for functional android malware detection. Sec Commun Netw 2018:10. Article ID 8087303
10. Talha KA, Alper DI, Aydin C (2015) APK auditor: permission-based android malware detection system. Digit Investig 13:1–14
11. Rahul P, Xiao X, Yang W, Enck W, Xie T (2013) WHYPER: towards automating risk assessment of mobile applications. In: Proceedings of the 22nd USENIX security symposium, Washington, DC, USA, August 2013, pp 527–542
12. Felt AP, Chin E, Hanna S, Song D, Wagner D (2011) Android permissions demystified. In: Proceedings of the 18th ACM conference on computer and communications security (CCS'11), Chicago, IL, USA, October 2011, pp 627–636
13. Arora A, Peddoju SK, Conti M (2020) PermPair: android malware detection using permission pairs. IEEE Trans Inf Forensics Secur 15:1968–1982
14. Wang W, Wei J, Zhang S, Luo X (2020) LSCDroid: malware detection based on local sensitive API invocation sequences. IEEE Trans Reliab 69(1):174–187
15. Mantoo B (2021) A hybrid approach with intrinsic feature-based android malware detection using LDA and machine learning. In: Proceedings of the forth international conference in computing, Central University of Jammu, March 2020, pp 295–306
16. Alzaylaee MK, Yerima SY, Sezer S (2017) Emulator vs real phone: android malware detection using machine learning. In: Proceedings of the 3rd ACM on international workshop on security and privacy analytics, 24 Mar 2017. ACM, Scottsdale, Arizona, USA, pp 65–72. https://doi.org/10.1145/3041008.3041010
17. Enck W, Gilbert P, Han S et al (2014) TaintDroid. ACM Trans Comput Syst 32(2):1–29
18. Fu H, Zheng Z, Bose S, Bishop M, Mohapatra P (2017) Leaksemantic: identifying abnormal sensitive network transmissions in mobile applications. In: Proceedings of the IEEE conference on computer communications (INFOCOM 2017), Atlanta, GA, USA, May 2017, pp 1–9
19. Yang W, Xiao X, Andow B, Li S, Xie T, Enck W (2015) Appcontext: differentiating malicious and benign mobile app behaviors using context. In: Proceedings of the international conference on software engineering (ICSE 2015), Florence, Italy, May 2015, pp 303–313
20. Kadri MA, Nassar M, Safa H (2019) Transfer learning for malware multi-classification. In: Proceedings of the 23rd international database applications and engineering symposium (IDEAS'19), Athens, Greece, June 2019, pp 1–7
21. Chen L, Hou S, Ye Y (2017) Securedroid: enhancing security of machine learning-based detection against adversarial android malware attacks. In: Proceedings of the 33rd annual computer security applications conference (ACSAC 2017), Orlando, FL, USA, December 2017, pp 362–372
22. Arora A, Peddoju SK, Chouhan V, Chaudhary A (2018) Hybrid android malware detection by combining supervised and unsupervised learning. In: Proceedings of the 24th annual international conference on mobile computing and networking (MobiCom 2018), October 2018, pp 798–800
23. Arora A, Peddoju SK (2018) NTPDroid: a hybrid android malware detector using network traffic and system permissions. In: Proceedings of the 2018 17th IEEE international conference on trust, security and privacy in computing and communications/12th IEEE international conference on big data science and engineering (TrustCom/BigDataSE), New York, NY, USA, August 2018, pp 808–813
24. Awad Y, Nassar M, Safa H (2018) Modeling malware as a language. In: Proceedings of the 2018 IEEE international conference on communications (ICC 2018), Kansas City, MO, USA, May 2018, pp 1–6

25. Alzaylaee MK, Yerima SY, Sezer S (2017) Emulator vs real phone: android malware detection using machine learning. In: Proceedings of the 3rd ACM on international workshop on security and privacy analytics. ACM, Scottsdale, Arizona, USA, 24 Mar 2017, pp 65–72. https://doi.org/10.1145/3041008.3041010
26. DroidBox: an android application sandbox for dynamic analysis. https://code.google.com/p/droidbox/
27. Yerima SY, Alzaylaee MK, Sezer S (2019) Machine learning-based dynamic analysis of android apps with improved code coverage. EURASIP J Inf Secur 2019(1):4. https://doi.org/10.1186/s13635-019-0087-1
28. Xiaio (2010) Principal component analysis for feature extraction of image sequence. In: Proceedings of international conference on computer and communication technologies in agriculture engineering
29. Yerima SY, Sezer S, Muttik I (2016) Android malware detection using parallel machine learning classifiers. arXiv:1607.08186
30. Hou S, Saas A, Chen L, Ye Y (2016) Deep4maldroid: a deep learning frame work for android malware detection based on linux kernel system call graphs. In: 2016 IEEE/WIC/ACM international conference on web intelligence workshops (WIW). IEEE, pp 104–111

A Survey on Different Security Frameworks and IDS in Internet of Things

Hiteshwari Sharma, Jatinder Manhas, and Vinod Sharma

Abstract Internet of things has significantly revamped the whole dynamics of communication. It consists of multiple heterogeneous devices which exchange and receive data over the Internet with phenomenal ubiquitous connection. The heavy penetration of these devices into everyone's life poses diverse nature of cyber security threats. IoT devices are susceptible to vast range of attacks due to their limited computation capabilities, low power and memory constraints. Even within one IoT standard, a device typically has multiple options to communicate with other devices. Various security measures have been proposed to counter these attacks from time to time. Intrusion detection system are excellent frameworks to save these devices from such vulnerabilities. This paper aims to conduct a deep, systematic and comprehensive survey on different IDS and security frameworks. Also, a novel security framework based on deep learning algorithms and blockchain platform has been proposed for this study.

Keywords Internet of Things · Intrusion detection system · Deep learning · Blockchain

1 Introduction

Internet of Things (IoT) incorporates multiple physical entities with pervasive Internet connection like sensors, actuators, microcontrollers and transceivers for facilitating communication between them. It is built with suitable protocol stacks and standards which interact with each other and communicating with the users, thus becoming the constitutive part of the Internet. IoT ushers a new contemporary era where every gadget or physical device has the potential to get connected to a network. Their effective utilization paves the path for rapid completion of multiple and time intensive complex tasks that requires a high degree of intellect and reasoning. Over the last few years, the popularity of Internet of Things (IoT) devices has exploded and

H. Sharma (✉) · J. Manhas · V. Sharma
Department of Computer Science and IT, University of Jammu, Bhaderwah Campus, Jammu, J&K, India
e-mail: hiteshwarisharma@gmail.com

Y. Singh et al. (eds.), *Proceedings of International Conference on Recent Innovations in Computing*, Lecture Notes in Electrical Engineering 1001,
https://doi.org/10.1007/978-981-19-9876-8_17

has increased to a greater extent. This advancement is due to their ubiquitous connection. It allows them to communicate and exchange data with other technologies to develop a system that is intelligent and capable of taking effective and efficient decisions. This results into a seamless user experience and improve people's daily lives. This is evident from the availability of such type of gadgets being used in today's world. These days the proliferation of smart devices is not only limited to domestic use, currently has penetrated into almost all the fields and acting as a driving force behind the development of a linked knowledge-based world [1].

The Internet of Things entails extending Internet connectivity beyond ordinary devices like desktops, laptops, smartphones, and tablets to any number as compared to the previously dumb or non-Internet-enabled physical gadgets and others objects being used on daily basis. The heavy penetration of these devices into everyone's life poses diverse nature of cyber security threat to the majority of our day-to-day activities. Cyber-attacks against essential infrastructure, such as power plants and transportation, have far-reaching effects for cities and countries. The incredible technology which is laden with so many benefits faces a lot of security challenges. Therefore, traditional security counter measures could not work efficiently for IoT networks due to different protocol stacks and standards followed by multiple heterogeneous entities like IEEE 802.15.4, Ipv6 over Low-power, Wireless Personal Area Network (6LoWPAN), IPv6 Routing Protocol for Low-power and Lossy Network (RPL), Constrained Application Protocol (CoAP), etc. Various methods that enhance IoT security include data confidentiality and authentication, access control within the IoT network, privacy and trust among users and IoT. Enforcement of security and privacy policies also adds to it in similar fashion. Even after implementing all the measures concerning IoT security, the networks still remain susceptible to a variety of assaults (multiple attacks) targeted to disrupt the networks. As a result, a second layer of protection tailored to identify attackers is required to be developed for its smooth functioning.

Intrusion detection systems (IDSs) are the frameworks that are being used to save these devices from vulnerabilities and attacks. Intrusion detection system (IDS) is an effective tool that collects network traffic as input data in order to detect intruders or malicious activities in the analyzed target, a network or a host. In IoT networks, these systems scan network packets and respond in real-time. To make them effective, these IDS must work in IoT-specific conditions like low energy, low process capacity, quick reaction and massive data processing. As a result, improving IoT embedded IDS is a continual and critical issue that necessitates thorough understanding of IoT system security vulnerabilities.

2 Literature Survey

A deep and thorough literature review has been undertaken to study the current research in the area of IDS performance for IoT devices. An extensive examination of different existing and possible cyber threats in IoT has been done. Different methodologies and proposed techniques has been carefully analyzed and various effective technologies are discussed in the security domain of IoT.

2.1 IDS and Its Types

Intrusion detection system constitutes software and hardware modules whose main purpose is to detect abnormal or malicious activities on the target. IDS are comprised of two main types:

1. Host-based IDS
2. Network-based IDS.

Host-based IDS are positioned in a host and they keep track on the traffic that are originating and coming to that particular hosts only. Attacks which are generated in any other part of the network will not be detected by this IDS. If there are attacks in any other part of the network, they will not be detected by the host-based IDS.

Network-based IDS are employed in a network to discover threats on the hosts of that network. It is imperative to place this IDS at the arrival and departure points of data from this network to the outside world to capture all the data passing through the network. Since a network-based IDS has to perform strong surveillance on all the data passing through the network, it needs to be very rapid and efficient to track the whole traffic and should drop as little packets as possible [2].

Based on their detection mechanism, IDS are classified into four main categories:

1. Signature based
2. Specification based
3. Anomaly based
4. Hybrid based.

Signature-based IDS compares the attack signatures with existing database and if matching occurs, it alerts the system for Intrusion. However, it is not effective in detecting new and advanced attacks. The main drawback is that they cannot detect unknown threats or attacks which have no signature in their pattern matching database. Specification-based IDS monitors the system according to the rules and threshold framed by the network administrators and if any deviations occurred within these rules, IDS will take an immediate action. Anomaly-based IDS are getting more popularity due to their innate capacity to detect unknown malicious traffic patterns using machine learning techniques. The biggest issue with this technology is the enormous number of false alarms as sometimes genuine traffic gets misclassified as cyber-attack. Hybrid-based IDS is the amalgamation of any of these IDS to improve the overall IDS performance.

Search Criteria

In this review paper, we intent to identify and review different types of IDS and security frameworks in the field of IoT. Table 1 lists the keywords and search parameters used to find and select the papers included in this review paper. The articles were searched from well-known electronic databases like Science Direct, IEEE Xplore, Mdpi, etc. A total of 35 research articles were selected according to our screening criteria. Redundant and similar articles were not considered as the same articles were published in different journals in more than one database. Also, some manuscripts were not found useful because they considered various attacks and threats in pure network domain and they differ from IoT due to its underlying architecture and different protocol stacks. Different search keywords were used to have a thorough insight into the electronic databases of research articles. Search keywords of articles and source journals are enlisted in Table 1 and the list of conference proceedings from which research articles have been retrieved is given in Table 2.

Previous Research for IDS on IoT

This section provides a detailed review of different techniques and frameworks used for IDS in IOT.

Hamid et al. [3] proposed an unsupervised and hybrid discovery model to detect two insider attacks, i.e., Sinkhole and Selective forwarding attack in IoTs. The whole

Table 1 Various resource databases and journals referred for the study

Resource	Keywords used for search	Journals from which papers are selected
Elsevier	Internet of Things, Intrusion Detection System, Deep learning, Attack detection, Blockchain, AI, constrained IoT devices	Computer Communications Future Generation Computer Systems International Journal of Information Management Mechanical System and Signal processing
IEEE	IoT, IDS, Security frameworks	IEEE Internet of Things Journal
Arxiv	IDS, Type of IDS, resource constraint devices	Cryptography and Security
SpringerOpen	Deep learning with Blockchain	Human Centric Computing and Information Sciences
IET	IoT, Blockchain, Authentication, Encryption	CAAI Transactions on Information Security

Table 2 List of different conference from which papers are collected

S. No.	Name of the conference
1	IEEE International Symposium on Electronics and Smart Devices (ISESD)
2	Elsevier, Third conference on Computing and Network Communication (CoConet)
3	ACM BlockSys'18: Proceedings of the 1st Workshop on Blockchain-enabled Networked Sensor Systems
4	2019 IEEE Global Communications Conference (GLOBECOM)

process of Intrusion detection is divided into different stages where a malicious node which can cause sinkhole and selective forwarding attacks is identified and threshold conditions for both the attacks are calculated using the proposed algorithm. The study followed by anomaly detection in 6BR which uses OPF with Map Reduce architecture to incorporate parallelism. Finally, the voting mechanism for anomaly detection mechanism is presented. The results based on specifications (local and global) and anomaly-based Intrusion detection agents, the root node makes a general decision about the occurred anomalies in the network using a voting mechanism.

Significant amount of advancement has been made in developing an intelligent IDS by incorporating deep learning algorithms. The study proposed by Diro and Chilamkurti [4] on distributed attack detection system using deep learning model. The whole architecture of IoT is divided into two main layers: IoT layer and Fog network layer which comprises of Sensors, Actuators and Fog nodes, Master nodes, respectively. The fog nodes are responsible in model training whereas master node perform parameter optimization. The experimentation performed on two class (normal and attack) and four class (Normal, DOS, Probe, R2L, U2R) categories. NSL-KDD dataset is used for model validation.

Next year, Roberto et al. [5] provided a hybrid IoT architecture based on blockchain technology for decentralized management of data. The proposed framework consists of IOT layer and blockchain layer. The edge computing paradigm provides computational advantage by reducing the amount of data to the network and eliminating centralized computing. The main objective of this paper is to improve the quality of data and false data detection and for that a game-based algorithm is developed that runs on the edge computing layer of new architecture on the data collected by the IoT nodes. Dinan et al. [6] created an IoT system with and without using blockchain and made a comparative analysis between the two systems. The communication protocol used in IoT systems without blockchain is MQTT. By simulating attacks and observing their security characteristics, both of these IoT systems were evaluated for their security degree. The results revealed that an IoT system based on blockchain technology is more robust and secure than an IoT system that does not use blockchain technology.

Integration of IoT and medical appliances improves the quality of healthcare services and provides an immediate check on patient condition. However, with all the advantages, IoT in medical domain is susceptible to wide variety of attacks due to sharing of sensitive patient data, complex standards, low awareness to security patches by manufacturers, etc. Notable work has been done in securing and protecting health records of patients using blockchain technology by secure management of keys by Huawei et al. [7]. Body sensor networks are integrated with blockchain technology to design a lightweight backup and efficient recovery scheme for keys of health blockchain to solve various issues like monopoly, vulnerability, integrity and privacy problems.

In the subsequent year, another significant work was carried out to develop an IoT architecture with integration of blockchain and artificial intelligence techniques to perform big data analysis by Sushil et al. [8]. The proposed BLOCKIoTINTELLIGENCE has four tiers namely Cloud Intelligence, Fog Intelligence, Edge Intelligence and Device Intelligence which clearly revealed secure and scalable IoT applications like smart health care, smart city and transportation. In every layer different aspect of blockchain and AI convergence is achieved like Bitcoin and Ethereum in physical layer, authentication and encryption in physical layer. The proposed architecture significantly outperformed existing IoT architecture in terms of accuracy and latency in object detection task.

Notable improvements have been made further by initiating the concept of migration learning for IoT data feature extraction and Intrusion detection system for smart cities by Daming et al. [9]. A deep migration learning model is proposed which includes instance and feature transfer, inductive migration learning algorithm, parameter transfer and Sybil Intrusion detection algorithm. KDD CUP 99 dataset is used for experimental analysis. The proposed algorithm resolved sample misclassifications and spatial constraints of data clustering problems resulting in effective IDS. The experimental results demonstrated higher detection efficiency and time as compared to traditional methods such as ELM and BP. In the same year, Eirini et al. [10] proposed a three-layer supervised IDS for detection of cyber-attacks in IOT networks. The presented system accomplished three main responsibilities, i.e., device profiling of IOT devices in a network, tracking malicious packets on the network during attack followed by classification of different cyber-attacks occurring in the network. The system is evaluated within a smart home test best consisting of 8 popular commercially available devices. The effectiveness of the proposed IDS architecture is evaluated by deploying 12 attacks. Nine classifiers were selected based on their ability to support multi-class classification in which J48 decision tree method achieved the best classification results among others supervised machine learning algorithms in all the proposed three tasks.

Later on, Aiden et al. [11] introduced distributive Generative Adversarial Networks (GANs) and come up with fully distributed IDS for IoTs. ANNs were trained using the TensorFlow library, a single NVIDIA P100 GPU with 20 Gigabits of memory. The training occurred till 5000 epochs and ten scenarios are surveyed in which a random Gaussian noise is added to every feature of the training data points, each with a different attack-to-signal power ratio. Internal attack detection is performed by inspecting IoTD own data and external attack detection is done by examining neighboring IoTDs. The proposed GAN IDS claimed to have tp 20% higher accuracy, 25% higher precision and 60% lower false positive rate compared to a standalone IDS.

A unique and innovative blockchain-based deep learning framework BLOCK-DEEPNET is proposed by Shailendra et al. [12] to achieve confidentiality and integrity of IOT data devices. The suggested work performed the integration of device intelligence and blockchain technique to support a collaborative, secure DL paradigm for IoT. Different entities/devices in the IoT network employs a deep learning model for training on the local data and finally uploads the local update which consist of gradient parameters, learning model weights, etc. to the Edge server. The proposed work guarantees confidentiality and secret sharing and mitigates the problem of single point of failure, privacy leak and data poisoning attack.

Junaid et al. [13] proposed an energy efficient IDS for IoT devices using collaborative approach between device sensor nodes and Edge Router. Each device level IDS executes signature-based methods for tracking packet traffic using predefined Intrusion detection signatures whereas Edge router perform correlation among different alerts transmitted by device level to evaluate the overall magnitude of a threat. The frequency of threat pattern generated is used to evaluate the overall impact of a threat which triggers a mechanism within edge router component to report the alert to a system administrator. The whole experimentation setup and implementation is performed using Contiki Operating system with Cooja Emulator and the proposed framework reduces the overall overhead in terms of energy consumption and memory and proven to be potent for resource constrained IoT devices.

In 2020, remarkable enhancements and developments have been made further when IOT is integrated with blockchain, deep learning and multi agent system technology for effective detection of cyber-attacks in network layer by Chao et al. [14]. The proposed SESS system consists of four main modules named as Collection module, Data processing module, Detection and analysis module and Response module. Each module interacts with each other using communication agents which are continuously updated and improved through interactive reinforcement learning.

Sai Kiran et al. [15] carried out a significant study have to build an IDS in IoT domain using machine learning approaches. The proposed IDS framework comprises of an IoT test bed simulator to detect cyber threats on IoT nodes. Think Speak Server platform is used as a gateway followed by building an adversarial system to initiate attacks. Packets are inspected using Wireshark and Kali Linux is used for attack initiation and penetration testing. Four machine learning algorithms are used namely Naive Bayes, SVM, Decision trees and Adaboost to classify attack and normal traffic.

A deep learning-based blockchain driven scheme is proposed in this study for Smart city by Sushil et al. [16]. The said framework is divided into five layers (device, edge, cyber, data analytics and application layer). Each layer utilized the functions of blockchain and deep learning methods to perform self-management, scalable and rapid development in smart manufacturing. The study specifies the offering of blockchain technology at the fog layer for securing communication and storing data on an immutable or tamper-proof ledger. A case study on smart vehicle manufacturing under the proposed framework and various challenges related to data storage and complexity are also discussed (Table 3).

3 Proposed Methodology

The framework for the development of IDS in IoT environment is proposed in different phases. The initial step deals with capturing of IoT network traffic by using simulation tools followed by injecting specific attacks pertaining to IoTs. The next phase focusses on extracting relevant features from the network traffic and on later stage more emphasis is on applying deep learning-based algorithms like LSTM, GAN, CNN, etc. for detection of malicious traffic patterns and cyber-attacks

Table 3 Different IDS methods and security frameworks in IoT

Authors	Methodology	Datasets	Security threat/type of attack
Hamid et al.	Unsupervised hybrid discovery model	Dataset generated from Waspmote Mote Runner and IoT devices	Sinkhole and selective forwarding attack
A. A. Diro et al.	Distributed deep learning-based attack detection system	NSL-KDD dataset	DoS, Probe, R2L and U2R
Dinan et al.	Blockchain-based secure system on Ethereum platform	Dataset generated from IoT devices	Sniffing attacks
Huawei et al.	Lightweight backup and efficient recovery scheme using health blockchain	Mathematical framework	General security like encryption, authentication of keys, etc.
Sushil et al.	BLOCKIOTINTELLIGENCE model	PASCAL VOC dataset	Security, privacy, scalability, etc.
Daming et al.	Deep migration learning model	KDD CUP 99 dataset	DoS, Probe, R2L and U2R
Eirini et al.	Three-layer supervised IDS using supervised learning algorithms	Dataset generated from testbed using 8 IoT devices	Reconnaissance, MITM, Replay, DoS, spoofing
Aiden et al.	Distributed IDS using Generative Adversarial Networks (GANs)	Dataset generated using ANN, Nvidia GPU	Both internal and external attacks
Shailendra et al.	BLOCKDEEPNET (Integration of blockchain and deep learning)	Data generated through IoT testbed	Single point of failure, privacy leak and data poisoning attack
Junaid et al.	Energy efficient IDS using collaborative approach between device sensor nodes and edge router	Data generated using Cooja emulator	Signature-based attacks
Chao et al.	Blockchain, deep learning and multi agent-based system technology for effective detection of cyber-attacks	NSL-KDD dataset	DoS, Probe, R2L and U2R

is given [17–19]. The next phase deals with deployment of blockchain technology for enhancing IDS capability by improving IoT device security in a decentralized manner [20–22]. At the end the IDS will be placed in the IoT network which is followed by training and testing of the proposed IDS with dataset gathered from real-time/online sources. Further the comparison of different deep learning and blockchain IDS will be performed using different standard performance measurement criteria [23, 24]. The entire loop continues till the best results are achieved and robust IDS is obtained which is capable of detecting wide variety of IoT attacks (Fig. 1).

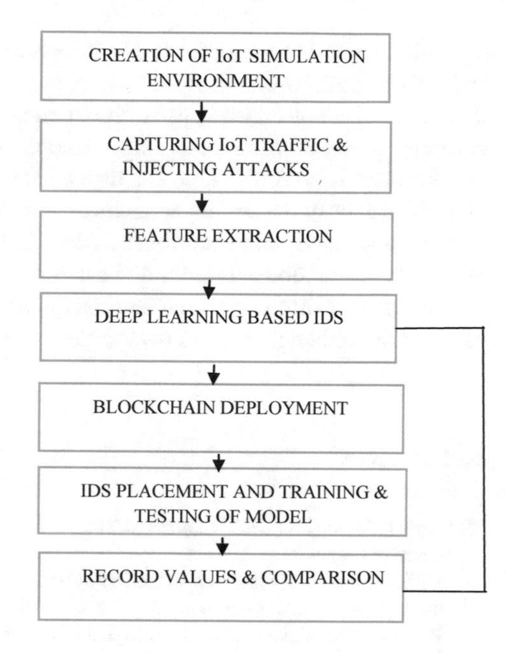

Fig. 1 Detailed flowchart of proposed methodology

4 Analysis and Discussion

The scope in developing security systems for IoT with less computational power, adaptable, minimal energy and provides good security and privacy is highly challenging. Artificial Intelligence has been proven to be a notable technology in diverse areas like robotics, manufacturing, medical sciences, etc. Implementing this technology in IoT networks shall surely add to its performance and includes multiple advantages like boosting operational efficiency, scalability, high security standards, etc. DL has the potential to enhance security measures and counter multiple attacks by discovering hidden patterns from the training data and can easily identify and discriminate attacks from routine traffic. Blockchain is another emerging area which can enhance the capabilities of IDS technology as it ensures decentralized and encrypted transactions between IoT devices and makes highly secure communication between them. It thus adds to the overall reliability of the system.

5 Conclusion

In this paper, a deep systematic review of different Intrusion detection system in IoT along with their techniques and frameworks is performed. A significant amount of research has been carried out to develop an enhanced and reliable security mechanism to protect IoT systems and considerable number of frameworks and methodologies have been proposed by different experts in this domain. IoT field always remain exposed to a wide range of security threats and are highly vulnerable to large-scale and remote attacks due to its autonomous nature. This field is highly dynamic due to its heterogeneous nature, use of independent protocol standards, utility of diverse communication methods, etc. Furthermore, coping with the IoT's diverse security architectures is critical. There is a dire need that during the design and development of an IDS, each and every layer needs to be addressed so that IDS framework shall be fully incorporated with security measures in holistic approach. Lightweight frameworks which are robust, reliable and can provide high resilient security solutions must be the focus of future research and design and development of such frameworks in IoT will prove beneficial and overcome several aspects of security issues.

References

1. Sethi P, Sarangi S (2017) Internet of Things: architectures, protocols, and applications. J Electr Comput Eng 1–25
2. Different types of intrusion detection systems (IDS). https://wisdomplexus.com/blogs/different-types-of-intrusion-detection-systems-ids/. Last accessed 2022/05/02
3. Bostani H, Sheikh M (2016) Hybrid of anomaly-based and specification-based IDS for Internet of Things using unsupervised OPF based on MapReduce approach. Comput Commun. Elsevier
4. Diro A, Chilamkurti N (2017) Distributed attack detection scheme using deep learning approach for Internet of Things. Future Gener Comput Syst. https://doi.org/10.1016/j.future.2017.08.043
5. CasadoVara R, Prieta F, Prieto J, Corchado J (2018) Blockchain framework for IoT data quality via edge computing. In: BlockSys'18: proceedings of the 1st workshop on blockchain-enabled networked sensor systems 2018. ACM, pp 19–24
6. Fakhri D, Mutijarsa K (2018) Secure IoT communication using blockchain technology. In: International symposium on electronics and smart devices 2018. IEEE, pp 1–6. https://doi.org/10.1109/ISESD.2018.8605485
7. Zhao H, Zhang Y, Peng Y, Xu R (2018) Efficient key management scheme for health blockchain keys. CAAI Trans Intell Technol 3
8. Singh S, Rathore S, Park J (2020) BlockIoTIntelligence: a blockchain-enabled Intelligent IoT architecture with artificial intelligence. Future Gener Comput Syst 110:721–743. https://doi.org/10.1016/j.future.2019.09.002
9. Li D, Deng L, Lee M, Wang H (2019) IoT data feature extraction and intrusion detection system for smart cities based on deep migration learning. Int J Inf Manage 49:533–545. https://doi.org/10.1016/j.ijinfomgt.2019.04.006
10. Anthi E, Williams L, Slowinska M, Theodorakopoulos G, Burnap P (2019) A supervised intrusion detection system for smart home IoT devices. IEEE Internet Things J 6:9042–9053. https://doi.org/10.1109/jiot.2019.2926365
11. Ferdowsi A, Saad W (2019) Generative adversarial networks for distributed intrusion detection in the Internet of Things. In: IEEE global communications conference 2019. IEEE, pp 1–6. https://doi.org/10.1109/GLOBECOM38437.2019.9014102
12. Rathore S, Pan Y, Park J (2019) BlockDeepNet: a blockchain-based secure deep learning for IoT network. Sustainability 11(14):3974. https://doi.org/10.3390/su11143974
13. Arshad J, Azad M, Abdeltaif M, Salah K (2020) An intrusion detection framework for energy constrained IoT devices. Mech Syst Sig Process 136:106436. https://doi.org/10.1016/j.ymssp.2019.106436
14. Liang C, Shanmugam B, Azam S, Karim A, Islam A, Zamani M, Kavianpour S, Bashah N (2020) Intrusion detection system for the Internet of Things based on blockchain and multi-agent systems. Electronics 9:1120. https://doi.org/10.3390/electronics9071120

15. Sai Kiran K, Devisetty R, Kalyan N, Mukundini K, Karthi R (2020) Building a intrusion detection system for IoT environment using machine learning techniques. Procedia Comput Sci 171:2372–2379. https://doi.org/10.1016/j.procs.2020.04.257
16. Singh S, Azzaoui A, Kim TW, Pan Y, Park JH (2021) DeepBlockScheme: a deep learning-based blockchain driven scheme for secure smart city. Sustain Cities Soc 11. https://doi.org/10.22967/HCIS.2021.11.012
17. Alzubaidi L, Zhang J, Humaidi A et al (2021) Review of deep learning: concepts, CNN architectures, challenges, applications, future directions. J Big Data. https://doi.org/10.1186/s40537-021-00444-8
18. Singh M, Aujla G, Bali R (2021) A deep learning-based blockchain mechanism for secure internet of drones environment. IEEE Trans Intell Transp Syst 22:4404–4413. https://doi.org/10.1109/tits.2020.2997469
19. Al-Garadi M, Mohamed A, Al-Ali A et al (2020) A survey of machine and deep learning methods for Internet of Things (IoT) security. IEEE Commun Surv 22:1646–1685. https://doi.org/10.1109/comst.2020.2988293
20. Alfandi O, Khanji S, Ahmad L, Khattak A (2021) A survey on boosting IoT security and privacy through blockchain. In: Cluster computing. Springer, pp 37–55. https://doi.org/10.1007/s10586-020-03137-8
21. Xu M, Chen X, Kou G (2019) A systematic review of blockchain. Financ Innov. https://doi.org/10.1186/s40854-019-0147-z
22. Zheng XR, Lu Y (2021) Blockchain technology—recent research and future trend. Enterprise Inform Syst. https://doi.org/10.1080/17517575.2021.1939895
23. Christidis K, Devetsikiotis M (2016) Blockchains and smart contracts for the Internet of Things. IEEE Access. 4:2292–2303. https://doi.org/10.1109/ACCESS.2016.2566339
24. Garg R, Gupta P, Kaur A (2021) Secure IoT via blockchain. IOP Conf Ser Mater Sci Eng

Emerging IoT Platforms Facilitate Entrepreneurship Businesses

Praveen Kumar Singh, Bindu Singh, Pulkit Parikh, and Mohil Joshi

Abstract In the current digital era, entrepreneurship has become a key differentiating reason in an extremely disruptive and competitive business environment to absorb constantly growing consumer demands. However, there is a need to comprehend the nuisances of digital entrepreneurship and its conceptual framework to navigate the associate businesses. IoT has a strong transformative potential to impact such entrepreneurship businesses. This paper discusses about the basic concept of IoT and briefly presents its security architecture. IoT connects diverse business entities and 'things' with real time data which is analyzed and combined with supplementary data in the cloud over varying platforms. This paper brings out briefly about some key platforms and their applications in entrepreneurship businesses. In this digital environment cognitive computing, artificial intelligence, machine learning and many other emerging technologies are driving new business capabilities with significant opportunities. Some prominent advantages of IoT platforms have been outlined in this paper. Different IoT application over distinct platforms are contributing to transform consumers' lives and changing their entrepreneurship business operations. Yet, these platforms are beleaguered with certain challenges which have been briefly discussed in this paper. Today, almost 40 billion 'things' are connected over the Internet. IoT appliances are likely to generate approximately 180 zettabytes data annually by 2025. Massive entrepreneurship business activities in future will need powerful and sophisticated analytical engines to support it. By highlighting some significant future prospects of IoT supported entrepreneurship businesses, the paper has been concluded.

P. K. Singh (✉) · B. Singh · P. Parikh · M. Joshi
IIIT, Lucknow, India
e-mail: praveen.197505@yahoo.com

B. Singh
e-mail: bindu@iiitl.ac.in

P. Parikh
e-mail: mdb21015@iiitl.ac.in

M. Joshi
e-mail: mdb21012@iiitl.ac.in

Y. Singh et al. (eds.), *Proceedings of International Conference on Recent Innovations in Computing*, Lecture Notes in Electrical Engineering 1001,
https://doi.org/10.1007/978-981-19-9876-8_18

Keywords Entrepreneurship · Digital · Business · IoT · Platforms · Data · Technologies · AIoT · Architecture · Security · Risks · Gateway · HTTP · Operation

1 Introduction

Proliferation of Internet of Things (IoT) platforms and their associated protocols have potential to transform the entrepreneurship businesses. A massive amount of IoT devices are connected in different sizes, shapes, standards, configurations and protocols. Accomplishing real entrepreneurship businesses outcome with IoT needs a resolute effort. IoT devices are likely to generate 180 zettabytes of annual data by 2025. Such mammoth data will need the most powerful analytical and sophisticated engines that the world would have ever witnessed. An IoT platform can be referred as a suite of apparatuses which enables installation of varying apps to monitor, manage and control the devices that are connected to a shared server. It also facilitates the connected equipment to share their data with one another. An enterprise oriented IoT platform permits entrepreneurship businesses to sync and streamline their devices to produce a business insight needs to acquire the craving output. In order to do so, an IoT platform aligns strategically to the business entrepreneurships current as well as future requirements.

Many entrepreneurship businesses have amended their way they used to operate by adopting IoT platforms. Weather we look at business-to-business (B2B) or business-to-consumer (B2C) to realize the decision alacrity, IoT technologies possess the ability to transform users' both social and personal aspects of lives and to answer precarious business queries with better flexibility and more profound data insights. Transformation of an entrepreneurship business through an IoT platform is not only about a mindset, it has also become a mandatory need on how we understand and perceive the business ecosystem as a whole. Deployment of IoT platforms warrant placement of a large number of high-resolution data sensors to enable new entrepreneurship business models which otherwise are primarily driven through the physical products in the industry. It also drives a tremendous pressure on entrepreneurs to contemplate their ventures on IoT platforms to enable their businesses on such business models which potentially hold better prospects for them. The orchestration of these entrepreneurship business ecosystems can leverage these IoT platforms to re-define their customer value.

2 Entrepreneurship Businesses

An entrepreneur is someone who organizes a business and then continues as an active contributor in that business operation. Entrepreneurs launch and carry on operating their own companies. Entrepreneurs finance their ventures through different sources like loans from investors, money from their close families or their own savings. It brings out the necessity and significance of a viable financial plan and a meticulous appreciation of the inherent risks involved in a successful entrepreneurship business. Entrepreneurship is about developing an idea for any exclusive or an occupied business. Learning about it and gaining experiences in varied business roles. It includes accounting, finance, marketing and management. It is about making a business plan, identifying and establishing sources for funding, recruiting talents with requisite skills of both workers and managers, testing, creating a support system, implementing and maintaining its company's products. It is also about devise business strategies to launch their products or services to attract new as well as retaining existing customers. After the entrepreneur has established the business, it's about seeking out different ways to raise the revenue. It could be also by indulging into new ventures as well as in product lines.

There are certain factors which assume significance in the path to accomplishments of the business entrepreneurs. These include political, legal considerations, taxation and the availability of desired capital to establish as well as to scale up the business entrepreneurships. Nation's political situations and law of the land become a critical factor to offer a conducive environment to facilitate the growth of such entrepreneurships. A heavy taxation imposed by the governments reduces the business profits and affects the business viability for the entrepreneurs. Likewise, depending upon the nature of entrepreneurships, a minimum quantum of capital becomes mandatory to support the business operations. In its absence, no entrepreneurship can thrive to its potential. Entrepreneurial Businesses generate employments and contribute in reduction of the nation's unemployment rate supporting indirectly the nation's economy. Besides, as their entrepreneurship business grow, additional positions build up and in a way that contribute significantly in poverty reduction as well. Entrepreneurial Businesses craft changes as they offer products which facilitate a solution to the people's routine lives. Entrepreneur's ideas and ambitions often become game-changer while carrying out business operations and impact these businesses all across the world. Entrepreneurial Businesses also assist in extending help to the societies.

3 The Internet of Things (IoT) and Its Platforms

The IoT facilitates real time information to the business entrepreneurs. It offers an insight about operations that when effectively responded can enable their business organizations more efficient. IT administrators, developers, investors, shareholders and the other associated stakeholders in the entrepreneurships must have an adequate understanding of IoT, how it operates, requirements, its usage, tradeoffs and also on how to implement its devices, data networks and infrastructures. Significance of IoT can be measured in different forms. It often involves data collection on processes, behavior and other related business environments. Many IoT devices undertake required actions to improve, correct or else to employ this data to endorse certain changes as required in business operations.

3.1 The Internet of Things (IoT)

Dedicated devices also referred as things and deployed to share the real time varying applications data in a network on Internet are known as Internet of Things (IoT). These devices conceptually can include everything right from speakers to office premise networks, cars, plethora of appliances in shopping malls, shipping labels, academic institutes, airports, etc. They may include smart sensors, security systems, electric bulbs, industrial machineries and all such items which can communicate through the Internet and can work along with it. IoT isn't a sole device, technology or software. It is a mix of networks, devices, computing resources, stacks and software tools. IoT terminology understanding generally begins with the IoT appliances themselves. Certain key IoT concepts which are applicable to business entrepreneurships are enumerated in succeeding paragraphs.

3.1.1 The Real Time Data

Business enterprises regularly deal with images, documents, spreadsheets, videos, Power Points and a lot of other types of digital information. IoT devices generate enormous data that usually reflects one or even more physical stipulations in the realistic scenario. A network connected with IoT devices enables an entrepreneur to acquaint with real time business operations. It can also facilitate to the business entrepreneurs to exercise their control on the business activities.

3.1.2 The Essential Need of Instant Business Operations

Where usual data transactions in business operations may subsist for even days or weeks with sometimes no utilization, IoT devices ought to deliver data instantly in most of the data processing cases without making any delay. It highlights the significance of the mandatory need of adequate availability of bandwidth in the associated data network. The connectivity assumes importance especially in IoT supported business environments where most of the data processing demands real time data transmission to prevent any substantial economic losses.

3.1.3 The Resultant Data

IoT supported business projects are mostly defined by data operational requirements for business purposes to necessitate the IoT network deployment at different scales.

In several cases, data of an IoT network becomes an element of the associated control loop to contribute in the source and outcome goal. For instance, a sensor conveys a homeowner about their front door gets unlocked, and thereby the homeowner could employ an actuator which is essentially an IoT device. Such IoT devices are meant to convert control signals in the data network into real time actions as a resultant to enable the homeowner to lock the door remotely.

Infrastructure demands in an IoT data networks are extensive. Processing and security add new complexities in the entrepreneurship businesses. IoT platforms address these issues with Software as Service (SaaS) data network architectures. Its adoption eradicates number of associated network issue which are usually required for edge computing gateways and some other IoT applications. A basic IoT Software as a Service (SaaS) data network has been illustrated in Fig. 1. IoT can support greatly and to the far reaching entrepreneur's business objectives. Millions of sensors in an IoT network can produce incredibly vast quantum of unprocessed data that may become far too much to the humans to evaluate and to act upon. Progressively, large IoT projects become the nucleus of big data operational initiatives, for example artificial intelligence (AI) and machine learning (ML) projects. The data accumulated from huge IoT device operations can be analyzed to process and initiate crucial business projections. It can also be employed to train AI-based systems for the real time business data transactions through the vast array of sensors. Simultaneously, back-end data operations analysis warrant considerable computing power and storage space. Such massive business data computation can be processed in centralized databases, public clouds or these may be distributed over numerous edge computing localities which are close to the places where data collection is done.

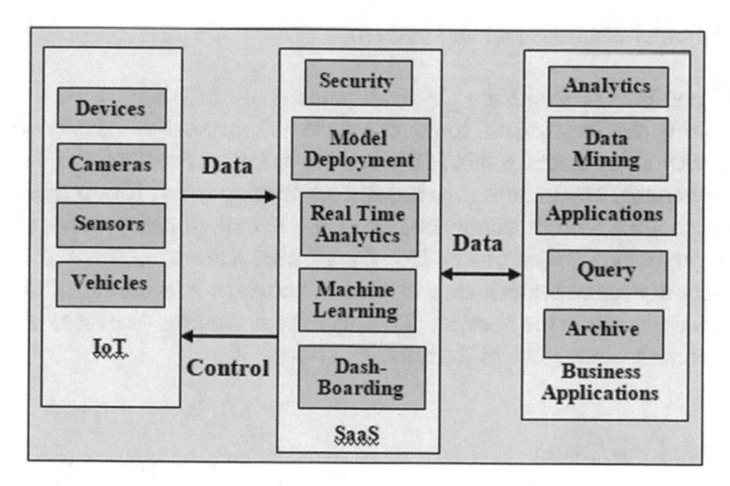

Fig. 1 A basic IoT SaaS data network architecture

3.2 *The Internet of Things (IoT) Platforms*

The proliferation of colossal amount of IT devices in business over the Internet of Things (IoT) platforms is likely to turn out a boon in entrepreneurships. It's not that just two devices are connected in the industrial ecosystem in it, rather a plethora of IT peripherals lead to a complexity which needs dedicated efforts to manage it out. These numerous IoT devices can be in all shapes, configurations, sizes, network protocols, standards, etc. It necessitates a viable platform that can facilitate its connectivity to the industrial ecosystem as well as within the ecosystem themselves. Though, there are lots of open source and free IoT platforms are available to the business startups, however there is a need to do a thorough examination by the entrepreneurs before the pick the most suitable one for their entrepreneurship venture. Some of the key industrial IoT platforms are being discussed in the succeeding paragraphs.

3.2.1 Link Management Platforms

This platform deals with the IoT system networking components and to facilitate its users the most suitable software and hardware for its effective functioning of the entire network linked devices in the online mode of Internet connectivity. All such data networks are established in the backdrop of already available infrastructure and it can have other wireless communication modes of Internet connectivity such as Wi-Fi to support the impending IoT infrastructure.

3.2.2 Apparatus Management Platforms

This IoT platform is about the physical connectivity of the associated hardware while ensuring that they remain connected in the data network and the security of the IoT network too is intact. In this IoT platform, users are constantly updated about the IoT apparatus, any changes in the network hardware and it also facilitates the equipment metrics with its continuous updating. In case of any network equipment failure, it provisions an alert to replace the affected hardware. This provision also assists users in routine maintenance of the IoT networks irrespective of the amount of devices connected in the network. This platform is therefore meant to upkeep and for appropriate maintenance of the network devices.

3.2.3 Cloud Platforms

Weather it is a data crunching or data storage, a data cloud is meant to be a data sharing facility to the users connected with that data cloud storage system. It signifies about the fact that the cloud holds all the benefits of a shared data facility. It does not require a physical connectivity and it facilitates the security of the data transmission and its further scaling if required, in the data network. Business entrepreneurs can establish the similar cloud system to generate the entire back-end connectivity of the IoT system which can take care of data processing and the required storage in the network.

3.2.4 Application Support Platforms

This platform enables users to acquire an IoT system in shortest feasible timeframe which is an integrated IoT ecosystem. It facilitates users inclusive data network access support system weather its software, an apparatus, network support system for the ease of network deployment against a quick turnaround. This platform id widely used by the new business entrepreneurs to support their digital ventures with a single window operation management. It offers a turnkey solution along with the requirement of manpower, network configuration hardware and maintenance.

One important thing needs to be understood here that all the above IoT platforms should not be seen in isolation. It's not only just the collection of data by employing these platforms; data management and its integration emanated from the different IoT platforms assume significance. In entrepreneurship, there may be a requirement of adding the business data from varying sources like websites, social media, mails, voice enabled devices, etc. It may increase the complexity of the IoT platforms to manage such volatile business data. Diverse business entrepreneurs collect and archive a huge quantum of business data. Figure 2 illustrates its utility in diverse business enterprises duly supported by IoT platforms. Latest technologies of data science and predictive analysis enhance the productive utility of such data on IoT platforms and bring efficiency, transparency, better system throughput and enrich the customer experiences. An IoT platform facilitates businesses with required infrastructure to link their assets, analyze and collect data in details with industry rating security protocols. The strength of an IoT platform is its ability to integrate data from innumerable sources to convert them in a coherent valuable business data.

4 The Key Benefits of IoT in Entrepreneurships

IoT has considerably impacted the business entrepreneurships. With an extensive increase in IoT devices, big data bases accessibility, sharing of a colossal amount of data enables business entrepreneurs to gain insight on their customer responses as well as on product performances. IoT facilitates the constant optimization in business practices and even influences employees' engagements. In certain entrepreneurships, IoT can facilitate to direct the systems in supply chains to independently execute transactions with certain pre-defined conditions. There are so many new exciting technologies which exhibit the IoT future as an incredibly versatile to meet the aspirations of both the business entrepreneurs as well as the consumers. Though,

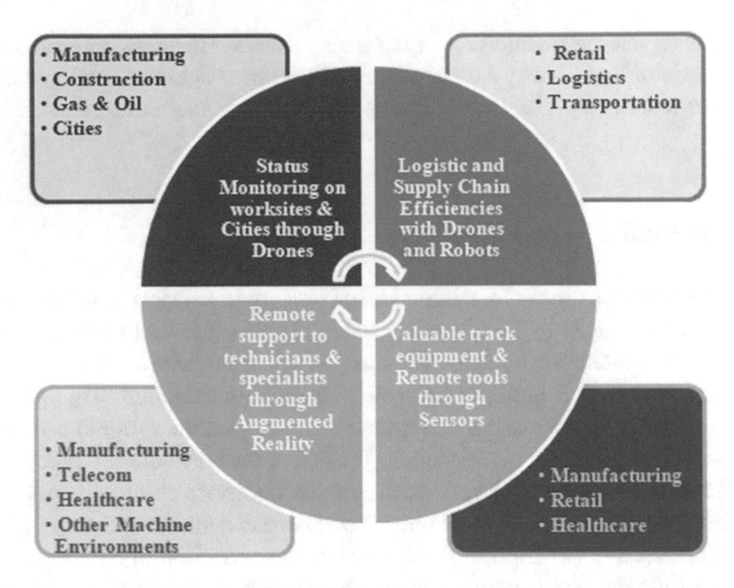

Fig. 2 Business enterprises supported by IoT platforms

there are innumerable business advantages and opportunities in entrepreneurship businesses, some of the key benefits are being enumerated in succeeding paragraphs.

4.1 Reduced Operative Cost

Establishing an IoT supported business model is a quite deliberate affair. If one intends to put entire required IoT infrastructure in an entrepreneurship, then the operative cost may go out of proportion to make it a viable business model. The best option in such case is to employ Software as a Service (SaaS) model which takes care of the upkeep and the maintenance issues inclusively.

4.2 Enhanced System Performance

In order to work effectively, a real time data sharing is the basic prerequisite for any organization. Based on the information received, an appropriate response is provided to the stakeholders. Availability of an IoT platform ensures the requisite predictive analysis can be performed by collecting the required data from the varying sources. It facilitates in optimization of the resources to accomplish relatively better maintenance to economize the project costs of the business entrepreneurships.

4.3 Augmented Security

In an IoT data network, data sharing devices hold only a limited security capability. However, once supported by the IoT ecosystem due to its peculiar network architecture and security provisions put in place, a required wherewithal is provided to these equipment. There is a procedure in place for identity management and a secure authentication mechanism to thwart any probable cyber security attacks and address the system security vulnerabilities.

4.4 Supports to New Business Models

Combination of a business and the IoT ecosystem leads to re-engineering the entire commerce activities. It develops and discovers many new revenue schemes and business models to support the new business ventures of the entrepreneurs. Partnering with different business ecosystems through the IoT platforms can open new avenues to business organizations. For instance, IBM partnering with General Motors offers a new transaction and marketing experience to automakers through the IoT supported mechanism in place. Analysis of the huge data from a vast sensor provides required insight to facilitate better optimization of resources and system performance.

4.5 Better Customer Experience

Uses of IoT transform the people experiences of the different services manifold offered through the business operations. For example employing IoT cloud through IBM platform by a global business leader in escalator and the elevator industry, KONE could optimize and monitor its management for millions of escalators, elevators, turnstiles and doors in cities worldwide. By connecting all their elevators with the cloud and by attending them carefully, by scrutinizing, KONE could do the due maintenance of all the required elevators. Likewise, there are numerous other sectors and business entrepreneurships in which IoT supported ecosystems offer much better experiences to customers to best of their satisfaction.

Combining IoT with some other emerging technologies such as 5G, AI, Robotics, ML and so on offer endless possibilities for entrepreneurship businesses. Entire commercial activities are gradually powered by IoT wherein security of the business ecosystem is being strengthened every day. Huge manufacturing productions are linked to a remote monitoring operational system through IoT. Utility enterprises can remotely accumulate data from the smart meters with associated infrastructure. Health care maneuvers can employ IoT to converse a patient's live status to the physicians remotely. Farmers can manage their harvest through analysis enabled through IoT. It has become an amazing asset to the widespread entrepreneurship businesses.

IoT allows entrepreneurs to assist better their customers as well as in managing their workforces to improve their services, products and business processes.

5 Proposed Security Architecture

IoT devices are prone to a multiple potentially harmful cyber-attacks which may include weak DNS systems, botnet attacks and many other such attacks. It can make way to ingress ransomware, malware, probable attack vectors, etc. due to unsecured and unauthorized devices over the data network and it can be the hazard for physical security. Security risks also carry analogous threats to an establishment's compliance posture. IoT is still under evolution. There are different design standards, configurations, operating as well as securing an IoT data infrastructure. Today, many business entrepreneurs pick IoT devices for their applications which follow existing technological standards like IPv6, Bluetooth Wi-Fi, Z-Wave, ZigBee and other data connectivity standards. In fact, further compliance standards emerge from industry like the IEEE 2413-2019 standards employed for an IoT data architectural framework. This standard is presently used for a widespread IoT architectural framework in diverse applications like healthcare, transportation, utility and other domains. It also conforms to the prevalent international data standard like ISO/IEC 42010:2011. An example of the proposed IoT system is illustrated in Fig. 3. Initially, data is converged to an IoT gateway emanated from number of sources through antennas, sensors microcontrollers, etc. Collected data is processed through the IoT hub for further data analysis. The desired data analytics is done through varying user interfaces like human, machine or even against diverse business application analytics to aggregate it with a viable business output data which can be commercially utilized in business applications.

However, there are some distinct IoT data network architectural issues which concerns primarily to the network security. In an IoT network, there are varying infrastructural requirements which may include diverse sensors, its locations and quantities, power connectivity, management tools, network configuration and interfaces. It demands adequate latency and bandwidth considerations. There may be a need to extensively deploy extra computing resources to handle additional processing or to employ add on resources like the cloud. Since, the data handled by the IoT network may be confidential and sensitive, there is a need to ensure adequate safeguards against data theft, snooping and hacking. Encryption may be a viable option for IoT data protection. A provision is also needed to prevent malicious alterations to machine configurations and hacking. Security entails diverse software tools and conventional security mechanisms like intrusion detection as well as prevention systems and firewalls. Figure 4 illustrates the proposed IoT architecture module which includes communication bus, data network, an analytics and the aggregation platform. In this proposal, management of different IoT agents, network managers and application requests are processed. There is a sensor which detects the transmitted data and communicates it through the data bus. A due data analysis is performed by

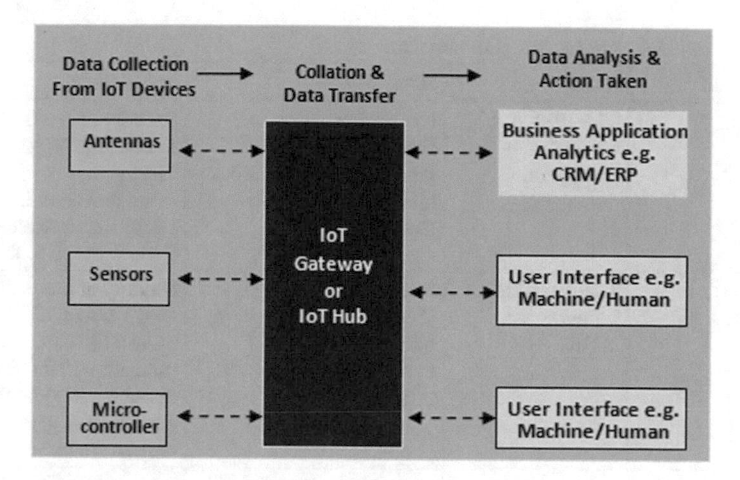

Fig. 3 An example of an IoT system

the network and the aggregated data ensures enhanced data security with seamless data integration in the IoT network.

The above mentioned data aggregation and the network integration ensures that network infrastructure, devices and tools are added to support interoperability with prevalent applications and systems like ERP and systems management. It will need a careful forecast and testing of proof of principle with appropriate IoT platforms and tools. The suggested IoT architecture will also entail a comprehensive understanding on how IoT data should be employed and analyzed. It will be done in an application layer with associated analytical tools. It may consist of training engines, AI, Robotics and ML modeling and rendering tools or visualization. There is a need to examine the various security threats, their mitigation and implementation. Table 1 illustrates the different security threats at device (Physical) level, varying risks, alleviation and their possible implementations. Proposed IoT architecture exhibits how to deal with the security threats identified in the entrepreneurship businesses. There are primarily four

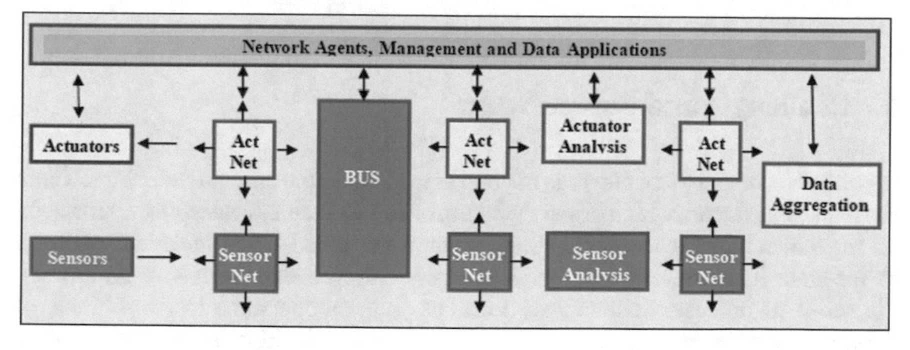

Fig. 4 Proposed IoT architecture modules

Table 1 Examination of proposed architecture

Components	Risks	Mitigation	Implementation
Apparatus IoT Hub	TLS [PSK/RSA] for traffic encryption	Eavesdropping or communication interference between the machine as well as the gateway	Security at protocol level to ensure protection. Mostly, the communication is established between the devices to IoT Hub
Machine to Machine	TLS [PSK/RSA] for encrypting traffic	Reading data during transportation. Tampering of data Machine overloading against new connections	Security at protocol level (MQTT/AMQP/HTTP) to protect the data. The threat mitigation to the DoS is through a field or cloud gateway and get them act only as clients
External Device	External entity pairing with the device	Eavesdropping and Interference to communication	Securely pairing of external entity with devices. Controlling operational (Physical) panel
Field and Cloud Gateway	TLS [PSK/RSA] for traffic encryption	Interfering communication or Eavesdropping	Security at protocol level (MQTT/AMQP/HTTP). Protection with custom protocols
Machine Cloud Gateway	TLS [PSK/RSA] for traffic encryption	Interfering communication or Eavesdropping	Security at protocol level (MQTT/AMQP/HTTP). Protection with custom protocols

focal areas wherein such security threats exist like data sources, data transmission, event processing and presentation. It is significant that the proposed IoT architecture segregates the gateway and device capabilities. These way users will have leverage with more secure gateway mechanisms by using secure protocols which demands greater dispensation overheads.

6 Challenges and Future Scope

Significance of an IoT platform is beyond any qualm in the entrepreneurships. There are numerous factors which prove the worth of IoT in business enterprises. However, its implementation is as challenging as any other cloud-based technical platform to facilitate its network solutions. A business entrepreneurship has to go through the series of processes which may turn out quite cumbersome to them. Scale of deployment, degree of security required at different terminals, numbers of user nodes, data network architectures and so on are some of the indicative parameters which will dictate the degree of challenges confronted in its implementation. Some of the key challenges faced in IoT implementation are enumerated in succeeding paragraphs.

6.1 Project Designing

Even though IoT appliances readily apply an array of standards, like 5G or Wi-Fi, currently there are no noteworthy international standards which can steer the implementation and design a foolproof IoT network architectures. There are no rulebooks to elucidate how to move toward in an IoT project. It permits a tremendous flexibility in project design; however it also permits for major design oversights and flaws. IoT projects should ideally be led through IT staff who are well conversed with IoT, although such expertise too is a self-evolving process. Eventually, there is no alternate for well measured design, careful and demonstrated presentation based on proof of projects and a committed testing.

6.2 Data Network Support

IoT data flows through an IP enabled network. While considering the IoT apparatus effect on data network bandwidth to ensure that sufficient, steadfast bandwidth becomes accessible, congested data networks with high latency and dropped data packets may result in delayed IoT data transmission. It may entail certain architectural alterations in the data. For instance, instead passing complete IoT data over the Internet, an entrepreneur may choose to deploy an edge data computing architecture which preprocesses and stores locally the raw data prior to passing only processed data over a centralized location for further analysis.

6.3 Retaining and Storage of Data

IoT devices generate colossal amounts of data that is simply multiplied based on number of network associated devices. This data becomes valuable asset for entrepreneurs to be stored and duly secured. Unlike conventional business data like contracts and emails, IoT data becomes extremely time sensitive. For instance, an automobile's road data state or speed reported on preceding day or earlier may not have any significance for today or next month. It means IoT data may possess a fundamentally diverse lifecycle than conventional business data. This needs considerable efforts in data security and storage capacity.

6.4 Data Security and Privacy

IoT projects warrant implementation of a secure network configuration. A suitable well planned IoT network security has also direct connotations for data regulatory compliances. While proliferation of IoT devices increase, the threat of IoT data network being compromised may also get proportionately enlarged. A lesser secure apparatus can endanger the entire IoT data network's ecosystem. With remote monitoring and sensors in a core IoT data network, there may always be an issue of privacy to ownership and access of data. Compliance becomes a complex issue especially in live applications.

IoT devices are likely to continue to grow. The future years will witness billions of added IoT devices over the Internet. It will be augmented by a blend of technologies supported by 5G connectivity. A countless new startups are emerging all across key industries. There is a tremendous future scope of IoT platforms as various technologies in IT domain and their associated applications are comparatively new and hold enormous growth prospects. In future, there is a likelihood of an enhanced and reevaluated IoT security commencing with primary device design by business implementation and selection. All such devices will include better security features which will be default enabled. Prevalent security tools like intrusion prevention and detection will incorporate strong support for IoT data architectures with ample active remediation and logging credentials. Concurrently, IoT device administrative tools are likely to progressively address IoT device security weaknesses and emphasize more on security auditing. In addition, certain aspects of IoT and AI are converging for a hybrid artificial intelligence of things (AIoT). It may create a platform which will be much more capable with human–machine interface and having an advanced learning capability. IoT data volumes are likely to continue to rise which translates into better revenue opportunities for entrepreneurship businesses.

7 Conclusion

Implementation of IoT is considered a way forward to the entrepreneur businesses. However, at the same time it also holds enormous challenges to deal with. An IoT enabled business facilitates any entrepreneur to connect with embedded tags and sensors, share their data, to control and monitor their business ecosystem remotely and to carry out many associated predictive analysis. The most heartening part of these IoT platforms enabled entrepreneurship businesses is about its accuracy and the processing speed at which the business data will be shared. IoT enabled solutions aim to assist in synergizing the use cases to proliferate economies of entrepreneurs to scale up and to provide security to the businesses. It can also identify the most appropriate technical solution in the core network to meet the constant growing business demands. In immediate future, there is a likelihood of a re-evaluation and enhancement of IoT security at varying segments to include the device designs of business entrepreneurs and their business implementations. These devices are likely to incorporate more resilient security features to augment their customer's trust.

Existing security tools on IoT platforms like an intrusion detection system includes support for prevalent IoT architectures which incorporate a comprehensive logging as well as an active remediation mechanism. Also, IoT platform management tools increasingly emphasize on periodic security audits to automatically address some inherent security vulnerabilities. In addition, certain aspects of IoT and AI are converge to demonstrate an artificial intelligence of things (AIoT) technique. This technology intends to combine the data-gathering abilities of IoT with other decision-making and computing capabilities of IoT platforms and AI in entrepreneurship businesses. AIoT has potential to create a platform which will be capable of machine-human interactions besides certain advanced learning capabilities. IoT data volumes are likely to continue to grow which translates into fresh businesses revenue opportunities for entrepreneurs. These business data will increasingly drive AI and ML initiatives for multiple entrepreneurships. It is also considered a promising transformation strategy in long-term for entrepreneurship businesses across many sectors. While implementation of IoT platforms in entrepreneurships may have challenging prospects, however with appropriate integration with suitable solution businesses partners, entrepreneurs can be benefited from the immense business potential through IoT platforms.

References

1. Metallo C, Agrifoglio R, Schiavone F, Mueller J (2018) Understanding business model in the Internet of Things industry. Technol Forecast Soc Chang 136(23):298–306
2. Ben-Hafaïedh C, Micozzi A, Pattitoni P (2022) Incorporating non-academics in academic spin-off entrepreneurial teams: the vertical diversity that can make the difference. R&D Manage 52(1):67–78
3. Ande R, Adebisi B, Hammoudeh M, Saleem J (2020) Internet of Things: evolution and technologies from a security perspective. Sustain Cities Soc 54:101728
4. Chalmers D, MacKenzie N, Carter S (2021) Artificial intelligence and entrepreneurship: implications for venture creation in the fourth industrial revolution. Entrep Theor Pract 45(5):1028–1053
5. Chen H, Tian Z (2022) Environmental uncertainty, resource orchestration and digital transformation: a fuzzy-set QCA approach. J Bus Res 139(24):184–193
6. Senyo PK, Liu K, Effah J (2019) Digital business ecosystem: literature review and a framework for future research. Int J Inf Manage 47(5):52–64
7. Corvello V, De Carolis M, Verteramo S, Steiber A (2021) The digital transformation of entrepreneurial work. Int J Entrep Behav Res 28(5):183–195. https://doi.org/10.1108/IJEBR-01-2021-0067
8. Del Giudice M (2016) Discovering the Internet of Things (IoT) within the business process management: a literature review on technological revitalization. Bus Process Manag J 22(2):263–270
9. Fossen F, Sorgner A (2021) Digitalization of work and entry into entrepreneurship. J Bus Res 125(4):548–563

10. Jabbari J, Roll S, Bufe S, Chun Y (2022) Cut me some slack! An exploration of slack resources and technology mediated human capital investments in entrepreneurship. Int J Entrep Behav Res 28(5):1310–1346. https://doi.org/10.1108/IJEBR-10-2020-0731
11. Kraus S, Palmer C, Kailer N, Kallinger F, Spitzer J (2019) Digital entrepreneurship: a research agenda on new business models for the twenty-first century. Int J Entrep Behav Res 25(2):353–375
12. Matricano D, Castaldi L, Sorrentino M, Candelo E (2021) The behavior of managers handling digital business transformations: theoretical issues and preliminary evidence from firms in the manufacturing industry. Int J Entrep Behav Res 28(5):1292–1309. https://doi.org/10.1108/IJEBR-01-2021-0077
13. Effiom L, Edet SE (2020) Financial innovation and the performance of small and medium scale enterprises in Nigeria. J Small Bus Entrep 34(7):1–34. https://doi.org/10.1080/08276331.2020.1779559
14. Laughlin C, Bradley L, Stephens S (2022) Exploring entrepreneurs business related social media typologies: a latent class analysis approach. Int J Entrep Behav Res 28(5):1245–1272
15. Trabucchi D, Buganza T (2021) Entrepreneurial dynamics in two-sided platforms: the influence of sides in the case of Friendz. Int J Entrep Behav Res 28(5):1184–1205. https://doi.org/10.1108/IJEBR-01-2021-0076
16. Masood T, Sonntag P (2020) Industry 4.0: adoption challenges and benefits for SMEs. Comput Ind 121(10):32–61
17. Troise C, Corvello V, Ghobadian A, O'Regan N (2022) SME's agility in the digital transformation era: antecedents and impact in VUCA environments. Technol Forecast Soc Chang 174(19):121–227. https://doi.org/10.1016/j.techfore.2021.121227
18. Upadhyay N, Upadhyay S, Dwivedi YK (2021) Theorizing artificial intelligence acceptance and digital entrepreneurship model. Int J Entrep Behav Res 28(5):1138–1166. https://doi.org/10.1108/IJEBR-01-2021-0052

Navigation and Cognitive Techniques for Humanoid Robots

Aqsa Sayeed, Zoltán Vámossy, Neerendra Kumar, Yash Paul, Yatish Bathla, and Neha Koul

Abstract A humanoid robot is a robot analogous to the human body in shape and structure. Navigation path planning requires the robot's actual state and the goal position to be within the same reference frame. Navigation is one of the indispensable functions of humanoids. This paper focuses on humanoid robot navigation and applied approaches. Robot navigation is illustrated using cascaded neural network structure, Mamdani fuzzy controller, and Petri-net controller. A combination of the three controllers, in the form of a hybrid controller, provides successful navigation. Humanoid robot uncertainty during multi-robot mobility is controlled by the hybrid of the three controllers. Cognitive and intelligent humanoid robotic software architectures have been discussed. Gaze control-based navigation strategy for path planning has been considered. Various challenges to humanoid robots are addressed in this article.

Keywords Humanoid robots · Robot controller (RC) · Robot control user interface (RCUI) · Cascade neural network (CNN) · Fuzzy controller · Petri-net controller (PNC) · Cognitive architecture

1 Introduction

Humanoid robots have become increasingly popular in recent years. Humanoid robotics is a new and complex topic of robotic research. Humanoid research has

A. Sayeed (✉) · N. Kumar · N. Koul
Department of Computer Science and IT, Central University of Jammu, Jammu, India
e-mail: aqsasyeed@gmail.com

N. Kumar
e-mail: neerendra.csit@cujammu.ac.in

Z. Vámossy · Y. Bathla
John von Neumann Faculty of Informatics, Óbuda University, Budapest, Hungary
e-mail: vamossy.zoltan@nik.uni-obuda.hu

Y. Paul
Faculty of Informatics, Eötvös Loránd University, Budapest 1117, Hungary

Y. Singh et al. (eds.), *Proceedings of International Conference on Recent Innovations in Computing*, Lecture Notes in Electrical Engineering 1001,
https://doi.org/10.1007/978-981-19-9876-8_19

progressed significantly, with a variety of humanoid robots capable of moving and performing sophisticated jobs. In the last decade, a significant range of science and technology has arisen in humanoid research, leading to the creation of highly sophisticated humanoid-embedded systems having a wide range of sensory abilities. Humanoid robots are meant to cohabit with people in the physical world. Humanoid robots must be able to deal with a broad range of tasks and objects in unstructured, dynamic environments. Humanoid robots must have a light body, great flexibility, a wide range of sensors, and a high degree of intelligence [1, 2, 5]. Humanoid robots are functional for various purposes, such as interacting with human tools and environments and bipedal locomotion. Humanoid robots have a body, a head, two arms, and two legs. Certain humanoid robots may just replicate a part of the body [3, 4]. For example, some humanoid robots have heads that are meant to mimic human facial characteristics like eyes and lips. The demand for robotics has increased as a result of technological developments. Acceptance of robotics has accelerated across a variety of applications. Humanoid robot research is divided into four categories: artificial intelligence, robot hardware development, biped or wheel drive movement, and human–robot interaction. Humanoid robots perform autonomous activities in various fields of life [5]. In health care, Pepper and NAO (humanoid robots), developed by SoftBank Robotics, assist the transportation of medicines and food from one place to another in hospitals and move freely through hospital corridors, elevators, and wards [6, 7]. In 2021, exclusively for the healthcare business, Hanson Robotics introduced Grace (a humanoid robot). Humanoid robots are programmed to do jobs that normally fall to caretakers, such as checking vital signs, providing medicine, aiding with feedings, and notifying healthcare experts in the event of an emergency. SoftBank Robotics, Kawada Robotics Corporation, Honda Motor Company, UBTECH Robotics Corp. Ltd., Hanson Robotics Ltd., PAL Robotics, and Toyota Motor Corporation are among the top companies in the humanoid business [8, 9]. The market growth of humanoids is due to the increasing development of humanoids with innovative features like the growing use of humanoids as educational robots. The high demand for humanoids in retail for personalized services and the rising need for humanoid robots in the medical sector are all driving market expansion. Between 2022 and 2027, the humanoid market is predicted to develop at a compound annual growth rate (CAGR) of 63.5%. The predicted increase in the United States dollar (USD) from 1.5 billion in 2022 to 17.3 billion in 2027 is shown in Fig. 1 [10].

1.1 Contributions

The following are the primary contributions of this work:

i. We discussed in-depth studies and analyses of several cognitive and humanoid techniques.
ii. We illustrate the precise navigation of humanoids.

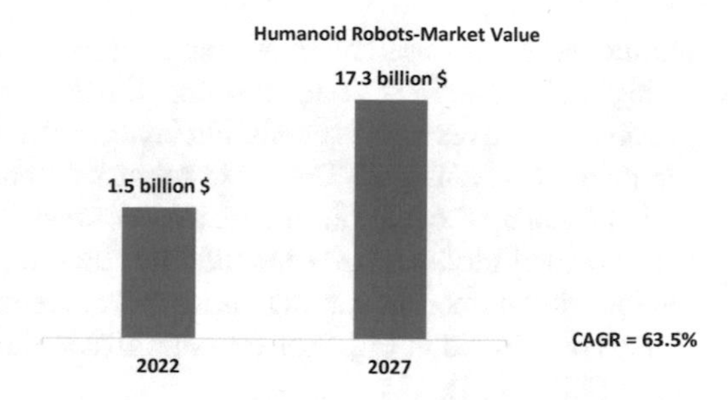

Fig. 1 Humanoid robot market value

iii. A brief illustration of the legged and wheeled humanoid robots.
iv. We highlight some open research problems as well as futuristic scope in this active field of research.

1.2 Organization of Paper

The organization of the article is arranged as follows: Sect. 2 provides a summary of the recent literature survey, and Sect. 3 provides an overview of humanoid robot types. The focus on the smooth navigation of humanoids is presented in Sect. 4 whereas Sect. 5 highlights some open research challenges in this active field of research, and Sect. 6 concludes the paper.

2 Related Work

2.1 Humanoid Robot Architectures

Cognitive and general software architectures for the humanoid robot are explained in Sects. 2.1.1 and 2.1.2, respectively.

2.1.1 Cognitive Architecture

Self-awareness, motivation, and emotions are big challenges for the cognitive system in robotics [11]. The early cognitive frameworks for robots were created using artificial intelligence. A cognitive robotic architecture supports perception, control, and task implementation at a low level [12]. Cognitive architecture helps in the recognition of complex contexts, planning intrinsic activities, and learning frequently processed behaviors at higher levels. Higher levels are associated with more complexity and comprehension [13]. To meet all of these criteria, a three-layered architecture is chosen to meet the needs of a humanoid. It is composed of parallel behavior-based modules that interact with one another and may be modified by higher-level entities. Quick response times to external factors, explicit assimilation of robot objectives in the scheduling layer, and a modular design strategy are among the primary advantages. The basic robot hardware includes sensors and actuators, which make up the interface in the robotic world [14, 15]. The single perceptual and task-oriented modules are separated by three layers of cognitive architecture. The database has numerous sub-databases that are required by all the layers' components. This is mentioned in Fig. 2 of the cognitive architecture of the Karlsruhe humanoid robot (KHR) [11].

The functionalities of the three layers of the cognitive architecture of KHR are illustrated in the following table (Table 1):

The cognitive architectural diagram (Fig. 2) illustrates various techniques in the system, such as active models for actual perceptions and actions are acquired from a universal data database and cached. Furthermore, they have a considerable degree of adaptability. The execution supervisor is in charge of the task and perceptual component prioritization [12]. Task knowledge is used to plan robot tasks on a more symbolic level. The task coordination on the mid-level receives the resultant sequence of actions and coordinates all running tasks. The low-level task execution executes the final, properly specified, and deadlock-free sequence of processes [11]. Interactions among cognitive modules at all levels are significantly relying on the

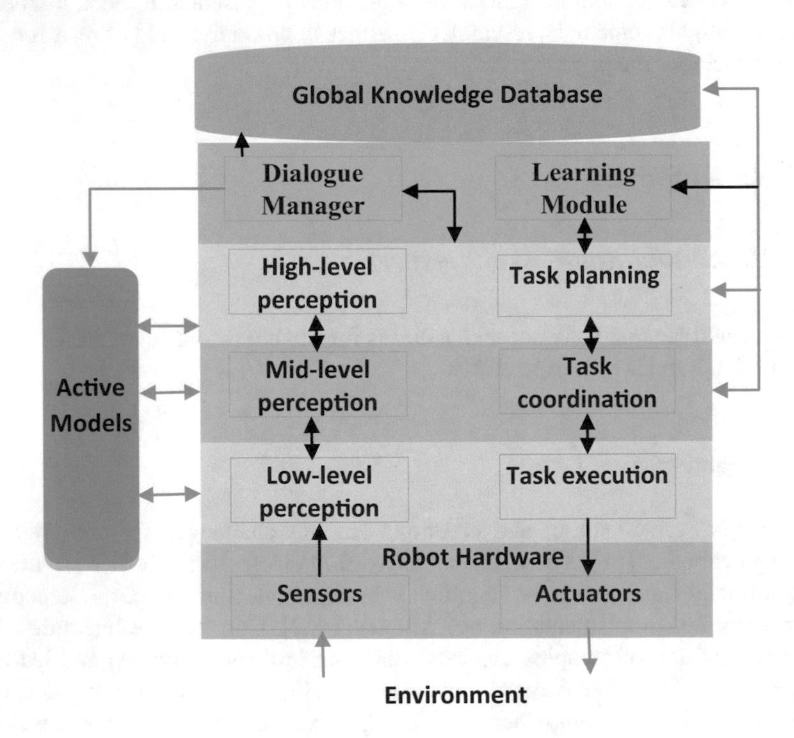

Fig. 2 Perceptive framework of Karlsruhe humanoid robot [1]

Table 1 Different layer functionalities of Karlsruhe humanoid robot

Serial No	Karlsruhe HR architecture levels	Functionalities
1	Low-level	Fast sensor data interpretation methods are used. The fast reactive components at low-level act in real time. Task execution is carried out at this level
2	Mid-level	It comprises many recognition components of the system, such as speech, acoustic, visual, and audiovisual speaker tracking. Task coordination is done here
3	Top-level	Organizes understanding components (single modality, multimodal fusion). This layer constitutes an interaction module, a task organizer, and a learning module

circumstances and behavior. This refers to the robotic system's focus of attention, the prioritization of necessary perceptual elements and activities, low-level elements' access to the database, interaction control, or intricate reflexes needing more intrinsic sensor analysis process than low-level perceptual elements can provide [16].

2.1.2 General Software Architecture for Humanoids

The architecture describes perceptual fusion, action automatic configuration, and high-level control. The architecture is meant to take advantage of current suitable software packages like ROS and Blender. The usage of the OpenCog software package to help operate humanoids made by Hanson Robotics is an initial survey stimulating this intelligent humanoid robotics software package. To achieve intelligent behavior in humanoids, a multipurpose cognitive framework is integrated with an adequate combination of robotics-specific software. A cognitive architecture is capable of processing perceptual input. This framework constructs a representation of a robot's current status and recommends relevant actions [6]. The software components of this architecture are given as follows (i–ix):

i. Physical layer
ii. Perception synthesizer (to generate audio signals) (PS)
iii. Body control
iv. Action orchestrator (automatic configuration) (AO)
v. Action creator (action generator) (AC)
vi. Action data source (data collecting process) (ADS)
vii. Robot controller (RC)
viii. Robot control user interface (RCUI)
ix. A gateway that can include more than the one source of data.

Figure 3 represents the software architecture for humanoid robots. Data from multiple sensors are taken and supplied to the system. The RC sends signals to the

AO that specifies which set of high-level actions to perform at a fixed given time and how to handle the robot to perform the tasks [17, 18]. Rather than being hard-coded, the AO's actions should be provided in a configuration file, enabling us to add new operations or adjust the current ones' settings. The AO might be paired with an actual or virtual robot depending on the simulation model specified in the configuration file. The orders delivered from the AO to the actual or virtual robots should ideally be similar. The AC converts externally generated movement scripts into robotic action patterns that may be executed in a robotic simulator or actual robot [6, 7]. The PS receives data from the robot and converts the data into a future-acceptable format. The PS processes and filters data before sending it to the RC. The PS may receive data from either the actual or virtual (simulated) robot. The data received in both situations would be similar in structure. For the time being, the AO and PS may collaborate in a future edition [19].

If ROS is employed as a platform, the AO, PS, and RC may be separate entities. Consider that the ROS nodes are in constant connection. The AO and PS would therefore interact with additional nodes that represented different parts of the robot or linked software [17]. An ADC is a part of extrinsic software that generates activity details that are transmitted to the action creator. The RC selects actions based on perceptions and user controls. It is unclear how to formalize the optimum action technique [20]. One method is to use OpenCog, either through the RCUI or other software tools. In a practical scenario, the mentioned software components (AC, ADS, and RCUI) may be deployed on several platforms (e.g., Windows, Mac, Android,

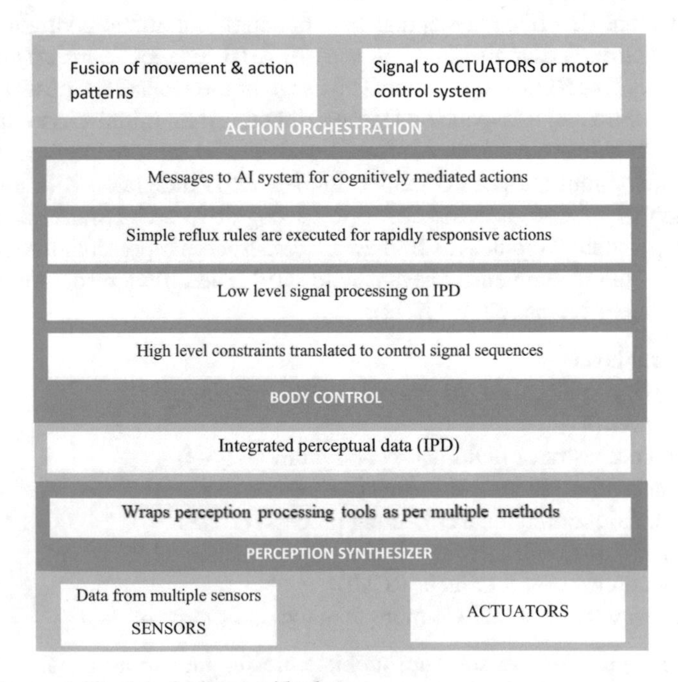

Fig. 3 Software architecture for humanoid robots

or iOS). The RC, PS, and AO, on the other hand, are expected to run on Linux in the future [6]. A summary of enabling technologies, architectures, limitations, and future tasks of Humanoid Robots is given in Table 2.

2.2 Navigation Based on Gaze Control

Humanoids gather information from various sources. Humanoids in dynamic environments determine gaze direction for data communication. A gaze direction is established by humanoid robots in the environment for data communication. A gaze control framework based on fuzzy integrals is required to establish the next gaze trajectory. The gaze direction points to a given position that meets the respective criterion. The robot's gaze direction can be chosen from a variety of directions based on the inclination of angles. The more gaze directions there are, the more computational time is required. A visibility check is required to reduce the time complexity [21, 22].

For ith gaze trajectory (α_i), a visibility test function ($M_l^j(\alpha_i)$) can be expressed using Eq. (1) [23].

$$M_l^j(\alpha_i) = \begin{cases} G_\alpha^j, \text{ if } \hat{\alpha}_l^j \in N_g(\alpha_i) \\ 0, \text{ otherwise} \end{cases} \tag{1}$$

where G_α^j is thc partial obtaincd valuc of $\hat{\alpha}_l^j$, and $N_g(\alpha_i)$ gives gaze area for α_i.

$\hat{\alpha}_l^j$ = gaze directions, for $j = 1, \ldots, n_l$, where l = gaze direction criteria, n_l = the number of gaze directions.

Mobile robots need this criterion in order to navigate safely and robustly in challenging surroundings. A normalizing function $w_\alpha(\alpha_i)$ adds and bounds (by 1.0) the partial obtained value for multiple gaze directions as mentioned in Eq. (2) [23, 24].

$$w_\alpha(\alpha_i) = \min\left(\sum_{j=1}^{n_l} M_l^j(\alpha_i), 1.0\right) \tag{2}$$

2.3 Human–Robot Interactions

Humanoids conduct complex tasks that necessitate higher consistency and precision. Humanoids imitate human social relationships autonomously without experiencing emotional trauma. As a result, humanoids are better suited for staff than humans [24, 25]. As a result, HRI enabling factors will give an overview of how humanoids can be developed to improve client satisfaction with HRI service. Example: The humanoid (Retail service robot) deploys artificial intelligence as its main asset to

Table 2 A summary of enabling technologies, architectures, limitations, and future tasks of humanoid robots

Author	Technologies	Humanoid robot	Summary	Architectures	Limitations	Future task
Burghart et al. [11]	Hidden Markov models, Kinematic models	ARMAR	A method for developing a cognitive framework for an intelligent humanoid robotic system	Cognitive architecture	A multimodal merger of speech and motions	Access to active models through tight integration
Nagarajan et al. [25]	Joystick controller, balancing controller, yaw drive mechanism	BALLBOT	A mechanism for physical HRI with the assistance of mobile robots	Control architectures (balancing, outer loop, leg control) [25]	One wheeled, limited but perpetual position displacements of ballbot	Laser range finders and stereo cameras are needed for accurate localization
Soyong et al. [26]	HRI, CFA models, SEM models	Retail service robots (RSRs): Sophia, NAO, Grace, and Pepper	In a survey on HRI in consumers, RSR helps customers in a store. RSR navigates and carries the products	Software architecture with sensors and abstract layers [27]	Depending on the service conditions, the work activities performed by robots and the kind of product/service customization of RSRs' design may be required	Future research should include a practical evaluation of the impacts of actual RSR physical designs
Yoo et al. [28]	Choquet fuzzy integral	Can be applied to any humanoid	A method allowing mobile robots to navigate safely and reliably in a dynamic environment using a gaze control-based localization framework	Fuzzy integral-based gaze architecture	The main issue is how to transmit and display various types of data at the same time [29]	The presented architecture will be expanded to deal with arbitrarily formed and equally sized objects traveling in peculiar ways

(continued)

Table 2 (continued)

Author	Technologies	Humanoid robot	Summary	Architectures	Limitations	Future task
Ariffin et al. [14]	Arduino microcontroller	NAO	New humanoid navigation using the mobile platform	Control system integration architecture	Limited detection area	To provide better detection by using a laser range sensor
Ariffin et al. [30]	Arduino controller interface (ACI)	NAO	ACI is used to build a humanoid-led navigation mobile platform within an obstacle in the surroundings by integrating exterior laser sensing with a humanoid	Control system integration architecture	Security is the issue	Path planning can be included to achieve optimal navigation [31]

provide personalized shopping assistance [26]. It is essential to ascertain whether the humanoid's appearance influences human activity toward it [27]. In a simple HRI, [28] the author matched members' impressions and actions toward two real humanoid robots. The test result showed a significant pattern regarding the spontaneity of the response range.

3 Humanoid Robots: An Overview

3.1 Legged and Wheeled Robots

Legged Humanoid Robots: A few examples are Sophia, NAO, Pepper, Petman, Atlas, Ocean One, etc. [10, 12]. Ocean One is an aquatic humanoid robot designed to explore coral reefs. Atlas is known as "the world's most dynamic humanoid." Atlas was designed to conduct search and rescue missions. It can use its range of senses, stereo vision, and other sensors to navigate its way through difficult terrain and obstacles in its path. NAO is made up of motors, sensors, and software. The Protection Ensemble Test Mannequin, or "Petman," was developed to test biological and chemical suits. The Pepper can recognize and respond to body languages such as facial expressions, tone of voice, and other signs [7]. Humanoid walking motions are 100 Hz controlled by accelerometers connected to the robot's hip. The best forward speed is 0.4 MPs. A humanoid robot cannot move in a diagonal path [11, 20, 34, 35]. Wheeled humanoid robots like M-Hubo assist humans in daily life tasks [36]. Wheeled robots consist of a degree of freedom for navigation [37–39]. Degree of freedom of a robot spatial procedure, f_i be the no. of freedoms supplied by joint i, and c_i be the number of constraints provided by joint i (it follows that $f_i + c_i = m_i$ for all i) [2, 12, 36]. Then, Grubler's formulas (3), (4), and (5) for the degrees of freedom (dof) of the robot are as follows:

$$dof = m(N-1) - \sum_{i=1}^{j} c_i \tag{3}$$

$$dof = m(N-1) - \sum_{i=1}^{j} (m - f_i) \tag{4}$$

$$dof = m(N-1-j) + \sum_{i=1}^{j} f_i \tag{5}$$

This formula is only retained if all joint constraints are autonomous. If they are not independent, then the formula gives a lower bound on the no. of degrees of freedom.

A homogeneous robot consists of identical types of sensors and actuators. The actuators and sensors differ slightly due to damage. Therefore, the ability to compete

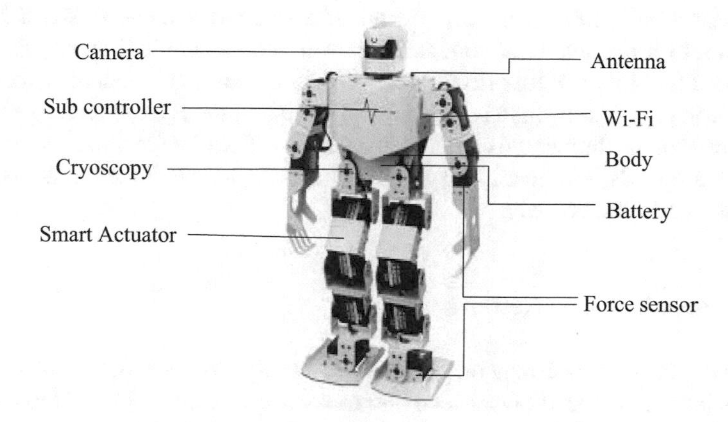

Fig. 4 A humanoid robot and its components [41]

in a wide range of tasks is limited. Heterogeneous robots usually have different sensing, motion, and computational power. Heterogeneous robots are categorized either as weakly or strongly based on their level of heterogeneity. A highly heterogeneous robot team has been used to develop an application such as aerial surveillance. Robot middleware, robot simulators, and multilevel testing are technologies for heterogeneous robots [4, 7, 30]. Figure 4 represents a robot diagram and mentions robotic components.

4 Navigation in Humanoid Robots

The ability of a robot to establish robotic position and orientation within a frame of reference is referred to as robot navigation. In this paper, we have used three methods consecutively to achieve the precise navigation of humanoids. The three methods are mentioned in Sect. 4.1.

4.1 Humanoid Robot Obstacle-Free Navigation Due to Hybrid Controller

Hybridized CNN with a fuzzy method is executed together with a PNC for obstacle-free mobility of humanoids in complex surroundings. The CNN is given three input parameters, a distance of humanoid from the front obstacle (DHRFO), a distance of humanoid from the left obstacle (DHRLO), and distance of humanoid from the right obstacle (DHRRO), and the target angle (TA), is generated as an output variable. The target angle, plus the three neural network input variables, is an input to the hybrid Mamdani fuzzy controller, and the effective target angle (ETA) is derived as an output for the humanoid robots. To prevent humanoid robot ambiguity during multi-robot mobility and for prioritized trajectory tracking of humanoids, the hybrid controller is required. If there is a humanoid collision, the ETA is given back to the PNC to prevent the humanoid collision [10, 28, 43]. The three methods in the hybrid approach are CNN, Mamdani fuzzy logic, and PNC, as mentioned in Sects. 4.1.1, 4.1.2, and 4.1.3, respectively.

4.1.1 Cascade Neural Network

CNN, like feed-forward networks, has connectivity between the input and each previous layer to the next layers. The output layer is directly connected to the input layer in CNN. The CNN layers, such as the input layer, a hidden layer, and the output layer, are well connected through neurons [44]. The CNN is being used to instruct the humanoid to acquire the TA. Thenceforth, TA is provided as an input to the fuzzy controller. Several training approaches are employed to instruct the neural network while receiving the TA. Several sensors are connected to the humanoid robot to identify obstacle distances and provide the data to the neural network. One layer is for I/O variables, and the other two are convolution layers for linking I/O variables with connection weights and neural bias [43, 45, 46]. The mean square method (ms) and root mean square method (rms) are applied for proving the potency of the CNN [43]. The formula for obtaining ms and rms is given in Eqs. (6) and (7), respectively:

$$\text{ms}(\%) = \left[\sum_{i}^{k} \left\{ \frac{(O)_{(A)} - (O)_{(P)}}{k} \right\}^{2} \right] * 100 \tag{6}$$

$$\text{rms}(\%) \sqrt{\frac{1}{k} \left[\sum_{1}^{k} \left\{ \frac{(O)_{(A)} - (O)_{(P)}}{(O)_{(A)}} \right\}^{2} \right]} * 100 \tag{7}$$

where

$(O)_{(A)}$ output (actual)
$(O)_{(P)}$ output TA (projected)
k no. of repetitions.

4.1.2 Hybrid Mamdani Fuzzy Method

The fuzzy control system is created by combining membership functions from triangles, quadrilaterals, and probabilistic distributions [43]. The fuzzy control system comprises three membership functions, TA, and an output, ETA. The three membership functions are grouped into five fuzzy variables: very very adjacent, very adjacent, adjacent, extreme, very extreme, and very very extreme. Similarly, the CNN output is the fourth input to the control system "TA," and the control system output "ETA" is classified into five fuzzy terms ranging from 30° to +30°, such as much negative, less negative, no turn, less positive, and much positive [42, 43, 47].

4.1.3 PNC

At a certain point, the presented hybrid CNN-fuzzy technique fails to convey critical information to the humanoids nearer the target. The hybrid method fails to stop the wreckage between the humanoids. To tackle the aforementioned circumstances, a PNC is aggregated with the hybrid CNN-fuzzy method to acquire the prioritized mobility. Figure 5 depicts the PNC utilized in this study [16, 32]. The PNC is discussed in detail in six plans [38, 43], which are listed below (plan 1–plan 6):

- Plan 1: The controller's first phase, in which all of the humanoids are randomly located in various locations in the environment. The humanoids in this scene have no notion of where the other humanoids are. When the hybridization procedure is triggered, the humanoids travel the course of their designated destinations.
- Plan 2: Humanoids continue to walk and may detect any obstacles in their path.
- Plan 3: This scene focuses on the humanoids, which detect any impediment.

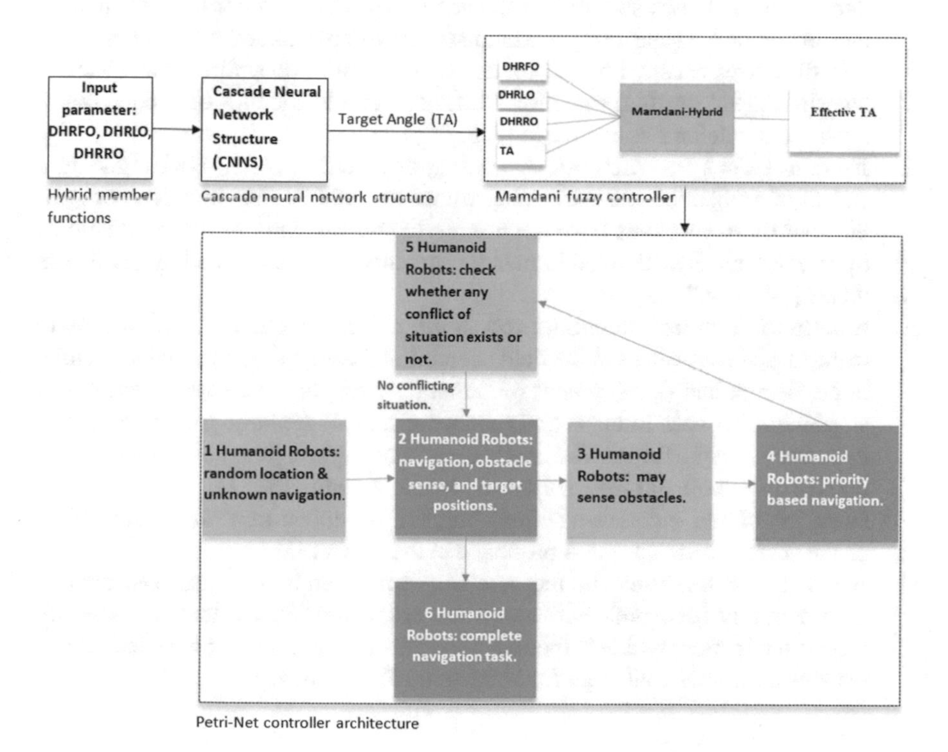

Fig. 5 A novel hybrid controller for precise navigation of humanoids [43, 49]

- Plan 4: Priority-based navigation begins at this stage. Humanoid nearer to the target is permitted to move with high priority. The rest humanoids remain static obstacles.
- Plan 5: The humanoid determines whether or not there is a conflict situation.
- Plan 6: This scene mirrors the behavior of stage 3, and scene 2 is followed by the remaining humanoids, who accomplish their navigational responsibilities.

The working of the hybrid controller is described in the flowchart as mentioned in Fig. 6.

5 Open Research Challenges

i. **Localization problems**: The navigational duties performed by humanoids remain restricted to motion modeling and position analysis, with little discussion of trajectory planning [46, 49, 51].

Ethical Issues:

ii. Robotics has been working on resolving the critical problem. Sir Isaac Asimov's three renowned laws should be followed in robotics. A robot may not harm a human person or cause injury to a human being through its actions. Except where such directives clash with the First Law, a robot must obey directions given by humans. As long as this protection does not clash with the First or Second Laws, a robot must defend its existence [7, 52].
iii. **Human–Robot Interactions**: According to recent research, HRI is facing a variety of problems in gaze tracking, voice interactions, and biological recognition, but these problems have not been tested yet and are mostly being studied by researchers. HRI-defined human movements must be adapted by intelligent robots [24, 25, 42, 43].
iv. **Emotional Robots**: Emotional robots bring their emotional relationships to reality. Recent advances in the field of emotional computing involve intervening in the design and development of "emotional robots" to create an emotional attachment between humans and robots. Still, there are huge gaps that need to be corrected in the future. The artificial software agents (bots) of "Pepper" pave the way for emotional interactions to become a reality [52].
v. **Noise problems**: Noise is a serious problem in robotic movement, depending on the surface resistance and pushback in the joints [53].
vi. **Accurate localization**: The measurements produced by the small sensors that are frequently used with humanoids are noisy and inconsistent. As a result, precise navigation, which is thought to be mostly addressed for wheeled robots, remains a difficult challenge for humanoid robots [38, 48].

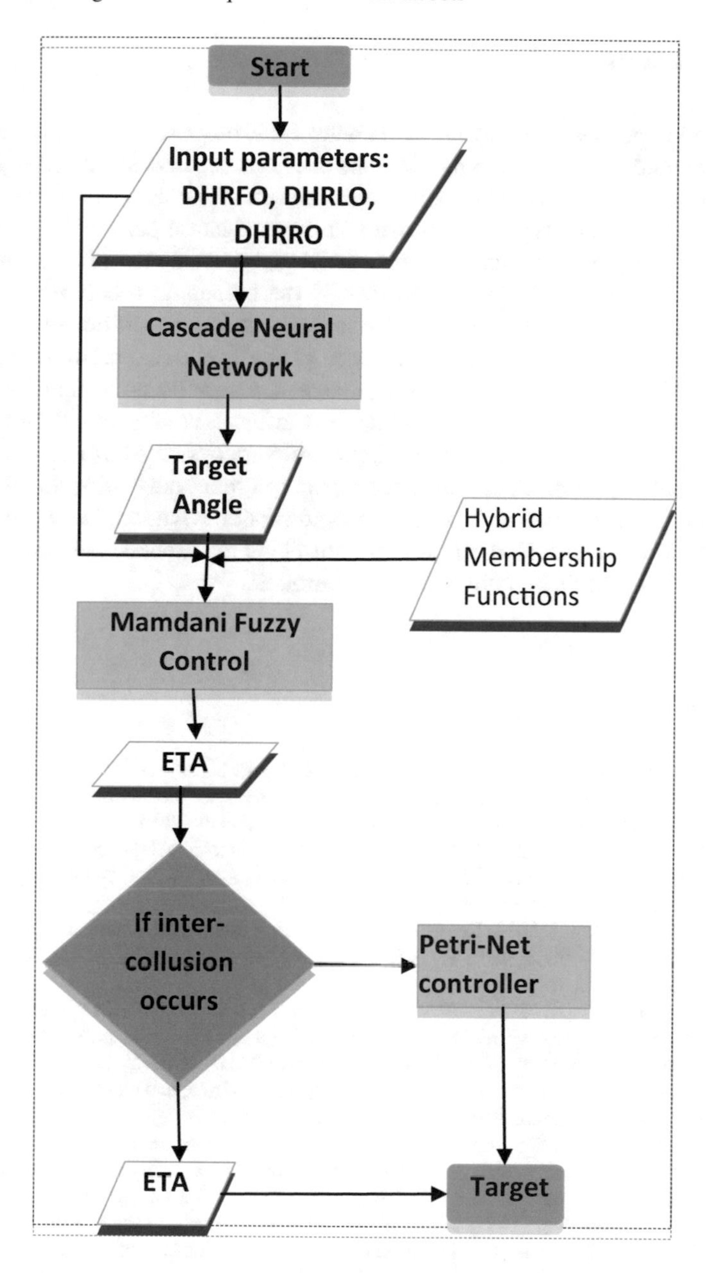

Fig. 6 Working flowchart of hybrid navigation controller [49]

6 Conclusion

For navigation purposes, a novel hybrid controller is proposed by combining CNN, fuzzy logic, and Petri-net. The hybrid controller can be used for the navigation of humanoid robots in static and dynamic environments. The navigation environment may consist of various types of obstacles in the navigation path. Firstly, the CNN is activated using barrier radii as input. CNN gives TA as output. Secondly, the fuzzy system is activated to yield the ETA to the humanoid robots while moving to the target position. The engineering development issues can be solved using a hybrid controller. For the cognitive purpose, a cognitive architecture for an intelligent humanoid robotic system has been presented. Cognitive architecture follows a hierarchical three-layer structure. In addition, a software architecture for interfacing between cognitive architectures and robotic software systems has been discussed. Different challenges related to humanoid robots and robot navigation are discussed. Path planning is the essential task of humanoid robot navigation. Precise collision-free navigation is a result of successful path planning. Robots can be accurately navigated using prominent path planning controllers.

References

1. Humanoid Robotics. www.humanoid-robotics.org/. Accessed 20 Apr 20 2022
2. Humanoids. https://www.ieee-ras.org/humanoid-robotics. Accessed 24 Mar 2022
3. Robots. https://en.wikipedia.org/wiki/Humanoid_robot. Accessed 11 Apr 2022
4. Kiener J, Von Stryk O (2010) Towards cooperation of heterogeneous, autonomous robots: a case study of humanoid and wheeled robots. Rob Auton Syst 58(7):921–929. https://doi.org/10.1016/j.robot.2010.03.013
5. Chen G, Wang H, Lin Z (2014) Determination of the identifiable parameters in robot calibration based on the POE formula. IEEE Trans Robot 30(5):1066–1077. https://doi.org/10.1109/TRO.2014.2319560
6. Munawar A et al (2018) MaestROB: a robotics framework for integrated orchestration of low-level control and high-level reasoning. In: Proceedings of IEEE international conference on robotics and automation, pp 527–534. https://doi.org/10.1109/ICRA.2018.8462870
7. Ray PP (2017) Internet of robotic things: concept, technologies, and challenges. IEEE Access 4:9489–9500. https://doi.org/10.1109/ACCESS.2017.2647747
8. Oh JH, Hanson D, Kim WS, Han IY, Kim JY, Park IW (2006) Design of android type humanoid robot Albert HUBO. In: IEEE international conference on intelligent robots and systems, November 2006, pp 1428–1433. https://doi.org/10.1109/IROS.2006.281935
9. Parviainen J, Coeckelbergh M (2021) The political choreography of the Sophia robot: beyond robot rights and citizenship to political performances for the social robotics market. AI Soc 36(3):715–724. https://doi.org/10.1007/s00146-020-01104-w
10. Markets in Focus. https://www.marketsandmarkets.com/. Accessed 20 Apr 2022
11. Burghart C et al (2005) A cognitive architecture for a humanoid robot: a first approach. In: Proceedings of 2005 5th IEEE-RAS international conference on humanoid robots, vol 2005, pp 357–362. https://doi.org/10.1109/ICHR.2005.1573593
12. Zaraki A et al (2017) Design and evaluation of a unique social perception system for human-robot interaction. IEEE Trans Cogn Dev Syst 9(4):341–355. https://doi.org/10.1109/TCDS.2016.2598423

13. Manca M et al (2021) The impact of serious games with humanoid robots on mild cognitive impairment older adults. Int J Hum Comput Stud 145:102509 (2021). https://doi.org/10.1016/j.ijhcs.2020.102509
14. Ariffin IM et al (2015) Sensor based mobile navigation using humanoid robot nao. Procedia Comput Sci 76(Iris):474–479. https://doi.org/10.1016/j.procs.2015.12.319
15. Vermesan O et al (2020) Internet of robotic things intelligent connectivity and platforms. 7:1–33. https://doi.org/10.3389/frobt.2020.00104
16. Le XT et al (2007) A study on obstacle avoidance and walking humanoid robots. In: ICCAS 2007—international conference on control, automation and systems, pp 2309–2312. https://doi.org/10.1109/ICCAS.2007.4406712
17. Goertzel B, Hanson D, Yu G (2014) A software architecture for generally intelligent humanoid robotics. Procedia Comput Sci 41:158–163. https://doi.org/10.1016/j.procs.2014.11.099
18. Park JH, Kim KO, Jeon JW (2006) PDA based user interface management system for remote control robots. In: 2006 SICE-ICASE international joint conference, pp 417–420. https://doi.org/10.1109/SICE.2006.315333
19. Kashyap AK, Parhi DR, Pandey A (2021) Multi-objective optimization technique for trajectory planning of multi-humanoid robots in cluttered terrain. ISA Trans. https://doi.org/10.1016/j.isatra.2021.06.017
20. Yi SJ, Lee DD (2016) Dynamic heel-strike toe-off walking controller for full-size modular humanoid robots. In: IEEE-RAS international conference on humanoid robots, pp 395–400. https://doi.org/10.1109/HUMANOIDS.2016.7803306
21. Adachi Y, Tsunenari H, Matsumoto Y, Ogasawara T (2004) Guide robot's navigation based on attention estimation using gaze information. In: 2004 IEEE/RSJ international conference on intelligent robots and systems, vol 1, pp 540–545.https://doi.org/10.1109/iros.2004.1389408
22. Oh HS, Lee CW, Mitsuru I (1991) Navigation control of a mobile robot based on active vision. In: IECON proceedings (industrial electronics conference), vol 2, pp 1122–1126. https://doi.org/10.1109/iecon.1991.239283
23. Yoo JK, Kim JH (2015) Gaze control-based navigation architecture with a situation-specific preference approach for humanoid robots. IEEE/ASME Trans Mechatron 20(5):2425–2436. https://doi.org/10.1109/TMECH.2014.2382633
24. Jeon HS et al (2015) Software architecture for humanoid robots with versatile task performing capabilities. In: 2015 12th international conference on ubiquitous robots and ambient intelligence (URAI 2015), pp 555–559.https://doi.org/10.1109/URAI.2015.7358829
25. Nagarajan U, Kantor G, Hollis R (2014) The ballbot: an omnidirectional balancing mobile robot. Int J Rob Res 33(6):917–930. https://doi.org/10.1177/0278364913509126
26. Soyoung C, Kim Y (2022) The role of the human-robot interaction in consumers' acceptance of humanoid retail service robots. 146:489–503
27. Hendrich N, Bistry H, Zhang J (2015) Architecture and software design for a service robot in an elderly-care scenario. Engineering 1(1):027–035. https://doi.org/10.15302/J-ENG-2015007
28. Jeong J, Yang J, Baltes J (2021) Robot magic show as testbed for humanoid robot interaction. Entertain Comput 40:100456. https://doi.org/10.1016/j.entcom.2021.100456
29. Kastner L, Lambrecht J (2019) Augmented-reality-based visualization of navigation data of mobile robots on the microsoft hololens–possibilities and limitations. In: Proceedings of 9th IEEE international conferences on cybernetics and intelligent systems (CIS), and robotics, automation and mechatronics (RAM), pp 344–349. https://doi.org/10.1109/CIS-RAM47153.2019.9095836
30. Ariffin IM, Baharuddin A, Atien AC, Yussof H (2016) Real-time obstacle avoidance for humanoid-controlled mobile platform navigation. Procedia Comput Sci 105:34–39. https://doi.org/10.1016/j.procs.2017.01.185
31. Zhang X, Wang J, Fang Y, Yuan J (2019) Multilevel humanlike motion planning for mobile robots in complex indoor environments. IEEE Trans Autom Sci Eng 16(3):1244–1258. https://doi.org/10.1109/TASE.2018.2880245
32. Adiprawita W, Ibrahim AR (2012) Service oriented architecture in robotic as a platform for cloud robotic (case study: Human gesture based teleoperation for upper part of humanoid robot).

In: Proceedings of international conference on cloud computing and social networking 2012: cloud computing and social networking for smart and productive society (ICCCSN 2012), pp 2–5. https://doi.org/10.1109/ICCCSN.2012.6215727
33. Kim MS, Kim I, Park S, Oh JH (2008) Realization of stretch-legged walking of the humanoid robot. In: 2008 8th IEEE-RAS international conference on humanoid robots, pp 118–124.https://doi.org/10.1109/ICHR.2008.4755941
34. Fedorov EA (2020) Humanoid bipedal balancing and locomotion. In: Proceedings of 2020 international Russian automation conference (RusAutoCon 2020), pp 35–41. https://doi.org/10.1109/RusAutoCon49822.2020.9208216
35. Park IW, Kim JY, Oh JH (2006) Online biped walking pattern generation for humanoid robot KHR-3 (KAIST humanoid robot-3: HUBO). In: Proceedings of 2006 6th IEEE-RAS international conference on humanoid robots (HUMANOIDS), vol 3, pp 398–403. https://doi.org/10.1109/ICHR.2006.321303
36. Chen F, Cao L, Tian M, Du G (2020) Research and Improvement of competitive double arm wheeled humanoid robot. In: Proceedings of 2020 IEEE 3rd international conference on information systems and computer aided education (ICISCAE 2020), pp 599–601. https://doi.org/10.1109/ICISCAE51034.2020.9236868
37. Nikita T, Prajwal KT (2021) PID controller based two wheeled self balancing robot. In: Proceedings of 5th international conference on trends in electronics and informatics (ICOEI 2021), pp 1–4. https://doi.org/10.1109/ICOEI51242.2021.9453091
38. Yanjie L, Zhenwei W, Hua Z (2009) The dynamic stability criterion of the wheel-based humanoid robot based on ZMP modeling. In: 2009 Chinese control and decision conference (CCDC 2009), pp 2349–2352.https://doi.org/10.1109/CCDC.2009.5192727
39. Li Y, Tan D, Wu Z, Zhong H, Zu D (2006) Dynamic stability analyses based on ZMP of a wheel-based humanoid robot. In: 2006 IEEE international conference on robotics and biomimetics (ROBIO 2006), pp 1565–1570. https://doi.org/10.1109/ROBIO.2006.340177
40. Taylor C, Ward C, Sofge D, Lofaro DM (2019) LPS: a local positioning system for homogeneous and heterogeneous robot-robot teams, robot-human teams, and swarms. In: 2019 16th international conference on ubiquitous robots (UR), (1), pp 200–207.https://doi.org/10.1109/URAI.2019.8768559
41. No Title. https://www.google.com/search?q=humanoid+robot+biped&tbm=isch&ved=2ahUKEwivovOPlKL3AhWDKbcAHUqYDg4Q2-cCegQIABAA&oq=humanoid+robot&gs_lcp=CgNpbWcQARgBMgcIIxDvAxAnMgcIIxDvAxAnMgUIABCABDIFCAAQgAQyBAgAEEMyBAgAEEMyBAgAEEMyBQgAEIAEMgUIABCABDIFCAAQgARQAFgAYI4jaA. Accessed 20 Apr 2022
42. Kumar N (2017) Robot navigation with obstacle avoidance in unknown environment. 5
43. Muni MK, Parhi DR, Kumar PB, Sahu C, Kumar S (2022) Towards motion planning of humanoids using a fuzzy embedded neural network approach. Appl Soft Comput 119:108588. https://doi.org/10.1016/j.asoc.2022.108588
44. Aved'yan ED, Barkan GV, Levin IK (1999) Synthesis of multi-layer neural networks architecture (for the case of cascaded NNs). In: Proceedings of international joint conference neural networks, vol 1, pp 379–382. https://doi.org/10.1109/ijcnn.1999.831523
45. Razafimandimby C, Loscri V, Vegni AM (2016) A neural network and IoT based scheme for performance assessment in internet of robotic things. https://doi.org/10.1109/IoTDI.2015.10
46. Aznan NKN, Connolly JD, Al Moubayed N, Breckon TP (2019) Using variable natural environment brain-computer interface stimuli for real-time humanoid robot navigation. In: Proceedings of IEEE international conference on robotics and automation, vol 2019-May, pp 4889–4895. https://doi.org/10.1109/ICRA.2019.8794060
47. Kumar N, Takács M, Vámossy Z (2017) Robot navigation in unknown environment using fuzzy logic, pp 279–284
48. Sabe K, Fukuchi M, Gutmann JS, Ohashi T, Kawamoto K, Yoshigahara T (2004) Obstacle avoidance and path planning for humanoid robots using stereo vision. In: Proceedings of IEEE international conference on robotics and automation, vol 2004(1), pp 592–597. https://doi.org/10.1109/robot.2004.1307213

49. Kashyap AK, Parhi DR, Muni MK, Pandey KK (2020) A hybrid technique for path planning of humanoid robot NAO in static and dynamic terrains. Appl Soft Comput J 96:106581. https://doi.org/10.1016/j.asoc.2020.106581
50. Huang K, Xian Y, Zhen S, Sun H (2021) Robust control design for a planar humanoid robot arm with high strength composite gear and experimental validation. Mech Syst Sig Process 155:107442. https://doi.org/10.1016/j.ymssp.2020.107442
51. Kumar N, Vámossy Z, Szabó ZM Robot path pursuit using probabilistic roadmap
52. Khalid S (2021) Internet of robotic things: a review. J Appl Sci Technol Trends 2(03):78–90. https://doi.org/10.38094/jastt203104
53. Hornung A, Wurm KM, Bennewitz M (200) Humanoid robot localization in complex indoor environments. In: 2010 IEEE/RSJ international conference on intelligent robots and systems (IROS 2010), pp 1690–1695. https://doi.org/10.1109/IROS.2010.5649751

An IoT-Based Approach for Visibility Enhancement and Fog Detection

Kapil Mehta, Vandana Mohindru Sood, Meenakshi Sharma, and Monika Dhiman

Abstract Nowadays, many camera-based advanced driver assistance systems have been created in recent years to assist drivers and assure their safety in a variety of driving situations. When driving in foggy circumstances, one of the issues drivers confront is decreased scene visibility and lower contrast. Aside from the visibility issues, the driver needs also to make a speed decision when driving. The major facts of fog are loss of contrast and color fading. Rain and snow are exceedingly disruptive to drivers, and glare from the sun or other road users can be quite dangerous, even if only for a short time. This paper aims to provide a novel approach that will let the users know about foggy weather conditions. The proposed system uses smart fog sensors by employing the Internet of Things (IoT). IoT is an emerging field that can be used for the intelligent transport system that can be used in vehicles due to zero visibility in fog. This paper highlights the approaches and systems that have been discovered and deemed important for estimating or even improving visibility in bad weather. This paper elaborates that the proposed technique can approximate and remove the corresponding fog functions for better visibility and safety.

Keywords Visibility enhancement · Internet of Things · Intelligent transport system · Smart fog sensors

1 Introduction

Rapid growth in the vehicle industry, along with an ever-increasing global population, results in increased traffic on the roads, which increases the number of accidents resulting in human death. Thousands of lives are lost every year in accidents caused by zero visibility in fog during the winter season. The goal of this study is to reduce

K. Mehta (✉) · M. Sharma · M. Dhiman
Department of Computer Science and Engineering, Chandigarh Group of Colleges, Landran, Mohali, Punjab, India
e-mail: kapilmehta5353@gmail.com

V. M. Sood
Chitkara University Institute of Engineering and Technology, Chitkara University, Chandigarh, Punjab, India

Y. Singh et al. (eds.), *Proceedings of International Conference on Recent Innovations in Computing*, Lecture Notes in Electrical Engineering 1001,
https://doi.org/10.1007/978-981-19-9876-8_20

the number of people killed in traffic accidents caused by fog. The goal is to propose an IoT-based built system in cars that can identify impediments in fog and avoid accidents, saving human lives. The major strategy to limit the number of accidents on public roads is to adapt vehicle speed to environmental circumstances [1]. Weather-related poor visibility when driving has been identified as the leading accident cause. Research studies in the previous decade produced a variety of features to aid drivers, such as revamping headlights with LED devices or enhancing the directivity of the real-time beam; with novel techniques, the output light is nature closer [2]. They too added a novel feature, auto-dimming technology, which is now standard on the majority of vehicles [3]. Unfortunately, during the event of fog, it is insufficient, with no dependable and stable system for installation on a commercial truck has been created to date. Picture processing methods such as marking of lanes, traffic light signs, or dangers like barriers [4], image dehazing and deblurring [5], image segmentation, or machine learning algorithms [6, 7] were used. Various approaches rely on studying scattering and beam dispersion to determine the direct transmission or backscattering optical power of a light source [8]. The world's largest corporations have been working for years to develop technology that can revolutionize running the autonomous car [9]. As technology is implemented in public areas, it is expected that crash rates will drop dramatically.

Consider the working of autonomous vehicles behave in the worst circumstances of weather: vehicle adherence loss, stability issues of vehicles, and perhaps decreased visibility: non-seen signs of traffic conditions and markings of lanes, unidentifiable pedestrians [10], objects in its path [11], reduced visibility due to sun glare [12], and so on [13]. Autonomous vehicles and human-driven vehicles will cohabit on public roads during a transitional period in the next decades. Because of drivers' unpredictable behaviour, these systems will need to be evaluated and react rapidly to prevent accidents. Based on reasoning, we conducted a review of the state of the research which uses image processing as estimating fog visibility situations, hence improving overall traffic safety.

Figure 1 depicts a field overview, beginning with state-of-the-art methods of visibility enhancement and fog detection, then sensor systems that employ the methods presented in the first two subsections for visibility detection in adverse or worst conditions of weather, and finally in such conditions, the human observer's reactions.

In general, the first group's methods rely on image processing, whereas the second group's methods rely on optical power measurements or image processing. The well-known and widely utilized approaches from major groups will be discussed in the following sections. The objective of this paper is to showcase the benefits as well as the drawbacks of each strategy to find new ways to improve it. Following that, as shown in the diagram underneath, we present a combination of strategies having the goal of compensating for the shortcomings of one method with the strengths of the other. The ending stage will be to determine whether the outcomes achieved by the system are valid for humans as well as autonomous vehicles.

The present invention relates to an intelligent vehicle system for determining fog level and changing the vehicle route in real time. A fog sensor is installed on the bonnet of the four-wheeler vehicle and will automatically measure the fog after every

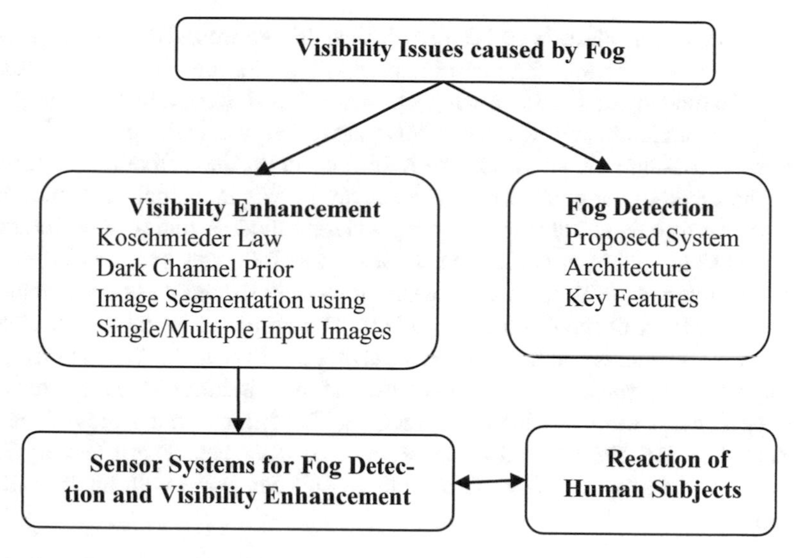

Fig. 1 Overall structure of intelligent vehicle system

five to ten minutes in case the engine of the four-wheeler vehicle is on. An Arduino is connected with the fog sensor for measuring the level of fog and sends the message to the four-wheeler display screen through bluetooth to aware a driving person. The Smart Fog Addition will display the four different types of dots on different routes on the map. A smart fog chip consists of bluetooth connected to the four-wheeler vehicle to display the level of fog on the display.

1.1 Organization of Paper

This paper is systematized as represented in Fig. 1. Section 2 characterizes the literature background. Section 3 elaborates on the visibility enhancement methods. Section 4 emphasizes the proposed system using IoT for fog detection along with the architecture and key features of the intelligent system. Section 5 concludes the paper.

2 Related Work

Various solutions to the aforementioned issues have been proposed throughout history. A summary of the work done to solve this challenge is offered below. Saraf et al. developed a method for distance calculations between vehicles and determining the density of traffic in a given location. His research attempts to improve the safety of roads by providing alerts in real time to vehicle drivers around their path obstacles and in their instant vicinity. The communication between the car and the roadside device is the fundamental mechanism [14]. Ismail et al. devised a method for using a wireless communication system for warning the driver while taking turns. To avoid collisions, a red-light signal is used to inform users on the different side lanes to restrict the speed of the other car [15]. There are intelligent vehicle system models which are the foundation of microscopic simulation models, and they're frequently used in the examination of vehicle behavior such as following speed and distances in different intervals of time. A car-following model [16] regulates the conduct of the vehicle in front of them in the same lane. The proposed invention will help in reducing road accidents and traffic jams occurring due to fog. The Smart Fog Sensor in the four-wheeler vehicle will help in measuring the fog through the fog sensor and sending a message through Bluetooth and GPS. The different intensities of fog will be detected by the Arduino and measured by the fog sensor. The following Table 1 shows how the proposed work is introduced and compared with other available solutions.

3 Visibility Enhancement Methods

There has been a lot of interest in enhancing visibility concerns during adverse weather conditions, also as foggy circumstances, over the previous decade. The image processing algorithms are categorized into two categories: single-image processing (one of the initiation approaches was offered by Tarel and Hautiere in [22]) and multiple-image processing [23]. In many real-world applications, taking numerous input photographs of the same scene is impractical; this is why uni-image haze reduction has gotten a lot of attention.

3.1 Basic Theoretical Assumptions

The primary strategies from the first group, visibility enhancement, will be mathematically described in this subsection: Koschmieder Law and dark channel prior.

(i) **Koschmieder Law**

Koschmieder discovered a link between the object luminance (L) detected at a distance × from the observer and the brightness L_0 close to the object during his work on the attenuation of luminance through the atmosphere:

$$L = L_0 e^{-\sigma x} + L_\infty\left(1 - e^{-\sigma x}\right)$$

Table 1 Comparison of proposed system with the existing state of art

Existing state of art	Drawbacks in the existing state of art	Benefits of proposed system
Automatic detection of fog and visibility distance estimation by using an on-board camera [17]	(i) This model is based on Koschmieder's model hypothesis (ii) It merely uses a camera for measuring fog (iii) It does not inform the people whether there is fog ahead or not	(i) It uses a fog sensor and Arduino for measuring fog content (ii) It uses Bluetooth and GPS for informing people about the fog content (iii) It helps in preventing accidents and traffic jams
Air Pollution and Detection of Fog using Vehicular Sensors [18]	(i) This model uses cameras and LIDAR for measuring fog (ii) The data collected on fog and pollution is used for improving weather conditions (iii) It will not inform the person sitting at home about the fog	(i) The proposed invention uses a fog sensor for measuring fog and Arduino for judging the level of fog (ii) The data collected by the fog sensor will be immediately sent to Smart Fog Chip and Smart Fog Addition (iii) It will report on the fog content at different locations
A New Technology of Fog Detection by using a Current (pA ~ nA) Path on the Vehicle's Interior Windshield and Application [19]	(i) This invention uses a current-based fog sensor for measuring fog (ii) It does not check if the fog content is high, moderate, or low (iii) It does not transmit the information about fog to the user via Bluetooth or GPS	(i) The proposed invention will use an infrared-based fog sensor to give the exact report of the fog (ii) It will use Arduino for measuring the fog levels as low, moderate, or high (iii) This will let the users know about the fog content through Bluetooth and GPS
Fog Intelligence for Real-time IoT Sensor Data Analytics [20]	(i) This model uses light and temperature as input signals (ii) It will only measure if the signal is normal or not (iii) It transmits signals to mobile phones through SMS only if it is abnormal	(i) The proposed invention uses a fog sensor and Arduino for fog content (ii) It will measure the four different levels of fog classified as low, moderate, high, or none (iii) It uses GPS for sending messages to Smart Fog Addition which displays the different dots in Google Map

(continued)

where L_0 is the luminance close to the object, L_∞ is the atmospheric luminance, and σ is extinction coefficient. In uniform luminance, this law is pertinent only in the daytime.

(ii) **Dark Channel Prior**

Table 1 (continued)

Existing state of art	Drawbacks in the existing state of art	Benefits of proposed system
Fog Detection and fog mapping using low-cost Meteosat-WEFAX transmission [21]	(i) This model uses Meteosat-WEFAX imagery for measuring the fog content (ii) The bi-dimensional threshold test is done in it to distinguish between fog and other surfaces (iii) It does not transmit the information about fog to the user via Bluetooth or GPS	(i) It uses a fog sensor for measuring fog and Arduino for judging fog level (ii) It uses Arduino for checking the different levels of fog as low, moderate, high, or none (iii) The proposed invention will let the users know about the fog content through Bluetooth and GPS

This is a statistical approach that employs an input as a single foggy image and is dependent on non-foggy photographs collected outdoors. It is built on the assertion that haze-free outdoor photographs have a minimum single channel with the lowest levels of intensity for a few pixels in practically all non-sky areas. In other words, the lowest intensity raises to zero. For a particular image J, formal can be demarcated as:

$$J^{\text{dark}}(x) = \min_{c \in \{R,G,B\}} \left(\min_{y \in \cap(x)} \left(J^c(y) \right) \right)$$

where J_c is the color channel of image J, and $W(x)$ is a local area or window centred in x.

Negru et al. offer a picture dehazing technique based on the same concepts, which calculates brightness attenuation through the environment [24, 25]. The project's purpose is to dehaze photos captured from a running vehicle and notify the driving person regarding the density of the fog and the appropriate rapidity under a few conditions. The initial stage in this method is to use a Canny–Deriche edge detector on a given image, followed by assessments of the image's horizon line and inflection point to see if fog exists.

(iii) **Image Segmentation using Single/Multiple Input Images**

In [26], Zhu et al. elaborated a mean shift-based algorithm that addresses several issues with traditional dehazing methods, including oversaturated images in [27], laborious and limited use of the method for gray-scale images in [28], and difficulties with sky images and computational complexity. Sky segmentation, re-refining, and restoring are three processes of the mean shift-based method. The algorithm begins by computing input image dark channels; next, a white balance adjustment is done to the same input image to lessen the color cast's influence (after dehazing, there is a significant increase in saturation).

Some image segmentations were already defined [29–31], where the sky was dissociated from the rest parts of the image for avoiding noise and boundary area distortions. Moving cars are segregated from the outer environment in various techniques, this time with numerous input photos, even in severe weather conditions like fog [32–34]. To distinguish between the foreground and backdrop, such a method necessitates a higher number of frames; hence, idea behind this work was to use the moving energy of the running vehicle for identifying them from the varying background.

4 Visibility Enhancement Methods

The proposed IoT-based smart fog detection system could be best used by the people who are either traveling or going to travel in further future. This system relates to a vehicle system for determining fog level and changing the vehicle route in real time. A fog sensor is installed on the bonnet of the four-wheeler vehicle and will automatically measure the fog after every five to ten minutes in case the engine of the four-wheeler vehicle is on. The Smart Fog Sensor will have four levels for measuring fog of different locations, depending upon the fog prevailing in the particular area. An Arduino is connected with the fog sensor for measuring the level of fog and sends the message to the four-wheeler display screen through bluetooth to aware a driving person. The Arduino is also used to send a message to the smart fog addition through the GPS installed in it. The smart fog addition will display the four different types of dots on different routes on the map. A smart fog chip consists of Bluetooth connected to the four-wheeler vehicle to display the level of fog on the display. The Smart Fog Sensor installed in the car will transmit the message to the Google Map through GPS according to which dots of different routes will change and help people decide which route to choose. Also, the signal will be transmitted to Smart Fog Chip through bluetooth which will display the fog condition on the display screen of the four-wheeler vehicle. Thus, a person by opening the Google Maps app could know the fog conditions of a particular route while sitting at home only.

4.1 Architecture

The proposed IoT-based system is shown in Fig. 2 which consists of the devices 'Smart Fog Sensor,' 'Smart Fog Chip' and 'Smart Fog Addition.' The fog sensor installed on the bonut of the four-wheeler vehicle will automatically measure the fog after every ten minutes in case the engine of the four-wheeler vehicle is on. Thus, whenever the 'Smart Fog Sensor' will find any of the four levels of the fog by measuring it through the fog sensor and detecting it through Arduino, it will send the message to the four-wheeler display screen through the bluetooth. This is done to aware the person driving the four-wheeler vehicle get aware and drive the four-wheeler vehicle carefully. Also, the Arduino in Smart Fog Sensor will send a message to the Smart Fog Addition through the GPS installed in it. Then, accordingly, the

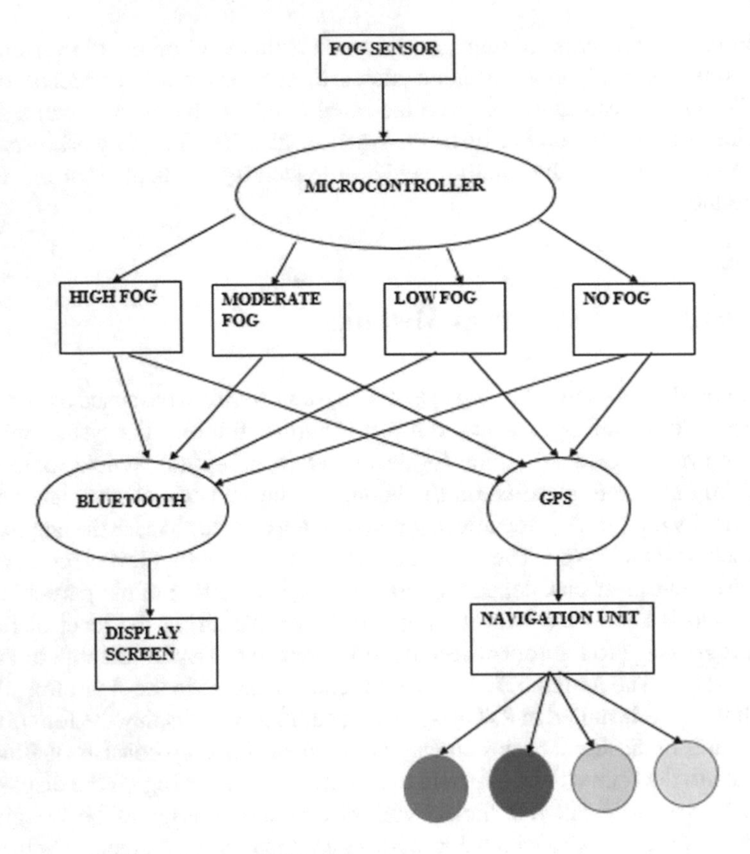

Fig. 2 Proposed IoT-based system architecture

Smart Fog Addition will display the four different types of dots on different routes on the map.

For example, a person A wants to go to a certain place through a particular route. He leaves the home and in the mid-way, he finds the fog, due to which there is less or no visibility. Thus, it can lead him to meet with an accident or stuck in a traffic jam. Also, a person B, who is ongoing his travel can get late if there is traffic on the street. In case, our invention is used, person A can check the Google map whether his route has fog or not. As a result, he can pick the right path before leaving the house and arrive at his destination on time. Also, through our invention, person B, who is traveling from a particular location and he is getting late, then he by seeing on the screen display can change his route according to the intensity of fog. Thus, our device will help the people know the appropriate location of the fog while sitting at home or while ongoing travel so that they can easily decide whether to continue the journey or abort it. Also, if to continue, then which route he should take depends upon the intensity of the fog.

4.2 Key Features

The present invention relates to a vehicle system for determining fog. More particularly, the present invention relates to a system to reduce the accidents which occur due to the low visibility and dense fog in a particular area. It comprises:

(i) A fog sensor installed on the bonnet of the four-wheeler vehicle will automatically measure the fog after every five to ten minutes in case the engine of the four-wheeler vehicle is on. The Smart Fog Sensor will have four levels (no fog with clear visibility, less fog with fine visibility, moderate fog with moderate visibility, and dense fog with no visibility) for measuring fog of different locations, depending upon the fog prevailing in the particular area.
(ii) An Arduino is connected with the fog sensor for measuring the level of fog and sends the message to the four-wheeler display screen through Bluetooth to aware the driving person. The Arduino is also used to send a message to the smart fog addition through the GPS installed in it.
(iii) The Smart Fog Addition will display the four different types of dots on different routes on the map. These four conditions would be denoted by four colors orange, brown, green, and yellow in the app. The orange color would denote the very high fog and extremely low visibility case, the brown color would denote a moderate level of fog and moderate visibility case, while the green color will denote the condition of low fog or a good amount of visibility, and the yellow color will denote no fog and clear visibility.
(iv) A smart fog chip consists of bluetooth connected to the four-wheeler vehicle to display the level of fog on the display.

4.3 Technical Design Functions

The external feature used in the proposed system is GPS and Google Map support. As GPS is used to transmit the signal from the 'Smart Fog Sensor' to 'Smart Fog Addition,' thus it is necessary for knowing the appropriate location. The Google Map support is used here to represent different colors on the different routes for helping the person decide which route to take depending on the intensity of fog. The Smart Fog Sensor contains the devices Arduino, GPS, fog sensor, and Bluetooth. The fog sensor will measure the fog of the area after every five minutes when the engine is ON. After checking, it will send the results to Arduino, which will measure the different intensities of fog, i.e., no fog, low fog, moderate fog, and dense fog. Once measure the fog level, it will transmit the message through bluetooth to Smart Fog Chip and to Smart Fog Addition through GPS.

5 Conclusion

This research described approaches and classifications from the scientific literature that were used for fog detection and to improve vision in foggy situations that seemed in recent years. The development of autonomous vehicles will be the primary emphasis of automotive firms in the coming years, and visibility requirements in adverse conditions of weather will be critical. The current approaches are based on the Internet of Things, image processing, optical power measurements, or a variety of sensing devices, few of which are currently accessible on commercial vehicles but are utilized for different purposes. Some approaches described in the literature only function during the day, rendering them unsuitable for automotive applications that demand systems that can provide dependable outcomes in real-time and typical scenarios 24×7. The proposed approach consists of dynamically implementing Koschmieder's Law for the computations of climatological visibility distance. The proposed solution features the benefit of utilizing a single camera and challenging the road availability as well as sky scenes. In opposition to various methods which demand the explicit extraction of the road, this method provides fewer constraints because it can be used with nothing more than the extraction of a homogenous surface inside the image that contains a section of the road and sky.

In the future, the plan is to leverage this concept to construct more complex frameworks for weather removal vision systems based on deep neural networks in the future. When a motorist needs to slow down to prevent a collision, we can consider various subject vehicles' deceleration levels in the future. We might also explore learning drivers' habits and adjusting the collision alert level threshold based on different users.

References

1. U.S. Department of Transportation (2015) Traffic safety facts—critical reasons for crashes investigated in the national motor vehicle crash causation survey. National Highway Traffic Safety Administration (NHTSA), Washington, DC, USA
2. OSRAM Automotive. Available online https://www.osram.com/am/specials/trends-in-automotive-lighting/index.jsp. Accessed on 30 June 2020
3. The Car Connection, 12 Oct 2018. Available online https://www.thecarconnection.com/news/1119327_u-s-to-allowbrighter-self-dimming-headlights-on-new-cars. Accessed on 30 June 2020
4. Aubert D, Boucher V, Bremond R, Charbonnier P, Cord A, Dumont E, Foucher P, Fournela F, Greffier F, Gruyer D et al (2014) Digital imaging for assessing and improving highway visibility. Transport Research Arena, Paris, France
5. Rajagopalan AN, Chellappa R (eds) (2014) Motion deblurring algorithms and systems. Cambridge University Press, Cambridge, UK
6. Palvanov A, Giyenko A, Cho YI (2018) Development of visibility expectation system based on machine learning. In: Computer information systems and industrial management. Springer, Berlin/Heidelberg, Germany, pp 140–153
7. Yang L, Muresan R, Al-Dweik A, Hadjileontiadis LJ (2018) Image-based visibility estimation algorithm for intelligent transportation systems. IEEE Access 6:76728–76740

8. Ioan S, Razvan-Catalin M, Florin A (2016) System for visibility distance estimation in fog conditions based on light sources and visual acuity. In: Proceedings of the 2016 IEEE international conference on automation, quality, and testing, robotics (AQTR), Cluj-Napoca, Romania, 19–21 May 2016
9. Levinson J, Askeland J, Becker J, Dolson J, Held D, Kammel S, Kolter JZ, Langer D, Pink O, Pratt V et al (2011) Towards fully autonomous driving: systems and algorithms. In: Proceedings of the 2011 IEEE intelligent vehicles symposium (IV), Baden-Baden, Germany, 5–9 June 2011, pp 163–168
10. Jegham I, Khalifa AB (2017) Pedestrian detection in poor weather conditions using moving camera. In: Proceedings of the IEEE/ACS 14th international conference on computer systems and applications (AICCSA), Hammamet, Tunisia, 30 Oct–3 Nov 2017
11. Dai X, Yuan X, Zhang J, Zhang L (2016) Improving the performance of vehicle detection system in bad weathers. In: Proceedings of the 2016 IEEE advanced information management, communicates, electronic and automation control conference (IMCEC), Xi'an, China, 3–5 Oct 2016
12. Miclea R-C, Silea I, Sandru F (2017) Digital sunshade using head-up display. In: Advances in intelligent systems and computing, vol 633. Springer, Cham, Switzerland, pp 3–11
13. Garg M, Chadha A, Mehta K, Garg R (2019) Brain gate technology—an analysis. Int J Adv Sci Technol 28(19):890–93. http://sersc.org/journals/index.php/IJAST/article/view/2676
14. Saraf PD, Chavan NA (2013) Pre-crash sensing and warning on curves: a review. Int J Latest Trends Eng Technol (IJLTET) 2(1)
15. Ismail M et al (2014) Intersection cross-traffic warning system for vehicle collision avoidance. Int J Adv Res Electr Electron Instrum Eng 3:13155–13160. https://doi.org/10.15662/ijareeie.2014.0311031
16. Wu Z, Liu Y, Pan G (2009) A smart car control model for brake comfort based on car following. IEEE Trans Intell Transp Syst 10:42–46
17. Lo WL, Chung HSH, Fu H (2021) Experimental evaluation of PSO based transfer learning method for meteorological visibility estimation. Atmosphere 12(7):828
18. Mehta K, Kumar Y (2020) Implementation of efficient clock synchronization using elastic timer technique in IoT. Adv Math Sci J 9(6):4025–4030
19. Sajjad F, Ali A, Ahmad SR (2021) Development of anti-fog agent for the reduction of potential occupational visual hazards at workplaces. Pak J Sci 73(1):144
20. Sharma M, Singh H (2021) Substrate integrated waveguide-based leaky-wave antenna for high-frequency applications and IoT. Int J Sens Wirel Commun Control 11(1):5–13
21. Arora T, Dhir R (2020) Geometric feature based classification of segmented human chromosomes. Int J Image Graph 20(01):2050006
22. Tarel J-P, Hautiere N (2009) Fast visibility restoration from a single color or gray level image. In: Proceedings of the 2009 IEEE 12th international conference on computer vision, Kyoto, Japan, 27 Sept–4 Oct 2009, pp 2201–2208
23. Narasimhan SG, Nayar SK (2003) Contrast restoration of weather degraded images. IEEE Trans Pattern Anal Mach Intell 25:713–724
24. Negru M, Nedevschi S (2013) Image-based fog detection and visibility estimation for driving assistance systems. In: Proceedings of 2013 IEEE 9th international conference on intelligent computer communication and processing (ICCP), Cluj-Napoca, Romania, 5–7 Sept 2013, pp 163–168
25. Rai V et al (2022) Cloud computing in healthcare industries: opportunities and challenges. In: Singh PK, Singh Y, Chhabra JK, Illés Z, Verma C (eds) recent innovations in computing. Lecture notes in electrical engineering, vol 855. Springer, Singapore. https://doi.org/10.1007/978-981-16-8892-8_53
26. Mehta K, Kumar Y, Aayushi A (2022) Enhancing time synchronization for home automation systems. ECS Trans 107(1):6197
27. Mehta K et al (2022) Enhancement of smart agriculture using Internet of Things. ECS Trans 107(1):7047

28. Bansal A, Mehta K, Arora S (2012) Face recognition using PCA and LDA algorithm. In: 2012 second international conference on advanced computing and communication technologies. IEEE
29. Das D, Roy K, Basak S, Chaudhury SS (2015) Visibility enhancement in a foggy road along with road boundary detection. In: Proceedings of the blockchain technology and innovations in business processes, New Delhi, India, 8 Oct 2015, pp 125–135
30. Yuan H, Liu C, Guo Z, Sun Z (2017) A region-wised medium transmission based image Dehazing method. IEEE Access 5:1735–1742
31. Mohindru V, Vashishth S, Bathija D (2022) Internet of Things (IoT) for healthcare systems: a comprehensive survey. In: Singh PK, Singh Y, Kolekar MH, Kar AK, Gonçalves PJS (eds) Recent innovations in computing. Lecture notes in electrical engineering, vol 832. Springer, Singapore. https://doi.org/10.1007/978-981-16-8248-3_18
32. Mohindru V, Garg A (2021) Security attacks in Internet of Things: a review. In: Singh PK, Singh Y, Kolekar MH, Kar AK, Chhabra JK, Sen A (eds) Recent innovations in computing (ICRIC 2020). Lecture notes in electrical engineering, vol 701. Springer, Singapore. https://doi.org/10.1007/978-981-15-8297-4_54
33. Rai V et al (2022) Cloud computing in healthcare industries: opportunities and challenges. Recent Innovations in computing: Proceedings of ICRIC 2021 2:695–707
34. Mehta, K et al (2022) Machine learning based intelligent system for safeguarding specially abled people. In: 2022 7th International conference on communication and electronics systems (ICCES). IEEE

Internet of Medical Things (IoMT) Application for Detection of Replication Attacks Using Deep Graph Neural Network

Amit Sharma, Pradeep Kumar Singh, Alexey Tselykh, and Alexander Bozhenyuk

Abstract The fast developments in micro-computing, mini-hardware manufacturing, and machine-to-machine communications have paved the way for innovative Internet of Things (IoT) solutions that are reshaping a great deal of networking software. The Internet of Things (IoT) has introduced a new branch of IoT that is known as the Internet of Medical Things (IoMT) systems. One of the applications that have been transformed by IoT is the healthcare system. Remote monitoring of patients suffering from chronic conditions is made possible by IoMT devices. As a result, it can deliver rapid diagnostics for patients, which in the event of an emergency may save their life. However, ensuring the safety of these vital systems is one of the primary obstacles standing in the way of their widespread use. The objective of this paper is to detect the replication attack, i.e., DoS (Denial-of-service) attack by using multilayer perceptron (MLP) classification algorithm in graph neural network. The proposed approach achieves better detection accuracy of 98.4% when compared with existing state-of-the-art classification models.

Keywords Attack · Internet of medical things (IoMT) · Deep Graph Neural Network · Security · Healthcare system

---

A. Sharma
Institute of Computer Technologies and Informational Security, Southern Federal University, Rostov-on-Don 344006, Russia

P. K. Singh (✉)
School of Technology Management and Engineering (STME), Narsee Monjee Institute of Management Studies (NMIMS), Chandigarh, India
e-mail: pradeep_84cs@yahoo.com

A. Tselykh
Department of Information and Analytical Security Systems, Southern Federal University, Rostov-on-Don 344006, Russia
e-mail: tselykh@sfedu.ru

A. Bozhenyuk
Southern Federal University, 44 Nekrasovsky Street, Taganrog 347922, Russia

A. Sharma
Chitkara University Institute of Engineering and Technology, Chitkara University, Punjab, India

Y. Singh et al. (eds.), *Proceedings of International Conference on Recent Innovations in Computing*, Lecture Notes in Electrical Engineering 1001,
https://doi.org/10.1007/978-981-19-9876-8_21

1 Introduction

The applications and services that are based on the Internet of Things (IoT) include sensor networks, healthcare systems, transportation systems, smart industrial systems, communication systems, smart cities, and manufacturing [1]. It has been suggested that the Industrial Internet of Things (IIoT) may significantly improve the characteristics of conventional industries, overcome geographical restrictions to accomplish remote monitoring, carry out autonomous production, and provide consumers with real-time information. By the year 2025, the Internet of Things (IoT) is expected to be responsible for the delivery of about 85% of all IoT devices used in health care. It has been predicted by the clever firm Tractia that the yearly revenues in this area employing blockchain technology would reach 9 billion USD by the year 2025 [2]. In the healthcare industry, Internet of Things devices are frequently employed to provide real-time services to patients and doctors [3]. Applications of IoMT in medical device technology may be found in both medical institutions and companies. On the other hand, the number of Internet-connected medical devices is expected to continue growing, which will result in a rise in both the amount and variability of the data produced. Handling considerable data traffic in the Internet of Things (IoT) has now become a serious challenge and a cause for worry. This is due to the centralized cloud-based features of IoT. As a consequence of this, issues over patient safety and confidentiality have increased, and data collecting, data ownership, location privacy, and other aspects of privacy will be put in danger. Intruders and hackers may easily attack the 5G-enabled IoMT network by making copies of data and altering the identity of healthcare equipment. As can be seen in Fig. 1, IoMT-Cloud presently suffers from a vulnerability known as a single point of failure, as well as assaults from hostile actors and privacy breaches. Trust, device identification, and user authentication are all necessities for data transfer between IoMT and the cloud if one demands to maintain network security and safely send PHR files. The typical Central Cloud service, on the other hand, is susceptible to a variety of security flaws since the nodes that make up this IoT network are connected around the clock [4]. These flaws include message manipulation, eavesdropping, and denial-of-service attacks. This poses substantial security concerns in the industrial sector since the improper use of data might lead to an inaccurate diagnosis, which in turn can lead to potentially life-threatening situations for the patients who are being observed [5].

However, there is a significant obstacle regarding the safety of IoMT devices and healthcare systems in general. It is essential that the data about patients' health care that is collected, transmitted, and stored by IoMT systems be safeguarded at every step of the process. According to the analysis that was published by 2020 CyberMDX [6], over half of all IoMT devices are susceptible to being exploited. IoMT systems are distinct from other types of systems in that they have the potential to influence patients' lives and raise privacy problems if patient names are disclosed. On top of that, the average cost of healthcare data is fifty times more than the cost of credit card information, which means that it is very valuable on the black market [7]. As a

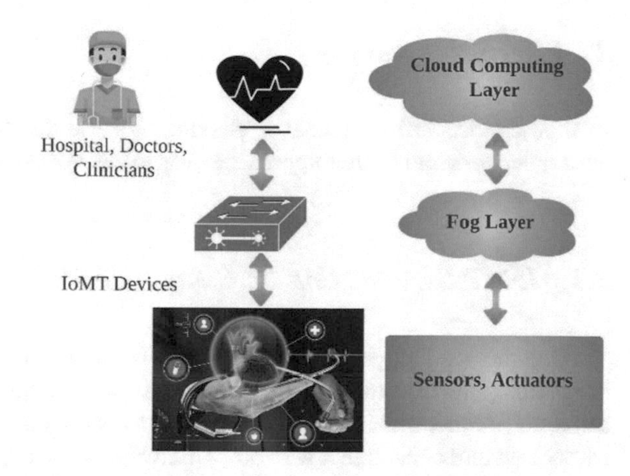

Fig. 1 Application of Fog computing

result, security is one of the primary prerequisites for the achievement of the goals of the IoMT system. These systems come with 11 different layers of security standards designed to ensure the security, integrity, and availability of data; no repudiation and authentication are terms that are often used interchangeably referred to as CIANA [8]. The conventional methods can fulfill these needs. Options for security are also available. Nevertheless, because of the power they possess usage, together with the needs of other systems, conventional solutions may fall short of providing adequate assurances of security. Researchers have instead put out several different strategies that are aimed specifically at IoMT and IoT system implementations.

The conventional method of providing medical treatment is being rapidly supplanted by IoMTs. However, when it comes to the development of IoMT devices and systems, their security needs have received comparatively little amount of attention. It may be difficult to tune conventional security solutions to work with the IoMT schemes, which may be one of the primary causes. The method of detecting attacks and mitigating their effects has been improved with the use of the Deep Graph Neural Network (DGNN). An advanced DGNN method may also be a potential solution to addressing the ongoing and upcoming concerns about the attacks, privacy, and security of IoMT [9, 10]. However, because of the problems that are already present with the IoMT system, it is of the utmost importance to learn how such techniques can be efficiently implemented to fulfill the requirements of security and privacy without negatively disturbing the quality of services of IoMT systems or the lifespan of the device. In this paper, proposed DGNN is applied in the IoMT devices to prevent attacks. The rest of the article is organized as follows: Sect. 2 consists of the literature survey and the proposed methodology of DGNN is discussed in Sect. 3. The result and discussion are shown in Sect. 4 which is followed by the conclusion in Sect. 5.

2 Literature Survey

In this section, various recent works done for the detection of attacks using graph neural networks and other approaches for IoMT are discussed.

2.1 IoMT Systems and Its Categorization

The Internet of Things (IoT) is a rapidly developing science that enables organizations, computerized machineries, smart physical things, several industrial applications, and connecting individuals to communicate with each other, capture data from each other, and exchange it with other individuals and applications through the use of networking [11]. Therefore, the use of IoT in the field of medicine and the healthcare business is recognized as the Internet of Medical Things (IoMT) [12]. It is estimated that with the successful execution of the IoMT schemes, a significant improvement will be achieved in both the effectiveness and standard of treatment as a result of the consistent innovations in IoT. These innovations include the growth of microchips, evolving 5G technologies, and biosensor architecture [13]. Despite this, it is difficult to keep a certain design as a baseline owing to the wide range of devices and the many ways in which they are used. As a result of this, several different strategies for the Internet of Things architectures and layers have been presented in the research literature [14]. These strategies include protocol-based architecture [15] and data processing phases which are recognized as cloud-base, edge, and fog strategies [16]. In this discussion, two different architectures are used by the IoMT. The initial architecture of IoMT is viewed as being made up of three primary components which are defined as:

(i) **Device layer**: It is a system/network of wireless body sensors.
(ii) **Fog layer**: Smart access points collect information and connected through Internet.
(iii) **Cloud layer**: Platform for analysis of data provides big data services and is also known as Cloud computing [17].

The sensing/device, network, and server (personal/medical) layers are the three components that make up the second architecture [18]. However, to meet the needs of a particular IoMT application, several researchers have subdivided the architecture of the IoMT into more than three layers [19]. In addition, the IoMT includes four distinct varieties of smart medical devices that may be differentiated from one another according to the area of the human body in which they are implanted [20]. The following is a quick overview of the various IoMT device types and their respective attributes, which may be found in Table 1. The Food and Drug Administration (FDA) in the United States suggests that the IoMT devices may also be categorized depending on the level of risk that they present. High-risk medical procedures include the use of implantable devices like ECG, EMG, and EEG. Implantable cardioverter-defibrillator (ICD) is an acronym for "implantable cardioverter and defibrillator."

Table 1 The categories and features of medical devices used in the IoMT system

	Category	Implementation	Illustration	Prediction of risk
Santagati et al. [22]	Implantable	Inside tissues of human	Insulin pump, deep brain implants, and pacemaker (heart)	High
Tseng et al. [23]	Wearable	On the human body	Smart watches, fitness devices	Low
Pandey and Litoriya [24]	Ambient	Outside the human body	Monitoring health of elderly people in smart home	Low
Xu et al. [25]	Stationary	Present in hospitals	MRI and CT scan (medical image processing)	Low

Because of this, they are subject to regulation and certification by the FDA. On the other hand, non-implantable gadgets like fitness trackers and smartwatches fall into the category of having minimal risk. As a result, the Food and Drug Administration does not need to certify them or regulate them [20].

2.2 Security and Privacy in the IoMT

Both conventional and zero-day attacks are capable of successfully targeting the IoMT devices. This is mostly the result of the production of the devices not including any pre-existing security processes or precautions, which is an addition to the nature of the IoT network and devices themselves. These smart devices are very small, and as a result, their processing resources and batteries are unable to do the calculation required by present stringent security measures. In addition, the IoMT network is heterogeneous, which means that it is built on diverse protocols at each layer. Because of this, a singular security solution cannot be applied to all of the devices. The number of smart medicinal IoT devices in the European Union (EU) might reach 26 million devices by 2025. Additionally, the continuous rise in the number of intelligent medical devices and the advantages brought about by the decrease in the cost of wireless sensors and concerns about privacy and safety have emerged as the primary area of focus [26]. In addition, the number of Internet-connected gadgets will ultimately lead to a rise in the amount of data created by such devices [27]. Statista estimated that IoT devices would generate around 79.4 zettabytes (ZB) by the year 2025 [28]. It is common knowledge that the data stored on IoMT devices, in addition to the devices themselves, are at significant risk of being compromised by a cyber-attack. Data exposure and privacy concerns are the two most pressing concerns in the IoMT infrastructure right now [29]. The requirements for security and privacy in IoMT are distinct from those of conventional networks, which are often summarized by the acronym CIA-triad and consist of three components that are availability, confidentiality, and data integrity. Other criteria, such as privacy and non-repudiation, are just as important for the IoMT schemes [30]. Nowadays, preventing IoMT devices from attacks is an important concern. So, to solve this problem authors used DGNN to detection of replication attacks on the Internet of Medical Things (IoMT).

3 Proposed Methodology

In Sect. 3, the proposed methodology of DGNN is discussed which further used IoMT devices for its security purpose.

3.1 Proposed Security Framework for IoMT

No one method can guarantee a safe setting for IoMT systems. As a result, a framework is suggested that is capable of safeguarding IoMT systems. Additionally, all of the security standards that are necessary for IoMT systems are met by the framework. As can be seen in Fig. 2, the framework that is based on the phases of the IoMT security model is composed of three distinct sections.

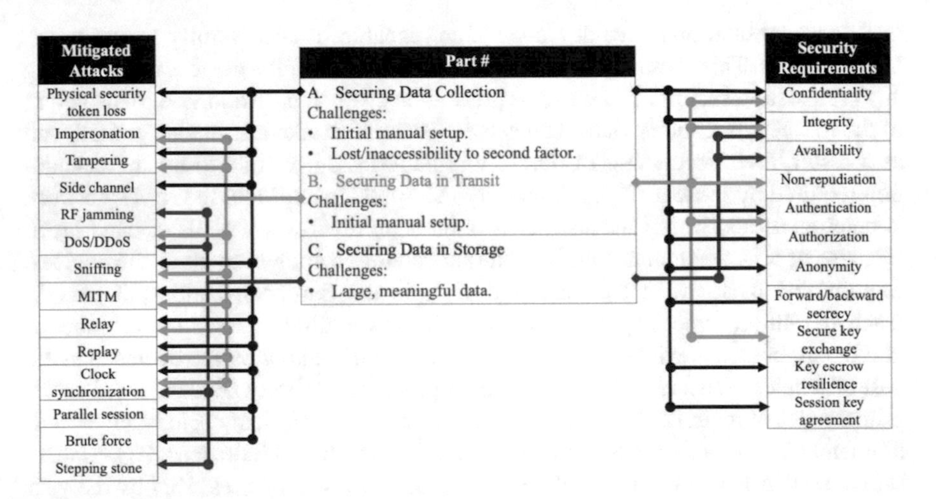

Fig. 2 Framework for security features

3.1.1 Securing Data Collection

The first thing that must be done to safeguard the patient's data collecting stage when protecting IoMT systems is to secure how other systems interface with them. Authentication methods that require a response from the user in both directions, often known as two-factor authentication, are excellent solutions that may give such protection and resistance to some of the threats outlined. Even if one of the two elements is breached, the whole protection provided by the other may still not be compromised.

Adopting the hierarchical access strategy with DGNN is an ideal way to ensure the safety of data sharing with other members of the medical staff depending on the roles they play. This method has been used in other industries, such as smart homes [31, 32]. This method calls for KGS, which may be found in the cloud layer (see Fig. 3 for an illustration of this). As a secondary consideration, biometric sensors are now the most often used method owing to the ease with which they may be used in both day-to-day lives and in times of crisis [33]. As seen in Fig. 3, these sensors are what are utilized to verify the patient's identity before allowing them access to the sensor layer nodes. Proxy-based strategies may be used to give security to existing unprotected sensors at the sensor layer [34].

If a software attack occurs while the data is being collected, DGNN and biometrics will be able to safeguard the system. However, the system requires an additional method to inform both the patient and the medical personnel in the event of a hardware assault. This will allow for the impacts of such an attack to be mitigated or eliminated. If the device in the gateway layer is unable to establish a connection to the IoMT sensor for a certain amount of time, the user as well as the attending physician should be notified (i.e., 1 h). Edge computing, also known as EC, has lately garnered a lot of interest in IoT systems since it cuts down on latency and offers strong resources to the sensors of these systems [35]. As seen in Fig. 3, the EC, which is often found at the gateway layer, has the capability of performing the duties of either acting as a gateway for IoMT smart sensors or a principal gateway for a group of secondary gateways. To meet the criteria for the patient's data integrity, its security, and confidentiality,

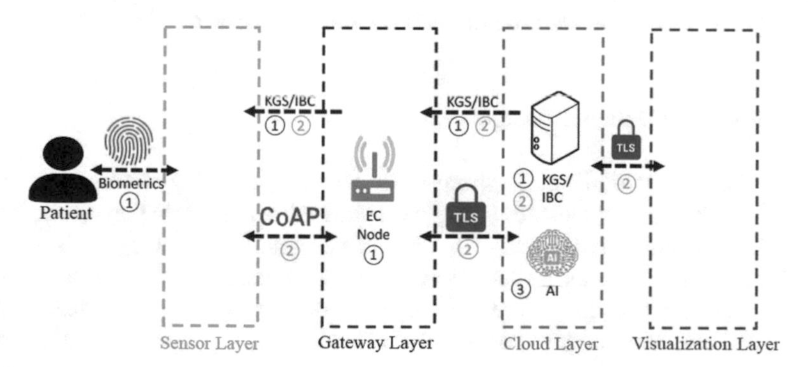

Fig. 3 IoMT secure system architecture

this model may be used to monitor the fluctuations in the readings provided by the sensors as an initial study. If one of the requirements is not met, the EC can provide an early warning to the patient of the breach of these requirements. If the patient does not answer, the system can promptly send a notification to the patient's doctor. These methods may provide the system with several benefits, including anonymity, privacy at forward–backward, backward–forward, and session-key agreement. Guaranteeing these criteria can make the system resistant to attacks. Other needs include clock synchronization, parallel sessions, and physical security tokens. Nevertheless, the approaches described in this section assume pre-shared keys or starting limitations, and the following difficulties are faced as a result.

(i) The KGS has to be manually set up at the beginning to be ready for the classified access approach.
(ii) Unfeasible in the event that the 2nd element is misplaced or cannot be accessed, particularly in times of crisis.

3.1.2 Securing Data in Transit

Adopting specific security protocols, such as the constrained application protocol (CoAP) can improve the level of security afforded by the IoMT systems when they are linked to other devices over the network [36]. As can be seen in Fig. 3, CoAP is an application protocol that was developed specifically for resource-constrained Internet of Things applications such as IoMT systems. Its purpose is to facilitate communication between the sensor and gateway levels. The other levels may be connected by using transport-layer security (TLS) version 1.3 or Secure HTTP (HTTPS) [37]. As a result, its use in IoMT systems is made easier by this trait. As seen in Fig. 3, certificate-less cryptography, which is a subset of ID-based cryptography (IBC), may be used to cut down on the administrative workload associated with managing certificates at the cloud layer [38, 39]. In certificate-less cryptography, the process of key creation is carried out by using the KGS public key in conjunction with certain initial settings to assist the nodes of the IoMT systems in the development of their keys. After that, authentication is carried out via the use of certificate-less authenticated encryption (CLAE), which does not need the administration of central keys [40].

3.1.3 Securing Data in Storage

Attacks such as DoS/DDoS, stepping-stone attacks, and RF jamming attacks are some of the kinds of attacks that may be launched against IoMT systems. These kinds of attacks aim to compromise the system's availability and integrity. The use of DGNN algorithms enables the detection of these assaults. Approaches from the field of DGNN are used in the construction of detection frameworks, with improvement techniques being superimposed on topmost of such models. For instance, DNN (deep neural network) might be used in the construction of intrusion detection models. As

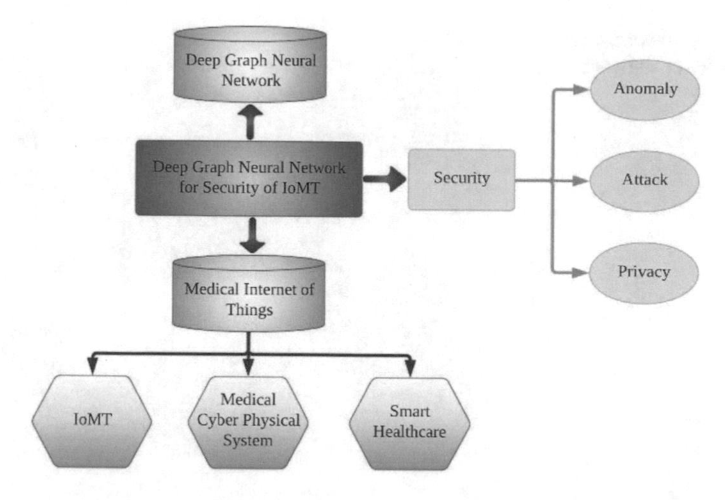

Fig. 4 Methodology for implementation of DGNN in IoMT

soon as this model identifies potentially malicious behavior, it forces the disconnection of the compromised connection to defend against the assault. When such assaults take place, the adoption of these intrusion detection/finding models at cloud layer, as illustrated in Fig. 3, may offer a caution to the network administrator, which provides early alerts from the EC nodes at gateway layer. It is highly important for DGNN approaches to collect data that is both sufficient and relevant. This is a tough phase that needs to be completed before the error rate can be reduced using these methods. By maintaining records of the existence of linked gateways or ECs, the cloud can identify any kind of breach that occurred. It is also able to provide a backup gateway, which allows it to locate alternate pathways to IoMT sensors if the primary gateway is compromised, damaged, or destroyed.

Figure 4 shows the proposed methodology for the implementation of the DGNN model for the detection/finding of replication attacks in IoMT.

3.2 *Different Categories of Attacks in IoMT Network*

Various cybercriminals employ feasible security flaws for triggering different attacks and attempts to gain access to system for stealing confidential information and to affect the device operations. These kind of attacks and attempts give rise to a large amount of resource-constrained medical smart devices. These devices are connected to IoMT-based network wirelessly and which further leads to several security vulnerabilities. This section provides a short overview of broad sorts of assaults that might be possible attacks against IoMT networks. These attacks can be classified as either physical or cyber-attacks. Some of the attacks in the IoMT network are depicted in Fig. 5.

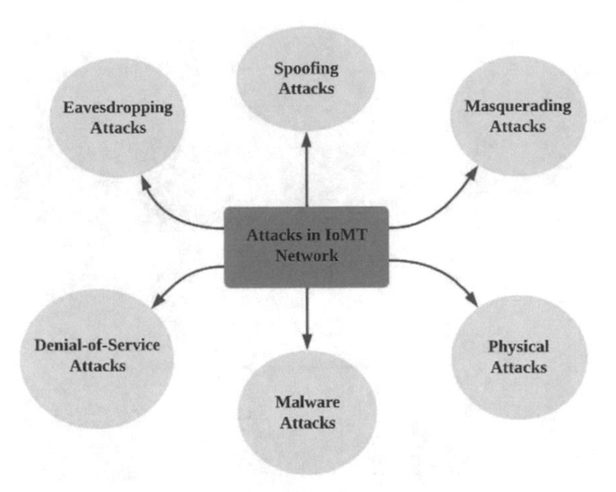

Fig. 5 Categories of several attacks in IoMT network

In eavesdropping attack, an attacker disrupts the connection among two entities like smartphones or any other connected smart devices without their permission by taking advantage of unprotected network communications and using such to their advantage. The adversary eavesdrops on the conversation to pick up any relevant information that they may store away for later use in their ruse to impersonate the claimant. Since they do not produce any irregularities in the transmission operations of the network, eavesdropping assaults are typically difficult to detect. Spoofing attack is the cautious stimulation of a process, resource, or an entity to behave inappropriately. For example, to unlawfully infiltrate a protected system, the invader can fake sender's address of the data transfer to steal its contents. It has been suggested that spoofing may also take the form of imitation and piggybacking. Masquerading attacks are also known as an impersonation attack, which is a kind of active attack where illegal entities fraudulently masquerade as legal entities for obtaining higher privilege as administrator of the system. In addition, the attacker may carry out a harmful operation by fraudulently impersonating an authorized organization to carry it out. For instance, the intruder may attain illegal rights to access confidential health information by posing as a genuine user and stealing the login credentials for the user's device. Physical attacks deal with the operations that focus on the physical layer, and therefore, the devices are under threat of compromising themselves and are known as physical attacks. For example, an adversary might modify the behavior or structure of devices that are a part of the IoMT network, which can then cause hardware failure in the system. The intrusive machine hardware attacks, device capture, side-channel attack, tampering, and reverse engineering are all examples of physical attacks. Other types of physical attacks include side-channel attacks. In malware attacks, to compromise the integrity of a system's security, an adversary will build and run malicious software or firmware on the system. It is common practice for firmware or software to be secretly inserted into alternative program to destroy data, run destructive-intrusive programs, or otherwise jeopardize the reliability, privacy and accuracy of the system's data, and applications. Malicious mobile programs, worms, root-kits, virus codes, trojan horses, and other program-based harmful entities which successfully infect a system are the example of common types of malwares that may be used to attack on a computer or smart device. The DoS (Denial-of-service) attack is a kind of attack which prevent time-critical functionality from being provisioned or prevent authorized users from accessing assets and facilities. Depending on the kind of service that is being given, time-sensitive might refer to either milliseconds or hours. This may be accomplished by overwhelming the resource-limited IoMT network with a large number of requests, which would result in congestion of the network's bandwidth.

4 Results Analysis

The outcomes that were accomplished by DGNN are detailed in Table 2, which shows the test accuracy as an average calculated throughout the first tenfold of the cross-validation process (standard deviations are detailed on the folds). The graphical presentation of the test accuracy comparison of DGNN methodology with the state-of-the-art methods is provided in Fig. 6.

Additionally, the performance attained by DGNN settings limited to only a single hidden layer, i.e., (L = 1), is also provided in the evaluation. In the same table, the accuracy observed through the literature models for several graphs is also provided. These include several different types of neural network for graphs, such as (Graph neural network) GNN [31], (Relational neural network) RelNN [41], (Deep graph

Table 2 Test accuracy of DGNN and comparison with state-of-the-art technique

	MUTAG	PTC	COX2	PROTEINS	NCCII
DGNN	$88.89_{\pm 5.24}$	$64.85_{\pm 6.73}$	$83.89_{\pm 6.54}$	$77.80_{\pm 2.94}$	$78.16_{\pm 3.94}$
$\text{DGNN}_{(L=1)}$	$87.83_{\pm 7.33}$	$64.85_{\pm 6.73}$	$82.56_{\pm 3.12}$	$77.80_{\pm 2.94}$	$78.03_{\pm 3.54}$
Scarselli et al. [31]	$80.49_{\pm 0.81}$	–	–	–	–
Blockeel and Bruynooghe [41]	$87.77_{\pm 2.48}$	–	–	–	–
Zhang et al. [42]	$85.83_{\pm 1.66}$	$58.59_{\pm 2.47}$	–	$75.54_{\pm 0.94}$	$74.44_{\pm 0.47}$
Tran et al. [43]	$87.22_{\pm 1.43}$	$61.06_{\pm 1.83}$	–	$76.45_{\pm 1.02}$	$76.13_{\pm 0.73}$
Atwood and Towsley [44]	–	–	–	$61.29_{\pm 1.60}$	$56.61_{\pm 1.04}$
Niepert et al. [45]	–	–	–	$75.00_{\pm 2.51}$	$76.34_{\pm 1.68}$
Shervashidze et al. [46]	$81.39_{\pm 1.74}$	$55.65_{\pm 0.46}$	–	$71.39_{\pm 0.31}$	$62.49_{\pm 0.27}$
Yanardag and Vishwanathan [47]	$82.66_{\pm 1.45}$	$57.32_{\pm 1.13}$	–	$71.68_{\pm 0.50}$	$62.48_{\pm 0.25}$
Vishwanathan et al. [48]	$79.17_{\pm 2.07}$	$55.91_{\pm 0.32}$	–	$59.57_{\pm 0.09}$	-
Neumann et al. [49]	$76.00_{\pm 2.69}$	$59.50_{\pm 2.44}$	$81.00_{\pm 0.20}$	$73.68_{\pm 0.68}$	$82.54_{\pm 0.47}$
Shervashidze et al. [50]	$84.11_{\pm 1.91}$	$57.97_{\pm 2.49}$	$83.20_{\pm 0.20}$	$74.68_{\pm 0.49}$	$84.46_{\pm 0.45}$

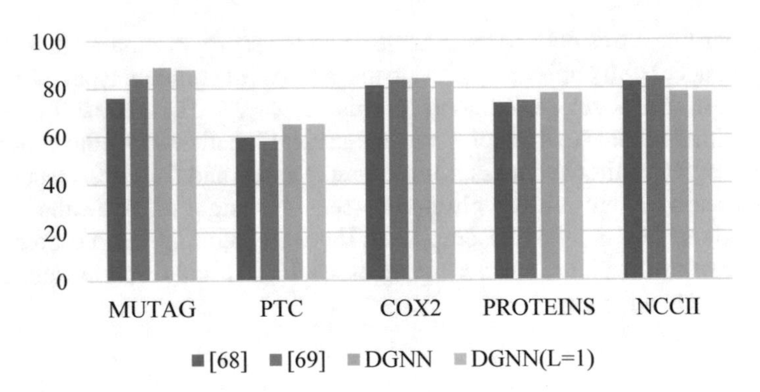

Fig. 6 Comparison of DGNN with state-of-the-art approaches in terms of accuracy

convolutional neural network) DGCNN [42], and (Parametric Graph Convolution DGCNN) PGC-DGCNN [43–45].

Several relevant models from state-of-the-art studies that use graph kernels are also taken into consideration, including the Graphlet Kernel (GK) [46], the Deep GK (DGK) [47], the Random-walk Kernel (RW) [48], the Propagation Kernel (PK) [49], and Weisfeiler-Lehman Kernel (WL) [50]. Table 2 shows the result of test accuracy of DGNN model for the detection of replication attack in application of IoMT. In Table 2, the proposed DGNN test accuracy results are compared with the state-of-the-art methods.

In order to carry out any DoS attack on a target web server, the attacking clients should have high HTTP flow all the time. The attack period should be at least 100 s, and because of the entropy rate of specified IP address is low, there are only small variations in the attack rate during the duration of the attack. The training sets obtained from the CAIDA datasets were used to train the MLP classification model. This model was tested against standard datasets such as CAIDA Ex1 [51], DARPA Ex2 generated datasets [52], and an experimentally generated dataset (see in Fig. 7). The detection accuracy is compared in these three standard datasets by implementing MLP for a sample size of 1000, as shown in Fig. 7. The experimentally produced dataset (DATA Ex3) is a grouping of several attack tools and network traffic from multiple sources. The graph shows that the accuracy of DoS attack detection increases as the sample size increases. It is observed from the experimentation that the detection mechanism is effective for large datasets for detecting DoS attacks. Blockchain technology can further be explored for securing the communication in IoMT network as implemented by one study for securing network [53]. This work can be considered for future development by using integration approaches of Artificial Intelligence and Machine learning as studied from several studies [54–56].

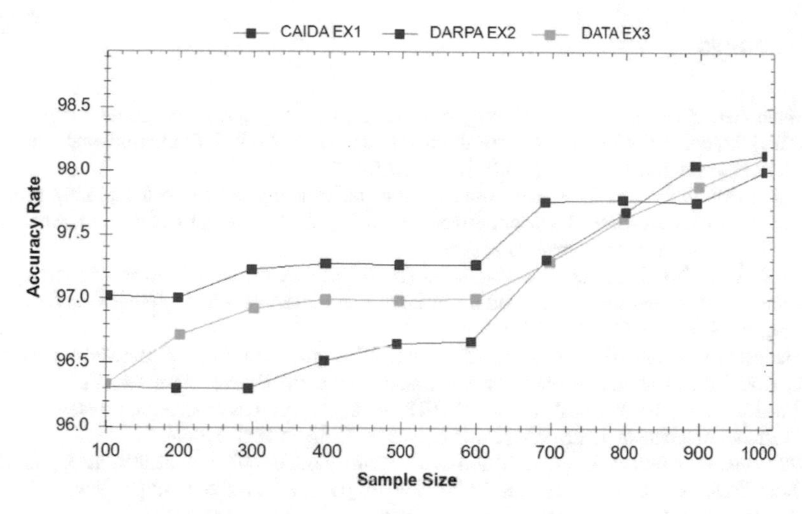

Fig. 7 Accuracy calculation of proposed method

5 Conclusion

Securing IoMT devices has become an incredibly critical concern as a direct result of the growing demand for the use of IoMT sensors to lower healthcare costs and improve the quality of treatment provided to patients. IoMT sensors, on the other hand, often have limited resources/elements, and some of those which ate already implanted need additional devices to maintain their safety. In this piece, a high-level look at the security needs is taken into consideration, covering some of the most cutting-edge security measures, and going through some of the newest forms of cyber-attacks. The authors presented a DGNN in IoMT devices to prevent attacks and secure the data in IoMT devices. This paradigm addresses all aspects of data and device security, beginning with the gathering of data and continuing through its storage and dissemination. This article analyzed the impacts of replication attack, i.e., DoS attack, and compared the important factors that influence the attack. The trained MLP along with GA algorithm is used for the detection of replication attack on the basis of several requests such as HTTP GET, entropy, and entropy variance. It is analyzed from the experimentation that there exists not much variation in the creation of HTTP GET requests during attack duration. The proposed model can detect the replication attack (DoS) with specificity rate of 57.1%, sensitivity rate of 99.7%, and accuracy of 98.4%. The proposed work is limited for the detection of replication attack only, which in future is considered for discovering a prevention mechanism for the replication attack.

Acknowledgements The research is supported by postdoctoral fellowship granted by the Institute of Computer Technologies and Information Security, Southern Federal University, project No PD/22-02-KT.

References

1. Shah AA, Piro G, Grieco LA, Boggia G (2019, June) A qualitative cross-comparison of emerging technologies for software-defined systems. In: 2019 sixth international conference on software defined systems (SDS). IEEE, pp 138–145
2. Ali A, Mehboob M (2018, September) Comparative analysis of selected routing protocols for WLAN based wireless sensor networks (WSNS). In: Proceedings of the 2nd international multi-disciplinary conference, vol 19, p 20
3. Shah AA, Piro G, Grieco LA, Boggia G (2020, July) A review of forwarding strategies in transport software-defined networks. In: 2020 22nd international conference on transparent optical networks (ICTON). IEEE, pp 1–4
4. Gatteschi V, Lamberti F, Demartini C, Pranteda C, Santamaría V (2018) Blockchain and smart contracts for insurance: is the technology mature enough? Future Internet 10(2):20
5. Jia B, Zhou T, Li W, Liu Z, Zhang J (2018) A blockchain-based location privacy protection incentive mechanism in crowd sensing networks. Sensors 18(11):3894
6. 2020 vision: a review of major IT and cyber security issues affecting healthcare. CyberMDX, New York, NY, USA. Accessed: 18 Nov. 2020. [Online]. Available https://www.cybermdx.com/resources/2020-visionreview-major-healthcare-it-cybersec-issues
7. Maddox W (2020) Why medical data is 50 times more valuable than a credit card. Accessed: 18 Nov. 2020. [Online]. Available https://www.dmagazine.com/healthcare-business/2019/10/why-medicaldata-is-50-times-more-valuable-than-a-credit-card/
8. Information Assurance, United States Naval Acad., Annapolis, MD, USA. Accessed: 18 Nov. 2020. [Online]. Available https://www.usna.edu/Users/cs/wcbrown/courses/si110AY13S/lec/l21/lec.html
9. He H, Wang J, Zhang Z, Wu F (2022) Compressing deep graph neural networks via adversarial knowledge distillation. arXiv preprint arXiv:2205.11678
10. Zhuang Y, Lyu L, Shi C, Yang C, Sun L (2022) Data-free adversarial knowledge distillation for graph neural networks. arXiv preprint arXiv:2205.03811
11. Farahani B, Firouzi F, Chang V, Badaroglu M, Constant N, Mankodiya K (2018) Towards fog-driven IoT eHealth: promises and challenges of IoT in medicine and healthcare. Futur Gener Comput Syst 78:659–676
12. Alsubaei F, Abuhussein A, Shiva S (2018, November) A framework for ranking IoMT solutions based on measuring security and privacy. In: Proceedings of the future technologies conference. Springer, Cham, pp 205–224
13. Ahad A, Tahir M, Yau KLA (2019) 5G-based smart healthcare network: architecture, taxonomy, challenges and future research directions. IEEE Access 7:100747–100762
14. Sethi P, Sarangi SR (2017) Internet of things: architectures, protocols, and applications. J Electr Comput Eng
15. Burhan M, Rehman RA, Khan B, Kim BS (2018) IoT elements, layered architectures and security issues: a comprehensive survey. Sensors 18(9):2796
16. Escamilla-Ambrosio PJ, Rodríguez-Mota A, Aguirre-Anaya E, Acosta-Bermejo R, Salinas-Rosales M (2018) Distributing computing in the internet of things: cloud, fog and edge computing overview. In: NEO 2016. Springer, Cham, pp 87–115
17. Rahmani AM, Gia TN, Negash B, Anzanpour A, Azimi I, Jiang M, Liljeberg P (2018) Exploiting smart e-Health gateways at the edge of healthcare Internet-of-Things: a fog computing approach. Futur Gener Comput Syst 78:641–658
18. Sun Y, Lo FPW, Lo B (2019) Security and privacy for the internet of medical things enabled healthcare systems: a survey. IEEE Access 7:183339–183355
19. Elrawy MF, Awad AI, Hamed HF (2018) Intrusion detection systems for IoT-based smart environments: a survey. J Cloud Comput 7(1):1–20
20. Nanayakkara M, Halgamuge M, Syed A (2019) Security and privacy of internet of medical things (IoMT) based healthcare applications: a review. In: International conference on advances in business management and information technology, pp 1–18

21. Jaigirdar FT, Rudolph C, Bain C (2019, January) Can I trust the data I see? A physician's concern on medical data in IoT health architectures. In: Proceedings of the Australasian computer science week multiconference, pp 1–10
22. Santagati GE, Dave N, Melodia T (2020) Design and performance evaluation of an implantable ultrasonic networking platform for the internet of medical things. IEEE/ACM Trans Netw 28(1):29–42
23. Tseng TW, Wu CT, Lai F (2019) Threat analysis for wearable health devices and environment monitoring internet of things integration system. IEEE Access 7:144983–144994
24. Pandey P, Litoriya R (2019) Elderly care through unusual behavior detection: a disaster management approach using IoT and intelligence. IBM J Res Dev 64(1/2):15–21
25. Xu G, Lan Y, Zhou W, Huang C, Li W, Zhang W, Che W (2019) An IoT-based framework of webvr visualization for medical big data in connected health. IEEE Access 7:173866–173874
26. Yang Y, Wu L, Yin G, Li L, Zhao H (2017) A survey on security and privacy issues in Internet-of-Things. IEEE Internet Things J 4(5):1250–1258
27. Dimitrov DV (2016) Medical internet of things and big data in healthcare. Healthcare Inform Res 22(3):156–163
28. O'Dea S (2020) Data volume of IoT connected devices worldwide 2018 and 2025. Statistica. Available at https://www.statista.com/statistics/1017863/worldwide-iot-connected-devices-datasize/
29. Gupta S, Venugopal V, Mahajan V, Gaur S, Barnwal M, Mahajan H (2020, January) HIPAA, GDPR and best practice guidelines for preserving data security and privacy—what radiologists should know. In: European congress of radiology (ECR 2020)
30. Spiekermann S (2015) Ethical IT innovation: a value-based system design approach. CRC Press
31. Scarselli F, Gori M, Tsoi AC, Hagenbuchner M, Monfardini G (2008) The graph neural network model. IEEE Trans Neural Netw 20(1):61–80
32. Xu K, Hu W, Leskovec J, Jegelka S (2018) How powerful are graph neural networks? arXiv preprint arXiv:1810.00826
33. Zheng G, Yang W, Valli C, Qiao L, Shankaran R, Orgun MA, Mukhopadhyay SC (2018) Finger-to-heart (F2H): authentication for wireless implantable medical devices. IEEE J Biomed Health Inform 23(4):1546–1557
34. Kulaç S (2019) A new externally worn proxy-based protector for non-secure wireless implantable medical devices: security jacket. IEEE Access 7:55358–55366
35. Almajali S, Salameh HB, Ayyash M, Elgala H (2018, April) A framework for efficient and secured mobility of IoT devices in mobile edge computing. In: 2018 third international conference on fog and mobile edge computing (FMEC). IEEE, pp 58–62
36. Constrained Application Protocol. Wikipedia. Accessed: 18 Nov 2020. [Online]. Available https://en.wikipedia.org/wiki/Constrained_Application_Protocol
37. Salowey JA, Turner S, Wood CA.TLS 1.3. Accessed: 18 Nov 2020. [Online]. Available https://www.ietf.org/blog/tls13/
38. ID-Based Cryptography. Wikipedia. Accessed: 18 Nov 2020. [Online]. Available https://en.wikipedia.org/wiki/ID-based_cryptography
39. Certificateless Cryptography. Wikipedia. Accessed: 18 Nov 2020. [Online]. Available https://en.wikipedia.org/wiki/Certificateless_Cryptography
40. Certificate-Less Authenticated Encryption, Wikipedia. Accessed: 18 Nov 2020. [Online]. Available https://en.wikipedia.org/wiki/Certificate-less_authenticated_encryption
41. Blockeel H, Bruynooghe M (2003) Aggregation versus selection bias, and relational neural networks. In: IJCAI-2003 workshop on learning statistical models from relational data. Date: 2003/08/11–2003/08/11, Location: Acapulco, Mexico
42. Zhang M, Cui Z, Neumann M, Chen Y (2018, April) An end-to-end deep learning architecture for graph classification. In: Proceedings of the AAAI conference on artificial intelligence, vol 32(1)
43. Tran DV, Navarin N, Sperduti A (2018, November) On filter size in graph convolutional networks. In: 2018 IEEE symposium series on computational intelligence (SSCI). IEEE, pp 1534–1541

44. Atwood J, Towsley D (2016) Diffusion-convolutional neural networks. In: Advances in neural information processing systems, vol 29
45. Niepert M, Ahmed M, Kutzkov K (2016, June) Learning convolutional neural networks for graphs. In: International conference on machine learning. PMLR, pp 2014–2023
46. Shervashidze N, Vishwanathan SVN, Petri T, Mehlhorn K, Borgwardt K (2009, April) Efficient graphlet kernels for large graph comparison. In: Artificial intelligence and statistics. PMLR, pp 488–495
47. Yanardag P, Vishwanathan SVN (2015, August) Deep graph kernels. In: Proceedings of the 21th ACM SIGKDD international conference on knowledge discovery and data mining, pp 1365–1374
48. Vishwanathan SVN, Schraudolph NN, Kondor R, Borgwardt KM (2010) Graph kernels. J Mach Learn Res 11:1201–1242
49. Neumann M, Garnett R, Bauckhage C, Kersting K (2016) Propagation kernels: efficient graph kernels from propagated information. Mach Learn 102(2):209–245
50. Shervashidze N, Schweitzer P, Van Leeuwen EJ, Mehlhorn K, Borgwardt KM (2011) Weisfeiler-lehman graph kernels. J Mach Learn Res 12(9)
51. The CAIDA "DDoS Attack 2007" Dataset. Accessed: 18 Nov 2020. [Online]. Available online https://www.caida.org/data/passive/ddos20070804_dataset.xml
52. LANDER: Los Angeles network data exchange and repository. Accessed: 18 Nov 2020. [Online]. Available online http://www.isi.edu/ant/lander
53. Zeng H, Dhiman G, Sharma A, Sharma A, Tselykh A (2021) An IoT and blockchain-based approach for the smart water management system in agriculture. Exp Syst e12892
54. Wang H, Hao L, Sharma A, Kukkar A (2022) Automatic control of computer application data processing system based on artificial intelligence. J Intell Syst 31(1):177–192
55. Sun L, Gupta RK, Sharma A (2022) Review and potential for artificial intelligence in healthcare. Int J Syst Assur Eng Manage 13(1):54–62
56. Cai Y, Sharma A (2021) Swarm intelligence optimization: an exploration and application of machine learning technology. J Intell Syst 30(1):460–469

Image Processing and Computer Vision

A Review on GIF-Based Steganography Techniques

Abhinav Chola, Lalit Kumar Awasthi, and Samayveer Singh

Abstract Data privacy has been a significant concern since the beginning of the Internet, and in the current scenario, the need and importance of data privacy are increasing exponentially. Today, a multitude of media formats including text, images, videos, Graphics Interchange Format (GIFs), etc., are available for exchanging data, and techniques of cryptography and steganography play a key role in maintaining data secrecy during such transmissions. This paper elaborates on the GIF-based techniques of steganography. It provides a detailed comparison of the working, advantages and limitations of the major approaches developed in this field in the last three decades. It provides a clear and chronological understanding regarding the evolution of GIF-based steganography methods and discusses future directions to enhance the existing techniques.

Keywords GIF · Steganography · Colour palette · Pixels

1 Introduction

The last few decades have seen a significant rise in data and its transmissions. Internet users all over the world are exchanging information through numerous media formats such as texts, videos, GIFs, audio and much more. However, the Internet is full of attackers waiting to intercept such transmissions and exploit the valuable information for their benefit [1]. To counter such threats, developers leverage cryptography and steganography techniques to provide a safe way for transferring sensitive information

A. Chola (✉) · S. Singh
Department of Computer Science and Engineering, Dr B R Ambedkar National Institute of Technology, Jalandhar, Punjab, India
e-mail: abhinavchola1@gmail.com

S. Singh
e-mail: samays@nitj.ac.in

L. K. Awasthi
Department of Computer Science and Engineering, National Institute of Technology, Srinagar, Uttarakhand, India

Y. Singh et al. (eds.), *Proceedings of International Conference on Recent Innovations in Computing*, Lecture Notes in Electrical Engineering 1001,
https://doi.org/10.1007/978-981-19-9876-8_22

over a network. The goal of both of these techniques is to ensure that only the authorized receiver must be able to access the message. Cryptography and steganography both operate to secure data but rely on different concepts of data hiding.

Cryptography requires the user to convert the given data into a form that is difficult to understand and decipher [4]. This changes the original data's appearance but ensures that valuable information in it is safe from middlemen. However, in steganography, users embed valuable information within another form of data in such a way as to keep the fundamental properties of the cover data relatively unchanged. Moreover, the data hidden inside also remain secure from attackers.

Steganography, on the other hand, is a technique that requires us to hide information (original data) within another form of data (cover data), for instance, hiding an image or text within a different image [4]. Even if steganography sounds similar to cryptography, stark differences exist when it comes to the implementation and functioning of both methodologies.

Today, numerous steganography approaches are used in different fields [10]. Some of the most popular types of steganography are as follows:

- ***Video-based Steganography***: It is a technique used for transmitting secret data securely by hiding it in a video file before transmission. This technique is gaining popularity with the increase in the number of video editing software.
- ***Text-based Steganography***: This technique uses a text file as the cover to hide and transmit data security. Essentially, this technique hides data within data and has various implementation types such as format-based, random and linguistic.
- ***GIF-based Steganography***: It involves embedding data in a GIF file for secure transmission. The sender can either hide data in the colour palette of the GIF or embed it in the GIF itself. Since GIFs are dynamic, the changes in their appearance are negligible even after hiding the data.
- ***Image-based Steganography***: This is the most common steganography technique in which an image is used to hold secret data during transmission. The sender can hide data in an image's frame and safeguard it from potential threats. Furthermore, the amount of data that a sender can hide depends on the format of the cover image.

The popularity and demand for animated GIFs have been on the rise in the last decade because of the social networking platforms and new-age online advertisements. A GIF provides you with a portable collection of image frames that together produce an animated effect. GIFs are a creative alternative for the dull and repetitive images used for the advertisement of shoes, bags and many more products. Moreover, these GIF files are also a pleasant way for people to interact on social media platforms today. The objective of this paper is to provide a review of the existing GIF-based steganography techniques.

The high-speed Internet and freely available encoders allow users to create and send GIF files within seconds. There are even dedicated websites that teach users about, sharing, searching and creating animated GIFs [8, 9]. This paper will cover the aspect of applying steganography to such GIF files.

The rest of the paper is structured as follows. Section 2 elaborates on the fundamental concepts of GIF files including the format of its storage. Afterwards, Sect. 3

lists down the performance metrics that can help in comparing techniques in the field of GIF-based steganography. Section 4 explains in detail the work done by researchers in the past to embed secret messages in GIFs. Section 5 covers the future directions to explore in the field of GIF-based steganography.

2 Structure of GIF Files

To understand the work done on Graphics Interchange Format (GIF) steganography, it is essential to first understand how a GIF (containing a single or multiple frames) is structured and saved on computer. A GIF frame is a digital image containing pixels that store its colours. Digital images are represented by using a bitmap. A bitmap (raster image) is stored as a two-dimensional array in the memory, and it can be described using the following three key characteristics:

- ***Number of Pixels***: The pixel quantity corresponds to the dimensions of the image and decides the space that an image will occupy.
- ***Resolution***: The resolution of an image depends on the size of pixels and decides the quality of the image.
- ***Colour Depth***: Colour depth represents the number of bits used to encode the colour of each pixel. A high depth implies a higher amount of colour in an image.

The above mentioned characteristics decide the bitmap for a given GIF. Moreover, a GIF file's bitmap can be designed either as a true colour image or as an indexed image.

2.1 True Colour Images

Each pixel in an image is assigned a number that represents its colour. If the image is stored in such a way that each pixel's colour is saved separately, then the image is known as a true colour image. This image has multiple pixels of the same colours and thus the same bits. Therefore, storing a true colour image contains a lot of redundant information and requires a huge space. Now, assuming that for a high-quality image 24 bits per pixel is being used, $2^{24} = 16$ million colours are possible. Storing data for each colour is not feasible. An optimal way to store an image is by using the RGB format. An RGB image is a combination of 3-bit maps (Red, Green and Blue) containing different intensities of these colours for each pixel in the original image as shown in Fig. 1.

A 24 bit pixel of an RGB Image = 8 Bits of Red + 8 Bits of Green + 8 Bits of Blue. This implies the original pixel represented as 00C8FF is 00 (Red) C8 (Green) and FF (Blue).

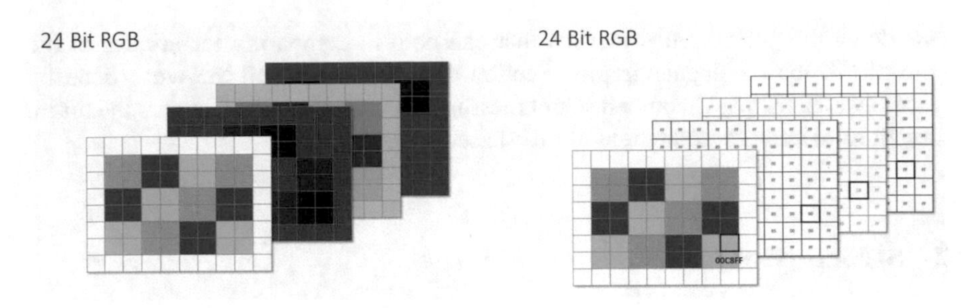

Fig. 1 RGB channels of a bitmap [7]

2.2 Indexed Images

However, image can also be saved as a bitmap using an index. This index or colour table is known as palette which contains information about colour and has an index value associated with it. This way in the image bitmap, instead of saving each colour, storing the index is sufficient. The main reason why a palette works is that human eyes cannot differentiate between colours that are close in terms of intensity. Therefore, an image does not need 16 million colours, it can have the same effect using only 256 colours. This type of image shown in Fig. 2 and most of the steganography techniques that this paper discuss is implemented on this representation only. Now since each image pixel is simply a reference to an entry in the palette (colour table), the number of bits required to represent each pixel in the image is comparatively less.

This implies a high-quality image that is available with only 256 colours instead of 16 million in the palette and only 8 bits are sufficient for its pixel representation.

Fig. 2 Image in indexed form [7]

Indexed colour

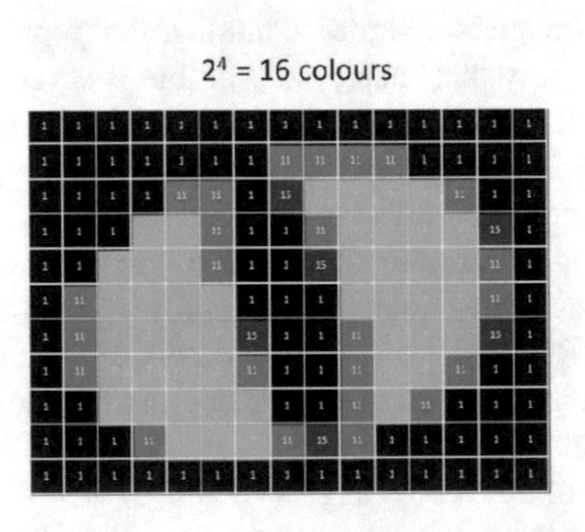

Palette

1		000000
2		0000FF
3		A2D9FB
4		87CEFA
5		F86C0F
6		74B9D8
7		00FF00
8		C9E9FD
9		FFFF00
10		5684FC
11		008C00
12		C3D895
13		FFFFFF
14		586C67
15		0C4317
16		9ED7CA

Furthermore, it will be feasible to save the palette and the image together in the memory. Today, the browsers in most computers provide the facility of easily saving image in the indexed form. They use a set of 216 colours called the web safe palette. This set of colours look exactly the same in every browser and anyone can create an image using them.

3 Performance Metrics

There are no fixed performance metrics that can be used to compare all GIF-based steganography techniques. Generally, the best measure of a technique's performance is obtained by subjective observations.

When it comes to objective measurement, different approaches of GIF steganography incorporate different metrics. However, the following two values can give great insights into a GIF-based steganography technique's performance.

3.1 Root Mean Squared Error (RMSE)

It is one of the most widely used objective metrics, and it is an ideal general purpose tool to quantify numerical predictions [9]. The Root Mean Squared Error works on the following formula:

$$\text{RMSE} = \frac{r}{\frac{1}{n}\sum_{i=1}^{n}\frac{d_i - f_i^{\prime 2}}{\sigma_i}} \tag{1}$$

Here, a_i is the general representation of observations, and n is the total number of observations used for analysis.

This formula finds the mean (average) of all errors and then calculate the square root of the resulting value. RMS error predicts accuracy and is good for cases where multiple models are to be compared on a particular variable.

3.2 Peak Signal-to-Noise Ratio (PSNR)

The peak signal-to-noise ratio (PSNR) computes the value for the ratio of the maximum value (power) possible of a signal to the value (power) of distortion affecting the quality of signal representation [15].

$$\text{PSNR} = 10\log_{10}\frac{(2^d - 1)^2 WH}{\sum_{i=1}^{W}\sum_{j=1}^{H}(p[i,j] - p'[i,j])^2} \tag{2}$$

In the above formula, *i* and *j* represent the rows and columns, respectively, from the input image. This metric is useful on the same tests images that are being subjected to different image enhancement algorithms. This way, a systematic comparison is possible, and it is easy to identify which algorithm generates the best results.

4 Literature Review

In 1998, Wang et al. proposed a novel data hiding approach through colour quantization which has shown substantially better performance as compared to the preceding techniques [14]. The technique utilizes palette-based representation of images due to the fact that natural photos can be generated at high quality using a very small percentage of the total RGB colour space. In this approach, two similar colours are merged to create a new colour to reduce the size of the colour palette. The approach uses a risk function that can select a colour with minimum distance and large capacity. This method is beneficial as the cover image's distortion is not dependent on the content material of the message. This implies that one can pre-determine the amount of rate-distortion behaviour of the Stego image even before embedding. Now, since the process of selecting the colour for quantization has a direct impact on the stereo image's quality, this paper relies on a greedy method. This greedy method generates a risk function that provides near-optimal results. However, this technique depends on colour quantization which is prone to errors. Furthermore, the technique rearranges the whole palette even if the data to be embedded is small. Rinaldi in 2016 also proposed a modified palette-based data hiding approach which uses chaos theory to embed data into GIFs [13]. It selects random frames and hides encrypted text in its pixels but not in a sequential order. Instead, the technique uses a key to select the order of embedding data to enhance security.

Fridrich in 1999 gave a new technique to hide information in the GIF using parity bits [8]. The message is hidden in binary form with one bit being placed in one pixel (its pointer to the palette). The pixels that carry message bits are selected using a random number generator called PRNG that operates using a secret key. Once a pixel is chosen at random, the palette is searched to find colours closest to that pixel. The closest colour containing the same parity (as that of the chosen pixel) is then used instead of the original colour. This technique works on the general principle that colours arranged close to each other in a luminance-ordered palette are also found close in the actual colour space. This method's advantage lies in the fact that it ensures a minimum overall change in colours of pixels as compared to its predecessor like the Least Significant Bit (LSB) of indices to a luminance-sorted palette method. Moreover, experiments proved that the new technique generates four times less distortion to the cover image compared to the EZ Stego method. The maximal colour change is 45 times smaller for the new technique than that of EZ Stego.

A similar approach relying on random pixel selection was proposed by Imran et al. in [2]. The technique uses a hashing key to select random pixels of a GIF to hide data. The approach used perfect hashing function and can work with both large- and small-sized GIFs. However, both of these techniques are constrained by the size of input image.

Wul et al., in 2004, proposed a message embedding technique that requires slicing the animated GIF into frames [9]. Then using the colour palette of the individual frame, the technique decides which bits will be chosen as the message carriers. The frame selection process can be both random and sequential, and the technique also relies on a threshold while deciding the multiple bits. The colour of the cover GIF is maintained by ensuring the prefix values of the pixel (that carries message bits) is similar to its neighbouring colours. The advantage of this technique lies in the fact that the pixels can carry more than one bit in every colour. Moreover, the multibit assignment technique is capable of handling colour changes in the GIF as it considers the distance colour value of neighbours while assigning multi bits. This technique, however, is complex to implement and is efficient only in the case of RMS error. Similarly, in 2018, Ratan et al. proposed a Pixel Value Differencing (PVD) technique, which extracts frames and hides data in the bits of overlapping pixels [6]. The clear advantage of the PVD-based method was to utilize maximum pixels and increase the size of secret text that can be hidden. However, the technique is complex in implementation and is applicable only on black and white GIFs.

Larbi et al., in 2014, devised a method which operates on the GIF image palette and the intermediate frames [12]. Moreover, this method uses an entropy-based function to decide the message bits to be embedded for each frame. The method decomposes a dynamic GIF into static frames and then deploys new distortion functions for each of them. These functions are then used to compute the initial cost of embedding message bits in each frame. After that, an adjustment factor is calculated that takes into consideration the GIF image palette and the adjacent frames. Using this function, the message is partitioned and embedded into each frame. This new method for dynamic GIFs operates on the STC framework. The approach rearranges the palette and computes the amount of modification required to adjust each RGB colour channel. Moreover, due to the strong correlation among interframes, the method utilizes the difference in motion of adjacent frames calculates the embedding costs. This technique, however, modifies the GIF file irreversibly.

In 2020, Sheik et al. proposed a technique to manipulate the transparent pixels in animated GIFs to embed messages secretly [15]. The technique involves inserting a new frame between two GIF frames and modify certain pixel bits to hide data. An important aspect is the animation duration, which should not increase due to the new frame. Therefore, the delay duration of each frame is adjusted so that the cover GIF and the Stego GIF have the same duration. This technique has an advantage in terms of reversibility. This implies that even after adding the extra frame, it can restore the GIF to its original state. Moreover, this method is scalable and also suited for applications of data hiding like fragile watermarks. When a new frame is added, GIF's size increases and thus the reverse zero-run length (RZL) encoding technique is implemented for compression. This technique excels only when used under small thresholds.

Generally, data hiding finds its main applications in security issues, such as copyright, protection and authentication. However, in 2020, Lin et al. provided a different perspective by using the data hiding mechanism to transmit audio using a GIF [11]. This new technique presented an approach to insert audio tracks into dynamic GIFs via data hiding concepts. It is important to note that the key principle of using data hiding is to ensure that the GIF's bitmap format does not lose its portability after adding sound. Moreover, the inserted bits are not a part of any secret message, and the method just borrows data hiding technique to use the GIF as a carrier. The Fridrich palette-based algorithm is modified to achieve the desired goal in this paper. In this approach, the sender embeds sound bits in the pixels of GIF frames using Fridrich's data hiding method. The receiver simply extracts the audio data from parity bits of the RGB channels of the GIF frames' pixels. This technique is original and works on low complexity. Moreover, it can add audio data to animated GIFs without causing a noticeable increase in the file size. Its another advantage is that it avoids the development of any data containers which means the cover GIF is not affected much. However, the method is expensive in terms of time and complexity.

Wong et al., in 2020, suggested an iterative method that works to minimize the RMS error in GIFs created by the embedded message [16]. The method first iteratively transforms both the palette and image pixels of the original dynamic GIF. Afterwards, it uses Fridrich's method for embedding the secret text in the modified GIF. The key principle behind this technique is to substitute the "less important" colours in the image with the adjacent ones to create empty entries in the palette. These entries are filled with the required "more important" colours. This in turn greatly diminishes the Root Mean Square (RMS) error in the embedding process. The advantage of this approach is that it can act as the pre-processing process for many steganography methods. Moreover, it can also work as a replacement process that removes a specified palette entry and replaces it with a new and more important colour. The method also calculates the effect of this colour transition on the GIF. Only when the substitution decreases the RMS value, does the method complete the colour switch. However, the technique does not exploit its parameters to their full extent.

Basak et al., in 2021, proposed a hybrid approach of embedding encrypted data in a GIF file using SHA-1 (secure hash algorithm) and least significant bits [5]. The technique uses the ASCII values of the input data and encrypts it using SHA-1 before initiating the embedding process. It splits a GIF into frames which are further converted into overlapping pixel blocks in which data is finally stored. During the data extraction phase, SHA code comparison verifies whether the data extracted is the same as the original data. The technique is beneficial for hiding a large stream of data. Moreover, its Pixel Differencing Histogram (PDH) provides a comprehensive overview of its performance. The proposed method can find applications in the modern-day promo code sharing tools. Its highly secure SHA-1 can cater to the needs of businesses trying to share classified information via animated promo codes and gift coupons. The proposed technique offers a higher bits per pixel (bpp) than its peers but is complex in implementation.

Balagyozyan et al., in 2021, presented a novel approach to hiding secret data in frames by randomizing the pixel selection process [3]. The technique deploys a mathematical function to choose random pixels from GIF frames with the objective of deceiving third party attackers. The technique hides data in the least significant bits of the mathematically selected pixel and uses overlaying to optimize data hiding. The proposed technique is fit for use in storing geo-location information of minerals. However, the security of this approach depends on the complexity of its mathematical function. This implies, to increase its security, the implementation complexity must also increase.

Table 1 provide a detailed comparison on the current techniques of GIF-based steganography.

5 Future Direction

The previous section discussed the working and benefits of the research done in GIF-based steganography. The following future directions can potentially enhance the capabilities of the current techniques to hide data in GIF files:

- The method provided by Fridrich can be improved by pushing two similar colours close enough so that the change is not visible [8]. Moreover, a better risk function can be derived using the optimal solution approach rather than a greedy method.
- Instead of using random pixels in the technique proposed by Wang et al., a function can be developed that may select the pixels that will have better probabilities of retaining the cover image's outlook [14]. This can be achieved by assigning weights to pixels depending on their transformation once a message bit is embedded. Moreover, factors like local variance in neighbouring pixels can also be used to select the best possible pixels, making changes to pixels in areas of uniform colour.
- In future, the technique given by Lin et al. can be enhanced to work on ways to adaptively select separate threshold values for different regions [12]. It can also incorporate the magnitude of local fluctuation to provide a better imperceptibility.
- In the technique devised by Wong et al., the current payload allocation algorithm can be enhanced to achieve the same level of results but with lesser time and effort [15].
- In the technique proposed by Larbi et al., more sophisticated embedding strategies need to be developed so that image quality can be preserved better at the cost of higher algorithmic complexity [11].
- The proposed scheme by Wong et al. can be improved by enhancing the iterative process so that the same results can be achieved with less complexity [16].

Table 1 Comparison of several GIF-based steganography techniques

S. No	Author	Method	Limitation	Results
1	Wang et al. [14]	The technique implemented data hiding in the GIF pixels and merged two similar colours to create a new colour to reduce the size of the palette. The technique used a risk function to decide which colours should be merged together	Quantizing two colours into one if not done correctly can raise suspicion from an attacker's point of view. If the secret message has a smaller cover image's capacity, it creates a unique situation. For example, if only a 1-bit long-secret message has to be embedded in an image that can accommodate hundreds of bits this approach will still rearrange the whole palette even when such reordering is not required	PSNR 36.07%
2	Fridrich et al. [8]	The technique is known as the "Fridrich" which selects random pixels and then finds parity bits of close colours. These parity bits then act as the message carriers. The technique works on the principle that colours arranged close to each other in a luminance-ordered palette are also found close in the actual colour space	The method selects random pixels to implant the secret message. This may not offer the best possible pixels at all times. Moreover, it is constrained by the number of pixels. This approach can hide a message in a GIF only if its size is less than the number of pixels in the image. Embedding larger messages will therefore create visible changes to the image	PSNR 27.31%
3	Wu et al. [9]	This modified Fridrich's method embeds secret text in a GIF. It also minimized the RMS error in the GIF file caused due to data hiding	The method only caters to removing the RMS error and ignores other errors that may be generated due to the data embedding process. Moreover, the cost (complexity) of this method is high	RMS error 2.22%
4	Larbi et al. [12]	This technique implemented a modified Fridrich palette-based algorithm to hide an audio message inside a GIF file	This algorithm relies on a sound encoder to compress the sound file before embedding. This is necessary to ensure that the Stego GIF has not exceeded the maximal embedding capacity and to keep the image distortion minimal. Moreover, the embedding procedure modifies the GIF irreversibly	PSNR 36.00%

(continued)

Table 1 (continued)

S. No	Author	Method	Limitation	Results
5	Renaldi [13]	The technique uses EZ Stego approach which works by sorting the GIF pixels in the order of luminance and hiding data in the LSB bits. It further adds a logistic map-based encryption	The technique attains extra security by adding an encryption mechanism which adds extra over-head. Moreover, the approach only works for static GIF and may not be feasible for dynamic GIFs	PSNR 38.95%
6	Ratan et al. [6]	This technique leverages pixel value differencing and hides data in pixel pairs in a GIF. It implements hashing to identify the number of message bits that the GIF can store. Its advantage lies in its capability to hide longer messages in a GIF	The method only works for black and white GIFs and since it operates using hash function, the complexity of its implementation is high	NA
7	Lin et al. [11]	The approach rearranges the palette and computes the amount of modification required to adjust each RGB colour channel using STC frameworks	The proposed method involves complex calculations regarding distortion function design. Moreover, for high-quality images, the current method's execution time will be high as it operates on each channel separately	PSNR 42.13%
8	Wong et al. [16]	This technique Inserts new frames in the GIF file and hides data using transparent pixels in the new frames	This process is complex in implementation. Moreover, the delay parameters between frames have not been utilized fully in the current approach	PSNR 30.02%
9	Basak et al. [5]	The technique uses the ASCII values of the input data and encrypts it using SHA-1 before initiating the embedding process	The proposed technique offers a higher bits per pixel (bpp) than its peers but is complex in implementation	PSNR 41.00%
10	Balagyozyan et al. [3]	This technique relies on overlaying and mathematical pixel selection to hide data securely	Its complexity increases with an increase in its security	NA

6 Conclusion

Maintaining data privacy over transmissions is a major area of study and practical application in the current information age. The fields of cryptography and steganography are constantly evolving to overcome the threat posed by attackers and safeguard the data carried in images, GIFs and other such media. This paper provided some insight to the fundamental concepts of GIF-based steganography. It also provided a detailed discussion on the significant research work that has been done to embed data into GIF files without changing its appearance. Furthermore, the tabular comparison summarized the key aspects of the existing research work with the aim to offer a means to analyse these techniques and understand their shortcomings. The paper also elaborated on the possible future directions that researches can explore to further enhance the existing techniques. The world is changing at a fast pace and concerns about data security and privacy are prevalent. In such a scenario, an optimal way of hiding data in GIF files can provide users with a convenient and secure way of exchanging information.

References

1. Anton AI, Earp JB, Young JD (2010) How internet users' privacy concerns have evolved since 2002. IEEE Secur Priv 8(1):21–27
2. Bajwa IS, Riasat R (2011) A new perfect hashing based approach for secure stegnograph. In: 2011 sixth international conference on digital information management. IEEE, pp 174–178
3. Balagyozyan L, Hakobyan R (2021) Steganography in frames of graphical animation. In: E3S web of conferences, vol 266. EDP Sciences, p 09006
4. Barakat M, Eder C, Hanke T (2018) An introduction to cryptography. Timo Hanke at RWTH Aachen University, pp 1–145
5. Basak RK, Chatterjee R, Dutta P, Dasgupta K (2022) Steganography in color animated image sequence for secret data sharing using secure hash algorithm. Wirel Pers Commun 122(2):1891–1920
6. Basak RK, Dasgupta K, Dutta P (2018) Steganography in grey scale animated GIF using hash based pixel value differencing. In: 2018 fourth international conference on research in computational intelligence and communication networks (ICRCICN). IEEE, pp 248–252
7. Drumm K True colour and indexed colour bitmaps. https://youtu.be/cwHPuU3sHOk?t=309
8. Fridrich J (1999) A new steganographic method for palette-based images. In: PICS, pp 285–289
9. Juzar MT, Munir R (2016) Message hiding in animated GIF using multibit assignment method. In: 2016 international symposium on electronics and smart devices (ISESD). IEEE, pp 225–229
10. Kaur N, Behal S (2014) A survey on various types of steganography and analysis of hiding techniques. Int J Eng Trends Technol 11(8):388–392
11. Larbi SD, Zaien A, Sevestre-Ghalila S (2016) Voicing of animated GIF by data hiding—a technique to add sound to the gif format. Multim Tools Appl 75(8):4559–4575
12. Lin J, Qian Z, Wang Z, Zhang X, Feng G (2020) A new steganography method for dynamic GIF images based on palette sort. Wirel Commun Mob Comput 2020
13. Munir R (2016) Application of the modified ezstego algorithm for hiding secret messages in the animated GIF images. In: 2016 2nd international conference on science in information technology (ICSITech). IEEE, pp 58–62

14. Wang X, Yao T, Li CT (2005) A palette-based image steganographic method using colour quantisation. In: IEEE international conference on image processing 2005, vol 2. IEEE, pp II–1090
15. Wong K, Nazeeb MN, Dugelay JL (2020) Complete quality preserving data hiding in animated GIF with reversibility and scalable capacity functionalities. In: International workshop on digital watermarking. Springer, pp 125–135
16. Wu MY, Ho YK, Lee JH. (2003) A method to improve the stego-image quality for palette-based image steganography. In: International workshop on digital water-marking. Springer, pp 483–496

Optimization and Design of Efficient D Flip-Flops Using QCA Technology

Naira Nafees, Suhaib Ahmed, and Vipan Kakkar

Abstract Digital circuits designed at a nano-scale with the growing use of technologies like quantum-dot cellular automata (QCA) are beneficial over the orthodox CMOS regime in context with certain parameters like lower power consumption, high speed and density of the device. This paper presents a literature survey of the D flip-flop designs in QCA. This is followed by proposed optimized circuits for these existing D flip-flop designs. The comparison is done between existing designs in literature and their optimized designs on the basis of parameters like number of cells used, area occupied by the cells and latency. After analysing the performance of the proposed design, it is found that it achieves performance improvements up to 62.71% over previous designs.

Keywords Flip-flop · Quantum computing · Nanoelectronics · Quantum dot cellular automata · Energy dissipation

1 Introduction

The technology for the fabrication and manufacturing of semiconductor devices has gone through an immense change in the last few decades. There are some applications which require functionalities like high speed and ultra-low power thus causing CMOS scaling to reach an impasse [1, 2]. This inspires researchers to move their focus

N. Nafees · S. Ahmed · V. Kakkar
School of Electronics and Communication Engineering, Shri Mata Vaishno Devi University, Katra, India
e-mail: vipan.kakar@smvdu.ac.in

S. Ahmed
Department of Electronics and Communication Engineering, Baba Ghulam Shah Badshah University, Rajouri, India

Present Address:
S. Ahmed (✉)
Department of Computer Science and Engineering, Chitkara University Institute of Engineering and Technology, Chitkara University, Rajpura, Punjab, India
e-mail: sabatt@outlook.com

Y. Singh et al. (eds.), *Proceedings of International Conference on Recent Innovations in Computing*, Lecture Notes in Electrical Engineering 1001,
https://doi.org/10.1007/978-981-19-9876-8_23

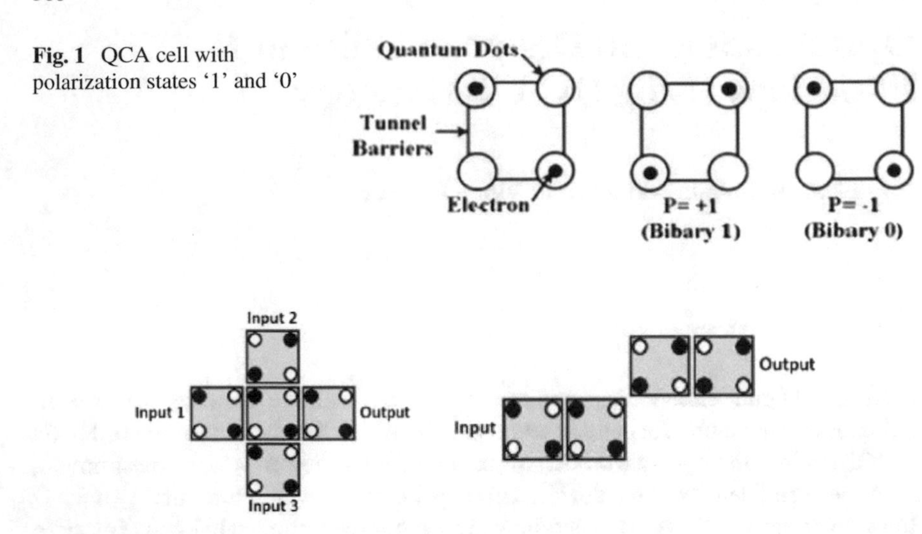

Fig. 1 QCA cell with polarization states '1' and '0'

Fig. 2 QCA designs of **a** three input majority voter and **b** inverter

towards the substitute technologies, and QCA nanotechnology is one of the promising technologies in this area [3, 4]. The fundamental principle for the flow of information in QCA technology is quantum mechanical tunnelling [5–8]. The QCA paradigm has some basic buildings which are incorporated in every circuit designed in it. QCA cell consists of four quantum dots which contain two mobile electrons [9–12]. The electrons try to occupy the antipodal sites due to the mutual columbic repulsion [13–15]. A QCA cell is defined by its polarization state. Therefore, two polarization states exist which define either logic '0' or logic '1' [16–19]. A basic QCA cell along with two polarization states is shown in Fig. 1. Apart from a basic QCA cell, a majority voter gate and an inverter are some fundamental structures to form circuit designs in QCA as shown in Fig. 2.

In QCA, clocking plays a very substantial part for the proper working of the circuits. It is not only responsible for the control of flow of data/information but also for supplying power to the cells [20–22]. The clocking is applied in four zones to a circuit in QCA. Each of these zones is represented by a different colour. Each clock zone has four distinct phases, i.e. switch, hold, release and relax [4, 12, 23, 24]. These different clocking zones in QCA are shown in Fig. 3.

The key contributions of this work are as follows:

- A survey of current D flip-flops based on QCA.
- Optimization and design of existing D flip-flops.
- An improved D flip-flop and its QCA circuit has been designed.

The subsequent sections of this paper are organized follows: the literature survey of current D flip-flop QCA designs is given in Sect. 2 followed by the optimized designs and their performance comparison in Sect. 3. The proposed improved design

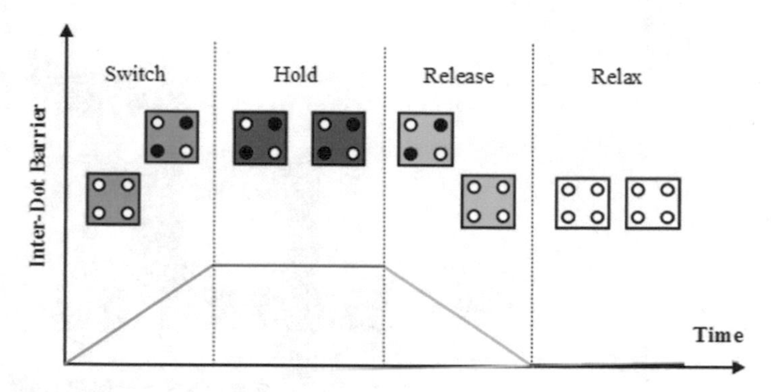

Fig. 3 Graphical representation of QCA clock phases

of D flip-flop along with its QCA circuit implementation is presented in Sect. 4. The concluding remarks are presented in Sect. 5.

2 Existing D Flip-Flop Designs

A flip-flop is a sequential circuit with output dependency on present input as well as past output. It is a memory unit which has the capability of storing 1 bit of binary information [12, 25]. For the realization of more complex sequential circuits, flip-flops and memory cells act as the characteristic building blocks. Figure 4 shows the design discussed in [26]. This D flip-flop design uses 59 cells with an area of 0.075 μm^2. It has a delay of 1.75.

Figure 5 shows the D flip-flop design proposed in [27]. It has 56 number of cells and 0.06 μm^2 area occupancy with latency of 2.5. Figure 6 shows the DFF design put forward in [28]. This design consists of 48 cells in its QCA layout and has a delay of 0.75, 0.014 μm^2 cell area and 0.03 μm^2 as total area. The flip-flop is being reset with input 'clear'. When both the enable inputs 'clock' and 'clear' are 1, then only input will be reflected at the output of flip-flop. Further, 1343 cells are utilized in 4-bit shift register circuit while occupying approximately 1.67 μm^2 total area with the use of their D flip-flop design. For the utilization of QCA arrangement of 8-bit universal shift register, 2801 cells are required, and it occupies 3.52 μm^2 area. Both these shift registers use four and eight 4 to 1 multiplexers, respectively. The proposed 4 to 1 multiplexer has 101 cells in its QCA design and occupies an area of 0.10 μm^2. It has a delay of 1.25.

In [29], a phase-frequency detector is devised using a new DFF which has a reset pin. First, the design of the latch is put forward which is then easily converted to a flip-flop. This is done using NNI gate and inverter circuit. QCA design for the latch comprises of 22 cells with area 0.018 μm^2, and latency is 0.5. Further D flip-flop which has the capability to reset is devised and a design which lacks reset ability is also purported. D flip-flop without reset ability has cell count of 44, area 0.046 μm^2

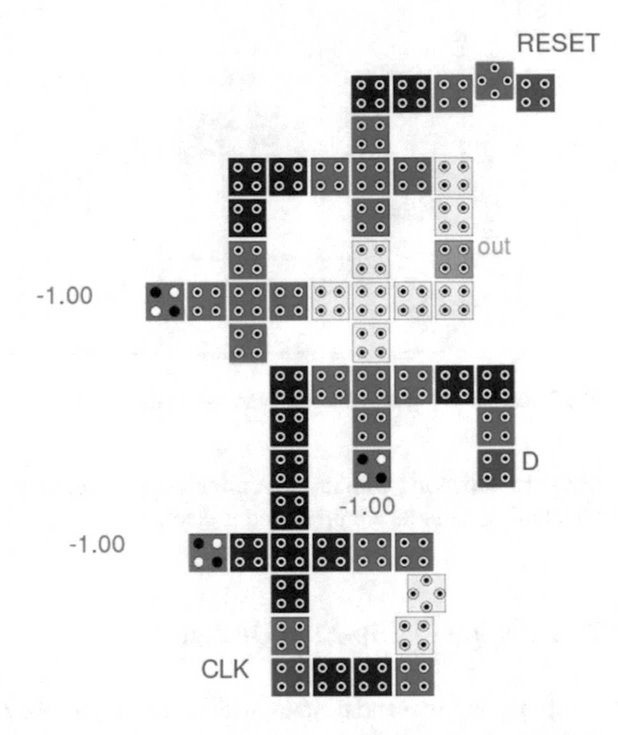

Fig. 4 QCA design 1 [26]

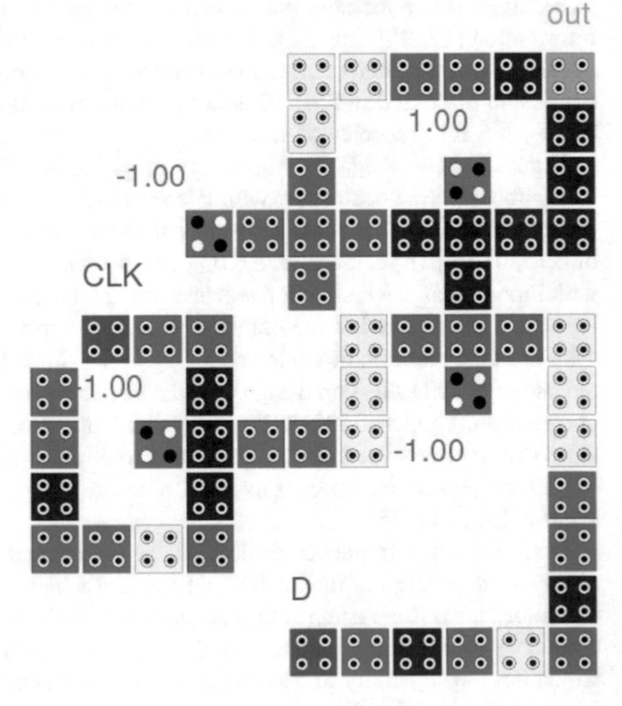

Fig. 5 QCA design 2 [27]

Fig. 6 QCA design 3 [28]

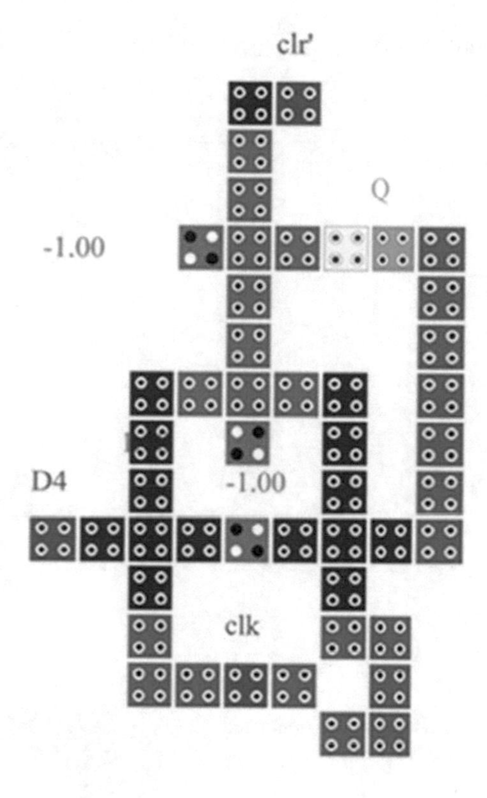

and latency as 1. D flip-flop which has reset ability has 48 cells with area 0.05 μm^2 and latency 1. Further, the proposed QCA layout of PFD utilizes 141 cells with a delay of 2 and occupies an area of about 0.17 μm^2. Figure 7 shows the design which is presented in [29].

3 Optimized Designs and Performance Comparison

The optimized design of D flip-flop design 1 is shown in Fig. 8. The number of cells is reduced to 46 from 59 along with the reduction in latency from 1.75 to 1.25. Figure 9 shows the improved version of second D flip-flop design. The number of cells required in this design is reduced to 46 from 56 along with the significant decrease in latency from 2.5 to 1.5. Figure 10 shows optimized design for existing design 3. It consists of 31 cells with significant reduction in cell area and total area occupied in comparison to its previously existing counterpart. Another optimized design of D flip-flop is presented in Fig. 11. It is the improved version of existing design 4. The number of cells required in this design is reduced to 33 from 44 resulting in the significant decrease in area and increase in overall performance.

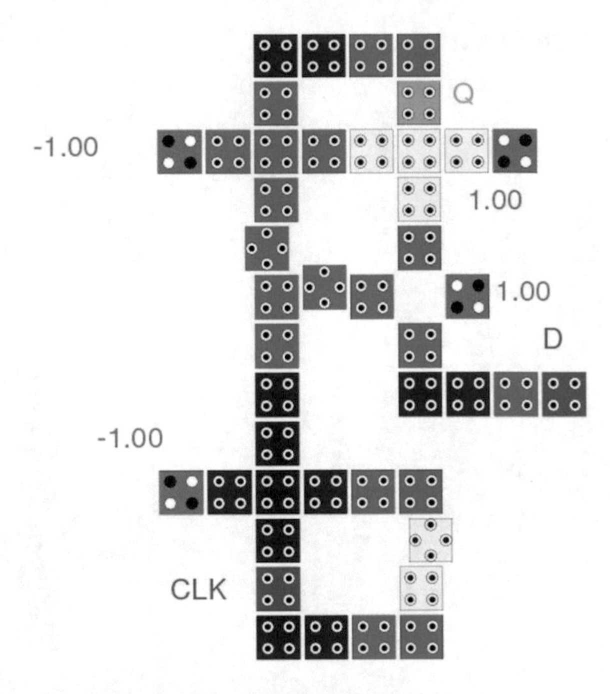

Fig. 7 QCA design 4 [29]

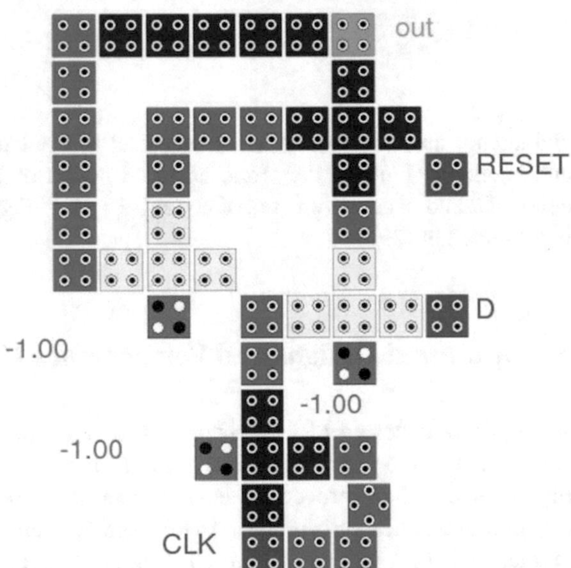

Fig. 8 Proposed optimized design 1

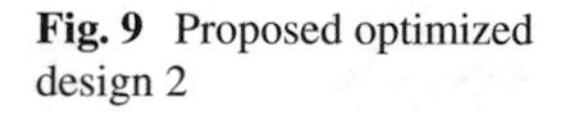

Fig. 9 Proposed optimized design 2

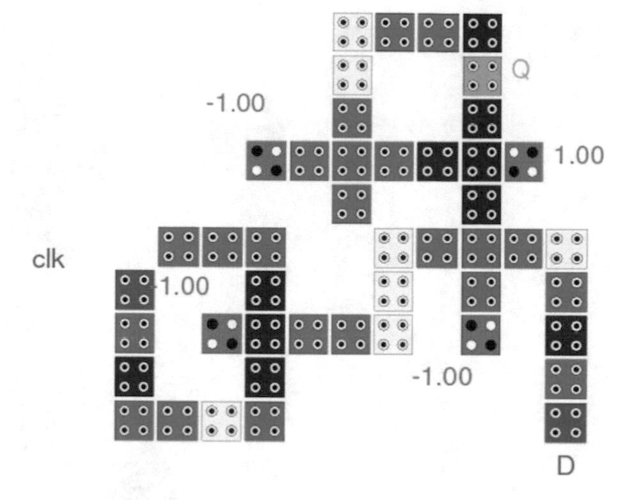

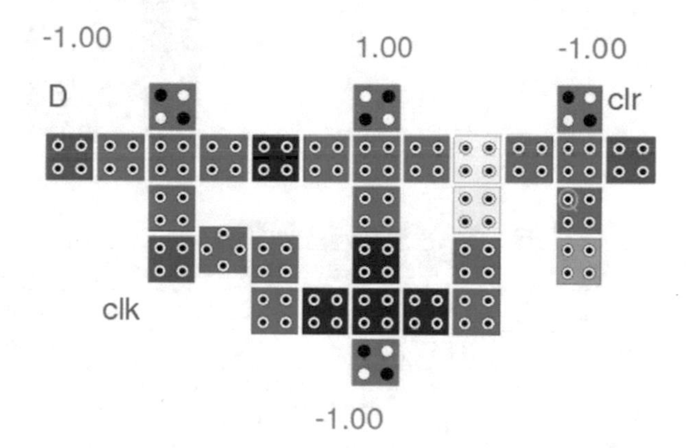

Fig. 10 Proposed optimized design 3

Table 1 gives performance analysis of various existing D flip-flops in the literature as discussed in previous section. Table 2 gives performance analysis of optimized D flip-flop designs and their overall improvement in performance. The anticipated designs are compared in terms of number of cells used, latency and area usage.

4 Proposed Design of D Flip-Flop

Figure 12 shows the logic diagram of proposed design of D flip-flop. This design follows the characteristic equation given as

$$Q(n) = D.\text{Clock} + Q(n-1).\text{Clock} \tag{1}$$

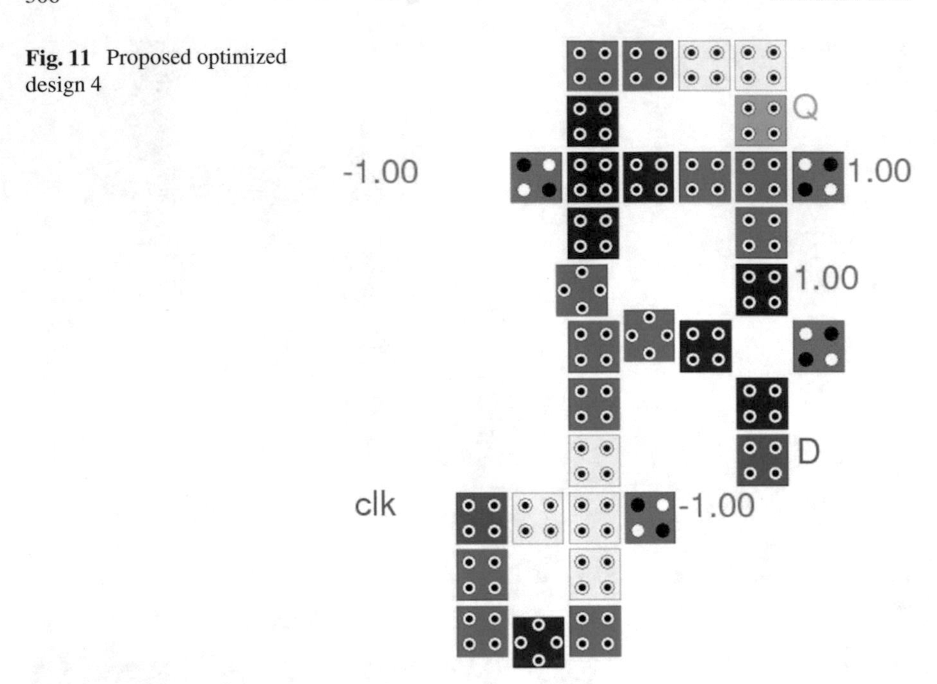

Fig. 11 Proposed optimized design 4

Table 1 Performance of existing D flip-flops

D flip flop	Cells utilized	Total area (μm^2)	Cell area (μm^2)	Latency
Design 1 [26]	59	0.075	0.0191	1.75
Design 2 [27]	56	0.06	0.0181	2.5
Design 3 [28]	48	0.05	0.0155	0.75
Design 4 [29]	44	0.046	0.0142	1.0

Table 2 Performance of optimized D flip-flops designs

D flip flop	Cells utilized	Total area (μm^2)	Cell area (μm^2)	Latency	Improvement from existing design (%)
Design 1 [26]	46	0.044	0.0149	2.25	22.03
Design 2 [27]	46	0.043	0.0149	1.5	17.85
Design 3 [28]	31	0.028	0.01	1.25	35.41
Design 4 [29]	33	0.036	0.0106	1.75	25

where D and clock are inputs, $Q(n)$ is the present output and $Q(n-1)$ is the previous output. The design uses 3 majority voter gates QCA design. The output of first majority voter Maj1 is 'D.Clock' and the output of the second majority voter is

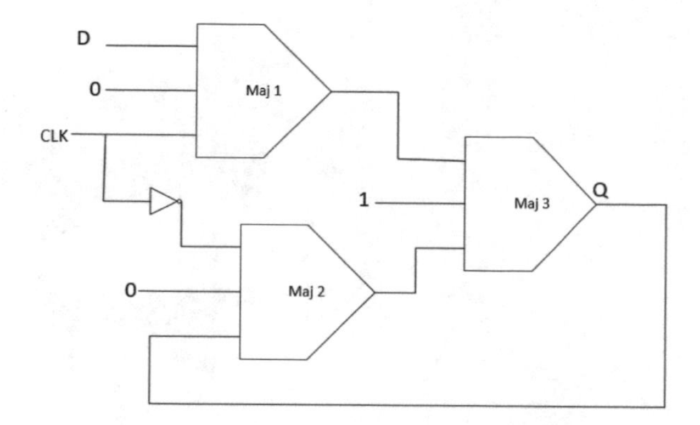

Fig. 12 Logic diagram of proposed D flip-flop

'$Q(n-1)$.Clock'. The QCA implementation of the proposed design and its simulation graph is shown in Fig. 13. The rotated cell inverter is used to obtain the complementary clock signal. It utilizes 22 cells in its QCA design and occupies 0.016 μm^2 as total area. It has a latency of 0.75 with cell area of 0.0071 μm^2. The simulation is done using QCADesigner tool [30].

5 Conclusion

As the established approaches of circuit designing have begun to show their limitations, therefore, alternate approaches are to be looked forward to. QCA technology has abundant advantages over conventional technology such as low power dissipation, operating at high switching frequencies. A survey of existing designs of D flip-flop in QCA technology is done, and these designs are compared on the basis of various parameters like cell count, area and delay. Then, these designs are optimized, and a performance comparison with their already existing counterparts is done on the basis of cell count and total area utilized. Additionally, we propose an improved D flip-flop and its implementation using QCA. The proposed design has been found to be better when it comes to parameters like the number of cells and the total area, as well as latency and latency time. In this way, it is possible to design complex sequential circuits of higher order more efficiently.

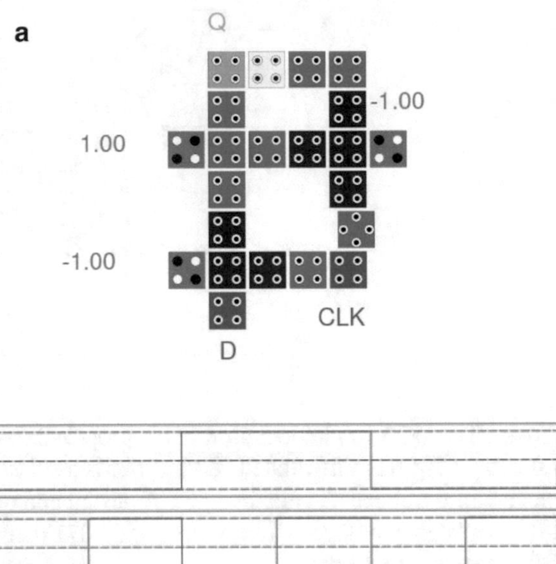

b

max: 1.00e+000
CLK
min: -1.00e+000

max: 1.00e+000
D
min: -1.00e+000

max: 9.85e-001
Q
min: -9.85e-001

max: 9.80e-022
CLOCK 0
min: 3.80e-023

max: 9.80e-022
CLOCK 1
min: 3.80e-023

max: 9.80e-022
CLOCK 2
min: 3.80e-023

max: 9.80e-022
CLOCK 3
min: 3.80e-023

0 1000 2000 3000 4000 5000 6000 7000 8000 9000 10000 11000

Fig. 13 **a** Proposed QCA circuit of D flip-flop. **b** Input–output waveforms of the proposed design

References

1. Compano R, Molenkamp L, Paul D (2000) Roadmap for nanoelectronics. European Commission IST Programme. Future and Emerging Technologies, pp 1–81
2. Wolkow RA, Livadaru L, Pitters J, Taucer M, Piva P, Salomons M, Cloutier M, Martins BV (2014) Silicon atomic quantum dots enable beyond-CMOS electronics. In: Field-coupled nanocomputing. Springer, pp 33–58
3. Naz SF, Ahmed S, Ko SB, Shah AP, Sharma S (2022) Fields: QCA based cost efficient coplanar 1 × 4 RAM design with set/reset ability. Int J Numer Model Electron Netw Dev 35:e2946
4. Naz SF, Shah AP, Ahmed S, Girard P, Waltl M (2021) Design of fault-tolerant and thermally

stable XOR gate in quantum dot cellular automata. In: 2021 IEEE European test symposium (ETS). IEEE, pp 1–2

5. Ahmed S, Baba MI, Bhat SM, Manzoor I, Nafees N, Ko S-B (2020) Design of reversible universal and multifunctional gate-based 1-bit full adder and full subtractor in quantum-dot cellular automata nanocomputing. J Nanophotonics 14:036002
6. Ganesh E, Kishore L, Rangachar MJ (2008) Implementation of Quantum cellular automata combinational and sequential circuits using Majority logic reduction method. Int J Nanotechnol Appl 2:89–106
7. Raj M, Ahmed S, Gopalakrishnan L (2020) Subtractor circuits using different wire crossing techniques in quantum-dot cellular automata. J Nanophotonics 14:026007
8. Safoev N, Ahmed S, Tashev K, Naz SF (2021) Design of fault tolerant bifunctional parity generator and scalable code converters based on QCA technology. Int J Inform Technol 1–8
9. Bilal B, Ahmed S, Kakkar V (2017) Multifunction reversbile logic gate: logic synthesis and design implementation in QCA. In: 2017 international conference on computing, communication and automation (ICCCA). IEEE, pp 1385–1390
10. Manzoor I, Nafees N, Baba MI, Bhat SM, Puri V, Ahmed S (2019) Logic design and modeling of an ultraefficient 3×3 reversible gate for nanoscale applications. In: International conference on intelligent computing and smart communication 2019. Springer, pp 1433–1442
11. Nafees N, Manzoor I, Baba MI, Bhat SM, Puri V, Ahmed S (2019) Modeling and logic synthesis of multifunctional and universal 3×3 reversible gate for nanoscale applications. In: International conference on intelligent computing and smart communication 2019. Springer, pp 1423–1431
12. Yaqoob S, Ahmed S, Naz SF, Bashir S, Sharma S (2021) Design of efficient N-bit shift register using optimized D flip flop in quantum dot cellular automata technology. IET Quantum Commun 2:32–41
13. Lent CS, Tougaw PD (1997) A device architecture for computing with quantum dots. Proc IEEE 85:541–557
14. Lent CS, Tougaw PDJ (1993) Lines of interacting quantum-dot cells: a binary wire. J Appl Phys 74:6227–6233
15. Tougaw PD, Lent CSJ (1994) Logical devices implemented using quantum cellular automata. J Appl Phys 75:1818–1825
16. Ahmad F, Ahmed S, Kakkar V, Bhat GM, Bahar AN, Wani S (2018) Modular design of ultra-efficient reversible full adder-subtractor in QCA with power dissipation analysis. Int J Theor Phys 57:2863–2880
17. Ahmed S, Naz SF, Bhat SM (2020) Design of quantum-dot cellular automata technology based cost-efficient polar encoder for nanocommunication systems. Int J Commun Syst 33:e4630
18. Ahmed S, Naz SF, Sharma S, Ko SB (2021) Design of quantum-dot cellular automata-based communication system using modular N-bit binary to gray and gray to binary converters. Int J Commun Syst 34:e4702
19. Ajitha D, Ahmed S, Ahmad F, Rajini G (2021) Design of area efficient shift register and scan flip-flop based on QCA technology. In: 2021 international conference on emerging smart computing and informatics (ESCI). IEEE, pp 716–719
20. Bilal B, Ahmed S, Kakkar V (2018) An insight into beyond CMOS next generation computing using quantum-dot cellular automata nanotechnology. Int J Eng Manuf 8:25
21. Bilal B, Ahmed S, Kakkar V (2018) Quantum dot cellular automata: a new paradigm for digital design. Int J Nanoelectron Mater 11:87–98
22. Senthilnathan S, Kumaravel SJC (2020) Power-efficient implementation of pseudo-random number generator using quantum dot cellular automata-based D flip flop. Comput Electr Eng 85:106658
23. Naz SF, Ahmed S, Sharma S, Ahmad F, Ajitha D (2021) Fredkin gate based energy efficient reversible D flip flop design in quantum dot cellular automata. Mater Today Proc 46:5248–5255
24. Sasamal TN, Singh AK, Ghanekar UJ (2019) Systems: design and implementation of QCA D-flip-flops and RAM cell using majority gates. J Circ Syst Comput 28:1950079

25. Huang J, Momenzadeh M, Lombardi F (2007) Design of sequential circuits by quantum-dot cellular automata. Microelectron J 38:525–537
26. Alamdar H, Ardeshir G, Gholami M (2020) Phase-frequency detector in QCA nanotechnology using novel flip-flop with reset terminal. Int Nano Lett 10:111–118
27. Binaei R, Gholami M (2019) Design of multiplexer-based D flip-flop with set and reset ability in quantum dot cellular automata nanotechnology. Int J Theor Phys 58:687–699
28. Purkayastha T, De D, Chattopadhyay T (2018) Universal shift register implementation using quantum dot cellular automata. Ain Shams Eng J 9:291–310
29. Alamdar H, Ardeshir G, Gholami M (2021) Novel quantum-dot cellular automata implementation of flip-flop and phase-frequency detector based on nand-nor-inverter gates. Int J Circ Theor Appl 49:196–212
30. Walus K, Dysart TJ, Jullien GA, Budiman RA (2004) QCADesigner: a rapid design and simulation tool for quantum-dot cellular automata. IEEE Trans Nanotechnol 3:26–31

Analysis of Breast Cancer Prediction Using Machine Learning Techniques: Review Paper

Rashika Pandita and Deo Prakash

Abstract Now-a-days machine learning plays an important role as there are a big number of human diseases detected which should be cured within the right time. So that the appropriate treatment or medicine is provided at the right time to save the life of a person. The result should be accurate and consume less time as well. Machine learning is a child of AI and it includes deep learning which mainly focuses on images. Consequently, these provide ultimate performance at the prediction/detection time. Machine learning provides the best solution in predicting or detecting breast cancer and evaluates the performance. Machine learning (ML) has been slowly entering every sector of our lives and its positive impact has been amazing. Today cancer is although a curable disease in early stages but most of the cancer patients are recognized in late stages. With the help of high-level specialists and machine learning techniques cancer can be detected. The aim of this review paper is to have an overview of thirteen studies predicting and then further classifying breast cancer. Through this article various machine learning algorithms and methodologies are being surveyed for breast cancer diagnosis. This paper presents theoretical and non-theoretical papers that are executed.

Keywords Breast cancer · Machine learning · Computer aided design

1 Introduction

Breast cancer is a main warning for today's women throughout the world and currently this is the second most intimidating reason of death in women. Cancer sets up when cells establish to grow out of control. Symptoms of this type of cancer incorporate a lump in the breast, discharge of blood from the nipple and changes in the shape or texture of the breast. Breast cancer stems from the tissue of breast, commonly from the interior lining of the milk ducts or lobules furnishing the ducts with milk. Basically, genes control how our cells work. They are composed of an actinic called

R. Pandita (✉) · D. Prakash
Shri Mata Vaishno Devi University, Katra, India
e-mail: rashikapandita8@gmail.com

Y. Singh et al. (eds.), *Proceedings of International Conference on Recent Innovations in Computing*, Lecture Notes in Electrical Engineering 1001,
https://doi.org/10.1007/978-981-19-9876-8_24

DNA. Cancer cells occur when genes start transforming and multiplying, an oncogene is formed which over time form cancer cells.

Breast cancer takes place in both men and women, but sparse in men. Even with magnified action toward, the lack of early detection settled females at more risk of not living from this disease. Several women expire due to this breast cancer.

Breast cancer is of two disparate types depending on the part of the breast it develops on. They are broadly divided as benign and malignant. Figure 1 shows the breast tumor, Fig. 2 shows the benign and Fig. 3 shows malignant tumors.

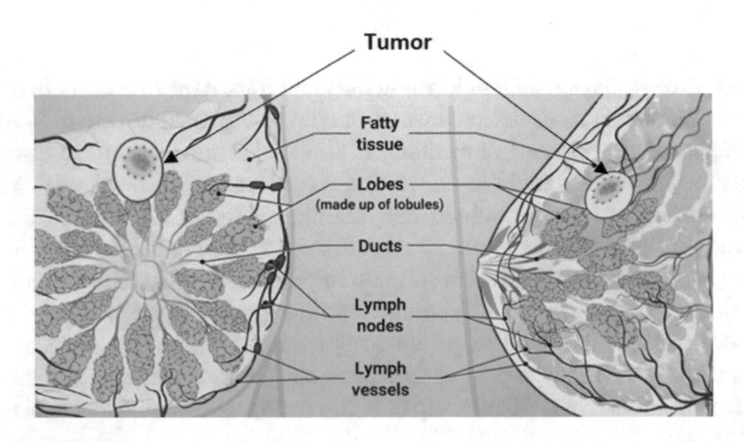

Fig. 1 Breast tumor [15]

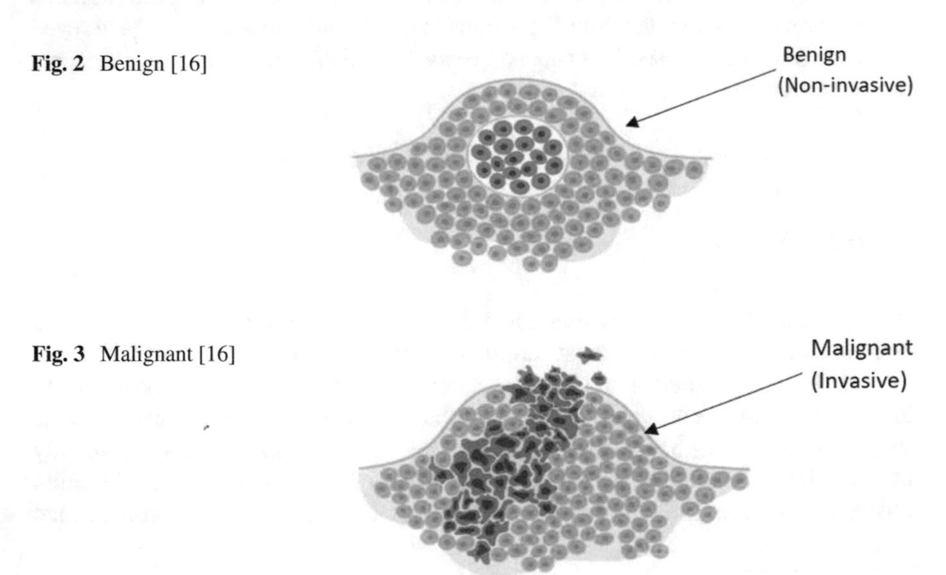

Fig. 2 Benign [16]

Fig. 3 Malignant [16]

1.1 Benign

Benign indicates a situation of tumor or growth which is not deadly. This means that it does not lay out to some other parts of the body. It does not invade nearby tissue. Sometimes, a condition is called benign to advise it is not deadly or something momentous. In general form, a benign tumor grows slowly and is not damaging.

1.2 Malignant

Malignant tumors are deadly. If they are kept without checking, malignant cells eventually may lay out beyond the real tumor to other parts of the body. The phrase "breast cancer" indicates to the malignant tumor that has been composed from cells in the breast.

Primal diagnosis of breast cancer is the only way to ensure a long survival of the patients. So, if primal the tumor detects before it lays out then there is more hope it gets fine. Therefore, exact diagnosis of breast cancer has become one of the mandatory as well as acute problems in areas of medicine. Primal and timely detection and prevention can definitely increase chances of being alive. An important actuality regarding breast cancer prognosis is to enhance the probability of cancer occurring again and again. If cancer is being found in its primal stages, there is a 30% possibility that the cancer can be treated productively. Also, detection of advanced-stage tumors behind time leads the treatment harder and demanding. Breast cancer is far simpler to treat if it is recognized earlier rather than behind time. Early recognition of breast cancer is a big issue; Moreover, breast cancer usually arrives without producing symptoms. If a patient has clear symptoms which only a doctor can determine. Careful examination and testing can bring unseen cancer to gleam.

The reception of cancer is based on its stages. It is composed of chemotherapy, radiation, hormone therapy and surgery. Breast cancer can be found in an individual if a person has specific signs. However, breast cancer patients often have no specific signs. Hence, daily base check-up plays a crucial role.

There are so many ways by which doctors check breast cancer. Some general information of the most commonly used tests is described as follows:

- **Clinical Breast Examination**

The examiner checks the cancer by watching the changes in the breasts, changes in structure or change in size and by feeling the armpits and by manually checking.

- **Mammogram**

A mammogram is an X-ray image of the very inner tissues as well lumps. The examining procedures start with the specialist uncovering the part of the breast to be examined and the breast is kept between plates proceeded with X-rays.

- **Axillary Dissection and Biopsy**

Breast cancer starts with small abnormal growth and spreads in the whole body. If the patient is suspected of cancer. The dissection method is used for removing abnormal cells from the armpit and then tissues are examined as malignant or benign.

- **Hormone receptor testing**

In this step doctors examine the tumor samples as biopsy testing gives samples and if cancer is detected at this stage with the help of medications chances of survival can be increased.

Breast Cancer Symptoms are shown in Fig. 4.

- Pain
- Change in nipple
- Skin dimpling
- Breast lump
- Blood stain discharge
- Change in skin color.

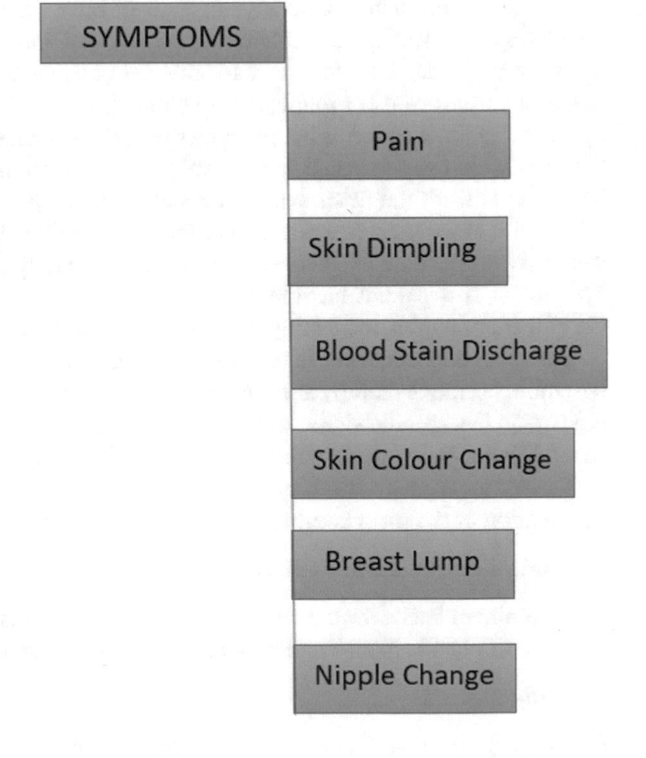

Fig. 4 Shows symptoms of breast cancer [17]

1.3 Breast Cancer Detection

Cancer or specifically cancer in breast is a deadly ailment. Breast cancer is although very complex in all aspects but primal check-up would make its reception a little easy. Finding cancer or tumor before time has reached to brim can rescue the lives of a number of people or may increase the longevity of a patient's life.

There are lots of algorithms or we can say techniques in the areas of machine learning which are working and are mentioned in the literature portion that comprises various portrayals as well as their framework which are hence fit for so many applications. Lots of machine learning algorithms and techniques aid in pinpointing breast cancer and show successful outcomes and promote other people to work in the same fields as well.

1.4 Breast Cancer Detection

Cancer or specifically cancer in breast is a deadly ailment. Breast cancer is although very complex in all aspects but primal check-up would make its reception a little easy. Finding cancer or tumor before time has reached to brim can rescue the lives of a number of people or may increase the longevity of a patient's life.

There are lots of algorithms or we can say techniques in the areas of machine learning which are working and are mentioned in the literature portion that comprises various portrayals as well as their framework which are hence fit for so many applications. Lots of machine learning algorithms and techniques aid in pinpointing breast cancer and show successful outcomes and promote other people to work in the same fields as well.

1.5 Levels of Breast Cancer Detection

The first method is to detect questionable patches (images) that are likely to contain a lymph or any other wired cell, and then confirming or denying the presence of a cancer. The second method, tries to detect and classify cancer from a dataset, history of patient which may include new occurring symptoms.

1.6 Medical Imaging Modalities

Images that are all mentioned in the above diagram in Fig. 5 are all examples of medical imaging modalities. There are also a number of scanning techniques for seeing the human body for diagnostic and therapeutic purposes. First of all, let us

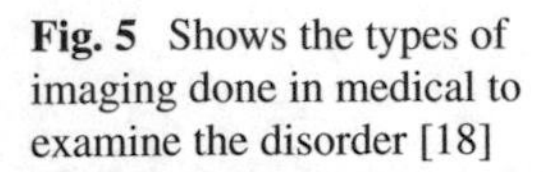

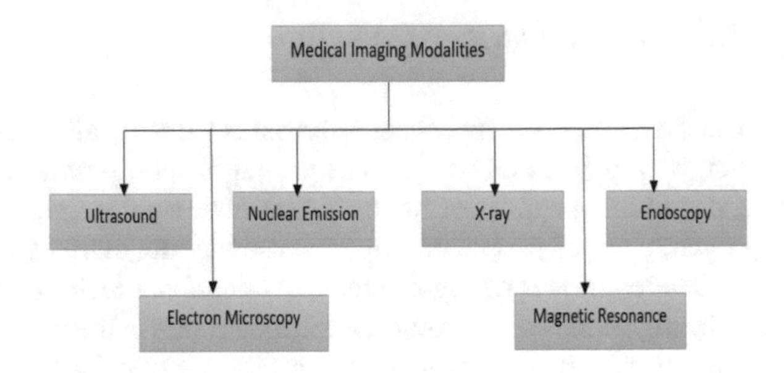

Fig. 5 Shows the types of imaging done in medical to examine the disorder [18]

understand what Image modality is. Imaging helps to detect an individual's disorder. These types of imaging are medical images and are invasive to the human body.

In an ideal world, we'd be able to diagnose and cure patients without causing them any harm. Medical imaging, which allows us to view what's going on within the body without the need for surgery or other invasive procedures, remains one of the greatest ways to accomplish this. Medical imaging is one of the most powerful tools we have for efficiently caring for our patients because it may be utilized for both diagnosis and treatment. The main purpose of these imaging is to get a complete view of abnormalities that are aroused into the human body from small disorders to cancer like disorders. Furthermore, next-generation healthcare systems are expected to include completely automated intelligent medical picture diagnosis systems.

2 Literature Review

Ravi Kumar et al. [1] considered the individuals out of which 499 were used for training the data and another two hundred for testing purposes. This outputs 241 cancerous and left over that is 458 were non-cancerous. K-fold cross over validation method was used to lower down error. They implemented the NB approach which successfully shows 94.5%. SVM produced almost the same results. Kim et al. [2] experimented on 679 patients that had breast cancer. The data was collected from medical laboratories. SVM was implemented which successfully resulted in 99% exactness. Hazara et al. [13] in which the WBCD dataset is considered. It had 569 datasets of which 300 were used for training and 269 for testing. NB, SVM, Ensemble were used to obtain maximum accuracy with less time consumption. NB successfully showed 97.3978% results. Kim et al. [4] took out data from a data bank of 679 individuals who already had breast cancer and now they wanted to check the probability of getting breast cancer again. Nomogram outputted 80% exactness. So, in conclusion this can be used for checking for recurrence. Rodrigues et al. [9] presented two distinct machine learning approaches. Two algorithms: first is the Bayesian network and the other is the decision tree. They successfully resulted in about 98% and 96%, respectively, when treated with different datasets. It should be noted that cleaning up of the data is very necessary because it replaces the missing values that are into the dataset.

Saabith et al. [10] studied the three machine learning algorithms or techniques such as J48, MLP and Rough set were used. A dataset from UCI that is a library for machine learning was pulled off. Experiments were conducted in different ways: feature selection included and excluded. Feature selection showed the best outcome. Irfan et al. [7], a very simple method was followed in which a dataset of patients was experimented with the application of different algorithms. These algorithms were collated and SVM successfully showed the best output over all other approaches.

Chang et al. [3] experimented by taking an XG-Boost classifier for classification of breast cancer. This experiment was done by taking 2964 breast cancer samples collected from the Chung Shan Medical University Hospital, Jen-Ai Hospital and Far Eastern Memorial Hospital. The results showed that using XG-Boost alone had a high exactness of 94.00%. Gómez-Flores and Hernández-López [5] presented a CAD which could be used by radiologists to classify breast cancer. The proposed CAD is based on 39 morphological features which is basically trained by numerous shapes of tumors that can differentiate between benign and malignant tumors. Dataset of Ultrasound Images about 2054 were taken out from INC and mammogram images about 892 were taken out from BCDR were used to experiment. The AUC was reported to be 82.0% in both databases.

Liu et al. [6] presented a model based on CAD which could classify breast cancer on the basis of extraction of edge features. The morphological features were extracted from the ROI such as, roughness, regularity, aspect ratio, ellipticity and roundness. After that the SVM approach was applied to distinguish the benign or malignant tumors. The ultrasound images were about 192 among which 71 were malignant and 121 were benign tumors. The method outputted in 67.31% accuracy. Zeebaree et al. [8] developed a CAD that is based on machine learning and segmentation. The ultrasound images were used to experiment this. The combination of more than two models was used. Ultrasound images about 250 in total were considered, among which benign were 100 and malignant were 150. Artificial neural network was applied which then outputted an exactness of 93.1 and 90.4% for malignant and benign, respectively. Asri et al. [11] used WBCD dataset and applied tenfold cross validation to replace missing values then applied SVM which resulted 97.13 accuracy.

Al Bataineh [12] analyzed five machine learning approaches which are non-linear, those are Multilayer Perception (MLP), K-NN, Classification and Regression Trees (CART), NB, SVM on WBCD. Among all approaches MLP outputted best accuracy with 96.7%. Summary of all papers is provided in Table 1.

3 Machine Learning Algorithms

Machine learning can play an important role in detecting and classifying cancer. These techniques can be utilized in the primal diagnosis of cancer and these are

Table 1 Summarizes the literature survey [14]

References	Dataset	ML algorithm	Accuracy (%)
Kumar et al. [1]	WBC	NB, SVM	94.5%
Kim et al. [2]	UCI (Medical laboratories)	SVM	99%
Hazara et al. [13]	WBDC	NB SVM Ensemble	NB: 97.39%
Kim et al. [4]	Breast Cancer Center	Nanogram	80%
Rodrigues et al. [9]	WBDC	Bayesian network and decision tree	98% 96%
Lab et al. [10]	UCI	J48, MLP and rough set	79, 75 and 71%
Irfan et al. [7]	Ultrasonic images	SVM	98%
Chang et al. [3]	Medical labs	XG-Boost classifier	94%
Gómez-Flores and Hernández-López [5]	Ultrasonic images and mammography	CAD	82%
Liu et al. [6]	Ultrasonic images	SVM	67%
Zeebaree et al. [8]	Ultrasonic images	ANN	93.1% (Malignant) 90.4% (Benign)
Al Bataineh [12]	WBDC	NB, SVM, K-NN, CART, MLP	MLP: 97%
Asri et al. [11]	WBDC	SVM	93.13%

shown in Fig. 6. Various techniques like ANN, SVM and Decision Trees. Besides machine learning there is one more area that is of deep learning, deep learning techniques can also aid in the detection of cancer.

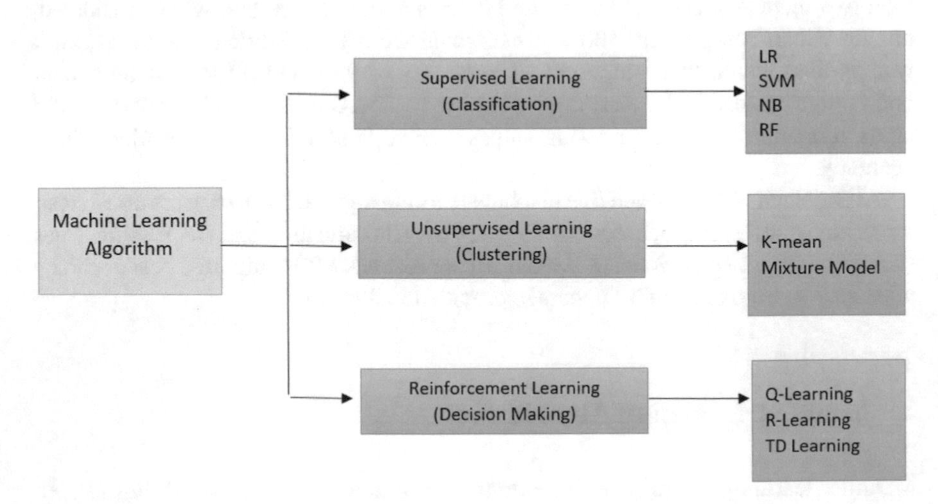

Fig. 6 Shows flowchart of machine learning techniques classification [19]

Table 2 Accuracy of various machine learning algorithms for cancer detection [21]

Approaches	Training set (%) (Accu.)	Testing set (%) (Accu.)
SVM	98.4	97.2
RF	99.8	96.5
LR	95.5	95.8
DT	98.8	95.1
K-NN	94.6	93.7

Support Vector Machine (SVM) is a morpheme that carves up the databanks in two subclasses one of abnormal class and another of normal class. It draws a line between normal and non-normal categories. SVM transforms into higher dimensions from the actual training data. Not only there can be a single line but a number of lines can be pulled out.

Random forest decision is a combinational method which is a combination of random forest and decision tree that is here random forest and decision tree are applied. So many decision trees are implemented at the time of training and outputting. When there is trouble of overfitting it is solved with the random forest technique.

K-Nearest Neighbors (K-NN) is an algorithm which comes under the class of supervised learning in which training is given. A collection of data points that are labeled are utilized to label other points according to the nearest points which are not labeled.

Logistic regression is an approach which considers two variables one is input and another is output variable. LR is basically used for foretelling binary (of two) variables.

Decision Tree is a Foretelling model that is helpful in many aspects of the automation world. It further splits the database and further sub-splits that data and slits it again until we reach a smaller unit. Machine learning approaches can make a significant contribution to the overall activity of breast cancer foretelling and primal diagnosis as well which have now-a-days become a research platform, and have been proven to be a powerful technique. Table 2 shows accuracies of different algorithms.

4 Suggested Methodology

The dataset is collected from Kaggle which includes numerical data and images data. Then data then is pre-processed by removing missing values in the numerical type dataset and removing noise from images dataset. Feature selection is used select some main features which influence prediction mostly. Then we construct a model to predict breast cancer with maximum accuracy. In this study we will compare three machine learning algorithms in terms of accuracy. The steps involved in methodology are provided in Fig. 7.

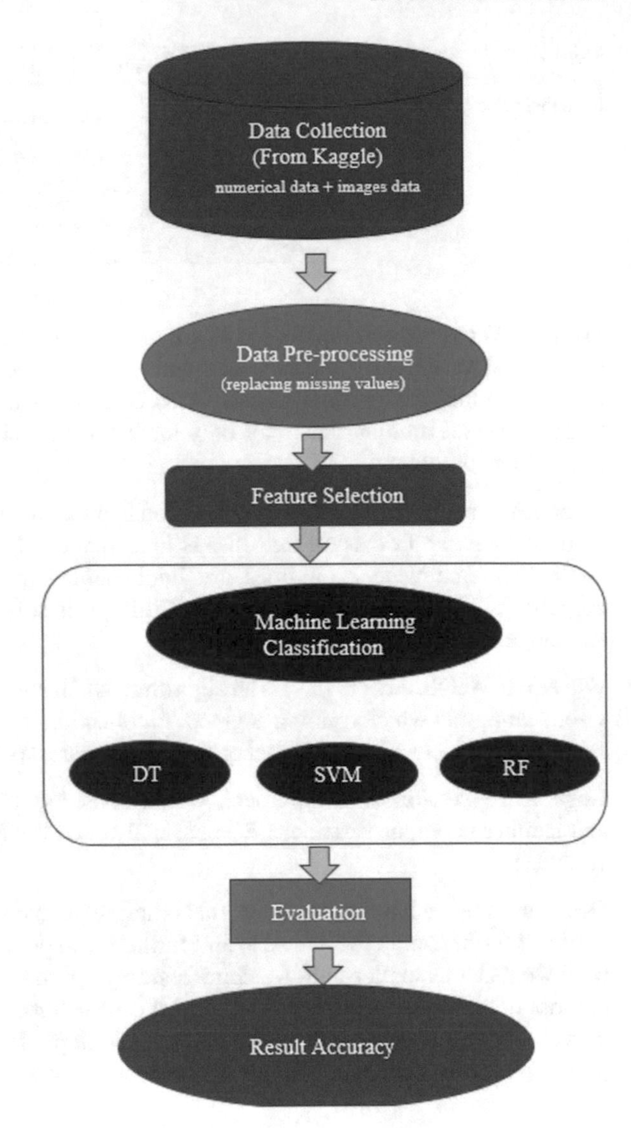

Fig. 7 Flowchart of proposed methodology [20]

5 Conclusion

In this study, we have gone through a number of papers which have used different algorithms to detect breast cancer such as Support Vector Machine (SVM), Decision Tree (J48), Naïve Bayes (NB), K-NN, Ensemble Techniques and many others. Based on the data and parameter selection, each method performs differently. It should be mentioned that all of the findings acquired are limited to the classification database, which can be seen as a restriction of our work. It is therefore essential to reflect on future works to implement these very same approaches and techniques to other datasets to confirm the output acquired through this dataset, as well as to implement our as well as other ML algorithms to use new parameters on massive data sets with more illness classes to obtain. As a result, every time a new dataset is explored, one must experiment with multiple feature selection methods or other algorithms to determine the optimal one to improve accuracy rather than relying on the previously proven one in the same domain. Once the optimal approach for a given dataset has been discovered, it can be utilized to improve cancer detection accuracy. This work can also be improved by altering the Support Vector Machine, which provides the highest level of accuracy. Constructing appropriate and computational efficiency classifiers for medical uses is a difficult task in machine learning. It is extremely hard to go through the many medical diseases of a cancer patient using computational methods, and the prediction of conditions is also more crucial in nature. Furthermore, in the not-too-distant future, the diagnosis of a specific stage of cancer will be possible.

References

1. Ravi Kumar G, Ramachandra GA, Nagamani K (2013) An efficient prediction of breast cancer data using data mining techniques. Int J Innov Eng Technol (IJIET) 2(4):139
2. Kim W, Kim KS, Lee JE, Noh DY, Kim SW, Jung YS, Park MY, Park RW (2012) Development of novel breast cancer recurrence prediction model using support vector machine. J Breast Cancer 15(2):230–238
3. Chang C-C, Chen S-H (2019) Developing a novel machine learning-based classification scheme for predicting SPCs in breast cancer survivors. Frontiers Genet 10:1–6
4. Kim W, Kim KS, Park RW (2015) Nomogram of Naive Bayesian model for recurrence prediction of breast cancer medicine. 206 World cup-ro, Yeongtong-gu, Received 29 Oct 2015
5. Gómez-Flores W, Hernández-López J (2020) Assessment of the invariance and discriminant power of morphological features under geometric transformations for breast tumor classification. Comput Methods Prog Biomed 185. Article 105173
6. Liu Y, Ren L, Cao X, Tong Y (2020) Breast tumors recognition based on edge feature extraction using support vector machine. Biomed Sig Process Control 58(101825):1–8
7. Irfan R, Al Mazroui AA, Rauf HT, Damaševičius R, Nasr EA, Abdelgawad AE (2021) Dilated semantic segmentation for breast ultrasonic lesion detection using parallel feature fusion. Diagnostics 11(7):1212
8. Zeebaree DQ, Haron H, Abdulazeez AM, Zebari DA (2019) Machine learning and region growing for breast cancer segmentation. In: 2019 international conference on advanced science and engineering (ICOASE). IEEE, Zakho–Duhok, Iraq, pp 88–93
9. Borges, Lucas Rodrigues Union College. Analysis of the Wisconsin breast cancer dataset and machine learning for breast cancer detection, October 05th–07th 2015
10. Saabith ALS et al (2014) Comparative study on different classification techniques for breast cancer dataset. Int J Comput Sci Mob Comput 3(10)
11. Asri H, Mousannif H, Moatassime HA, Noel T (2016) Using machine learning algorithms for breast cancer risk prediction and diagnosis. In: Proceeding's 6th international symposium on frontiers in ambient and mobile systems (FAMS)
12. Al Bataineh A (2019) A comparative analysis of nonlinear machine learning algorithms for breast cancer detection. Int J Mach Learn Comput 9(3). https://doi.org/10.18178/ijmlc.2019.9.3.794248
13. Hazara A (2016) Study and analysis of breast cancer cell detection using Naïve Bayes, SVM and Ensemble algorithms. Int J Comput Appl 145(2):39. ISSN 0975-8887

14. Naji MA, El Filali S, Aarika K, Abdelouhahid RA (2021) Table1 in "Machine learning algorithms for breast cancer prediction and diagnosis". In: International workshop on edge IA-IoT for smart agriculture (SA2IOT), 9–12 Aug 2021
15. Breast tumor. Edited version. Source https://www.youtube.com/watch?v=KyeiZJrWrys&t=238s
16. Types of cancer, Benign and Malignant. Source https://www.miskawaanhealth.com/cancer/different-tumor-types/
17. Symptoms of Brest Cancer. [2022]. Source https://www.google.com/search?q=symptoms+of+cancer&rlz=1C1CHBD_enIN920__920&oq=symptoms+of+cancer+&aqs=chrome..69i57j0i512l9.7184j0j7&sourceid=chrome&ie=UTF-8
18. Medical imaging modulities, ways to detect the disorder. Source https://www.google.com/search?q=medical+imaging+technology&source=lmns&bih=552&biw=1263&rlz=1C1CHBD_enIN920__920&hl=en&sa=X&ved=2ahUKEwjWz5Gxz5H7AhXtnNgFHTBqCrEQ_AUoAHoECAEQ\
19. Machine learning techniques classification. Source https://www.analyticsvidhya.com/blog/2021/03/everything-you-need-to-know-about-machine-learning/

Soil Classification Using Machine Learning, Deep Learning, and Computer Vision: A Review

Ashish Kumar and Jagdeep Kaur

Abstract The goal of this work is to examine the primary soil variables that impact crop development, such as organic matter, vital plant nutrients, and micronutrients, and to determine the appropriate connection proportion among those qualities using ML and DL models. Agriculture relies heavily on the soil. Different kinds of dirt can be found in various locations. Each type of soil can have unique characteristics to develop crops. To cultivate crops in diverse soil types, we need to know which kind of soil is best. Accurate soil moisture and temperature forecast are beneficial to agriculture planting factors. According to the review articles, machine learning approaches, deep learning prediction models, and computer vision are helpful in this instance. The external temperature and humidity and soil moisture and temperature are used to train and test to anticipate soil moisture and temperature. A deep learning model was developed based on the extended short-term memory network (LSTM). Machine learning models such as k-nearest neighbor, SVM, and RF methods are utilized for soil classification. Several contributions of computer vision models in fields like planting, harvesting, enhanced weather analysis, weeding, and plant health detection and monitoring have been shown in the application of computer vision in soil categorization. This paper presents a review of soil classification and various challenges associated with it using machine learning, deep learning, and computer vision.

Keywords Soil classification · Machine learning (ML) · Deep learning (DL) · Computer vision

A. Kumar (✉) · J. Kaur
Dr B R Ambedkar National Institute of Technology Jalandhar, Grand Trunk Road, Barnala-Amritsar Bypass Rd, Jalandhar, Punjab 144011, India
e-mail: ashishkumar291996@gmail.com

J. Kaur
e-mail: kaurj@nitj.ac.in

Y. Singh et al. (eds.), *Proceedings of International Conference on Recent Innovations in Computing*, Lecture Notes in Electrical Engineering 1001,
https://doi.org/10.1007/978-981-19-9876-8_25

1 Introduction

The potential of soil to promote plant development by providing crucial plant nutrients as well as desired chemical, physical, and biological qualities as fill out environment is mentioned as soil fertility. Soil fertility is one of the majority needs soil qualities for crop growth [1]. To get bigger and yield properly, crops need the right proportions of nitrogen, phosphorus, potassium, and more nutrients. A substantial portion of the Indian economy is based on agriculture, and to enhance agricultural practices, it is required to properly forecast crop output responses, which may be accomplished using ML. Agricultural soil quality is determined by the macro and micro nutrient contents of the soil, such as S, K, pH, C, Mg, P, Ca, B, and so on. Our primary goal is to investigate, adapt, and formulate soil qualities and crop development variables. This difficulty may be met by boosting soil fertility by providing the essential nutrients to the plant at the appropriate time and in an acceptable amount. Increased agricultural output necessitates soil management, which may be accomplished by enhancing and preserving soil nutrients. Choosing the right soil management approach can help boost crop yields. Experts can more easily make predictions and make judgments about the best soil resource. Soil fertile has average to high nutrients required for plant development and productivity. India has many natural and human resources, and its economy is rapidly expanding. A substantial portion of the Indian economy is based on agriculture. To enhance agricultural practices, it is required to correctly forecast crop output responses, which may be done using ML, DL, and computer vision. The main goal of this paper is to investigate macro and micro soil parameters that impact crop output, similarly organic content, and essential plant nutrients and to determine the rank of a particular soil based on previously graded soils using supervised learning. Agriculture involves the domestication of plants. While the ground is frequently referred to as a fertile substrate, not all soil is the same. The importance of soil parameters in soil analysis, environmental analysis, and soil humidity and temperature is recent critical advances in proximate sensor technologies, such as portable X-ray fluorescence (pXRF) spectroscopy [2] or visible and near-infrared (Vis–NIR) spectroscopy [3] provide quick and non-destructive ways to quantify data from soil profiles. This work is broken into two parts and discusses several computer-based soil categorization approaches. The first is computer vision-based soil classification technologies, which use standard computer vision algorithms and methodologies to categorize soil based on texture, color, and particle size. Another is DL- and ML-based soil classifications. To predict soil qualities from portable X-ray fluorescence devices and Vis–NIR data, most (ML) and (DL) approaches are applied. The most popular machine learning methods are used, such as random forest, support vector machine, decision tree, MLR, and deep learning methods such as multiple perceptron neural network and convolutional neural network. Agriculture is a vital part of the global economy and provides for one of humanity's most fundamental needs, namely food [4]. It is regarded as the primary source of employment in most countries. Many nations, such as India, still practice traditional farming. Farmers are hesitant to adopt new technology in their fields due to a lack of knowledge, high costs, or a lack of awareness of the benefits of these technologies. Lack of understanding of soil types, yields, crops, weather, poor pesticide application, irrigation issues, incorrect harvesting, and a lack of market information has resulted in the loss of farmers or increased costs.

2 Literature Review

Here, the authors surveyed a few papers related to soil analysis [5] using various ML, DL, and computer vision techniques as given in Table 1. Soil is a necessary component of life, and soil qualities are critical in determining the health of the soil. Recent advances in proximate sensor technologies, such as portable X-ray fluorescence (pXRF) spectroscopy or visible and near-infrared (Vis–NIR) spectroscopy, provide quick and non-destructive ways to quantify data from soil profiles. A model that incorporates machine learning methods (random forest, SVM, and ANN) to predict the crop is now available in this industry.

Research Questions:

RQ1: What is the role of machine learning in soil classification?
RQ2: What is the role of deep learning in soil classification?
RQ3: What is the role of computer vision in soil classification?
RQ4: What is the relation between machine learning, deep learning, and computer vision?

3 Findings

Research Questions are answered in this section.

RQ1: What is the role of machine learning on soil classification?

Throughout the whole growing and harvesting cycle, machine learning is present. It starts with a seed being placed in the soil. It continues with soil preparation, seed breeding, and water feed measurement until the harvest is picked up by robots who use computer vision to determine ripeness. Crop disease diagnosis, pest detection, plant species recognition, crop production prediction, exact fertilization, smart agricultural IoT, food material supply chain security tracking, crop security, and other significant challenges in smart agriculture are all improved using machine learning approaches. Several machine learning methods are utilized for soil categorization, including weighted k-nearest neighbor (KNN) and random forest-based support vector machines (SVM).

Table 1 Literature review of soil classification

Year	Title	Technique	Objective	Limitations
2021 [3]	To analyze soil properties using ML and DL	Random forest PLSR, CNN	Develop DL and ML techniques with an emphasis on predicting soil qualities	PLSR does not perform well on a large number of a heterogeneous sample
2021 [1]	Machine learning was used to estimate the soil quality of the Bhimtal block	ANN, RELU TAN-GENT, SVM	The research aims to classify Bhimtal's soil fertility by the village	This block is located in hilly areas, so there are more possibilities of soil erosion
2021 [7]	A hybrid process for crop yield prediction using machine learning	SVM, LSTM, RNN	Finds difficult to estimate soil properties and weather conditions	SVM does not perform large data set
2018 [9]	Data analytics in agriculture	MapReduce	Enhance the accuracy by using the Bayesian classification algorithm	MapReduce is very complex than SVM, another algorithm, it does not work well in response
2018 [4]	Using a variety of regression approaches in village level soil fertility for many nutrients in India was made	Lasso and Ridge regression, random forest, SVM	Collect more data for those nutrients, which is available to predict a geographic mapping	Lasso is not suitable for interpreting the data when two or more variables are collinear
2021 [5]	Soil analysis and crop fertility prediction using machine learning	SVM, random forest, Naive Bayes, ANN	CNN models are used to recognize the soil fertility images	ANN does not give the best accuracy to identify soil fertility images evaluation

(continued)

Table 1 (continued)

Year	Title	Technique	Objective	Limitations
2014 [6]	Using back propagation neural network in soil fertility and plant nutrient management	Back propagation	VPNs can act as an automated method to forecast crop growth rates based on soil parameters	Extended time for training, less interpretability for humans point of view

RQ2: What is the role of deep learning in soil classification?

Deep learning is most useful for weed classification, soil classification, and plant recognition. Using a combination of CNN and long short-term memory, as well as back propagation DL-based techniques, weed plants are identified and classified (LSTM) [6, 7]. Convolutional neural networks (CNNs) have a distinct structure for extracting discriminative features from input pictures, while LSTM allows for combined classification optimization. Seed classification categorizes various types of seeds into different groups based on their physical characteristics. The goal is to do an automated variety of seed quality. The convolutional neural network is one approach that may be employed since it has powerful classification accuracy results, but its disadvantage is that it takes a long time to train.

RQ3: What is the role of computer vision on soil classification?

Computer vision in the soil aids in the detection of product problems and the sorting of produce by weight, color, size, maturity, and a variety of other variables. They may save a lot of time and shorten time-to-market when used with the proper mechanical equipment—a branch of artificial intelligence in computer vision. Machines can now perceive and understand the visual world in the same way humans do. In soil classification, computer vision algorithms combined with picture gathering through remote cameras provide non-contact and scalable sensing systems.

RQ4: What is the relation between machine learning, deep learning, and Computer vision?

ML used some techniques to classify the soil, i.e., SVM, KNN, and deep learning models are used to plant recognition. Computer vision is used to identify color, texture, and image acquisition. Table 2 compares ML, DL, and computer vision used in soil classification.

Table 2 Compare ML, DL, and computer vision

ML	DL	Computer vision
ML is a subfield of AI that builds intelligent systems using statistical learning methods	This branch of AI is a strategy based on how the human brain filters data	It gives computers the ability to process, analyze, and interpret visual data
ML is a subset of artificial intelligence that grants software to improve its prediction accuracy in the absence of explicitly coded	DL is a field of artificial intelligence that employs artificial neural networks to mimic the human brain's operation	Computer vision is a branch of machine learning concerned with teaching computers or machines to interpret human movements, behaviors, and languages in the same way that people do
ML is used in crop management, yield management, soil properties	DL is used in plant recognition, seed classification, weed classification	Computer vision techniques combined with picture collecting through remote cameras provide non-contact and scalable sensing systems
Farmers may now use machine learning to analyze complicated patterns and correctly identify related plant and weed species	DL algorithms are used to detect the temperature and water level of the crops	Improved autonomous flying capabilities allow drones to obtain crucial data using computer vision-enabled cameras

4 Different ML, DL, and Computer Vision Techniques Are Used for Soil Classification

Table 3 defines the models used in this paper to categorize the soil with the help of ML, DL, and computer vision techniques used to predict the soil properties. Because of the time and equipment required to undertake laboratory tests regularly before earthwork design and construction, the prediction has become vital. These methods are used to categorize dirt.

5 Machine and Deep Learning Application in Soil Classification

Farmers rely on traditional agricultural methods based on the trustworthiness of the elderly advice and expertise. Farmers are left at the compassion of unpredictable meteorological circumstances, which are before becoming more unpredictable caused by global warming and irregular rainfall patterns. The manual pesticide spraying approach wastes resources and is harmful to the environment. The authors shall cover the state-of-the-art techniques put forward by many scholars and exponents worldwide in this part.

Table 3 Techniques of ML, DL, and computer vision

ML, DL and computer vision	Algorithm description
Random forest	It is a supervised learning approach ML algorithm. It is built on the ensemble learning concept. It is a classifier that uses an average to improve the accuracy of a dataset by containing more trees on different criteria. When more trees are taken into account, more precision is expected, which helps to avoid the problem of overfitting
SVM	It is a supervised learning approach used for regression and categorizing problems. This technique generates a best-fit line, also known as a decision boundary, that can separate in the future
ANN	ANN effort to reproduce the network of neurons that build up a human brain so that a computer may learn and make choices like a human brain
KNN	It is an ML algorithm based on supervised learning. This works on the alike between the new data and put the new data into the category most similar to the existing type. It stores all current data and breaks down the latest based on similarity because new data can easily classify the best category
CNN	CNN is a neural network that is good at processing data with a grid-like architecture, such as images
RNN	RNN is a neural network that may be handed down to model data in sequences. RNNs, which are derived from feed-forward networks, behave like human brains
Regression algorithm	Regression algorithms expect the output values to depend on the input feature from the data given into the system. The conventional method is for the algorithm to produce a model based on the attributes of training data and then use that model to predict the value of new data
Image classification	Viewpoint variation, size variation, intra-class variance, picture distortion, image occlusion, lighting circumstances, and background clutter are all issues in image clarity
Object detection	Identifying things inside photographs generally entails producing bounding boxes and labels for each item

5.1 Prediction of Soil Properties and Weather

Soil attributes primarily include predicting ground nutrients, the soil outside humidity, and meteorological state over the crop's lifespan. Human pursuit has had a remarkable effect on soil qualities and, as an outcome, our range to develop crops. As given in Table 2, 16 essential components are required for plant development. Crop growth is effect by the nutrients available in a particular soil. Electric and electromagnetic sensors are used to monitor soil nutrients. Farmers construct informed conclusions about which crop is best for the land based on the nutrients available. Although needed for measuring soil characteristics, soil nutrients, soil moisture, and pH [8] are also required. The experimental findings revealed a low root mean square error (RMSE) of 2.17% when applying the sine kernel function.

5.2 Soil Management

Soil is a various natural resource with a variety of processes and procedures. Its single temperature can provide details on the implications of climate change on the regional output. ML algorithms look into evaporation processes, soil moisture, and temperature to recognize the dynamics of ecosystems and their effect on agriculture.

5.3 Weed Classification

Using a mix of CNN and LSTM, deep learning (DL)-based algorithms are used to identify and classify weed plants (LSTM). Convolutional neural networks (CNNs) have a distinct structure for extracting discriminative features from input pictures, while LSTM allows for combined classification optimization.

5.4 Seed Classification

Seed classification categorizes various types of seeds into distinct groups based on their physical characteristics. The goal is to do an automated variety of seed quality. The CNN is one approach that may be employed since it has powerful classification accuracy results. Still, its disadvantage is that it takes more time to train. The remaining network and custom layer architecture was chosen. The usage of a custom layer to shorten training time, the inclusion of a proposal region to sharp more on corn seeds, and augmentation to enhance the variety in the quantity of data have all been included in the approach.

6 Comparison of Machine Learning-Based Soil Classification Methods

Table 4 presents the papers using the ML algorithm giving the best accuracy to find soil properties. Soil moisture is crucial for understanding the connection between the ground and the atmosphere, and it has an impact on hydrological and agricultural processes including drought and crop productivity. Seed germination is impacted by the temperature of the soil. The temperature of the soil has a direct effect on plant

Table 4 Prediction of soil characteristics using various ML techniques

Reference	Features	Experimental place	ML algorithm	Accuracy
Park et al. [10]	Soil moisture	AMSR2 soil moisture data from MODIS	Random forest	RMSE of 0.06 and R^2 of 0.96
Mohammadi et al. [11]	Soil temperature	Maize field located in Shouyang Country	ELM, RF	ELM is best predicted with RMSE of 0.56 and R^2 of 0.99
Lazri et al. [12]	Estimate of precipitation	North region of Algeria	Random forest, ANN, SVM, KNN, k-means algorithm	R^2 of 0.93
Acar et al. [13]	Soil outside humidity	Polarimetric measurement utilizes Radarset-2 for a field at Dicle University	ML-based regression model	RMSE of 2.19%
Mahmoudzadeh et al. [14]	Soil organic carbon	Kurdistan province of Iran	KNN, XGBoost, SVM	RMSE of 0.35% and R^2 of 0.60

development. The major part of soil organisms does gain the advantage when the soil temperature is optimal. Organic carbon in the soil enhances the tilth of the soil, making it more physically stable. This improves soil aeration, water drainage, and retention, as well as reduces erosion and nutrient leaching.

7 Comparison of Deep Learning-Based Soil Classification Methods

Table 5 presents the papers using the DL algorithm giving the best accuracy to find soil properties. Rainfall has an important effect on soil. Nutrients in the soil can move and not arrive at the roots of plants if the weather is too moist or too dry, resulting in very low growth and general health. Because all plants need at least some water to shoot up, agriculture relies heavily on rain.

8 Comparison of Computer Vision-Based Soil Classification Methods

Table 6 presents the papers using the DL algorithm giving the best accuracy to find soil properties. Traditional image processing techniques and procedures to categorize

Table 5 Prediction of soil characteristics using various deep learning techniques

Reference	Features	Experimental place	DL algorithm	Accuracy
Sierra and jesus [15]	Rainfall prediction	Tenerife Spain	Random forest, neural network	Neural networks predict the rainfall incident with an F score close to 0.3 and R 0.1–0.7
Racon et al. [16]	Plant disease identification	Soil data from France	CNN	Accuracy of 90.7%
Tseng et al. [17]	Obstacle detection	Image net and COCO dataset	CNN and FCN	Accuracy of 70.81%
Duarte et al. [18]	Soil segmentation	Image net dataset	CNN	Accuracy of 0.57
Younis et al. [19]	Plant recognition	Soil data from African	CNN	Accuracy of 82.4%

Table 6 Prediction of soil characteristics using various Computer-Vision techniques

Reference	Features	Experimental place	Computer vision techniques	Accuracy
Kevin et al. [20]	Color	Northcentral Missouri	Image processing	Accuracy of 99.6%
Sudarsan Texture et al. [21]	Texture	Locations from Field86	Image acquisition	Accuracy of 87%
Barman et al. [22]	Color	Soil pH laboratory	K-means clustering for image segmentation	pH of 0.859
Wu et al. [23]	Texture	Southwest China	SVM, ANN, image acquisition	Accuracy of 94.3%

soil using diverse characteristics such as texture, color, and particle size are included in computer vision-based soil classification methodologies. One of the most essential qualities of soil is its texture. The "feel" of the soil material, which can be rough and gritty or smooth and silky, is referred to as soil texture.

9 Challenges

Challenges in the forecast of soil properties using Machine learning

- Nonlinear dataset—These forecasts frequently include a nonlinear dataset that may be used by regression algorithms like ELM, RF, and SVR to make correct predictions.

- Accurate prediction—The RMSE and R^2 error indices for parameter prediction using ML are low, common indicators of statistical analysis accuracy.
- The universal design of prediction algorithms is complicated by varying parameters and complex datasets.

Challenges in the forecast of soil properties using Deep learning

- CNN with a deeper structure outperforms those with a shallow system. Deeper networks better represent high-level information from the input, which improves prediction accuracy.
- By enabling the end-to-end process, deep learning applications reduce the need for spatial-form designs and preprocessing approaches.

Challenges in the forecast of soil properties using Computer Vision

- Traditional computer vision techniques and methods to classify soil using diverse characteristics such as texture, color, and particle size are included in computer vision-based soil classification methodologies.
- On soil photographs, computer vision techniques may be used to classify them. Because color is a broad indicator of a soil's physical qualities and chemical compositions, studying the color of the ground can provide helpful information on soil statistics.

10 Conclusions and Future Work

The number of papers in this review was 27 in total from the analysis of these papers. Using ML and DL techniques to decrease the farmer's issue of getting losses on their farms due to a lack of knowledge of growing in different soil and weather conditions, this model is presented to forecast soil fertility and the sorts of crops grown on fertile soil. The study was based on a global database of soil data. This paper focuses on using computer vision, ML, and DL in soil classification. DL approaches are used for improved crop selection based on soil categorization. Machines interact among themselves to determine which soil is ideal for a crop using datasets developed by researchers and applied to the system. Color, texture, and particle size are some of the essential qualities of soil employed by the system which incorporates supervised machine learning approaches such as linear regression multi-variate based on error analysis, SVM, and RF classifier to produce the best results. As a result, this technique will assist farmers in reducing their struggles. We are trading with the grading of the soil and the prediction of crops suited for the land based on an analysis of the essential soil parameters. This will serve as a simple way to provide farmers with the information they require to build a high harvest, thereby increasing their surplus and reducing their troubles. More soil data can be added to the dataset that has been analyzed in the future as an improvement. Other classification methods, such as the Bayesian classification algorithm, ANN, and others, can be used to improve this application. Other data visualization tools, like Tableau, Sap Lumira, and others, can also be used to visualize data.

References

1. Pant J, Pant P, Pant RP, Bhatt A, Pant D, Juyal A (2021) Soil quality prediction for determining soil fertility in Bhimtal Block of Uttarakhand (India) using machine learning. Int J Anal Appl 19(1):91–109
2. Andrade R, Faria WM, Silva SHG, Chakraborty S, Weindorf DC, Mesquita LF, Curi N et al (2020) Prediction of soil fertility via portable X-ray fluorescence (pXRF) spectrometry and soil texture in the Brazilian coastal plains. Geoderma 357:113960
3. Pham V, Weindorf DC, Dang T (2021) Soil profile analysis using interactive visualizations, machine learning, and deep learning. Comput Electron Agric 191:106539
4. Sirsat MS, Cernadas E, Fernández-Delgado M, Barro S (2018) Automatic prediction of village-wise soil fertility for several nutrients in India using a wide range of regression methods. Comput Electron Agric 154:120–133
5. Yadav J, Chopra S, Vijayalakshmi M (2021) Soil analysis and crop fertility prediction using machine learning. Mach Learn 8(03)
6. Ghosh S, Koley S (2014) Machine learning for soil fertility and plant nutrient management using back propagation neural networks. Int J Rec Innov Trends Comput Commun 2(2):292–297
7. Agarwal S, Tarar S (2021) A hybrid approach for crop yield prediction using machine learning and deep learning algorithms. J Phys Conf Ser 1714(1):012012. IOP Publishing
8. Wang A, Li D, Huang B, Lu Y (2019) A brief study on using pH_{H_2O} to predict pH_{KCL} for acid soils. Agric Sci 10(02):142
9. Lata K, Chaudhari B (2019) Crop yield prediction using data mining techniques and machine learning models for decision support system. J Emerg Technol Innov Res (JETIR)
10. Park S, Im J, Park S, Rhee J (2015) AMSR2 soil moisture downscaling using multisensor products through machine learning approach. In: 2015 IEEE international geoscience and remote sensing symposium (IGARSS). IEEE, pp 1984–1987
11. Mohammadi K, Shamshirband S, Motamedi S, Petković D, Hashim R, Gotic M (2015) Extreme learning machine based prediction of daily dew point temperature. Comput Electron Agric 117:214–225
12. Lazri M, Labadi K, Brucker JM, Ameur S (2020) Improving satellite rainfall estimation from MSG data in Northern Algeria by using a multi-classifier model based on machine learning. J Hydrol 584:124705
13. Acar E, Ozerdem MS, Ustundag BB (2019) Machine learning-based regression model for prediction of soil surface humidity over moderately vegetated fields. In: 2019 8th international conference on agro-geoinformatics (Agro-Geoinformatics). IEEE, pp 1–4
14. Mahmoudzadeh H, Matinfar HR, Taghizadeh-Mehrjardi R, Kerry R (2020) Spatial prediction of soil organic carbon using machine learning techniques in western Iran. Geoderma Regional 21:e00260
15. Diez-Sierra J, del Jesus M (2020) Long-term rainfall prediction using atmospheric synoptic patterns in semi-arid climates with statistical and machine learning methods. J Hydrol 586:124789
16. Rancon F, Bombrun L, Keresztes B, Germain C (2018) Comparison of SIFT encoded and deep learning features for the classification and detection of Esca disease in Bordeaux vineyards. Rem Sens 11(1)
17. Tseng D, Wang D, Chen C, Miller L, Song W, Viers J, Goldberg K et al (2018) Towards automating precision irrigation: deep learning to infer local soil moisture conditions from synthetic aerial agricultural images. In: 2018 IEEE 14th international conference on automation science and engineering (CASE). IEEE, pp 284–291

18. Duarte C, Schiele R, Friedel C, Gerth S, Rousseau D (2018) Transfer learning from synthetic data applied to soil–root segmentation in X-ray tomography images. J Imaging 4(5):65
19. Younis S, Weiland C, Hoehndorf R, Dressler S, Hickler T, Seeger B, Schmidt M (2018) Taxon and trait recognition from digitized herbarium specimens using deep convolutional neural networks. Botany Lett 165(3–4):377–383
20. O'Donnell TK et al (2010) Identification and quantification of redoximorphic soil features by digital image processing. Geoderma 157:3–4
21. Sudarsan B et al (2018) Characterizing soil particle sizes using wavelet analysis of microscope images. Comput Electron Agric 148:217–225
22. Barman U, Choudhury RD, Uddin I (2019) Predication of Soil pH using K mean segmentation and HSV color image processing. In: 2019 6th international conference on computing for sustainable global development (INDIACom). IEEE, pp 31–36
23. Wu W, Li AD, He XH, Ma R, Liu HB, Lv JK (2018) A comparison of support vector machines, artificial neural network and classification tree for identifying soil texture classes in southwest China. Comput Electron Agric 144:86–93
24. Keerthan Kumar TG, Shubha C, Sushma SA (2019) Random forest algorithm for soil fertility prediction and grading using machine learning. Int J Innov Technol Explore Eng (IJITEE) 9(1)
25. Sharma A, Jain A, Gupta P, Chowdary V (2020) Machine learning applications for precision agriculture: a comprehensive review. IEEE Access 9
26. Sidhu KS, Singh R, Singh S, Singh G (2021) Data science and analytics in agricultural development. Environ Conserv J 22(SE):9–19
27. Veres M, Lacey G, Taylor GW (2015) Deep learning architectures for soil property prediction. In: 2015 12th conference on computer and robot vision. IEEE, pp 8–15

Robust Color Texture Descriptors for Color Face Recognition

Shahbaz Majeed and Chandan Singh

Abstract Extraction of the local color texture has been one of the most desired and challenging research areas in face image recognition due to the ever-increasing use of color face images in numerous applications including biometric pattern matching, surveillance, and security. In this paper, we propose two novel color local texture descriptors called the local binary pattern of correlation of adjacent pixels (LBPC_A) and the local binary pattern of correlation of neighborhood pixels with the center pixel (LBPC_C). These two operators provide different local characteristics of an image. The operator LBPC_A provides the low-order texture characteristics while the operator LBPC_C provides the high-order texture characteristics. When these two operators are fused at the feature level, they provide high recognition rates with a small feature size for the face images. The resulting approach is referred to as LBPC_A + LBPC_C. The performance of the proposed operators has been compared with the state-of-the-art LBP-like operators on three color face databases-FERET, CMU-PIE, and AR. Experiments conducted on FERET, CMP-PIE, and AR demonstrate that the proposed descriptors outperform the state-of-the-art color texture descriptors.

Keywords Local binary pattern (LBP) · Local binary pattern for color images (LBPC) · Local color texture · Color face recognition

1 Introduction

Face recognition is an active research area in computer vision and pattern recognition applications. A well-designed face recognition system depends heavily on the feature extraction process. Over the last four decades, many descriptors have been developed which can broadly be categorized as the holistic and the local descriptors. Most commonly used holistic methods, such as eigenfaces [1] and Fisherfaces [2] built on PCA and LDA, respectively, have been found very successful for face recognition.

S. Majeed (✉) · C. Singh
Department of Computer Science, Punjabi University, Patiala 147002, India
e-mail: Shahbaz.nengroo858@gmail.com

Y. Singh et al. (eds.), *Proceedings of International Conference on Recent Innovations in Computing*, Lecture Notes in Electrical Engineering 1001,
https://doi.org/10.1007/978-981-19-9876-8_26

They are comparatively simple and fast to compute. Most of them can deal effectively with the global changes in the image such as rotation, translation, scale, and noise. However, their performance is not as high as the local descriptors. As a consequence, the local descriptors have gained much popularity in face recognition systems. Local descriptors are robust to illumination and pose variations by generating distinguishable features to efficiently represent the spatial structural information of an image. However, their performance is affected by noise in the image.

The local binary pattern (LBP) is one of the most successful local descriptors. The LBP operator was originally designed as a texture descriptor [3] for grayscale images. It has been used successfully in feature classification [4–6], face recognition and face detection [7–10], facial expression recognition [11–13], and image retrieval [14]. The LBP operator is a non-directional first-order circular descriptor that concatenates binary results for generating the micropatterns. A higher-order local derivative pattern (LDP) derived from an LBP image is proposed by Zhang et al. [15] for face recognition which can capture much more successfully the discriminative features of a face image. Murala et al. [16] proposed local tetra patterns (LTrP) to generate more distinguishable information by using four distinct values from two high-order derivative direction patterns. Fan et al. [17] proposed another higher-order descriptor called local vector pattern (LVP).

With the increasing use of color images over the Internet and with the enhancement in the processing speed and memory of the current computing devices, color face recognition has attracted considerable attention in the recent past. This motivates the researchers to develop descriptors that can represent color texture patterns as effectively as the LBP does for grayscale images. A simple but effective method to derive texture from color images is to extend the existing grayscale descriptors to each channel of the color image. Maenpaa et al. [18] used this strategy for color texture descriptors. But, it suffers from information redundancy as observed by Maenpaa et al. [19] who have used only three sets of opponent colors for representing the correlation between different color channels. Choi et al. [20] used YC_bC_r color space and derived an LBP histogram for each channel and three sets of opponent channels. After concatenating LBP histograms from different regions of a face image, feature dimension reduction techniques, viz. PCA [1], FLDA [2], RLDA [21], and ERR [22], were used. Lee et al. [23] proposed a local color vector binary pattern (LCVBP) that uses color norm patterns and color angular patterns for color face images. Lan et al. [24] proposed a quaternion local ranking binary pattern (QLRBP) for image retrieval by combining color information using quaternion representation of color images. Its recognition rate is less as compared to other similar operators as reported by [25]. Li et al. [26] proposed color local similarity patterns (CLSP) for representing color images as the co-occurrence of its image pixels color quantization information and the local color image texture information. Recently, Dubey et al. [27] proposed two descriptors multichannel adder local LBP (MALBP) and multichannel decoder LBP (MDLBP) and applied them to the problems of image retrieval. Singh et al. [25] proposed a new descriptor, the local binary pattern for color images (LBPC) for image retrieval.

In this paper, we propose two operators for deriving texture features of color images. The two operators are referred to as the local binary patterns of correlation of adjacent pixels (LBPC_A) and correlation of neighborhood pixels with the center pixel (LBPC_C). These operators are the result of the projection of the color pixels on the reference unit vector. The reference unit vector also represents a color pixel which is obtained adaptively in the neighborhood of each pixel of the image. These two patterns represent two types of characteristics of the image in the local regions. The low-order texture characteristics are represented by the LBPC_A operator and the high-order characteristics are represented by the LBPC_C operator. When the features of these two operators are fused, they provide a high recognition rate.

The rest of the paper is organized as follows. The proposed operators based on the local correlation of pixels are presented in Sect. 2. In Sect. 3, detailed experimental results on face recognition have been provided for the face database: CMU-PIE, FERET, and AR. Finally, results and conclusions are given in Sect. 4.

2 The Proposed Methods: Local Correlation of Pixels

The proposed methods for deriving correlation among pixels in a neighborhood consist of two operators—local binary patterns of correlation of adjacent pixels (LBPC_A) and correlation of neighborhood pixels with the center pixel (LBPC_C). The two operators are then fused to obtain discriminative texture features of color images.

2.1 *The LBPC_A Operator*

Let a vector $\boldsymbol{I}(,y) = (r(x,y), g(x,y), b(x,y))$ or simply $\boldsymbol{I} = (r,g,b)$ represent a color pixel in the (R,G,B) color space. We define a local window W of size $(2R+1)\times(2R+1)$, $R \geq 1$, centered at a pixel c with a color vector $\boldsymbol{I_c} = (r_c, g_c, b_c)$. To understand the steps of the derivation of the operator, we set $R = 1$ and $P = 8$ circularly enumerate the neighborhood pixels. However, the discussions are carried on for a general value of P. Let $\boldsymbol{I_p} = \left(r_p, g_p, b_p\right)$, $p = 0, 1, 2, \ldots, P-1$ be the neighborhood of the center c. To determine the correlation of the adjacent pixels we take a reference unit vector $\boldsymbol{I_f} = \left(r_f, g_f, b_f\right)$ that represents the local characteristics of the image and determines its response to the pixels $\boldsymbol{I_p}$, $p = 0, 1, \ldots, P-1$ by taking the dot product of the two vectors as follows:

$$l_p = \boldsymbol{I_p}.\boldsymbol{I_f}, \quad p = 0, 1, 2, \ldots, P-1, \tag{1}$$

where P is the total number of neighborhood pixels. It is noted that the response l_p is the cousin measure of the angle between the vectors $\boldsymbol{I_f}$ and $\boldsymbol{I_p}$ manifested by the magnitude of the vector $\boldsymbol{I_p}$. A cosine response, instead, does not reflect the behavior of intensity of pixel p which can be an important trait for representing the local characteristic of the color image. Alternatively, l_p maybe regarded as the projection of the vector $\boldsymbol{I}_p$ on the unit vector $\boldsymbol{I}_f$. Keeping this aspect in view, we do not, therefore, use directly the angle between the two vectors. Having obtained the value of l_p, we threshold these values to 1 or 0 depending on whether $l_p \geq l_{p+1}$ in a circular order. Mathematically, we determine a binary value of 1 or 0 as:

$$E(l_p) = \begin{cases} 1, \text{ if } l_p \geq l_{p+1}, \; p = 0, \ldots, P-1, \text{ with } l_{P+1} = l_0, \\ 0, \text{ otherwise.} \end{cases} \tag{2}$$

The binary string of P pixels is converted into a decimal value as

$$D = \sum_{p=0}^{P-1} E(l_p) 2^p. \tag{3}$$

Now, the matter rests on the determination of the reference unit vector $\boldsymbol{I}_f$. As shown in [25], among the several choices for a reference vector, the vector obtained by averaging the color pixels of the window W provides an effective proposition for the said task. Although in [25], the averaging of local pixels is carried out to find the normal to a hyperplane and the context of the problem there is entirely different, the normal vector provides a role similar to that provided by the reference vector in the proposed operator. Mathematically, the components of the reference vector $\boldsymbol{I}_f$ are obtained by averaging the components of the neighborhood pixels

$$r_f = \frac{1}{(2R+1)^2} \sum_{x=0}^{2R} \sum_{y=0}^{2R} r(x, y), \; g_f = \frac{1}{(2R+1)^2} \sum_{x=0}^{2R} \sum_{y=0}^{2R} g(x, y),$$

$$b_f = \frac{1}{(2R+1)^2} \sum_{x=0}^{2R} \sum_{y=0}^{2R} b(x, y). \tag{4}$$

To provide cross-correlation between the reference vector and the pixel $\boldsymbol{I}_p$, we provide a circular shift on the components r_f, g_f and b_f and chose the reference vector as

$$\boldsymbol{I}_f = (b_f, r_f, g_f) / \left(\sqrt{b_f^2 + r_f^2 + g_f^2} \right). \tag{5}$$

There are four other combinations of the circular shift (r_f, g_f, b_f) but their response to the performance of the resulting operator is more or less the same. However, all five combinations provide better results than the choice $\boldsymbol{I}_f = (r_f, g_f, b_f)$. Therefore, we select the reference vector given in Eq. (5). With this choice of the reference vector, the explicit form of the response l_p in Eq. (1) is given by

$$l_p = r_p b_f + g_p r_f + b_p g_f, p = 0, 1, 2, \ldots, P - 1 \quad (6)$$

It may be noted that the concept of cross-correlation among color channels is also used by CLBP [20] and LCVBP [23] who have used opponent colors to take the advantage of cross-correlation. Their approach to using cross-correlation is, however, different from what we have proposed here.

It is observed that the value of the LBPC_A operator is obtained as a result of the interaction of a neighborhood pixel with its adjacent pixel. The role of the center pixel comes into play in obtaining the reference vector $\boldsymbol{I}_f$. The interaction of the neighborhood pixels with the center pixel is obtained using the second operator LBPC_C which is explained as follows.

2.2 The LBPC_C Operator

As opposed to the LBPC_A operator which finds the local binary patterns by comparing the response of the adjacent pixels with a reference vector, the LBPC_C operator compares the response of the neighborhood pixels with the response of the center pixel. Thus, Eq. (6) responds to the neighborhood pixels with the response unit vector given by $l_p, p = 0, 1, \ldots, P - 1$. The response of the center pixel with the reference unit vector is given by

$$l_c = \boldsymbol{I}_c.\boldsymbol{I}_{f,} = r_C b_f + g_C r_f + b_C g_f \quad (7)$$

Next, we compare the neighborhood response values l_p with the center response value l_c to obtain a binary value of 1 or 0, depending on whether l_p is greater or equal to a scaled value of l_c or otherwise, i.e.,

$$E(l_p) = \begin{cases} 1, \text{ if } l_p \geq kl_c, p = 0, 1, \ldots, P - 1, \\ 0, \text{ otherwise}, \end{cases} \quad (8)$$

where $k \geq 1$ represents the scaling parameter which plays an important role in determining the local characteristics of the image which will be discussed a little later. The process of obtaining a binary string of length P and its conversion to the decimal value is the same as for the LBPC_A.

The main objective of using the LBPC_C operator is to locate the regions with high details such as edges, corners, and other fine structures. It may be recalled that the LBP is a non-directional derivative operator, and hence, it is used here to find out the high-order gradient region in the image. It is observed that for constant regions where the variation in the intensity value of a neighborhood pixel is not significantly different from the center pixel, the operator will provide a binary string with all zeros, whereas the response will be the opposite when the difference is large. This phenomenon helps in identifying key regions of a face image such as the eyes, nose, mouth, etc. For the case, $k = 1$, the operator reduces to the LBPC operator [25].

3 Experimental Analysis

This section provides several experimental results to demonstrate the effectiveness of the proposed methods and compare their results with the closely related existing color texture operators—CLBP [20], LCVBP [23], MDLBP [27], LBPC [25], and LBPC + LBPH + CH [25]. The number of features used by all methods is also mentioned in the results. In the following subsections, we describe in detail the databases we use in our experiments, the evaluation methodology, and the experiment results.

3.1 Experimental Setup

The performance of a face recognition system depends not only on effective features but also on effective similarity measures or distance metrics. Several similar measures are very successful in face recognition systems. We have used non-training-based distance measures which are very fast and commonly used for face recognition systems. The most prominent distance measures are chi square, square-chord, and histogram intersection.

Three publicly available face databases, CMU-PIE [28], AR [29], and FERET [30] are used in our experiments. To extract the face region from the images, all facial images used in our experiments are cropped manually. Each image is cropped to the size 120×120 using coordinates of the eyes. Each image is then divided into 64 blocks, each block of size 15×15.

To validate the robustness of the proposed LBPC_A and LBPC_C feature descriptors against variation due to illumination, expression, and occlusion, we have constructed six datasets for experimental analysis. The first dataset is constructed from the CMU-PIE database to analyze illumination variations. For this purpose, we have used 1428 frontal images of 68 persons with 21 images per person. All the images were captured with room lights off condition. The second dataset is prepared from the FERET database with expression variation. For this purpose, we take fa subset of the FERET dataset as the gallery database which consists of 847 images. The subset fb of FERET is used for expression variation which also consists of 847 images.

The other four datasets are constructed from the AR database which represents four variations: illumination, expression, occlusion due to sunglasses, and scarf. We randomly selected 100 persons from the full AR database of 4000 images. Two natural images of each of these 100 persons represent two successive sessions of photography resulting in a dataset of 200 images which serves as the gallery dataset for all experiments. The first four variations are used to construct four datasets AR-I,

AR-II, AR-III, and AR-IV which consist of 600 images each. These datasets have three images for each person with two sessions of photography resulting in six images per person.

3.2 Selection of Parameters

Several parameters affect the performance of the proposed operators. These are the size of the window (W), number of neighborhood points (P), number of blocks into which the face image is divided, and the scale parameter k. As suggested in [25], we take $R = 2$ which yields W of the size 5×5. A bilinear interpolation is performed to find equally spaced neighborhood points with coordinated $x_p = R\cos\left(\frac{2\pi p}{P}\right)$ and $y_p = R\sin\left(\frac{2\pi p}{P}\right)$, $p = 0, 1, \ldots, P-1$, and the current pixel is the center of the circle. In this paper, we have taken $P = 8$. As considered by several authors [7–9, 20], we divide a face image into 64 blocks (8×8). The selection of the scaling parameter k for the effective use of the operator LBPC_C which encodes the high-order details of an image is very crucial for the high performance of the proposed system. We determine its value empirically by conducting face recognition experiments on the AR database which represents the four variations: illumination, expression, and occlusion due to sunglass and scarf. For this purpose, we take 200 natural images of 100 persons in the gallery dataset and 2400 images are taken in the probe dataset. The average recognition rates obtained by the LBPC_A and LBPC_C are shown in Fig. 1a for $k = 1.0$ to 2.0 with an increment of 0.1. The recognition rate of LBPC_A is 89.94% which is represented by the horizontal line as it is independent of k. The recognition rate of LBPC_C is 87.34% for $k = 1$. When we concatenate the features of LBPC_A and LBPC_C, the average recognition rate increases from 89.72% ($k = 1.0$) to 92.04% ($k = 1.4$) as shown in Fig. 1b. To compare the recognition rates of LBPC_A + LBPC_C, the recognition rate of 89.94% of LBPC_A is also plotted as the horizontal line. Note that for $k = 1.0$ the performance of the individual operator LBPC_A and LBPC_C are 89.94% and 87.34%, respectively as shown in Fig. 1a. The recognition rate of the fused operators (LBPC_A + LBPC_C) is 89.72% which is lower than the 89.94% achieved by LBPC individually. Therefore, the gain in the recognition rate achieved by LBPC_A + LBPC_C over the value 89.94% is 2.10%, for $k = 1.4$. Therefore, for all experiments conducted in this paper, we take a value of $k = 1.4$. This shows that the operator LBPC_C represents complementary features to LPC_A features for $K > 1.0$ and around $k = 1.4$ its effect is the most prominent.

Tables 1 and 2 show the recognition rate (%) on AR-1 and CMU-PIE datasets, respectively.

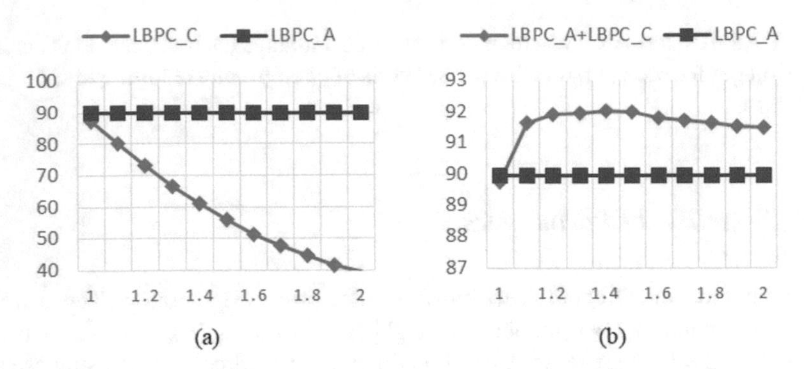

Fig. 1 Average recognition rate as a function of scale parameter k: **a** LBPC_A and LBPC_C, and **b** LBPCA + LBPC_C. Experiments are conducted on the AR dataset

Table 1 Recognition rate (%) achieved by various methods on AR-I dataset of AR face database under illumination variation using various distance measures

Methods	Recognition rate (%)			
	Square chord	Chi square	Histogram intersection	Average
CLBP [20]	88.83	87.50	80.83	85.72 (6)
LCVBP [23]	96.50	97.50	97.83	97.28 (4)
MDLBP [27]	56.43	58.71	56.57	57.24 (7)
LBPC [25]	99.50	99.17	99.00	99.22 (3)
LBPC + LBPH + CH [25]	83.17	86.83	90.00	86.67 (5)
LBPC_A (proposed)	98.83	99.50	99.67	99.33 (2)
LBPC_A + LBPC_C (proposed)	99.50	99.50	99.67	99.56 (1)

Table 2 Recognition rate (%) achieved by various methods on CMU-PIE dataset under illumination variation using various distance measures

Methods	Recognition rate (%)			
	Square chord	Chi square	Histogram intersection	Average
CLBP [20]	92.13	92.43	93.46	92.67 (5)
LCVBP [23]	95.81	95.37	94.12	95.10 (2)
MDLBP [27]	94.85	94.85	93.97	94.56 (4)
LBPC [25]	92.65	92.13	87.94	90.91 (6)
LBPC + LBPH + CH [25]	57.72	59.34	62.50	59.85 (7)
LBPC_A (proposed)	94.56	94.93	94.71	94.73 (3)
LBPC_A + LBPC_C (proposed)	97.87	97.94	97.43	97.75 (1)

3.3 Performance Evaluation Under Illumination Variation

To analyze the effect of the illumination variation, we perform experiments on AR-I and CMU-PIE datasets whose constructions are explained in Sect. 3.1. The recognition rates achieved by the various existing methods and the proposed methods along with the size of feature vectors are given in Table 1 for the AR-I dataset across the various distance measures. For a glance, the ranking of the methods is written within brackets along with the average recognition rates. The highest average recognition rate is achieved by the proposed method LBPC_A + LBPC_C which is 99.56% followed by LBPC_A (99.33%), LBPC (99.22%), and LCVBP (97.28%), LBPC + LBPCH + CH (86.67%), CLBP (85.72%), and lastly, by MDLBP (57.24%). It is observed that the performance of LBPC and LCVBP is very close. However, LBPC not only provides a better recognition rate than LCVBP but also the size of its feature vector is one-fourth of the LCVBP. The recognition rates on CMU-PIE obtained by various methods are given in Table 2. Here too, the proposed method LBPC_A + LBPC_C provides the best average recognition rate of 97.75% which is 2.65% more than the next best average recognition rate of 95.10% achieved by LCVBP. Not only does the method LBPC_A + LBPC_C achieves a higher recognition rate, but also its feature vector is half of the size of the feature vector of LCVBP. The performance of our second approach LBPC_A (94.73%) is only 0.37% less than that of LCVBP, whereas the size of the feature vector of LCVBP is four times the feature size of LBPC_A. The performance of LBPC (94.56%) is also comparable to LBPC_A and LCVBP and this operator also uses a feature vector whose size is one-fourth of the LCVBP.

3.4 Performance Evaluation Under Expression Variation

For this purpose, we use the AR-II dataset of the AR database and fa fb datasets of the color FERET database. The FERET dataset fa is used as the gallery and fb is used as the probe. The results of all seven methods are given in Table 3 for the AR-II dataset. It is observed that the proposed method LBPC_A + LBPC_C provides the highest average recognition rate of 91.67%, followed by LBPC + LBPH + CH (90.44%), CLBP (89.39%), LCVBP (87.61%), LBPC_A (87.505), MDLBP (86.17%), and lastly, by LBPC (85.22%). When look for the performance of the various methods on the FERET dataset, here too, the method LBPC + LBPH + CH provides the best average recognition rate of 88.21% as shown in Table 4. Here, LCVBP has stood second with an average recognition rate of 86.97% followed by MDLBP (86.60%), the proposed methods LBPC_C + LBPC_A (84.78%), and LBPC_A (84.78%), CLBP (84.16%), and LBPC (82.8%).

Tables 3 and 4 show the recognition rate (%) on AR-II and FERT database, respectively.

Table 3 Recognition rate (%) achieved by various methods on AR-II database under expression variation using various distance measures

Methods	Recognition rate (%)			
	Square chord	Chi square	Histogram intersection	Average
CLBP [20]	89.00	89.17	90.00	89.39 (3)
LCVBP [23]	88.33	87.50	87.00	87.61 (4)
MDLBP [27]	85.50	87.33	85.67	86.17 (6)
LBPC [25]	84.17	85.33	86.17	85.22 (7)
LBPC + LBPH + CH [25]	91.33	90.50	89.50	90.44 (2)
LBPC_A (proposed)	87.83	87.00	87.67	87.50 (5)
LBPC_A + LBPC_C (proposed)	90.00	92.75	92.25	91.67 (1)

Table 4 Recognition rate (%) achieved by various methods on $fafb$ dataset of FERET database under expression variation using various distance measures

Methods	Recognition rate (%)			
	Square chord	Chi square	Histogram intersection	Average
CLBP [20]	84.00	84.24	84.24	84.16 (6)
LCVBP [23]	86.48	87.10	87.34	86.97 (2)
MDLBP [27]	86.23	86.35	87.22	86.60 (3)
LBPC [25]	81.64	82.75	84.00	82.80 (7)
LBPC + LBPH + CH [25]	88.34	88.21	88.09	88.21 (1)
LBPC_A (proposed)	84.62	84.99	84.74	84.78 (5)
LBPC_A + LBPC_C (proposed)	84.74	84.99	84.62	84.78 (4)

3.5 Experimental Results Under Occlusion Variation

For this purpose, we use AR-III and AR-IV datasets of the AR database. The results are given in Tables 5 and 6, respectively. As given in Table 5, the proposed method LBPC_A provides the best recognition rate of 94.78%. The performance of LBPC_A + LBPC_C is lower than LBPC_A because the operator LBPC_C which is supposed to encode higher-order details fails in the present case to do so. It is because the sunglasses occlude a significant part of the face image where the high details exist in the large amount and the high-order details remain ineffective in the occluded region. A notable difference between LBPC and LBPC_A is that the LBPC_A provides a significantly high average recognition rate as compared to LBPC, although their feature size and the derivation process are the same. This kind of difference is observed only in the illumination variation on the CMU-PIE dataset. In other experiments, the difference is between 0.11 and 2.28%. Table 6 displays the average recognition rates achieved by various methods on the AR-IV dataset which represents occlusion due to scarf. Here, the best average recognition rate of 84.44% is achieved by LBPC_A + LBPC_C followed by LBPC_A (77.78%), LCVBP (76.67%), LBPC (68.78%), LBPC + LBPH + CH (66.50%), CLBP (66.22%), and MDLBP (54.00%). An important difference between the recognition rates achieved by LBPC_A + LBPC_C on AR-III (occlusion due to sunglasses) and AR-IV (occlusion due to scarf) is that the LBPC_A + LBPC_C method not only provides the best recognition rate but its recognition rate is higher by 6.6% than the second-best recognition rate achieved by LBPC_A. In the AR-IV dataset, the occlusion due to the scarf does not block the eyes and their surrounding regions which represent the high-order details, unlike the AR-III dataset where these regions are blocked by sunglasses.

Table 5 Recognition rate (%) achieved by various methods on AR-III (occlusion due to sunglasses) dataset of AR face database under occlusion variation using various distance measures

Methods	Recognition rate (%)			
	Square chord	Chi square	Histogram intersection	Average
CLBP [20]	85.00	86.83	85.83	85.89 (6)
LCVBP [23]	90.50	91.00	91.17	90.89 (3)
MDLBP [27]	72.50	75.50	75.33	74.44 (7)
LBPC [25]	86.83	89.00	92.00	89.28 (4)
LBPC + LBPH + CH [25]	85.17	88.33	91.00	88.17 (5)
LBPC_A (proposed)	94.00	94.67	95.67	94.78 (1)
LBPC_A + LBPC_C (proposed)	93.00	93.83	95.00	93.94 (2)

Table 6 Recognition rate (%) achieved by various methods on AR-IV (occlusion due to scarf) dataset of AR face database under occlusion variation using various distance measures

Methods	Recognition rate (%)			
	Square chord	Chi square	Histogram intersection	Average
CLBP [20]	68.50	66.83	63.33	66.22 (6)
LCVBP [23]	76.83	76.50	76.67	76.67 (3)
MDLBP [27]	53.17	54.17	54.67	54.00 (7)
LBPC [25]	63.00	68.33	75.00	68.78 (4)
LBPC + LBPH + CH [25]	56.50	67.67	75.33	66.50 (5)
LBPC_A (proposed)	73.00	77.33	83.00	77.78 (2)
LBPC_A + LBPC_C (proposed)	81.67	84.33	87.33	84.44 (1)

Tables 5 and 6 show the recognition rate (%) on AR-III and AR-IV database, respectively.

4 Results and Conclusion

The results are summarized as follows:

1. The operator LBPC_A provides an effective color texture descriptor with the small size of the feature vector. This represents the low-order variations in texture patterns. Its performance is better than a recently developed color texture descriptor LBPC [25] with the same size of features.
2. The operator LBPC_C represents high-order variations in color texture. To provide both the low and high-order variations in color texture patterns, the features of LBPC_A and LBPC_C are fused. The resulting method LBPC_A + LBPC_C provides a high recognition rate for face images and surpasses the performance of the existing state-of-the-art color texture descriptors.
3. Among the existing state-of-the-art methods, LCVBP provides the best overall recognition rate of 86.89% which is less than that obtained by LBPC_A + LBPC_C and LBPC_A by 3.07% and 1.33%, respectively. Not only do the proposed methods provide higher recognition rates, but also their feature size are smaller, i.e., half and one-fourth of the feature size of LCVBP.

The proposed two operators—local binary patterns of correlation of adjacent pixels (LBPC_A) and correlation of neighborhood pixels with the center pixel (LBPC_C)—provide effective texture descriptors for color images. The operator LBPC_A encodes the low-order local characteristics of the texture features and the operator LBPC_C encodes the high-order local characteristics. When the local binary patterns of these two operators are fused at the feature level the resulting method, called LBPC_A + LBPC_C, provides a very high recognition rate with the small size of feature vectors. The performance of the proposed methods and the state-of-the-art color texture descriptors for the recognition of the face images has been carried out on the CMU-PIE, FERET $fafb$, and AR color face databases. We have considered five variations in face images: illumination, expression, occlusion due to sunglasses, and occlusion due to scarf.

References

1. Turk MA, Pentland AP (1991) Face recognition using eigenfaces. In: IEEE computer society conference on computer vision and pattern recognition. IEEE Computer Society, pp 586–587
2. Belhumeur PN, Hespanha JP, Kriegman DJ (1997) Eigenfaces vs. fisher faces: recognition using class specific linear projection. In: European conference on computer vision. Springer, Berlin, Heidelberg, pp 43–58

3. Topi M, Matti P, Timo O (2000) Texture classification by multi-predicate local binary pattern operators. In: 15th international conference on pattern recognition (ICPR-2000), pp 939–942
4. Topi M, Timo O, Matti P, Maricor S (2000) Robust texture classification by subsets of local binary patterns. In: 15th international conference on pattern recognition (ICPR-2000), pp 935–938
5. Pietikäinen M, Mäenpää T, Viertola J, Pietikainen M, Maenpaa T, Viertola J (2002) Color texture classification with color histograms and local binary patterns. In: Workshop on texture analysis in machine vision, pp 1–4
6. Guo Y, Zhao G, Pietikäinen M (2012) Discriminative features for texture description. Pattern Recogn 45(10):3834–3843
7. Ahonen T, Hadid A, Pietikäinen M, Pietik M (2004) Face recognition with local binary patterns. In: European conference on computer vision. Springer, Berlin, Heidelberg, pp 469–481
8. Ojala T, Pietikäinen M, Mäenpää T (2002) Multiresolution gray-scale and rotation invariant texture classification with local binary patterns. IEEE Trans Pattern Anal Mach Intell 24(7):971–987
9. Ahonen T, Hadid A, Pietikäinen M (2006) Face description with local binary patterns: application to face recognition. IEEE Trans Pattern Anal Mach Intell 28(12):2037–2041
10. Huang D, Shan C, Ardabilian M, Wang Y, Chen L (2011) Local binary patterns and its application to facial image analysis: a survey. IEEE Trans Syst Man Cybern Part C (Applications and Reviews) 41(6):765–781
11. Shan C, Gong S, McOwan PW (2005) Robust facial expression recognition using local binary patterns. In: IEEE international conference on image processing. IEEE, Genova, Italy, pp II-370
12. Zhao G, Pietikainen M (2007) Dynamic texture recognition using local binary patterns with an application to facial expressions. IEEE Trans Pattern Anal Mach Intell 29(6):915–928
13. Shan C, Gong S, McOwan PW (2009) Facial expression recognition based on local binary patterns: a comprehensive study. Image Vis Comput 27(6):803–816
14. Takala V, Ahonen T, Pietikainen M (2005) Block-based methods for image retrieval using local binary patterns. In: Scandinavian conference on image analysis. Springer, Berlin, Heiledlberg, pp 882–891
15. Zhang B, Gao Y, Zhao S, Liu J (2009) Local derivative pattern versus local binary pattern: face recognition with high-order local pattern descriptor. IEEE Trans Image Process 19(2):533–544
16. Murala S, Maheshwari RP, Balasubramanian R (2012) Local tetra patterns: a new feature descriptor for content-based image retrieval. IEEE Trans Image Process 21(5):2874–2886
17. Fan KC, Hung TY (2014) A novel local pattern descriptor-local vector pattern in high-order derivative space for face recognition. IEEE Trans Image Process 23(7):2877–2891
18. Maenpaa T, Pietikainen M, Viertola J (2002) Separating color and pattern information for color texture discrimination. In: Object recognition supported by user interaction for service robots. IEEE, Quebec, Canada, pp 668–671
19. Mäenpää T, Pietikäinen M (2004) Classification with color and texture: jointly or separately? Pattern Recogn 37(8):1629–1640
20. Choi JY, Ro YM, Member S, Plataniotis KN (2012) Color local texture features for color face recognition. IEEE Trans Image Process 21(3):1366–1380
21. Lu J, Plataniotis KN, Venetsanopoulos AN (2003) Regularized discriminant analysis for the small sample size problem in face recognition. Pattern Recogn Lett 24(16):3079–3087
22. Jiang X, Mandal B, Kot A (2008) Eigenfeature regularization and extraction in face recognition. IEEE Trans Pattern Anal Mach Intell 30(3):383–394
23. Lee SH, Choi JY, Ro YM, Plataniotis KN (2012) Local color vector binary patterns from multichannel face images for face recognition. IEEE Trans Imge Process 21(4):2347–2353
24. Lan R, Zhou Y, Member S, Tang YY (2016) Quaternionic local ranking binary pattern: a local descriptor of color images. IEEE Trans Image Process 25(2):566–579
25. Singh C, Walia E, Kaur KP (2018) Color texture description with novel local binary patterns for effective image retrieval. Pattern Recogn 76:50–68
26. Li J, Sang N, Gao C (2016) Completed local similarity pattern for color image recognition. Neurocomputing 182:111–117

27. Dubey SR, Singh SK, Singh RK (2016) Multichannel decoded local binary patterns for content-based image retrieval. IEEE Trans Image Process 25(9):4018–4032
28. Sim T, Baker S, Bsat M (2003) The CMU pose, illuminlation, and expression database. IEEE Trans Pattern Anal Mach Intell 25(12):1615–1618
29. Martinez AM, Benavente R (1998) The AR face database. CVC Technical Report 24
30. Phillips PJ, Moon H, Rizvi SA, Rauss PJ (2000) The FERET evaluation methodology for face-recognition algorithms. IEEE Trans Pattern Anal Mach Intell 22(10):1090–1104

Comparative Analysis of Transfer Learning Models in Classification of Histopathological Whole Slide Images

Javaid Ahmad Wani and Nonita Sharma

Abstract Histopathology is essential in medical imaging. Consequently, automatic histopathological computer vision significantly influences the overall affordability, reliability, and accessibility of healthcare. Histopathological examination is the diagnostic and research of tissue disorders. An integrated suite of histopathological samples has greatly aided doctors and researchers in the world of clinical research. Identifying cancerous tissues is critical for clinicians in providing appropriate oncology treatments. A whole slide image (WSI) is a digitized scanning of the tissues on the glass slide that allows the samples taken to be stored digitally on the computer system in the form of a digital picture. India is witnessing more than one million cases of breast cancer per year. WSI processing and storage have greatly aided professionals while encouraging researchers to develop more reliable and efficient fully automated analysis diagnosing models. ML, specifically DL-based models. DL models have outperformed in various disciplines, along with clinical applications and profound features in healthcare. In this study, we have exemplary tunned four models with a transfer learning approach to classify histopathical images. The validation accuracies achieved after fine-tuning these models are 91% for MobileNet V2, 90% for NasNet Large, 96% for Xception 94.43%, and a benchmark accuracy of 96.95% for EfficientNetV2L which is the latest model in Keras. We have plotted the confusion matrix for each model.

Keywords Histopathology · Deep learning · MobileNet V2 · NasNet large · Xception · EfficientNetV2L

J. A. Wani (✉)
Department of Computer Science and Engineering, Dr. BR Ambedkar National Institute of Technology Jalandhar, Jalandhar, Punjab, India
e-mail: Javaidcse14@gmail.com

N. Sharma
Department of Information Technology, Indira Gandhi Delhi Technical University for Women, Delhi, India
e-mail: Nonita@nitj.ac.in

Y. Singh et al. (eds.), *Proceedings of International Conference on Recent Innovations in Computing*, Lecture Notes in Electrical Engineering 1001,
https://doi.org/10.1007/978-981-19-9876-8_27

1 Introduction

Histopathology is the identification and analysis of tissue pathogens that includes the examination of body tissue under a magnifying glass. Histopathologists are capable of designing tissue diagnosis and treatment and assisting healthcare professionals in managing a healthcare service. Histopathological analysis refers to the procedure of pathologists that use a magnifying glass to locate, analyze, and classify deadly diseases such as cancer.

Substantial technological advancements have resulted in the implementation of novel digital imaging potential solutions in histopathology. Whole slide imaging (WSI), which refers explicitly to scanning traditional slides to create digital slides, is now the latest imaging technique used by pathologists worldwide. WSI is gaining popularity among many pathologists for diagnosing, academic, and scientific research [1].

Automatic analysis of histopathology image data has greatly aided doctors and researchers in medicine. Primarily because of the abundance of labeled data and technology that really can be used for analysis purposes, specialists from various fields of computer.

Science and computer vision are capable of contributing to medical advances. WSI is a digital scanning of the tissues which allows the representative sample to be stored as digital files on the computer system in the form of a colored image. WSI can be analyzed remotely by specialists for comprehensive diagnosis and treatment or as regard for future forecasts.

WSI processing and storage have greatly aided professionals while encouraging researchers to develop more reliable and efficient fully automated analyses diagnosing car models. ML, specifically DL-based models, strength medical imaging analysis software solutions. DL with a CNN is a rapidly growing area of histopathologic image analysis [2].

The continuous technological advancements in digital scanners, image visualization methods, and the incorporation of AI-powered methodologies with these kinds of systems available offer opportunities for new and emerging technologies. Its advantages are numerous, including easy accessibility via the web, exclusion of physical storage, and no risk of smudging depletion or slide breakdown, to name a few. Several hurdles, including heavy price, technical difficulties, and specialist reluctance to adopt an innovation, have been used in pathology [3].

Deep neural networks are used to perceive complex images. Convolutional networks are transforming many areas of medical imaging, yet the medical transformation of these advanced technologies is still in its early stages. One reason for the discrepancy is that CNNs, by definition, require massive labeled training data, which are not easily accessible from the perspective of histopathological examination. Another main factor is that CNN risk assessment requires validation in clinically defined external validation cohorts [4]. The main contribution of this paper is a comparative analysis of four fine-tunned transfer learning models in the classification of various diseases in the histopathological images. This paper consists of a literature review of previous proposed studies in Sect. 2, various approaches for classification of histopathological images using Artificial Intelligence in Sect. 3, proposed methodology in Sect. 4, models in Sect. 5, result and analysis in Sect. 6, and conclusion in Sect. 7.

2 Literature Review

DL models have outperformed in various disciplines, along with clinical applications and profound features in healthcare. Considering a dataset and a neural network that has already been specially trained to try to differentiate on massive vague datasets, there are indeed pre-trained alternative solutions. This method entails fine-tuning compared to pre-trained CNN as a classification method by fine-tuning only the last layers of most of the pre-trained network or by training algorithms from scratch. The most efficient and accurate deep learning techniques were used to classify colorectal cancer tissue using the ResNet50 achieved 79% accuracy reported the best accuracy rate, and the VGG-16 achieved 77% accuracy on the NCT-HE-100K data set of 100,000 histological [5] Osteosarcoma tumor has been slightly growing over the past two decades. It mainly affects long bones.

In [4], Kather et al. evaluated various DL models, fine-tuned them, and trained on the NCT-HE-100K, which was composed of 100,000 HE patches, and tested on the CRC-VAL-HE-7K, which contains about 7000 HE patches. VGG-19 has shown the outstanding performance of 94.3% in classifying the 9 different classes of cancer. In [6], Rashika et al. have carried out a comparative study of various deep learning models and a self-build model to classify the Osteosarcoma tumor images. The main aim was to develop an efficient and accurate classification model. The self-build model has shown a magnificent performance than the fine-tune models. The dataset utilized contains about 64k image samples, which were annotated manually with the help of experienced pathologists. The accuracics of VGGNet, LeNet, AlexNet, and the self-build model were recorded as 67%, 67%, 73%, and 92%, respectively [7]. Morteza et al. have introduced a dataset under the title of kimina path24, which has 24 different types of tissue classes selected on the basics of their texture pattern. Patches are the Three other methods employed to retrieve and classify the patches. The Bag-of-words method did not perform well; the local binary pattern (LBP) and CNN-based process have shown relatively good accuracy 41.33% and 41.80%, respectively.

In [8], Hira et al. performed patch-based DL-based were implemented for the detection and classification of breast cancer. The relatively small dataset consists of about 300 images of four different classes of breast cancer, which were taken from the publicly available were utilized, and 70k patches were created from the same dataset for the training and validation of the methods and achieved the accuracy of 86%. In [9], Kumar et al. introduced a dataset that is freely accessible from the KIMIA Lab official site. The dataset is composed of about 960 histopathological image samples of 20 different types of tissues. The LBP method showed a slightly good performance of 90.62%, BoVW achieved better accuracy, followed by CNN 96.50%, and 94.72%, respectively.

In [10], Tsai et al. collected colorectal cancer (CRC) histopathological samples and utilized them as the exploratory dataset to validate optimized parameters, and the potential of the 5 most widely DL models was used to accurately classify colorectal cancer tissues was evaluated by comparing performance on CRC-VAL-HE-7K, and CRC-VAL-HE-100K datasets and achieved an accuracy of 77 and 79%. In [11], Shubham et al. proposed the CNN model for the detection and classification of breast cancer histopathology images, which have reported 97.05% accuracy.

In [12], Shafieii et al. used the Kimia-Path 24C dataset, containing 24 WSIs from various tissue classes. The whole dataset is designed to resemble retrieval work activities in clinical practice. Color is a vital feature in histopathology. The color was completely disregarded in the Kimia-Path24 dataset, with all patches stored as gray-scale since retrieved from colored WSIs in the Kimia-Path24C dataset; the color feature has great significance. K-means clustering and the Gaussian mixture model (GMM) segmentation algorithms were employed to extract the interesting patches. VGG-16, Inception, and DenseNet models were used as feature extraction to provide further initial findings for setting a benchmark and have achieved accuracies of 92%, 92.45%, and 95.92%, respectively. The details of models and dataset along with accuracy for the selected papers is provided in Table 1.

Table 1 Literature review [4, 7, 9–16]

References	Model	Dataset	Accuracy (%)
2017 [7]	LBP CNN	Kather-texture-2016-image	41.33 41.80
2017 [14]	VGG-16, Inception-V	Kimia Path24	42.29, 56.98
2017 [9]	LBP CNN BoVW	KIMIA Path960	90.62 94.72 96.50
2019 [13]	CNN	MICCAI 2014	96
2018 [15]	CNN	AGATA	97
2019 [4]	VGG-19	NCT-HE-100K, CRC-VAL-HE-7K	994.3
2021 [10]	VGG-16, Inception-v3	Kimia Path24	77 79
2021[11]	CNN	Breast cancer Histopathological	97.05
2021 [16]	DenseNet, KamiaNet	TCG	71, 99
	VGG-19, KamiaNet	Endometrial cancer dataset	76.38, 81.41
	VGG-19, KamiaNet	Colorectal data	94.90, 96.80
2021 [12]	VGG-16 InceptionV3 DenseNet	Kimia Path24C	92 92.45 95.92

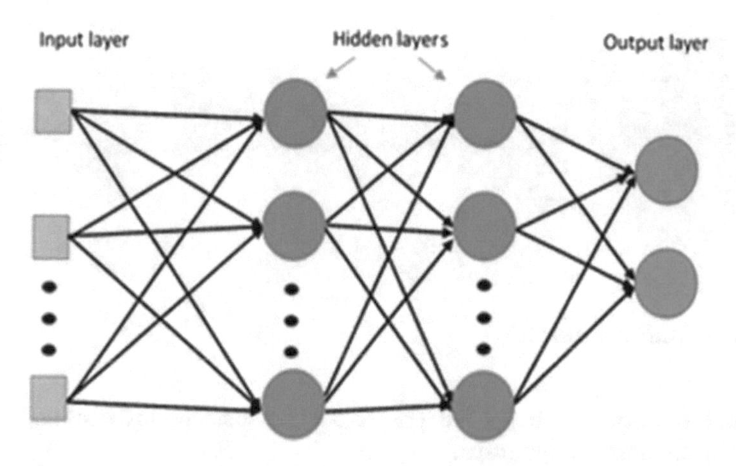

Fig. 1 Neural network [1]

3 Approaches in AI for Classification of Histopathological Images

3.1 Neural Networks

Also known as ANNs are a subset of ML and form the core of DL algorithms. They mimic the human brain working. ANNs are composed of multiple layers of nodes. The neurons are connected and have a certain weight and threshold. When the output of any particular node is greater than the threshold value, we say that the node is active, and it sends the data to the next layer node/neuron of the network. The structure of a basic neural network is shown in Fig. 1.

3.2 Deep Learning

DL is an ML subsection that uses Artificial neural networks (ANN). DL reduces the work required to generate a new feature extraction for each new challenge. ANN is the underlying methodology of DL. DL model directly accepts images as an input layer, feature extraction is carried in hidden layers, by convolution layer, pooling layer, every layer uses corresponding activation functions, final fully connected layer is added at output layer for actual classification of classes as deprecated in Fig. 2. Without human involvement in extracting the features, CNN-based classifications may be taught directly by employing raw images. Engineers have considerably improved the precision of disease diagnosis and categorization models in recent developments in hardware technology and DL. In agribusiness, Military, automation technology, speech synthesis, drug discovery and toxicology, Recommendation

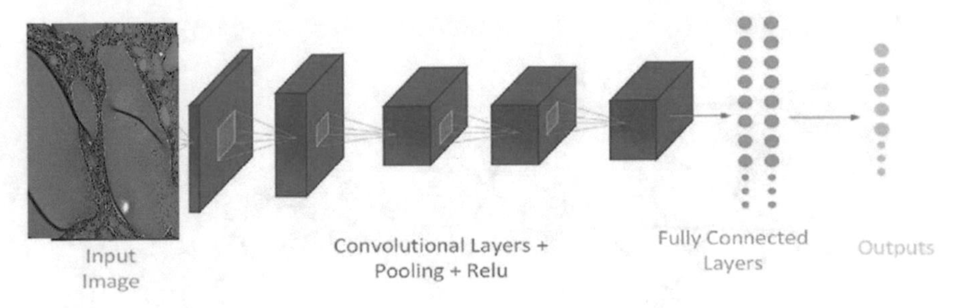

Fig. 2 Deep learning model [10]

systems, Bioinformatics, image recognition, and Financial fraud detection, CNN has demonstrated outstanding results.

CNN is a deep learning calculation that can catch symbolism, give esteem (intelligible and prejudicial instruments) to the different components/components in the image, and have the option to recognize one from the other. The underlying preparation needed for ConvNet is contrasted with other stage calculations. While the old channels are hand-made, ConvNets can peruse these channels/highlights with good practice. Singular neurons react to upgrades just in the limited area of the review field known as the Reception Field. The assortment of such fields penetrates to cover the whole apparent surface. Without much of a stretch, CNN, which can recognize and arrange objects with negligible pre-preparing, prevails in breaking down visual pictures. Without much time, it can remember the necessary high lights by its numerous lines structure.

The DL model can be broadly divided into two components:

(a) Feature extraction phase: In this phase, we train deep architectures on an extensive dataset and extract a feature using the cascade of different layers. We simply input the images and then feed them to other layers.
(b) Disease classification: In this phase, the images of the various diseases are classified into the respected class.

3.3 Transfer Learning (TL)

TL is an ML optimization technique that emphasizes transferring knowledge acquired while solving one task to another but a similar study. Using transfer learning minimizes the training cost, improves accuracy, and obtains low generalization error.

To obtain the best performance compared to existing CNN models in different implementations, building and training a CNN architecture from scratch is a complex procedure. Therefore, other models may be retrained according to the applications used for feature engineering. Machine learning and Knowledge Discovery in Data (KDD) have made massive progress in numerous knowledge engineering areas, including classification, regression, and clustering [17]. The model's pre-trained model or wanted segment can be incorporated straightforwardly into another CNN model. The weight of the pre-trained models might be freezing in some applications. During the development of the new models, weight can be updated with slower learning rates, which allows the pre-trained model to behave like weight initialization when the new model is learned. The pre-trained model may also be utilized as a weight initialization, classifier, and extractor.

3.4 Fine-Tunned CNN Model

Building a deep learning model from scratch is no easy task. We can make some changes in the architecture of the deep learning model as per the problem to be solved. The generalized flowchart for fine-tuning any deep learning model is demonstrated in Fig. 3. Fine-tuning a deep learning model is crucial to improving the precision of anticipated outcomes. We can adjust the deep learning model to our interesting data by freezing the inclinations of convolutional layers as we train. We gradually update the weights beginning with the most minimal level layers and working our way to the top. It learns a lot from pre-trained consequences while training fine-tuned models. In the wake of training and testing, we can contrast our networks. The work flow diagram for the fine-tuned CNN model is shown in Fig. 3.

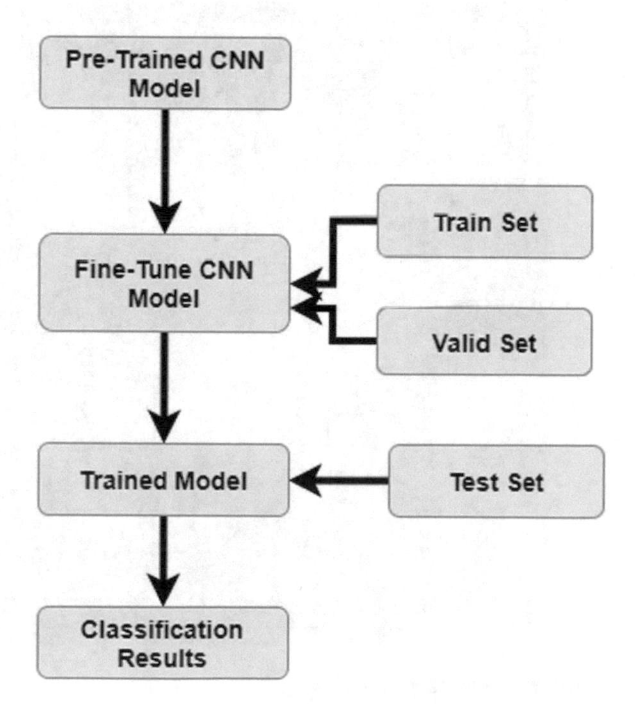

Fig. 3 Work flow diagram [10]

4 Proposed Methodology

Proposed methodology stages are shown in Fig. 4.

Dataset: In this study, we have used a new dataset, KimiaPath24C, for the classification of digitized pathological images. This dataset consists of 24 WSI images of various tissues. The background pixels were blurred to white. The whole dataset is designed to resemble retrieval work activities in clinical practice. Color is a vital feature in histopathology. The color was completely disregarded in the Kimia-Path24 dataset, with all patches stored as gray-scale since retrieved from colored WSIs in the Kimia-Path24C dataset; the color feature has great significance. The patch size is set to 1000 × 1000 selected by the manual process by marking the differences. The selected patches were processed with removal and selection processes, thereby further saving them as testing patches, and the remaining patches were kept for the training set. The different types of images of the dataset are shown in Fig. 5.

- **Preprocessing**: In this step, we loaded the data from a directory. We split the data 20% for validation and 80% for training. We set our batch size to 32, and the image is 224 × 224. Since the images are histopathical, so the color mode is kept as RGB. The shuffle size for images is actual by default, but for creating the confusion matrix, we set the value to false in that same step.

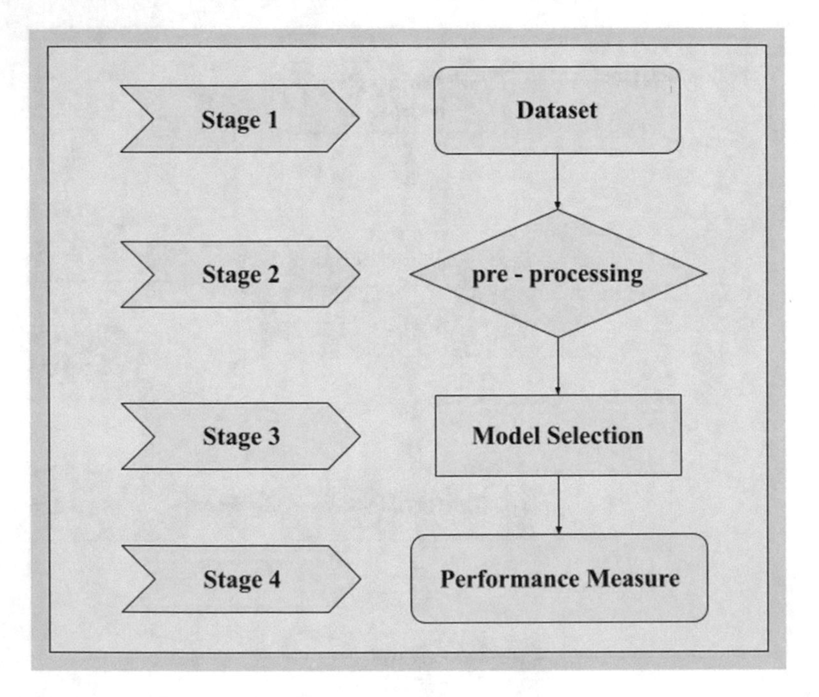

Fig. 4 Proposed methodology [10]

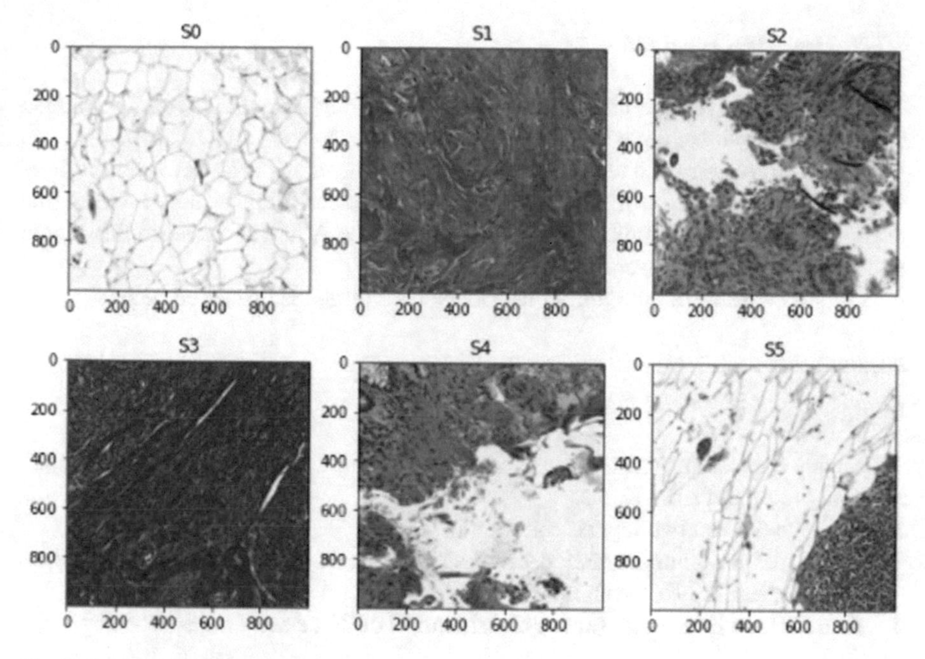

Fig. 5 Sample from dataset

5 Models Used

5.1 Mobile Net V2

The Mobile Net V1 was introduced to decrease complexities, and it was developed so light that it was able to use on mobile devices. It reduces required memory. Mobile Net V1 uses Depth wise Separable Convolution. Mobilc Net V2 uses three structures in it.

5.1.1 Depth Wise Separable Convolution

This operation has two parts.

Depth wise in which filter is applied per channel, and a pointwise filter is used to an output of the previous phase.

5.1.2 Linear Bottleneck

Pointwise is combined with bottleneck, and linear activation is used.

5.1.3 Inverted Residual

The expansion layer is added at the block input's beginning, and output is added together as output for the whole block.

The two blocks in it have three layers the first layer is in the convolution layer with ReLu proceeding with the next with depthwise convolution, and the third one does convolution with nonlinearity. We have used the Adam compiler in this model and the categorical_crossentropy loss function.

The step wise representation of MobileNet V2 model is discussed below.

Algorithm

Begin

1. Divide dataset into the train, test, and validate.
2. Call the model from Keras.
3. Remove the last two layers from model.
4. Used Adam optimizer and decreased learning to 0.0001 and used categorical_crossentropy loss function.
4. Add a dense layer class for the classification of 24 classes in the images.

End

The accuracy achieved from MobileNet V2 is shown in Fig. 6.

We have achieved a training accuracy of

- 91.5% training accuracy
- 91.35% testing accuracy, and
- 99.65% validation accuracy.

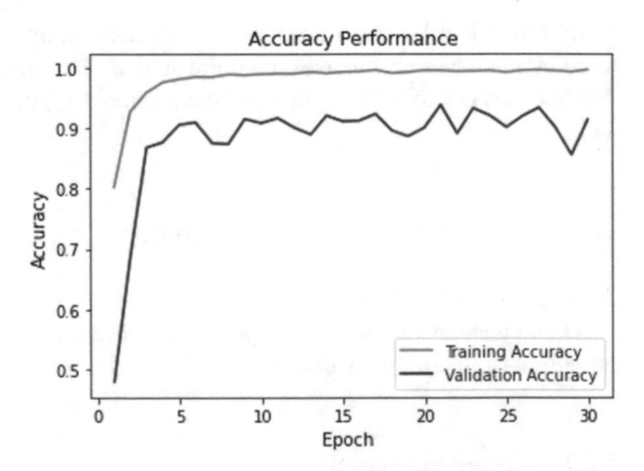

Fig. 6 Mobile Net V2 performance

5.2 Xception Model

It was developed by Google scholars and is inspired by the inception model and ResNet. The convolution operations are depthwise and pointwise in it. It performs shortcut operations like ResNet. We have used the activation function as softmax in this model. We have used categorical_crossentropy as a loss function in it. In this model, there are 84,936,983 total parameters out of which 84,740,315 are trainable. The step wise representation of Xception model is discussed below.

Algorithm

Begin

1. Divide dataset into the train, test, and validate.
2. Call the model from Keras.
3. Remove the last two layers from model.
4. Used Adam optimizer and decreased learning to 0.0001 and used categorical_crossentropy loss function.
4. Add a dense layer class for the classification of 24 classes in the images.

End

The accuracy achieved after successful completion of Xception model is presented in Fig. 7.

In this model, we have achieved the accuracy of

- 94.44% training accuracy
- 96.87% testing accuracy, and
- 99.64% validation accuracy.

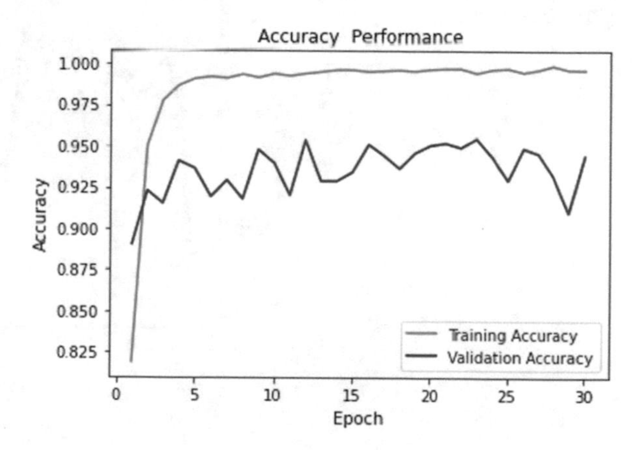

Fig. 7 Xception model performance

5.3 Efficient Net V2L

This family of CNNs has high speed in training than previous ones. EfficientNet V2 uses a 3×3 kernel size. In this model, we have used the global average pooling 2D layer and activation function as SOFTMAX and compiler as SGD. The accuracy achieved from Xception model is presented in Fig. 8.

In this model, we have achieved an accuracy of

- 98.12% training accuracy
- 95.16% testing accuracy, and
- 98.65% validation accuracy.

The step wise representation of EfficientNet V2L model is discussed below.

Algorithm
Begin

1. Divide dataset into the train, test, and validate.
2. Call the model from Keras.
3. Add a global average pooling layer, a drop out layer before the last two layers.
3. Remove the last two layers from model.
4. Used Adam optimizer and decreased learning to 0.0001 and used categorical_crossentropy loss function.
4. Add a dense layer class for the classification of 24 classes in the images.

End

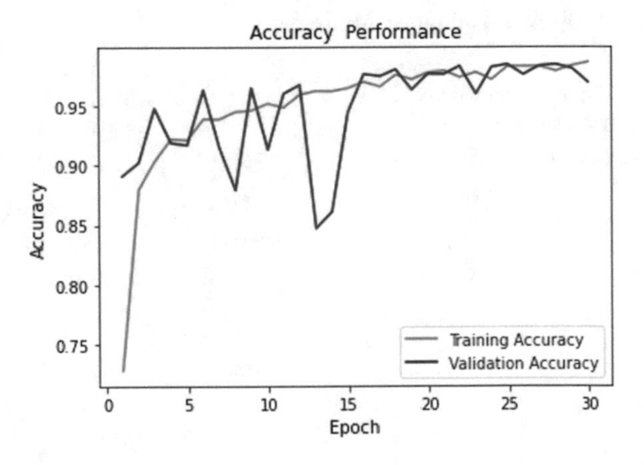

Fig. 8 Efficient Net V2L performance

5.4 NasNet Large

This model is one of the finest models of the CNN family, which has been trained on millions of images. The model is acquainted with classifying thousands of images of large datasets, and it reuses this approach in the classification of different problems. The input shape is (331, 331, 3) in this model by default. In this model, we have used SOFTMAX as an activation function. There are 84,936,983 total parameters out of which 84,740,315 are trainable. The step wise representation of NasNet Large model is discussed below.

Algorithm
Begin

1. Divide dataset into the train, test, and validate.
2. Call the model from Keras.
3. Remove the last two layers from model.
4. Used SGD optimizer and decreased learning to 0.0001 and used categorical_crossentropy loss function.
4. Add a dense layer class for the classification of 24 classes in the images.

End

The accuracy achieved from NasNet large model is presented in Fig. 9.

In this model, we have achieved an accuracy of

- 99.95% trainig accuracy
- 90% testing accuracy, and
- 90.53% validation accuracy.

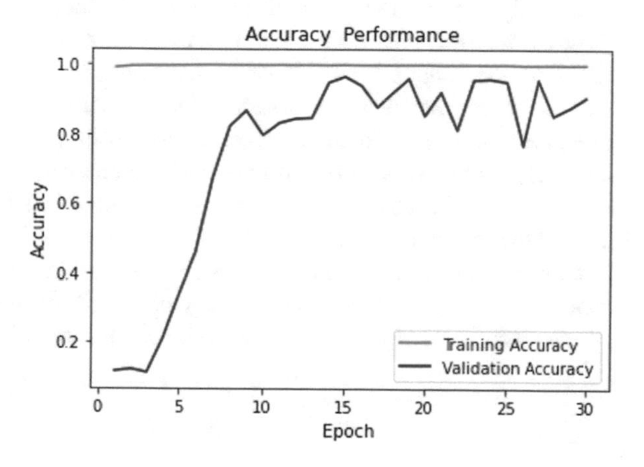

Fig. 9 NasNet large performance

6 Result and Analysis

In this study we have done fine-tuning in CNN models on the KimiaPath24C which contains 23,915 images, We have given 70% of data for training, 20% of data for validation, and 10% of data for testing. We have trained our models in 30 epochs with a learning rate of 0.001, we have used Adam and SGDM optimizers in our models. We have plotted the confusion matrix for the models used in this. We have evaluated our models based on precision, recall, and f1 score. We have analyzed that we have achieved the highest validation accuracy in the EfficientNet V2L at 98.2% followed by Xception at 94%.

These confusion matrix have been plotted for the validation data of MobileNet V2, NasNet Large, Xception, and EfficientNet V2L, as show in Figs. 10, 11, 12, and 13, respectively. We have un_freezing all weights and have trained our models from scratch. We have .hd5 files for each model which have architecture and weights saved in it. We can provide them on request via mail.

There is a total of 1325 images and 24 classes of tissues present in this dataset. Out of which MobileNet V2 has predicted 1185 correct.

This model has predicted 1185 images as correct.

This model has predicted 1299 images as correct.

This model has predicted 1300 images as correct. Analysis of different models in shown in Table 2.

7 Conclusion

This study explored the four different CNN models trained from the scratch for the classification of various tissues in the KimiaPath24C dataset which consists of histopathical images. We have used cross-entropy loss functions for our models. We will like to use the other optimizers and loss functions as well in our future work. We intend to improve our research work by using segmentation and comparing various datasets as well. This work requires high-speed GPU because these models require hours to train. Based on the literature survey our models were able to achieve better accuracies in comparison to [9, 12, 14] on the same dataset. Our methodology and models can help in classifying various diseases and can assist doctors in the diagnostic process. We have analyzed that we have achieved the highest validation accuracy in EfficientNet V2L of 96.95% followed by Xception with 94.43%.

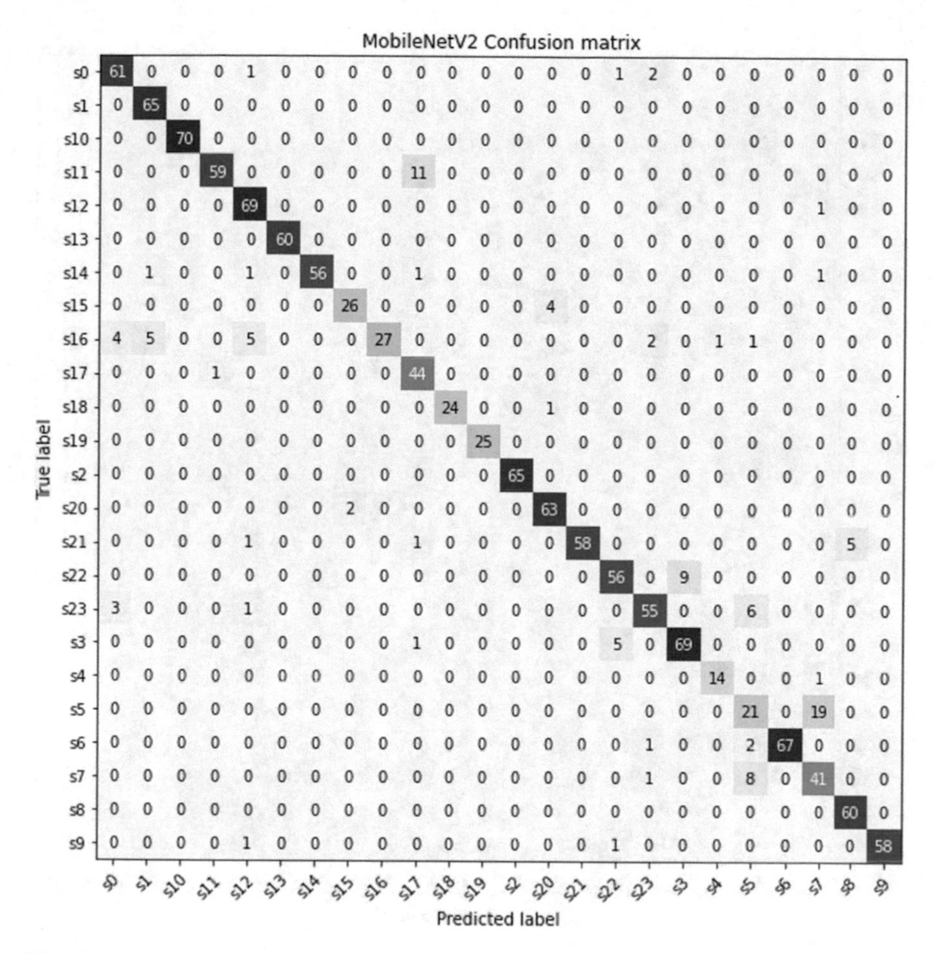

Fig. 10 Confusion matrix of MobileNet V2

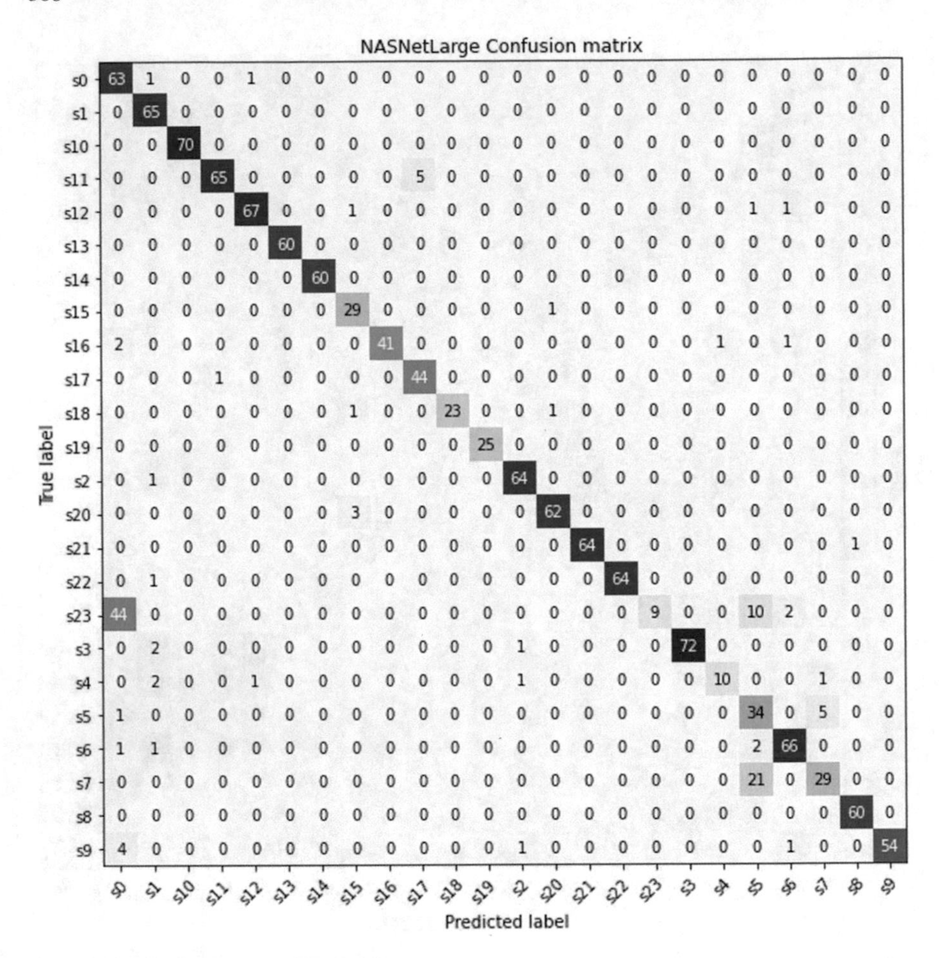

Fig. 11 Confusion matrix of NasNet large

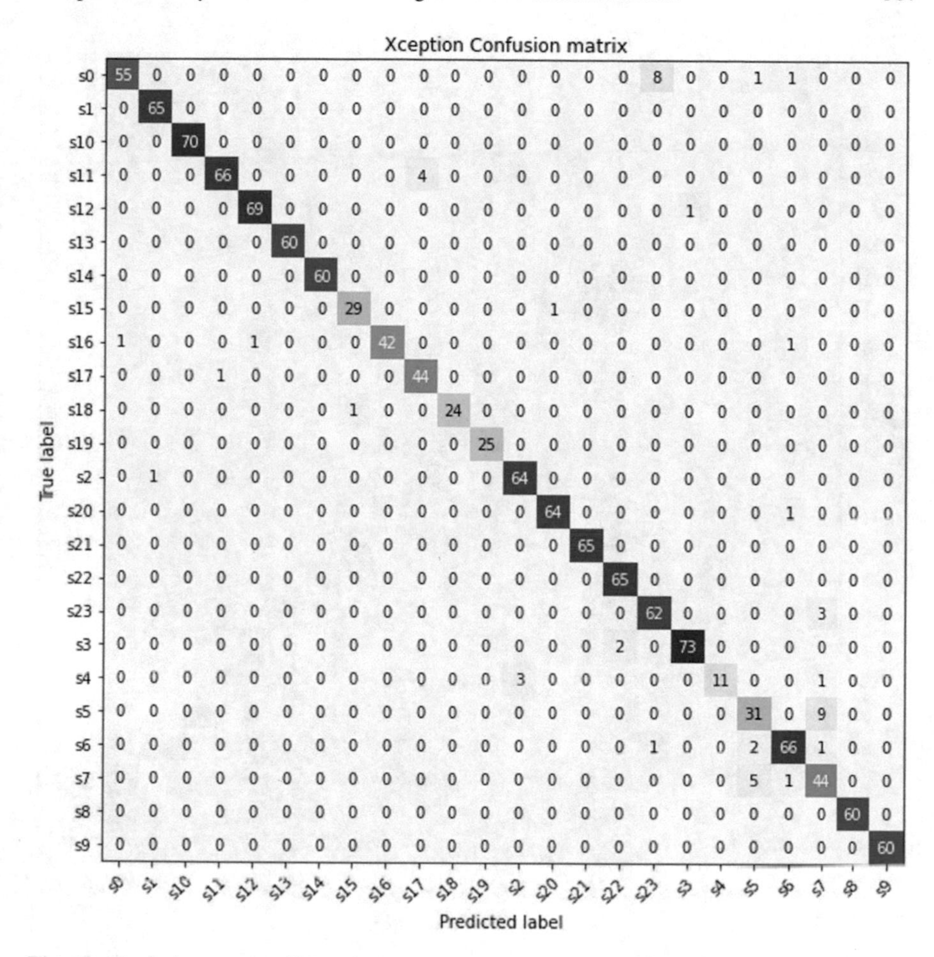

Fig. 12 Confusion matrix of Xception

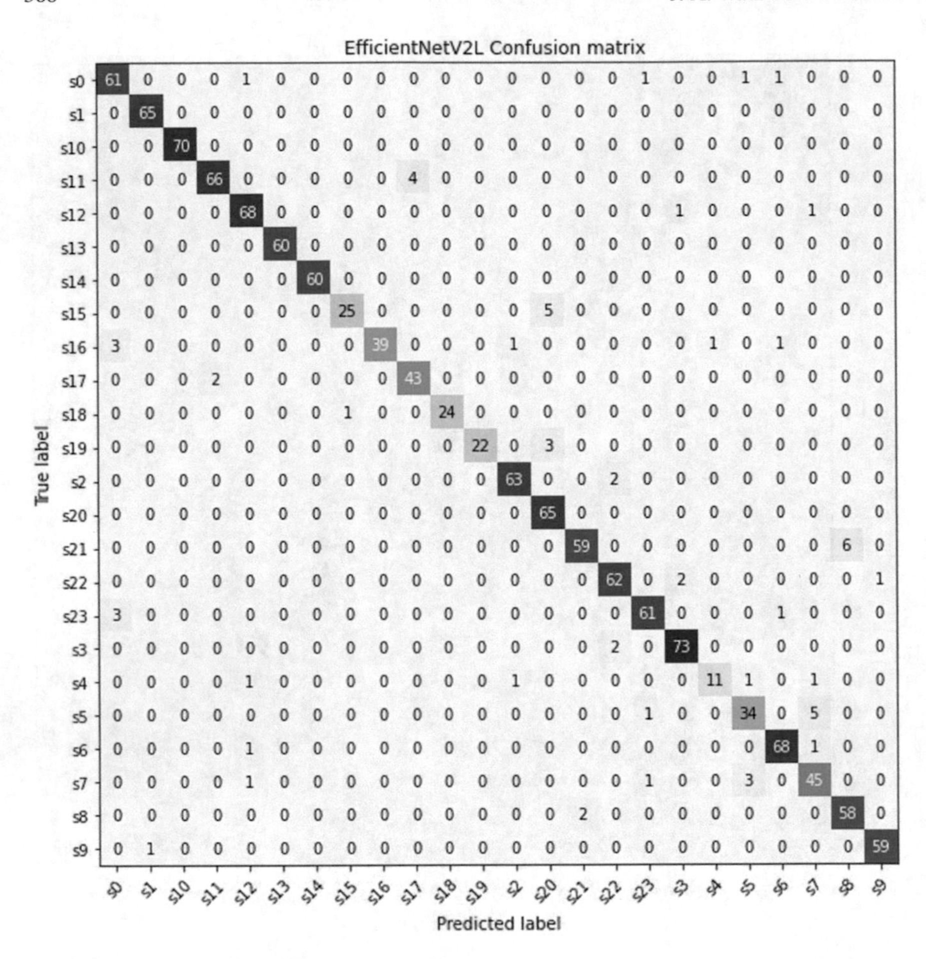

Fig. 13 Confusion matrix of EfficientNet V2L

Table 2 Analysis

Model	Training accuracy (%)	Validation accuracy (%)	Test accuracy		
			Precision	Recall	F1 score
MobileNet V2	99.65	91.34	0.92	0.91	0.92
NasaNetLarge	99.95	90.53	0.92	0.91	0.92
Xception	99.64	94.43	0.96	0.95	0.96
EfficientNet V2L	98.64	96.95	0.95	0.94	0.95

References

1. Farahani N et al (2015) Whole slide imaging in pathology: advantages, limitations, and emerging perspectives. **7**:23–33
2. Dimitriou N, Arandjelović O, Caie PD (2019) Case, deep learning for whole slide image analysis: an overview. **6**:264
3. Kumar N, Gupta R, Gupta S (2020) Whole slide imaging (WSI) in pathology: current perspectives and future directions. J Digit Imaging 33:1034–1040
4. Kather JN et al (2019) Predicting survival from colorectal cancer histology slides using deep learning: a retrospective multicenter study. 16(1):e1002730
5. Ahmed S et al (2021) Transfer learning approach for classification of histopathology whole slide images. 21(16):5361
6. Mishra R et al (2018) Convolutional neural network for histopathological analysis of osteosarcoma. 25(3):313–325
7. Babaie M et al (2017) Classification and retrieval of digital pathology scans: a new dataset. In: Proceedings of the IEEE conference on computer vision and pattern recognition workshops
8. Hirra I et al (2021) Breast cancer classification from histopathological images using patch-based deep learning modeling. 9:24273–24287
9. Kumar MD et al (2017) A comparative study of CNN, BoVW, and LBP for classification of histopathological images. In: 2017 IEEE symposium series on computational intelligence (SSCI). IEEE
10. Tsai M-J, Tao Y-HJE (2021) Deep learning techniques for the classification of colorectal cancer tissue. 10(14):1662
11. Kushwaha S et al (2021) Deep learning-based model for breast cancer histopathology image classification. In: 2021 2nd international conference on intelligent engineering and management (ICIEM). IEEE
12. Shafiei S et al (2021) Colored Kimia Path24 dataset: configurations and benchmarks with deep embeddings
13. Xu Y et al (2017) Large scale tissue histopathology image classification, segmentation, and visualization via deep convolutional activation features. 18(1):1–17
14. Kieffer B et al (2017) Convolutional neural networks for histopathology image classification: training vs. using pre-trained networks. In: 2017 seventh international conference on image processing theory, tools and applications (IPTA). IEEE
15. Bejnordi BE et al (2018) Using deep convolutional neural networks to identify and classify tumor-associated stroma in diagnostic breast biopsies. 31(10):1502–1512
16. Riasatian A et al (2021) Fine-tuning and training of densenet for histopathology image representation using TCGA diagnostic slides. 70:102032
17. Kavakiotis I et al (2017) Machine learning, and data mining methods in diabetes research. 15:104–116

Attention-Deficit Hyperactivity Disorder Spectrum Using ADHD_sfMRI

Faisal Firdous, Deepti Malhotra, and Mehak Mengi

Abstract The attention-deficit hyperactivity disorder also known as ADHD is a collective mental health syndrome in young groups. Efficacious involuntary analysis of ADHD which is based on features that are extracted from magnetic resonance imaging data that would give the reference knowledge for treatment (Wang and Kamata, Classification of structural MRI images in ADHD using 3D fractal dimension complexity map, pp. 215–219, 2019) [1]. It affects children and adolescents and is categorized by distractibility, poor-concentration, or weak self-control. Traditional-based diagnosis is based on a behavioral reliant approach, which is slightly subjective, and now most of the research has been carried out by considering biomarkers where the human's eye, writings have been analyzed. A lot of research has been performed with the help of various artificial intelligence techniques using biomarker (MRI, EEG) with a unimodal approach. However, a multi-modal approach can diagnose early, which will be cost-effective and can give more accuracy than existing models. The prime objective of the research is to propose an intelligent ADHD_sfMRI framework that can be used very efficiently and effectively by health practitioners to predict ADHD at an early stage (Riaz et al., J Neurosci Methods 335:108506, 2020) [2].

Keywords Machine learning · Magnetic resonance imaging · Attention-deficit hyperactivity disorder structural · MRI · Functional MRI · Multimodal · Prediction · etc.

1 Introduction

Neuro-developmental disorders (NDDs) are related to brain disorders that impede the neurological system's development, resulting in amend brain function. A person

F. Firdous (✉) · D. Malhotra · M. Mengi
Department of Computer Science and Information Technology, Central University of Jammu, Jammu 181143, India
e-mail: faisalparray39@gmail.com

D. Malhotra
e-mail: deepti.csit@cujammu.ac.in

Y. Singh et al. (eds.), *Proceedings of International Conference on Recent Innovations in Computing*, Lecture Notes in Electrical Engineering 1001,
https://doi.org/10.1007/978-981-19-9876-8_28

pretentious with ADHD is usually intertwined with learning complications that lean toward frustration once they influence adulthood [3]. ADHD is the most complicated disorder since its symptoms overlap with those of other NDDs, making the diagnosis task a challenging one. When overlapping is high, symptoms of other disorders among these ADHD is more predominant NDD among kids. The majority of children with ADHD will continue to experience clinical symptoms throughout adulthood, displaying pernicious elements since there are inadequate therapies available [4]. Diagnosis of ADHD is a challenging task as traditional-based tools rely on subjective measures (scale evaluation and interview-based observations) which may lead to deferred diagnosis. Biomarkers used for ADHD detection could be found before birth or after birth, and some of them might be used to anticipate or diagnose specific illnesses that require early medication. The early prediction may be possible with the help of the artificial intelligence technique by considering objective measures (MRI, EEG, Gait, Kinesics). Magnetic resonance imaging has been identified as a possible biomarker for tracing the evolving route for the very system as a whole or a section of it. The anatomical and functional abnormalities in autistic patient's brain have been extensively studied using magnetic resonance imaging (MRI). The assessment of functional changes in the brain, such as changes in neuronal function, blood circulation, and connections among various nervous system areas, is done using functional MRI.

1.1 Structural MRI

It also known as structural MRI anatomizing for neuro-developmental dysfunctions materialized in mid of 1990s. Voxel-based morphometry and surface-based morphometry are the two additional categories that structural MRI is further divided into for more fully defining the brain's structure.

1.2 Functional MRI

It is a type of MRI that assesses physical, cognitive, and psychological skills in addition to neuron activity. In functional MRI, there are two subcategories: task-based fMRI and resting-state fMRI. When a person is executing any activity, then task-based fMRI is incorporated to measure link among neuronal processes in the brain. While the patient is not performing any activities, the rs-fMRI analyzes the functional affinity of different brain areas [5].

Recently, various AI-based models using biomarkers have been developed for early ADHD prediction. However, until now, no such model has been deployed in clinical practices. Previously, most of the ADHD models employed unimodality which sometimes gives less accuracy. Keeping this essence, now the researchers are moving toward developing a multimodal system that gives more comprehensive information and yields better accuracy.

1.3 Organization of Paper

There are several sections of paper: Sect. 2 discusses numerous ADHD detection strategies being developed by various researchers. Relative analysis of many ADHD techniques is illustrated in Sect. 3. Section 4 confers the existing challenges in research of ADHD field of diagnostic. Moreover, Sect. 5 presents intelligent multimodal approach for early ADHD diagnosis. Finally, Sect. 6 includes conclusion.

2 Literature Review

For determination of methodologies that have been used in past studies of the similar topics, it was necessary to do a literature surveys.

- Khullar et al. [5] developed a method for classifying people with ADHD and TD using resting-state fMRI data. This model was determined as having better outcomes when compared to both models on several parameters.
- Anitha and Thomas Geroge [4] focused on diagnosis, deep learning with hidden features is used to represent ADHD data. For the first time, it considered deep learning capable of evaluating MRI data and established two-stage algorithms for detecting ADHD condition.
- Ariyarathne et al. [3] focused on the categorization of ADHD, the influence of the key brain areas in the default mode network. By examining the key area of several parameters, the suggested method was applied for several DMN in the view of seed correlation.
- Zhang et al. [6] presented a multi-site rs-fMRI dataset, an attention-based separated channel CNN for recognizing both ADHD and HCs was developed. It is a one-of-a-kind deep learning architecture that feeds signals straight into a separate CNN model to learn feature abstraction and weights them to classify with the attention-based network.
- Aradhya et al. [7] a novel MST-based classifier has been developed. MST uses a non-deterministic technique to descend distinguishable features. Only, rs-fMRI data were utilized to classify ADHD using MST latent space features and PBL-McRBFN.
- Riaz et al. [2] proposed a cutting-edge model for the categorization of ADHD using fMRI data. The proposed model learns to predict the categorization label using preprocessed time series information from fMRI.

- Shao et al. [8] proposed to distinguish ADHD and control participants; a redesigned gcForest approach was used. A multigrained structure fuse to ALFF and FC characteristics was proposed to merge FC and ALFF features.
- Eslami and Saeed [9] proposed the technique is robust in terms of sensitivity, with greater than 50% sensitivity for each of the three datasets studied. It looked at the issue from the standpoint of multivariate time series. It also treated each brain as a multivariate time series and classified the test set using the k-NN method.
- Qureshi et al. [10] proposed a model with an ADHD neuroimaging data in which an RFE feature selection was incorporated to ELM and other classifiers, with H-ELM achieving superior accuracy for both classifications.

3 Comparative Analysis

Figure 1 shows comprehensive overview of existing ADHD. From this figure, it has been observed that the CNN classifier yields the best accuracy in comparison to other approaches. The reason behind this is that CNN can automatically learn features from data (images) and obtain scores from its output of it (Table 1).

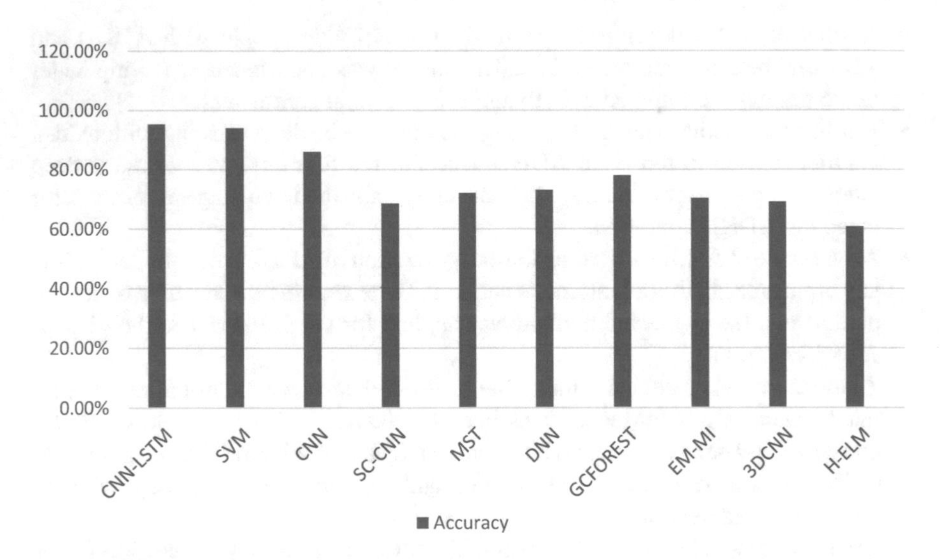

Fig. 1 Performance percentage accuracy of existing AI techniques for ADHD detection

Table 1 Comprehensive overview of the review of recent ADHD papers

Authors	Objectives	Limitations	Method	Dataset	Biomarker Used	Results (accuracy)
Wang and Kamata [1]	Classification of ADHD and healthy controls using DL	This technique is compared with an existing model. Other techniques can be implemented to give better results	2D (CNN) algorithm and hybrid (2D CNN–LSTM)	ADHD-200	fMRI	2D CNN-95.32% 2D CNN–LSTM 95.12%
Riaz et al. [2]	Using ML techniques for the documentation and classification of ADHD	Helpful in treatment of affected region and prevention from long-term impacts	SoftMax regression and SVM	ADHD	fMRI	94%
Ariyarathne et al. [3]	Classification of ADHD in view of seed correlation using convolution neural networks	Based on the seed correlation technique, the solution may potentially be refined to categorize different subtypes of ADHD	CNN	ADHD-200	fMRI	Between 84 and 86%
Anitha and Thomas Geroge [4]	Multi-site rs-fMRI data, SC-CNN used to categorize ADHD and HCs	Replicating the findings across bigger, more varied datasets and generalizing them to the real-world clinical setting remained a difficulty	Separated channel convolutional neural network	ADHD-200	fMRI	68.6%

(continued)

Table 1 (continued)

Authors	Objectives	Limitations	Method	Dataset	Biomarker Used	Results (accuracy)
Khullar et al. [5]	The spatial filter design issue is converted into a restraint optimization problem utilizing a hybrid genetic algorithm employing the metaheuristic spatial transformation (MST) technique	The findings show that mathematical modeling-based strategies are ineffective for a dataset with significant intra-class variations but low inter-class variance	Metaheuristic spatial transformation (MST	ADHD-200	fMRI	72.10%
Zhang et al. [6]	The deep learning model for ADHD classification uses preprocessed fMRI time series data as input and predicts a label (1 for ADHD and 0 for healthy control) as output	Retrain the feature extractor network for each imaging site independently, keeping the classification network and similarity measure network parameters constant	Deep neural network	ADHD-200	fMRI	73.1%
Aradhya et al. [7]	ADHD discrimination by gcForest against normal controls	Functional PCA approach for fMRI data processing	gcForest	ADHD-200	fMRI	Peking-78.05%, KKI-90.91%, NYU-78.05%, NI-80%

(continued)

Table 1 (continued)

Authors	Objectives	Limitations	Method	Dataset	Biomarker Used	Results (accuracy)
Shao et al. [8]	A short-time fMRI analysis approach for ADHD that is ideal for edge computing and can successfully assist doctors to perform remote consultations	The threshold is utilized to distinguish between healthy persons and sick; however, the threshold-based classifier is insufficient to reflect the more complicated statistical distribution	Threshold-based EM-MI algorithm	ADHD-200	fMRI	70.4%
Eslami and Saeed [9]	ADHD automated diagnosis utilizing a unique classification technique based on a 3D fractal dimension complexity map (FDCM)	To increase classifier performance and detect additional mental diseases, FD complexity maps of fMRI features were used	3D convolutional neural network	ADHD-200	sMRI	69.01%
Qureshi et al. [10]	Machine learning technique, to classify healthy versus ADHD children	This method is robust in terms of sensitivity, since it attained more than 50% sensitivity for each of the three datasets studied	K-nearest neighbor classifier	ADHD-200	fMRI	NA

Table 2 Relative comparison of ADHD detection techniques

Parameters	Biomarkers	
	sMRI	fMRI
Hardware requirement (HR)	Yes	Yes
Software requirement (SR)	Yes	Yes
Early diagnosis (ER)	Yes	Yes
Human intervention (HI)	Yes	Yes
Data collection	Easy (already available in ADHD-200)	Easy (already available in ADHD-200)
Hardware costs	High	High
Performance rate	Low	Low
Data accuracy	Low	High

3.1 Relative Comparison of Existing MRI-Based Attention-Deficit Hyperactivity Disorder Techniques

There are several approaches for detecting attention-deficit hyperactivity disorder. However, a selection of acceptable and efficient approaches is made by taking into account the relevance, limitations, performance efficiency, and features. For relative comparison, some parameters are used. The success rate of ADHD detection is generally determined by several essential elements. Relative comparison of MRI-based attention-deficit hyperactivity disorder techniques has been discussed in Table 2.

Investigating a substantial number of ADHD screening approaches has a slow efficiency rate. To resolve this issue, ADHD detection biomarkers and their performance analysis are generalized in the form of graph taking sMRI, fMRI as shown in Fig. 2.

From Sects. 2 and 3, it is apparent that the proficiency of various AI techniques and biomarkers utilized for ADHD diagnosis vary widely. The fact is that these techniques are insufficient for the determinate diagnosis at an earlier age. When ADHD is overlying by other disorders and unimodal diagnostic systems, ADHD is not much sufficient to give precise results. Thus, to identify the disorder, there is a need to develop an efficient and effective framework that yields more comprehensive information and reduces the misdiagnosis rate to a greater extent by combining the knowledge across the different modalities. By consolidating the fMRI data into one, it will be possible to improve the efficiency and accuracy of existing ADHD detection methods.

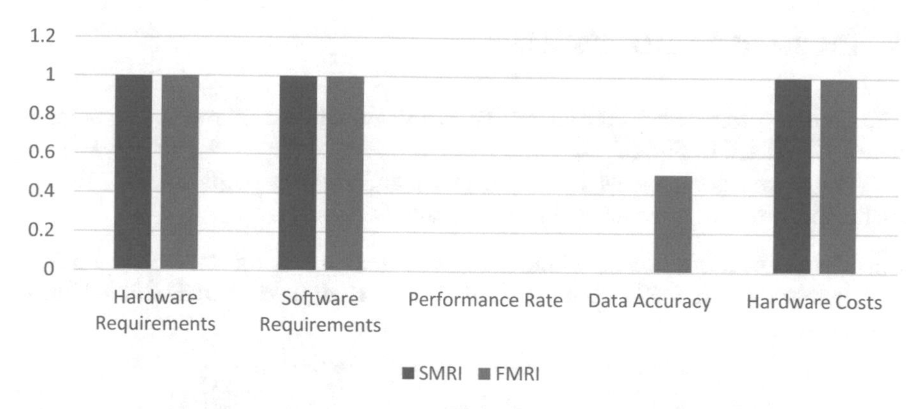

Fig. 2 Relative comparison of existing ADHD detection techniques based on MRI data

4 Research Gaps and Challenges

- **Early detection is not possible**: There is a meagerness of proper research on the early diagnosis of ADHD using biomarkers. Therefore, the proposed work needs to be implemented, which will be a vital aid to the person suffering from ADHD.
- **Less competence for unimodality**: Traditional classification models and feature extraction approaches often rely on a single channel model and static data. Therefore, applying the multimodal method takes less computational time and gives better results than existing methods.
- **Reliability subsides by unimodal**: The late fusion is a significant contributor to improved performance. The use of late fusion has been demonstrated to improve picture classification accuracy and decrease variation. As a result, using the multimodal technique makes data categorization networks more reliable.
- **Classifier exert are not ample to give complex distribution**: Threshold is used to distinguish between healthy persons and sick; however, the threshold-based classifier is unable to characterize the more composite statistical distribution; thus, the application field is unsatisfactory. It is possible to continue to research the learning algorithm based on a more complicated statistical model, intending to expand the algorithm's application scenarios and improve the model's accuracy.
- **Less sample size**: More sample numbers and other aggregate samples would be required to verify the deep network. New metrics are therefore needed to not only grasp the impact of various brain regions for a classification task but also to improve the model's causal inference and support clinical applications in the future.

5 Proposed ADHD_sfMRI

The proposed framework ADHD_sfMRI is divided into the following key components as depicted in Fig. 3.

Various key components involve data collection, data preprocessing, feature extraction, prediction model, and results as follows (a–e):

(a) **Data collection**: Most scathing for every research is to assemble data. For this framework, an MRI dataset is needed. This study uses ADHD-200—Global

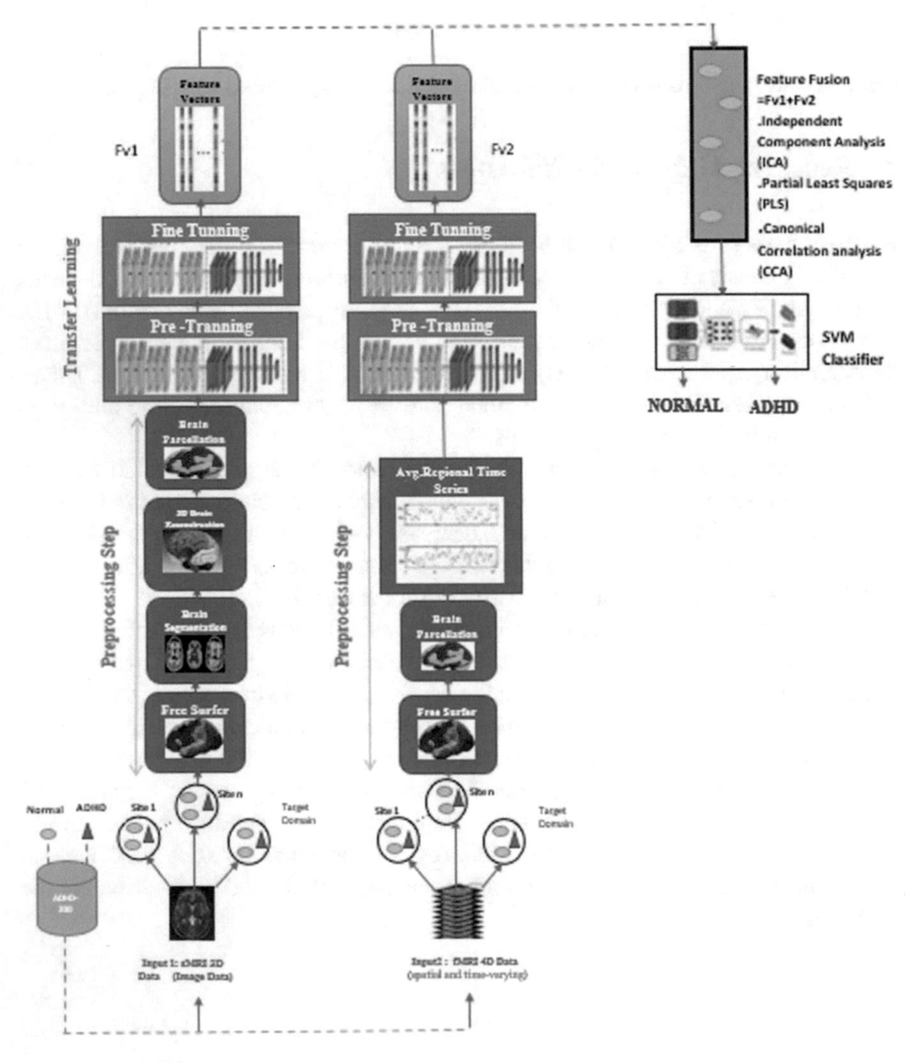

Fig. 3 Components of proposed framework

Table 3 Overview of the dataset used in the study

	Train dataset		Test dataset	
	Healthy controls	ADHD	Healthy controls	ADHD
NYU	98	118	12	29
NI	23	25	14	11
Peking	61	24	24	27
KKI	61	22	08	03
OSHU	42	37	28	06

Competition dataset provided by the Neuro Bureau (https://www.nitrc.org/frs/?group_id=383). The dataset comes from five datasets, namely Peking (Peking University), KKI (Kennedy Krieger Institute), NYU, NI (Neuro Imaging), and OSHU. All imaging locations have a different number of subjects. Table 3 defines the outline of the data used in this study.

(b) **Data preprocessing**: Neuroimaging data have a complex structure that, if not appropriately preprocessed, might alter the final diagnosis. Both the data are preprocessed discretely.

sMRI: Preprocessing of this data usually comprises several standard procedures carried out by various tools. In the following section, preprocessing steps are briefly explained for sMRI data:

Brain Segmentation: Brain segmentation is the process of division of an image into its constituent regions. It involves the segmentation of structural MRI images by using FreeSurfer, an open-source software.

3D Brain Reconstruction: It encompasses reconstructing brain MRI images using deep learning (Convolutional autoencoder). There are two parts to the autoencoder: an encoder and a decoder. The encoder has three convolution blocks, each with a convolution layer and a batch normalization layer. After the first and second convolution blocks, the max-pooling layer is employed. The decoder consists of two convolution blocks, each with a convolution layer and a batch normalization layer. After the first and second convolution blocks, an upsampling layer is utilized.

Brain Parcellation: It allows to be more specific about the brain areas.

fMRI: Preprocessing the data is the initial stage in fMRI analysis. FreeSurfer, an open-source program, was used to perform the preprocessing in this experiment. The preprocessing steps applied in this study are as follows:

Data Parcellation: It involves pyClusterROI a collection of Python scripts for generating whole-brain parcellations from functional magnetic resonance imaging (fMRI) data. The produced areas are adequate for use as regions of interest (ROIs) in the analysis of fMRI data. The outcome subsequently parcellation is average regional time series.

(c) **Feature Extraction**: The most significant characteristics should be chosen at the feature extraction step. These are the classifier's inputs. In our proposed model, transfer learning method is employed for feature extraction. Transfer learning is a well-known technique in which a network is trained on a large labeled source dataset and then fine-tuned on a smaller target dataset, thereby transferring the gained knowledge from the source to the target dataset.

(d) **sMRI and fMRI fusion**: FMRI data are used to determine brain function, whereas sMRI data are used to determine brain anatomy. Subsequently feature extraction of sMRI and fMRI data, the feature vectors are attained by assimilation of the SMRI and FMRI data features based on some resemblances through the process of fusion. First, compared to PCA, another well-liked component analysis technique, ICA might occasionally be unorganized as a constituent analysis tool. The first difference between ICA and PCA is that whereas ICA's components are autonomous, PCAs are orthogonal. Second, the only statistical properties of the observed signal that PCA can mention are those of second order. However, ICA is able to extract the signal's high-order statistic unique walloping. Furthermore, the original data should be satisfied with a normal distribution when using PCA for signal analysis.

(e) **Prediction model**: The classification stage of our research involves using a support vector machine (SVM) classifier to assess the discriminative capabilities of the features chosen in the previous steps. SVM is a widely used machine learning classification technique that has shown to be effective in a variety of neuroimaging research. The classifier is supplied with labeled training data during the training phase (for healthy control and ADHD subjects). The training and testing data ration will be 3:1.

6 Conclusion and Future Scope

In this study, we present a multimodal approach, which can more accurately predict the ADHD-based complexity map extracted from brain structural MRI scans and brain functional MRI, than existing unimodal approaches. This paper mainly focuses on providing a brief idea about already prevailing ADHD detection techniques. Most of the proposed methods are based on fMRI brain imaging that provides the brain's functional anatomy, and some of the methods are based on sMRI brain imaging that results in anatomical structure. However, no study has been carried out by considering two or more potential biomarkers (sMRI and fMRI) in ADHD diagnostic field. Hence, this paper presents a multimodal approach (combination of sMRI and fMRI) which will be cost-effective and can give more accuracy than existing models. The future work includes the proposed framework to be incorporated in coming era for early analysis of ADHD.

References

1. Wang T, Kamata SI (2019) Classification of structural MRI images in ADHD using 3D fractal dimension complexity map. In: Proceedings of international conference image process (ICIP), vol. 2019-Septe, pp 215–219. https://doi.org/10.1109/ICIP.2019.8802930
2. Riaz A, Asad M, Alonso E, Slabaugh G (2020) DeepFMRI: end-to-end deep learning for functional connectivity and classification of ADHD using fMRI. J Neurosci Methods 335:108506. https://doi.org/10.1016/j.jneumeth.2019.108506
3. Ariyarathne G, De Silva S, Dayarathna S, Meedeniya D, Jayarathne S (2020) ADHD identification using convolutional neural network with seed-based approach for fMRI data. In: Pervasive computing technologies for healthcare, pp 31–35. https://doi.org/10.1145/3384544.3384552
4. Anitha S, Thomas Geroge S (2021) ADHD classification from FMRI data using fine tunining in SVM. J Phys Conf Ser 1937(1). https://doi.org/10.1088/1742-6596/1937/1/012014
5. Khullar V, Salgotra K, Singh HP, Sharma DP (2021) Deep learning-based binary classification of ADHD using resting state MR images. Augment Hum Res 6(1). https://doi.org/10.1007/s41133-020-00042-y
6. Zhang T et al (2020) Separated channel attention convolutional neural network (SC-CNN-attention) to identify ADHD in multi site Rs-fMRI dataset. Entropy 22(8):1–10. https://doi.org/10.3390/E22080893
7. Aradhya AMS, Sundaram S, Pratama M (2020) Metaheuristic spatial transformation (MST) for accurate detection of attention deficit hyperactivity disorder (ADHD) using rs-fMRI. In: Proceedings of annual international conference of the IEEE engineering in medicine and biology society (EMBS), vol 2020-July, pp 2829–2832. https://doi.org/10.1109/EMBC44109.2020.9176547
8. Shao L, Zhang D, Du H, Fu D (2019) Deep forest in ADHD data classification. IEEE Access 7(2017):137913–137919. https://doi.org/10.1109/ACCESS.2019.2941515
9. Eslami T, Saeed F (2018) Similarity based classification of ADHD using singular value decomposition. In: Proceedings of 2018 ACM international conference computing frontiers (CF 2018), pp 19–25. https://doi.org/10.1145/3203217.3203239
10. Qureshi MNI, Min B, Jo HJ, Lee B (2016) Multiclass classification for the differential diagnosis on the ADHD subtypes using recursive feature elimination and hierarchical extreme learning machine: structural MRI study. PLoS ONE 11(8):1–20. https://doi.org/10.1371/journal.pone.0160697
11. Hanson E, Cerban BM, Slater CM, Caccamo LM, Bacic J, Eugenia C (2013) Brief report: prevalence of attention-deficit/hyperactivity disorder among individuals with an autism spectrum disorder. J Autism Dev Disord 43(6):1459–1464
12. Mayes SD, Calhoun SL, Mayes RD, Molitoris S (2012) Autism and ADHD: overlapping and discriminating symptoms. Res Autism Spectr Disord 6(1):277–285
13. Kushki A, Anagnostou E, Hammill C, Duez P, Brian J, Iaboni A, Schachar R, Crosbie J, Arnold P, Lerch JP (2019) Examining overlap and homogeneity in ASD, ADHD, and OCD: a data-driven, diagnosis-agnostic approach. Transl Psychiatry 9(1)
14. Yael L (2014) The co-occurrence of autism and attention deficit hyperactivity disorder in children—what do we know? Front Human Neurosci 8:1–8
15. Ramtekkar U (2017) DSM-5 changes in attention deficit hyperactivity disorder and autism spectrum disorder: implications for comorbid sleep issues. Children 4(8):62
16. Antshel KM, Yanli Z-J, Faraone SV (2013) The comorbidity of ADHD and autism spectrum disorder. Exp Rev Neurother 13(10):1117–1128
17. Thabtah F, Peebles D (2019) Early autism screening: a comprehensive review. Int J Environ Res Public Health 16(18)
18. Biswas SD, Chakraborty R, Pramanik A (2020) A brief survey on various prediction models for detection of ADHD from brain-MRI images. In: Proceedings of the 21st international conference on distributed computing and networking. Association for Computing Machinery, New York

19. Romiti S, Vinciguerra M, Saade W, Cortajarena IA, Greco E (2020) Artifcial intelligence (AI) and cardiovascular diseases: an unexpected alliance. Cardiol Res Prac (Ml)
20. Le Berre C, Sandborn WJ, Aridhi S, Devignes M-D, Fournier L, Smaïl-Tabbone M, Danese S, Peyrin-Biroulet L (2020) Application of artifcial intelligence to gastroenterology and hepatology. Gastroenterology 158(1):76–94.e2
21. Hirschmann A, Cyriac J, Stieltjes B, Kober T, Richiardi J, Omoumi P (2019) Artificial intelligence in musculoskeletal imaging: a review of current literature, challenges, and trends. Semin Musculoskel Radiol 23(3):304–311
22. Fakhoury M (2019) Artificial intelligence in psychiatry, vol 1192. Springer, Singapore
23. Tary JB, Herrera RH, Van Der Mirko B (2018) Analysis of time-varying signals using continuous wavelet and synchrosqueezed transforms. Philosoph Trans Royal Soc A Mathem Phys Eng Sci 376(2126):2017025
24. Jones R (2014) NIH public access. Bone 23(1):1–7
25. Voineagu I, Yoo H (2013) Current progress and challenges in the search for autism biomarkers. Dis Mark 35(1):55–65
26. Mayeux R (2004) Biomarkers: potential uses and limitations. NeuroRx 1(2):182–188
27. Mengi M, Malhotra D (2022) Anatomy of various biomarkers for diagnosis of socio-behavioral disorders. In: Singh PK, Singh Y, Kolekar MH, Kar AK, Gonçalves PJS (eds) Recent innovations in computing. Lecture notes in electrical engineering, vol 832. Springer, Singapore. https://doi.org/10.1007/978-981-16-8248-3_7

Applications of Deep Learning in Healthcare: A Systematic Analysis

Ishani Kathuria, Madhulika Bhatia, Anchal Garg, and Ashish Grover

Abstract Deep learning (DL) is a subfield of artificial intelligence (AI) that deals with the recognition of patterns. It learns from the input provided to it to predict an output according to the features it evaluates. With the extensive increase in unstructured data in the past few years, the ability to train machines to predict outcomes became much more difficult but the development of artificial neural networks (ANNs) and DL techniques changed that. One of the biggest advancements made with DL is in the field of healthcare. The objective of this research is to provide a comprehensive analysis of the vast applications of DL techniques used in the healthcare system, specifically in the domains of drug discovery, medical imaging, and electronic health records (EHRs). Due to the past epidemics and the current situation of the ongoing pandemic disease, i.e., COVID-19, the application of AI, ML, and DL in this field has become even more critical. Such work has become even more significant, and these techniques can help make timely predictions to combat the situation. The result showed a lot of research is ongoing to continuously tackle the limitations and improve upon the advantages. Many important advancements have been made in the field and will continue to grow and make our quality of life more efficient, cost-effective, and effortless.

Keywords Artificial intelligence · Machine learning · Deep learning · Drug discovery · Drug-target interaction · De novo drug design · Medical imaging · Electronic health records

---

I. Kathuria (✉) · M. Bhatia
Amity School of Engineering and Technology, Amity University, Noida, Uttar Pradesh, India
e-mail: ishani@kathuria.net

M. Bhatia
e-mail: mbhadauria@amity.edu

A. Garg
School of Creative Technologies, University of Bolton, Bolton, UK
e-mail: a.garg@bolton.ac.uk

A. Grover
Manav Rachna International Institute of Research and Studies, Faridabad, India
e-mail: ashishgrover.fet@mriu.edu.in

Y. Singh et al. (eds.), *Proceedings of International Conference on Recent Innovations in Computing*, Lecture Notes in Electrical Engineering 1001,
https://doi.org/10.1007/978-981-19-9876-8_29

1 Introduction

Artificial intelligence (AI) is described as the science of providing machines with "intelligence" so that machines can think like humans. It covers everything from Good Old Fashioned Artificial Intelligence (GOFAI or symbolic AI) like decision trees to connectionist architectures like deep learning (DL) as explained thoroughly in [1]. Under AI lies machine learning (ML) that permits machines to comprehend information directly from practicing with already known data. Instead of being given rules as such was the case in GOFAI, ML models are given training data as examples of how to approach the problem. ML and DL are different terms but are often used interchangeably. DL is a system based around artificial neural networks (ANNs). ANNs are a simplified version of human biological neural networks [2] that can replicate the functionalities of a human brain within computer programs, whereas ML focuses on simple mathematical formulations to discover patterns from structured data. So, ML is a part of a wider AI, while DL is a subset of ML. This relationship between the three terms is shown in Fig. 1.

These neural networks contain hundreds of layers as compared to previously used shallow networks containing just two to three layers. With more layers, the capabilities of the network have been extended extensively popularizing DL in recent times. One of the major contrasts between ML and DL is that DL models do not require human intervention. They work on their own, figuring out what patterns and connections exist between the items and performs classifications/predictions according to the features extracted making them suitable for unstructured data (e.g., videos, images, audio, etc.) which is what exists in a majority in today's date. On the other hand, ML requires formatted or structured data that requires human assistance. Figure 2 explains this concept in a more simplified manner with a classification problem that requires a model to take in an image and classify it as an ambulance or not ambulance. The top half of the figure shows the workflow of a ML model where the training images need their features extracted with the help of a human before being fed into the model, while the bottom half shows a DL model that can extract the features on its own.

Fig. 1 Relationship between AI, ML, and DL [1]

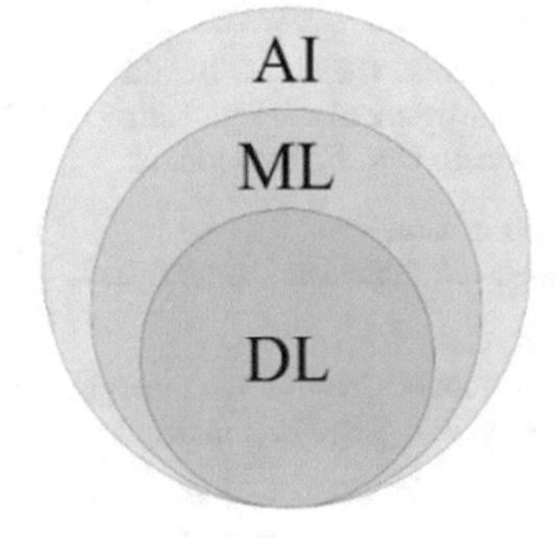

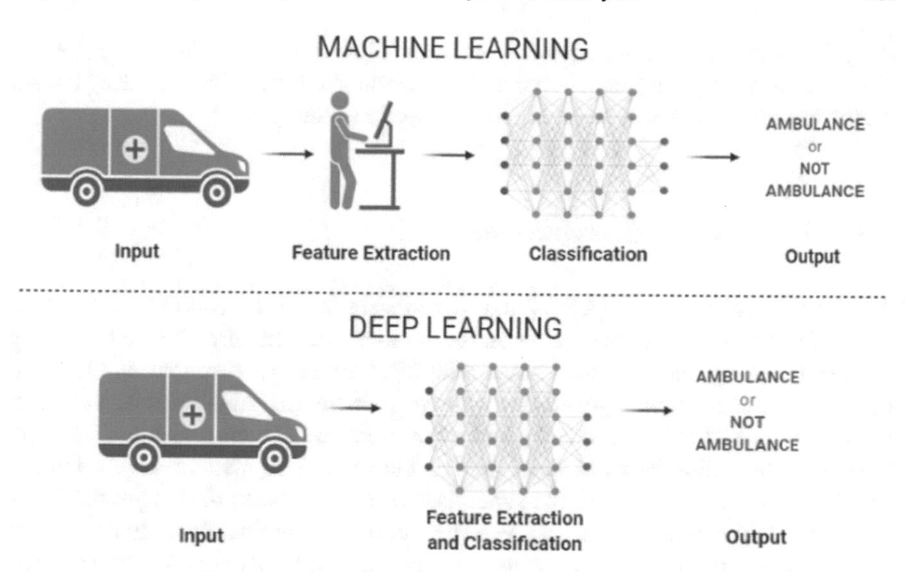

Fig. 2 Example showing the difference between DL and ML

DL is being widely utilized to overcome issues like natural language processing, pattern recognition, and computer vision. The applications of these concepts can be seen in our everyday life in voice recognition systems employed by virtual assistants, for instance, Siri or Google, image recognition systems used in different social media platforms, or recommendation systems used by video streaming platforms like Netflix or online shopping websites such as Amazon. These technologies are also readily improving fields like transport through self-driving cars or in public services to help combat homelessness. With this abundance of ongoing research in DL, we felt a need to recapitulate the research in this area, with a special focus on healthcare. The healthcare industry ensures that the appropriate care is provided to the patient at the appropriate time, but such is not always the case due to several reasons such as limitations in technology, knowledge, or simply human error. DL techniques are being employed to deliver valuable insights and combat the challenges through the evaluation of intricate biomedical data [3]. The current worldwide pandemic of COVID-19 has also reminded everyone that healthcare needs to be a major focus of any nation to ensure the well-being of its citizens. COVID-19 is a novel coronavirus which emerged in Wuhan, China in 2020 and has since spread all over the world. It is very contagious and presents as an acute respiratory disease. It was declared as a pandemic in March 2020 by the World Health Organization (WHO). With DL's success in many domains, researchers have applied these concepts to combat this deadly disease.

Some of the most common DL applications in healthcare include drug discovery, medical imaging, genome analysis, robotic surgeries, electronic health records, etc. This ever-expanding field has applications we could not even comprehend a decade ago but are now becoming a part of our daily lives. We have given special emphasis on drug discovery, medical imaging, and electronic health records. These applications were chosen as they have had the most development in recent years and a deep impact on the healthcare industry in saving lives as well as money.

1.1 Deep Learning Architectures

Artificial neural networks (ANNs) aim to replicate the inner workings of a human brain. The research has been going on for decades, and although it is an extremely hard undertaking, the field has shown a lot of progress [4]. An input, hidden, and output layers make up the general structure of an ANN [5]. ANNs can be trained in three broad ways: (1) supervised, (2) unsupervised, and (3) reinforcement learning. The main distinction between supervised and unsupervised methods is whether the training data is labeled or not. For supervised learning, labeled data is provided that facilitates inferencing information. With unsupervised learning, the training data set does not have any labels, it is up to the network to find patterns and make sense of the data. The third method of reinforcement learning falls between these two categories by not giving the labels to the model directly but telling the model while it is learning whether it is going in the right direction or not [6]. There are lots of ANN architectures that differ in the number of layers or learning techniques [7]. We will be discussing recurrent neural network (RNN), long short-term memory (LSTM), and convolutional neural network (CNN).

Recurrent Neural Networks (RNNs). These are neural networks where the input to each layer is the output from the previous layer that is with recurrent connections that help them recognize patterns as shown in Fig. 3, hence being called recurrent neural networks [8]. These networks have hidden states which make it possible for them to memorize previous information [9]. Figure 3 shows the architecture of a simple RNN where $\boldsymbol{x}$ is the input layer, $\boldsymbol{y}$ is the output layer, and $\boldsymbol{h}$ represents the hidden layers. Also, the variable parameters to improve the model, called weights, are represented by $\boldsymbol{w}_x$, $\boldsymbol{w}_y$, and $\boldsymbol{w}_h$ for input, output, and hidden layers, respectively.

Long Short-Term Memory (LSTM). A major problem with earlier RNN models was the “vanishing gradient problem” as explained by Sepp Hochreiter in his 1991 thesis which was fixed by the introduction of a new type of RNN termed long short-term memory (LSTM) in 1997 by Hochreiter [10]. LSTM introduced the concept of a memory cell as a replacement for the neuron-based neural network. This architecture can forget unnecessary information at each step with the help of a forget gate. By only keeping the important information in the current state, it can retain more information and work better than a simple RNN. Figure 4 shows the architecture of a LSTM memory cell where $\boldsymbol{c}_t$ represents cell state, $\boldsymbol{x}_t$ represents input vector, and $\boldsymbol{h}_t$ represents new state, all at time $\boldsymbol{t}$. Also, $+$ stands for addition, $\times$ represents multiplication, and σ and $\boldsymbol{tanh}$ are the $\boldsymbol{sigmoid}$ and $\boldsymbol{tanh}$ activation functions.

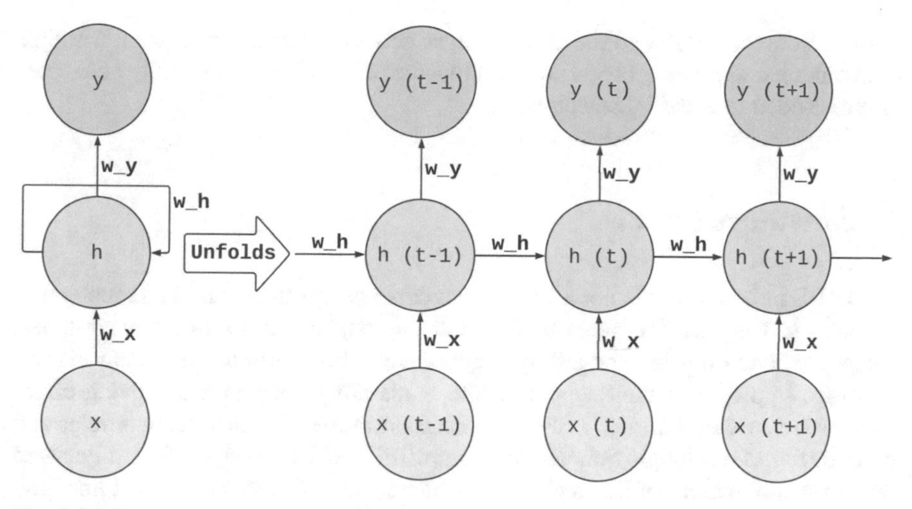

Fig. 3 RNN architecture

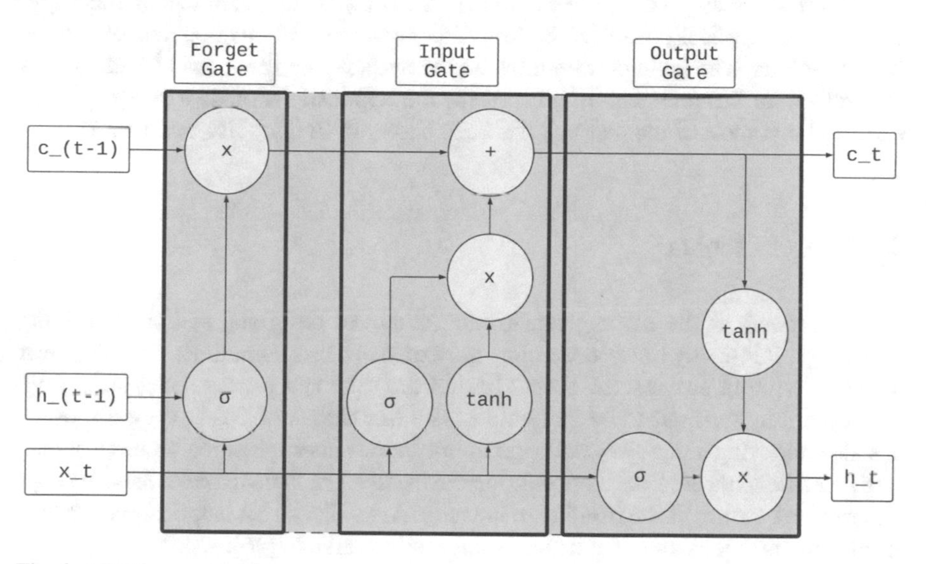

Fig. 4 LSTM memory cell

Convolutional Neural Networks (CNNs). Convolutional neural networks were developed with inspiration from cells found in an animal's visual cortex making them the indomitable DL architecture for working with images as seen in our review as well. All the medical imaging papers reviewed have used CNN architectures (Table 1). 1989 saw LeCun establishing the initial framework for CNN and improving on this for the next few years and finally achieving the modern framework in 1998 [11]. The basic CNN architecture consists of a convolutional layer followed by pooling layers and fully connected layers as shown in Fig. 5. The convolution layer extracts feature from an image and the pooling layer acts as a dimension reducer. Multiple convolution and pooling layers are used together after which a final fully connected layer is added to classify the output.

2 Literature Review

Using DL in healthcare is not a very recent concept, but there has been much more research in the past five years with healthcare establishments of all magnitudes, categories, and domains are getting progressively more curious about how AI can reinforce improved patient support while moderating budgets and refining effectiveness. In further sections, a review is done on recent literature using articles and journals found on Google Scholar, Springer, IEEE, arXiv, ACM, CAS, and PubMed related to applications of DL in healthcare, namely drug discovery, medical imaging, and electronic health records as well as COVID-19 in these fields.

Table 1 summarizes all the research papers reviewed and mentioned in this paper based on the applications of DL in drug discovery, medical imaging, and electronic health records. The table concentrates on the authors and the focus of their papers as well as the DL architecture utilized for the applications of their model. It was observed that most of the research has been based on CNN, RNN, and LSTM.

3 Drug Discovery

Drug discovery is the name given to the domain of designing and/or identifying new drugs. It is a very time-consuming and expensive task sometimes spanning over 10 years without any results. Even after all the time and money spent on a drug, the result may turn out to be ineffective. The limitations of DL in drug discovery are also imposing a massive challenge in this field. This application requires people with specific skillset to develop and operate these AI models to make these models a permanent fixture of the healthcare industry. Also, like all DL applications, data is needed in enormous amounts which is not easily available [36].

3.1 Drug Properties Prediction

DL has also been very useful in predicting the properties of the drugs, which makes the basis for drug discovery. Various deep neural network applications have been used for classification of drugs based on their properties. Reference [12] recently demonstrated how their model based on a CNN architecture showed improvements over state-of-the-art methods in predicting membrane permeability, lipophilicity, and aqueous solubility. These methods can help improve and speed up the drug discovery

Table 1 Reviewed research papers

Domain	Application	Architecture	References
Drug discovery	Prediction of properties of a given molecule	CNN	Wang et al. [12]
	Presented DeepPurpose, for DTI prediction	CNN	Huang et al. [13]
	8-layer DL framework for identifying and predicting a drug's performances in different stages	ANN	Jamshidi et al. [14]
	Proposed a platform to forecast and accelerate vaccine designs for COVID-19	LSTM	Jamshidi et al. [15]
	Presented DeepNovo, to use in de novo peptide sequencing	CNN + LSTM	Tran et al. [16]
	Generation of new molecules starting from a seed compound, its three-dimensional (3D) shape, and its pharmacophoric features	CNN + LSTM	Li et al. [17]
	DL-based multi-scale feature fusion method for creating structural similarity features for drugs and proteins called DeepFusion	CNN	Song et al. [18]
	AI model for to discover/design new therapeutic agents for SARS-CoV-2	RNN	[19]
Medical imaging	Prediction of onset of DR using tomography scans	CNN	De Fauw et al. [20]
	Detection and classification of DR from color fundus images	CNN	Shankar et al. [21]
	Classification of enhancing lesions as benign or malignant at multiparametric breast MRI	CNN	Truhn et al. [22]
	Breast cancer risk detection based on mammography	CNN	Yala et al. [23]
	Detection of breast cancer on screening mammograms	CNN	Shen et al. [24]

(continued)

Table 1 (continued)

Domain	Application	Architecture	References
	An accurate computer-aided method to assist in identification of COVID-19-infected patients using CT images	CNN	Ying et al. [25]
	Assisting radiologists to automatically diagnose COVID-19 in X-ray images	CNN	Hemdan et al. [26]
	Detect coronavirus 2019 and differentiate it from community-acquired pneumonia and other lung conditions	CNN	Li et al. [27]
	COV19-CT-DB database developed and used to train a CNN-RNN model	CNN + RNN	Kollias et al. [28]
EHR	Reads medical records, stores previous illness history, infers current illness states, and predicts future medical outcomes	LSTM	Pham et al. [29]
	A framework for disease onset prediction that combines free-text medical notes as well as structured information	CNN + LSTM	Liu et al. [30]
	Accurately predicting multiple medical events from multiple centers without site-specific data harmonization	LSTM	Rajkomar et al. [31]
	A cost-sensitive formulation of LSTM using expert features and contextual embedding of clinical concepts	LSTM	Ashfaq et al. [32]
	Prediction of the number of COVID-19 positive cases in Indian states	LSTM	Arora et al. [33]
	Proposed a model to give mortality prediction based on confirmation of COVID-19	RNN	Sankaranarayanan et al. [34]
	Word embedding-based CNN for predicting whether a patient has COVID-19 based on their reported symptoms	CNN	Obeid et al. [35]

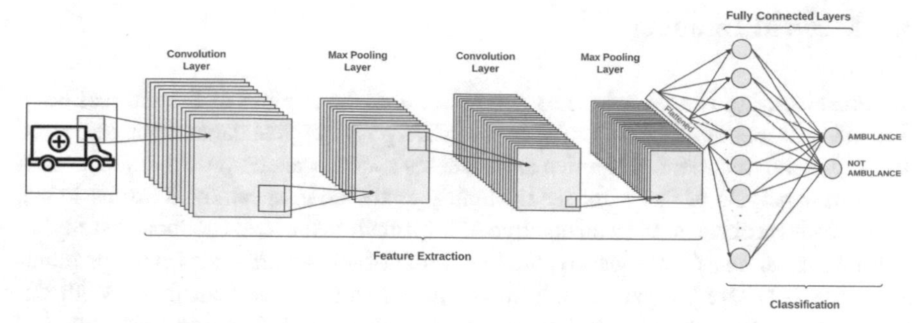

Fig. 5 CNN architecture

process by predicting which drugs would be most useful in the development based on the properties required.

3.2 Drug-Target Interaction (DTI)

One of the biggest challenges in drug discovery is to comprehend how a drug will interact with its target. Drugs are sequences of proteins and even though the sequence can be found, their shapes are highly unpredictable due to the enormous number of possibilities, making it very difficult to figure out DTI. Google's DeepMind applied DL to achieve just this in their project titled AlphaFold [37] which was able to anticipate the three-dimensional configuration of proteins at a trailblazing speed and accuracy, besting some of the world's best biologists and researchers of the field. Further, studies proved to be even more successful with their models [13]. These methods have also been employed to develop COVID-19 vaccines or predicting how the virus will react with specific drugs [14, 15].

3.3 De Novo Drugs

The process of designing a specific drug with certain properties that can interact with the precise drug you want is called de novo drug design. The concept of using DL for de novo drug design is relatively newer but has the potential to reach great heights. Reference [16] shows how DL can be used for de novo peptide sequence. They designed a DL system, DeepNovo, with an architecture taking advantage of CNNs and LSTM networks working together. Another model presented in [17] showcases DeepScaffold, a 20 layer CNN model capable of molecule generation. Reference [18] proposed another CNN-based model, DeepFusion for creating structural similarity features for drugs and proteins called DeepFusion. Similarly, [19] showcases how this concept can be used to develop therapeutic agents for SARS-CoV-2.

4 Medical Imaging

Medical imaging is a technique to get an accurate description of the internal body including tissues and organs for the creation of a physiological database as well as be helpful in the treatment of injuries and diseases such as cancer, pneumonia, or brain wounds. Analysis of these images is slightly restricted to specialized professionals due to their intricacies. With an upsurge of information, the load for doctors studying this data escalates. Consequently, the likelihood of human error may perhaps intensify. This is where DL can be helpful and act as an unbiased consultant. With DL come new challenges such as images used in the medical imaging fields like CT scans, X-rays, MRIs, etc., might be of low resolution which would be hard to convert to a proper resolution for training a DL model. Also, finding large, good quality labeled data sets is very hard due to data being unlabeled and unstructured [38]. These challenges are also being continuously being improved upon to further the research in the field.

4.1 Diabetic Retinopathy

Diabetic retinopathy (DR) is a primary cause of preventable vision loss globally. It presents in patients with severe cases of diabetes and results in complete vision loss if not found at an earlier stage which is usually not possible as it is hard to diagnose. Using DL methods to diagnose DR at earlier stages and prevent it has shown immense promise. Reference [20] proposed a model that performed better than the top specialists of the field. Another recent study [21] was able to attain extraordinarily high results with a maximum accuracy of 99.49% in diagnosing DR with their HPTI-v4 model. These studies show how useful DL can be to help diagnose preventable blindness.

4.2 Breast Cancer

Breast cancer is one of the most common types of cancers, killing upwards of 600,000 people every year. Using DL can help speed up the diagnosis process and help save lives. Reference [22] used a CNN algorithm in comparison with radiomic analysis (RA). CNN had a higher performance in comparison with RA and although neither of them could outperform radiologists, the CNN can be improved with more training data. Another study [23] used a mammography-based CNN model that improved risk determination compared to the current clinical standards. Reference [24] had some limitations due to less training data and low GPU memory but still showed high potential with another CNN-based DL model. Overcoming these limitations will help DL reach new heights in the future.

4.3 COVID-19

Since COVID-19 presents as an acute respiratory disease like pneumonia, it caused multiple misdiagnosis which is why many research studies have focused on developing medical imaging models to diagnose COVID-19. Reference [25] used CT images to develop a CNN-based DL model, DeepPneumonia, to identify accurately the patients with COVID-19. The study was conducted using CT scans of patients from 2 provinces in China and was able to achieve high accuracy in diagnosing patients carrying the virus. Another study developed COVIDX-Net [26] using X-ray images. Reference [27] was a similar study using CT scans but with a larger data set that naturally led to a higher performance. Reference [28] created an annotated database of over 7000 3D CT scans called COVID-19-CT-DB database and used it to train and validate their model to detect COVID-19 and its severity.

5 Electronic Health Records

Electronic health records (EHRs) are databases collecting heterogeneous patient data in the form of clinical notes, medical images, prescriptions, laboratory tests, and patient demographic information. With the extensive amount of data being collected in today's date, using DL techniques can be majorly beneficial. With DL models, by recognizing patterns, many preventable diseases can be predicted before a diagnosis. Money can be saved with predictions about a patient's length of stay or readmission probabilities.

DeepCare [29] was proposed based on the LSTM architecture to predict future diagnoses based on previous illness history. Similarly, [30] is another model based on a combined LSTM and CNN architecture used to predict future diagnoses and act as a tool for efficient healthcare resource allocation. In 2018, Google developed a model [31] capable of predicting 30-day unexpected readmission, extended length of stay, in-hospital fatality, and a patient's ultimate discharge diagnoses. This model was able to perform much better than traditional models. Google has also developed a unified EHR tool developed with AI technologies [38]. A 2019 study [32] presented a cost-effective LSTM-based DL model with the capability of predicting unscheduled readmissions probabilities. The advantages of such a system are not only monetary but also capable of saving lives. Reference [33] has developed a model to predict the increase in cases in India. With research like this, better plans can be created to control the spread of such deadly diseases and reduce the positivity rates. A similar model [34] was designed to predict the mortality rate of a person diagnosed with having COVID-19. Another study [35] used text-analysis techniques along with screening algorithms to develop more robust predictive models to diagnose COVID-19. In a pandemic situation like COVID, tools such as these can be valuable.

There are a few problems with DL research in EHR as well. Medical text data is very unstructured and hard to mine properly. There might be missing information in the records or ambiguity with generic words causing a big challenge. Also, to maintain privacy, de-identification is required which is a massive task with ways to improve it still being developed [39].

6 Challenges

1. **Availability of data and privacy concerns**: DL models require enormous amounts of training and testing data, which is not always easily available, especially in a domain like the healthcare industry as it does have major privacy concerns. The data might also be spread across multiple systems as people continuously change hospitals or clinics, they visit which makes the availability of proper data sets scarce.
2. **Region-specific models**: Research is conducted in specific regions with data sets of those areas. Since DL, models rely heavily on these data sets to learn, they will only be able to obtain certain information. In the medical line of work, information varies from region to region, race, gender, etc. Therefore, these models though accurate for their regions, may not be applicable for others without more data.
3. **Bias in the system**: It has also been observed that there may be some bias or unfairness that creeps into the DL systems. This could be because the data set used to train the algorithm itself is biased or maybe the way a user interacts with the algorithm leads to the development of this bias. This is a big challenge that should be taken into consideration seriously. It is explained in wonderful detail in [40].

7 Opportunities

1. **Automation**: AI systems will be able to automate the menial work and be able to save a physician's precious time. The burden of menial tasks such as keeping records, typing on keyboards, etc., will be taken away and the doctor will get more time to focus on the patient.
2. **Reach new heights**: Research into this field also helps push human boundaries and to be able to do things we cannot yet achieve on our own. For example, the research into diabetic retinopathy (DR), which is the primary cause of preventable vision loss globally. As discussed earlier, DR is not usually diagnosable until its later stages where the treatment becomes minimal and the procedures expensive. With the application of DL in this domain, it can be caught earlier with an actual chance of recovery without any loss of vision.
3. **Cost-effective and timesaving**: As seen in our review, DL models can reduce the cost of research by being able to predict which resources will be needed. In a field like drug discovery, where the development of drugs can take up to 10 years and still fail at clinical trials, DL models will be very beneficial. Also as seen

with DL applications in EHR, allocation of resources like beds and medicines can be done economically.

8 Conclusion

The medical industry is one of the most important, if not the most important domain of society. As humans, we are limited to some extent to which machines can exceed. This is exactly what DL helps us achieve. With unprecedented amounts of ever-increasing data, now is the best time to utilize such concepts. The concepts of DL and the architectures of ANNs were discussed in detail. We reviewed 24 research papers from the past 5 years which explored the applications of DL in the topics of drug discovery, medical imaging, and electronic health records, as well as a special focus of COVID-19 applications in these topics. DL has been proven to be an immensely advantageous tool, even though it does have its limitations as discussed, which are continuously being worked upon and improved. CNNs and LSTMs are some of the oldest established architectures but seem to be the preferred choice for developing models in the medical field.

The ongoing global pandemic has made everyone more aware of how important the medical industry is and how much we, as humans, need to develop. With concepts like DL, human capabilities can be extended to achieve greater heights at faster paces than ever before. This is just the beginning of AI, ML, and DL applications in healthcare, with more research being done every day, these concepts will continue to advance and enhance human life.

References

1. Tiwari T, Tiwari T, Tiwari S (2018) How artificial intelligence, machine learning and deep learning are radically different? Int J Adv Res Comput Sci Softw Eng 8:1. https://doi.org/10.23956/ijarcsse.v8i2.569
2. Yegnanarayana B (2009) Artificial neural networks. PHI Learning Pvt. Ltd.
3. Sharma K, Bhatia M (2020) Deep learning in pandemic states: Portrayal. https://www.semanticscholar.org/paper/Deep-Learning-in-Pandemic-States%3A-Portrayal-Sharma-Bhatia/f14a49a82a3e28f29c8c0f1a310fbd4c9bb0d7e5. Accessed 13 Jul 2022
4. Yao X, Liu Y (1997) A new evolutionary system for evolving artificial neural networks. IEEE Trans Neural Netw 8:694–713. https://doi.org/10.1109/72.572107
5. da Silva IN, Hernane Spatti D, Andrade Flauzino R et al (2017) Artificial neural network architectures and training processes. In: da Silva IN, Hernane Spatti D, Andrade Flauzino R et al (eds) Artificial neural networks: a practical course. Springer International Publishing, Cham, pp 21–28
6. Alloghani M, Al-Jumeily D, Mustafina J et al (2020) A systematic review on supervised and unsupervised machine learning algorithms for data science. In: Berry MW, Mohamed A, Yap BW (eds) Supervised and unsupervised learning for data science. Springer International Publishing, Cham, pp 3–21
7. Applications of artificial neural networks: a review. https://indjst.org/articles/applications-of-artificial-neural-networks-a-review. Accessed 13 Jul 2022

8. Bengio Y, Simard P, Frasconi P (1994) Learning long-term dependencies with gradient descent is difficult. IEEE Trans Neural Netw 5:157–166. https://doi.org/10.1109/72.279181
9. Sutskever I, Martens J, Hinton GE (2011) Generating text with recurrent neural networks
10. Hochreiter S, Schmidhuber J (1997) Long short-term memory. Neural Comput 9:1735–1780. https://doi.org/10.1162/neco.1997.9.8.1735
11. Lecun Y, Bottou L, Bengio Y, Haffner P (1998) Gradient-based learning applied to document recognition. Proc IEEE 86:2278–2324. https://doi.org/10.1109/5.726791
12. Wang X, Liu M, Zhang L et al (2020) Optimizing pharmacokinetic property prediction based on integrated datasets and a deep learning approach. J Chem Inf Model 60:4603–4613. https://doi.org/10.1021/acs.jcim.0c00568
13. Huang K, Fu T, Glass LM et al (2020) DeepPurpose: a deep learning library for drug–target interaction prediction. Bioinformatics. https://doi.org/10.1093/bioinformatics/btaa1005
14. Jamshidi MB, Talla J, Lalbakhsh A et al (2021) A conceptual deep learning framework for COVID-19 drug discovery. In: 2021 IEEE 12th annual ubiquitous computing, electronics and mobile communication conference (UEMCON), pp 00030–00034
15. Jamshidi MB, Lalbakhsh A, Talla J, et al (2021) Deep learning techniques and COVID-19 drug discovery: fundamentals, state-of-the-art and future directions. In: Arpaci I, Al-Emran M, Al-Sharafi MA, Marques G (eds) Emerging technologies during the era of COVID-19 pandemic. Springer International Publishing, Cham, pp 9–31
16. Tran NH, Zhang X, Xin L et al (2017) De novo peptide sequencing by deep learning. Proc Natl Acad Sci 114:8247–8252. https://doi.org/10.1073/pnas.1705691114
17. Li Y, Hu J, Wang Y et al (2020) DeepScaffold: a comprehensive tool for scaffold-based de novo drug discovery using deep learning. J Chem Inf Model 60:77–91. https://doi.org/10.1021/acs.jcim.9b00727
18. Song T, Zhang X, Ding M et al (2022) DeepFusion: a deep learning based multi-scale feature fusion method for predicting drug-target interactions. Methods 204:269–277. https://doi.org/10.1016/j.ymeth.2022.02.007
19. Artificial intelligence-guided de novo molecular design targeting COVID-19. ACS Omega. https://pubs.acs.org/doi/full/10.1021/acsomega.1c00477. Accessed 24 Jul 2022
20. De Fauw J, Ledsam JR, Romera-Paredes B et al (2018) Clinically applicable deep learning for diagnosis and referral in retinal disease. Nat Med 24:1342–1350. https://doi.org/10.1038/s41591-018-0107-6
21. Shankar K, Zhang Y, Liu Y et al (2020) Hyperparameter tuning deep learning for diabetic retinopathy fundus image classification. IEEE Access 8:118164–118173. https://doi.org/10.1109/ACCESS.2020.3005152
22. Truhn D, Schrading S, Haarburger C et al (2019) Radiomic versus convolutional neural networks analysis for classification of contrast-enhancing lesions at multiparametric breast MRI. Radiology 290:290–297. https://doi.org/10.1148/radiol.2018181352
23. Yala A, Lehman C, Schuster T et al (2019) A deep learning mammography-based model for improved breast cancer risk prediction. Radiology 292:60–66. https://doi.org/10.1148/radiol.2019182716
24. Shen L, Margolies LR, Rothstein JH et al (2019) Deep learning to improve breast cancer detection on screening mammography. Sci Rep 9:12495. https://doi.org/10.1038/s41598-019-48995-4
25. Ying S, Zheng S, Li L et al (2020) Deep learning enables accurate diagnosis of novel coronavirus (COVID-19) with CT images. Radiol Imaging
26. Hemdan EE-D, Shouman MA, Karar ME (2020) COVIDX-Net: a framework of deep learning classifiers to diagnose COVID-19 in X-ray images. ArXiv200311055 Cs Eess
27. Li L, Qin L, Xu Z et al (2020) Using artificial intelligence to detect COVID-19 and community-acquired pneumonia based on pulmonary CT: evaluation of the diagnostic accuracy. Radiology 296:E65–E71. https://doi.org/10.1148/radiol.2020200905
28. Kollias D, Arsenos A, Kollias S (2022) AI-MIA: COVID-19 detection and severity analysis through medical imaging

29. Pham T, Tran T, Phung D, Venkatesh S (2017) Predicting healthcare trajectories from medical records: a deep learning approach. J Biomed Inform 69:218–229. https://doi.org/10.1016/j.jbi.2017.04.001
30. Liu J, Zhang Z, Razavian N (2018) Deep EHR: chronic disease prediction using medical notes. ArXiv180804928 Cs Stat
31. Rajkomar A, Oren E, Chen K et al (2018) Scalable and accurate deep learning with electronic health records. NPJ Digit Med 1:18. https://doi.org/10.1038/s41746-018-0029-1
32. Ashfaq A, Sant'Anna A, Lingman M, Nowaczyk S (2019) Readmission prediction using deep learning on electronic health records. J Biomed Inform 97:103256.https://doi.org/10.1016/j.jbi.2019.103256
33. Arora P, Kumar H, Panigrahi BK (2020) Prediction and analysis of COVID-19 positive cases using deep learning models: a descriptive case study of India. Chaos Solitons Fractals 139:110017. https://doi.org/10.1016/j.chaos.2020.110017
34. Sankaranarayanan S, Balan J, Walsh JR et al (2021) COVID-19 mortality prediction from deep learning in a large multistate electronic health record and laboratory information system data set: algorithm development and validation. J Med Internet Res 23:e30157. https://doi.org/10.2196/30157
35. Obeid JS, Davis M, Turner M et al (2020) An artificial intelligence approach to COVID-19 infection risk assessment in virtual visits: a case report. J Am Med Inform Assoc 27:1321–1325. https://doi.org/10.1093/jamia/ocaa105
36. Selvaraj C, Chandra I, Singh SK (2022) Artificial intelligence and machine learning approaches for drug design: challenges and opportunities for the pharmaceutical industries. Mol Divers 26:1893–1913. https://doi.org/10.1007/s11030-021-10326-z
37. Senior AW, Evans R, Jumper J et al (2020) Improved protein structure prediction using potentials from deep learning. Nature 577:706–710. https://doi.org/10.1038/s41586-019-1923-7
38. Care studio: clinical software—google health. https://health.google/caregivers/care-studio/. Accessed 27 Jul 2022
39. Pandey B, Kumar Pandey D, Pratap Mishra B, Rhmann W (2021) A comprehensive survey of deep learning in the field of medical imaging and medical natural language processing: challenges and research directions. J King Saud Univ Comput Inf Sci. https://doi.org/10.1016/j.jksuci.2021.01.007
40. Mehrabi N, Morstatter F, Saxena N et al (2019) A survey on bias and fairness in machine learning. ArXiv190809635 Cs

Real-Time Grade Prediction for Students Using Machine Learning

Tameem Alshomari and Chaman Verma

Abstract One of the most crucial aspects is predicting students' performance in the educational domains such as schools and universities, because it aids in the creation of effective mechanisms that, among other things, enhance academic performance and prevent dropout. There are several mechanisms that are involved in typical student activities that handle a large amount of data gathered via software programs for technology-enhanced learning that are aided by automation. We can therefore learn suitable information about their knowledge and their connection to academic activities by properly reviewing and analyzing this data. This data is used to feed promising algorithms and approaches that can predict students' success. In addition to the objectives that must be met in this field, this research will reveal numerous contemporary methodologies that are widely used for predicting performance such as machine learning. The study's mission is to implement and compare existing approaches to come up with an optimized solution and an accurate model to predict the performance as well. The regression task involved modeling the fundamental classes. Additionally, different selections (such as with and without prior grades) and DM models were investigated. The findings demonstrate that, given access to grades from the first and/or second school periods, high predictive accuracy can be attained. Although prior ratings have a significant impact on student progress, other factors (such as the frequency of absences, the employment and educational status of parents, and alcohol use) have also been found to be significant, and better tools can be developed, which will improve the administration of school resources as well as the standard of instruction.

Keywords Machine learning in education · Deep learning · Regression · Decision trees · Random forest · XGBoost

T. Alshomari (✉) · C. Verma
Eötvös Loránd University, Budapest, Hungary
e-mail: h9whav@inf.elte.hu

C. Verma
e-mail: chaman@inf.elte.hu

Y. Singh et al. (eds.), *Proceedings of International Conference on Recent Innovations in Computing*, Lecture Notes in Electrical Engineering 1001,
https://doi.org/10.1007/978-981-19-9876-8_30

1 Introduction

In most educational entities and institutes, predicting student performance becomes a crucial need to provide exceptional learning resources and experiences, as well as to improve the university's rating and reputation [1]. However, for universities, especially those that concentrate on graduate and postgraduate programs and have few students' records to analyze [1], this may be challenging. As a result, the goal of this research is to demonstrate the feasibility of training and modeling a tiny dataset and constructing a prediction model with a respectable accuracy rate. This study also explores the likelihood of finding important indicators in a dataset [1]. Predictive analytic applications have become increasingly popular in higher education [2]. Predictive analytics used sophisticated analytics, which included ML deployment, to generate high-quality performance and meaningful data for students at all levels of education [2]. Most individuals are aware that one of the important indicators instructors may use to monitor students' academic development is their grade [3]. Many different ML algorithms have been proposed in the education arena during the last decade. As a result [3], this study provides a thorough examination of ML algorithms for predicting final students' grades [4]. It is frequently necessary to predict future student behavior. DM is favorable in this scenario [4]. DM approaches examine datasets and extract data to convert it into usable forms for later use [4]. The most important computer algorithms that use this data to predict the performance, grades, or the risk of leaving out are CF, ML, ANN, and RS. There is a significant number of studies that follow the lines of forecasting student behavior, as well as other associated educational themes of interest. Many works on this subject have been published in journals and presented at conferences. Furthermore, in most classification or prediction studies, researchers used to put in a lot of effort merely to locate the key indications that may be more effective in building realistic, precise forecasting models. This research aims to fill in any spaces with answering the subsequent research inquiries in which ML model is the most effective with a realistic and significant accuracy rate, using a limited dataset size. What are the most important key indications for developing a method for forecasting the results? Is it possible to forecast the performance in any subject with a reasonable and significant accuracy rate utilizing their prediction records, personal, social, and academic features?

2 Related Work

The authors showed how former school grades can be used to predict secondary student grades in two essential classes. SVM, NN, DT, and RF were used to test three different DM goals [5]. It is important to mention that this research is also based on the dataset used in this work. Table 1 shows the binary classification results.

Table 1 Binary classification results [5]

Input setup	NV	NN	SVM	DT	RF
A	91.9†±0.0	88.3±0.7	86.3±0.6	90.7±0.3	91.2±0.2
B	83.8†±0.0	81.3±0.5	80.5±0.5	83.1±0.5	83.0±0.4
C	67.1±0.0	66.3±1.0	70.6*±0.4	65.3±0.8	70.5±0.5

Table 2 Algorithms and accuracy [6]

Algorithm	Accuracy
LR	0.653
LDA	0.812
KNN	0.786
CART	0.685
NB	0.759
SVM	0.804

In this work [6], the authors used K-nearest neighbors, classification and regression trees, Gaussian distribution, logistic regression based on previous data regarding students' achievement in the courses, Naive Bayes, and SVM were used to construct a model that anticipates students' grades. The investigations revealed that examination of linear analysis is the most efficient method for precisely predicting exam results. Out of 54 records, the model correctly predicted 49, giving it an accuracy of 90.74% [6]. Table 2 shows how different algorithms are compared in terms of accuracy. Table 2 shows the algorithms and their corresponding accuracies.

In this work [7], the authors used simple baselines, predict grades at random from a distribution that ranges from 0 to 4 using the unified random method, Total Mean: Use the mean of all previously observed grades to predict grades, and Mean of Means: Add the global, per student, and per course together, standard rank k singular value decomposition, SVD-kNN, SVD post-processed with kNN ,FM: Factorization machine model. Table 3 shows the cold and none cold start prediction, student grades were predicted here using cooperative filtering based on user input. Elective course prediction accuracy of 96.4% is promising for assisting students in selecting the proper area of expertise [8]. The final predictions are generated by applying the mean centering equation, resulting in possible suggestions [8]. The results are compared in a binary sense, BB were regarded as indicative of successful students. Rather than basic accuracy, the evaluation was based on a variety of measures. Various similarity algorithms were examined using various performance measurements which were developed and divided into categories: coverage, prediction, and error [8]. Ceyhan et al. [8] while the authors applied DL approaches that can be utilized to forecast student performance vs typical ML techniques. This study used feedforward NNs and recurrent NNs to create a model that can accurately predict student's GPA. It compared various recurrent neural architectures and found that recurrent NNs were more accurate than feedforward NNs. The root mean square error was used as a

Table 3 Cold and none cold start prediction [7]

Method	RMSE	MAE
FM	0.7751	0.5301
Mean of means	0.8591	0.5661
SVD with KNN	0.9341	0.6281
SVD	0.9581	0.6541
Total mean	0.9581	0.6112
Unified random	1.8671	1.0541

Table 4 Metrics [2]

Metrics	RF	NB	KNN	LR	SVM	J48
Accuracy	0.986	0.975	0.984	0.983	0.981	0.982
Precisions	0.986	0.976	0.984	0.983	0.981	0.982
Recalls	0.986	0.975	0.984	0.983	0.981	0.992
F-measures	0.986	0.975	0.984	0.983	0.972	0.981

comparison metric, and the bidirectional long short-term memory network had the lowest error rate of 8.2 [3]. The single-layered LSTM NN's RMSE was found to be around 24. This indicates that the model is around 76% correct [3]. The RMSE was adjusted to 15 using numerous layers, five to be exact. This means that the model is 85% accurate [3]. This study also looks into the idea of employing visualization and clustering techniques to discover significant indicators in a small dataset in order to be employed to construct a prediction model. The study's key findings showed that SVM and learning discriminant analysis algorithms are effective in training small datasets and delivering acceptable classification accuracy and reliability test rates [1]. It functioned well with nominal attributes in comparison with the baseline, the prediction rates for dissertation grade and all courses grade classes are 76.3% and 69.7%, respectively. Because SVM with radial kernel function can train and cope with imbalanced datasets, it was chosen. Furthermore, kappa results revealed that LDA is 44.7% accurate in predicting a student's dissertation grade. However, SVM was the greatest for all course categorization (i.e., grade's attribute) and was better than the baseline recording a 41.7% accuracy rate [1].

In this paper [2], several models were employed by the authors. In order to counteract the impacts of unbalanced that result in overfitting, they developed a model based on oversampling with feature selection. The gathered data demonstrates that the proposed model's integration with RF results in a considerable improvement, with an F-measure maximum of 99.5. Results for the proposed models are comparable and encouraging [2]. The findings show that RF has the best score of 0.989, then KNN which scored 0.985. Meanwhile, the precision of SVM was 0.981, with a score of 0.978, not to mention that NB has the lowest model [2]. Table 4 shows the used metrics with their corresponding accuracies.

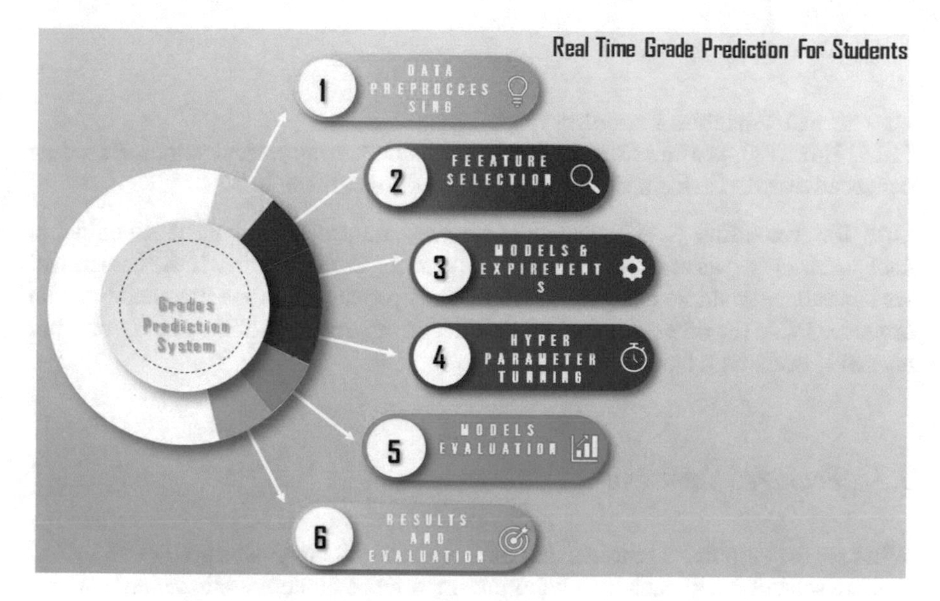

Fig. 1 Workflow of the real-time grade predictions

3 Methodology and Experimental Design

This chapter comprises the practical aspect of the work, in which the experimental environment is described, datasets, oversampling technique settings, and findings, and a full assessment of the advantages and disadvantages of each method used. It also describes the tests that were carried out. To expedite the study process, a range of software applications were used. Figure 1 displays the workflow of the real-time grade predictions.

3.1 Dataset Description

It relates to students performance at two Portuguese secondary schools. The characteristics were obtained from surveys and reports from the schools. For performance in two separate subjects, there are two datasets available. Regression task was used to model the dataset [9].

3.2 Dataset Preprocess

Categorical Variables Encoding
Label Encoding is a common encoding technique for categorical values. Based on consecutive order, a distinct number is assigned to each label.

One Hot Encoding One hot encoding solves the problem of label encoding. It uses just binary data and creates a new column for each category. If the categorical variables have multiple categories, the features produced by one hot encoding can explode. PCA (or other dimensional reduction approaches) followed by one hot encoding deals with this issue.

3.3 Features Selection

What are the significant features that played a vital role in grades prediction
Feature selection approaches like feature importance and correlation matrix with a heat map to decrease overfitting, enhance accuracy, and reduce training time were used. This will provide an insight into which characteristics affect the model's performance and have the greatest impact on accuracy. By doing so, there are other factors to consider (e.g., the number of absences, the work and education of the parents, and alcohol usage). Some features were deleted during the prepossessing stage due to their lack of discriminate significance. For example, few students responded to questions concerning their family income (owing to privacy concerns) even though most pupils reside with parents and have a computer at home. Defined the random forest (RF) as an ensemble of T unpruned DT. A random feature pick from bootstrap training samples is used to build each tree. Furthermore, given that the DT and RF algorithms specifically undertake a set of feature selections. In comparison with the DT/RF approaches, the NN/SVM are impacted by irrelevant inputs. For an A input selection example, G2 is the most crucial attribute, whereas G1 is very important for the B setup. In addition, the number of previous failures, when there are no student scores available, the most essential factor is previous student performance. Other important factors include social, demographic, and school-relevant factors, such as the number of absences, additional school aid, or travel time. School-relevant characteristics (such as absence rates, and the justification for selecting a school), as well as demographic, social, and behavioral aspects (such as student age, and alcohol usage) should be addressed too. Considering that just a tiny portion of the examined input variables seem to be significant, automatic feature selection techniques (such as filtering or wrapping) will be researched.

Grid Search for Hyperparameter Tuning Grid search for hyperparameter tuning was applied to the different models for optimization to get the most out of them. Grid search is the most basic hyperparameter tuning approach. In a nutshell, the hyperparameters domain is divided into a discrete grid. Then, using cross-validation, every possible combination of values in this grid was attempted, computing various performance measures. The ideal combination of values for the hyperparameter is the point on the grid that maximizes the average value in cross-validation.

3.4 Machine and Deep Learning Techniques

This section focused on the implementation of machine learning algorithms. The authors applied ML and DL models to identify the student's academic grades and compared the accuracy. After applying feature extracting, and hyperparameter tuning the authors achieved good results in all algorithms. These models were SVM, RF, Lasso Regression, LR, RG, DT, XGB, GaussianNB, and LSTM.

3.5 Evaluation Metrics

Absolute Mean Error (MAE): This is a metric that evaluates the average size of forecasting errors without addressing the orientation of the losses. Over the test sample, it stands for the estimated value of the absolute disparities between the forecast and the observed data. The weight of each individual deviation is equal. Below x and y are D dimensional vectors, and x_i denotes the value on the ith dimension of x

$$\sum_{i=1}^{D} |x_i - y_i|. \tag{1}$$

Mean Square Error (MSE): This indicates how near a regression line is to a group of points. In other words, a calculation is made to determine the average squared error between the estimated and original data. A regression metric is in use.

$$\sum_{i=1}^{D} (x_i - y_i)^2. \tag{2}$$

Root mean square error (RMSE): This is a popular way for a model or estimator to measure the differences between estimated and original data.

$$\sqrt{\frac{1}{n}\sum_{i=1}^{n}\left(\frac{d_i - f_i}{\sigma_i}\right)^2}. \tag{3}$$

Table 5 Models evaluation and experiments

Model	Accuracy	RMSE	MAE
SVM	0.994	0.006	0.00355
RF	0.991	0.009	0.00655
RG	0.981	0.019	0.01655
LGR	0.801	0.199	0.1745
LR	0.992	0.008	0.00555
DT	0.994	0.006	0.00355
XGB	0.989	0.011	0.00855
LSTM	0.983	0.017	0.01455
GAUS	0.989	0.011	0.00855

4 Models Evaluation

The authors showed that achieving promising results with high accuracy in comparison with related works was possible, and the methods scores were improved. Got a deep insight into the relationship between the features, and showed what are the best features which impact the accuracy and have a critical effect on the student's final grades, not to mention that after finding the most correlated features which impact the accuracy and have the most effect on the models, and with the hyperparameter tuning, the accuracy was improved to the highest. Table 5 shows the models evaluation and experiments.

The authors found that DT and SVM went up to high accuracy with the help of the significant features and based on their importance such as G2 is the most crucial attribute, whereas G1 is very important for the setup. In addition, the number of previous failures, when there are no student scores available, the most essential factor is previous student performance. Other important factors include socioeconomic, demographic, the frequency of absences, extra assistance, or transport time. Absenteeism rate, familial situation, and educational level, and the justification for selecting a school, as well as behavioral aspects should be addressed too.

5 Conclusion and Future Directions

This investigation has offered a thorough examination of student performance prediction in real time. While in modern society, education is really important. BI and DM techniques, which allow for high-level information extraction from raw data, have the intriguing potential for the education sector. Former school grades, socioeconomic, demographic, and others to predict secondary student grades were used in this work. Decision trees (DT), random forests, SVM, and NNs were used to test different DM goals. Different input options were also investigated. The acquired results show that if period grades are known, a high prediction accuracy is possible. Other relevant features include school-relevant, demographic, and social variables, based on a study of the information supplied by the most accurate forecasting models. Promising results with high accuracy with a comparison with related works were achieved, and the methods scores were improved and got a deep insight into the relationship between the features. And showed what is the best feature which impacts the accuracy and has a critical effect on the students final grades, this study showed that utilizing a student prediction in a school administration support system, will have the potential for the learning environment. This will enable the collection of extra information (e.g., grades from prior school years) as well as valuable feedback from school personnel. However, expanding the studies includes more schools to enrich the student databases. Because just a tiny percentage of the variables analyzed and appear to be significant, automatic feature selection approaches will also be investigated that are responsive to unrelated inputs and should benefit from this. More research is also needed to understand certain characteristics which affect student performance. In addition to that the presented machine learning models can be deployed on university learning management system.

Acknowledgements The work of Chaman Verma and Tameem Alshomari was supported by the Faculty of informatics, Eötvös Loránd University, Budapest, Hungary.

References

1. Abu Zohair LM (2019) Prediction of student's performance by modelling small dataset size. Int J Educ Technol High Educ 16:27. https://doi.org/10.1186/s41239-019-0160-3
2. Bujang SDA et al (2021) Multiclass prediction model for student grade prediction using ML. IEEE Access 9:95608–95621. https://doi.org/10.1109/ACCESS.2021.3093563
3. Patil AP, Ganesan K, Kanavalli A (2017) Effective deep learning model to predict student grade point averages. In: IEEE international conference on computational intelligence and computing research (ICCIC), pp 1–6. https://doi.org/10.1109/ICCIC.2017.8524317
4. Ramesh V, Parkavi P, Ramar K (2013) Predicting student performance: a statistical and data mining approach. Int J Comput Appl 63(8):35–39. https://doi.org/10.1504/IJTEL.2012.051816
5. Cortez P, Silva AMG (2008) Using data mining to predict secondary school student performance. In Brito A, Teixeira J (eds) Proceedings of 5th annual future business technology conference, Porto, pp 5–12
6. Gull H, Saqib M, Iqbal SZ, Saeed S (2020) Improving learning experience of students by early prediction of student performance using ML. In: IEEE international conference for innovation in technology (INOCON), pp 1–4. https://doi.org/10.1109/INOCON50539.2020.9298266
7. Sweeney M, Lester J, Rangwala H (2015) Next-term student grade prediction. In: IEEE international conference on big data (Big Data), pp 970–975. https://doi.org/10.1109/BigData.2015.363847
8. Ceyhan M, Okyay S, Kartal Y, Adar N (2021) The prediction of student grades using collaborative filtering in a course recommender system. In: 2021 5th international symposium on multidisciplinary studies and innovative technologies (ISMSIT), pp 177–181. https://doi.org/10.1109/ISMSIT52890.2021.9604562
9. https://archive.ics.uci.edu/ml/datasets/student+performance

10. Nabil A, Seyam M, Abou-Elfetouh A (2021) Prediction of students' academic performance based on courses' grades using deep NNs. IEEE Access 9:140731–140746. https://doi.org/10.1109/ACCESS.2021.3119596
11. Li H, Li W, Zhang X, Yuan H, Wan Y (2021) Machine learning analysis and inference of student performance and visualization of data results based on a small dataset of student information. In: 2021 3rd international conference on machine learning, big data and business intelligence (MLBDBI), pp 117–122. https://doi.org/10.1109/MLBDBI54094.2021.00031
12. Hongthong T, Temdee P (2022) The classification-based machine learning algorithm to predict students' knowledge levels. In: 2022 joint international conference on digital arts, media and technology with ECTI northern section conference on electrical, electronics, computer and telecommunications engineering (ECTI DAMT & NCON), pp 501–504. https://doi.org/10.1109/ECTIDAMTNCON53731.2022.9720334
13. Nguyen-Huy T et al (2022) Student performance predictions for advanced engineering mathematics course with new multivariate copula models. IEEE Access 10:45112–45136. https://doi.org/10.1109/ACCESS.2022.3168322
14. Nabizadeh Rafsanjani AH, Goncalves D, Gama S, Jorge J Early prediction of students final grades in a gamified course. IEEE Trans Learn Technol. https://doi.org/10.1109/TLT.2022.3170494
15. Hassan YMI, Elkorany A, Wassif K (2022) Utilizing social clustering-based regression model for predicting student's GPA. IEEE Access 10:48948–48963. https://doi.org/10.1109/ACCESS.2022.3172438
16. Suleiman R, Anane R (2022) Institutional data analysis and machine learning prediction of student performance. In: 2022 IEEE 25th international conference on computer supported cooperative work in design (CSCWD), pp 1480–1485. https://doi.org/10.1109/CSCWD54268.2022.9776102

A Comparative Analysis on Machine Learning Techniques for Driver Distraction Detection

Garima Srivastava and Shikha Singh

Abstract Driving any transport vehicle carries a significant amount of risk. Driving, on the other hand, is a necessity that cannot be ignored or replaced with something safer. The only thing that can be done in this concern is by taking precautionary measures to make driving safer and reducing the injuries and death to almost 0. The distracted driving model was built with the aim of achieving accident-free roads, with no threat and danger of any accident. This model accomplishes this task by detecting the behavior of the driver while driving the vehicle and inform the driver of any distraction which can possibly result in an accident. For this purpose, the machine learning model is trained to categorize the real-time received images of the driver into various classes of distractions. The distracted driver model has been developed on various different models. The accuracy of the predictions made by each of the models is compared to finally achieve the model which can best suit the requirement and provide highly accurate results. After performing a validation check on all the algorithms, KNN model has best accuracy and found out that KNN over LDA had ~99% accuracy which will give near about accurate detection of drivers' distraction. We also validated it on PCA and PCA over LDA too where PCA gave the accuracy of ~81% PCA over LDA gave ~76%, but LDA gave us the best accuracy, i.e., ~99%. The authors can assist the government in detecting drivers who engage in practices such as drunk driving, rash driving, texting, calling, eating, drinking, and many more by properly implementing this model. The laws can be implemented successfully and easily using this paradigm because no human intervention is necessary. The paper discusses about building a machine learning model for detecting distractions of the driver while driving. The paper comprehensively discusses the steps and methodology required for creating such model. Finally, results were calculated for respective algorithms and based on those selections can be made as per the requirement of the user. It is difficult to choose the best algorithm or the technique that will fit as per the user's requirement. Therefore, this paper tried to draw

G. Srivastava (✉) · S. Singh
Department of Computer Science and Engineering, ASET, Amity University, Lucknow, Uttar Pradesh, India
e-mail: gsrivastava1@lko.amity.edu

S. Singh
e-mail: ssingh8@lko.amity.edu

Y. Singh et al. (eds.), *Proceedings of International Conference on Recent Innovations in Computing*, Lecture Notes in Electrical Engineering 1001,
https://doi.org/10.1007/978-981-19-9876-8_31

a comparison among the various techniques by calculating the values of different parameters and helps the users to select the best technique as per their requirement. Selection of the technique would vary for different users.

Keywords Machine learning · SVM · Decision tree · KNN · Driver distraction

1 Introduction

Road accidents are very common in today's world, and most of the deaths take place due to harsh driving or sleepiness of the person who is driving the vehicle. According to surveys taken place worldwide, the US National Highway Traffic Safety has estimated that nearly 100,000 vehicles crash every year, and the reason is the driver drowsiness which leads to 1550 deaths and 71,000 injuries and a great money loss [1]. The development of various technologies has made the prevention of accidents at wheel a big challenge. Distracted driving has become one of the major reasons for the cause of accidents these days. As per another statistics of the year 2015, the number of deaths caused by distractions while driving was approximately 3477, and the number of injuries was 391,000 [2]. Because of this reason, the prevention of the driver and vehicle has become important. The reasons are different but lead to the same disaster. Therefore, a proper system is needed to alarm the person and reduce car crashes. Machine learning models for distracted drivers are trained using data that is currently accessible. It enables computers to make decisions and learn from their mistakes. The distracted driver model is designed to warn the driver ahead of time if the model detects any distractions that might lead to an accident. Distractions that occur while driving can be categorized as follows: visual, manual, and cognitive.

The paper starts with the introduction followed by a rigorous literature survey on the various machine learning techniques for detection of distraction. Further Sect. 3 of the paper discusses about the methodology adopted for building the model using various machine learning techniques. Section 4 presents the steps involved in model development; Sect. 5 presents the results of the implementation followed by the conclusion and the future scope.

2 Literature Survey

Monitoring and analyzing the distracted driver behavior and hence addressing the problem using machine learning has been popular and attracted the researchers as an emerging solution nowadays [3]. When the statistics of vehicles for the number of accidents equipped with distracted driver feature and the ones not equipped with this feature were compared the former category of vehicles registered a very low rate of accidents [4]. By the proper implementation of this model, researchers can help the government in the detection of drivers indulged in practices like drunk driving, rash driving, texting, calling, eating, drinking, and many more. Implementation of the laws using this model can be done effectively and easily as no human intervention is required at all. Although no model for the distracted driver has yet been established that will guarantee the driver's and other passengers' safety 100% [5]. Controlling the number of vehicles on the road is a very difficult task. With each passing day, the number of vehicles is increasing on the roads [6]. This increases the risk of even more accidents. In such a scenario, it is very necessary to develop a model which can help in controlling the number of accidents. Various implementations have been done for the distracted driver detection model using different techniques. The main challenge for its implementation is that the processing of driver's images by the model must be fast to provide real-time data with almost no time lag. Alexandra Elbakyan researched in the driver detection field and compared well-known CNN models to traditional handcrafted features for automatically detecting if the drivers are engaging in distracting behaviors [7]. The images for the model were collected using a high-quality dashboard camera. He also highlighted the fact that though this model is very helpful in preventing accidents, but on other hand, this can intrude on the privacy of the driver and other passengers in the car. Abouelnaga and Moustafa implemented a model called real-time distracted driver posture classification. In this, they aimed to identify the distractions based on the postures and positioning of the driver while driving the vehicle [8]. Omerustaoglu in his research classified the driver distraction detection techniques into three categories, namely vision-based, sensor-based, and the combination of vision and sensor-based techniques. He also concluded that combining the sensor data with vision data increased the accuracy of the model [9]. Alotaibi and Alotaibi suggested that data required for the distracted driver detection model can also be collected using wearable devices that can help track real-time data of driver's heartrate, oxygen level, stress [10]. There are two popular approaches using which researchers can build our models train them and further use them to automate the process [11]. These two approaches are machine learning techniques and deep learning techniques. Feng and Yue built a machine learning model for detecting distractions of the driver and alert them before hand to avoid any accidents [12]. The model was built by them using multiple machine learning algorithms such as KNN, SVM, Naïve Bayes, and decision tree. Another technique for model building is deep learning. Earlier researchers only used machine learning techniques, but in recent times, many have shifted their focus toward deep learning techniques for model building purposes [13]. Abouelnaga, Eraqi, and Moustafa developed a model which could detect the stress levels of the driver. As input, the model fetched real-time images of the driver and analyzed their postures and body movements. This deep analysis was used by model to further predict the stress levels of the driver [14]. Higher the stress levels of the driver higher will be the risk associated with driving [15]. Their model classified the drivers among four different stress levels. Jain, Singh, Koppula, Soh, and Saxena, 2016, used recurrent neural networks (RNNs) and sensory fusion architecture for building their model [16]. It is found that deep learning algorithms are more complex than machine learning models [17]. Although deep learning techniques require more time in modeling process, then machine learning techniques, but deep learning techniques are more accurate and flexible in nature [18]. Kim,

Choi, Jang, and Lim 2017 built a model using deep learning techniques. They used RestNet50 and MobileNet CNN models for building their model which classified the images of the driver in two separate classes. These classes were driver looking in front and driver looking at the back [19]. Although the model was accurate and had high precision, but it was not comprehensive. Since the model only classified images into two classes [20].

3 Proposed Methodology

The workflow of our model is clearly visible in Fig. 1, where first we are taking the data or the dataset and performing feature extraction on it using various methods, then pre-processing it, reducing its dimensionality, applying different ML algorithms, and finally plotting and providing various evaluation metrics.

Algorithm for steps involved in the model development:

Step 1: Perform feature extraction on the dataset images.
Step 2: Preprocessing the images by resizing them and splitting them into classes.
Step 3: Apply dimensionality techniques on images to reduce bulkiness of data.
Step 4: Finally, classify the images into pre-defined classes using various algorithms such as decision tree, SVM, and KNN.

State Farm Distracted Driver Detection is the dataset that was used to train and test the model. The model's dataset was sourced from https://www.kaggle.com/c/state-farm-distracted-driver-detection/data. The dataset used to create the model included 22424 photos for training and 79727 images for testing. The photos were 640 × 480 pixels in size. The dataset was divided into nine sub-classes. Each of these classes represented a distinct category. The photos in the dataset were then divided into ten classes, from C0 to C9. The classes and their description is as follows:c0: safe driving, c1: texting-right, c2: talking on the phone-right, c3: texting-left, c4: talking on phone-left, c5: operating the radio, c6: drinking, c7: reaching behind, c8: talking to a passenger.

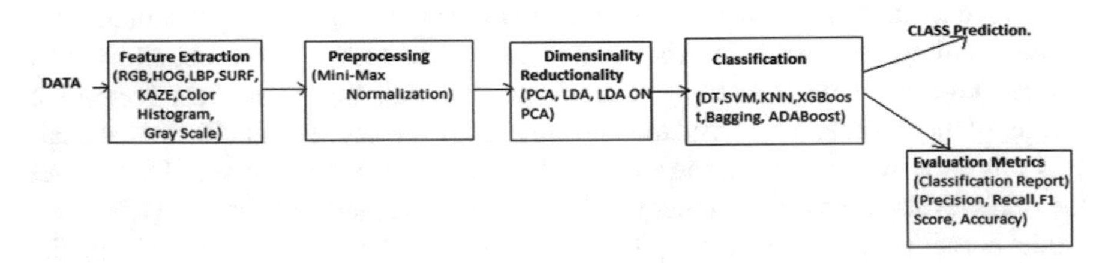

Fig. 1 Workflow of the model

3.1 Feature Extraction

Humans can interpret and comprehend visual or image data just by looking at it and applying their prior knowledge [21]. Feature extraction entails breaking down the image collection into smaller groupings. To provide accurate and desirable outcomes, the extracted features from the image must still represent the original data, and no changes to the current data must be made. On the resized images, various feature extraction techniques have been applied, like histogram of oriented gradients (HOGs), local binary pattern (LBP), color histogram, speeded up robust feature (SURF).

3.2 Data Preprocessing

This step's major goal is to make the picture categorization process more efficient, accurate, and quick. The photos were downsized to 64 × 64 × 3 in order to achieve these goals. Then, the dataset was split into training and testing dataset—80:20 ratio and then further splitting of training dataset into 90:10 ratio using stratified splitting to training and validation sets. After stratified partitioning the data into a training-testing ratio, the training set has 16145 photos; the invalidation set has 1794 images, and the testing set has 4485 images.

3.3 Dimensionality Reduction

The data we are usually given for modeling is large and contains a lot of information. The dimensionality-reduced dataset contains only the information needed for the actual processing. Before the actual modeling, dimensionality reduction techniques are considered an important stage in data preprocessing. [22], in the driver distraction model, dimensionality reduction was implemented using following methods:

3.4 PCA

PCA, or principal component analysis, aims to emphasize important trends in a dataset by accommodating as much volatility as feasible. The three axes are entirely independent of one another. All of the axes are at a 90-degree angle to one another [23]. PCA method is based on projections as the dataset after reduction is projected on the perpendicular axes. The final result will accommodate maximum possible variations so as to represent the whole dataset within reduces space.

3.5 LDA

LDA stands for linear discriminant analysis. LDA is an algorithm that is used to classify a dataset into multiple categories. The dataset is divided into several pre-defined classes based on similarity criteria in multiclass classification. It assumes that an item can only belong to one class at a time. An object cannot be a member of more than one class at the same time.

3.6 LDA Over PCA

This methodology was created to generate even better outcomes than the two methods previously stated, namely LDA and PCA. It is more efficient since it combines the two approaches mentioned above to create the dimensionally reduced dataset. The dimensionally reduced dataset produced by PCA is sent into LDA as an input. Then, LDA after further processing the already processed dataset LDA will produce the result [24].

4 Model Building

4.1 Grid Search

Grid search is a tuning technique that attempts to compute the optimum values of hyperparameters. It is an exhaustive search that is performed on a specific parameter values of a model. So, for our model, grid search was applied to find the optimal parameters for modeling so that we don't need to manually adjust the parameters. The library will automatically find the parameters, and we can use them to model our data. The models which we will be using are as follows: decision tree, SVM, KNN, ADA, bagging, XGBOOST.

4.2 Model Training

After creating grid search functions, we can proceed with model training to find the optimal parameters while also saving the model at the same time. The modeling is done for PCA, similarly done for LDA, LDA over PCA also. From this, we can get the exact optimal parameter required for each model to be trained on. From implementation, we can interpret that the optimal n-neighbors required in KNN is 5, and accuracy will be near to 98% which is really good and similarly for other algorithms

Table 1 Hyperparameter tuning

Model	Optimal hyperparameters
DT	Criterion = 'entropy', max-depth = 20
SVM	C = 10 and kernel = 'rbf'
KNN	n-neighbors = 5
XGB	Max-depth = 6, eta = 0.5
Bagging	n-estimators = 40
AdaBoost	n-estimators = 200

also. Same procedure is repeated for all the algorithms as well as dimensionality reduction methods.

4.3 Model Analysis

In this step, loading of all the models that were trained was done to make predictions of trained datasets using validation sets, and finding the precision, recall, F1-score, and accuracy is also done; the different scores for each algorithm used in PCA, LDA, LDA over PCA were also found.

4.4 Validation Score

Hyperparameter manages the accuracy of the model. They needs to be explicitly defined and set by the user. Table 1 shows the optimal hyperparameters that are set for creating functions for applying grid search on different models.

Tables 2, 3, and 4 show the validation scores for PCA, LDA, and LDA over PCA.

Table 2 Validation score for PCA

Model	Precision	Recall	F1-Score	Accuracy
DT	0.8221	0.8213	0.8214	0.822
SVM	0.9973	0.9973	0.9973	0.997
KNN	0.9872	0.9870	0.9870	0.987
XGB	0.9856	0.9849	0.9852	0.985
Bagging	0.7927	0.7848	0.7861	0.789
AdaBoost	0.7197	0.6957	0.7010	0.693

Table 3 Validation score for LDA

Model	Precision	Recall	F1-Score	Accuracy
DT	0.9753	0.9752	0.9751	0.974
SVM	0.9881	0.9876	0.9876	0.987
KNN	0.9922	0.9924	0.9923	0.992
XGB	0.9912	0.9912	0.9912	0.991
Bagging	0.9825	0.9826	0.9825	0.982
AdaBoost	0.5169	0.5785	0.5191	0.574

Table 4 Validation score for LDA over PCA

Model	Precision	Recall	F1-Score	Accuracy
DT	0.9638	0.9639	0.9638	0.964
SVM	0.9799	0.9791	0.9795	0.979
KNN	0.9806	0.9793	0.9799	0.979
XGB	0.9757	0.9756	0.9756	0.976
Bagging	0.9720	0.9710	0.9714	0.971
AdaBoost	0.6880	0.6634	0.6304	0.658

5 Performance Measure

An ROC curve (receiver operating characteristic curve) is a graph showing the performance of a classification model at all classification thresholds [25]. This curve plots two parameters:

- **True Positive Rate**
- **False Positive Rate**

True positive rate (TPR) is a synonym for recall and is therefore defined as follows: $\text{TPR} = \frac{\text{TP}}{\text{TP+FN}}$.

False positive rate (FPR) is defined as follows: $\text{FPR} = \frac{\text{FP}}{\text{FP+TN}}$.

An ROC curve plots TPR vs. FPR at distinct type thresholds. Lowering the classification threshold classifies extra gadgets as positive, as a consequence increasing each false positives and true positives.

Figure 2 shows ROC curve showing TP rate versus FP rate at different classification thresholds, also TP versus FP rate at different classification thresholds. To compute the factors in an ROC curve, the authors could evaluate a logistic regression version commonly with one of a kind classification threshold, and however, this would be inefficient. Figures 3, 4, and 5 show ROC curve for PCA, LDA, LDA on PCA.

Fig. 2 ROC curve

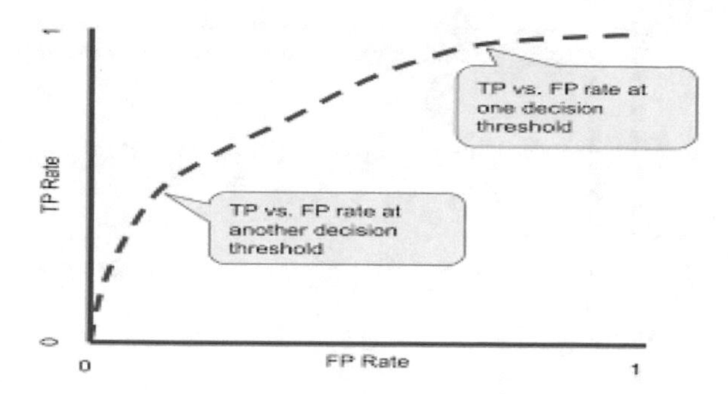

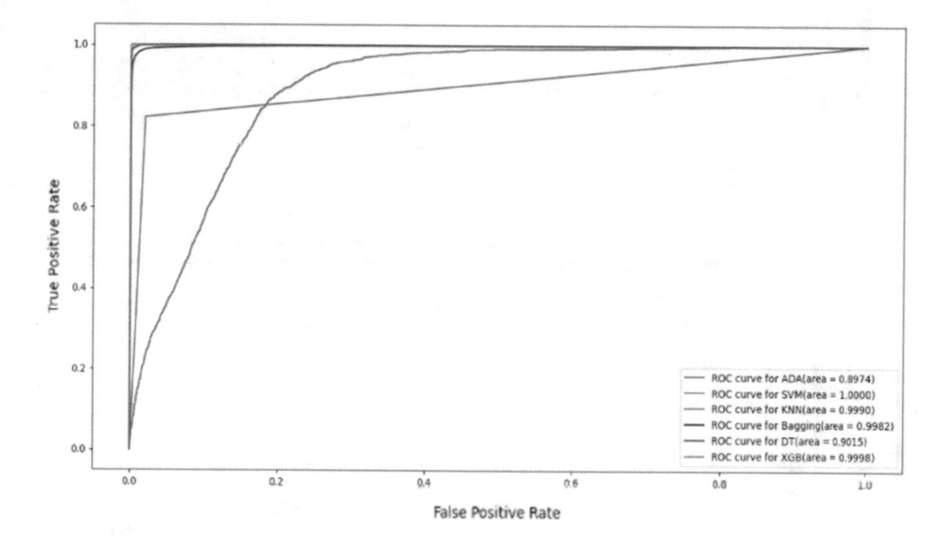

Fig. 3 ROC curve for PCA

6 Result and Conclusion

Modeling was successful and the development of different ML models with varying parameters but optimal for each algorithm. We also performed a validation check for KNN model as it had the best accuracy and found out that KNN over LDA had ~99% accuracy which is very accurate for a ML model. We also validated it on PCA and LDA on PCA too where PCA gave the accuracy of ~81% LDA on PCA gave ~76%, but LDA gave us the best accuracy, i.e., ~99% as shown in Fig. 6. The vehicles can be equipped with this distracted driver detection model along with the other required hardware such as dashboard camera. This model does not guarantee to completely resolve the problem of accidents, but it assures to mitigate this problem and significantly reduce the number of accidents.

In future work, the authors will try to implement the same model using deep learning to achieve better accuracy of results. Using deep learning, the results are

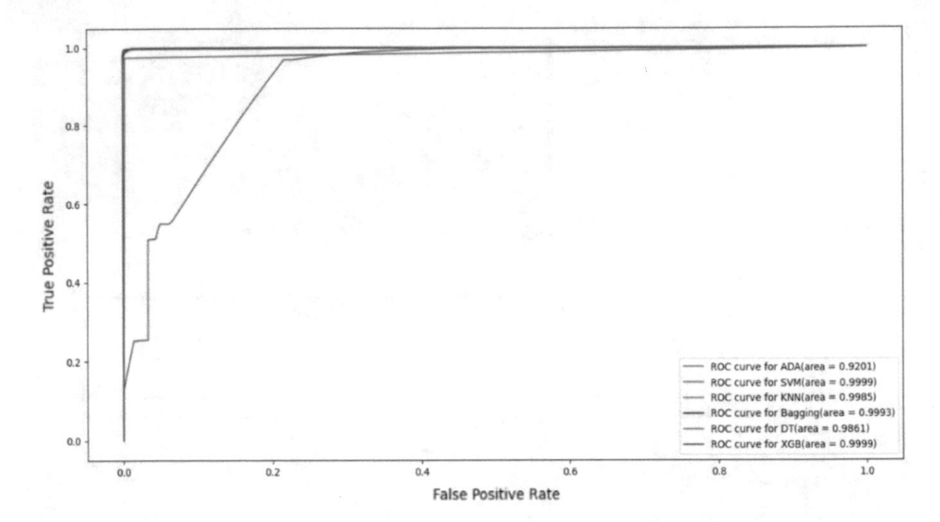

Fig. 4 ROC curve for LDA

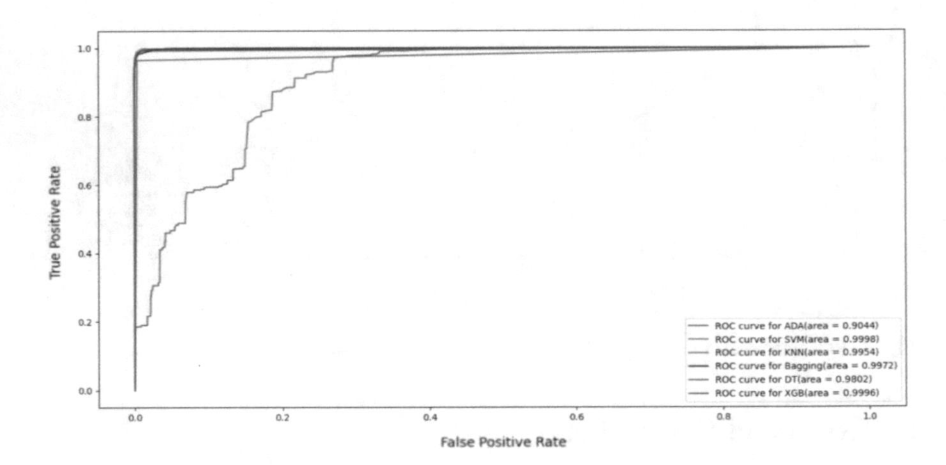

Fig. 5 ROC curve for LDA on PCA

produced faster and with no time lag. Also, in order to further increase the efficiency of the model, the authors will capture the image of the driver from various angles rather than just the dashboard. Driver detection model can be clubbed with drowsy driver detection classification model for better results.

```
features_train_pca = np.load('cache/features_train_pca.npy')
features_test_pca = np.load('cache/features_test_pca.npy')
print(features_train_pca.shape, features_test_pca.shape)
comp_model = train_model(features_train_pca, train_y, model_name='KNN')
pickle.dump(comp_model, open('cache/final_model_1.pkl', 'wb'))
y_hat = comp_model.predict(features_test_pca)
np.save('cache/final_predictions_1.npy', y_hat)
acc = metrics.accuracy_score(val_y, y_hat)
print(acc*100)

(16145, 600) (1794, 600)
81.66109253065775
```

```
features_train_lda = np.load('cache/features_train_lda.npy')
features_test_lda = np.load('cache/features_test_lda.npy')
print(features_train_lda.shape, features_test_lda.shape)
comp_model = train_model(features_train_lda, train_y, model_name='KNN')
pickle.dump(comp_model, open('cache/final_model_2.pkl', 'wb'))
y_hat = comp_model.predict(features_test_lda)
np.save('cache/final_predictions_2.npy', y_hat)
acc = metrics.accuracy_score(val_y, y_hat)
print(acc*100)

(16145, 54) (1794, 54)
99.66555183946488
```

```
lda_pca_train = np.load('cache/lda_pca_train.npy')
lda_pca_val = np.load('cache/lda_pca_val.npy')
print(lda_pca_train.shape, lda_pca_val.shape)
comp_model = train_model(lda_pca_train, train_y, model_name='KNN')
pickle.dump(comp_model, open('cache/final_model_3.pkl', 'wb'))
y_hat = comp_model.predict(lda_pca_val)
np.save('cache/final_predictions_3.npy', y_hat)
acc = metrics.accuracy_score(val_y, y_hat)
print(acc*100)

(16145, 9) (1794, 9)
76.25418060200067
```

Fig. 6 Validation check for KNN on PCA, LDA, and LDA on PCA

References

1. Rau P (2005) Drowsy driver detection and warning system for commercial vehicle drivers: field operational test design, analysis, and progress. National Highway Traffic Safety Administration, Washington, DC, USA (Google Scholar)
2. Research note on "Traffic safety facts" by U.S. Department of transportation national highway Traffic safety Administration, March 2017
3. Hossain MU et al (2022) Automatic driver distraction detection using deep convolutional neural networks. Intell Syst Appl 14:200075
4. The Government of India (2016) Road accidents in India. [Online]. Available: https://morth.nic.in/sites/default/files/Road_Accidents_in_India_2016.pdf. Accessed 15 Jan 2020
5. Federal Communication Commission (2016) The dangers of distracted driving
6. National Highway Traffic Safety Administration (2017) Distracted driving
7. Tian R, Li L, Chen M, Chen Y, Witt GJ (2013) Studying the effects of driver distraction and traffic density on the probability of crash and near-crash events in naturalistic driving environment. IEEE Trans Intell Transp Syst 14(3):1547–1555
8. Regan MA, Hallett C, Gordon CP (2011) Driver distraction and driver inattention: definition, relationship and taxonomy. Accid Anal Prev 43:1771–1781
9. Willis S (2002) Shorter oxford english dictionary on historical principles, 5th edn. Oxford University Press, New York, NY, USA
10. Kaplan S, Guvensan MA, Yavuz AG, Karalurt Y (2015) Driver behavior analysis for safe driving: a survey. IEEE Trans Intell Transp Syst 16(6):3017–3032
11. National Institutes of Health Research (2014) Distracted driving raises crash risk
12. Center for Disease Control and Prevention (2020) Distracted driving. 15 Jan 2020
13. National Highway Traffic Safety Administration (2020) Policy Statement and Compiled FAQs on Distracted Driving. U.S. Department of Transportation, Washington, DC (Online). 25 Jan 2020
14. Li Z, Bao S, Kolmanovsky IV, Yin X (2018) Visual-manual distraction detection using driving performance indicators with naturalistic driving data. IEEE Trans Intell Transp Syst 19(8):2528–2535
15. Iranmanesh SM, Mahjoub HN, Kazemi H, Fallah YP (2018) An adaptive forward collision warning framework design based on driver distraction. IEEE Trans Intell Transp Syst 19(12):3925–3934

16. Business Insider (2017) Tesla fatal crash. (Online). Available: https://www.businessinsider.in/New-details-about-the-fatal-Tesla-Autopilotaccident-reveal-the-drivers-last-minutes/articleshow/59238933.cms. Accessed 15 Jan 2020
17. The Guardian (2018) Uber fatal crash (Online). Available: https://www.theguardian.com/technology/2018/jun/22/driver-wasstreaming-the-voice-when-uber-self-driving-car-crashed-say-police. Accessed 15 Jan 2020
18. Zhang X, Zheng N, Wang F, He Y (2011) Visual recognition of driver hand-held cell phone use based on hidden CRF. In: Proceedings of IEEE international conference on vehicular electronics and safety, July 2011, pp 248–251
19. Das N, Ohn-Bar E, Trivedi MM (2015) On performance evaluation of driver hand detection algorithms: challenges, dataset, and metrics. In: Proceedings of IEEE 18th international conference on intelligent transportation systems, Sep 2015, pp 2953–2958
20. Seshadri K, Juefei-Xu F, Pal DK, Savvides M, Thor CP (2015) Driver cell phone usage detection on strategic highway research program (SHRP2) face view videos. In: Proceedings of IEEE Conference on Computer Vision and Pattern Recognition Workshops, June 2015, pp 35–43
21. Le THN, Zheng Y, Zhu C, Luu K, Savvides M (2016) Multiple scale faster-RCNN approach to driver cell-phone usage and hands on steering wheel detection. In: Proceedings of IEEE Conference on Computer Vision and Pattern Recognition Workshops, June 2016, pp 46–53
22. Ohn-Bar E, Martin S, Tawari A, Trivedi MM (2014) Head, eye, and hand patterns for driver activity recognition. In: Proceedings of 22nd international conference on pattern recognition, Aug 2014, pp 660–665
23. Zhao CH, Zhang BL, He J, Lian J (2012) Recognition of driving postures by contourlet transform and random forests. IET Intell Trans Syst 6(2):161–168
24. Zhao C, Zhang B, Zhang X, Zhao S, Li H (2012) Recognition of driving postures by combined features and random subspace ensemble of multilayer perceptron classifiers. Neural Comput Appl 22:175–184
25. State Farm (2016) Distracted driver detection competition (Online). Available: https://www.kaggle.com/c/state-farm-distracted-driverdetection. Accessed: 15 Jan 2020

Machine Learning Recognition Mechanism Based on WI-FI Signal Optimization in the Detection of Driver's Emotional Fluctuations

Zhu Jinnuo and S. B. Goyal

Abstract With the development of the automobile industry, the use of vehicles has become the most basic means of transportation in people's daily life. However, due to the rapid increase in the number of vehicles, more and more drivers are causing traffic accidents due to emotional fluctuations during driving. In order to reduce the occurrence of this problem, the artificial intelligence machine learning mechanism under Industry 4.0 technology can currently capture and detect the instantaneous facial emotions of drivers during driving. Through the literature survey, it is found that in the machine learning mode, the existing mechanism mainly relies on human vision and biological signals sensors to identify the driver's emotional fluctuations during driving. However, due to the frequent distortion of visual detection methods and the problems of invasiveness and privacy invasion of biological signals, the investment cost is relatively high. In order to solve the problems under the existing algorithm, this paper firstly formulates the corresponding emotional fluctuation recognition model according to the existing WI-FI signal detection principle and designs the antenna position according to the Fresnel zone to achieve the best signal acquisition effect. In addition, the driver's action status while using the brake and accelerator is collected. The emotion recognition coefficient is calculated by the collected data, and the fluctuation recognition is performed by using the recognition wash and LSTM emotion discriminator. Finally, according to the evaluation data, the recognition rate of about 85% in the real scene is achieved.

Keywords Machine learning · Emotion detection · LSTM · Biological signals

Z. Jinnuo
Nanchang Institute of Science & Technology, Nanchang, China

Z. Jinnuo · S. B. Goyal (✉)
City University, Petaling Jaya, Malaysia
e-mail: drsbgoyal@gmail.com

Y. Singh et al. (eds.), *Proceedings of International Conference on Recent Innovations in Computing*, Lecture Notes in Electrical Engineering 1001,
https://doi.org/10.1007/978-981-19-9876-8_32

1 Introduction

In today's society, vehicles have become our most common means of transportation [1]. With the continuous increase in the number of vehicles, the number of traffic accidents caused by excessive emotional fluctuations in the driver's driving process is also increasing. Because when human beings have excessive emotions, our thinking, judgment, and operation of the target will be greatly affected, and it is very easy to cause traffic accidents due to mistakes. In the process of driving, it is particularly important to predict the driver's dangerous emotions in advance and reduce the occurrence of traffic accidents through effective detection methods [2]. To reduce the occurrence of this phenomenon, the key point is to suppress the road rage and impatience of drivers. By reviewing the cashback, psychologist Novaco [3] once said that when human beings encounter negative things such as troubles, failures, and psychological trauma, they will be most prone to anger. Only, by improving the driving environment of the driver, we can effectively control the generation of irritable emotions dangerous behavior. The total emotion recognition of the driver's driving process needs to be detected with the help of equipment. Through the research on effective recognition algorithms such as CNN and LSTM on the driver's face, the driver can timely understand the appearance of road rage and impatience during the driving process through the intervention of equipment [3]. The research focus of this paper is to use machine learning algorithms to operate under Industry 4.0. Because machine learning is the ideal state of human–computer interaction at present, when humans implant algorithms into the mechanism, the mechanism can provide intelligent services according to human consciousness, which is accurate and fast [4]. The core work of this paper is to establish an effective WI-FI wireless network environment, transmit effective test data through the WI-FI network, so that machine learning can identify the driver's biological signals while driving under an effective algorithm and feedback whether it is risky behavior with road rage. Finally, the human–computer interaction between the driver and the detection equipment is realized.

2 Background

2.1 Research Status Analysis

Through the data analysis of driving accidents at the current stage, the frequency of dangerous driving processes or traffic accidents is proportional to the number of road rage in the driver's driving [5]. At present, in the field of driver emotion detection, there is no specific work on the detection and research of driver's dangerous emotion in driving. From the literature collection, it is known that there are usually two most common ways to detect emotions during driving under the existing algorithms: One is visual perception through traffic scenes. In the visual detection of the target, the detection is carried out by the rotation change of the eyeball. If the driver does not move his eyes for a long time, he is judged to be tired. If the pupil of the eyeball is enlarged, it may be identified as road rage. At the same time, in the implementation process of obtaining road rage emotion through visual detection, it is very likely that there will be concerns about the invasion of user privacy. At the same time, wearing special equipment for testing will result in high cost and discomfort to the testing personnel. The other is sensor-based biosignal capture of drivers. The sensor uses the biological signals caused by the target's emotional fluctuations to reflect the emotional characteristics of the driver at that time. Some scholars in my country have used a method of inducing road rage in participants in the "real sense traffic" mode to obtain biological signals that appear in EEG signals and have identified and analyzed the signals. However, this method may also easily lead to the existence of problems such as user privacy violation. At present, according to the latest data in recent years, we found that WI-FI network has been used as an effective medium for detecting human emotions because WI-FI signals can send and receive human body data through unique wireless characteristics and are not limited to the length of wires. In the 5G era, some scholars specialize in research on human body signal monitoring for wireless networks. Based on the above findings [6], we plan to use the wireless network as the platform architecture to carry out various methods such as accelerator and brake use, driver emotional coefficient, and LSTM algorithm, to study a set of research work based on the wireless network to realize whether the driver has road rage. To address the issue of high costs and the effect on user privacy in current research.

2.2 *The Research Focus of This Paper*

This paper starts with the way researchers in related fields capture human facial emotions, combined with our current research. We need to solve the following parts in the research process:

The key research questions are provided above in Table 1. The question we discuss in this paper is how to conduct preliminary research on the current driver's road rage detection mechanism and develop a more accurate detection mechanism by improving the algorithm based on the existing theoretical and experimental basis. In the research method of the third chapter, we will select the research method through the detailed architecture, using the global architecture diagram, the research process algorithm design, and the output.

3 Methodology

By detecting the driver's driving emotion in the traditional mode, we expect to use a special WI-FI signal architecture in the Fresnel zone mode to implement the whole

Table 1 List of research questions

Serial number	Research problem
1	What is the significance of road rage detection for drivers?
2	What is the capture method in the process of road rage emotion?
3	What is the process of driver road rage under the existing algorithm?
4	How to optimize existing algorithms to improve detection of driver road rage emotion?

process from capture to detection of the driver's facial features [7]. The advantages of Wi-Fi signal under this architecture are airtight, small in scope, strong in signal, and easy to install. The method can effectively obtain the driver's biological signal through the signal. In the early stage of the experimental study, we will collect all the required data. First, we will use the model architecture of the Fresnel zone with a special antenna device, which can increase the characteristics of WI-FI transmission signals compared with the traditional mode [8] and enhance the acquisition of the driver's facial data by the mechanism; secondly, we use a kind of driver's use frequency of the vehicle's accelerator and brake during driving to obtain the driver's emotional fluctuation (biological signal performance); thirdly, by using the LSTM algorithm to build the emotional recognition mechanism measurement part, the Wi-Fi algorithm is applied to the real driving. The performance test is carried out in the environment; finally, the WI-FI detection mechanism in the machine learning mode is used to obtain the actual emotional fluctuation of the driver during driving [9]. We divide the final driver emotional results into three types according to actual needs: cautious driving, normal driving, and road rage driving. The following is the detailed research method.

3.1 Global Architecture

According to the research steps proposed in this paper, this paper proposes the following global process architecture, as shown in Fig. 1.

From the above flowchart, we can see that the system implementation process is divided into three steps: system preparation, execution, and detection. In the first step, data are communicated using the wireless signal in the first step, which involves sending a wireless signal across a WI-FI environment. In the second step, the collected sample data, the usage data of the accelerator and the brake during the driver's driving process, and the LSTM feature data are extracted from three algorithms. The third step is the detection operation under the final machine learning [10]. The machine learning output includes two ways of valid recognition and invalid recognition. There are no feedback results for invalid identification, and there are two methods for valid identification: result return and unmatched data.

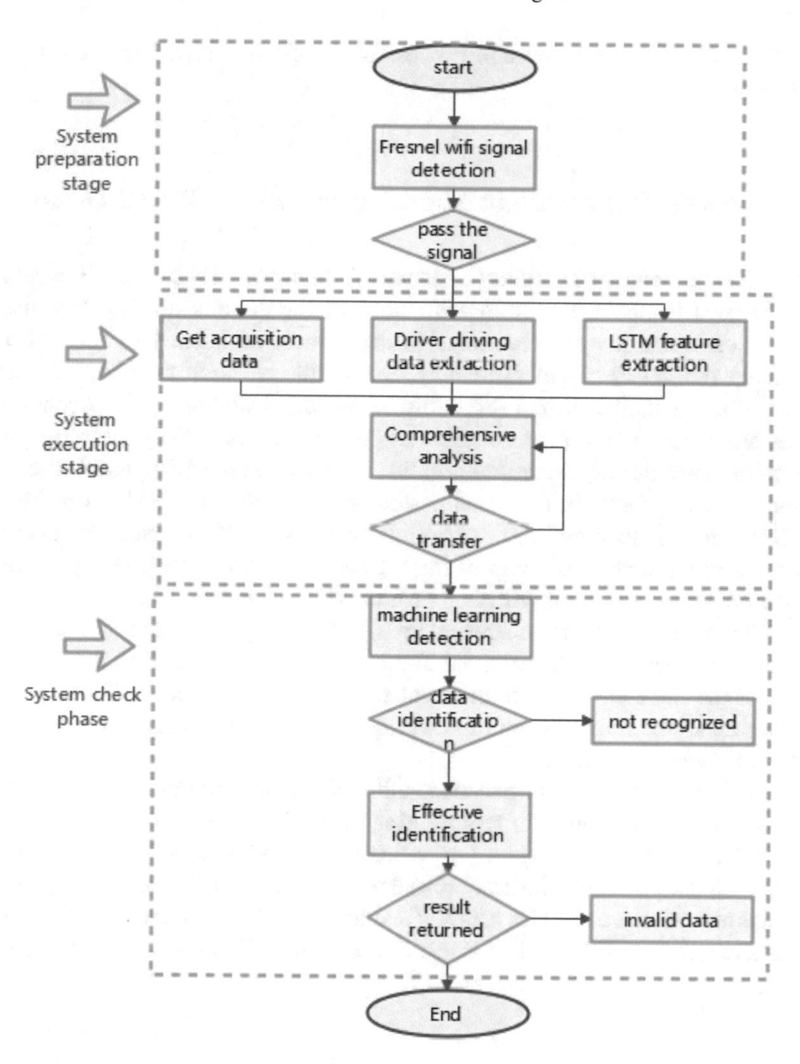

Fig. 1 Global flowchart

3.2 *Data Collection Work*

In the process of data collection, this paper selects the classic driving scene. The sample data in this paper is realized by the actions of drivers when their emotions fluctuate. Under normal circumstances, the driver will slam on the brakes and slow-down in the event of an emergency. For example, a driver in a normal mood has moderate accelerator force and moderate time while driving [11]. When a driver with road rage is driving, he may be more aggressive and last for a long time. We will collect effective data through the strength, frequency, action duration, and other characteristics of the transmission process of the driver's signal to the brake and accelerator.

3.3 Antenna Identification Signal Algorithm of WI-FI Device

In this paper, a method of building a special WI-FI signal based on the Fresnel zone model is used to obtain the biological signal of the driver's driving. Because the transmission technique and transmission angle must be taken into account when calculating the WI-FI signal's intensity. During the research process, we need to select the best installation area according to the actual situation [12]. According to the literature survey, the mode of the Fresnel zone is formed by the reflection of the electric wave through the refraction line between the receiving and transmitting antennas. Among them, in the transmission process, the focus position is the position of the method antenna, and the overall transmission signal presents an ellipse. According to the model structure of the Fresnel area, this paper proposes a special installation setting for the installation position of the antenna, where Qn represents the boundary of the outermost Fresnel area, and per represents the first area of the Fresnel area projected to the X-axis. The x value corresponding to the y value, S represents the calculation area of the x value at the boundary points TX and per [13]. The specific signal transmission model diagram and algorithm are shown in Fig. 2 and the following formulas.

Figure 2 is divided into 4 concentric ellipses, where n represents the nth ellipse from the inside to the outside, representing the wavelength of the signal λ in this process. The most central region represents $n = 1$ and is also the region with the strongest signal, becoming the first Fresnel region. There are boundary points T_x, R_X in the figure, which are used for sending and receiving [14]. The separation between the vehicle's driver and the data transmission point. We choose the strongest 5G

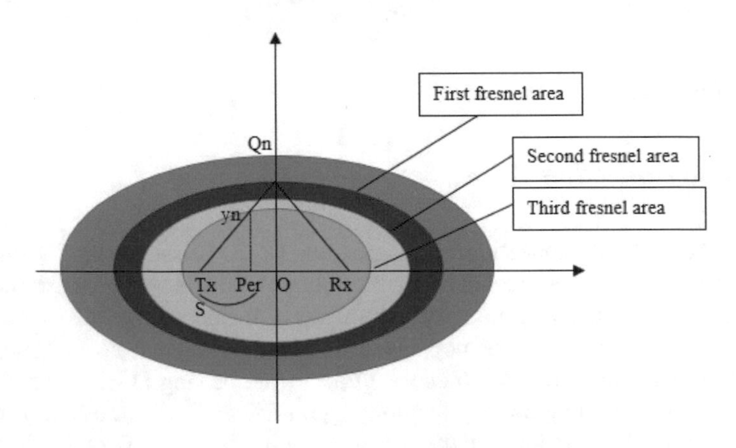

Fig. 2 Antenna identification signal algorithm for WI-FI devices

Wi-Fi signal for testing during the calculation. With the area of the ellipse, we finally give the following algorithm:

$$y_n = \sqrt{1 - \frac{16\left|\frac{T_X R_X}{2} - S\right|}{4T_X R_X + 4n\lambda|T_X R_X| + n^2\lambda^2}}$$

y_n represents the value of y on the Y axis when the value of x on the X-axis is $\left(\frac{T_X R_X}{2} - S\right)$ the value. This architecture is based on theoretical calculation laws. As the initial value, we set the distance of $T_X R$ as 1 m, and the entire Fresnel area in the figure from the first area to the third area is the range of the enhanced driver's operating signal [15].

3.4 *Brake and Accelerator Signal Recognition Algorithm During Driving*

3.4.1 Displacement Calculation While Driving

In this paper, through the literature review, Mruphey's expert [16] team once pointed out that the speed or acceleration of the vehicle can be quantified while the vehicle is running and proposed a very authoritative R_{driver} algorithm. At the same time, the Ding expert team pointed out that there is a relationship between the acceleration and the distance of the emotion recognition coefficient and proposed the R_{de} algorithm [17]. Combining the content of the two algorithms, this paper adds the driver's acceleration measurement and braking distance to the driving distance and carries out the braking and accelerator signal recognition algorithm. First, we set the brake displacement as S, and the specific displacement formula is as follows:

$$s = v_0 t + \frac{at^2}{2}$$

3.4.2 Distance Coefficient *P* Value Calculation

In the above formula, V_0 represents the initial speed of the vehicle; a represents the acceleration, and t represents the acceleration duration. It can be seen from the above formula that a and V_0 are proportional to the braking distance. At the same time, the braking time is also a key factor affecting the braking distance. Through comprehensive analysis, the braking signal duration is used to replace the deceleration duration [18]. According to the emotion recognition coefficient R_{de} in the previous case, the P-parameter algorithm used in this paper is defined as follows:

$$p = \frac{A_{\mathbf{break}} t_{\mathbf{break}}^2}{A_{N-\mathbf{break}} t_{N-\mathbf{break}}^2}$$

In the above definitions, A_{break} and t_{break} represent the amplitude and duration of the braking signal, respectively, while $A_{N\text{-break}}$ and $t_{N\text{-break}}$ represent the amplitude and duration of the braking signal under normal emotions, respectively [19]. Judging by the ratio.

3.4.3 Acceleration Coefficient Calculation

After completing the amplitude and duration, the next thing we need to define is the coefficient of acceleration. Usually, when the vehicle accelerates too much, the depth of the accelerator pedal is also larger [20]. In this paper, the average value of the amplitude of the throttle signal generation process and the standard deviation of the throttle signal are used to calculate. The final result is the ratio between the standard deviation of the throttle signal and the average value of the throttle signal amplitude to reflect the acceleration identification coefficient r_{driver}, the algorithm is as follows:

$$r\_\text{driver} = \frac{\sigma_{\text{driver}}}{A_{\text{driver}}}$$

In the above algorithm, A_{driver} represents the average value of the throttle signal amplitude, and σ_{driver} represents the standard deviation of the throttle signal [21].

3.4.4 Emotion Recognition Coefficient Calculation

Finally, we will determine the emotion recognition coefficient according to the algorithm already given. After analyzing the driving data, we give the driver's emotion recognition coefficient algorithm R_{de} as follows:

$$R_{\text{de}} = \frac{\sigma_{\text{driver}}}{A_{\text{driver}}} * \frac{A_{\text{break}} t_{\text{break}}^2}{A_{N_{\text{break}}} t_{N_{\text{break}}}^2}$$

3.5 *Mood Fluctuations Under the LSTM Algorithm*

Because every driver has a unique driving style and because emotional shifts and time series are closely related. Ultimately, our biggest research challenge is to use the features of LSTM based on the summary and architecture of all the previous algorithms and use the previously calculated driver emotion recognition coefficients

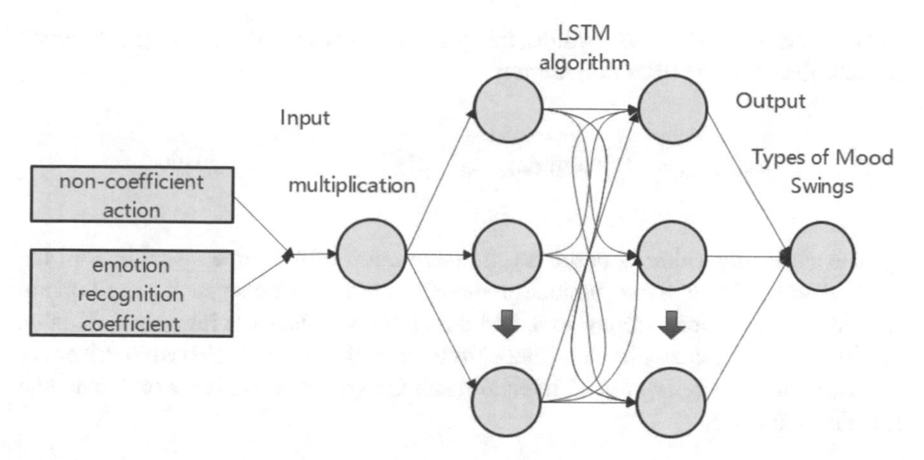

Fig. 3 Flowchart of emotional fluctuations under LSTM algorithm

and non-coefficient action data other than the coefficients to define the original data of emotional fluctuations [22]. The emotion recognition coefficient obtained from the characteristics of the accelerator and brake signals has been obtained, and the emotion fluctuation algorithm is obtained.

$$R_{\mathrm{b}} = (R_{\mathrm{de}} + t_{\mathrm{d}}) * R_{\mathrm{LSTM}}$$

In the above algorithm, this paper uses the R parameter as the target parameter of the whole process, uses R_{LSTM} to represent the LSTM feature of the corresponding driver; td represents the non-coefficient action value, and the final output emotional fluctuation value is replaced by R_{b} [23]. The advantage of this algorithm is that through the characteristics of the LSTM algorithm, it is possible to control the fusion of the emotional coefficient and the value under the time series. The output value after multiplying the feature value of the two can be calculated more accurately, and it is also beneficial to our output in the later machine learning algorithm [24]. Another advantage is that the LSTM algorithm can effectively store data features of different lengths for a long time. The model diagram of the execution process of the LSTM algorithm is shown in Fig. 3.

3.6 *Machine Learning Algorithm Output*

According to the previous sample collection work, the antenna area identification algorithm, the accelerator and brake signal algorithm, and the output method of the L STM mood fluctuation algorithm under the Frenier zone model [25], combined with the machine learning algorithm that needs to be implemented in this paper, the driver's road rage the output process of emotion detection. Through the calculation and analysis of the above algorithms, this paper summarizes the following algorithms for machine learning detection output:

$$m(n)_{\text{sum}} = \int_{1}^{n} \left[\text{sum}\left(Y_{n(1,\ldots,n)} + R_{\text{de}(1,\ldots,n)} + R_{\text{b}(1,\ldots,n)}\right)\right]\text{d}n$$

The above algorithm is based on the integration of the three algorithms in the methodology. Through the summation method, the values under each algorithm are weighted and summed from 1 to *n*, and the obtained values are integrated. Finally, the corresponding characteristic values from the first driver data to the nth driver are calculated by integrating [26]. The final result is used as the basis for later machine learning detection.

4 Result

4.1 Antenna Distance Radius Statistics Under Fresnel Zone Model

WI-FI signal installation method and the application of the algorithm under the previous Fresnel zone model, we divided the necessary radius area of the wireless signal corresponding to the Fresnel zone according to the actual situation. According to the needs of the experiment, we have here the data of the gradual diffusion from the first radius to the thirteenth radius. Under the current Wi-Fi test conditions, the signal-allowed area is within a 1.3 radius under the Fresnel zone, and beyond the 1.3 radius, there is no Wi-Fi signal area. The specific statistics are shown in Table 2.

The above data is obtained from the antenna algorithm under the Fresnel zone model. According to the actual calculation results, this paper selects the first, eleventh,

Table 2 Fresnel zone model statistics

$T_X R_X$ length (m)	The first radius of the Fresnel zone	Eleventh radius of the Fresnel zone	Twelfth radius of the Fresnel Zone	Thirteenth radius of the Fresnel zone
0.65	0.974	0.3601	0.3891	0.3987
0.75	0.1035	0.3802	0.3999	0.4189
0.85	0.1081	0.3997	0.4186	0.4388
0.95	0.1166	0.4099	0.4301	0.4508
1.05	0.1228	0.4198	0.4450	0.4655
1.15	0.1301	0.4296	0.4533	0.4768
1.25	0.1366	0.4396	0.4598	0.4844

Table 3 Statistical table of driver's road rage emotion under machine learning

Serial number	Classification	Number of detections	Detection rate (%)
1	Drive cautiously	27	85.88
2	Normal driving	32	82.01
3	Road rage	29	84.33

twelfth, and thirteenth Fresnel zones except for the most important parts of the head and tail parts radius.

4.2 *The Algorithm After Machine Learning Outputs the Detection Result*

Through the output process under the machine learning algorithm, this paper completes the detection of the driver's road rage during the driving process. According to the output of the algorithm, the biosignal of the driver is obtained and detected. Sentiment statistics were carried out according to the three categories before the experiment. According to the experimental situation, the number of drivers with the three emotional expressions is basically the same, and the detailed detection data is shown in Table 3.

It can be seen from the above data that we used the features of the algorithm fusion in the research method to carry out machine learning, and the overall detection rate is basically the same among the three types, among which the cautious type has the highest detection rate, which currently exceeds 85%. The overall situation has achieved the expected effect. At the same time, in the experiment process, this paper also uses the recurrent neural network RNN algorithm and the convolutional neural network CNN algorithm to compare the same feature data, but the detection result is much lower than the detection rate based on the Fresnel zone model WI-FI. The main reason is that the overall algorithm of convolutional neural networks and recurrent neural networks focuses on the form of vector sets and matrices. In the process of using matrix vectors for algorithm testing, the matching accuracy in vector calculation and transmission cannot be fully integrated with algorithms such as emotion recognition coefficients and acceleration, resulting in a low detection rate [27–29]. Because the radiation range of the WI-FI signal antenna using the Fresnel zone model, architecture can basically meet the integration of the accelerator and brake algorithms and the LSTM algorithm through our experiments, especially the characteristics of the LSTM algorithm are also suitable for feature data extraction with different lengths to measure the changing values in the driver's emotion recognition. At present, there are still some target data that has not been detected by the mechanism during the experiment. Through our analysis of the data, the main reason at present is that some of the WI-FI signals for the Fresnel zone model under the preliminary machine learning are blocked by objects in the car, such as the driver's body or other on-board items, and there are also some reasons. This is because the sample algorithm collected in machine learning has an incomplete area, and we need to further analyze the machine learning process. It is believed that the output process of the machine learning algorithm will continue to be optimized in the later stage, and the machine recognition rate under emotional fluctuations will be improved further improvement.

5 Conclusion

It can be seen from the above data that we used the features of the algorithm fusion in the research method to carry out machine learning, and the overall detection rate is basically the same among the three types, among which the cautious type has the highest detection rate, which currently exceeds 85%. The overall situation has achieved the expected effect. At the same time, in the experiment process, this paper also uses the recurrent neural network (RNN) algorithm and the convolutional neural network CNN algorithm to compare the same feature data, but the detection result is much lower than the detection rate based on the Fresnel zone model WI-FI. It shows that the road rage emotion detection process under the framework of this paper is more suitable for the machine learning detection method in the WI-FI environment. At present, there are still some data not recognized by the mechanism during the experiment process. It is believed that the output process of the machine learning algorithm will continue to be optimized in the later stage, and the machine recognition rate under emotional fluctuations will be further improved.

References

1. Qu J et al (2020) Convolutional neural network for human behavior recognition based on smart bracelet. J Intell Fuzzy Syst 38(5). https://doi.org/10.3233/JIFS-179651
2. Wang X et al (2020) Driver emotion recognition of multiple-ECG feature fusion based on BP network and D–S evidence. IET Intell Transp Syst 14(8). https://doi.org/10.1049/iet-its.2019.0499
3. Chen L et al (2019) Cross-subject driver status detection from physiological signals based on hybrid feature selection and transfer learning. Expert Syst Appl 137. https://doi.org/10.1016/j.eswa.2019.02.005
4. Yongdeok Y, Hyungseok O, Rohae M (2019) The effect of takeover lead time on driver workload. In: Proceedings of the human factors and ergonomics society annual meeting, vol 63, no 1. https://doi.org/10.1177/1071181319631523
5. Jaeger SR et al (2019) Using the emotion circumplex to uncover sensory drivers of emotional associations to products: six case studies. Food Qual Prefer, vol 77. https://doi.org/10.1016/j.foodqual.2019.04.009
6. Kowalczuk Z et al (2019) Emotion monitoring system for drivers. IFAC PapersOnLine 52(8). https://doi.org/10.1016/j.ifacol.2019.08.071

7. Zhang J et al (2019) Prediction method of driver's propensity adapted to driver's dynamic feature extraction of affection. Advances in Mechanical Engineering, vol 5. Pt. 6. https://doi.org/10.1155/2013/658103
8. Fan X et al (2019) A personalized traffic simulation integrating emotion using a driving simulator. The Visual Computer, vol 36. prepublish. https://doi.org/10.1007/s00371-019-01732-4
9. Zhang M et al (2019) Discriminating drivers' emotions through the dimension of power: evidence from facial infrared thermography and peripheral physiological measurements. Transp Res Part F: Psychol Behav, vol 63. https://doi.org/10.1016/j.trf.2019.04.003
10. Wang X et al (2019) Feature extraction and dynamic identification of drivers' emotions. Transp Res Part F: Psychol Behav, vol 62. https://doi.org/10.1016/j.trf.2019.01.002
11. Wang L (2018) Three-dimensional convolutional restricted Boltzmann machine for human behavior recognition from RGB-D video. EURASIP J Image Video Process 2018(1). https://doi.org/10.1186/s13640-018-0365-8
12. Steinhauser K et al (2018) Effects of emotions on driving behavior. Transp Res Part F: Psychol Behav, vol 59. https://doi.org/10.1016/j.trf.2018.08.012
13. Youwen H, Chaolun W (2018) Human recognition behavior algorithm based on deep learning. Dianzi Jishu Yingyong 44(10). https://doi.org/10.16157/j.issn.0258-7998.182201
14. Dolinski D, Odachowska E (2018) Beware when danger on the road has passed. The state of relief impairs a driver's ability to avoid accidents. Accid Anal Prev, vol 115. https://doi.org/10.1016/j.aap.2018.03.007
15. Lafont A et al (2018) Driver's emotional state and detection of vulnerable road users: towards a better understanding of how emotions affect drivers' perception using cardiac and ocular metrics. Transp Res Part F: Psychol Behav, vol 55. https://doi.org/10.1016/j.trf.2018.02.032
16. Zhao L et al (2018) Driver drowsiness detection using facial dynamic fusion information and a DBN. IET Intell Transp Syst 12(2). https://doi.org/10.1049/iet-its.2017.0183
17. Izquierdo Reyes J et al (2018) Advanced driver monitoring for assistance system (ADMAS): based on emotions. Int J Interact Des Manuf (IJIDeM) 12(1). https://doi.org/10.1007/s12008-016-0349-9
18. Ming Y et al (2017) Uniform local binary pattern based texture-edge feature for 3D human behavior recognition. PLoS One 10(5). https://doi.org/10.1371/journal.pone.0124640
19. Van Lissa Caspar J et al (2017) The cost of empathy: parent-adolescent conflict predicts emotion dysregulation for highly empathic youth. Dev Psychol 53(9). https://doi.org/10.1037/dev0000361
20. Batchuluun G et al (2017) Fuzzy system based human behavior recognition by combining behavior prediction and recognition. Expert Syst Appl, vol 81. https://doi.org/10.1016/j.eswa.2017.03.052
21. Ooi JSK et al (2017) A conceptual emotion recognition framework: stress and anger analysis for car accidents. Int J Veh Saf 9(3). https://doi.org/10.1504/IJVS.2017.085188
22. Ding I-J, Liu J-T (2016) Three-layered hierarchical scheme with a Kinect sensor microphone array for audio-based human behavior recognition. Comput Electr Eng, vol 49. https://doi.org/10.1016/j.compeleceng.2015.03.032
23. Ye Q et al (2015) 3D Human behavior recognition based on binocular vision and face–hand feature. Optik—Int J Light Electron Opt 126(23). https://doi.org/10.1016/j.ijleo.2015.08.103
24. Yao B et al (2015) A fuzzy logic-based system for the automation of human behavior recognition using machine vision in intelligent environments. Soft Comput 19(2). https://doi.org/10.1007/s00500-014-1270-4
25. Zhang T, Chan AHS (2014) How appraisals shape driver emotions: a study from discrete and dimensional emotion perspectives. Transp Res Part F: Psychol Behav, vol 27. https://doi.org/10.1016/j.trf.2014.09.012
26. Li H et al (2014) Multi-feature hierarchical topic models for human behavior recognition. Sci China Inf Sci 57(9). https://doi.org/10.1007/s11432-013-4794-9
27. Yagil D (2001) Interpersonal antecedents of drivers' aggression. Transp Res Part F: Psychol Behav 4(2). https://doi.org/10.1016/S1369-8478(01)00018-3

28. Jinnuo Z, Goyal SB, Tesfayohanis M, Omar Y (2022) Implementation of artificial intelligence image emotion detection mechanism based on python architecture for industry 4.0. J Nanomater 2022(5293248):13. https://doi.org/10.1155/2022/5293248
29. Diwan TD, Choubey S, Hota HS, Goyal SB, Jamal SS, Shukla PK, Tiwari B (2021) Feature entropy estimation (FEE) for malicious IoT traffic and detection using machine learning. Mobile Inf Syst 2021(8091363):13. https://doi.org/10.1155/2021/8091363

Dark Channel Prior-Based Image Dehazing Algorithm for Single Outdoor Image

Lalita Kumari, Vimal Bibhu, and Shashi Bhushan

Abstract Hazed image is a critical issue for scene understanding which need to be mitigated properly by applying an image dehazing algorithm. Single image dehazing is a challenging preprocessing task of image vision for object detection. This paper presents a dark channel prior (DCP)-based image dehazing method to remove haze from a single outdoor input image. The DCP receives great attention in the field of dehazing/defogging because it takes advantage of the property that some pixels in at least one of the RGB channels have the lowest intensity which tends to zero resulting in the diversion toward dark or black. The presented algorithm divides a scene image into multiple regions guided by brightness estimation. All regions are processed separately for dehazing which includes normalization, transmission map estimation, depth estimation, and radiance recovery. All the regions are processed separately and combined using adaptive region merge to generate a haze-free image. The efficiency of the proposed method is evaluated in terms of MSE, PSNR, and SSIM and compared same with the result of the methods in the literature. The presented method is very much effective in fog removal from an input image captured outdoor containing trees, mountains, buildings, sky, etc. Experimental results indicate that the proposed method gives better results on Dense-Haze CVPR 2019, NH-HAZE, and O-HAZE datasets.

Keywords Dark channel prior (DCP) · Image restoration · Single image dehazing · Defogging

---

L. Kumari (✉) · S. Bhushan
Amity School of Engineering & Technology, Amity University, Patna, India
e-mail: kumaril2003@yahoo.co.in

V. Bibhu
CSE Department, Amity University, Greater Noida, India

Y. Singh et al. (eds.), *Proceedings of International Conference on Recent Innovations in Computing*, Lecture Notes in Electrical Engineering 1001,
https://doi.org/10.1007/978-981-19-9876-8_33

1 Introduction

Capturing a good quality image is the primary requirement for the various task under image processing such as text extraction, object recognition, feature detection, satellite remote sensing system, aerial photosystem, and outdoor monitoring. Sometimes the captured image seems to be not of good quality and visibility goes down. The quality of the outdoor image varies to various natural conditions, and it gets degraded with its presence such as fog, smoke, dust, mist, and haze at the time of capturing the image [1, 2]. Objects and features in the hazed image become difficult to identify. The degrading of the image happens due to the scattering of bright light at the line of sight in presence of the conditions mentioned above. This degrading results in poor visibility, loss of contrast in the image, and faded image output. Removing haze from an input image restores the color shift, brightness, and contrast and finally increases the visibility of a scene image. Haze is a nonlinear noise to a scene that degrades the image. Previous work [3] has discussed the negative impact of haze in the path of computer vision and proposed dehazing as preprocessing step in computer vision tasks.

There exist various methodologies in the literature to remove haze. Haze removal methods may be categorized primarily into two groups: single input images and multiple input images. In a single input image, there is no additional source of information for the purpose of haze removal whereas the multiple input image-based haze removal methods use multiple images captured from different degrees of polarization [4]. Haze removal problem from a single input image requires stronger assumptions such as higher contrast [5] and locally uncorrelated surface [6].

Hsu [7] proposed a wavelet-based decomposition of the hazed image and reconstruction of the haze-free image. Dark channel prior (DCP)-based image dehazing methods proposed by He et al. [8], Chia-chi et al. [9], Li [10] have been chosen as the base of our proposed algorithm. This paper presents a DCP-based method for haze removal from a single hazy outdoor input image. The presented algorithm divides a scene image into multiple regions guided by brightness estimation. The DCP works very efficiently on outdoor single input images as it is based on the key feature observation in a haze-free image.

2 Related Work

2.1 Hazed Image Model

An optical model of hazy image formation is shown in Fig. 1 which is generated as per the widely adapted model of McCartney [11] and is described globally in the field of computer vision by Eqs. (1), (2), and (3) where I is the hazed image, A is atmospheric light, and J is the scene radiance. The non-scattered light that reaches the camera is described by the medium transmission t. A is the global atmospheric

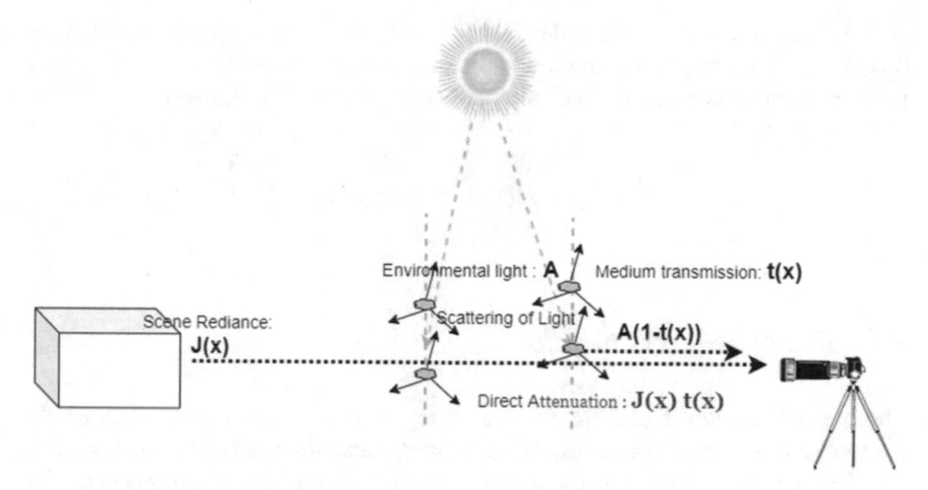

Fig. 1 Optical model for light scattering and hazy image formation

light and t is the medium transmission describing the portion of the light that is non-scattered and reaches the camera. J is the scene radiance and t is described by Eq. (2) where scene depth and the scattering coefficient is represented by d and β, respectively.

$$I(x) = J(x)t(x) + A(1 - t(x)) \quad (1)$$

$$t(x) = \mathrm{e}^{-\beta d(x)} \quad (2)$$

2.2 *Dark Channel Prior*

DCP was first observed by He et al. in [4] through statistical observation of haze-free outdoor image. He described that at least one of the three channels at some pixel contains very low intensity close to zero [4] if image contains the non-sky patches [9]. Splitting a colored image into its basic channels: red, green, and blue channels and selecting pixels with minimum intensity from all three channels will result in a kind of dark image. Again, the dark channel can be referenced by selecting the pixels with the lowest intensity from the three channels of the RGB image. The process of selecting the least intense pixels from the RGB image can be denoted by the term—previous, and we call it the dark channel before (DCP). The min{minimum} operator is used to select the minimum pixels from the RGB color channels. Dark channel I^{dark} of an image I can be given by Eq. (3) where I^{ch} is one of the available three (RGB) color channels of input image. $\Omega(x)$ is a local patch anchored at x. Che et al. in [12] observed that the value of I^{dark} is low and tends to zero except for very bright object or image part. In outdoor natural image dark channels of tree, grass, flowers, mountains, shadows, etc., are really dark (except sky regions).

$$I^{\text{dark}}(x) = \min_{y \in \Omega(x)} \left(\min_{\text{ch} \in \{r, g, b\}} I^{\text{ch}}(y) \right) \tag{3}$$

2.3 Transmission Estimation

The generalized target of haze removal is to estimate A and t and subsequently compute J using I as shown in Eq. (1). The patch transmittance is denoted by $t \sim (x)$, and the local patch transmittance $\Omega(x)$ is assumed as constant. Equation (1) can be rewritten as Eq. (4).

$$\frac{I(x)}{A} = t(x)\frac{J(x)}{A} + 1 - t(x) \tag{4}$$

Computing dark channel from Eq. (4), we get Eq. (5).

$$\min_{y \in \Omega(x)} \left(\min_{\text{ch}} \frac{I^{\text{ch}}(x)}{A} \right) = t \sim (x) \min_{y \in \Omega(x)} \left(\min_{\text{ch}} \frac{J^{\text{ch}}}{A} \right) + 1 - t \sim (X) \tag{5}$$

As the dark channel of haze-free image tends to zero, Eq. (5) may be reduced to Eq. (6)

$$t \sim (x) = 1 - \min_{y \in \Omega(x)} \left(\min_{\text{ch}} \frac{I^{\text{ch}}(x)}{A} \right) \tag{6}$$

Incorrect estimation of transmission map may results to false texture and blocked artifacts. To handle such issue, further sharpening of transmission map has been proposed in the literature [7–9, 12, 13]. Post filtering methods such as Gaussian, bilateral, soft matting, cross-bilateral filter, guided filter, and wavelets have been used in the literature [13] for this purpose.

3 Proposed Method

The flow chart of the proposed dehazing model is shown in Fig. 2. The input image is first segmented into multiple regions for further processing separately and parallel in line with haze removal. The dark channel is computed for each parallel assembly line after the normalization of the regions. The dark channel is computed as presented by

Fig. 2 Input hazed image (CVPR 2019) and its computed dark channel

Eq. (3), and further transmission map estimation and atmospheric light estimation are performed. Figure 3a shows a hazed input image, and Fig. 3b–f shows the computed dark channel computed from Fig. 3a with patch sizes 3×3, 5×5, 9×9, 11×11, and 15×15, respectively. The atmospheric light A is estimated initially from 0.1% of the brightest region [12] to obtain the transmission map $t \sim (x)$. Once the dark channels are computed, the atmospheric light is estimated using each dark channel. Finally, all regions are merged using adaptive region merge to obtain the final haze removed output image.

4 Experiments and Result

The experiment has been conducted on CVPR 2019, NH-HAZE, and O-HAZE foggy image datasets. These datasets contain outdoor foggy images. The experiment setup has been made using Python-3.10 with the help of open CV, numpy, and SciPy modules. Figure 4 shows the obtained result from the hazed input image using the proposed algorithm. Figure 4a, f shows the input image from CPVR 2019 dataset and obtained dehazed image, respectively. Figure 4b–c, g–h shows the input images from NH-HAZE dataset and obtained dehazed images, respectively. Figure 4d–e, i–j shows the input images from the O-HAZE dataset and obtained dehazed images, respectively. Figure 6a, b shows the histogram of an input image (taken from Fig. 4d) and its dehazed image, respectively. For result evaluation of the proposed algorithm, structural similarity index metric (SSIM), peak signal-to-noise ratio (PSNR), and mean square error (MSE) have been computed for the dataset under experiment.

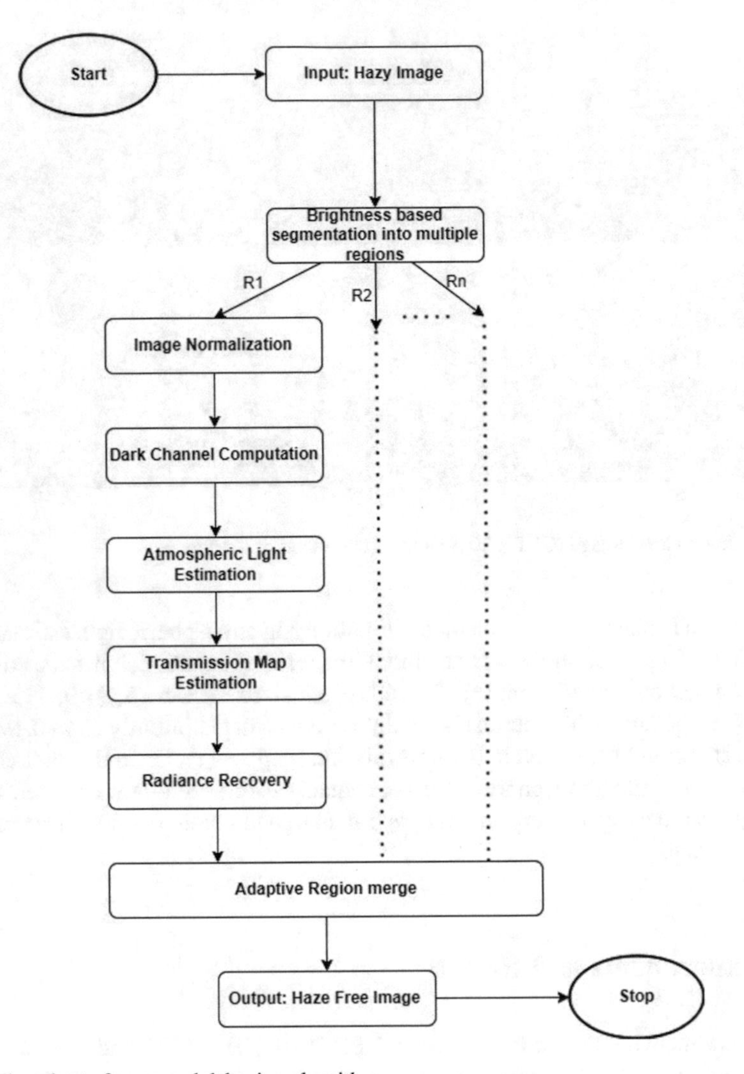

Fig. 3 Flowchart of proposed dehazing algorithm

Mean square error (MSE) is the mean squared difference between the estimated and actual values and is represented by

$$\text{MSE} = \frac{1}{m \times n} \sum_{r=1}^{m} \sum_{c=1}^{n} \left[\text{IGT}_{r,c} - \text{IOB}_{r,c}\right]^2 \tag{7}$$

where IGT and IOB are ground truth image and actually obtained image, respectively. Size of images is $m \times n$.

Fig. 4 Dehazed image using proposed algorithm. Figure a, b, c, d, e as input image (taken from CVPR 2019, NH-HAZE, and O-HAZE datasets), and figure f, g, h, i, j as their respective dehazed image

Peak signal-to-noise ratio (PSNR) is proportional to logarithmic representation ratio between the maximum possible intensity and the mean square error, represented by

$$\text{PSNR} = 10 \log \frac{(I_{\max})^2}{\text{MSE}} = 10 \log \frac{(255)^2}{\text{MSE}} \tag{8}$$

where $I_{\max}$ is the maximum intensity of image and MSE is the mean square error represented by Eq. (9). $I_{\max}$ in our case is 255 (maximum possible value of a pixel).

Structural similarity index (SSIM) is another performance evaluation method for an image. It evaluates the similarity between ground truth image and image under evaluation in terms of brightness, contrast, and structure. SSIM is represented by Eq. (11). Here, μ_{gt} and μ_{i} are the local mean of ground truth (actual haze-free image) and dehazed image, respectively. σ_{gt}, σ_{i}, and σ are the standard deviation of ground truth image, standard deviation of dehazed image, and their cross-covariance, respectively. θ_1 and θ_2 are two variables to stabilize the division with weak denominator.

$$\text{SSIM} = \frac{\left(2\mu_{\text{gt}}\mu_{\text{i}} + \theta_1\right)(2\sigma + \theta_2)}{\left(\mu_{\text{gt}}^2 + \mu_{\text{i}}^2 + \theta_1\right)\left(\sigma_{\text{gt}}^2 + \sigma_{\text{i}}^2 + \theta_2\right)} \tag{9}$$

5 Conclusion

In this paper, we proposed a dark channel prior (DCP)-based haze removal method from a single outdoor input image. The presented algorithm divided the scene image into multiple regions guided by brightness estimation and performed dehazing separately which later combined back using adaptive region merge to reconstruct the dehazed image. The dataset used for performance evaluation was a combination of

Table 1 Performance evaluation of proposed algorithm in terms of MSE, PSNR, and SSIM for datasets: CPVR, NH-HAZE, and O-HAZE

Dataset	Image name	MSE	PSNR	SSIM
CVPR-2019	A	58	30.49	0.88
NH-HAZE	B	118	27.41	0.81
NH-HAZE	C	448	21.61	0.68
O-HAZE	D	183	25.50	0.79
O-HAZE	E	246	24.22	0.76

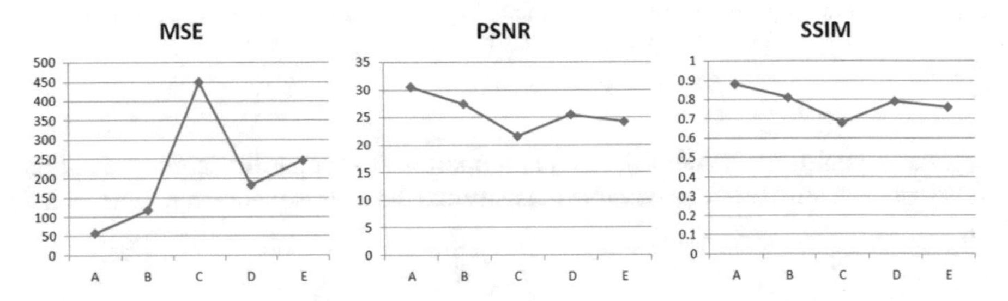

Fig. 5 Graphical representation of MSE, PSNR, and SSIM obtained from image under experiments

CVPR, NH-HAZE, and O-HAZE. Table 1 shows the performance evaluation of the proposed method in terms of MSE, SSIM, and PSNR. Figure 5 shows the graphical representation of MSE, SSIM, and PSNR, obtained from the image under experiments and shown in Table 1. The proposed method has been evaluated and compared with available methodologies in the literature and which is cited in the reference. Table 2 shows the comparative result of performance evaluation on the O-HAZE dataset in which the proposed method is compared with the methods in [6–8, 14]. The graphical representation of comparative results in terms of PSNR and SSIM is shown in Fig. 7. The performance result as shown in Fig. 4, Tables 1, and 2 confirms the usability of the proposed method.

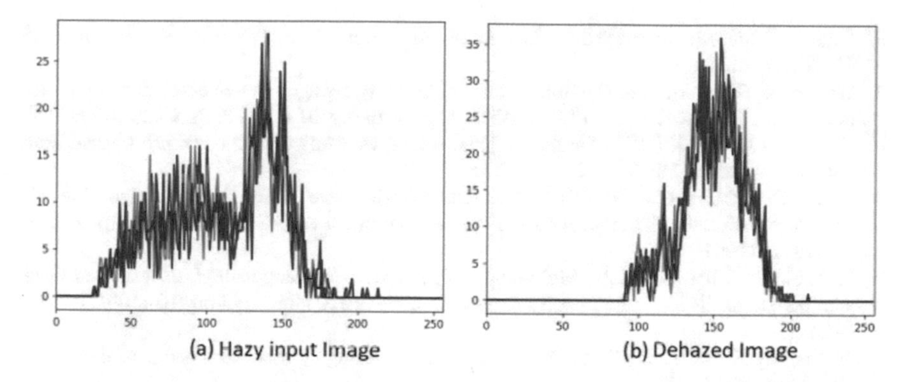

Fig. 6 Histogram of **a** hazy input image and **b** dehazed image of source image Fig. 4d

Table 2 Comparative performance evaluation in terms of average value of PSNR and SSIM for O-HAZE dataset

Method	Hsu et al. [7]	Cai et al. [14]	He et al. [8]	Fattal [6]	Ours
PSNR	20.97	16.207	16.59	15.64	24.86
SSIM	0.761	0.666	0.735	0.707	0.775

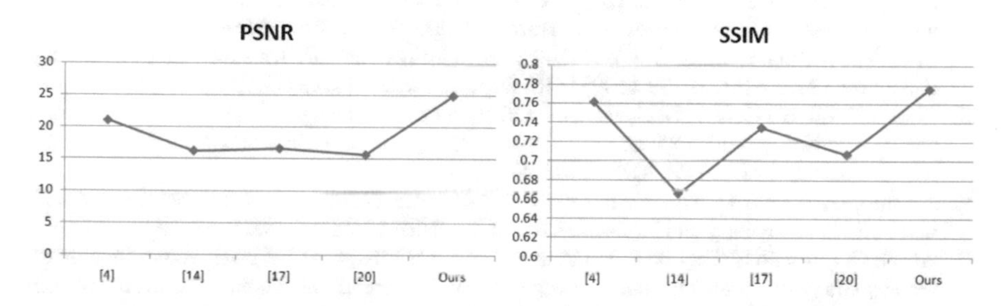

Fig. 7 Histogram graphical representation of comparative results in terms of PSNR and SSIM

References

1. Narasimhan SG, Nayar SK (2002) Vision and the atmosphere. Int J Comput Vis 48(3):233–254
2. Oakley JP, Satherley BL (1998) Improving image quality in poor visibility conditions using a physical model for contrast degradation. IEEE Trans Image Process 7(2):167–179
3. Li B, Peng X, Wang Z, Xu J, Feng D (2017) AOD-Net: all-in-one dehazing network. In: 2017 IEEE international conference on computer vision (ICCV), pp 4780–4788. https://doi.org/10.1109/ICCV.2017.511
4. He K, Sun J, Tang X (2009) Single image haze removal using dark channel prior. In: 2009 IEEE conference on computer vision and pattern recognition, pp 1956–1963. https://doi.org/10.1109/CVPR.2009.5206515
5. Tan R (2008) Visibility in bad weather from a single image. In: Proceedings of IEEE conference on computer vision and pattern recognition, June 2008

6. Fattal R (2008) Single image dehazing. ACM Trans Graph 27(3):1–9. https://doi.org/10.1145/1360612.1360671
7. Hsu W-Y, Chen Y-S (2021) Single image dehazing using wavelet-based haze-lines and denoising. IEEE Access 9:104547–104559. https://doi.org/10.1109/ACCESS.2021.3099224
8. He K, Sun J, Tang X (2011) Single image haze removal using dark channel prior. IEEE Trans Pattern Anal Mach Intell 33(12):2341–2353
9. Tsai C-C, Lin C-Y, Guo J-I (2019) Dark channel prior based video dehazing algorithm with sky preservation and its embedded system realization for ADAS applications. Opt Express 27:11877–11901
10. Li Z, Shu H, Zheng C (2021) Multi-scale single image dehazing using Laplacian and Gaussian pyramids. IEEE Trans Image Process 30:9270–9279. https://doi.org/10.1109/TIP.2021.3123551
11. McCartney EJ (1976) Optics of the atmosphere: scattering by molecules and particles. Wiley, New York, NY, USA
12. Chen W-T et al (2021) All snow removed: single image desnowing algorithm using hierarchical dual-tree complex wavelet representation and contradict channel loss. In: 2021 IEEE/CVF international conference on computer vision (ICCV), pp 4176–4185. https://doi.org/10.1109/ICCV48922.2021.00416
13. Tarel JP, Hautière N, Caraffa L, Cord A, Halmaoui H, Gruyer D (2012) Vision enhancement in homogeneous and heterogeneous fog. IEEE Intell Transp Syst Mag 4(2):6–20
14. Cai B, Xu X, Jia K, Qing C, Tao D (2016) Dehazenet: an end-to-end system for single image haze removal. IEEE Trans Image Process
15. Kumar R, Bhandari AK, Kumar M (2022) Haze elimination model-based color saturation adjustment with contrast correction. IEEE Trans Instrum Meas 71:1–10
16. Huang Z, Li J, Hua Z, Fan L (2022) Underwater image enhancement via adaptive group attention-based multiscale cascade transformer. IEEE Trans Instrum Meas 71:1–18
17. Li P, Tian J, Tang Y, Wang G, Wu C (2021) Deep retinex network for single image dehazing. IEEE Trans Image Process 30:1100–1115. https://doi.org/10.1109/TIP.2020.3040075
18. Zhu Z, Wei H, Hu G, Li Y, Qi G, Mazur N (2021) A novel fast single image dehazing algorithm based on artificial multiexposure image fusion. IEEE Trans Instrum Meas 70(5001523):1–23. https://doi.org/10.1109/TIM.2020.3024335
19. Ancuti CO, Ancuti C, Timofte R (2020) NH-HAZE: an image dehazing benchmark with non-homogeneous hazy and haze-free images. In: IEEE CVPR NTIRE workshop
20. Ancuti CO, Ancuti C, Timofte R, De Vleeschouwer C (2018) O-HAZE: a dehazing benchmark with real hazy and haze-free outdoor images. IEEE Conference on computer vision and pattern recognition, NTIRE workshop, NTIRE CVPR'18
21. Song Y, Luo H, Lu R, Ma J (2017) Dehazed image quality assessment by haze-line theory. J Phys Conf Ser 844:012045. https://doi.org/10.1088/1742-6596/844/1/012045
22. Lee S, Yun S, Nam JH et al (2016) A review on dark channel prior based image dehazing algorithms. J Image Video Proc 2016:4. https://doi.org/10.1186/s13640-016-0104-y
23. Zhu Q, Mai J, Shao L (2015) A fast single image haze removal algorithm using color attenuation prior. IEEE Trans Image Process 24(11):3522–3533
24. Narasimhan SG, Nayar SK (2003) Contrast restoration of weather degraded images. IEEE Trans Pattern Anal Mach Learn 25(6):713–724

Real-Time Driver Sleepiness Detection and Classification Using Fusion Deep Learning Algorithm

Anand Singh Rajawat, S. B. Goyal, Pawan Bhaladhare, Pradeep Bedi, Chaman Verma, Ţurcanu Florin-Emilian, and Mihaltan Traian Candin

Abstract In the past few years, driving while tired has become one of the main causes of car accidents. This kind of dangerous driving can cause serious injuries or even death and huge financial losses. The face, a significant part of the body, can send many different messages. When a person is tired, their facial expressions, like how often they blink their eyes, differ from when they are alert. This is very clear when it comes to drivers. In this study, a fusion deep learning algorithm (FDLA) is used to determine if a driver is tired and how tired they are in real time. This algorithm uses video footage to determine how tired a driver is based on how often they blink and how long their eyes are closed. We are making a new face-tracking algorithm because the ones that are already out there aren't good enough. This will make face tracking more accurate. Also, a new way to find problems in facility areas has been made, and it is based on seventy essential spots. After that, these parts of the driver's face can be used to determine how he or she is feeling. When the eyes and lips are used together, a driver care solution can tell if the driver is getting tired.

A. S. Rajawat · P. Bhaladhare
School of Computer Science and Engineering, Sandip University, Nashik, India
e-mail: pawan.bhaladhare@sandipuniversity.edu.in

S. B. Goyal (✉)
Faculty of Information Technology, City University Malaysia, Petaling Jaya, Malaysia
e-mail: sb.goyal@city.edu.my

P. Bedi
Galgotias University, Greater Noida, India
e-mail: pradeepbedi@galgotiasuniversity.edu.in

C. Verma
Faculty of Informatics, University of Eötvös Loránd, Budapest 1053, Hungary
e-mail: chaman@inf.elte.hu

Ţ. Florin-Emilian
Department of Building Services, Faculty of Civil Engineering and Building Services, Gheorghe Asachi Technical University of Iasi, 700050 Jassy, Romania
e-mail: florin-emilian.turcanu@academic.tuiasi.ro

M. T. Candin
Faculty of Building Services Cluj-Napoca, Technical University of Cluj-Napoca, 400114 Cluj-Napoca, Romania

Y. Singh et al. (eds.), *Proceedings of International Conference on Recent Innovations in Computing*, Lecture Notes in Electrical Engineering 1001,
https://doi.org/10.1007/978-981-19-9876-8_34

Keywords Drivers fatigue detection · Driver care solution · Alarm · Drowsiness detection

1 Introduction

People are increasingly relying on public transportation systems to get around their cities. As we look at the pros and cons of public transportation, we also look at how transit traveler information systems [1] might be able to help people deal with these problems as public transit becomes more popular. People who can't or won't drive themselves to places like work, school, or the doctor can use public transportation. Researchers have found that having access to a car is directly linked to a person's ability to find and keep a job. This shows how important mobility is as a leading indicator of getting a job. By encouraging people to take public transportation instead of driving single-occupant cars, communities might be able to cut down on traffic congestion and the damage it does to the environment [2]. The main goals of this section are for people who use public transportation to be happy and for the number of people who do so to grow. Mobile devices can now figure out where they are on their own. It is helpful for programs that run on the device itself and for programs that run elsewhere [3] and need to track the device. Also, the tracking system needs to be able to give accurate updates on the location even when conditions change. These conditions include delays caused by positioning and communication and different levels of positioning accuracy. The system in place keeps a close eye on people on the street who are carrying GPS-enabled targets. Using a variety of user interfaces, people with mobile devices can find out how flights are doing now. The information here will be helpful for first-timers and people who have been to this event. Users could get to information by looking at a list of the stops along a specific route. Even though the full Web interface showed information about stops and routes on a map, you still had to look for stops by number, route, or location. This was because of how the information was shown on the map. As a solution to this problem, I have made a native app that knows where it is. The program is set up so that it works best on phones. A bus that uses the ability of modern mobile devices to figure out where it makes it easy for passengers to find nearby bus stops and get search results that are more relevant to their current situation. In recent years, there has been more demand for modern mobility, which has led to more parking lots and garages being built faster. The National Highway Traffic Safety Administration sent out a report that said people had been hurt. 20–30% of these accidents were caused by drivers too tired to drive safely. Drowsy driving is one of the leading causes of car accidents because it poses a hidden risk. In recent years, one of the most researched topics has been how to tell when a driver is tired. In the process of finding something, both subjective and objective methods are used. A driver must take part in an evaluation using the unique detection method. The driver has to ask himself questions and keep an eye on himself. This assessment is based on what the driver thought about the situation. Gustavo Olague helped the main editor with the article [4]. He oversaw the review and gave the final go-ahead for the article to be published. Evaluation and the use of questionnaires are both parts of the method. In the end, the data are used to estimate how many cars are being driven by people who are too tired to do so. This gives the drivers a way to organize their schedules more effectively. On the other hand, the objective detection method doesn't depend on drivers' feedback because it constantly checks the driver's physical state and how they drive. The information gathered is used to figure out how tired the driver is. Accurate detection comes in two forms: touch and non-contact. The device can be used in more cars if it doesn't need computer vision technology or a high-tech camera. This makes it cheaper and easier to use than the contact method, which requires the device and the car to be in direct physical contact. The non-contact method has often been used to find tired drivers because it is easy to set up and doesn't cost much. With the help of technologies like attention technology and SmartEye, which track the driver's eye and head movements, it is possible to tell if the driver is getting tired or not. Based on what we found in our study, we propose a method called real-time driver sleepiness detection and classification. This method doesn't involve touching the driver in any way. Our system only uses the camera on the car, so the driver doesn't have to wear or carry any extra gear. By looking at each video frame taken inside the car, it is possible to figure out what is going on with the driver.

2 Related Work

Using a GPS-enabled phone, you can do the following to find out where a bus is: A look at what is going on bus tracking systems that use Global Positioning System (GPS) combine automatic vehicle location [5] with fleet data collection software to get a full picture of where the vehicles being tracked are. This picture was made by putting together information from each car in the fleet. If you have the right software or can get on the Internet, you can use electronic maps to find out about bus routes. In bus tracking systems of today, the Global Positioning System (GPS) is often used to pinpoint the exact location of the bus. When it comes to big cities, these strategies are beneficial. Operators often use this technology for many things, such as tracking, routing, dispatching, collecting data on board, and keeping things safe. These are just some of the ways it can be used. These methods can be constructive for dealing with problems that often arise, like traffic jams, unplanned delays, vehicles leaving at different times, and many others. Because it shows where buses are in real time, it makes bus delays less likely. For the proposed system [6] to work, the ellipse needs to be an open-source software development environment. The ellipse comprises an integrated development environment (IDE) and a system for adding plug-ins (Integrated Development Environment). An integrated development environment, or IDE, can help speed up the making of android apps. It is a system that can do more than one thing simultaneously. Students and workers need to know how important it is to be on time if they want to get the most out of their daily routines.

GPS tracking systems are becoming more and more popular among consumers as a way to find lost or stolen items.

The face, a substantial body part, can say a lot. When a person is tired, their facial expressions, like how often they yawn and blink, differ from when they are not tired. DriCare is the system we built during our investigation to handle examples. A number of things, like how long a driver has his or her eyes closed, how often they yawn, and how often they blink [7], are taken into account when figuring out how tired a driver is in real time. A new face-tracking algorithm is being made to fix the problems with accuracy caused by the algorithms already in use. Our team also devised a new way to recognize the parts of the face based on 68 key points. Then, we use how these different parts of the drivers' faces look to figure out their general health. The DriCare [8] system uses the driver's eyes, lips, and its own sensors to tell the driver early on if they are getting sleepy. During the testing, it was found that DriCare was accurate about 92% of the time. Based on machine learning, a system to tell if someone is getting sleepy. An event is called an accident when it happens suddenly, unexpectedly, and by chance in circumstances that were not expected. Every day, a large number of people around the world are seriously hurt or killed in accidents involving motor vehicles. More than a quarter of all highway deaths each year are caused by drivers who are so tired they can't drive safely. This shows that driving while fatigued is a much bigger risk for an accident than driving while drunk. Long-haul truck drivers are more likely to fall asleep at the wheel if they drive for long periods without stopping, and they are also less likely to notice a problem early on. When you use the [9], using computer vision technologies, a multidisciplinary field that looks at how computers can be made smarter by making techniques that help them get high-level understanding from digital images or videos, a way can be found to find and warn drowsy drivers. This would be a way to find and wake up drivers who are falling asleep at the wheel. With this technology, the driver will get a warning [10] if they are about to fall asleep at the wheel. The goal of this project is to use OpenCV to make a computer vision system that can find signs of driver fatigue in real-time video without making the user's experience worse. If the system notices that the driver is starting to nod off, it will sound an alarm to wake them up [11]. To use this technology, a driver would need to use an algorithm that can figure out where the visual landmarks he or she sees are. Based on this information, the system can tell if the driver is tired or not. If it finds that they are, it will play a warning tone to let them.

3 Proposed Methodology

Fusion deep learning algorithm (FDLA)

The data are prepared in such a way that the fusion deep learning method may be used to perform both deep learning and clustering simultaneously. Because it employs

CNN's arranged in a stack [12], this strategy is most effective when applied to real-time images. This recurrent architecture uses agglomerative clustering, symbolized by the forwarding pass [13], and CNN learning (shown in the backward pass). When optimizing the recurrent framework, which includes both CNNs and large-scale clustering, a single loss function is utilized instead of numerous loss functions. This is done so that the framework may be optimized more effectively.

$$y^t = f_0\left(h^t\right) = h^t$$

$$h^t = f_m\left(X^t, h^{t-1}\right)$$

$$X^t = f_r\left(I|\theta^t\right)$$

f_r is the transformation function that will be used to get X^t features from the deep model. This function will be used in conjunction with the input I. This function will allow you to use the t value from the training session [14] before this one. That is an element of it. One of the fascinating aspects of agglomeration clustering is how it can join two clusters with each iteration [15] up until it reaches convergence. At every one of these stages, the f_m merging function is put to use. Because of the effect of X^t, a state concealed in the past is brought into view, resulting in the production[16] of the state of interest, h^t (also known as the label of the cluster at time step t). In order to keep the CNN learning variables up to date, backpropagation is applied with the agglomerative clustering parameters in play. The objective function of the process is typically referred to as a "loss function" in common parlance.

$$\left(\{y^1, \ldots y^T\}, \{\theta^1, \ldots \theta^T\}|I\right) = \sum_{t=1}^{T} L^t, \left(y^t, \theta^t|y^{t-1}, I\right)$$

The picture label sequence for the period p can be stated in this illustration: y^p. The agglomerative clustering process [17] will start as soon as the merger of two neighboring clusters is finished. Subsequent processes then follow the first step. During the forward pass [18], the standard agglomerative clustering technique parameters will be modified.

$$L^p\left(y^p|\theta^p, I\right) = \sum_{t=t_p^s}^{t_p^e} L^t\left(y^t|\theta^p, y^{t-1}.I\right)$$

To create an ideal sequence label, the approach first locates [19] the clusters and then combines them to make a perfect sequence label with the image clusters. This is represented by the formula "$y^p = y^t$" which means "forward pass equals forward pass." With the backward pass, you are aiming to minimize and reduce the loss caused by the front pass, which can be expressed mathematically as L. It is done this way so students can learn independently without being constantly monitored.

$$L\big(\theta|\{y_*^1 \ldots, y_*^p\}, I\big) = \sum_{k=1}^{p} L^k\big(\theta|y_*^k, I\big)$$

The remaining (x, y)-coordinates for each eye can be found by moving your cursor clockwise around the area while maintaining eye contact with the person. Because of this, you can duplicate the eye's appearance as if you were staring directly at it. From looking at this graphic [20], the most crucial thing one can determine is that the width and height coordinates are linked in some way. "Real-Time Eye Blink Detection using Facial Landmarks: A Case Study" serves as the equation's basis for eye aspect ratio (EAR). For the 2D face landmarks, which have been labeled as "p1," "...," and "...," these are the locations of their coordinates. There are a total of six of these historical sites. As the numerator and denominator [21], the distance from vertical eye landmarks is employed because there are only two sets of vertical points whereas there is only one set of horizontal points as shown in Fig. 1.

Blinking quickly approaches zero and remains there until the eye is opened again, while the aspect ratio remains somewhat stable when the eye is closed. Eventually, this will become evident [22]. We do not need to resort to picture processing if we use this basic equation to determine whether someone is blinking. We can immediately detect if someone is blinking by looking at their eyes. To compare ratios, we don't have to look at the distances between the landmarks in our eyes one at a time. This is possible because the ratios can be compared [23]. The figure offered by Soukupová and Each to shed some light on the situation [24] may be found below. The top left

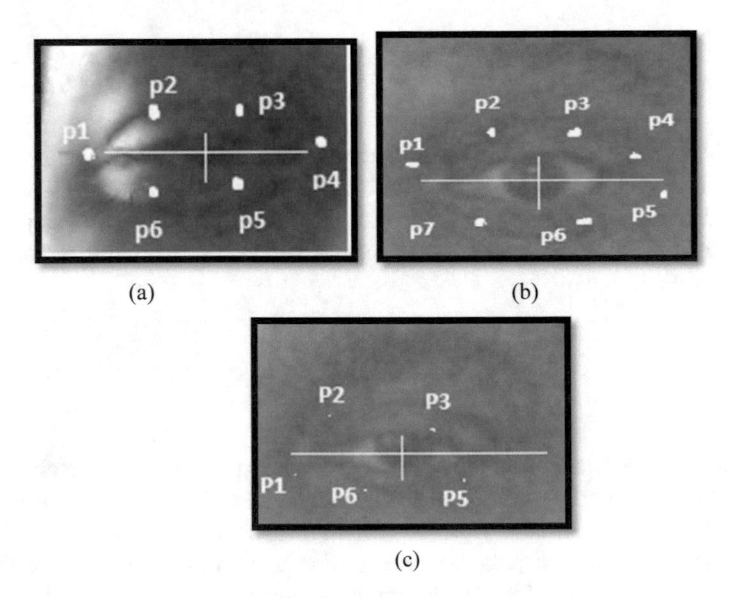

Fig. 1 **a–c** Eye representation in points and graph

corner of the photograph shows what looks to be a wide open eye. Because of this, the eye's aspect ratio is high and stable in the situation we are discussing. Consequently [25], when someone blinks, their eye aspect ratio decreases dramatically (top right). Observe how the ratio varies over time by looking at the graph at the very bottom of the page. The eye's aspect ratio retains [26] its existing value during a single blink, before unexpectedly dipping below zero and then rising back to its original value.

Proposed Algorithm

Input: Video Frames
excerpts from various motion pictures
The tiredness of the driver as an output
Step 1: It is recommended that still images from a video be used as the input
Step 2: Extract and evaluates the properties of the mouth and eyes in their current states
Step 3: Compute the eye-closure frame ratio for the 60 s and the duration of eye-closure, which is denoted by the letter e
Step 4: Calculate r and y, which represent, respectively, the number of times you blinked and the total number of times you yawned in the 60 s
Step 5: In the event where e is more than 25%, then
De equals 1
Step 6: therefore, this brings our discussion to a close
Step 7: if the time has passed more than 2.5 s and the subject has not yawned,
Step 8: The value of Dt is equal to 1
Step 8: if this is the final act…
Step 9: If r is more than 28 or r 7, then
Step 10: One is equivalent to Dr
Step 11: if this is the final act…
Step 12: If y is greater than 2, then x must be greater than 2 as well
Step 13: The value of dy is equal to one
Step 14: if this is the final act…
Step 15: D, you have to compute the total weight of everything
Step 16: It is equal to the quotient obtained by dividing the sum of (De + Dr + Dy) and (De + Dt)
Step 17: In this specific instance, T2
Step 18: Drowsiness in the driver was indicated
else
Step 19: The motorist is awake and aware of his surroundings
Step 20: if this is the final act…

In Fig. 2, driver sleepiness detection and classification using fusion deep learning algorithm FDLA, which stands for multiple convolutional neural networks [27], will be used to do more work on the data before it is used. From the video stream that was coming in, the face of the driver could be seen. In network infrastructure [28], the kernelized correlation filter algorithm is used. The kernelized correlation filter algorithm (KCF) will be used to track [29] the infrared (IR) markers during this technique. Using a method called "feature extraction," it was found out which parts of the face, like the eye and the mouth, are the most important. After that, people were watched as they closed their eyes, yawned, and blinked. In Fig. 3, real-time driver sleepiness detection and classification.

First, let's take a look at the camera that comes with android. Once the photo has been taken, the next step is to find the person whose face is in it. The face is used to get rid of the eyes. There are seventy different places to look in the picture. After figuring out how often each eye blinks, we check to see if the person is sleepy. If we saw you start to fall asleep [30], we could call your phone number to wake you

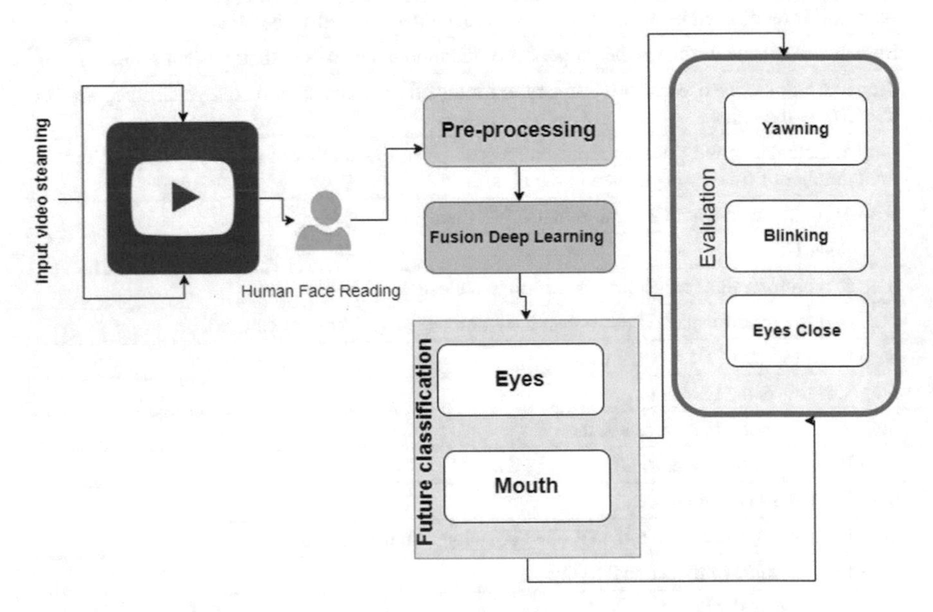

Fig. 2 Driver sleepiness detection and classification using fusion deep learning algorithm

Fig. 3 Real-time driver sleepiness detection and classification

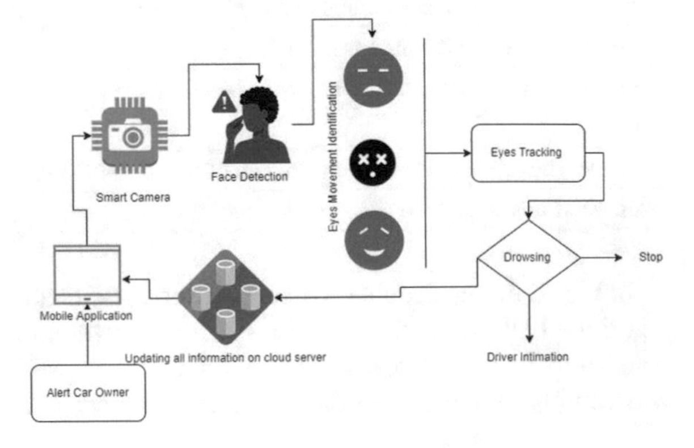

up. After using accelerometers to determine where an accident-damaged car is, we can send a text message to the car's owner. This information is gathered by GPS and updated on the vehicle's firebase cloud and owner site, so we can see where it is right now.

4 Conclusion

Face tracking and identifying facial key points are the foundations of our new method for figuring out how tired a driver is. This system will figure out how tired the driver is. The fusion deep learning algorithm [32–35] is a new way we came up with to make the CNN algorithm even better (FDLA). This algorithm uses both CNN and FDLA to track how the driver feels. Face critical points are used to figure out which parts of the face are used to identify people. We also came up with a new way to judge how tired someone is by looking at their eyelids and lips. Driver care solution works like a real-time system because it can do things quickly. The tests show that the driver care solution can handle many situations and still give reliable results. Face recognition and face tracking determine how tired the driver is. The results can be used to figure out how tired the driver is. This system can tell whether or not the driver is tired. To improve the original CNN algorithm, we devised a new method called the FDLA algorithm. These two algorithms are used to make up the FDLA algorithm. Face essential points help us figure out which parts of the face are used for detecting. A unique way to tell if a person is sleepy is to look at how their eyes and mouth are moving. Driver care solution has a fast processing speed that works like a real-time system. The tests show that the driver care solution can be used in many different situations and can deliver consistent results.

Acknowledgements This paper was financed by the Postdoctoral Advanced Research Program within IOSUD-TUIASI: Performance and excellence in postdoctoral research—2022, granted by the "Gheorghe Asachi" Technical University of Iasi, Romania. This paper was partially supported by UEFISCDI Romania and MCI through projects ALPHA, ADCATER, iPSUS, Inno4Health, AICOM4Health, SmarTravel, Mad@Work, DAFCC, PREVENTION, NGI-UAV-AGRO, ArtiPred, SOLID-B5G, 5G-SAFE+, F4itech and by European Union's Horizon 2020 research and innovation program under grant agreements No. 883522 (S4ALLCITIES) and No. 883441 (STAMINA)

References

1. Umut İ et al (2017) Detection of driver sleepiness and warning the driver in real-time using image processing and machine learning techniques. Adv Sci Technol Res J 11(2):95–102. https://doi.org/10.12913/22998624/69149. Accessed 26 May 2021
2. Shrestha A, Dang J (2020) Deep learning-based real-time auto classification of smartphone measured bridge vibration data. Sensors 20(9):2710. https://doi.org/10.3390/s20092710

3. Chandraprabha S et al (2020) Real time LDR data prediction using IoT and deep learning algorithm. Innovations in Information and Communication Technology Series, 30 Dec 2020, pp 158–161. https://doi.org/10.46532/978-81-950008-1-4_033
4. Haider KZ et al (2017) Deepgender: real-time gender classification using deep learning for smartphones. J Real-Time Image Process 16(1):15–29. https://doi.org/10.1007/s11554-017-0714-3
5. Kuyuk HS, Susumu O (2018) Real-time classification of earthquake using deep learning. Procedia Comput Sci, vol 140, pp 298–305. https://doi.org/10.1016/j.procs.2018.10.316
6. Sivakumar K et al (2019) Real time objects recognition and classification using deep learning algorithm for blind peoples. IJARCCE 8(2):289–292. https://doi.org/10.17148/ijarcce.2019.8256
7. Zhu J, Xu W (2020) Real-time data filling and automatic retrieval algorithm of road traffic based on deep-learning method. Symmetry 13(1):1. https://doi.org/10.3390/sym13010001
8. Real-time object detection and recognition using deep learning with YOLO algorithm for visually impaired people. J Xidian Univ, vol. 14, no. 4, 2020. https://doi.org/10.37896/jxu14.4/261
9. Kekong PE et al (2021) Real time drowsy driver monitoring and detection system using deep learning based behavioural approach. Int J Comput Sci Eng 9(1):11–21. https://doi.org/10.26438/ijcse/v9i1.1121
10. Neupane B et al (2022) Real-time vehicle classification and tracking using a transfer learning-improved deep learning network. Sensors 22(10):3813. https://doi.org/10.3390/s22103813
11. Chan-Hon-Tong A (2018) An algorithm for generating invisible data poisoning using adversarial noise that breaks image classification deep learning. Mach Learn Knowl Extr 1(1):192–204. https://doi.org/10.3390/make1010011
12. Mohan A, Meenakshi Sundaram V (2020) V3O2: hybrid deep learning model for hyperspectral image classification using vanilla-3D and octave-2D convolution. J Real-Time Image Process. https://doi.org/10.1007/s11554-020-00966-z
13. Cevik KK (2020) Deep learning based real-time body condition score classification system. IEEE Access 8:213950–213957. https://doi.org/10.1109/access.2020.3040805
14. Lu K, Karlsson J, Dahlman AS, Sjöqvist BA, Candefjord S (2022) Detecting driver sleepiness using consumer wearable devices in manual and partial automated real-road driving. IEEE Trans Intell Transp Syst 23(5):4801–4810. https://doi.org/10.1109/TITS.2021.3127944
15. Yehia A et al (2021) Using an imbalanced classification algorithm and floating car data for predicting real-time traffic crash risk on expressways. SSRN Electron J. https://doi.org/10.2139/ssrn.3994300
16. Mazhar M. et al (2022) Real-time defect detection and classification on wood surfaces using deep learning. Electron Imaging 34(10):382–1, 382–6. https://doi.org/10.2352/ei.2022.34.10.ipas-382
17. Mårtensson H, Keelan O, Ahlström C (2019) Driver sleepiness classification based on physiological data and driving performance from real road driving. IEEE Trans Intell Transp Syst 20(2):421–430. https://doi.org/10.1109/TITS.2018.2814207
18. Persson A, Jonasson H, Fredriksson I, Wiklund U, Ahlström C (2021) Heart rate variability for classification of alert versus sleep deprived drivers in real road driving conditions. IEEE Trans Intell Transp Syst 22(6):3316–3325. https://doi.org/10.1109/TITS.2020.2981941
19. Balandong RP, Ahmad RF, Saad MN, Malik AS (2018) A review on EEG-based automatic sleepiness detection systems for driver. IEEE Access 6:22908–22919. https://doi.org/10.1109/ACCESS.2018.2811723
20. Bakker B et al (2022) A multi-stage, multi-feature machine learning approach to detect driver sleepiness in naturalistic road driving conditions. IEEE Trans Intell Transp Syst 23(5):4791–4800. https://doi.org/10.1109/TITS.2021.3090272
21. Zhao Y, Xie K, Zou Z, He J-B (2020) Intelligent recognition of fatigue and sleepiness based on inceptionV3-LSTM via multi-feature fusion. IEEE Access 8:144205–144217. https://doi.org/10.1109/ACCESS.2020.3014508

22. Kamran MA, Mannan MMN, Jeong MY (2019) Drowsiness, fatigue and poor sleep's causes and detection: a comprehensive study. IEEE Access 7:167172–167186. https://doi.org/10.1109/ACCESS.2019.2951028
23. Maheswari VU, Aluvalu R, Kantipudi MP, Chennam KK, Kotecha K, Saini JR (2022) Driver drowsiness prediction based on multiple aspects using image processing techniques. IEEE Access 10:54980–54990. https://doi.org/10.1109/ACCESS.2022.3176451
24. Ed-Doughmi Y, Idrissi N, Hbali Y (2020) Real-time system for driver fatigue detection based on a recurrent neuronal network. J Imaging 6(3):8. https://doi.org/10.3390/jimaging6030008. PMID: 34460605; PMCID: PMC8321037
25. Husain SS, Mir J, Anwar SM et al (2022) Development and validation of a deep learning-based algorithm for drowsiness detection in facial photographs. Multimed Tools Appl 81:20425–20441. https://doi.org/10.1007/s11042-022-12433-x
26. Pandey NN, Muppalaneni NB (2022) A survey on visual and non-visual features in driver's drowsiness detection. Multimed Tools Appl. https://doi.org/10.1007/s11042-022-13150-1
27. Min J, Cai M, Gou C et al (2022) Fusion of forehead EEG with machine vision for real-time fatigue detection in an automatic processing pipeline. Neural Comput Appl. https://doi.org/10.1007/s00521-022-07466-0
28. Gumaei A, Al-Rakhami M, Hassan MM et al (2020) A deep learning-based driver distraction identification framework over edge cloud. Neural Comput Appl. https://doi.org/10.1007/s00521-020-05328-1
29. Guo JM, Markoni H (2019) Driver drowsiness detection using hybrid convolutional neural network and long short-term memory. Multimed Tools Appl 78:29059–29087. https://doi.org/10.1007/s11042-018-6378-6
30. Wijnands JS, Thompson J, Nice KA et al (2020) Real-time monitoring of driver drowsiness on mobile platforms using 3D neural networks. Neural Comput Appl 32:9731–9743. https://doi.org/10.1007/s00521-019-04506-0
31. Rajawat AS, Barhanpurkar K, Goyal SB, Bedi P, Shaw RN, Ghosh A (2022) Efficient deep learning for reforming authentic content searching on big data. In: Bianchini M, Piuri V, Das S, Shaw RN (eds) Advanced computing and intelligent technologies. Lecture Notes in Networks and Systems, vol 218. Springer, Singapore. https://doi.org/10.1007/978-981-16-2164-2_26
32. Rajawat AS, Bedi P, Goyal SB, Alharbi AR, Aljaedi A, Jamal SS, Shukla PK (2021) Fog big data analysis for IoT sensor application using fusion deep learning. Math Probl Eng 2021(6876688):16. https://doi.org/10.1155/2021/6876688
33. Goyal SB, Bedi P, Kumar J et al (2021) Deep learning application for sensing available spectrum for cognitive radio: an ECRNN approach. Peer-to-Peer Netw Appl 14:3235–3249. https://doi.org/10.1007/s12083-021-01169-4
34. Goyal SB, Bedi P, Kumar J (2022) Realtime accident detection and alarm generation system over IoT. In: Kumar R, Sharma R, Pattnaik PK (eds) Multimedia technologies in the internet of things environment, vol 2. Studies in Big Data, vol 93. Springer, Singapore. https://doi.org/10.1007/978-981-16-3828-2_6
35. Bedi P, Goyal SB, Sharma R, Yadav DK, Sharma M (2021) Smart model for big data classification using deep learning in wireless body area networks. In: Sharma DK, Son LH, Sharma R, Cengiz K (eds) Micro-electronics and telecommunication engineering. Lecture Notes in Networks and Systems, vol 179. Springer, Singapore. https://doi.org/10.1007/978-981-33-4687-1_21

E-Learning Cloud and Big Data

Tutorial on Service-Oriented Architecture: Vision, Key Enablers, and Case Study

Darshan Savaliya, Sakshi Sanghavi, Vivek Kumar Prasad, Pronaya Bhattacharya, and Sudeep Tanwar

Abstract In modern web applications, service-oriented architecture (SOA) allows enterprises to use reusable functional components and form interoperable services. From an applicative viewpoint, the services use standard interfaces and communication protocols, where the associated service and operations are decoupled via microservices. This removes the redundancy in task development and provides interoperability with back-end legacy frameworks. With the advent of Web 3.0, the requirement is even more critical as services communicate over open wireless channels, and an adversary may gain access to confidential information through associated application programming interface (API) points. Until now, limited research has been carried out to understand the critical visions of SOA architecture and its associated enablers. Thus, motivated by the research gap, we discuss the SOA vision, its key components, and enabling technologies in this article. We present a discussion of SOA with the web and the critical communication protocols to support the case. Next, we discuss the security viewpoint of SOA and address the critical security principles. Research challenges are suggested, and a case study is presented that integrates blockchain (BC) and Web 3.0 with SOA architecture in healthcare ecosystems. The tutorial aims to let the readers gain valuable insights into SOA integration into web-based applicative frameworks.

D. Savaliya · S. Sanghavi · V. K. Prasad · P. Bhattacharya (✉) · S. Tanwar
Institute of Technology, Nirma University, Ahmedabad, Gujarat, India
e-mail: pbhattacharya@kol.amity.edu

D. Savaliya
e-mail: 19bce237@nirmauni.ac.in

S. Sanghavi
e-mail: 19bce241@nirmauni.ac.in

V. K. Prasad
e-mail: vivek.prasad@nirmauni.ac.in

S. Tanwar
e-mail: sudeep.tanwar@nirmauni.ac.in

P. Bhattacharya
Amity School of Engineering and Technology, Amity University, Kolkata, India

Y. Singh et al. (eds.), *Proceedings of International Conference on Recent Innovations in Computing*, Lecture Notes in Electrical Engineering 1001,
https://doi.org/10.1007/978-981-19-9876-8_35

Keywords Service-Oriented Architecture · Microservices · Web 3.0 · Quality of Security Service · Security Protocols · Blockchain

1 Introduction

In modern enterprises, the applications have become data-driven, and thus diverse service orientations are required to be integrated in different enterprise applications. In such cases, Service-Oriented Architecture (SOA) is a design technique that enables us to seamlessly integrate various applications into a single platform. The concept of SOA allows us to interact with each other without affecting the existing software. A service can interact with each other and execute contracts without affecting the rest of the software. There is a plethora of services deployed and can interact with each other seamlessly using contract-based messages and policies [8]. Some of the general characteristics of service according to numerous descriptions of SOA are presented in [10, 16]. It basically represents recurring task which produces a predefined output, it follows black box approach, it is made up of multiple distinct services, it is reusable, and scaling is easier.

SOA is applied to large distributed systems for which scalability and fault tolerance are crucial in upkeeping such systems. Figure 1a presents a generic enterprise SOA architecture that involves customers interacting with enterprise services like customer relationship management (CRM) service, billing, and registry services. A service registry is set up to manage the services that interact bidirectionally with users and enterprise applications. The current self-seeking world demands fast, flexible, cheap, and problem-free systems. Fast and cheap solutions do not go hand in hand with errorless solutions. Hence, it is necessary to prune the repercussions of faults occurring in the system. The concept applied to handle the needs of scalability, flexibility, and fault tolerance is called loose coupling [11]. Loose coupling focuses on reducing the system dependencies, so that alterations or errors in a system will have minimal effects on others. Businesses that use SOA have a lot more flexibility in adapting and changing processes. Telecommunications firms, for example, might introduce new services or change their current costs.

From an applicative viewpoint, microservice architecture is regarded as a progression of SOA because its activities are finer-grained and act independently of one another. As a result, if one of the services within an operation fails, the app will still work because each service serves a separate role. The fundamental difference between the SOA and microservice is their extent. Generally said, service-oriented architecture (SOA) is focused on the enterprise, whereas microservices design is focused on the application. Figure 1b presents the global microservice market, and it is expected to reach 32.01 billion by 2023. It is growing at a compounded annual growth rate (CAGR) of 16.17%, and the forecast is presented till 2030 [19]. Although microservices have a high market cap, an effective realization for large-scale enterprises is not possible without the inclusion of SOA [4]. Thus, the existence of the SOA is still on, and its utility will last for a long time.

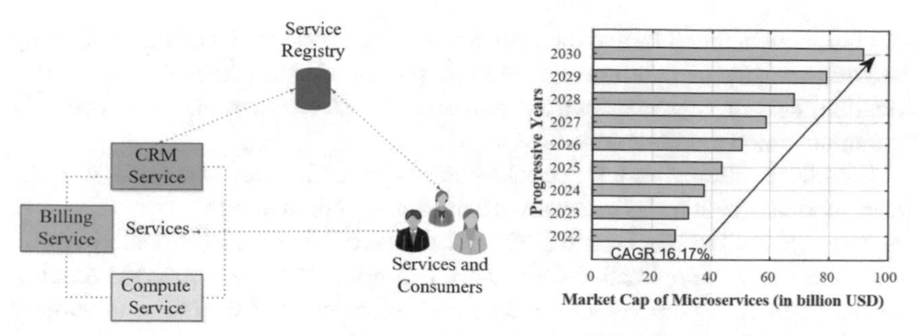

Fig. 1 Generic SOA architecture for an enterprise solution, and microservice expected market cap by 2030

Within the cloud architecture, SOA provides translations, communication, and management layer that lowers the barrier to cloud customers accessing requested services. Networking and messaging protocols can be built and utilized to interface with one another using SOA clients and components.

1.1 Motivations

Business processes and services require components, codes, and architectures to be integrated cohesively. As the industry has shifted toward continuous and agile development approaches, there is a requirement for loosely coupled services and associated actions. SOA defines reusable modules to synergize business applications, where the context remains independent. SOA is easily integrated with cloud environments, which improves efficiency and monitoring at large scales. In case data is shared among distributed applications, SOA can support microservices to address the applicative needs. Recent studies have not suggested the architectural visions of SOA, interplay with web and communication protocols and associated security concerns. Thus, in this article, we propose a tutorial approach to discuss SOA concepts and how SOA can be extended to support microservice integration in an application domain.

1.2 Research Contributions

Following are the research contributions of the paper:

1. An interplay of the SOA and the web protocols are presented, and the underlying secured transfer of data through SOA is presented that preserves the authentication, session key distribution, perfect forward secrecy, and anonymity in communication are maintained.
2. A SOA testing framework for contemporary applications are presented, forming a secured interaction and control with third-party applications.
3. A case study of the blockchain (BC) assisted Web 3.0 and SOA-oriented healthcare microservice applicative framework is proposed that discusses the requirements of SOA for enterprise services and microservice integration to support web-based APIs that fetches data from different healthcare setups.

1.3 Article Layout

The article is divided into eight sections. Section 2 presents the key enablers of SOA and discusses the existing state-of-the-art approaches in SOA design.

Section 3 presents the interplay of SOA with the Web. Section 4 is oriented toward the security aspects in SOA design regarding authentication, session key distribution, forward secrecy, and anonymity. Section 5 presents an SOA testing framework that supports contemporary enterprise applications. Section 6 presents the key challenges and future research directions. Section 7 presents the case study that integrates blockchain and Web 3.0 with SOA in healthcare application setups, and finally, Sect. 8 concludes the article.

2 SOA Enablers and Existing Schemes

This section discusses the design principles and enablers of SOA and summarizes the applicative findings of existing research works on SOA. The details are presented as follows.

2.1 SOA Enablers

There are several design principles to keep in mind while creating an SOA service, and these principles provide a foundation for the design and development of the service [7]. Firstly, there must be a description of the service that would provide the details about the functionalities offered by the service, which means there should be a *standardized service contract*. The design concepts used in defining the contracts should be the same, making the interaction between different services easier. In addition, services which are defined in a system should be least reliant on the others, i.e., services should be *loosely coupled*. The reusability of the services will increase due to loose coupling. Furthermore, there should be *service abstraction* meaning that the consumer knows only the essential details about the service by hiding the logical application of the service. This makes the service easily usable as the client does not have to worry about how and why the service works.

Moreover, *service autonomy* is essential because of the reusability principle mentioned above. Service autonomy means the logic of the service is entirely under the control of the service itself. Additionally, the services defined ought to be stateless, i.e., the services should be transparent between the states. This principle improves the scalability, due to which more stateless service calls can be easily made in the current environment. Essentially, *service discoverability* is an important principle which means that services should become available quickly and effortlessly by the use of a service registry. Besides, the services for each small problem are created and then integrated to solve a significant problem for which the *service composability* should be effortless. One of the most important principles is *service interoperability*. This ensures that one service is compatible with all the other services.

2.2 Existing Schemes

This section summarizes the findings and evaluation of existing surveys and research articles that have discussed SOA. Benatallah et al. [1] signify the need for SOA in everyday life and have proposed a framework for integrating different applications through SOA. Canfora et al. [3] overview SOA testing and discuss different issues faced in different testing levels, and provide new ways to improve service testability. Liu et al. [17] propose a unified testing framework to enable Continuous Integration Testing (CIT) of SOA solutions. Kontogogos et al. [14] present the approaches of SOA in various fields like e-Commerce, eLearning, etc. The paper compares the characteristics of different SOA solutions and their applications in diverse fields.

Rumez et al. [21] elaborate different security measures for a hybrid architecture integrating signal and service-oriented world. Ouda et al. [20] propose a complete Quality of Security Service (QoSS) model for addressing security in SOA. This model incorporates different techniques like symmetric keys, public keys, etc. Tan et al. [24] discuss the integration of SOA with web services API like Simple Object Access Protocol (SOAP), Representational State Transfer (REST), and others. Hui Chen [5] introduces novel technologies like artificial immunity and cloud security into the SOA.

3 Interplay of SOA and the Web: Key Protocols

In the web framework, the clients can access the data that other applications can retrieve through the use of web services [15]. In web services, three major protocols are used, which are listed as follows [22].

3.1 *Simple Object Access Protocol*

Simple Object Access Protocol (SOAP) is a protocol that is designed to transfer information in a distributed system. SOAP messages can be transferred whenever and whatever the client requires. SOAP is a perfect fit for the Internet as it improves the interoperability of the services. SOAP is based on eXtensible markup language (XML) [23] and uses HTTP to transfer the SOAP messages over the Internet. The documents are structured around the root elements, while the child element contains the information. A SOAP message in XML format is to be sent, which contains the request for some information from the server. This document is analyzed, and then the server responds to the request by sending the information in the form of a SOAP message.

3.2 *Web Services Description Language*

Web Services Description Language (WSDL) describes a web service in XML format to inform the client about the service, how to send a request to the server, and what to write in the request. The logical and physical part of WSDL is separated in the definition. The physical details comprise the information regarding reaching the server, e.g., HTTP port number, etc. When a WSDL is generated, it is validated using the WS-I basic profile; the validation errors generated during this should be resolved before deploying the WSDL.

3.3 *Universal Description, Discovery, and Integration*

The service discoverability principle mentioned in Sect. 2 is implemented using Universal Description, Discovery, and Integration (UDDI). UDDI makes it possible for the services to be available to other organizations. UDDI is a registry for the services, and companies providing the services can list themselves on this registry. UDDI aims to make it easy for companies to find other companies for a seamless transaction system over the Internet. UDDI permits companies to list themselves based on name, product, location, or the company's service.

4 Security in SOA Architecture: Handshake-based Scenarios

To assure security over SOA exchanges, a few protocols are defined to make the services secure over an untrusted medium of data transfer. A majorly used model

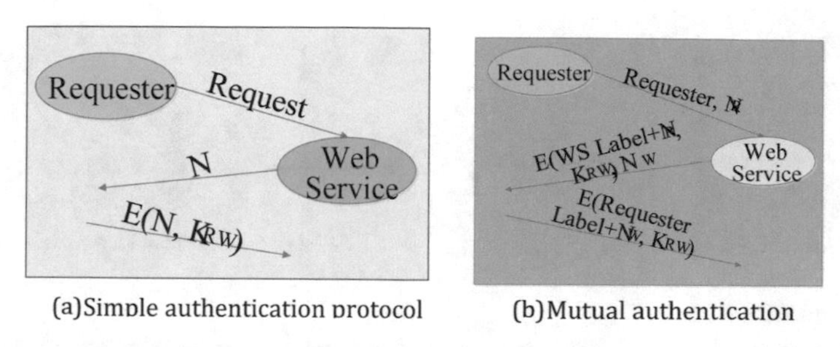

Fig. 2 Handshake scenarios of simple and mutual authentication

used for securing this data transfer using a web service is Quality of Security Service (QoSS) [20].

QoSS defines many protocols to secure web service interaction, as it is impossible to define a single communication protocol that would best fit the need. QoSS focuses on the vulnerability analysis of the data, the requirements to set up authentication, and the distribution of session keys to different clients. Further, it focuses on the anonymity of the user. Based on the vision, we present the key security principles of QoSS as follows.

4.1 *Simple Authentication Protocol*

Considering that there is a need for authentication between the service and the client, a symmetric key is necessary for the authentication protocol. When the client sends a request to the service, a nonce (N) is sent to the client by the service. Then the client needs to encrypt the nonce using a key and pass the encrypted nonce to the service. If the decrypted form of the message received by the server matches the nonce, the data requested is sent. This process is illustrated in Fig. 2a.

4.2 *Mutual Authentication*

This protocol is incorporated when a mutual authentication between the client and service is necessary. Here, a nonce as well as the request for the data are sent by the client. The service needs to encrypt the nonce using a key and send it to the client along with another nonce. Then if the decrypted form of the message matches the nonce sent, the client then needs to encrypt the received nonce using that same key and send that back to the service. If the decrypted form of the received message matches the nonce sent by the service, the requested data is sent to the client. This protocol overcomes the possibility of man-in-the-middle (MiM) attack [18]. This process is illustrated in Fig. 2b.

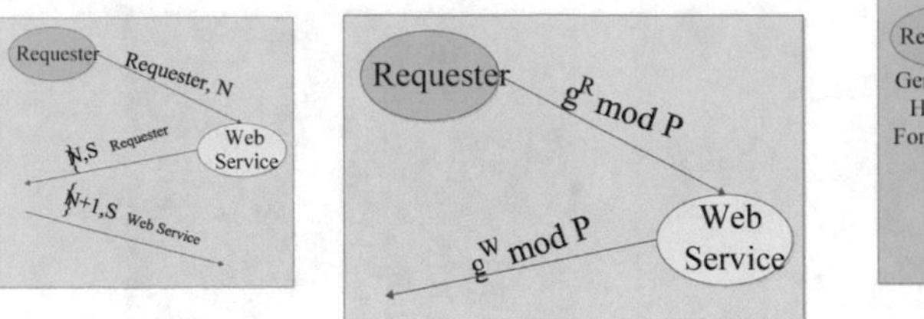

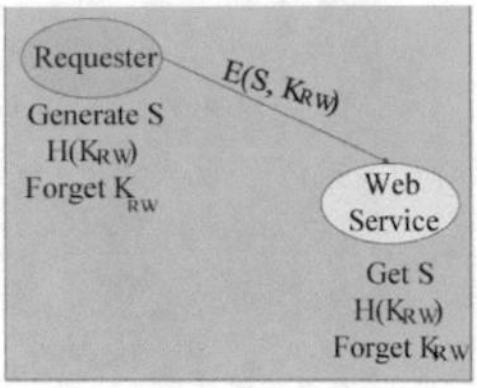

(a) Authentication and session keys (b) Perfect forward secrecy (c) Anonymity communication

Fig. 3 Handshake scenarios of authentication, session key distribution, perfect forward secrecy, and anonymity preservation

4.3 Authentication and Session Keys

Here, the request and a nonce are sent to the service by the client. The service uses a key to encrypt the nonce and a session key and sends the encrypted message to the customer. The client decrypts the message, increments the nonce, and encrypts the nonce. The client sends this encrypted nonce and session key using the same key. The service sends the requested data if the decrypted message matches the incremented nonce. This process is illustrated in Fig. 3a.

4.4 Perfect Forward Secrecy

The issue with keeping the same key is that a third party might get access to the key, and all the data from the service can be decrypted. This problem can be solved by using Diffie–Hellman [6], which creates a secret between the client and the server by changing the key frequently. The difficulty with the protocol stated above is that it is vulnerable to MiM attacks. To avoid this problem, the values of the key that is frequently changed can be communicated in encrypted form using the long-term key which is available to both the client and the service. This process is illustrated in Fig. 3b.

4.5 Anonymity Communication

The client generates a session key, and then that key, along with a hash of the private key, is encrypted by using the service's public key. The service receives and decrypts the message using the private key, then the hash of the long-term key is calculated and verified with the hash sent in the message. The session key is stored if the

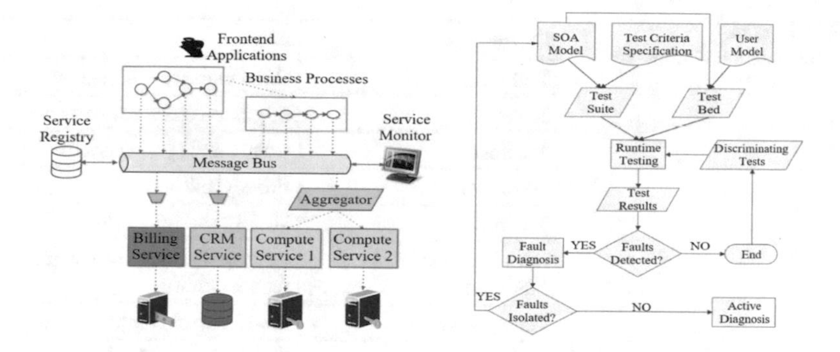

(a) A high-level overview of contemporary SOA (b) A flowchart depicting the SOA testing framework

Fig. 4 High-level SOA testing view and flow steps

hashes match, and the requested data is sent to the anonymous client. This process is illustrated in Fig. 3c.

5 A SOA Testing Framework

Due to the rapid growth of SOAs, the underlying complexity of the systems using this architecture has also shown rapid growth. The systems using SOA no more comprises only the services and the clients. These more complex SOAs are commonly known as contemporary SOAs. The structure of these contemporary SOAs is shown in Fig. 4a.

As a result of this high complexity, the chances of errors occurring in the system increase exponentially. Thus, to make these systems robust, there is a need for extensive testing. Testing these contemporary SOAs is a challenging task because it has many interconnections between systems. The higher-level overview of the SOA testing for these contemporary SOAs is shown in Fig. 4b. As the services contain many levels, it is a challenging task to isolate the error.

Tools like Genesis 2 [12], PUPPET [2], and many others can be used for creating a testbed [9]. The tests defined at the design time are executed at the runtime using a testbed, an abstraction, and a staging environment for the whole SOA.

6 Key Challenges

The various challenges of the SOA can be categorized as technical, management, and governance. The present section aims to showcase the challenges while implementing the SOA. These challenges are given in Table 1. Communication between the services

Table 1 SOA implementation challenges

Category (s)	Issues
Technical	Communication in real time
Technical	Migration of legacy software
Technical	Compatibility of services
Technical	Privacy and security
Technical	Testing and reliability
Both Technical and Management	Getting services to cooperate (cooperation)
Management	Requirement comprehension
Management	Standards must be defined
Governance	Managing the complexity of SOA (SOA overhead)
Governance	Services for monitoring

in real time is a challenge, as SOA is a collection of services, and to fulfill its objectives, the services have to communicate among themselves. The migration of the legacy software into the SOA architecture is yet to be optimized. The compatibility of the services and their security is of utmost importance as these services are exposed to other services also. The integration testing in the SOA is an issue, and this needs to be improvised, as many services will be communicated to each other. The monitoring of the services is also required to get the overall performance of the SOA design architecture.

7 Case Study: BC-Assisted Web 3.0 and SOA-Oriented Healthcare Microservice Applicative Framework

This section outlines a case study that mitigates the limitations of centralized SOA control through a proposed distributed file system, like Interplanetary File Systems (IPFS), where the data from heterogeneous nodes are stored. We have proposed a BC-assisted Web 3.0 and SOA-oriented healthcare microservice framework. Figure 5 outlines the proposed framework. This framework has three basic sections: (1) distributed file systems, (2) blockchain, and (3) enterprise SOA and Web 3.0.

7.1 Distributed File Systems

The medical records for a single patient may exist at different setups like medical labs, hospitals, and wireless body area networks (WBAN) [13]. Firstly, all those records from different setups are aggregated and sent to aggregator and analytics.

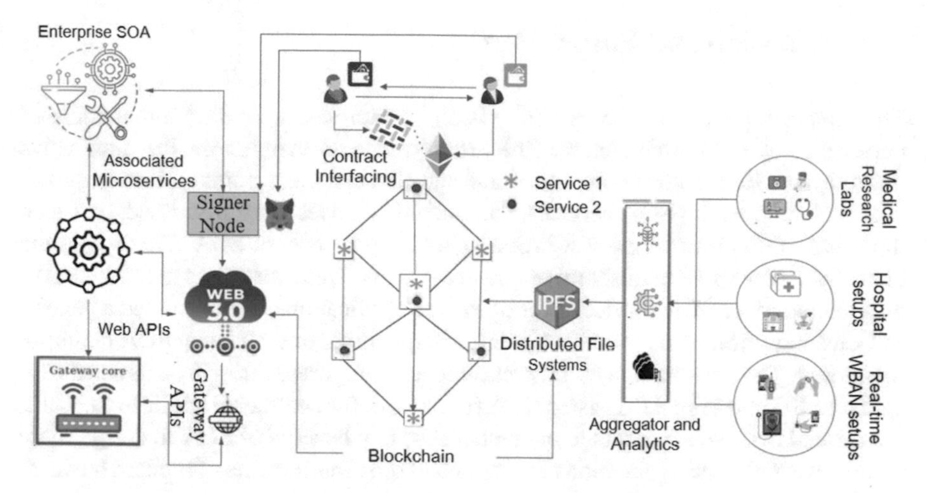

Fig. 5 A BC-assisted Web 3.0 and SOA-oriented healthcare microservice application framework

The analytics are performed on the cloud, and once completed, the records are sent to IPFS, from which data can be retrieved through the user authorization and the IPFS content key.

7.2 Blockchain

For accessing the file system mentioned above, the services (S_1, S_2) required will be distributed on different nodes (N_1, N_2, ..., N_n) over a consortium blockchain (BC). The stakeholders of this BC are various healthcare institutes that provide the necessary medical history of patients. Service S_1 is accessed by nodes (N_1, N_3, and N_5), while service S_2 is accessed by nodes (N_2, N_4, and N_6). The central node is the host for both the services S_1 and S_2. The users interact with nodes via a contract interface on Ethereum BC.

7.3 Enterprise SOA and Web 3.0

The BC mentioned above is connected with Web 3.0, which helps in communication between users and services provided by the nodes of BC. Furthermore, the signer node is connected with Web 3.0, as it is necessary to authenticate any user to maintain the privacy of healthcare data. User when tries to access a service, the request is handled by a service gateway that exists on every node of BC. The gateway core handles this request, which provides the nearest node containing the service requested. The request is then forwarded to the node where it is handled.

8 Conclusion and Future Work

This tutorial presented the critical vision, principles, and different associated paradigms of SOA architecture. The article presents insights on the applicative interplay of SOA with web architecture and discusses the communication protocols that support the SOA services. The technical aspects are discussed, and then the article shifts toward the discussion of security aspects in SOA. We present the handshake scenarios for different security scenarios. This establishes the addressal of data sharing over SOA services over open web applications. We discussed a generic SOA testing framework, possible test criteria specifications, and how SOA faults are diagnosed. The key challenges are presented and supported through an assisted case study on BC and Web 3.0 assisted SOA framework for healthcare application. Thus, the tutorial orients its readers to understand the key benefits of SOA to design more responsive and agile applications that can meet the requirements of modern business applications.

As part of the future scope, the authors would investigate the integration of microservices in Web 3.0 architecture. The authors would propose a memory effective and secure scheme design in healthcare ecosystems.

References

1. Benatallah B, Motahari Nezhad HR (2008) Service oriented architecture: overview and directions, pp 116–130. Springer, Berlin, Heidelberg. https://doi.org/10.1007/978-3-540-89762-04, https://doi.org/10.1007/978-3-89762-0_4
2. Bertolino A, De Angelis G, Frantzen L, Polini A (2008) Model-based generation of testbeds for web services. In: Suzuki K, Higashino T, Ulrich A, Hasegawa T (eds) Testing of software and communicating systems. Springer, Berlin Heidelberg, pp 266–282
3. Canfora G, Di Penta M (2009) Service-oriented architectures testing: a survey, pp 78–105. Springer, Berlin, Heidelberg. https://doi.org/10.1007/978-3-540-95888-84, https://doi.org/10.1007/978-3-95888-8_4
4. Cerny T, Donahoo MJ, Pechanec J (2017) Disambiguation and comparison of SOA, microservices and self-contained systems. In: Proceedings of the international conference on research in adaptive and convergent systems, pp 228–235. RACS'17, Association for Computing Machinery, New York, NY, USA. https://doi.org/10.1145/3129676.3129682
5. Chen H (2021) SOA security strategy based on cloud immune protection. In: 2021 13th International conference on measuring technology and mechatronics automation (ICMTMA), Beihai, China. pp 514–517. https://doi.org/10.1109/ICMTMA52658.2021.00118
6. Diffie W, Hellman M (2021) New Directions in cryptography (1976). In: Ideas that created the future: classic papers of computer science. The MIT Press. https://doi.org/10.7551/mitpress/12274.003.0044
7. Erl T (2008) SOA: principles of service design prentice hall. Upper Saddle River, NJ
8. Evdemon J (2007) Service oriented architecture (SOA) in the real world
9. Hirzalla M, Cleland-Huang J, Arsanjani A (2009) A metrics suite for evaluating flexibility and complexity in service oriented architectures. In: Feuerlicht G, Lamersdorf W (eds) Service-oriented computing—ICSOC 2008 workshops. pp 41–52. Springer, Berlin, Heidelberg
10. Hurwitz JS, Bloor R, Kaufman M, Halper F (2009) Service oriented architecture (SOA) for dummies. Wiley

11. Josuttis NM (2007) SOA in practice: the art of distributed system design. O'Reilly Media, Inc.
12. Juszczyk L, Dustdar S (2010) Programmable fault injection testbeds for complex soa. In: Maglio PP, Weske M, Yang J, Fantinato M (eds) Service-oriented computing. Springer, Berlin, Heidelberg, pp 411–425
13. Khan JY, Yuce MR, Bulger G, Harding B (2012) Wireless body area network (WBAN) design techniques and performance evaluation. J Med Syst 36(3):1441–1457
14. Kontogogos A, Avgeriou P (2009) An overview of software engineering approaches to service oriented architectures in various fields. In: 2009 18th IEEE international workshops on enabling technologies: infrastructures for collaborative enterprises, Groningen, Netherlands. pp 254–259. https://doi.org/10.1109/WETICE.2009.44
15. Kreger H et al (2001) Web services conceptual architecture (WSCA 1.0). IBM Softw Group **5**(1):6–7
16. Laskey KB, Laskey K (2009) Service oriented architecture. Wiley Interdisc Rev Comput Stat 1(1):101–105
17. Liu H, Li Z, Zhu J, Tan H, Huang H (2009) A unified test framework for continuous integration testing of SOA solutions. In: 2009 IEEE international conference on web services, Los Angeles, CA, USA. pp 880–887. https://doi.org/10.1109/ICWS.2009.28
18. Mallik A (2019) Man-in-the-middle-attack: understanding in simple words. Cyberspace: Jurnal Pendidikan Teknologi Informasi 2(2):109–134
19. Ocean TNR (2022) Microservice architecture market share, size 2022 to 2030 consumption analysis by application, future demand, leading players, competitive situation and emerging trends. https://www.taiwannews.com.tw/en/news/4400883. Accessed: 06 June 2022
20. Ouda AH, Allison DS, Capretz MA (2010) Security protocols in service-oriented architecture. In: 2010 6th world congress on services, Miami, FL, USA, pp 185–186. https://doi.org/10.1109/SERVICES.2010.44
21. Rumez M, Grimm D, Kriesten R, Sax E (2020) An overview of automotive service oriented architectures and implications for security countermeasures. IEEE Access 8:221852–221870. https://doi.org/10.1109/ACCESS.2020.3043070
22. Shaghayegh B (2011) Using service oriented architecture in a new anonymous mobile payment system. In: 2011 IEEE 2nd international conference on software engineering and service science, Beijing, China, pp 393–396. https://doi.org/10.1109/ICSESS.2011.5982335
23. Skonnard A, Gudgin M (2001) Essential XML quick reference: a programmer's reference to XML, Xpath, XSLT, XML Schema, SOAP, and more. Addison-Wesley Longman Publishing Co., Inc.
24. Tan W, Fan Y, Ghoneim A, Hossain MA, Dustdar S (2016) From the service oriented architecture to the web API economy. IEEE Internet Comput 20(4):64–68. https://doi.org/10.1109/MIC.2016.74

Container Division and Optimization (CDO) Algorithm for Secured Job Processing in Cloud

S. Muthakshi and K. Mahesh

Abstract Generally, container resource allocation and job scheduling are the major concern in cloud computing. To produce proper cloud virtualization and to perform an exact allocation, migration process is opted. Several preceding methods opted for migration techniques to produce an appropriate cloud container allocation and resource scheduling. But, only migration is not enough to process a job scheduling and resource allocation. The system proposes a container division and optimization (CDO) algorithm that follows a three set of stage in job scheduling and processing. (1) 'Entry-level/Bottleneck container' to handle and processes allocation for user data. (2) 'Processing container' to perform the job scheduling and maintain the secrecy of the user job (3) 'Exit-level container' to cross-check both the container and user performance. Further, the system proposes a reduced migration model that performs migration only if highly needed (lack of space). Hence, the experimental results proves that the proposed algorithm is effectively performing a secured optimized job scheduling in the cloud.

Keywords Bottleneck · Container · Resource allocation · Migration model · Optimization · Security

1 Introduction

Cloud computing can provide container resources on-demand adapting themselves to varying service demands. In cloud computing, virtualization plays the main role to achieve high resource utilization and job scheduling. User job rendering through several cloud service providers (CSPs) is not sufficient for providing secured and viable cloud service. Hence, stream processing is used for properly organized inlet and outlet of data involves techniques similar to batch and stream processing of big

S. Muthakshi (✉) · K. Mahesh
Department of Computer Applications, Alagappa University, Karaikudi, India
e-mail: muthakshi.researchscholar@gmail.com

K. Mahesh
e-mail: maheshk@alagappauniversity.ac.in

Y. Singh et al. (eds.), *Proceedings of International Conference on Recent Innovations in Computing*, Lecture Notes in Electrical Engineering 1001,
https://doi.org/10.1007/978-981-19-9876-8_36

data [1]. Generally, in job scheduling, the job processing time, battery usage, and energy consumption of the resource-limited devices are the main concern. Based on the demand and utilization of the cloud container, some of the cloud-based engines such as Docker, Mesosphere, and Kubernetes are provided for resource allocation.

When the data are not produced to a suitable container for allocation, then the system seeks a 'migration process' (moving the data to a bigger size container). Container allocation and job management authorize the revenue providers to utilize limits of resources. Some conventional cloud computing provisions follow a guided performance in resource allocation and task scheduling that helps in managing important problems. Thus, researchers are reasoning the studies in motivating the job scheduling and container allocation of cloud computing resources.

There are many algorithmic techniques followed for processing a perfect container resource allocation. The proposed system aims to provide secured container resource allocation providing a secured storage and transaction in the cloud [2].

The contributions of the method are as follows:

1. Initially, the system uses a novel container division and optimization (CDO) algorithm for performing allocation. The algorithm undergoes three stages to accomplish a proper resource allocation strategy. The system first analyzes the incoming data to initiate a proper container allocation. Then, using the processing container, the suitable container is allocated, and the user resources are maintained securely. The unwanted connections are blocked and blacklisted if detected.
2. Further, a cross-verification is made to know the satisfaction of the container provider and the user in resource allocation. The main aim of the system is to provide a container for resource allocation and to maintain the resources confidentially. Before retrieval of resources, the data are exchanged with the fake data and then deleted to enhance the security.
3. Further, the system proposes an optimal migration technique to handle the volatile data. The use of the migration process saves computational time and consumes energy. Further, the algorithm reduces the processing time and improves reliability and energy consumption.

The rest of the paper is as follows. In Sect. 2, previous research on container allocation, migration process, and some security measures are discussed. The proposal implementation on container division and optimization resource allocation with some security measures are discussed and implemented in Sect. 3. Then, following Sect. 3, the overall methodology used in this work is detailed. In Sect. 4, the analysis of the results is presented and discussed. Finally, in Sect. 5, the paper is concluded.

2 Literature Survey

Ouyang et al. in [3] deliberated a new job management model named Band-area Application Container. BAC includes processing of users, documents, fine-grained application services or micro-services, messages, and a deposit of related function rules. The theory also used the fish swarm algorithm to predict the node and categorize them.

Saxena et al. in [4] proposed an adaptive learning-based neural NW model for forecasting the load balancing in cloud data centers. The learning-based approaches also help in learning the cloud resource allocation and produce an adaptive model. Some attacks may occur during the resource allocation operation; several defenders and defeating algorithms are produced and discussed.

Li et al. in [5] introduced a DDos mitigation strategy to stop DDos attacks. The author proposed a queuing theory to study the strengths and weaknesses of the container-cloud environment to defeat the DDoS attack.

Further in [6], Bai et al. deliberated an ant colony algorithm to perform multi-objective optimization in cloud-based container scheduling and resource allocation.

In [7], Kaur et al. produced a multi-objective scheduling model implementing a fuzzy particle swarm optimization for resource scheduling. The system provides time consumption, energy consumption, and power consumption for processing to maximise resource utilization.

Some of the migration techniques to handle the suitable and adaptable resource allocation are introduced by the researchers. Kim et al. in [8] introduced an Optimal Container Migration to process allocation in edge computing. The migration system reduces the network traffic that occurs during allocation and solves the container size utilization issue.

2.1 Container Scheduling Techniques

The scheduling-related strategies and techniques used in the previous researches are discussed.

Artificial neural network (ANN): *I*n [9], Garg et al. produced a boosting utilization algorithm for resource allocation. Further, the system uses artificial neural network (ANN) with a multi-objective optimization theory to improve the profit of virtual machines in the CDC.

Boosting Algorithm: Adhikari et al. in [10] fulfills the QoS necessities for users in the service-level agreement in cloud through implementing the boosting algorithm. A new resource allocation strategy along with job optimization is proposed with help of the BAT theory and k-means algorithm for the VM. The utilization of BAT algorithm and k-means minimizes the computational time and task execution model.

Multi-objective scheduling model: Kaur et al. [7] produced a fuzzy particle swarm optimization (FSO) for providing a multi-objective optimized scheduling in cloud. Fuzzy particle swarm optimization model reduces the computational time and energy utilizing large virtual machine (VM).

Automation Model: The system in [11] produced a reinforcement learning based on the traditional learning automata model. A job scheduling-related learning automaton technique used in this model. Also, for more optimized utilization, a self-adapting job scheduling technique is proposed in [11].

Heuristic Algorithm: For a proper scheduling, optimization in [14] focuses on the three algorithms (enhancement of ant colony algorithm, particle swarm algorithm, and genetic model), based on heuristic algorithm. The model focuses mainly on fulfilling the requirements of the task scheduling scenarios in cloud environment.

Integer Linear Programming Model: Xu and Wang in [12] proposed an effective adapting scheduling algorithm by modeling the scheduling problem as integer linear programming. Moreover, the existing system in [13] proposed a linear scheduling model for task scheduling and job scheduling. An application scenario of 5G edge computing based on resource utilization to calculate the theory is used.

3 Methodology Implementation

The system implies a set of container categorization/classification models to provide enhanced security in container allocation and job scheduling.

In the proposed algorithm, the system follows three divisions to perform processing. The three container categorizations for job scheduling are as follows:

1. Entry-level/bottleneck container
2. Processing container
3. Exit-level container

The processing and the operation of the proposed theory container division and optimization (CDO) algorithm are deliberated in the below algorithm. Here is the input jobs for container allocation ($D_{s)}$. Then, the algorithm container division and optimization (CDO) follows a set of rules for container allocation. Initially, the jobs are assigned to (EBC_{ra}) the entry-level/bottleneck container. The EBC checks for available containers (Ca) in the processing container (Pc). The authorization to access in processing container (PC) and job scheduling is done via (EBC_{ra}) the entry-level/bottleneck container. The next level is processing container (PC_{ra}) which handles job allocation ($J_{a)}$, secures the resources (Rs), and performs deletion after retrieval ($R_{del)}$. Then, the process is moved to the exit-level container ($EC_{ra)}$. The exit-level container (EC) performs cross-verification ($C_{CV)}$, checks for the container clearance ($C_{Clr)}$, and initiates Fake Data Deletion ($D_{fd)}$ in place of the original data to enhance the security. Optimal migration ($M_{optm)}$ container limit exceeds or immense data for job scheduling is detected. Hence, by performing the proposed operation, an optimized container scheduling is produced for resource allocation.

Algorithm1. Algorithm for secured resource allocation using **container division and optimization (CDO) algorithm**

Input: D_s // Jobs for Container allocation

(continued)

(continued)

Algorithm1. Algorithm for secured resource allocation using **container division and optimization (CDO) algorithm**
Output: MA_r // Secured Job scheduling and resource allocation
Initialization: Resources for allocation
While (EBC_{ra}) //Entry-level/Bottle-neck Container
$C_a \leftarrow$ Checks for available containers
$Ar_{pc} \leftarrow$ Authorization to access Processing container
If (PC_{ra}) **then** // Processing Container
$J_a \leftarrow$ Allocated user job
$R_s \leftarrow$ security for Resources
$R_{del} \leftarrow$ Performs deletion after retrieval
$C_{blck} \leftarrow$ blacklist unwanted connection
End if
If (EC_{ra}) **then** // Exit-level Container
$C_{CV} \leftarrow$ Cross-verification container
$C_{Clr} \leftarrow$ Container clearance
$D_{Fd} \leftarrow$ Fake Data Deletion
End if
If (M_{ce}) **then** // Container limit exceed performs Migration
$M_{optm} \leftarrow$ Optimal migration
Else
If (PC_a) **then** // perform allocation in PC
End if
End while

3.1 *Container Division and Optimization (CDO) Algorithm*

The system architecture shown in (Fig. 1) illustrates how the proposed method works.

Entry-level/Bottleneck container

An entry-level container is a container that initially analyzes the user job and performs allocation accordingly.

Data Preprocessing: Initially, when a job enters the type of the job, i.e., users, application services, documents, and messages are analyzed. The data are preprocessed where the size, format, and originality of the job are checked thoroughly.

Scheduling/Allocation as a Bottleneck container: The entry-level container is also called as a bottleneck container that takes and processes the bottleneck job first. When

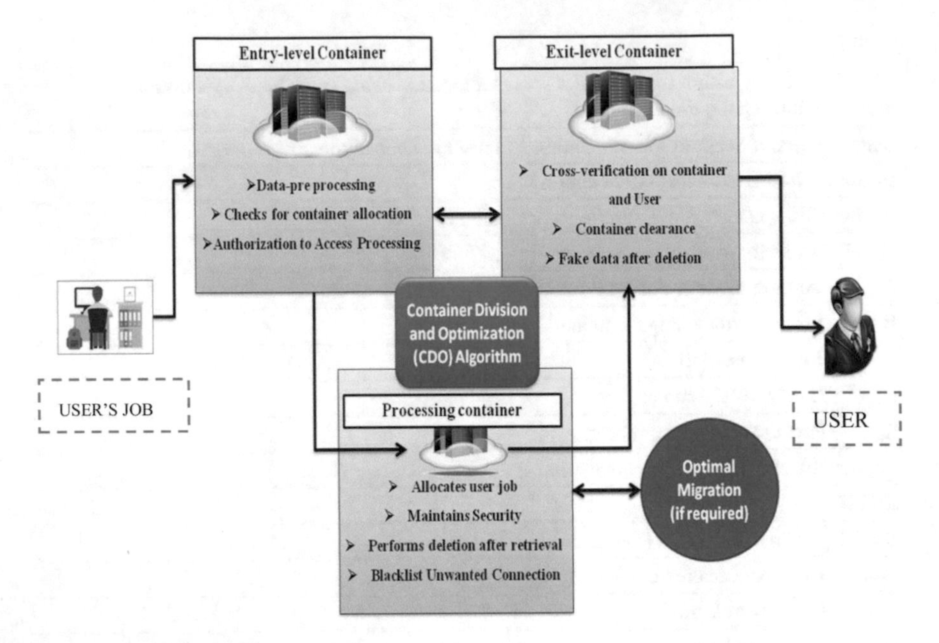

Fig. 1 System architecture diagram

a user job enters, the entry-level container immediately connects to the processing container to check the availability of the resources to allocate them. The container performs allocation immediately when a job is assigned.

Authorization to access processing: Only, the entry-level container has authorized access to operate the processing container. In case a job is set for allocation, the entry-level container will check for the available space and suitable space for resource allocation in the processing container. If a suitable and sustainable container is found in the processing container, then immediately the authority to access the storage in the processing container is provided to the user.

Processing Container

The processing container is a container that has to be maintained with high security. All the user-related job scheduling and resource allocations are handled by the processing container. The processing container acts as a secret resource handler between the entry-level container and exit-level container.

Security Maintenance: As the processing container contains all the original content of the user, the container should be maintained with high security. The PC contains a bundle of resources maintaining the user's job; all the users' confidential resources are processed by the PC only. The container should not have the authority or accessibility to outsource the resources. PC has provided authority to the link between the entry-level and exit-level containers only to maintain the secrecy of the resources.

Container Clearance: When the user resources are retrieved for the container space, then immediately the processing containers are instructed to clear the allocation space.

Fake data after deletion: There are chances of attackers to recovering even the deleted data. Further to provide enhanced security, the system initializes the fake data concept. Here, a container before clearance is exchanged with the fake values instead of the original values. Then, the fake data are deleted. Even if the authorized user accesses the data, again, the person will get only the deleted fake data instead of the original data. This concept is an added advantage to prevent the resources from unwanted intruders.

Exit-Level Container

The exit-level container is a container that performs a proper exit of a resource at the user retrieval time. When the user requests the resources, then the resources of that particular user are collected from the processing container and produced to the user with verification.

Cross-verification: The exit-level container also performs a cross-verification on the performance of the container and the user.

Container Feedback: Feedback from the container side is taken regarding the resource allocation, space utility, and user cost settlement. The user can retrieve the data only after a complete cross-verification of the exit-level container. If any issues in payment or dissatisfactory from the user side are detected, then the resources are held and sent only after fulfilling the containers requirements.

User Feedback: Same as that the user feedback is also collected. The users are queried for the quality, security, and authorization of the job scheduling and resource allocation. If the user faces any discomfort during allocation, then immediately the alert insisting a proper allocation is sent, and the problem is rectified at once. This way both the user and the container allocator are mutually benefitted.

Security check: The exit-level container also checks and conforms to the implementation of fake data instead of original data. Here, all the resource-related security is monitored then and thereby the exit-level container. If any unwanted connection is established, then immediately the intrusion is blocked/blacklisted.

After the final cross-verification, the user's confidential resources are securely produced to the user.

3.2 Optimal Migration

At the time, the user resources may exceed the container limit during the resource allocation process. In that case, the optimal migration process has opted for only if highly recommended. By using the migration process, a suitable container can be chosen, and the resources can be shifted to the exact container. The migration process can opt only if immense data are detected. This migration process is used mainly for time and energy consumption.

4 Result Analysis

To evaluate the QoS parameters and validate the performance for the proposed resource scheduling strategy and job allocation, a dataset is taken. The system adopted the dataset cluster-trace V2018 provided by Alibaba [15]. The dataset was analyzed with parameter settings and the setup of the experiment. Alibaba used Github tool to analyze the data using the random function of the synthetic datasets for testing.

1. The execution container parameters set in 100 container instance datasets are set for container execution in randomly.
2. 10 synthetic datasets are randomly generated with different task numbers. The number of execution container instances is 100. The task allocation is set as 100–1000.

Synthetic dataset for task processing time and energy: The time and energy required for processing each job to be analyzed and uploaded to each execution container. From the below graph (Fig. 2), a QOS, performance, and task-offloading analysis is made, and comparison of all the existing algorithms with the proposed model is done.

From the below graph (Fig. 3), a comparison of all the existing algorithms with the proposed model is done. The comparison of existing algorithms such as the Band-area Application approach, fuzzy particle swarm optimization, PSO, and Adaptive migration model is done with the proposed container division and optimization (CDO) algorithm. In the comparison, the proposed algorithm is proved to be the best in task scheduling, security maintenance, accuracy, and resource allocation.

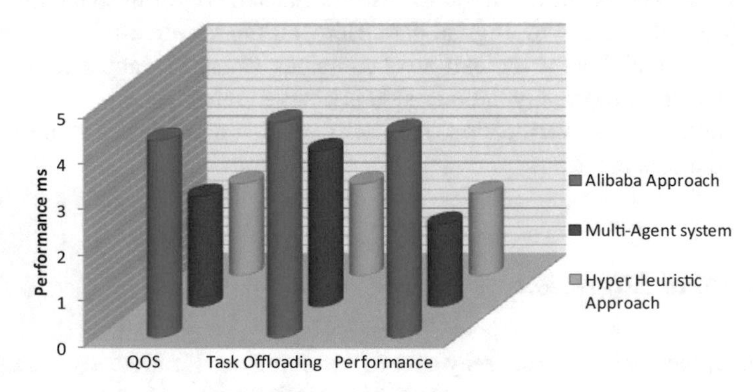

Fig. 2 Performance existing versus Alibaba approach

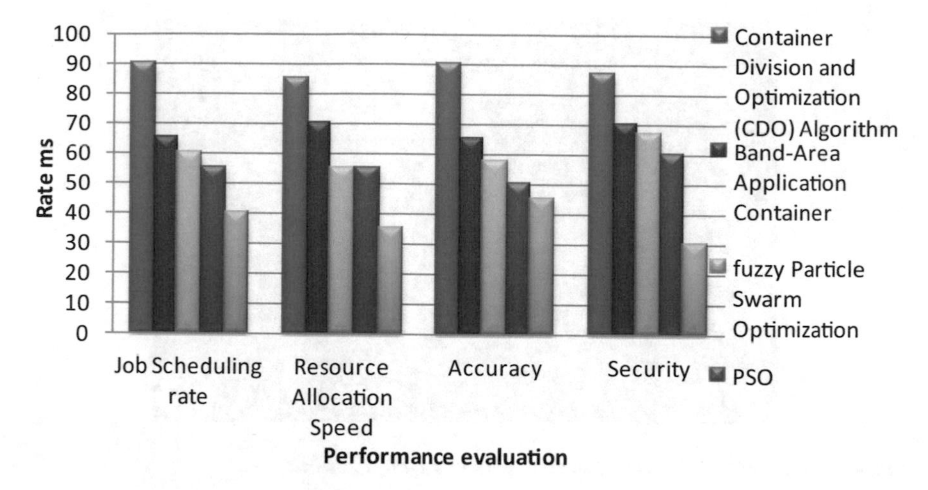

Fig. 3 Existing versus proposed algorithm

Table 1 proposed algorithm is compared with existing algorithms. The job scheduling rate, speed, accuracy, and security shows the proposed algorithm has best in performance.

In the result analysis, a comparison of the existing migration models with the proposed optimal migration model is done in the below graph (Fig. 4). The comparison of existing migration algorithms such as BAC migration approach and container migration model is done with the proposed optimal migration model. In the comparison, optimal migration model is made with the BAC migration approach, container migration model, integer linear programming model, and reinforcement learning on automata. The proposed system is proved to be the best in task scheduling and resource allocation providing speed, reliability, minimal time and energy.

Table 1 Comparative analysis of the proposed algorithm

	Container Division and Optimization (CDO) Algorithm	Band-Area Application Container	Fuzzy Particle Swarm Optimization	PSO	Adaptive migration model
Job scheduling rate	90	65	60	55	40
Resource allocation speed	85	70	55	55	35
Accuracy	90	65	58	50	45
Security	87	70	67	60	30

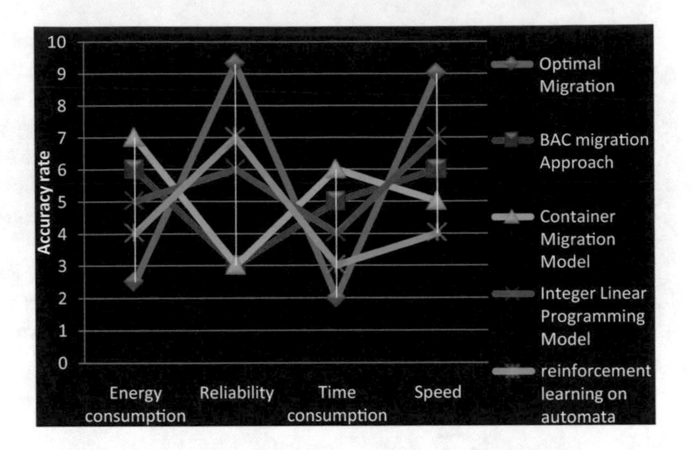

Fig. 4 Performance evaluation comparing various migration models

Table 2 Comparative performance analysis

	Optimal migration	BAC migration approach	Container migration model	Integer Linear Programming Model	Reinforcement learning on automata
Energy consumption	2.5	6	7	5	4
Reliability	9.3	3	3	6	7
Time consumption	2	5	6	4	3
Speed	9	6	5	7	4

Table 2 a comparative analysis of energy consumption, reliability, time consumption, and speed of various models is made. In these comparative results, the proposed optimal migration model has proven to be the best in performance.

5 Conclusion

The main aim of the proposed system is to perform an appropriate resource allocation in the cloud. The paper used a container division and optimization (CDO) algorithm following a set of container-based theories for secured container scheduling. The system uses an optimal migration process is used to shift and allocate the immense data in the container. The system further performs a fake data replacement and cross-verification to enhance the security of the container storage. The implementation model maintains the efficiency, speed, accuracy, security of task scheduling, and resource allocation in cloud containers. Hence, the user resources are allocated in the processing container confidentially, and the original data are retrieved to the user securely.

Future Enhancement

In future, a deep neural learning can be opted to resolve the performance degradation. All the containers are studied, and the less expensive and suitable containers are focused in future.

Acknowledgements This research work has been supported by UGC Junior Research Fellowship from New Delhi.

References

1. Dos Anjos JCS et al (2020) Data processing model to perform big data analytics in hybrid infrastructures. IEEE Access 8:170281–170294. https://doi.org/10.1109/ACCESS.2020.3023344
2. Zhang X, Wu T, Chen M, Wei T, Zhou J, Hu S, Buyya R (2019) Energy-aware virtual machine allocation for cloud with resource reservation. J Syst Softw 147:147–161
3. Ouyang M, Xi J, Bai W, Li K (2022) Band-area application container and artificial fish swarm algorithm for multi-objective optimization in internet-of-things cloud. IEEE Access 10:16408–16423. https://doi.org/10.1109/ACCESS.2022.3150326
4. Kumar J, Saxena D, Singh AK et al (2020) BiPhase adaptive learning based neural network model for cloud datacenter workload forecasting. Soft Comput 24:14593–14610. https://doi.org/10.1007/s00500-020-04808-9
5. Li Z, Jin H, Zou D, Yuan B (2020) Exploring new opportunities to defeat low-rate DDoS attack in container-based cloud environment. IEEE Trans Parall Distrib Syst 31(3):695–706. https://doi.org/10.1109/TPDS.2019.2942591
6. Lin M, Xi J, Bai W, Wu J (2019) Ant colony algorithm for multiobjective optimization of container-based microservice scheduling in cloud. IEEE Access 7:83088–83100. https://doi.org/10.1109/ACCESS.2019.2924414
7. Kaur M, Kadam S (2018) A novel multi-objective bacteria foraging optimization algorithm (MOBFOA) for multi-objective scheduling. Appl Soft Comput J 66:183–195
8. Kim T, Al-Tarazi M, Lin J-W, Choi W (2021) Optimal container migration for mobile edge computing: algorithm, system design and implementation. IEEE Access 9:158074–158090. https://doi.org/10.1109/ACCESS.2021.3131643
9. Garg SK, Toosi AN, Gopalaiyengar SK, Buyya R (2014) SLA-based virtual machine management for heterogeneous workloads in a cloud datacenter. J Netw Comput Appl 45:108–120
10. Adhikari M, Nandy S, Amgoth T (2019) Meta heuristic-based task deployment mechanism for load balancing in IaaS cloud. J Netw Comput Appl 128:64–77
11. Zhu L, Huang K, Hu Y, Tai X (2021) A self-adapting task scheduling algorithm for container cloud using learning automata. IEEE Access 9:81236–81252. https://doi.org/10.1109/ACCESS.2021.3078773
12. Xu R, Wang W, Gong X, Que X (2018) Delay-aware resource scheduling optimization in network function virtualization. J Comput Res Dev 55(4):738–747
13. Kong DJ (2017) Kubernetes resource scheduling strategy for 5G multi-access edge computing. Comput Eng 44(3):89–97

14. Shi XL, Xu K (2014) Utility maximization model of virtual machine scheduling in cloud environment. Chin J Comput 36(2):252–262
15. Alibaba Corp (2021) Alibaba Cluster Trace V2018. Accessed: 26 Sep 2021. (Online). Available: https://github.com/alibaba/clusterdata

An Optimized Search-Enabled Hotel Recommender System

Mudita Sandesara, Prithvi Sharan, Deepti Saraswat, and Rupal A. Kapdi

Abstract Nowadays, it is very challenging and confusing for the users to identify a well-suited hotel according to their requirements. Hotels are also increasing tremendously with a broader range of amenities. Many applications exist that recommend hotels based on various criteria such as location, rating, best offer and prices, amenities. In this paper, to minimize time and cost, we propose an optimized recommendation system using machine learning model based on multiple criterion and various filtering methods. We simulate the performance of different machine learning (ML) models and select the best model for hotel recommendation based on multiple performance metrics.

Keywords Recommender system · Hotel recommenders · Machine learning · Silhouette score · Dunn index · Davies Bouldin · MiniBatch · K-means · Birch · Hierarchical · Clustering

1 Introduction

The last decade has witnessed a boost in the tourism industry, hospitality and luxury sectors, considering the rise in hotel bookings. But as a result of COVID-19 pandemic, in recent times, the world has experienced a tough time, hampering the growth of tourism. A recent report by [1] indicates that the hotel sector has improved post-pandemic scenario and is expected to rise by a compound annual growth rate (CAGR)

M. Sandesara · P. Sharan · D. Saraswat (✉) · R. A. Kapdi
Department of Computer Science and Engineering, Institute of Technology, Nirma University, Ahmedabad, Gujarat, India
e-mail: deepti.saraswat@nirmauni.ac.in

M. Sandesara
e-mail: 18bce187@nirmauni.ac.in

P. Sharan
e-mail: 18bce207@nirmauni.ac.in

R. A. Kapdi
e-mail: rupal.kapdi@nirmauni.ac.in

Y. Singh et al. (eds.), *Proceedings of International Conference on Recent Innovations in Computing*, Lecture Notes in Electrical Engineering 1001,
https://doi.org/10.1007/978-981-19-9876-8_37

of 4.73% by 2026. Figure 1 shows the market trend of hospitality management up to the year 2027. As the number of tourists in any location increases, there is a proportionate increase in the demand for hotel stays. Today, various online hotel booking website exists wherein any user can look for their choice of hotel and book their stay online. These websites are a common decision-making place, containing ratings based on various aspects [17]. With the rise in the number of hotel booking websites, users have to search through numerous websites, compare them according to their preferences, and then book their stay. This is a very tedious and time consuming activity in today's modern world. Hotel management system wants their products to be unique and act in the desired manner [19], hotels being no exception. Hence, a user-focused approach is required for the said industry. The approach will enable an "open user model", a well-known method in user-adaptive systems to increase customization, transparency, and user-controllable [21]. This requires a one-stop solution for comparing hotels and websites as per user requirements (preferences) and recommending the list directly to users.

Recommender systems aim to actively propose products to a user by estimating the user's rating or preference for various products and showing the ones that the user prefers [17]. A recommender system gets access to data using Information Retrieval (IR) [2, 15, 18] takes into account two factors: ratings and features that affect aspects. It involves machine intelligence (supervised and unsupervised machine learning, deep learning, and reinforcement learning) to come up with relevant and useful recommendations. Being a numeric data, ratings have to be processed and interpreted based on mathematical models such as regression and dimensionality reduction. Features data needs to be cleaned and improvised based on Part-of-Speech

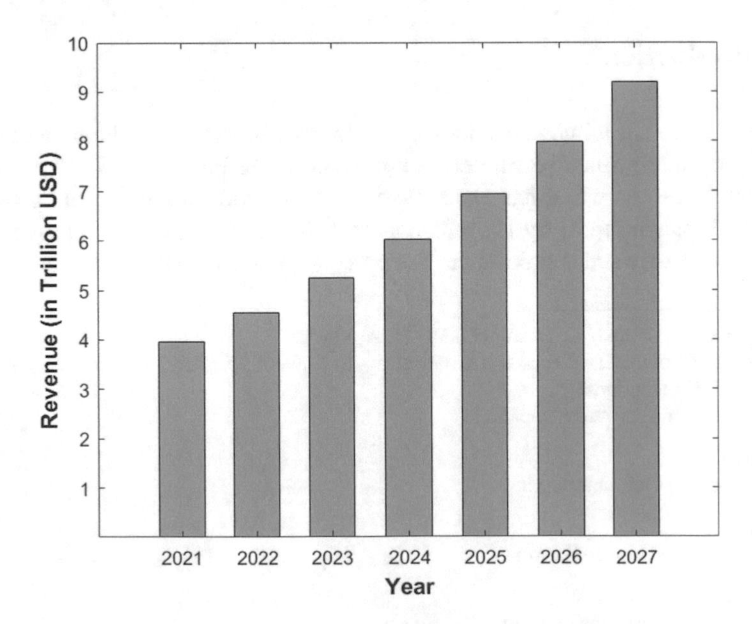

Fig. 1 Rise of market trend in hospitality management [1]

(POS) tagging [15] and other natural language processing (NLP) techniques. The text obtained can be interpreted based on models such as classification and clustering improving time efficiency [23]. One of the most acceptable uses of recommendation systems is recommending hotels to a traveler given a particular tourist spot [15]. Generally, there are three types of recommender system algorithms, namely, Collaborative Filtering (CF), Content-based Filtering (CBF), and Hybrid Filtering [6].

The paper aims to provide an easy-to-use, simple search-enabled hotel recommender system built with state-of-the-art techniques. The system will take user input such as location/city of hotel, amenities (free Wi-Fi connectivity, free parking), ratings, distance of hotel from main post office, and other criteria explicitly mentioned by the user. The system utilizes data of the top 3 pages according to user preference from TripAdvisor, Goibibo, OYO, Booking.com; the obtained data is processed, and finally the recommender engine gives the required recommendation list as the output. We consider various recommender algorithms and calculate the performance parametric like weighted average, Dunn index, and Silhouette score. Based on the score calculations, the best model is selected for the recommender system.

1.1 Recommender Systems

Table 1 presents the existing recommender engines utilized generally for hotel recommender system. Some of the other recommender engines are as follows:

- *Knowledge-based recommender systems*

It is based on feature grouping, preferences of user, and criteria for recommendation. Each time the user inputs a query, a relevant analysis is performed to recommend a combination of items or services, taking into account user specifications, item attributes, and domain knowledge [4]. The suggestions of content-based and collaborative systems [5] are generally based on historical data, whereas recommendations from knowledge-based systems are based on direct requests from users [4]. There are generally two types of knowledge-based recommender systems: Constraint-based and Case-based.

Table 1 Recommender engines

S. no	Filtering	Description
1	CF	Collaborative filtering [8, 23] utilizes data, social relations and proposals, and assessments with a similar background [11]. It calculates similarities between users or items based on collected user ratings for items in a given domain to reproduce relevant recommendations
2	CBF	Content-based filtering [5] utilizes apriori knowledge of content and user [11]. It solely considers the involved user's activity (history, rating, feedback, etc.)
3	Hybrid	It is a combination of CF and CBF. It is widely used in real-life applications

- *Candidate Generation Network*

It is an initial phase of recommendation where activity is considered to recommend a small subset from a large corpus. The mechanism generates a collection of candidates (matching the user preference). Candidate generation can be addressed using separate networks, RNNs, autoencoders, and hybrid methods [22]. Deep learning via candidate generation network can have particular applications by using DNN for YouTube, RNNs for question–answer (Q&A) systems, autoencoders for word embedding, and deep Key phrase generation.

2 Novelty of Work

The main contribution of the research work are as follows:

- Most of the current work focuses on the approach of direct scoring based on user ratings. However, we incorporate user preferences to provide the more customized result according to user requirements.
- We utilize clustering methodology which would provide robust performance compared to existing cosine similarity [6] generally used for the recommendation approach.
- We compare the different recommender engines and select the best based on performance score.

3 Literature Review

This section presents the existing state-of-the-art for hotel recommendation system. Authors in [15] describe a complex recommendation system based on the similarity between user item ratings and reviews. Using Naive Bayesian and k-nearest neighbor (KNN) classification algorithms, sentiment analysis is executed on hotel reviews. Uses ensemble models, using weighted vote-based classifier ensemble technique to extract user's opinions. The results of sentiment analysis can be used as feedback for recommendations. The authors in [23] utilize a combination of lexical, syntax, and semantic analysis for personalized hotel recommendations based on sentiment or hotel features and profiles of guests. In [21], the user input data is used to extract knowledge via an expert-curated list. A knowledge graph is then created to store semantic relationships among knowledge components, based on key-phrases. Based on the user's priority, a relevance score is calculated and ranking for recommendation is obtained. The authors in [9] proposes scoring based on similarity index and builds a useful Point Of Interest (POI) for the system. Authors in [27] utilizes hypercuboids that represents users to capture complex and diverse interests. They provide recommendations on the basis of relationship between user hypercuboids and items, provide the recommendation score through calculation of compositional distance between the user hypercuboid and the item and propose an LSTM-based neural architecture to facilitate user hypercuboid learning by capturing the activity sequences (like buying and rate) of the users. In [2], ensemble models are used to extract user opinions. Ensemble modeling is done using a weighted vote-based classifier ensemble approach. Data is collected using web crawlers. For sentiment analysis, text normalization and ensemble method approaches were used. The authors in [13] perform text mining on hotel reviews viz. data cleaning and dictionary-based NLP utilizing supervised algorithms like Probit, binomial logistic, and decision tree to predict explicit recommendations. In [11], an iteration formula is used to optimize the recommendation utility function throughout the adaption phase. Reference [16] utilizes the user location to build a hotel recommendation system based on a supervised link prediction method by incorporating a customer-hotel bipartite network. The authors in [10] propose a tourism recommendation system using an artificial bee colony (ABC) algorithm and a fuzzy TOPSIS model based on the preference type. They utilize an online database to acquire a dataset and apply the above algorithm to recommend the best tourist spot to users. The authors in [24] presents a survey on a recommender system based on Location-Based Social Networks (LBSN) that explores the recommendation techniques and information sources. The authors in [3] propose a scheme CAFOB, a context-aware tourism recommendation system that utilizes a fuzzy-enabled ontology and novel sentiment score scheme. The scheme also utilizes weather, location, and time parameters to enhance the accuracy. The figure of merits such as F-score, NDCG, and MRR results shows robust performance compared to state-of-the-art tourism recommendation systems. Table 2 presents the discussion of state-of-art approaches that are currently existing.

4 Proposed Methodology

This section explains the methodology proposed for recommendation process. The scheme is divided into various phases as explained in below subsections.

4.1 Data and Its Extraction

Out of the various online datasets available for hotels and their details present, certain features are missing or the data present is only restricted to a particular region only. To find a solution to this problem, we have implemented web scraping without the constraints of the regional area.

Web scraping and web information extraction gathers a large amount of data in an automated manner from an open source site. Web scratching utilizes principle instances to accommodate value checking, value knowledge, news observing, and statistical surveying. Individuals and organizations use web scraping to incorporate

Table 2 Comparison of existing state-of-the-art approaches

Author refs.	Year	Objective	User input	NLP	Scoring	Model(s)	Limitation(s)
[25]	2016	Recommendation system based on keyword extraction and ranking and then similarity calculation for recommendation	✓	✓	Contingency coefficient	–	–
[15]	2017	Parallel transformation of the user item ratings using hybrid collaborative filtering, item recommender and user reviews using sentiment analyzer	✓	✓	Weighted vote-based	Naive Bayesian, K-Nearest Neighbor (KNN)	Generic system with only basic classifiers
[20]	2018	Ranking-based clustering modeling post regression, dimensionality reduction for final recommendation output	×	✓	Sorting	Clustering	–
[23]	2019	An opinion-based sentiment analysis method for achieving hotel feature matrix via polarity identification	✓	✓	Polarity identification	Fuzzy logic	Dynamic automatic data updation missing
[13]	2020	Machine learning over cleaned data	×	✓	Probit algorithm, binomial logistic algorithm, CHAID, CART, and Random Forest algorithm	Limited to text-based systems	–

(continued)

Table 2 (continued)

Author refs.	Year	Objective	User input	NLP	Scoring	Model(s)	Limitation(s)
[12]	2020	Preprocessing and sentiment analysis-based recommendation based on weighted scoring on the web data	×	✓	Weighted scoring	–	–
[21]	2021	Demonstrates a customized framework with an open client model and knowledge graph to form user-controllable recommender system	✓	×	Cosine similarity	Knowledge graph	Basic model created, recommendations ranked solely based on relevance score
[27]	2021	Performs collaborative filtering using hypercuboids	×	×	Density-based	LSTM along with hypercuboid	Limitedscope and datasets considered
[2]	2021	Collaborative filtering recommender involving sentiment analysis and ensemble modeling	✓	✓	Weighted votes	Ensemble classifiers, KNN, linear regression, random forest, SVR	Utilizes basic classifiers without considering time factor
[26]	2021	Preprocessing and sentiment analysis-based analysis and recommendation on web extracted data	✓	✓	Sentiment and model scoring	–	–
Proposed	2022	Recommendation based on clustering and performance evaluation based on performance score	✓	✓	Weighted average, Boolean-based TF-IDF matrix	MiniBatch K-means, Hierarchical, DBSCAN, BIRCH, clustering	–

intelligent data handling in the end-to-end ML process. There are two parts of web scraping.

1. *Crawler:* It is also known as "spider". It is an AI-based algorithm that extracts the web data through the creation of linkage connection and investigation of searched information.
2. *Scraper:* It is a specific tool that precisely and immediately removes information from a page. Each piece of scrubber is represented as data finders or selectors and utilized to extract information from the HTML record. Widely used scraper tools are XPath, CSS selectors, regex, JSON, etc.

Various online libraries are available for web scraping, viz., requests, lxml, beautifulsoup, selenium, scrapy, and many more. We have explored the implementation of recommendations, scrapy, and beautifulsoup for the data extraction. Taking into consideration the on-time performance constraints, and available CPU power, we propose scrapy, a full-fledged solution as the web scraper. Scrapy provides advantages like low CPU and memory usage, well-designed architecture, and provides the flexibility of adding various plugins.

Based on the location [6] (country/state/city) entered by the user, the data of hotels in the particular area will be fetched from four websites: TripAdvisor, booking, goibibo, and Oyo. The required fields are queried using scrapy spider and response scraper. From each website, three pages are scraped, assuming that each website displays the most suggested hotels for the entered location.

4.2 Preprocessing and Feature Extraction

The data extracted via web scraping is stored in a.csv file. The data is cleaned and preprocessed using Pandas Python package, which provides fast, flexible, and expressive data structures language by performing below tasks.

- Performs information retrieval system methods and natural language processing (NLP) [9] on text fields, viz., stopword removal, tokenization, stemming, lemmatization, etc.
- Performs mathematical computations on numeric fields, viz., weighted rating (a weighted average based on ratings and number of ratings obtained from each website) and normalization scaling [14].
- Calculates new fields based on extracted data, i.e., distance of the hotel (based on its extracted address) from the main city.

4.3 Machine Learning (ML)

The queried data has attributes, viz., Hotel Name, Weighted Rating, Total number of ratings, Distance from main city, Amenities, Amenities' preferences added by the user (Boolean matrix's values), and Booking Link from extracted website. Here, the obtained data is multidimensional unlabeled data. Hence, it is required to find data points with similar feature coordinates and are closest to the ideal data point as per the entered user preferences. The user is also provided the leverage to choose the weightage to the ratings and amenities offered by the hotel. Also, the ranking of the hotel recommendations can be done according to the distance from the main city so that the commute time for a traveler in the place can be reduced. For a better understanding, computations and visualization, data reduction has to be performed. There are various methods of performing data reduction for better understanding, computations and visualization: Principal Component Analysis (PCA) and Singular Value Decomposition [19]. We have selected PCA for our analysis. PCA is a widely used linear algebraic approach that automatically perform dimensionality reduction. PCA is a projection-based transformation approach that changes data by projecting it onto a collection of orthogonal axes.

The collected data is classified using machine learning (ML) model using clustering approach [19] to find logical similarities and rankings based on the distances. A cluster is a collection of data points that have been grouped together due to particular similarities. Clustering alleviates traveler pain-points. It also elevates traveler's satisfaction, based on numerous hotel attributes and criteria [7]. Various clustering models used in the scheme are K-means, MiniBatch K-means, Hierarchical Clustering, DBSCAN, and BIRCH.

- *K-means Clustering*

It is a centroid-based and one of the widely utilized unsupervised machine learning techniques. It splits an unlabeled dataset into *k* clusters; each dataset belongs to one group with similar qualities. The process is repeated in an iterative manner. The technique reduces the sum of distances between data points and the clusters to which they belong.

- *MiniBatch K-means Clustering*

The MiniBatch K-means is a variation of the K-means which utilizes small scale clusters to lessen the calculation time. Smaller than usual bunches are subsets of the information, haphazardly examined in each preparation cycle. These smaller than usual bunches radically diminish the measure of calculation needed to merge to a neighborhood arrangement. As opposed to different estimates that decrease the union season of k-means, small cluster k-implies produce results that are by and large just somewhat more terrible than the standard calculation.

- *Hierarchical Clustering*

Hierarchical clustering forms nested clusters by combining or parting them progressively. This pecking order of groups is addressed as a tree or dendrogram (a visual portrayal of the compound connection information. The singular mixtures are organized along the lower part of the dendrogram and alluded to as leaf hubs). The base of the tree is the unique bunch that assembles every one of the samples, the leaves being the clusters with just one sample. The agglomerative clustering object

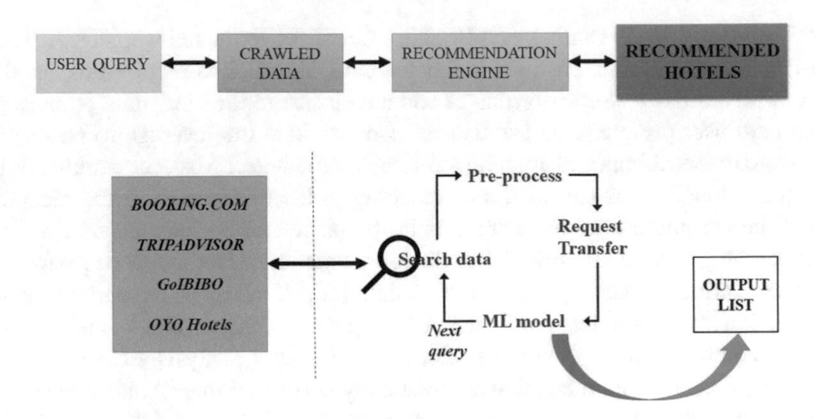

Fig. 2 Flow diagram of the proposed scheme

of sklearn plays out a progressive bunching utilizing a granular perspective: every perception begins in its group, and bunches are progressively combined.

- *Density-Based Spatial Clustering of Applications with Noise (DBSCAN)*

The technique combines data points that are densely packed into a single cluster. Through examination of the local density of data points, the cluster can be spotted in a huge geographical dataset. The technique does not require prior knowledge of number of sets and is outlier resistant. It is sensitive to the parameter ϵ, the value of which is determined from the K-distance graph.

- *Balanced Iterative Reducing and Clustering Hierarchies (BIRCH)*

The technique generates a clustering feature tree (CFT) for the given dataset. A set of clustering feature (CF) nodes can be created utilizing lossy compression of the actual data which is additionally processed by a global cluster. Since the input dataset is reduced to a subcluster, the approach is also called an instance or data reduction method. BIRCH does not scale very well to high-dimensional data; hence PCA-passed data helps in this case.

Figure 2 shows the complete flow of the proposed methodology, and Table 3 shows the detailed description and comparison of various methods.

5 Experiment and Simulation Results

This section discusses the experiments conducted and simulation results. As mentioned earlier, we deal with the clustering approach for the recommendation. Various evaluation measures are used for clustering unlabeled data viz., Dunn's index, Silhouette coefficient/score, Calinski-Harabaz index, Davies Bouldin score, and many more. These measures are beneficial for model evaluation in absence of validation methods for clustering using quantities and characteristics inherent to the dataset. The proposed methodology aims at evaluation of the Silhouette score, as it provides robust performance for calculating the goodness to all clusters. Figure 4 represents the comparison of various measures for all the clustering algorithms.

We compare the performance of four clustering algorithms such as MiniBatch K-means, Hierarchical, DBSCAN, and BIRCH. Table 4 mentions the configuration used for building the recommender system. Figure 3a depicts the variation of the Silhouette score and DB score according to cluster size. For DB score, at initial values of the cluster, the score value decreases after a certain cluster size, score

Table 3 Comparison of various measurement scores

Measure	Value range	Value meaning	Optimal value	Advantages	Disadvantages
Silhouette score	[−1, 1]	‘1’ indicates clusters are well separated from one another and recognized: best grouping, ‘0’ indicates impassive clusters implies that the distance between groups is not critical, ‘ 1’ indicates clusters are allotted in the incorrect way	Maximized	Abides by standards concepts of clustering	Generally higher for convex clusters, high-computational complexity
Dunn Index (DI)	(0, ∞)	The number of groups that expands Dunn Index is taken as the ideal number of clusters k	Maximized	Provides a comprehensive model for validating cloud-like clustered partitions	Considerable time Complexity, sensitive to the presence of noise in datasets
Davies Bouldin Index (DBI)	(0, 1)	Regularly used to assess the split quality utilizing k-means clustering calculation for a given number of clusters	Minimized	Simpler computation than Silhouette score	Limited to Euclidean space, generally higher for convex clusters

(continued)

Table 3 (continued)

Measure	Value range	Value meaning	Optimal value	Advantages	Disadvantages
Calinski Harabasz index (CH)	(0, ∞)	CH indices on line plot to pick solution that produces a peak or, at the very least, an abrupt elbow. If the line is smooth, however, there is no reason to choose one option over another	Maximized	Fast to compute and abides by standards concepts of clustering	Generally higher for convex clusters

Table 4 System specifications

S. no	Feature	Specification
1	Processor used	11th Gen Intel(R) Core(TM) i5-1135G7 @ 2.40 GHz 1.38 GHz
2	RAM	8.00 GB
3	Memory	512 GB

is almost constant. In the case of Silhouette method, the value is initially constant and starts falling. In case of MiniBatch K-means method, the Silhouette score lies between 0.7 and 1.2 and DB score lies between 0.5 and 0.7. It shows that Silhouette score is higher than DB score. In Fig. 3b, Silhouette score lies between 0.7 and 1.1 and DB scan lies between 0.5 and 0.7 which indicates that Silhouette score is greater than DB scan score. The variation of the Silhouette and DB scan core varies similarly except for the range values, however, Silhouette score is greater than the DB score in DBSCAN. In the case of BIRCH, Fig. 3c, we see no variation in both scores. We prefer Silhouette score as a measuring factor for selecting the best clustering methodology for hotel recommendation. Figure 4 depicts that among all the four methodologies, MiniBatch K-means provides the best result for our search-enabled hotel recommender system.

6 Conclusion

The paper looks at various aspects like web scraping from various websites in real-time. Then, we worked on the preprocessing and feature extraction part of the recommendation system and discussed ML models and clustering algorithms to analyze

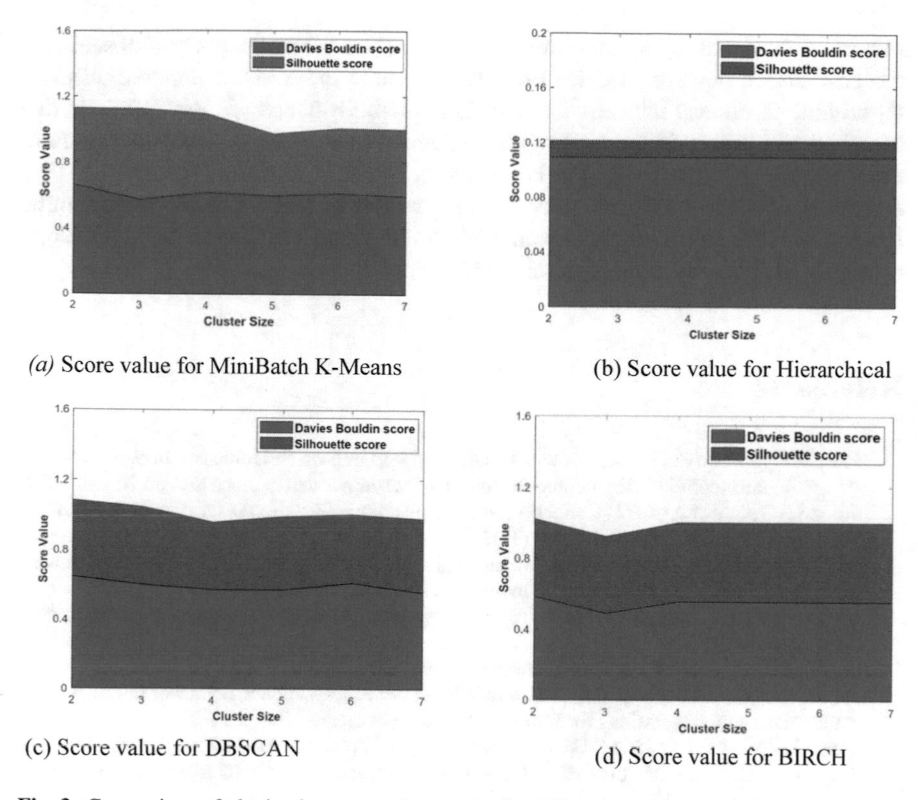

(a) Score value for MiniBatch K-Means

(b) Score value for Hierarchical

(c) Score value for DBSCAN

(d) Score value for BIRCH

Fig. 3 Comparison of obtained score vs. cluster size for different models

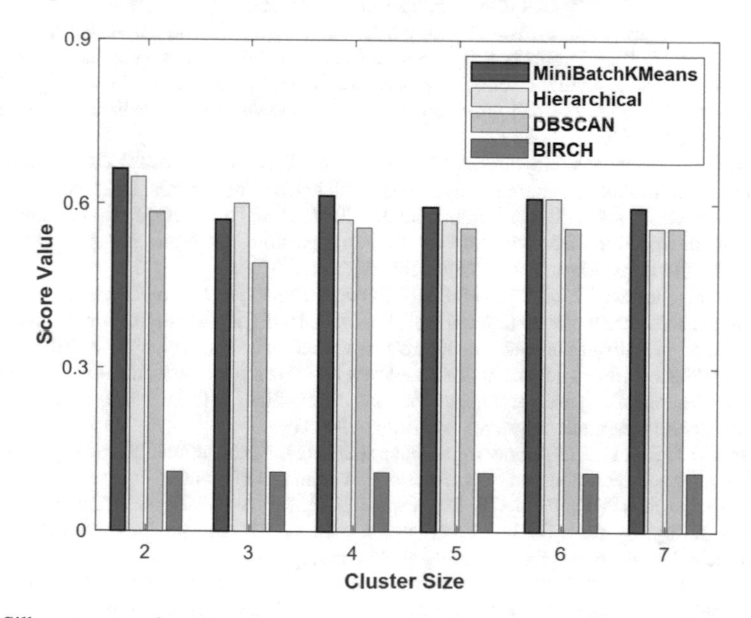

Fig. 4 Silhouette score for all models

recommender systems. We have used the Silhouette score as a major part of selecting the best model; however, Davies Bouldin and Dunn Index also compared different algorithms to choose the best from various methods analyzed. We conclude that MiniBatch K-means is the best model for recommendation of hotels that recommends based on distance of the hotel points from the center of the clusters. We have developed the complete recommendation system using Python programming language. In the future, authors would like to augment the proposed methodology through evaluation of more research parameters.

References

1. Hospitality industry in India—growth, trends, covid-19 impact, and forecasts. https://www.mordorintelligence.com/industry-reports/hospitality-industry-in-india. Accessed on 20 Feb 2022
2. Abbasi F, Khadivar A (2021) Collaborative filtering recommendation system through sentiment analysis. Turk J Comput Math Educ (TURCOMAT) 12(14):1843–1853
3. Abbasi-Moud Z, Hosseinabadi S, Kelarestaghi M, Eshghi F (2022) Cafob: context-aware fuzzy-ontology-based tourism recommendation system. Expert Syst Appl 199:116877. https://doi.org/10.1016/j.eswa.2022.116877. https://www.sciencedirect.com/science/article/pii/S0957417422003232
4. Aggarwal CC (2016) Knowledge-based recommender systems
5. Alyari F, Navimipour NJ (2018) Recommender systems: a systematic review of the state of the art literature and suggestions for future research. Kybernetes 47:985–1017
6. Chen CL, Wang CS, Chiang DJ (2019) Location-based hotel recommendation system. In: Chen JL, Pang AC, Deng DJ, Lin CC (eds) Wireless internet. Springer International Publishing, Cham, pp 225–234
7. Chen T (2020) A fuzzy ubiquitous traveler clustering and hotel recommendation system by differentiating travelers' decision-making behaviors. Appl Soft Comput 96:106585
8. Eirinaki M, Gao J, Varlamis I, Tserpes K (2018) Recommender systems for large-scale social networks: a review of challenges and solutions. Future Gener Comput Syst 78:413–418. https://doi.org/10.1016/j.future.2017.09.015. https://www.sciencedirect.com/science/article/pii/S0167739X17319684
9. Forhad MS, Arefin MS, Kayes AS, Ahmed K, Chowdhury MJ, Kumara I (2021) An effective hotel recommendation system through processing heterogeneous data. Electronics
10. Forouzandeh S, Rostami M, Berahmand K (2022) A hybrid method for recommendation systems based on tourism with an evolutionary algorithm and topsis model. Fuzzy Inf Eng 14(1):26–50. https://doi.org/10.1080/16168658.2021.2019430
11. Gluhih IN, Karyakin IY, Sizova LV (2016) Recommender system providing recommendations for unidentified users of a commerial website. In: 2016 IEEE 10th international conference on application of information and communication technologies (AICT). pp 1–3. IEEE
12. Godakandage M, Thelijjagoda S (2020) Aspect based sentiment oriented hotel recommendation model exploiting user preference learning. 2020 IEEE 15th international conference on industrial and information systems (ICIIS) pp 409–414
13. Guerreiro J, Rita P (2020) How to predict explicit recommendations in online reviews using text mining and sentiment analysis. J Hosp Tour Manag 43:269–272
14. Ifada N, Sophan MK, Putri NF, Setyawan GE (2021) A user-based normalization multi-criteria rating approach for hotel recommendation system. In: 6th International conference on sustainable information engineering and technology 2021
15. Jayashree R, Kulkarni D (2017) Recommendation system with sentiment analysis as feedback component. In: Proceedings of sixth international conference on soft computing for problem solving. pp 359–367. Springer

16. Kaya B (2019) A hotel recommendation system based on customer location: a link prediction approach. Multimedia Tools Appl 79:1745–1758
17. Khaleghi R, Cannon K, Srinivas RS (2018) A comparative evaluation of recommender systems for hotel reviews
18. Levi A, Mokryn O, Diot C, Taft N (2012) Finding a needle in a haystack of reviews: cold start context-based hotel recommender system. In: Proceedings of the sixth ACM conference on Recommender systems. pp 115–122
19. Mavalankar A, Gupta A, Gandotra C, Misra R (2019) Hotel recommendation system. https://doi.org/10.13140/RG.2.2.27394.22728/1
20. Nawrocka A, Kot A, Nawrocki M (2018) Application of machine learning in recommendation systems. In: 2018 19th International carpathian control conference (ICCC), pp 328–331
21. Rahdari B, Brusilovsky P, Javadian Sabet A (2021) Connecting students with research advisors through user-controlled recommendation. In: Fifteenth ACM conference on recommender systems, pp 745–748
22. Rama K, Kumar P, Bhasker B (2019) Deep learning to address candidate generation and cold start challenges in recommender systems: a research survey. ArXiv abs/1907.08674
23. Ramzan B, Bajwa IS, Jamil N, Amin RU, Ramzan S, Mirza F, Sarwar N (2019) An intelligent data analysis for recommendation systems using machine learning. Sci Program 2019:5941096. https://doi.org/10.1155/2019/5941096
24. S´anchez P, Bellog´ın A (2022) Point-of-interest recommender systems based on location-based social networks: a survey from an experimental perspective. ACM Comput Surv https://doi.org/10.1145/3510409. Just accepted
25. Shrote KR, Deorankar AV (2016) Review based service recommendation for big data. In: 2016 2nd International conference on advances in electrical, electronics, information, communication and bio-informatics (AEEICB), pp 470–474
26. Yu J, Shi J, Chen Y, Ji D, Liu W, Xie Z, Liu K, Feng X (2021) Collaborative filtering recommendation with fluctuations of user' preference. In: 2021 IEEE international conference on information communication and software engineering (ICICSE), pp 222–226
27. Zhang S, Liu H, Zhang A, Hu Y, Zhang C, Li Y, Zhu T, He S, Ou W (2021) Learning user representations with hypercuboids for recommender systems. In: Proceedings of the 14th ACM international conference on web search and data mining, pp 716–724

An Approach for Integrating a Prominent Conversational Chatbot with a Contact Center

Yatish Bathla and Neerendra Kumar

Abstract A conversational chatbot has plenty of advantages and the capability to grow a business. Telecommunication software like contact center has been increasingly collaborating with conversational chatbots. It uses artificial intelligence that helps increase customer satisfaction scores and reduce the customer waiting time and effort to request the service. Depending on the use cases and requirements of a business, a prominent chatbot with a certain set of features has been integrated into a contact center. The integration problem arises when use cases have been changed, and there is a need for a prominent chatbot that satisfies the specific set of requirements. Therefore, this research work proposes a Bot Bridge which integrates a prominent chatbot into a contact center based on the requirement of a business. It can resolve the integration issues of multiple conversational AI chatbots interacting with telecommunication systems.

Keywords Chatbot · Contact center · Artificial intelligence · Cloud computing · Google DialogFlow · Microsoft Azure Bot Service · Amazon Lex · IBM Watson Assistant

1 Introduction

For over a decade, companies purchase their call center solutions and add new software to expand their capabilities. This method has limitations for several reasons, including isolation, suffering usability issues without a unified, connected solution, data sharing, and collaboration. This leads to their effectiveness and speed, affecting customer service and satisfaction [1]. Therefore, the contact center era has started

Y. Bathla
Doctoral School of Applied Informatics and Applied Mathematics, Óbuda University, Budapest, Hungary

N. Kumar (✉)
Department of Computer Science and IT, Central University of Jammu, Jammu, India
e-mail: neerendra.csit@cujammu.ac.in

Y. Singh et al. (eds.), *Proceedings of International Conference on Recent Innovations in Computing*, Lecture Notes in Electrical Engineering 1001,
https://doi.org/10.1007/978-981-19-9876-8_38

where the customer journey extends the capability from telephone to Web technologies like chat, email, messaging apps, social media, text, fax, traditional mail, etc. It offers improved internal collaboration, customer journey management, and omnichannel capabilities. Then, there is a need for conversational artificial intelligence (AI) chatbot in the contact center due to excessive and repetitive task-related workload sharing for the human agent, 24×7 customer support, changing customer demands, reducing the average handling time (AHT), and increasing the customer satisfaction (CSAT). Large technology giant companies are investing in the development of smart conversational AI chatbots like Google DialogFlow [2], Microsoft Azure Bot Service [3], Amazon Lex [4], IBM Watson Assistant [5], etc. There is always tough competition between conversational AI chatbots about the offered features and services.

Then, the contact center software needs a dedicated integration strategy and security constraints with a conversational AI chatbot. But, business needs are always changing, and chatbots are chosen as per the requirements. It requires a different integration strategy and security constraints accordingly. As a result, it increases the handling cost of the contact center. Therefore, we have proposed a Bot Bridge between the contact center and conversational AI chatbots. It is the bridge that provides a standard interface between the conversational AI chatbots and contact centers deployed by the different vendors. This paper starts with the preliminary work of conversational chatbot, contact center, and artificial intelligence (AI) concepts in the contact center. Then, conversational chatbot integration with contact center AI is explained, and integration issues with the existing architecture are emphasized. After that, the proposed Bot Bridge, its concepts, integration with the contact center, and conversational AI are explained. Finally, the conclusion is discussed.

2 Preliminary Work

2.1 Conversational Chatbot

A chatbot is a software application that communicates with a human in text or spoken format. A conversational chatbot is also called conversational AI. It falls under the category of AI assistants. These assistants are full systems that use full conversational dialogue to accomplish one or more tasks [6]. Chatbots are frequently deployed in various domains like social media networks, healthcare systems, telecommunication systems, etc. The reason for its popularity is 24/7 customer support, no waiting time, providing support for frequently asked questions (FAQs) and knowledge base (KB) articles, lower cost, and boosted performance. Chatbots process their scripts by performing cognitive computations which initiate the identification of behavioral patterns and analyze data. The authors have considered the Google DialogFlow conversational chatbot [7] for explaining the research work.

2.2 *Contact Center*

A contact center is a business department within an organization that manages customer interactions. It handles inbound and outbound customer communication over multiple channels such as telephone, Web, chat, email, messaging apps, social media, text, fax, and traditional mail. It uses different types of advanced technology to help resolve customer problems faster, track customer engagement, and capture interaction and performance data [8]. Contact centers depend on multiple applications to accomplish their tasks like automatic contact distribution (ACD) application, interactive voice response (IVR) application, customer relationship management (CRM) applications, dashboard applications, etc. [9]. In the cloud technology era, contact center solutions are also being deployed from on-premise to the cloud. The key market leaders in the contact center solutions are Avaya [10], Genesys [11], Nice [9], and Five [12].

2.2.1 Contact Center Artificial Intelligence (CCAI)

Artificial intelligence (AI) is an evolving technological infrastructure whether smarter conversational chatbots or deployment of a successful omnichannel strategy or intelligent multichannel concepts in the contact center. The ideal conversational AI chatbot can interact with a customer in such a way that they are practically indistinguishable from a contact center human agent [13]. The greatest advantage of using artificial intelligence (AI) in the contact center is the cost reduction by increasing self-service and improving the customer experience. Moreover, human agents' workloads are also reduced as they can receive a larger percentage of calls that are not being resolved. There are two paths for contact centers and AI are explained in the article [14]:

- Use AI to assist agents

When a human agent provides support to a customer during the customer interaction, AI can run parallel and do some of the complex liftings of searching a knowledge base for answers.

- Use AI to run analytics in the background

Predictive contact routing analyzes the key pieces of information about the customer such as the customer's financial background, customer interest in the product types, and customer sales history. All the gathered information provides predictive data for its future behavior. It helps the human agent route calls to the business department with the best set of skills to successfully work with the customer and increase customer satisfaction.

3 Conversational Chatbot Integration with Contact Center AI

Contact center uses artificial intelligence (AI) to be integrated with conversational chatbot, for example, contact center giant companies like Avaya and Genesys have collaborated with Google Contact Center AI (CCAI) for enabling conversational chatbot [15, 16]. Figure 1 gives the architecture diagram of Google Conversational Chatbot to a contact center.

The elements of contact center AI (CCAI) are as follows:

- **Voice Portal**: The voiced equivalent of Web portals gives access to information through spoken commands and voice responses. It may be referred to as interactive voice response (IVR) systems, but also includes dual-tone multi-frequency (DTMF) services [17]. It supports the single point of orchestration of all automated voice and multimedia applications and services, for example, Avaya Experience Portal (AEP) [18].

- **AI (Artificial Intelligence) Voice Platform**: It interacts with the voice portal for voice automation installed on-premise and dedicated software for conversational AI chatbots. It handles customer requests via PSTN [19] to interact with the conversational AI chatbot. Then, the voice portal calls the REST API endpoints that are Voice Extensible Markup Language (VoiceXML) applications [20]. The VoiceXML applications call AI voice platform for escalation. The VoiceXML applications call AI voice platform for back-end service. It calls the adaptor to decide about the right vector directory number (VDN) for escalation.
- **AI (Artificial Intelligence) Configuration Platform**: It interacts with the AI voice platform and widgets via the widget configuration platform. Here, widget calls and provides REST endpoints. Then, the internal mechanism stores the conversational AI chatbot configuration and mapping information.
- **Contact Center**: It interacts with the customer via HTTPS and provides the necessary configuration to the widget workspace to handle the customer request. Further, the contact center has provided workspaces to agents, supervisors, and administrators to handle customer interactions. It could be a call, email, Web messenger, or Bring Your Own Channel (BYOC). Workspaces contain the widgets where most of the artificial intelligence (AI) tasks have been retrieved, for example, widget for converting the call into the transcript, customer sentiment analysis based on the conversation analysis, and the best possible article from the knowledge base (KB).
- **Dedicated Integration Software**: This software is coded to integrate the contact center with the conversation AI chatbot. It acts as a pipeline between both platforms. This is dedicated because AI components of conversational chatbots are not always compatible with the AI components of the contact center. Moreover, security is the other important factor from both ends. Therefore, the proper flow of information is done after the compatibility and security agreements.

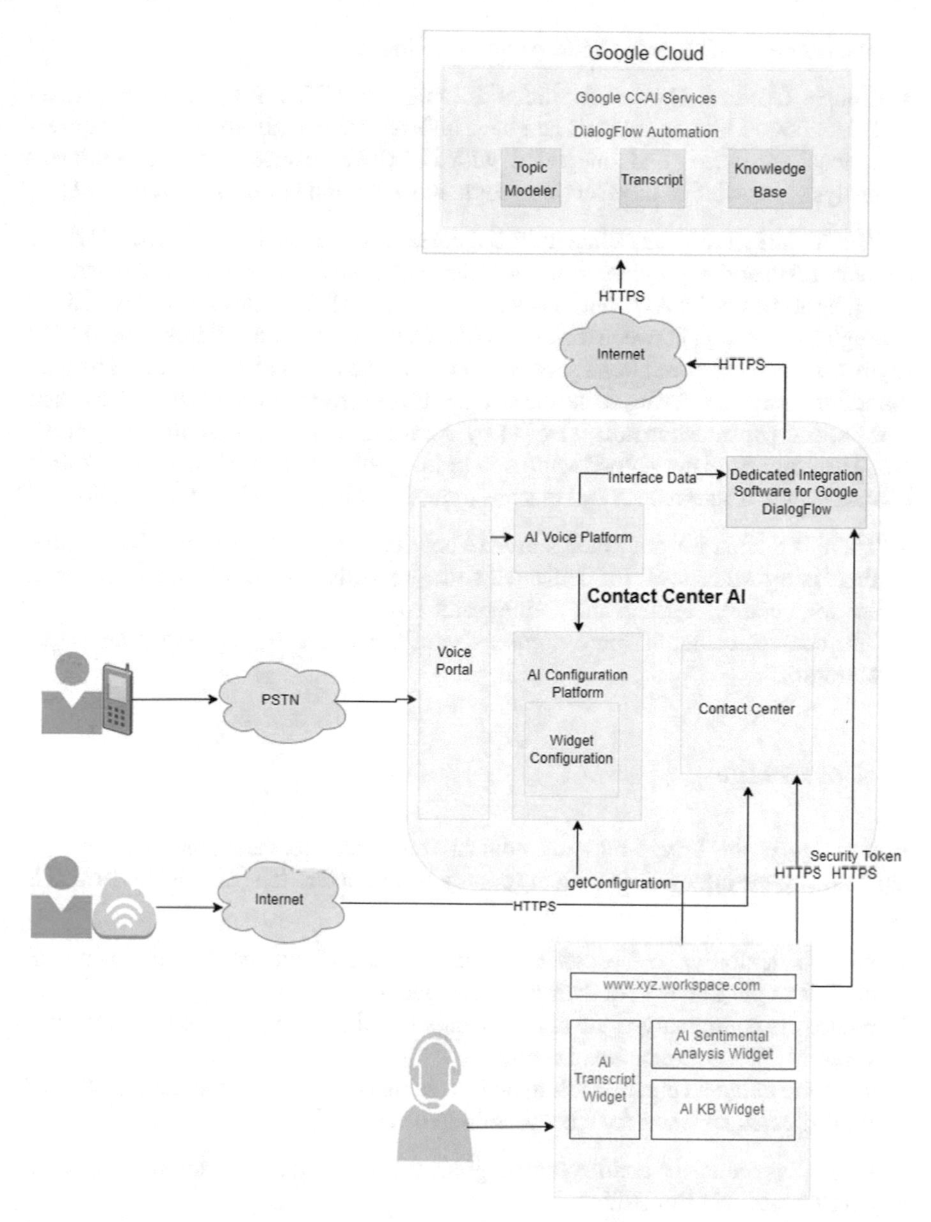

Fig. 1 Architecture diagram of Google Conversational Chatbot to a contact center. *Note* This image is the own work of the authors. It has been generated based on the concepts discussed in [7, 15–17]

The elements of Google Cloud AI are as follows:

- **Google Contact Center Artificial Intelligence (CCAI)**: Google has offered Dialogflow, which is a natural language understanding platform to design conversational interfaces and integrated with cognitive services such as sentiment analysis, knowledge base services, topic modeler, and voice to transcript [21].

This architecture works efficiently if a contact center has been integrated with a dedicated AI service provider. But, the requirements and needs of the customers are changing abruptly. An AI service provider can't meet all the criteria. Moreover, there is tough competition between tech firms like Google, Amazon, Microsoft, and IBM to provide the best possible AI services to satisfy the needs of the market. The tech giant firms that are offering AI services in terms of conversational chatbots are Azure Bot Service [3] by Microsoft, Lex [4] by Amazon, Watson Assistant [5] by IBM, etc. Therefore, it is not a good approach to integrate a dedicated AI service with a contact center. There are several reasons to justify this point which are as follows:

- There is a need for dedicated software and its integration. For each integration, there is a further need for dedicated software in the contact center. It results in software coding, testing, and maintenance cost.
- The contact center is more complex with the integration of every dedicated software.

4 Bot Bridge

Bot Bridge is the bridge between contact center and conversational chatbots by generalizing the interactions with the common parameters. The aim of Bot Bridge is as follows:

- provide flexibility to the contact center to use and change the conversational chatbot as per the need of their business model
- reduce the complexity of the contact center by eliminating the need for dedicated software for a conversational chatbot
- saves the contact center handling and maintenance costs by eliminating the need for dedicated software for a conversational chatbot

Figure 2 presents the architecture diagram of a standard conversational chatbot to a contact center via Bot Bridge.

4.1 Implementation of Bot Bridge with Cloud Contact Center

The similarity between the conversational chatbots is they are using the project id, conversational id, tenant id, and interaction id information while interacting with the contact center. Considering this point, the authors have proposed the Bot Bridge in

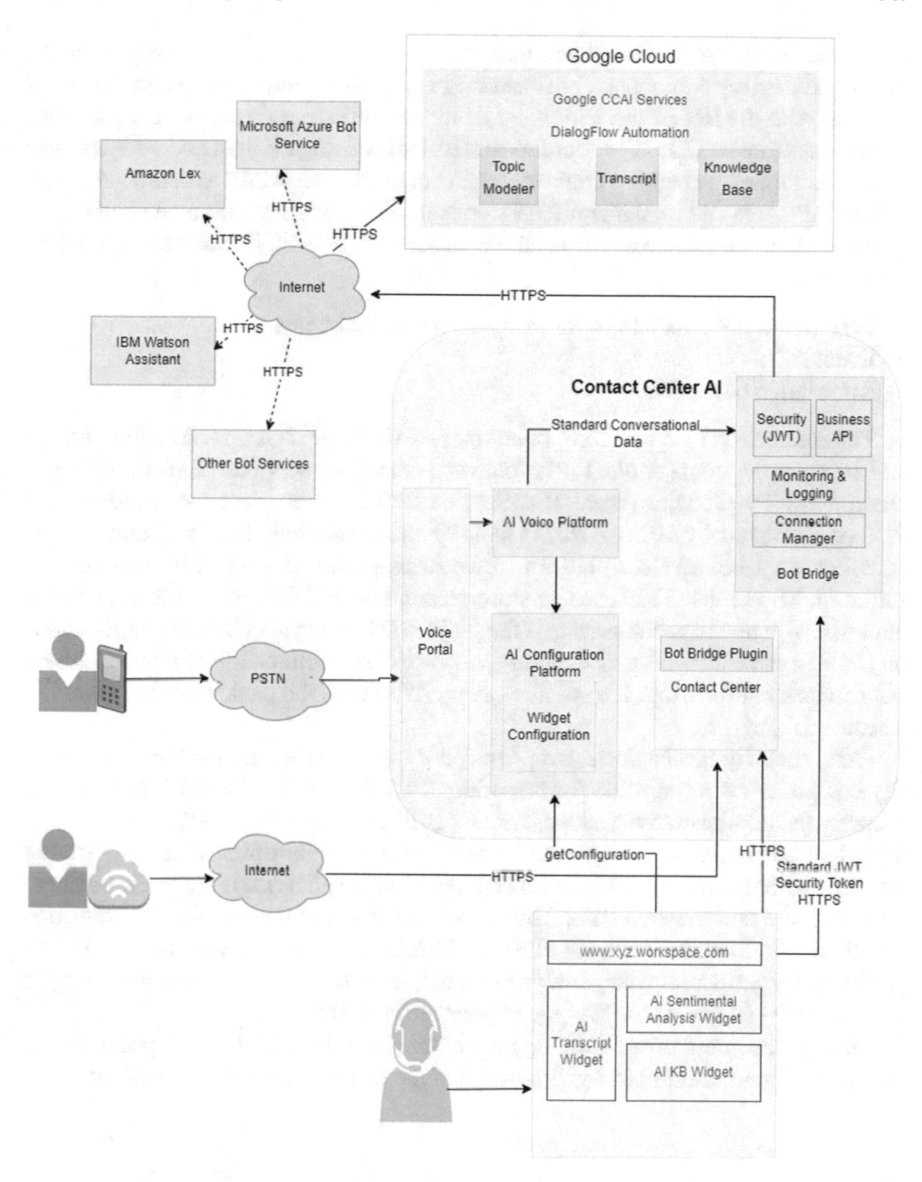

Fig. 2 Architecture diagram of a standard conversational chatbot to a contact center via Bot Bridge. *Note* This image is the own work of the authors. It has been generated based on the proposed Bot Bridge concepts

the contact center AI model. It contains the connection manager component, monitoring and logging component, business API component, and security component. It interacts with the Bot Bridge Plugin and widgets workspace. It provides representational state transfer (REST) endpoints and calls Bot Bridge Plugin and widgets. The connection manager handles connections to contact center AI for assisting an agent. All the APIs use and implement REST calls to APIs for conversational chatbots by Google, Microsoft, Amazon, or IBM. From the offered API, it analyzes the content for example:

- Conversation APIs that use transcripts, chats, and intents
- Event APIs
- Knowledge base APIs

The connection manager component plays a vital role. It is an automation bridge and handles the connections to the conversational bot services platform example DialogFlow by Google, Azure Bot Service by Microsoft, Lex by Amazon, and Watson Assistant by IBM. Also, it handles the credentials for the cloud and its conversational bot services platform. It receives tenant id, project id, and conversation id. The projectID is used to store credentials and configures the connection manager. You need a cache service if the projectID is not available to the applications on the first call. Internally, it handles the proper connections to the conversational bot services platform that is a session client. After that, the credentials are stored in a secure folder.

Here, security mechanisms are provided by the security component. The security component is responsible for providing the JSON Web Token (JWT) to handle securely the administrative tasks of the Bot Bridge. The business APIs interact with the AI voice platform and conversational bot services platforms. When a customer interacts with the voice portal via PSTN, the AI voice platform is using the business API to create conversation data. This conversation data interact with the conversational bot services platforms via HTTPS. Similarly, business APIs are used to add participants, get participants, analyze content, trigger event, get messages, suggest frequently asked questions (FAQs), suggest articles, etc.

Finally, the monitoring and logging component is used for the performance evaluation by monitoring and getting logs of the Bot Bridge and its interaction.

5 Conclusion

This research work proposes Bot Bridge which is an approach to integrate a prominent conversational chatbot with a contact center. It starts with the preliminary work of the contact centers and the conversational chatbots. Then, it focuses on the integration issues of multiple conversational chatbots to a contact center as each integration requires dedicated software. As a result, concepts of the Bot Bridge have been proposed in the contact center AI. It resolves the integration issue by using the common integration parameters from the multiple chatbots like integration Id, tenant id, project id, and conversation id. Also, it reduces the contact center software complexity as well as the overall cost when multiple conversational chatbots have been integrated. It is important to note that the Bot Bridge is on the conceptual level and needs to be tested for verification.

References

1. RingCentral, Inc (2022). https://www.ringcentral.com/us/en/blog/4-benefits-of-an-integrated-contact-center-solution/
2. Google Inc (2022). https://cloud.google.com/dialogflow/docs
3. Microsoft Inc (2022). https://azure.microsoft.com/en-us/services/bot-services/#documentation
4. Amazon Web Services Inc (2022). https://aws.amazon.com/lex/
5. IBM Inc (2022). https://www.ibm.com/products/watson-assistant
6. Freed AR (2021) Conversational AI chatbots that work. Manning Publications, pp 1–320
7. Boonstra L (2021) Dialogflow essentials concepts. In: The definitive guide to conversational AI with dialogflow and google cloud: build advanced enterprise chatbots, voice, and telephony agents on google cloud, pp 59–91. Apress
8. Amazon Web Services (2022). https://aws.amazon.com/what-is-a-contact-center/
9. Nice Inc (2022). https://www.nice.com/glossary/what-is-contact-center-software-application
10. Avaya Inc (2022). https://support.avaya.com/products/P0793/avaya-aura-contact-center
11. Genesys Inc (2022). https://www.genesys.com/capabilities
12. Five Inc (2022). https://www.five9.com/
13. Ruslan Makrushin: contact-centers and artificial intelligence, pp 1–5 (2021). https://doi.org/10.32370/IA_2021_12_26
14. Gerson Lehrman Group (2022). https://glginsights.com/articles/ai-impacts-the-contact-center/
15. Avaya & Google Cloud Contact Center AI (2022). https://www.avaya.com/en/avaya-google-cc-ai/
16. Orchestrate better experiences with genesys and google cloud (2022). https://www.genesys.com/googlecloud
17. Avaya Inc (2022). https://www.avaya.com/en/avaya-google-cc-ai/
18. Wikipedia Foundation (2022). https://en.wikipedia.org/wiki/Voice_portal
19. Bates J, Gallon C, Bocci M, Walker S, Taylor T (2006) The NGN and the PSTN. In: Converged multimedia networks. Wiley. https://doi.org/10.1002/9780470035177.ch4
20. Dunn MD (2007) Creating VoiceXML applications, pro microsoft speech server 2007: developing speech enabled applications with .NET, pp 129–154, Apress. https://doi.org/10.1007/978-1-4302-0272-1_5
21. Sabharwal N, Agrawal A (2020) Introduction to google dialogflow. In: Cognitive virtual assistants using google dialogflow, pp 13–54. Apress. https://doi.org/10.1007/978-1-4842-5741-8_2

IPFS: An Off-Chain Storage Solution for Blockchain

Manpreet Kaur, Shikha Gupta, Deepak Kumar, Maria Simona Raboaca, S. B. Goyal, and Chaman Verma

Abstract Due to intrinsic properties such as transparency, immutability, decentralization, and cryptographic security, blockchain has emerged as a revolutionary technology. Despite its widespread application, blockchain is not used to store huge data due to the restricted storage capacity offered by each block. Furthermore, because the data contained in a block must be duplicated on many other nodes on the network, storage space is wasted. Off-chain storage is therefore required to store significant volumes of data while preserving the efficiency and performance of the network. Additionally, off-chain solutions such as Inter-Planetary File System (IPFS) would reduce the cost requirements of blockchain nodes as data stored and processed by these nodes would be reduced. Hence, strong decentralized storage is provided by IPFS that is a P2P-based content-addressed file sharing system that uses cryptographic hashes to store data. As a result, by combining blockchain with IPFS, an efficient file sharing system would be created. This paper presents a comprehensive IPFS architecture, a simplified version of this integrated approach, and potential applications that might assist to maximize the worth of both systems.

M. Kaur (✉) · S. Gupta
Department of Computer Science and Engineering, Chandigarh University, Mohali, India
e-mail: preetmand@gmail.com

M. Kaur
Department of Computer Science and Engineering, Guru Nanak Dev Engineering College, Ludhiana, India

D. Kumar
Apex Institute of Technology, Chandigarh University, Mohali, India

M. S. Raboaca
National Research and Development Institute for Cryogenic and Isotopic Technologies-ICSI, Râmnicu Vâlcea, Romania
e-mail: simona.raboaca@icsi.ro

S. B. Goyal
Faculty of Information Technology, City University, London, Malaysia
e-mail: Sb.goyal@city.edu.my

C. Verma
Department of Media and Educational Informatics, Eötvös Loránd University, Budapest, Hungary
e-mail: chaman@inf.elte.hu

Y. Singh et al. (eds.), *Proceedings of International Conference on Recent Innovations in Computing*, Lecture Notes in Electrical Engineering 1001,
https://doi.org/10.1007/978-981-19-9876-8_39

Keywords Blockchain · IPFS · Off-chain storage

1 Introduction

Following the success of Bitcoin, blockchain has emerged as one of the most promising technologies. The blockchain is the technology behind Bitcoin. The necessity for a third-party verifier is eliminated, resulting in a safe and entirely decentralized system. The powers of cryptography have been used to offer reliable solutions to earn the trust of businesses [1]. Blockchain is a linked list of blocks, each of which stores several transactions. Block validation is done using cryptographic methods, and new blocks are appended to the existing chain using a specified consensus algorithm among the nodes to ensure immutability and consistency [2]. Due to the numerous features offered by blockchain such as transparency, immutability, absence of third party, and trustworthiness, it has been utilized in many industries.

Scalability and privacy issues arise when data is uploaded to a public blockchain. In practice, it is not advised to include large amounts of data inside a blockchain transaction due to the high costs involved and the latency problems posed when full nodes must obtain the entire ledger. Moreover, permanent nature of data available on the blockchain restricts their usage in applications involving personal or sensitive data [3].

Blockchain is often coupled with off-chain storage options to address its scalability and security challenges while retaining the advantages of decentralization. For these solutions, it is necessary to store the files themselves outside the chain, keeping only the timestamp and references to these files in the blockchain [3].

One such solution is IPFS—a peer-to-peer protocol and network that aims to establish a permanent, decentralized means of storing and distributing data that is both efficient and resilient. To provide a new means of delivering online content, IPFS links all computing devices to the same file structure. Although several attempts to introduce distributed file systems have been made in the past, but the first worldwide file system to achieve low-latency and decentralized large-scale file distribution is IPFS [4].

The major contributions of this research paper are listed as follows:

1. We explore the opportunities offered by IPFS to address blockchain storage requirements.
2. A thorough IPFS architecture enables researchers to comprehend the capabilities offered by IPFS as an effective offline storage solution.
3. This article offers insights toward integration of IPFS and blockchain to build a secure data storage system.

1.1 Blockchain

Blockchain has received increasing attention as a consequence of its unique properties [5]. It is currently acknowledged as one of the fastest-growing technologies [6]. "Blockchain is a decentralized network of nodes that provides immutability, anonymity, security, and an append-only data structure that is freely distributed across multiple non-trusting nodes and in which new blocks may only be appended to the end of the current chain without affecting the previous blocks" [7, 8].

Blocks, nodes, and miners are three important elements of blockchain. Each block is decomposed into two parts, block header and block body. Each block's unique identifier is specified in the block header. Primary information included in a block header is block number, block hash, previous block hash, timestamp, Merkle tree root and nonce (unique number used only once). While the block body consists of a series of transactions, a transaction counter and total block fee for all the transactions included in a block [5].

Any device with computational and storage capabilities can act as a node. These nodes are capable of storing, sharing, and maintaining the integrity of blockchain data, and so serve as a solid foundation for the blockchain operation. In blockchain, nodes can be categorized into two categories, full node and light node. A full node is capable to process and store all the transaction and blocks. In contrast, a light node is capable to perform limited set of operations with limited resources available. A miner node is specialized full node designated to add and validate new blocks to the blockchain.

Blockchain is an immutable data structure that allows data to be stored on a blockchain on permanent basis. Hence, blockchain not only stores the transactions but also maintains their entire history. When it comes to storing large files or documents on the blockchain, however, there are significant restrictions. In Bitcoin, the block size is limited up to 1 MB only; thus, a file size to be uploaded on bitcoin is restricted. As the amount of data to be stored in a block is fixed, it would make it a costly database for storing vast volumes of data. Decentralized storage media are developed to meet the needs of storing relatively large volumes of data. The Inter-Planetary File System (IPFS) is common example of a content-addressable distributed file system that runs analogous to blockchain [9].

1.2 Inter-planetary File System (IPFS)

IPFS is a peer-to-peer distributed file system that, unlike HTTP, allows users to retrieve files based on their content rather than their location. The file that must be saved on IPFS is divided into several fragments with a maximum data size of 256 KB called IPFS objects. The chunk size is chosen 256 KB as it is easily divisible by standard disk block size. Each of these objects is identifiable by a unique cryptographic hash value called "content identifier," or CID. The content of each file fragment is used to calculate CID. As a result, the hash value associated with an IPFS object not only permits access to the data contained in that chunk, but also assures its authenticity [10, 11]. The immutable IPFS objects connected to each file are stored in a Merkle Directed Acyclic Graph (Merkle DAG) [11]. The Merkle DAG's leaves would carry unique IPFS objects, while the Merkle DAG's root is given as a unique CID for each file [12]. Thus, Merkle DAG effectively identifies each file and ensures its contents are temper-resistant.

These IPFS objects are distributed over entire network. Every IPFS node only stores a selective data files in their local storage, i.e., cache, not the entire files published in network. Users who want to retrieve a particular object may search with hash of the required file. IPFS upon looking at the distributed hash table (DHT) will locate the nearest peer who either hold the data file or connected to the peer who may hold that file. In order to upload accessed resources to other nodes and help with load distribution for frequently accessed content, accessed resources are locally saved in cache inside an IPFS node [3].

A comparison among HTTP and IPFS has been depicted in Fig. 1; HTTP employs a location-based addressing to fetch a file from a central entity (location) which in turn slows down the process when multiple requests are being sent to same location. The requested data may be unavailable in the event of failure of the central server [13]. While in IPFS, a content-based addressing is involved to access the data from the closest peer who has the copy of the same. Hence, the resultant system eliminates the need of central entity, thus provides requested data or file relatively fast and is made available even if some may peers may fail or down.

IPFS is characterized as the "permanent web" because it provides content availability without relying on a single entity. If no change in the content is captured, the reference to the file remains unchanged [3]. Hence, a permanent reference to a file is provided regardless of its location.

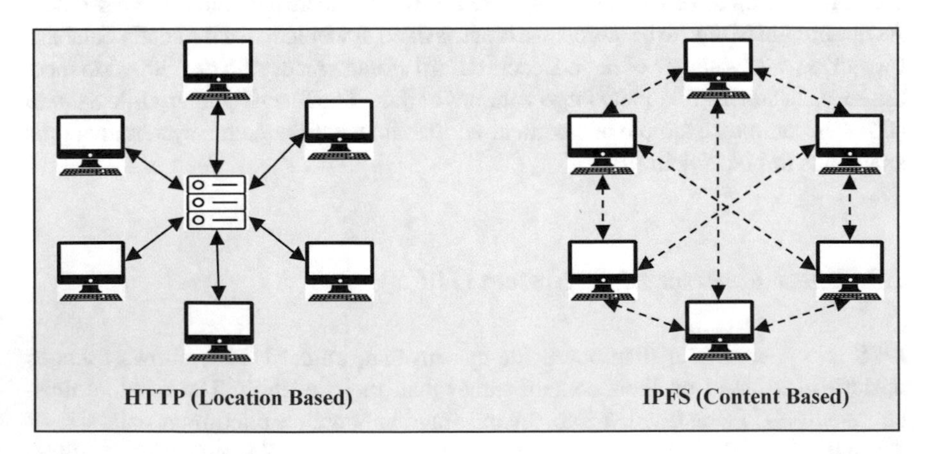

Fig. 1 HTTP versus IPFS

The remaining part of this paper includes related work in Sect. 2. Section 3 elaborates the IPFS architecture and detailed protocol stack. Section 4 focuses on blockchain-IPFS convergence. Section 5 concludes this work.

2 Related Work

This section highlights some recent research work that integrates blockchain and IPFS together to address the limitation and restrictions imposed so as to improve the system performance.

Nizamuddin et al. [1] proposed a system for document sharing and version control that combined blockchain and smart contracts with IPFS. The version control criteria and document sharing procedure were self-executed using smart contracts. To ensure system stability and security, the suggested system is implemented and evaluated.

As decentralized file storage and sharing systems become more widely used to store confidential information, their immutability and data persistence qualities generate significant uncertainties in terms of their compliance with privacy and data protection rights. Politou et al. [3] developed an IPFS-specific content erasure method that handles individual content erasure requests for personal data spread across many nodes. This erasure, however, is only permitted for the original content owner.

Huang et al. [9] discovered that a decentralized storage medium is necessary to alleviate the storage limitations of blockchain nodes when large files must be stored. As a solution to this challenge, IPFS seemed ideal. The security issues connected with IPFS, on the other hand, must be addressed. An IPFS proxy server was thus set up to offer decentralized access control with group key management.

Hasan et al. [14] noted that the streaming data created by IoT devices presents several issues owing to the devices' limited storage and processing capabilities. As a result, the cloud has been integrated with these streaming devices. Because of the risk of single-point failure and security issues imposed by central control, a blockchain-based decentralized secure solution was created. Furthermore, off-chain storage through IPFS has been deployed to store data in encrypted chunks, with hashes to these data chunks kept on blockchain to ensure immutability and transparency.

Jayabalan et al. [15] aimed to create a blockchain-enabled infrastructure with IPFS to allow hospitals to securely share medical information. AES-128 was utilized to encrypt patient data before putting it on IPFS, and RSA-1024 was used to transfer the symmetric key to the appropriate authorities. Scalability concerns related with blockchain have been addressed by leveraging IPFS off-chain storage. The intrinsic properties of blockchain have safeguarded the confidentiality of healthcare records.

Kaur et al. [16] purposed a novel consensus algorithm DPoAC that makes use of IPFS to store the encrypted parts of a secret produced to determine the block generation rights among network nodes.

3 IPFS Architecture

An innovative distributed file system called IPFS combines the best elements of earlier peer-to-peer networks including DHTs, Bit Torrent, and Git [4]. New mechanisms for sharing and versioning vast amounts of data, as well as a new platform for developing and deploying applications, are both offered by IPFS. IPFS has the potential to transform the Web itself [4].

IPFS is a peer-to-peer network with no privileged nodes. Local storage is used by IPFS nodes to keep IPFS objects. Nodes communicate with one another and exchange data objects. Files and other data structures are represented by these objects. The protocol stack of IPFS as shown in Fig. 2 consists of various sub-protocols discussed as follows:

3.1 Identities

This protocol layer of the stack is responsible for managing the node generation and verification operations and to identify each node on IPFS network. Each node has

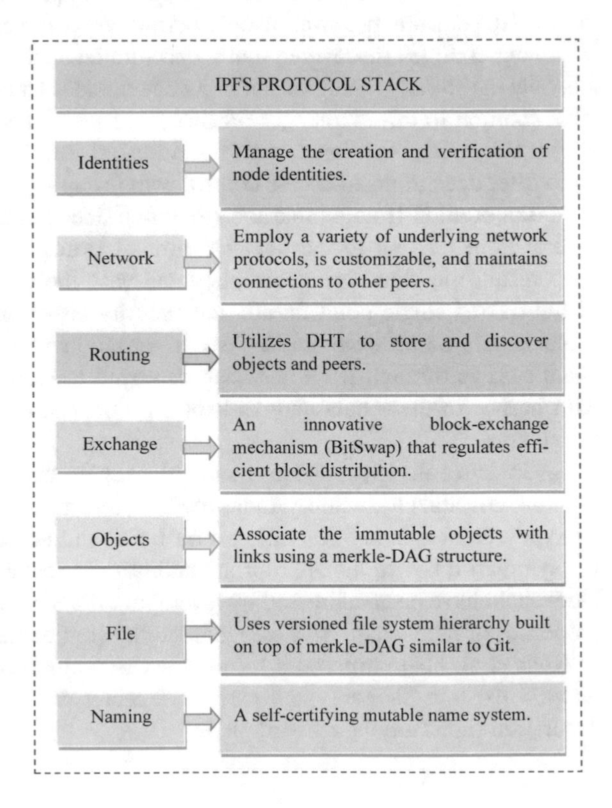

Fig. 2 IPFS protocol stock

a unique *NodeId* and a cryptographic pair of public–private key. Every node has the option to request a distinct *NodeId* each time it joins the network; however, if the same *NodeId* is utilized, nodes are eligible for incentives. Public key is used to decide whether to connect or disconnect while joining to other peers for the first time. Connection is established if the hash of the public key matches the *NodeId*; else, the connection is terminated.

The mechanism described in S/Kademlia is used by IPFS to enhance the expense of generating a new identity, hence escalating the cost and difficulty of Sybil attack [4].

3.2 Network

IPFS nodes need to interact with several other nodes across the Web. Because IPFS is not reliant on IP connectivity, any overlay network constructed on any other network can be utilized for communication. The crucial element implemented on network layer of IPFS is libp2p. Libp2p provides a modular network stack that can discover additional peers and networks without relying on centralized registries, allowing applications to execute offline also [4].

Libp2p Features: Numerous features have been provided by libp2p; some of them are explained as under:

Transport Module. An application can use any transport protocol that best suits its requirements. Depending on the runtime under which they are being used, these protocols change. The interface-transport standard defines how these transport protocols provide a clean interface for calling and monitoring.

No advance port assignments. Users used to allocate a listener to a port and then bind that port to certain protocols before libp2p was developed, so as to inform the port information to other nodes in advance. Users do not need to allocate ports ahead of time while using libp2p.

Encrypted communication. In order to prevent eavesdropping on user communication, libp2p establishes an encrypted channel to secure data and authenticates peer IDs.

Peer discovery and routing. Finding peers to connect to be made easier by a peer discovery module for libp2p. Peer routing searches the network for additional peers by deliberately sending out requests, which may be recursive, until a peer is located. To determine the location of content on the network, a content routing procedure is adopted.

3.3 Routing

IPFS nodes need a routing mechanism that can locate (a) the network addresses of other peers and (b) peers that can serve certain objects. IPFS performs routing using a distributed hash table (DHT) based on S/Kademlia [4]. DHTs are decentralized key-value pair storage systems. Data posted to the IPFS network serves as the value, while associated hashes serve as the key. Instead of the complete table being duplicated across all nodes, each P2P node in the network has a portion of the DHT table [15]. IPFS DHT saves data objects based on their size. Small data objects up to 1 KB are stored directly on DHT, while large objects need to be partitioned and stored individually on multiple peers. For large data objects, DHT holds references to those peers in the form of *NodeId* which will serve that data object [4].

When a request is received by a peer to retrieve a value associated with a given key, that peer will look up for the specified key in its own table. If corresponding key found, then value associated is returned to the requesting peer. Otherwise, the request is forwarded to peers of current node. This process continues until the value is found, and required value is passed through the same path in reverse direction until the value is received by initial requesting peer [15].

3.4 Exchange

BitSwap protocol is used by IPFS for data distribution via exchanging blocks among peers. BitSwap peers have two lists: a *want_list* of data blocks requested by that peer and a *have_list* of data blocks held by that peer and available to other peers [4, 15]. A node sends a *want_list* to peers to convey which blocks it need. When a peer receives a *want_ list*, it should check its *have_list* to see whether any blocks on the list are available and then send the matching blocks to the requestor. On the other hand, when requested blocks are received by a node, it must send out a notice known as a "Cancel" to inform its peers that block is not required by node because it has already been received.

3.5 Objects

Data blocks are stored and distributed using a Merkle Directed Acyclic Graph (DAG) to provide fast and reliable access. Merkle DAG associates the object with the cryptographic hash of target data file. The architecture of the Merkle DAG is similar to that of Git, and it offers various features as mentioned below [4]:

Content addressing. The multi-hash checksum uniquely identifies all content along with the corresponding links associated.

Tamper resistance. IPFS identifies tampered with or damaged data as content could be easily validated using checksum.

De-duplication. Objects with same content are treated as equal and need to store only once. This will reduce the likelihood to store same content repeatedly and optimize storage.

3.6 File

On top of the Merkle DAG, IPFS specifies a collection of objects for representing a versioned file system. This object model resembles with the one used by Git; therefore, it is easy to map from one system to the other. The types of objects that are supported are as follows [4]:

Block: It is a variable-size data block to store user data.

List: It is a large de-duplicated file composed of numerous blocks or lists.

A tree: It is a directory to map names to hashes of blocks, lists, or other trees.

Commit: It is a type of object reference that reflects a snapshot in a tree's version history. The changes between two versions of the file system are shown by comparing the objects of two separate commits. In IPFS, version control is implemented using commit object.

3.7 Naming

Self-certifying File System (SFS) is a distributed file system paradigm that "self-certifies" itself by requiring no special permissions to transport data between nodes [4, 15]. It enables users to safely gain remote file access privileges similar to local file access while keeping the transaction transparent. IPFS employs a public key cryptography mechanism to self-certify files released by network peers. Inter-planetary Name Space (IPNS) is the name given to the SFS integrated within IPFS. Nodes, like data/files, may be uniquely recognized in IPFS by using thc hash of their public key (*NodeId*) [4]. Nodes employ their secret keys to sign objects, while the receiver uses their public keys to verify their mutable state [15].

4 Blockchain-IPFS Convergence

The entire transaction information is kept in the network using blockchain. While this information is permanent and duplicated across numerous nodes, its magnitude

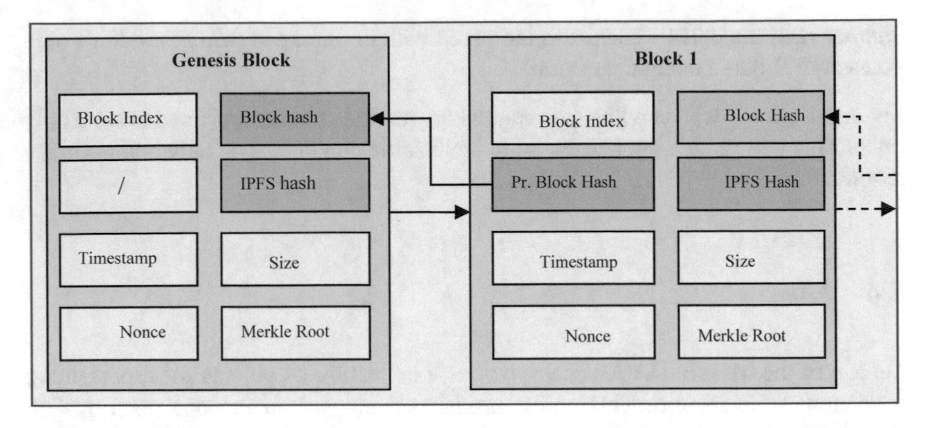

Fig. 3 Block header with IPFS hash

causes major concerns. Storing complete transactions along with metadata could elevate the storage requirements and has significant impact on blockchain speed. As a result, the process of duplicating transactional details across all nodes becomes time consuming and costly, especially when huge volumes of data must be stored.

The IPFS protocol offers a solution to the problem of limited data storage. It aids blockchain in overcoming storage constraints. The IPFS will store decentralized and immutable data outside the blockchain. Instead of storing the entire data as single file, it needs to fragment in multiple chunks. Blockchain stores the hash address of the data fragment seen in Fig. 3 as a field on the block header [8]. Data has to be saved in IPFS, and index to this data will be maintained on the blockchain. By taking the aforementioned strategy, IPFS can overcome data storage constraints in blockchain [17].

4.1 Features of Blockchain-IPFS Convergence

There are certain benefits of storing data off-chain in blockchain as mentioned below.

1. **Reduced Block Size**: Storing the index to the data on the chain while the data is stored on an off-chain media like as IPFS reduces the size of a block. It contributes to the development of lightweight blockchain, which improves system efficiency and speeds up the process of replicating data among network nodes.
2. **Data Immutability**: By storing the IPFS hash as a part of the block header, the data is kept intact and safe. To ensure data immutability, data retrieved from off-chain media is easily verified against the given IPFS hash.

As a result, IPFS would be an appealing concept to combine with blockchain to improve its efficiency and minimize restrictions. In Fig. 4, a simplified design of blockchain-IPFS integrated system has been demonstrated [15] in which the transaction data needs to put after encryption on IPFS while its hash is returned to the

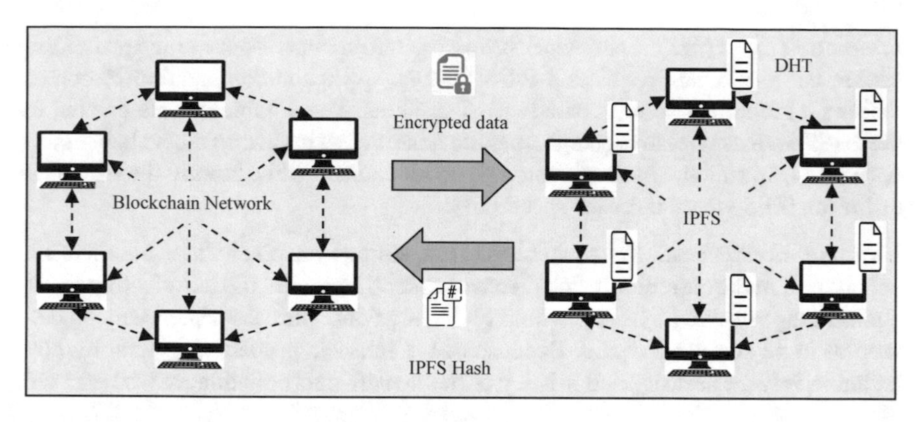

Fig. 4 A simplified blockchain-IPFS integrated system

blockchain network and stored as a part of block header. The use of the IPFS technology in blockchain architecture for off-chain storage removes the need for full nodes while maintaining network transparency. IPFS storage supports uploading of any kind of digital transaction, making it possible to utilize the system for numerous applications [15].

Another key characteristic of IPFS is to provide de-duplication. If the same file is repeatedly uploaded to the network, the same CID will be generated each time depending on its content, keeping the system consistent and preventing data duplication [15]. Thus, combination of decentralization and de-duplication algorithms of IPFS improves system efficiency and reduces storage space waste. For new network nodes or nodes rejoining the network, synchronization is simpler and faster since blockchain network nodes only use hash addresses of data in their transactions instead of the total volume of data. The incorporation of IPFS as off-chain storage addresses the major bottleneck in blockchain networks, namely scalability related to storage concerns [15].

4.2 Potential Applications of Blockchain-IPFS Convergence

Because of the multiple functionalities provided by the union of blockchain with IPFS, this integration benefits many applications by producing better efficiency and performance that would otherwise be impossible with blockchain alone. In this section, few main applications of this convergence are discussed.

Health Care. IPFS helps to solve the scalability issues with blockchain by enabling the storage of patient electronic health records (EHRs) on a blockchain-integrated infrastructure. Because of the scale of EHR, storing these massive documents on chain becomes challenging. As a result, it is recommended that these health records be stored off-chain via IPFS [15].

Internet of Things (IoT). IoT devices have a limited amount of storage and processing power, but blockchain requires a lot of storage space and computation resources, making it incompatible with existing IoT systems. The volume of data created by these IoT devices is relatively high, making it inefficient to save on blockchain nodes. As a result, storing the hash of such data on blockchain while storing the data itself offline on IPFS would increase speed [14].

Learning Management Systems: Large data storage on blockchain nodes is not economical in the current blockchain environment. However, the introduction of off-chain storage enabled by IPFS would allow huge data files associated with student records to be stored off-chain. Because of the intrinsic properties offered by both technologies, the security and privacy of this sensitive information may be ensured.

5 Conclusion and Future Work

In this article, we offered a solution to the blockchain's restricted storage capacities in the form of IPFS, a decentralized storage medium. There was a considerable demand for an offline storage medium since major blockchain applications are limited to store only a specified size of data in a block, making it difficult to employ blockchain to store massive amounts of data, such as IoT streaming data. As a result, a detailed architecture with IPFS protocol stack has been discussed. We offered an overview of the integrated system enabled by blockchain and IPFS in this work.

In the future, we want to assess the validity of our integrated solution in the aforementioned applications.

Acknowledgements This paper supported by Subprogram 1.1. Institutional performance-Projects to finance excellence in RDI, Contract No. 19PFE/30.12.2021 and a grant of the National Center for Hydrogen and Fuel Cells (CNHPC)—Installations and Special Objectives of National Interest (IOSIN). This paper was partially supported by UEFISCDI Romania and MCI through projects AISTOR, FinSESco, CREATE, I-DELTA, DEFRAUDIFY, Hydro3D, EREMI, SMARDY, STACK, ENTA, PREVENTION, SWAM, DISAVIT iPREMAS and by European Union's Horizon 2020 research and innovation program under grant agreements No. 872172 (TESTBED2) and No. 101037866 (ADMA TranS4MErs).

References

1. Nizamuddin N, Salah K, Azad MA, Arshad J, Rehman MH (2019) Decentralized document version control using ethereum blockchain and IPFS. Comput Electr Eng 76:183–197
2. Palaiokrassas G et al (2021) Combining blockchains, smart contracts, and complex sensors management platform for hyper-connected SmartCities: an IoT data marketplace use case. Computers 10(10):133
3. Politou E, Alepis E, Patsakis C, Casino F, Alazab M (2020) Delegated content erasure in IPFS. Future Gener Comput Syst 112:956–964

4. Benet J (2014) IPFS-content addressed, versioned, p2p file system. arXiv preprint arXiv: 1407.3561
5. Kaur M, Gupta S (2021) Blockchain technology for convergence: an overview, applications, and challenges. In: Blockchain and AI technology in the industrial internet of things, IGI Global, pp 1–17
6. Kaur M, Gupta S (2021) Blockchain consensus protocols: state-of-the-art and future directions. In: 2021 International conference on technological advancements and innovations (ICTAI), pp 446–453. https://doi.org/10.1109/ICTAI53825.2021.9673260
7. Kaur M, Khan MZ, Gupta S, Noorwali A, Chakraborty C, Pani SK (2021) MBCP: performance analysis of large scale mainstream blockchain consensus protocols. IEEE Access 9:80931–80944
8. Kaur M, Khan MZ, Gupta S, Alsaeedi A (2022) Adoption of blockchain with 5G networks for industrial IoT: recent advances, challenges, and potential solutions. IEEE Access 10:981–997
9. Huang HS, Chang TS, Wu JY (2020) A secure file sharing system based on IPFS and blockchain. In: Proceedings of the 2020 2nd international electronics communication conference
10. Steichen M, Fiz B, Norvill R, Shbair W, State R (2018) Blockchain-based, decentralized access control for IPFS. In: 2018 IEEE international conference on internet of things (iThings) and IEEE green computing and communications (GreenCom) and IEEE cyber, physical and social computing (CPSCom) and IEEE smart data (SmartData)
11. Zaabar B, Cheikhrouhou O, Jamil F, Ammi M, Abid M (2021) HealthBlock: a secure blockchain-based healthcare data management system. Comput Netw 200:108500
12. Doan TV et al (2022) Towards decentralised cloud storage with IPFS: opportunities, challenges, and future considerations. arXiv preprint arXiv: 2202.06315
13. Tenorio-Fornés Á, Hassan S, Pavón J (2021) Peer-to-peer system design trade-offs: a framework exploring the balance between blockchain and IPFS. Appl Sci 11(21):10012. https://doi.org/10.3390/app112110012
14. Hasan HR, Salah K, Yaqoob I, Jayaraman R, Pesic S, Omar M (2022) Trustworthy IoT data streaming using blockchain and IPFS. IEEE Access 10:17707–17721
15. Jayabalan J, Jeyanthi N (2022) Scalable blockchain model using off-chain IPFS storage for healthcare data security and privacy. J Parallel Distrib Comput 164:152–167
16. Kaur M, Gupta S, Kumar D, Verma C, Neagu B-C, Raboaca MS (2022) Delegated proof of accessibility (DPoAC): a novel consensus protocol for blockchain systems. Mathematics 10(13):2336. https://doi.org/10.3390/math10132336
17. Alizadeh M, Andersson K, Schelén O (2020) Efficient decentralized data storage based on public blockchain and IPFS. In: 2020 IEEE Asia-Pacific conference on computer science and data engineering (CSDE)

Hybrid Approach for Resource Allocation and Task Scheduling on Cloud Computing: A Review

Saraswati Narayan, Neerendra Kumar, Neha Koul, Chaman Verma, Florentina Magda Enescu, and Maria Simona Raboaca

Abstract Cloud computing is an accelerating technology in today's world. Cloud computing is a technology which changed the traditional ways of providing services that are deployed by either companies or individuals. Task scheduling and resource allocation are very crucial aspects of cloud computing. One of the fundamental factors in resource management, which often focuses on resource allocation, workload balance, resource provisioning, task scheduling, and QoS to achieve performance, applies to cloud-based IoT systems. This paper provides an overview of existing behavior-based load balancing method in the cloud system and also provides a hybrid approach of task scheduling and resource allocation using various techniques. Technologies like cuckoo search, harmony, genetic algorithm, random forest, resource optimization are discussed and hybrid method for allocation of resources and scheduling of tasks is proposed.

Keywords Cloud computing · Task scheduling · LAGA algorithm · Cuckoo harmony search algorithm · Resource allocation

S. Narayan · N. Kumar (✉) · N. Koul
Department of Computer Science and IT, Central University of Jammu, Jammu, India
e-mail: neerendra.csit@cujammu.ac.in

C. Verma
Department of Media and Educational Informatics, Eötvös Loránd University, Budapest, Hungary
e-mail: chaman@inf.elte.hu

F. M. Enescu
Department of Electronics, Communications and Computers, University of Pitesti, Pitesti, Romania

M. S. Raboaca
National Research and Development Institute for Cryogenic and Isotopic Technologies-ICSI, Râmnicu Vâlcea, Romania
e-mail: simona.raboaca@icsi.ro

Y. Singh et al. (eds.), *Proceedings of International Conference on Recent Innovations in Computing*, Lecture Notes in Electrical Engineering 1001,
https://doi.org/10.1007/978-981-19-9876-8_40

1 Introduction

In latest years, there has been an undeniable rapid enhance in cloud handling as it allows start-ups and small organizations to spend on a paid basis. With the exponential development of the cloud paradigm, there is a growing demand for multiple services from the diverse user community. According to [1], cloud service providers assist in the management and distribution of various cloud resources. Cloud resource management is a rooftop activity that covers the various stages of resources, including loading, scheduling, allocating cloud resources, and completing work. Two important factors are involved in the cloud paradigm, namely resource provision and resource scheduling. These steps involve discovering potential resources for a particular task that is heavily dependent on the QoS requirements set by cloud clients, called resource provisioning [2]. This research focuses on two main activities, namely task scheduling between cloud servers (virtual machines) and resource utilization in virtualization technology. The motivation for this paper is to find the limitations of existing algorithms and effectively use the proposed hybrid algorithm provided. The hostile nature of the cloud environment is the problem of efficient algorithm for resource allocation in achieving and efficient method for cloud computing. Cloud resources operate in a dynamic shared environment. The work schedule becomes ambiguous sometimes because the availability of resources when required the most is incorrect. Therefore, as part of this research work, a nature-inspired multi-purpose optimization model with load balancing was developed as an additional factor for scheduling work and resource utilization in the cloud paradigm.

1.1 Components of Cloud Computing

The system consists of three key components: clients, data centers, and circulated servers, each of which has a specific reason and performs a specific function [3]. This is well explained in Fig. 1.

Customers: Cloud information is processed when end users interact with customers and is divided into different categories such as mobile, laptop, and tablet computers. Customers control the interactions that drive data management across cloud servers.

Data Center: A set of servers that host heterogeneous applications that can be away from the client system and allow end users to connect to share different applications.

Distributed servers: These servers are part of the cloud space, which is a virtual server that hosts various applications on the Internet. These servers are not located in a geographical area.

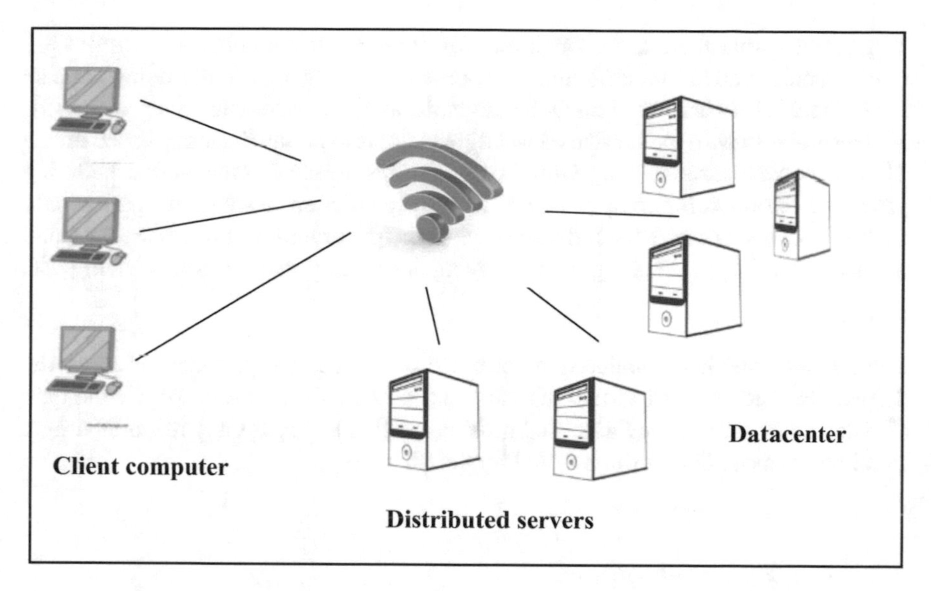

Fig. 1 Components of cloud system

1.2 Important Processes of Cloud Computing

- **Task scheduling**

A main problem with cloud computing simulations lies in job scheduling methodologies to adapt to some of the basic limitations with similar existing virtual machines. Thus, the problem arises due to differences in the allocation of effective resources and the lack of multiple functions due to a number of limitations [4].

- **Conflicting sources**

The use of available resources (virtual machines) is a major problem for work schedules due to the distribution of existing machines in different geographical locations due to specific configurations such as bandwidth, memory, computer power, and so on.

- **Recycling resources**

One of the main objectives of cloud computing is the efficient utilization of resources. It expands the services depending on the situation. Thus, the cloud environment works according to the automation method. Cloud service providers schedule tasks for the appropriate resources for efficient use of resources. This methodology involves the creation of an effective planning mechanism.

- **QoS service providers**

To guarantee that there are enough cloud resources for cloud users to fulfill QoS requirements such as execution time, response time, budgetary limits, delays, and so forth. Since SLAs are based on QoS standards, any violations can be penalized [5].

The problem with the service e level agreement is to define the terms based on the SLAs to achieve a satisfactory QoS threshold. This means that the work scheduling process in cloud computing can stop accurate prediction, testing, evaluation, and trading between complexity and clarity when the maximum user needs are met, and measuring, testing, evaluating, and resolving is very important. It's easy giving life.

- **Cost**

The market price is dynamically determined by the service provider based on the demand for each type of virtual machine and service. Cloud users specify the type of resource or service they need and must be willing to pay for any resource usage. In all other cases, fines will be levied by the SLA [6].

1.3 Paper Contributions

In this paper, we have reviewed high-tech works in cloud computing domain by considering the challenges and constraints in cloud computing. Following are the main contributions:

(a) The review paper's contributions emphasize the importance of hybrid approaches for resource allocation and task scheduling.
(b) A high-tech resource allocation mechanism and cloud resource mechanism is reviewed and analyzed in literature survey section of paper.
(c) Combined approaches for task scheduling and resource allocation on Cloud Computing along with their techniques are described in this paper.
(d) A hybrid approach for task scheduling and resource allocation is planned in this paper.

1.4 Paper Outline

Section 1 discusses introduction of cloud computing. In Sect. 2, a literature survey is done. In Sect. 3, hybrid approaches for resource allocation are discussed. In Sect. 4, hybrid approaches for task scheduling are discussed. The Sect. 5 presents the proposed hybrid approach for task scheduling and resource allocation of the paper. The paper concludes in the Sect. 6, as shown in Fig. 2.

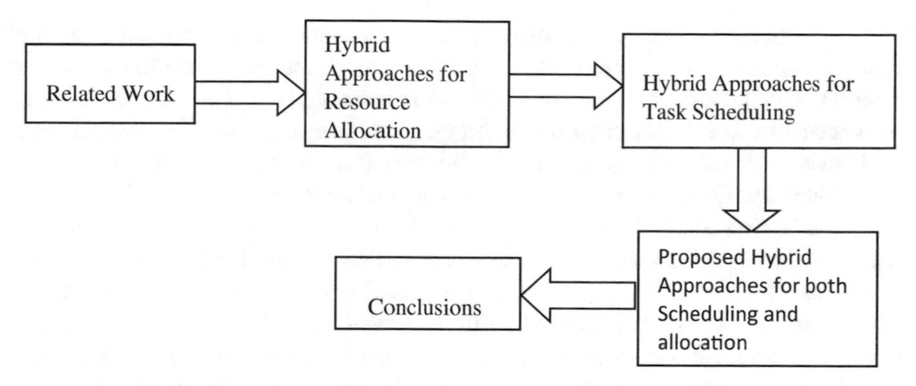

Fig. 2 An outline of the paper

2 Related Work

A high number of diverse workloads and resource leads to a big number of possible alternatives, which increases time overhead such issues have been categorized as an NP-hard issue in which the task scheduling in cloud computing is so challenging [7]. The author has demonstrated that some of the available approaches were unable to discover a workable solution to the complicated issues in a polynomial amount of time. Techniques based on metaheuristics have shown to produce the required result almost immediately. As a result, the referenced article will present a thorough research and roughly similar examination of several VM placement strategies based on cloud converged issues [8].

A research of postmodern scheduling problem has focused on three metaphysical tactics and two different techniques. The approaches include Optimization Algorithms, Linear Programming (GA), and Optimization (PSO). The programs involve League Title Algorithm (LCA), and Ant Colony optimization (ACO) (BA) (PSO), among others [9]. Convergence concerns, however, are not taken into account in the evaluation [10]. Carried undertaken a study of league championship algorithms based on a summary of the literature on LCA in conferences, peer-reviewed journals, and book chapters. The review highlighted certain LCA applications, including handling associated with adverse health effects single processor planning problem in a group also (JIT) system shipping fees and various constraints, arithmetical parameter optimization. The report also outlined some of the limitations of the method, such as its lack of robustness and generalizes ability. The method does not, however, concentrate on convergence problems. Tang et al. [11] Reviewed the load balancing methods that are currently in use. The review focuses on load balancing methods, which are common in cloud computing and enhance service delivery while enhancing overall network signaling and energy efficiency. The approach uses fair load balancing strategies to compete on cutting costs, latency, loss, execution order processing, and transmission inefficiency. The result is a dramatic drop in costs. Convergence difficulties, however, are not at all mentioned in the review [12].

Out a thorough analysis of resource allocation approaches, focusing on the many resource schemes employed by various scholars, the problems the techniques addressed, and the metrics by which they were assessed [13]. The evaluation takes into account capacity planning increased cloud infrastructure's entire performance with respect of the delivery of high-quality services and effective resource usage. The evaluation does not, however, specifically address convergence-related problems. Author in paper [14], have done a thorough analysis of computation offloading methods. The review focused on the concerns and challenges that come with current load balancing strategies, including geographical node distribution, single points complexity of the computational algorithms, probability of failing, virtual desktop migrating, homogeneous ports, offline caching, load-balancer scale. Reference [15] based on numerous criteria employed by the methodologies, the review similarly offers a comparative analysis of the articles utilized. The metrics comprise single goal, multiple objectives, and quality of service measurements. Convergence, however, is not viewed as a problem in the review. The reviewed and surveyed publications reveal that task scheduling methods based on convergence problems have not received the slightest amount of security [9].

2.1 Review of Task Scheduling and Resource Allocation Methodologies

Literature review on task scheduling and resource allocation methodologies has been presented in Table 1.

2.2 Review of Behavioral Inspired Algorithms-Based Scheduling and Load Balancing Methodologies

The major motivation of the study is to provide an existing behavior-base load balancing method in the cloud system, and this release provides an overview of the behavior-based load balancing algorithm and resources for help with load planning. The Behavioral Inspired Algorithms-Based Scheduling and Load Balancing Methodologies are presented in Table 2.

Table 1 describes task scheduling and resource allocation methodologies. Various methodologies and algorithms have been used in the existing research field of task scheduling and resources allocation. Centralized resource scheduling load balancing and QoS. Heuristic algorithm based on convex optimization and dynamic programming provides high QoS with minimized energy cost. To optimizes energy, bandwidth aware task scheduling (BATS) algorithm. Maximum utilization of resources with

Table 1 Task scheduling and resource allocation methodologies

Author	Method	Advantages	Limitations
Radojevic et al. [16]	Resource scheduling by centralized load balancing	Improves load balancing and QoS	Node abilities and configuration details are not determined effectively
Goudarzi et al. [6]	Heuristic algorithm based on convex optimization and dynamic programming	High QoS with minimized energy cost	Do not meet the SLA agreements
Lin et al. [17]	Bandwidth aware task scheduling (BATS) algorithm	Optimized energy consumption	Reliability in load balancing is missing
Zhu et al. [18]	Rolling horizon resource scheduling	Reduced energy consumption and maximized CPU	Resource availability not considered
Ghanbari et al. [19]	Workload scheduling algorithm	Maximum utilization of resources with optimal execution time	High total finish time
Mathew et al. [20]	New fuzzy resource scheduling algorithm (NFSA)	Minimized turnaround time missed deadline and response time	Resource utilization is not considered
Liu et al. [21]	Improvised social learning optimization (SLO) technique	Improve better convergence and global optimization	Reduced throughput
Dubey et al. [22]	Modified heterogeneous earlier finish time (HEFT)	Minimizes the makespan and load balancing problems	Higher cost
Gawali et al. [23]	Heuristic approach	Maximum utilization of CPU, memory, and bandwidth	High turnaround time and response time
Srichandan et al. [24]	Multi-purpose hybrid bacteria foraging algorithm (MHBFA)	Minimizes makespan and energy consumption	MHBFA incurs extra timing for crossover and transmutation and chemotaxis and reproduction process
Zhou et al. [25]	Deadline-constrained cost optimization for hybrid cloud (DCOH)	Minimizes the cost and makespan	Other QOA metrics are not considered

optimal execution time is performed by workload scheduling algorithm. NFS algorithm minimizes turnaround time and response time and HEFT algorithm minimizes the makespan and load balancing problems.

Table 2 describes various methodologies and algorithms for scheduling and load balancing methodologies. Algorithms like particle swarm optimization (PSO), K-means clustering, evolutionary approach-based load balancing helps in reducing the

Table 2 Behavioral inspired algorithms-based scheduling and load balancing methodologies

Author	Method	Advantages	Limitations
Rodriguez et al. [26]	Particle swarm optimization (PSO)	High throughput and low makespan	Resource starvation and vagueness are not considered
Malinen et al. [27]	K-means clustering	Reduced the make span and cost	Outliers and noisy tasks where not handled properly
Zhan et al. [17]	Evolutionary approach-based load balancing	Minimizes the makespan	High complexity in search process
Wang et al. [28]	Work spanning time along with load balancing oriented genetic algorithm	Enhanced throughput and decreased the makespan	Local throughput
He et al. [29]	Adaptive multi-objective job scheduling with particle swarm optimization	Minimized the value of processing time, energy, transmission time and cost	Availability of the resource is not effective
Alkayal et al. [26]	Teaching–Learning optimization algorithm	High throughput and less makespan	Not reliable and flexible
Reddy et al. [22]	Multi-objective task scheduling with whale optimization algorithm	Effective utilization of resource with minimized makespan	High searching time and overall execution time
Gupta et al. [30]	Ant colony optimization	Better makespan	Degraded QoS
Kaur et al. [31]	Multi-objective bacterial foraging optimization algorithm (MOBFOA)	Reduced resource usage cost, makespan and flow time	High computation complexity
Kaur et al. [32]	Context-aware load balancing genetic algorithm	Reduced the global solution search and convergence rate	Increased the cost
Bindu et al. [33]	Energy aware multi-objective genetic algorithm	Low energy utilization and less makespan	Short reliability

makespan and cost. Adaptive multi-objective job scheduling with particle swarm optimization helps in minimized the value of processing time, energy, transmission time and cost. Multi-objective bacterial foraging optimization algorithm reduced resource usage cost, makespan and flow time.

3 Hybrid Approaches for Resource Allocation

3.1 Load Balancing in Cloud Computing

The ability of the cloud to maintain optimally in the face of the dynamic nature of input constraints is highly dependent on an efficient load balancing algorithm. Load balancing also helps maintain resource tools efficiently, leading to the most efficient use of energy for green computing. Load balancing often initiates and executes a recovery strategy in the event of a server failure or malfunction that may occur at any time. Most load balancing algorithms can be started by a transmitter, a receiver, or grouping of both. Static and dynamic load balancing algorithms are the two main categories of load balancing algorithms. Static policies distribute the load evenly across as many servers as possible, ensuring that all servers receive the same amount of traffic. Dynamic methods distribute loads based on weights to the lowest node load, balance traffic, and minimize computational overhead. Increase the time spent in the distribution process of a low-load server to allocate server traffic. Ant colony optimization, particle swarm optimization clustering methods is just some of the existing algorithms for load balancing. Key criteria evaluated load balancing algorithms such as runtime, throughput, response time, load balance index, resource utilization factors, and QOA performance [22]. The resource tool determines the optimal use of resources for the immediate performance of a given task, while scalability is a calculation of how a cloud can perform a given task with a limited number of resources and processors. Operating power is defined as the ratio of tasks completed over a period of time, and error tolerance is defined as the ability of a cloud to perform a given task in the eventual system or network failure [16].

Search for a friendly combination creature strategy for cloud computing job scheduling has been put out in [34]. The method combined symbiotic organism search (SOS) with simulated annealing (SA) to address the issue of task assignment in a virtualized environment. Additionally, the combined approach improves SOS's convergence rate and solution quality, which further optimizes work scheduler. In addition, a fitness function is suggested to solve the issue of ineffective use of virtual machines (VMs). The author in paper [35] proposed employing priority queues based on an enhanced evolutionary method for platform for managing workflow scheduling. The proposed method, called N-GA, assigns processors to subtasks by combining a heuristic-based HEFT search with a Genetic algorithm (GA). A Linear Temporal Logic (LTL) version of the intended specification was also retrieved. Additionally, the integrated strategies improve execution speed while minimizing makespan. The author in paper suggested employing a hybridized mechanism for cloud-based job scheduling. The methods employed, Planned Annealing (PSO) or the Hill Climbers Automated process (HCA), are made to hasten the resolution of viable solutions in a demographically problem space. Moreover, the hybridized strategies might speed up the search function to obtain a quick work completion [36]. To address the issues of VM imbalance, makespan minimization, and VM migration reduction, an effective load balancing solution for cloud computing was presented. The technique was based on an improved bee colony algorithm. Additionally, the proposed approach employs honeybee foraging behavior to effectively load balance throughout the resource pool. To lessen the extent of the Vm' unbalance, the job with the lowest priority value is chosen for migration. As a result, activities are not needlessly delayed before processing [9].

3.2 Resource Allocation in Cloud Computing

The first and major task of a cloud computer is resource allocation. The allocation of resources to virtual machines (VMs) is carried out by analyzing the availability of memory, CPU, and disk space. Subsequently, Cloud Service Pro developer improves network throughput standards and error tolerance as value-added services. Resource allocation includes allocating and allocating cloud resources subject to constraints such as server performance, networking, and fault tolerance. An efficient resource allocation system must ensure that resources are available at the user's request and can operate under high load and stress conditions. Job scheduling algorithms are an NP-hard problem, which, along with many solutions, has a high computational complexity and takes a lot of time to find the best solution. Resource allocation is a complex issue in the IaaS hiring model because the reliability of cloud services and the provisioning and management of resources often depends on assigning tasks to users. Figure 3 illustrates that resource allocation is based on user requests. Each task spends a different execution time (task execution time) to execute the task on different virtual machines. Task scheduling is a big challenge as it is necessary to schedule tasks from different users to the respective IT resources in the shortest possible time. Effective planning helps you use resources and evaluate the best values, starting levels ,and costs in MaxPen. Thus, the efficiency of resource planning has a significant impact on the performance of the cloud environment [30, 37].

3.3 Cloud Resource Managers

The provision and scheduling of resources by cloud services with payment policy, if desired, is done with confidence and consistency in resource management. But the complex role that cloud consumers play in changing different needs makes service delivery more efficient. This is achieved by using different methods of cloud resource planning. Cloud consumers submit their workload with their QoS requirements and based on the resources provided from a set of sources, as shown in Fig. 4.

Providing services for workloads based on QoS parameters is a cloud application and reducing operating time is a major concern in this service. Providing the right resources for workloads is a complex task, and selecting the best resource-workload pair based on QoS standards is an important part of cloud-based research. Resource allocation or reservation is at the time of requesting resource allocation. Similarly,

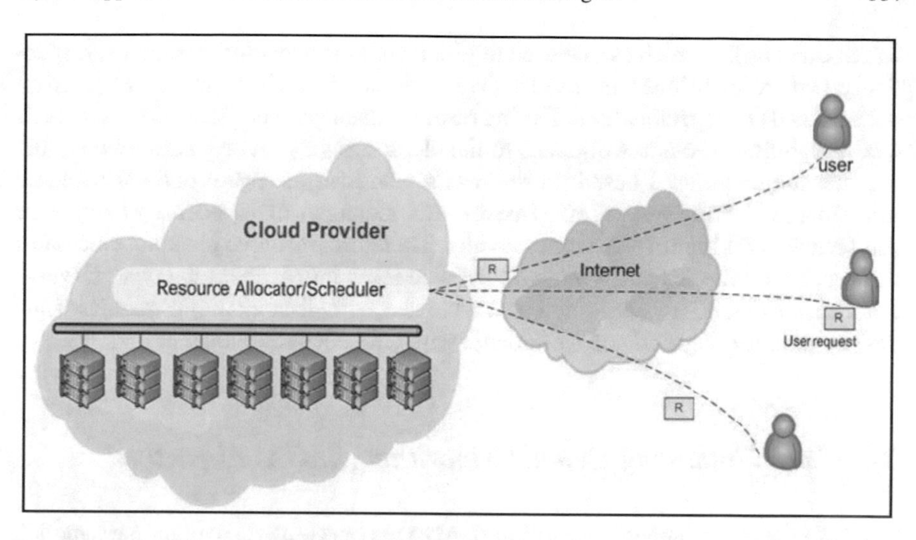

Fig. 3 Resource allocation in cloud computing [30, 37]

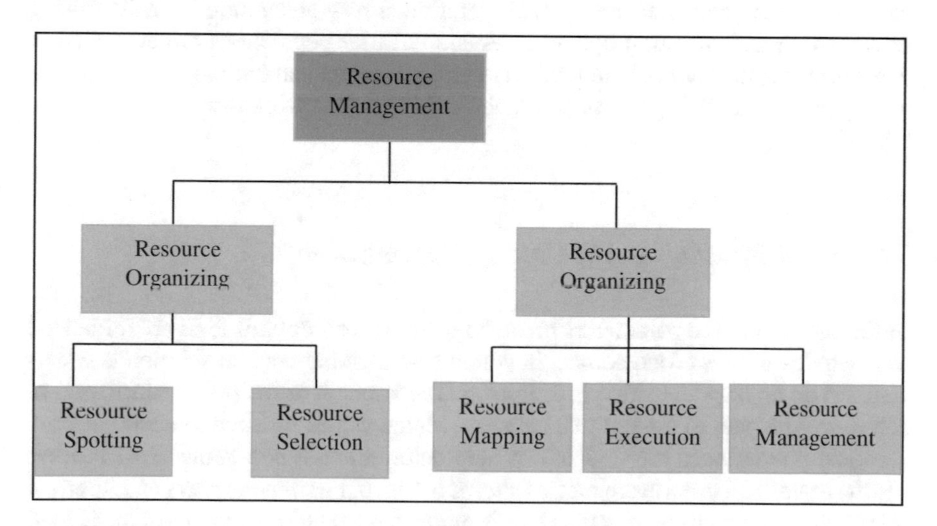

Fig. 4 Taxonomy of cloud resource management

allocation can provide resources such as CPU, network (power or cooling system) to support users. Schedule a request at a specific time. Coordinates development planning and processes and describes user planning based on Service Level Agreements (SLAs) [38].

In a cloud environment, the cloud resource management process is controlled by a central resource manager (CRM). CRM manages all workloads and manages resources efficiently. As shown in Fig. 4, various objects and interfaces are associated with CRM. Cloud resource managers manage resource containers such as accounts, folders, and tasks, which can be used to group and organize other resources, respectively. Dimension listeners are used to map workloads to available resources based on user-defined QoS specifications. During resource management, cloud users with their QoS capabilities provide workloads to the cloud service provider. Resource bundle services are configured based on the user's workload to ensure optimal resource utilization. In a cloud system, effective resource management is necessary to increase productivity and improve customer loyalty. Therefore, resource planning calculates the execution time of each workload, but also takes into account the type of workload, such as workloads (conflicting workloads) with dissimilar QoS requirements and workloads (single-mixed workloads) with QoS needs as shown in Fig. 7.

3.4 Load Balancing Genetic Algorithm (LAGA) Algorithm

The load balancing genetic algorithm (LAGA) is based on an evolutionary model. Min–max and max–min are used to initialize the population. Each person's score is based on fitness performance. In each iteration, a new generation is created using select, cross, and mutation operators. Again, all new populations are evaluated to determine the most suitable individuals during each replication; they are before the next generations. This process continues until the end. Workflow is shown in the Fig. 5.

3.5 Load Balancing Ant Colony Optimization

In this algorithm, the parameters involved in ant colony optimization are initialized to determine the best food source, in which case a global random solution is generated and an appropriate value is estimated. The strength of the pheromones used to determine the path to the optimal solution. Match values for each route are discovered and pheromones are updated to help determine the new route. This process continues until all criteria are met, as shown in Fig. 6. Pheromone plays an important role in choosing a virtual machine that is optimized to achieve a more balanced load distribution [22].

4 Hybrid Approaches for Task Scheduling

Multi-object-based task scheduling is hybrid of two algorithms:

(a) Cuckoo search algorithm.
(b) Harmony search algorithm.

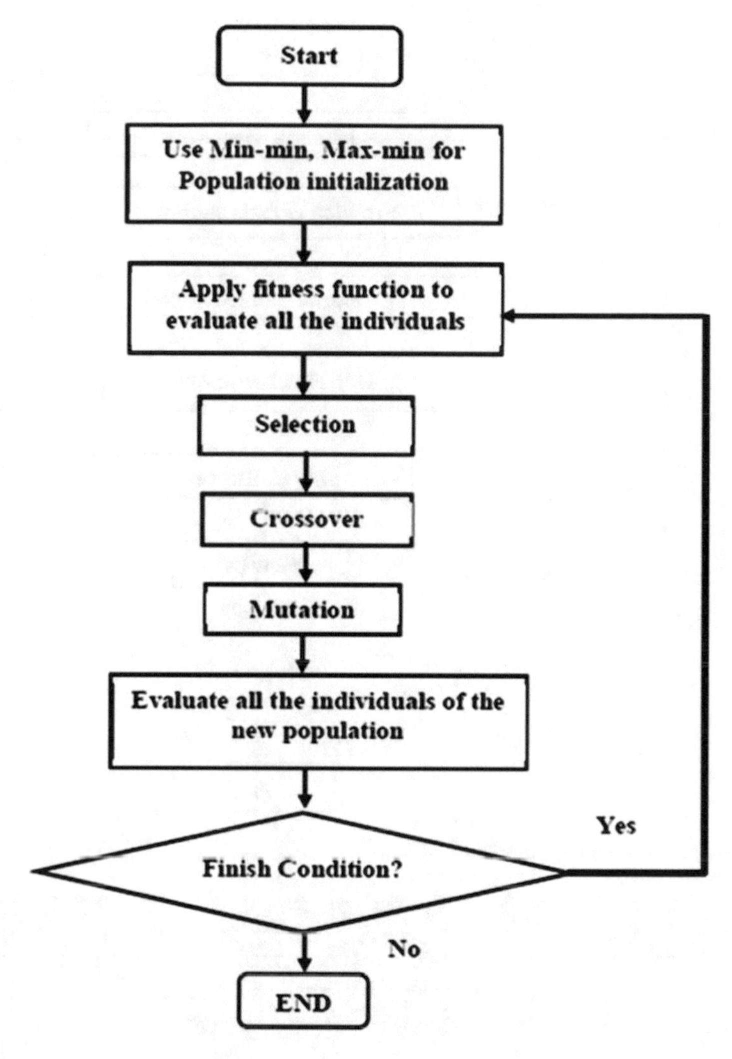

Fig. 5 Work flow of genetic algorithm

4.1 Cuckoo Search

This algorithm is based on swarm intelligence. The cuckoo's instinctive behavior served as the model for this program. Every one egg in a nest represents a potential solution within this algorithm. Each cuckoo can often only lay one egg in a nest with a certain form, yet each nest may include a variety of eggs that each stand for a variety of solutions. The fundamental goal of cuckoo search is to develop new solutions to change the poorest ones being used by the population. This can be understood from the steps below:

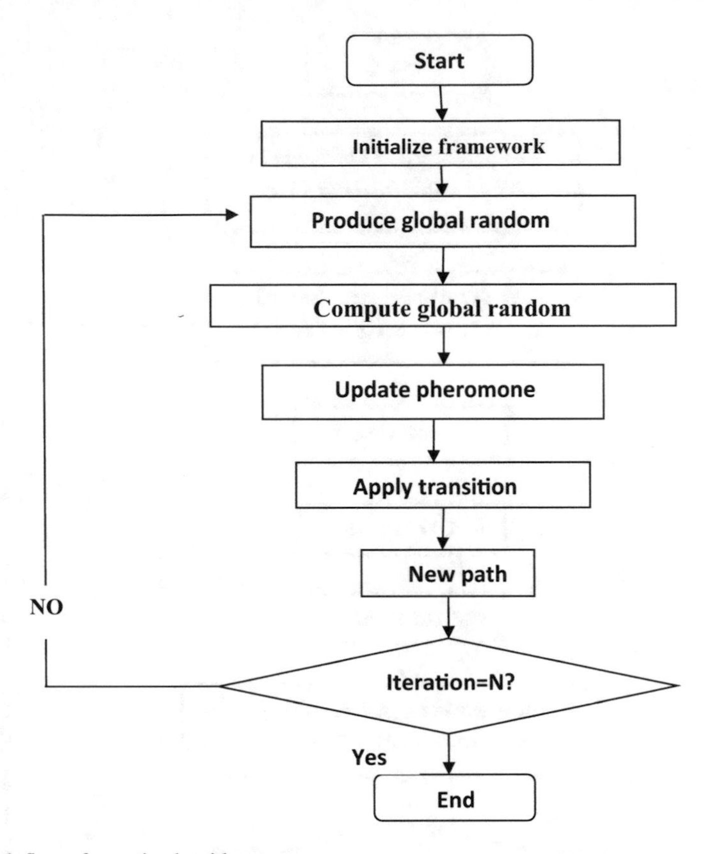

Fig. 6 Work flow of genetic algorithm

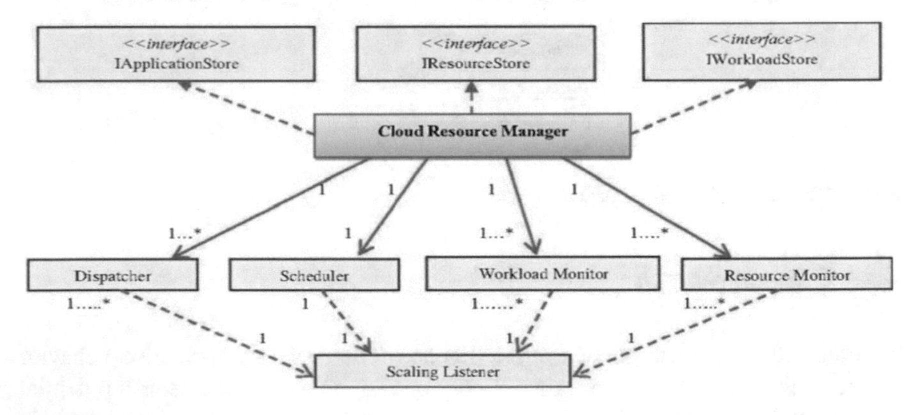

Fig. 7 Work of cloud resource manager

Step 1: initialization stage (Here a population of nest/solution is created randomly).
Step 2: initiate new cuckoo phase.
Step 3: fitness evaluation phase.
Step 4: updating phase.
Step 5: reject worst solution.
Step 6: phase of the stopping criterion.

4.2 Harmony Search

The improvisation of jazz musicians served as the basis for the Harmony Search algorithm. Its term is derived from the type of harmony that musicians create by experimenting with several combinations of the musical pitches they have learned. The method used by the musicians, who had never performed together swiftly improve their individual acts of impulsiveness via diversity, resulting in an artistic harmony; this HS calculation method looked to be particularly important in Figure, and its important steps are:

- Step 1: Initialization phase of harmony.
- Step 2: Memory initialization.
- Step 3: New harmony improvisation.
- Step 4: Updation of harmony memory.

4.3 Cuckoo Harmony Search Algorithm (CHSA)

Using a hybrid method for optimization, the primary aim of task scheduling is to plan the work. Cuckoo search and harmony search are both components of this hybrid optimization technique. The approach combines the optimization skills of harmony explore and cuckoo search to attain greater accuracy. The cuckoo search is a recognized optimization technique that aside from population size, executes the local search more effectively. Despite this, there are certain restrictions to this algorithm. In CS, continuous-valued points are where the solution is updated in the search space. The execution of the HS algorithm depends on parameter set. Therefore, static constraint characteristics are useless for modifying intensification and diversification where this algorithm is an alternative solution. By combining these algorithms, it will be possible to conquer the imperfection of the HS and CS's individual performances. It also has the compensation of accurately recognize and fast focusing, allowing this organizing strategy to obtain a idea or flawed course of exploit in the least amount of computational time. It is also used to enhance the suggested method's exploitation potential so that it can avoid becoming stuck in local optima. In light of the above, this way may discover the new exploration area using a hybrid harmony search administrator and engage the populace using cuckoo search, thoroughly abusing the advantages of both CS and HS.

4.4 Task Scheduling Using Hybrid CHSA

The scheduling procedure is based on five factors, including the cost of migration; energy use; memory utilization; recognition; and fine. The methodical approach of the planned CHSA-based job scheduling is clarified under steps.

- Step 1: solution encoding.
- Step 2: suitability encoding.
- Step 3: modernize base on cuckoo search algorithm.
- Step 4: hybridization.
- Step 5: termination criterion.

5 Analyzed Hybrid Approaches for Resource Allocation and Task Scheduling on Cloud Computing

This approach combines the optimization capabilities to obtain greater accuracy. By combining these two development algorithms, it will be possible to surmount the shortcoming of the harmony search and cuckoo search individual performances. It also has the reward of accurately recognize and fast focusing, allowing this organizing strategy to obtain flawed way of acting in the least amount of computational time. While the hybrid algorithm relies on five variables, including passage cost, energy utilization, memory utilize, recognition, and fine, to schedule a job [4]. In continuous optimization issues such as spring design and welding beams in engineering design applications, cuckoo search performs much better than other techniques. Large-scale issues are well suited to this technique. The longest-running cuckoo search is maximum, with a low success rate, is classified by an equation-based method, and the cuckoo bird's brooding parasitism forms the basis of the program. The best algorithm, according to this research, is cuckoo, followed by harmony search. The three methods behave well to solve the problem. Demonstrating that the methods used to address the problem performed well. The HS and CS performed better than the bat algorithm (BA) in the search for the best solutions for a large number of generators to identify, but the cuckoo search ultimately yielded the best results [39]. The proposed methods are evaluated using CloudSim, and performance is validated by a corresponding evaluation of other existing resource planning policies in terms of power consumption, response time, and index load balancing as key factors [31]. Hence, for better cloud computations like task scheduling and allocation of resources, cuckoo approach is considered as the best hybrid approach for cloud computation.

6 Conclusion

Convergence difficulties haven't been included in any cloud computing literature as one of the biggest obstacles to task scheduling. Most methods used to improve work scheduling focus on concepts that are based on efficiency, cost, or resource utilization. The task scheduling issues in the Np-hard minimization tasks are challenging. Incoming requests are arranged in a certain order for task scheduling in classify to create use of valuable resources. The ultimate goal of this article is to create versatile work schedule models using nature-inspired optimization algorithms with powerful resource utilization in the cloud environment. In this study, four multi-objective optimization models were analyzed and compared to solve task and resource scheduling problems in a cloud computing environment. This study suggested a multi-objective work scheduling solution based on a hybrid cuckoo harmony search algorithm (CHSA). By this hybridizing technique, it is possible to acquire scheduling explanation of higher quality. To improve the enormous cloud computations, further research is required in this active field of research.

Acknowledgements This paper supported by Subprogram 1.1. Institutional performance-Projects to finance excellence in RDI, Contract No. 19PFE/30.12.2021 and a grant of the National Center for Hydrogen and Fuel Cells (CNHPC)—Installations and Special Objectives of National Interest (IOSIN). This paper was partially supported by UEFISCDI Romania and MCI through projects AISTOR, FinSESco, CREATE, EREMI, STACK, ENTA, PREVENTION, DAFCC, RECICLARM, UPSIM, SmartDelta, ENRICH4ALL and by European Union's Horizon 2020 research and innovation program under grant agreements No. 872172 (TESTBED2), No. 883522 (S4ALLCITIES).

References

1. Chopra N, Singh S (2013) HEFT based workflow scheduling algorithm for cost optimization within deadline in hybrid clouds. pp 2–7
2. Alworafi MA, Dhari A, El-Booz SA, Nasr AA, Arpitha A, Mallappa S (2019) An enhanced task scheduling in cloud computing based on hybrid approach. Springer, Singapore. https://doi.org/10.1007/978-981-13-2514-4
3. Midya S, Roy A, Majumder K (2017) Author's accepted manuscript multi-objective optimization technique for resource allocation and task scheduling in vehicular cloud architecture: a hybrid adaptive nature inspired approach reference: phadikar, multi-objective optimization technique fo. J Netw Comput Appl. https://doi.org/10.1016/j.jnca.2017.11.016
4. Cheng M, Li J, Bogdan P, Nazarian S (2019) Resource and quality of service-aware task scheduling for warehouse-scale data centers : a hierarchical and hybrid online deep reinforcement learning-based framework. 14(8). https://doi.org/10.1109/TCAD.2019.2930575
5. Goudarzi H, Ghasemazar M, Pedram M (2012) SLA-based optimization of power and migration cost in cloud computing. https://doi.org/10.1109/CCGrid.2012.112
6. Liu T, Chen F, Ma Y, Xie Y (2016) An energy-efficient task scheduling for mobile devices based on cloud assistant. Futur Gener Comput Syst 61(2016):1–12. https://doi.org/10.1016/j.future.2016.02.004

7. Xiong Y, Wan S, She J, Wu M, He Y, Jiang K (2016) An energy-optimization-based method of task scheduling for a cloud video surveillance center. J Netw Comput Appl 59:63–73. https://doi.org/10.1016/j.jnca.2015.06.017
8. Wang X, Wang Y, Cui Y (2016) An energy-aware bi-level optimization model for multi-job scheduling problems under cloud computing. Soft Comput 20(1):303–317. https://doi.org/10.1007/s00500-014-1506-3
9. Wu K, Lu P, Zhu Z (2016) Distributed online scheduling and routing of multicast-oriented tasks for profit-driven cloud computing. IEEE Commun Lett 20(4):684–687. https://doi.org/10.1109/LCOMM.2016.2526001
10. Tang Z, Qi L, Cheng Z, Li K, Khan SU, Li K (2016) An energy-efficient task scheduling algorithm in DVFS-enabled cloud environment. J Grid Comput 14(1):55–74. https://doi.org/10.1007/s10723-015-9334-y
11. Li Z et al (2016) A security and cost aware scheduling algorithm for heterogeneous tasks of scientific workflow in clouds. Future Gener Comput Syst 65:140–152. https://doi.org/10.1016/j.future.2015.12.014
12. Zhu X, Chen C, Yang LT, Xiang Y (2015) ANGEL: agent-based scheduling for real-time tasks in virtualized clouds. 9340(c):1–14. https://doi.org/10.1109/TC.2015.2409864
13. He H, Xu G, Pang S, Zhao Z (2016) Strategies and schemes AMTS: adaptive multi-objective task scheduling strategy in cloud computing. pp 162–171
14. Ramezani F, Lu J, Taheri J, Hussain FK (2015) Evolutionary algorithm-based multi-objective task scheduling optimization model in cloud environments. pp 1737–1757. https://doi.org/10.1007/s11280-015-0335-3
15. Radojević B, Žagar M (2014) Analysis of issues with load balancing algorithms in hosted (Cloud) environments. No. Jan 2011
16. Zhan ZH, Zhang GY, Gong YJ, Zhang J (2014) Load balance aware genetic algorithm for task scheduling in cloud computing. Lect Notes Comput Sci (including Subser Lect Notes Artif Intell Lect Notes Bioinform), 8886(2013):644–655. https://doi.org/10.1007/978-3-319-13563-2_54
17. He C, Zhu X, Guo H, Qiu D, Jiang J (2012) Rolling-horizon scheduling for energy constrained distributed real-time embedded systems. J Syst Softw 85(4):780–794. https://doi.org/10.1016/j.jss.2011.10.008
18. Ghanbari S, Othman M (2012) A priority based job scheduling algorithm in cloud computing. Procedia Eng 50(Jan):778–785. https://doi.org/10.1016/j.proeng.2012.10.086
19. Ogedengbe MT, Agana MA (2017) New fuzzy techniques for real-time task scheduling on multiprocessor systems. Int J Comput Trends Technol 47(3):189–196. https://doi.org/10.14445/22312803/ijctt-v47p129
20. Liu Y, Yu FR, Li X, Ji H, Leung VC (2018) Hybrid computation offloading in fog and cloud networks with non-orthogonal multiple access. pp 154–159
21. Arunarani AR, Manjula D, Sugumaran V (2018) Task scheduling techniques in cloud computing : a literature survey. Future Gener Comput Syst. https://doi.org/10.1016/j.future.2018.09.014
22. Gawali MB, Shinde SK (2018) Task scheduling and resource allocation in cloud computing using a heuristic approach. J Cloud Comput 7(1). https://doi.org/10.1186/s13677-018-0105-8
23. Srichandan S, Kumar TA, Bibhudatta S (2018) Task scheduling for cloud computing using multi-objective hybrid bacteria foraging algorithm. Future Comput Inform J 3(2):210–230. https://doi.org/10.1016/j.fcij.2018.03.004
24. Schäfer D, Edinger J, Eckrich J, Breitbach M, Becker C (2018) Hybrid task scheduling for mobile devices in edge and cloud environments. pp 669–674
25. Beegom ASA, Rajasree MS (2019) Integer-PSO: a discrete PSO algorithm for task scheduling in cloud computing systems. Evol Intell 12(2):227–239. https://doi.org/10.1007/s12065-019-00216-7
26. Malinen MI, Fränti P (2014) Balanced k-means for clustering. Lect. Notes Comput. Sci. (including Subser Lect Notes Artif Intell Lect Notes Bioinform), vol 8621 LNCS, pp 32–41. https://doi.org/10.1007/978-3-662-44415-3_4

27. Ma R, Li J, Guan H, Wang BIN, Wei DSL (2015) WaSCO: a hybrid enterprise desktop wake-up system based on cloud infrastructure. IEEE Access 3:2000–2013. https://doi.org/10.1109/ACCESS.2015.2491959
28. Niu M, Cheng B, Chen JL (2020) GMAS : a geo-aware MAS-based workflow allocation approach on hybrid-edge-cloud environment. pp 574–581
29. Kumar T S, Mustapha SD, Gupta P, Tripathi RP (2021) Hybrid approach for resource allocation in cloud infrastructure using random forest and genetic algorithm. vol 2021
30. Kiruthiga G, Mary Vennila S (2020) Energy efficient load balancing aware task scheduling in cloud computing using multi-objective chaotic darwinian chicken swarm optimization. Int J Comput Networks Appl 7(3):82–92. https://doi.org/10.22247/ijcna/2020/196040
31. Vhatkar KN, Bhole GP (2022) Optimal container resource allocation in cloud architecture: a new hybrid model. J King Saud Univ Comput Inf Sci 34(5):1906–1918. https://doi.org/10.1016/j.jksuci.2019.10.009
32. Bindu GH, Ramani K, Bindu CS (2018) Energy aware multi objective genetic algorithm for task scheduling in cloud computing. Int J Internet Protoc Technol 11(4):242–249. https://doi.org/10.1504/IJIPT.2018.095408
33. Pop F, Dobre C, Cristea V, Bessis N, Xhafa F, Barolli L (2015) Deadline scheduling for aperiodic tasks in inter-cloud environments: a new approach to resource management. J Supercomput, pp 1754–1765. https://doi.org/10.1007/s11227-014-1285-8
34. Jiao H, Zhang J, Li JH, Shi J, Li J (2015) Immune optimization of task scheduling on multidimensional QoS constraints. Cluster Comput 18(2):909–918. https://doi.org/10.1007/s10586-015-0447-7
35. Zhang F, Cao J, Li K, Khan SU, Hwang K (2014) Multi-objective scheduling of many tasks in cloud platforms. Future Gener Comput Syst 37:309–320. https://doi.org/10.1016/j.future.2013.09.006
36. Ojha SK, Optimal load balancing in three level cloud computing using osmotic hybrid and firefly algorithm
37. Weingärtner R, Bräscher GB, Westphall CB (2015) Cloud resource management: a survey on forecasting and profiling models. J Netw Comput Appl 47:99–106. https://doi.org/10.1016/j.jnca.2014.09.018
38. T. scheduling algorithm Comparision, Cuckoo and harmony algorithnm comparison. https://www.academia.edu/RegisterToDownload/UserTaggingSurvey

Beginners Online Programming Course for Making Games in Construct 2

Veronika Gabal'ová, Mária Karpielová, and Veronika Stoffová

Abstract The article regards the game-making process as a method for teaching programming in primary school classrooms. The article analyses the key programming concepts that primary school pupils are expected to learn are describes the place of programming in the Slovak educational system. The authors of the article describe an alternative in a form of teaching programming through making games in an authentic game-making software. The article includes a preview of such a learning process and a brief description of a beginner's online programming course for making games in Construct 2 which was tested on primary school pupils during an extracurricular course of programming in selected schools. To analyze the effectiveness of the course, a pedagogical experiment was carried out using the questionnaire method to gather data and using the descriptive statistics to assess the results. Participants in the experiment were pupils aged 8–14 who responded to a short questionnaire at the end of each tutorial in the course. First of all, we wanted to find out and check the suitability and quality of video tutorials to gain practical experience in programming for elementary school students.

Keywords Game-making process · Constructionism · Programming in primary schools · Programming course

1 Introduction and Related Work

Computer games are a powerful device in the hands of educators. For one reason or another, teachers are looking for more ways to salvage their potential in all aspects of

---

V. Gabal'ová · M. Karpielová · V. Stoffová (✉)
Faculty of Education of Trnava University in Trnava, Trnava, Slovakia
e-mail: veronika.stoffa@gmail.com; veronika.stoffova@truni.sk

M. Karpielová
Faculty of Mathematics, Physics and Informatics, Comenius University in Bratislava, Bratislava, Slovakia

V. Stoffová
Faculty of Informatics, Eötvös Loránd University, Budapest, Hungary

Y. Singh et al. (eds.), *Proceedings of International Conference on Recent Innovations in Computing*, Lecture Notes in Electrical Engineering 1001,
https://doi.org/10.1007/978-981-19-9876-8_41

a lesson. While it is imperative in all school subjects, it is even more so in computer science, since a computer game manifests what an algorithm is and how it works. Therefore, we could say that for most children, computer games are the initial gate into programming, even though they might not know it yet.

Computer games, however, provide a different level of engagement, depending on how computer science teachers use them—whether the game presents an input or an output of the learning process. In this article, we would like to focus on the latter and present how the game-making process affected pupils' relationship to programming according to our survey.

The initial idea behind the research leading to the survey was the authors' shared belief that in the game-making process, there are incorporated, among other benefits further discussed later, all of the key programming concepts that pupils of primary schools are expected to learn during their education. The goal was to provide an alternative based on constructionism rather than instructionism, which has been prevalent in Slovak schools in recent years. This alternative took the form of a beginner's programming course for making computer games in an authentic game-making software. However, before the course could be designed, it was important to answer the question of suitability of programming environments for the game-making process. Authors chose Construct 2 as one of the suitable options, and this article discusses its effectiveness as a tool for teaching programming to primary school pupils.

2 Game-Making and Teaching Programming and Algorithmisation

The notion of creating a computer game is usually viewed as intricate, time-consuming, and more than anything else it denotes proficiency in programming and a thorough design thinking. That is more than enough reasons for teachers to be wary of including it in the educational process. The extent of what is to be learnt is the result of the thought processes behind the design and content of the didactic game. As Kafai [6] concluded, even in the process of making an instructional game, "the greatest learning benefit remains reserved for those engaged in the design process, the game designers, and not those at the receiving end, the game players."

The constructionist approach pioneers having learners consciously engaged in the process of creating their own public entities [10], which in the context of programming naturally includes the game-making process. Papert advocates that in education, doing is more than looking from a distance, "and that connectedness rather than separation are powerful means of gaining understanding" [1]. On closer inspection, in the game-making process, the focus is on the way a game, and its elements are constructed, similar to how we learn about a computer by assembling its pieces into a working machine. By putting together all of its pieces, we get a firmer grasp on their relations and their function, and it helps to bridge what we already know about them and new connections created in the process. The same applies for programming games.

Programming is considered a solitary activity, which combined with the amount of abstract concepts pupils need to learn and understand, and makes the quality of learning even more vital for gaining the skill. However, as a game is literally a manifestation of an algorithm, making a game makes for a beneficial learning environment. Moreover, according to Stoffová [11–13], Czakóová [2], Czakóová and Udvaros [3], a game itself might be viewed as a visualized simulation model of an algorithm, one pupils can engage with, and the creation process of which allows for deep learning as it creates meaningful context for learning, which is crucial for programming.

Another benefit of programming a game is the fact that it also highlights a part of the curriculum which is commonly overlooked and that is fixing errors and debugging. Weinthrop and Wilensky draw a connection between playing a game and programming one by pointing out how both create a low-stake environment for trying various strategies for success and learning from prior mistakes. "By aligning the construction of programs with the act of gameplay, players are situated in a context where early failures are expected and provide valuable learning experiences" [16]. This mind-set could potentially promote the freedom of discussion in the classroom and peer-learning as the goal of everyone included is to create "a working machine." However, for this to happen, Kolodner [7] advises, it would be imperative to give tasks divergent enough that pupils could have the freedom of how to complete them.

In our survey, we had attempted to have pupils learn to program computer games by working in an authentic software for making games. This choice was intentional, as we believe that the significance of the context for learning is great, especially for the constructionist approach, which is still mostly absent in Slovakia. We hoped to observe whether the unintuitive environment would boost their motivation in the process of creating their own games [14, 15].

3 How Making Games Teaches Pupils Key Programming Concepts

To ensure that the whole programming curriculum would be appropriately covered in the programming course, it was important to analyze which concepts pupils are supposed to learn.

In the first four years of primary school in Slovak republic, the main objective is to build a foundation for programming concepts as they are expected to learn in future. Pupils are introduced to what a program is, how to communicate with it and to understand sequence and its meaning for programming. To prepare them for more complex and abstract concepts such as loops or conditions, they learn to verify arguments. Eventually, by the end of the lower secondary education (grades 5–9 in Slovakia), pupils learn how to interact with programs and how to use programming language and its basic elements. They learn to use loops, conditions, variables, and command inputs and outputs. Since these concepts are more difficult to explain because of their abstract nature, the role of teachers and their pedagogical transformation is imperative. Based on authors' own experience, it is possible to, at least to some extent, explain any of the aforementioned concepts even to younger pupils, given the appropriate analogies from pupils' real life, and, coincidentally, from computer games too.

The conclusion is that by the end of their primary and lower secondary education, pupils are expected to learn verification of an argument/logic, sequence, loops, conditional statements, variables, and working with tools for interaction—input. Analyzing various computer game elements, this is how these programming concepts are incorporated in games.

Understanding sequence is simple, since it is prevalent in everyday life much as it is in a game. When an event happens in a game, be it the start of a new game, the collision of a player with an enemy or clearing the level, there is a list of actions that need to happen next. Whenever pupils program events, they are faced with a number of actions and the more games they make and debug the more they become aware that the order does make a difference. This is how they learn about sequence, even though they might not know the word for it yet.

Another key concept is using conditionals. Conditionals are used in the simplest of events, such as the detection of a collision between a two objects. The condition can contain a variable (when variable *score* is higher a certain value, open the door to the next level), key inputs (press W to jump) or even the two combined (when variable lives is higher than 0 and space key is pressed, player spawns an arrow).

Conditionals are also used with loops and since a loop needs an event to start it, even if the event is the start of a game, loops are exclusively used with conditionals.

A loop is active while the condition is true. The end of a game is one example. If a player has at least one life, he can continue and can be controlled by the input. If not, the game goes back to the game menu. Another example can be the use of boosts, e.g., to become invisible or to increase speed for a few seconds. When the player uses the boost and while *timer* is higher than 0, the spaceship spawns light particles to signal a boost being used.

Although variables can be difficult to explain due to their abstract character, in programming games, they play an important role and are very useful. Therefore, pupils intrinsically learn what they are by using them for various game elements. Games make use of them in keeping the count of score, lives, or health in general or the reached level of the player. However, even more important is their use in operating states of objects or the game itself. For example, it is common practice that Boolean is used to operate states of the game; true when the game is being played and false when the game ends. Moreover, when more than two states are needed, e. g., when programming a basic artificial intelligence for enemies to follow the player when the player gets in their line of sight and stopping when the player escapes, variables manage these states and help switch among them.

When we discuss a game as a result of an activity, Winthrop and Wilensky (2016) recognize two processes—the process of creating a game and a process of fixing a game, modifying it or adding new functionalities to it. Both of these processes fulfill different expectations of the innovative state educational program in Slovakia, and the latter is covered in the subsection, named fixing problems and debugging. According to the standards, pupils should learn to recognize when a program is not working well, and they should be able to discuss options for solutions [4, 5, 9]. Eventually, they should be able to navigate or even help finish other pupils' projects. For a teacher, to achieve this is not an easy task, however, from the authors' experience, the game-making process helps to naturally create safe space for these discussions to take place, which the authors of the article consider a great benefit.

This brief analysis is similar to the discussion that had taken place before the beginner's programming course was designed. It was one of the most important steps in the whole process, because the analysis allowed for more concise tutorials that would consistently build one on another. In the following section, the course and the chosen game-making software will be discussed in more detail.

The notion of creating a computer game is usually viewed as intricate, time-consuming, and more than anything else, it denotes proficiency in programming and a thorough design thinking. That is more than enough reasons for teachers to be wary of including.

4 Description of the Beginner's Programming Course for Making Games

Before the course was designed, two points had been agreed on. The first point was that the course would introduce beginners to programming games. And after finishing the course, beginners should be able to create three types of simple games. The other point was to design it so that it can become a sort of a guide through the software and the whole process of programming a game, one that can be used individually as the course would include all the information, a beginner would need for a successful completion of the course.

The game-making software Construct 2 was chosen as a suitable environment. It is a software created for game-makers, designers, and game enthusiasts who desire to create their own 2D games without programming skills. It works based on the event system. Instead of a programming language, game-makers use events, where each event consists of a condition and actions that would be carried out in a sequence they were ordered in. Moreover, since the software was created for the purpose of creating games, it includes specific features such as effects and various pre-programmed functions called *behaviors* which make the whole process more effective. The tutorials were designed to be completed with its free version, too.

5 Tutorials

The core of the programming course is eight tutorials that consistently introduce pupils to game programming. The idea behind the first six tutorials was to design them in a way that at the end of each, pupils would have playable games. This approach was to ensure a positive first experience using the unknown software. In these tutorials, pupils are shown how to create three basic types of games: a platformer game, a game for two players, and a simple space shooter game.

As the goal was to give pupils tools to make their own games and not just the three types of games that are presented in tutorials, eventually, the last two tutorials were designed. These tutorials do not instruct pupils to make games, rather they show how they can leverage the features of the game-making software to program interesting elements into their games which they can use in their next projects.

The process of creating the materials was not linear, and the tutorials themselves have been edited based on the feedback that was received from lectors from the courses where the tutorials had been tested. After the revision, the next step was to create supporting materials for the Website of the course.

Using tutorials for making games has three functions: It helps to gain basic skills in the software, it shows how to plan a game and its various parts, and it helps build confidence through positive experience. However, this is not the most difficult part [8] (Figs. 1 and 2).

The subsection *practice* presents six simple games which can be used as an inspiration for the next independent project. Each of the games shows a screenshot of the game, it is explained, and there is a hyperlink too in order for them to try it and see how it is operated. Moreover, to make the process of making the next project easier, with each game, there are hints as to how to program certain elements of the game in Construct 2, which behaviors would be useful or what they can add to the game to make it more interesting. Although the games are conceptually simple, they were

Fig. 1 Tutorial 1—your first platformer—simple basic game

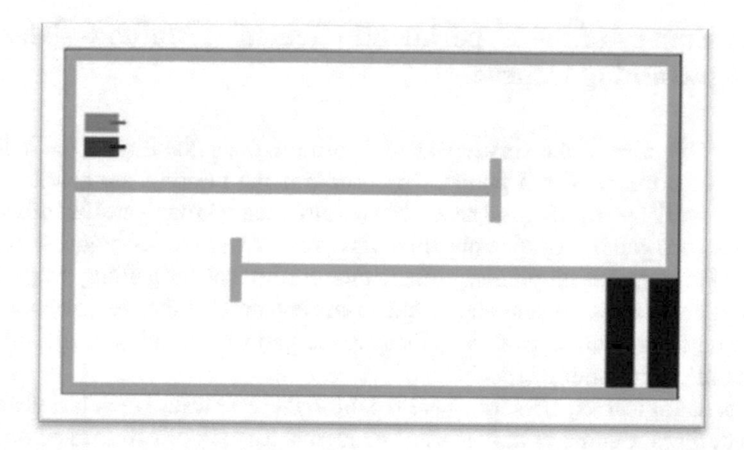

Fig. 2 Tutorial 2—car racing—example from the game

chosen on the basis of their difference from the games and elements in the tutorials to encourage pupils to try and learn new mechanics in Construct 2. The goal is to provide pupils with a repertoire of skills they could use later to make their own games in future (Fig. 3).

The sequence of images 1–3 only illustrates that the teaching progressed from simple to more complex—from the creation of a basic elementary platform with the gradual addition of new functions of the Construct 2 environment. Thus, the students gradually became familiar with a sufficient number of elements and discovered other possibilities, which they learned to use effectively during implementation their ideas and the implementation of their own program projects.

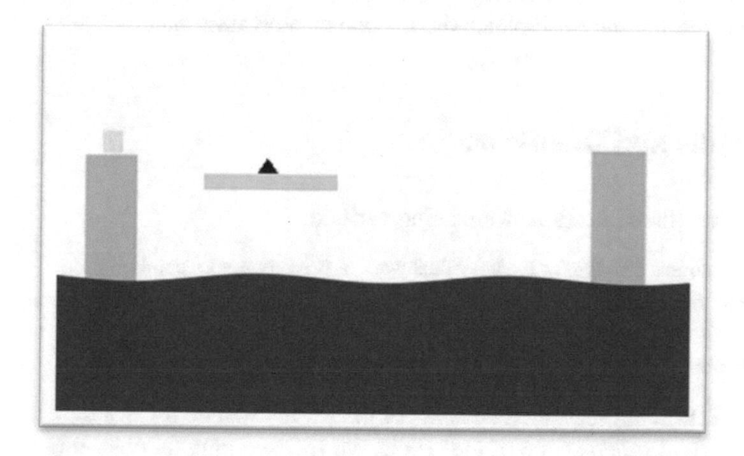

Fig. 3 Tutorial 8—obstacles in Construct 2—sample project

6 The Survey of the Experimental Teaching Method—The Programming Course

The main objective of the survey was to determine the suitability of the individual tutorials for primary school pupils and observe if the tutorials are able to support individual work by providing all needed materials to learn to navigate the software and make their own games. Another objective was to explore the motivational effect of the game-making process on pupils. Authors understand that for gaining programming skills, it is important to ensure pupils' autonomy and provide them with opportunities to study on their own, since it is an inseparable part of becoming proficient in any skill, but in programming especially.

The pedagogical experiment where the tutorials were tested took last years, with a break during the summer holidays and pandemic time. Pupils were aged 8–14, and they were following tutorials and instructions simultaneously, so when they finished the tutorial, they also had their project finished. At the end of each tutorial, they could choose whether they would do the optional challenge, and then, they filled in a short questionnaire for self-reflection.

Considering the sample, it was apparent that the questionnaire would have to be adjusted. Therefore, the number of items was minimized and divided in several sections. The first section collected data on pupils' grade, the extent of their previous experience in C2, and whether they needed assistance or not. The next three sections were inquiring about the three new introduced items, e. g., *would you be able to program that the blue tank loses life when a bullet shoots it?* And the options were *yes, without help, yes, with a small help, yes, but with the help* or *no*. The goal was to have pupils assess their skill and to provide information whether the tutorials were suitable to encourage autonomous learning. In the final part, the inquiry was about the optional challenge, whether pupils completed at least one of them or not, and if yes, then whether they completed them alone or with assistance.

7 Results and Discussion

By answers, three questions were being verified:

Was the course suitable for the individual work of primary school pupils?

How many elements would the primary school students be able to learn from those tutorials?

Would they spend more time and effort to complete optional challenges?

In the following section, the results of the survey will be reviewed.

Due to the limited scope of the article, we focused only on obtaining an answer to the first question. After each tutorial, we evaluated the students' answers to the question of how they worked with the instructions. As an answer, the student could

Table 1 Overview of responses from the questionnaire

Tutorial number	Answer			
	Alone	The lector helped me	Together with the lector	Number of respondents
1	50% (17)	47% (16)	3% (1)	34
2	62% (16)	38% (10)	0% (0)	26
3	75% (18)	25% (6)	0% (0)	24
4	82% (18)	18% (4)	0% (0)	22
5	65% (13)	35% (7)	0% (0)	20
6	70% (14)	30% (6)	0% (0)	20
7	71% (15)	29% (6)	0% (0)	21
8	71% (15)	29% (6)	0% (0)	21
Average	68.25%	31.375	0.375	23.5

choose from three options: alone, the lector helped me, together with the lector. An overview of the results is presented in the Table 1.

When surveying the suitability of the course and its efficiency to provide enough information for a successful completion of the course, the collected data assessed whether pupils needed help to complete a tutorial, therefore, whether the new information was presented in a way suitable for individual learning. Tutorials were designed to be concise, visually appealing, and containing short instructions and arrows to make the tutorial easier to follow. The analysis proved that on average, it was 68.28% of pupils who worked with tutorials individually and 31.375% of them needed a small assistance from their lector. Less than a percentage of pupils replied that they required a lector to complete the tutorial with them (Fig. 4). While the analysis looks promising, we understand the uniqueness of the survey sample and the fact that pupils at programming courses are generally led to work individually. Nonetheless, understanding that learning programming is specific because it requires processing of content on an intellectual, abstract level, and the results of almost three quarters of pupils working with tutorials alone which is viewed as positive feedback.

The last column in Table 1 shows that at the beginning, there were 34 students in the group, of which the number gradually decreased, and finally, there were 21 who successfully completed the course.

8 Conclusion

Schools face many challenges in teaching programming. There is a lack of qualified teachers, and more often teachers struggle with the curriculum and do not know how to approach programming and its concepts to their students in an appropriate way.

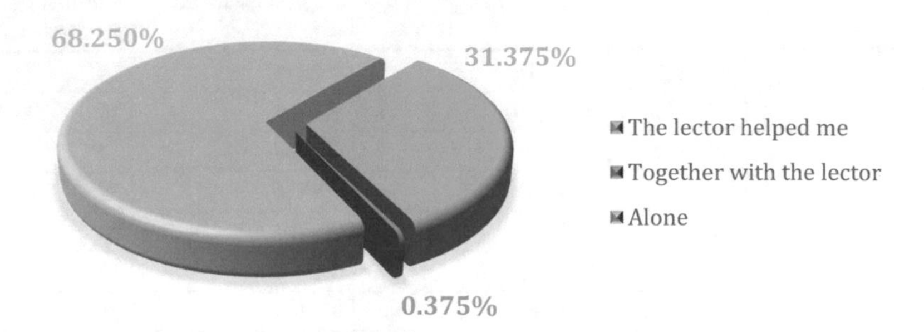

Fig. 4 Visualization of how the students worked with the tutorials in average

Such teachers need help. The number of methodologies for teaching programming through programming games in Slovakia is to this day very low.

The game-making process is, understandably, a scarecrow for some teachers, as they cannot imagine what it involves or how to approach it.

With the presented game programming course for beginners, we wanted to show that programming can be taught not only by creating games, but that teaching in Construct 2 enables discovery and supports students' creativity and their autonomy in the learning process, both of which are important in the optic of constructionism.

Programming games merge a meaningful context for learning and a natural curiosity as pupils often dedicate time to playing games in their free time and to be involved in the process of making rather than playing provides a new perspective and new opportunities for their learning.

The results of the survey created more questions than it had answered in the field of teaching programming games, even more so for using an authentic game-making software in the process. Although programming environment Construct 2 has not been used in Slovak schools so far, the positive feedback together with gathered data could fuel further, more insightful research. In our next research work, we would like to closely inspect the relationships between variables and a more homogenous research sample, how they would assess the course and their gained abilities, if the course as a unit would be suitable to support individual learning and later making their own games based on what they learnt from it. On teachers' part, it is important to gain a sort of designer's thinking to be able to analyze games and suitably use various game elements, so that no programming concept is underrepresented in the lessons.

We do not consider our research to be over. The results we obtained only created space for other questions in the field of game programming in the school preparation of pupils. We see it as a motivation for further, deeper research in this area. In future, we would like to focus on tracking the relationships between the variables in more detail with a research sample of complete beginners, how they would evaluate their knowledge after using individual tutorials as well as after completing the entire course. Another direction in which the research could be continue is the analysis and evaluation of the created own games after the completion of the game-making programming course and monitoring which mechanics the students use most often. We are also interested in whether any of the key programming concepts are neglected in the game creation process, so that appropriate prevention can be set. We truly believe that this work can serve as the first of the building blocks of a larger academic research that would bring valuable insights for modern programming didactics.

Acknowledgements The paper was supported by the national project, KEGA 013TTU4/2021 "interactive animation and simulation models for deep learning."

References

1. Ackermann E (2001) Piaget's constructivism, Papert's constructionism: what's the difference?
2. Czakóová K (2021) Game-based programming in primary school informatics. In: INTED 2021 Proceedings of the 15th international technology, education and development conference. IATED Academy, Valencia
3. Czakóová K, Udvaros J (2021) Applications and games for the development of algorithmic thinking in favor of experiential learning. In: EDULEARN21: Proceedings of the 13th international conference on education and new learning technologies. https://doi.org/10.21125/edulearn.2021.1389, pp 6873–6879, IATED Academy, Valencia. ISBN 978-84-09-31267-2. ISSN 2340-1117
4. Informatika – nižšie stredné vzdelávanie. https://www.statpedu.sk/files/articles/dokumenty/inovovany-statny-vzdelavaci-program/informatika_nsv_2014.pdf (2014)
5. Informatika - primárne vzdelávanie. https://www.statpedu.sk/files/articles/dokumenty/inovovany-statny-vzdelavaci-program/informatika_pv_2014.pdf (2014)
6. Kafai YB (2006) Playing and making games for learning: instructionist and constructionist perspectives for game studies. Games Culture 2:36–40
7. Kolodner JL (2006) Case-based reasoning. In: Sawyer K (ed) The Cambridge handbook of the learning sciences. Cambridge University Press, Massachusetts, pp 225–242
8. Karpielová M (2020) Creating of E-course as supporting material for teaching the course of programing in Construct 2 [Master Thesis], Trnava University in Trnava, Faculty of Education, Department of Mathematics and Computer Science, Supervisor of the Thesis PaedDr. Veronika Gabaľová, PhD, Degree of professional qualification. Master (Mgr) Trnava, TU FoE, 129p
9. Katyetova A (2022) Development of algorithmic and programming thinking at primary school in state educational programs. In: Klement M, Částková P, Šaloun P, Dostál J, Sedláček M, Kubrický J (eds) Trends in education & DidMatTech 2022 (abstracts) (eds.), Univerzita Palackého v Olomouci, p 14. https://doi.org/10.5507/pdf.22.24461243 ISBN 978-80-244-6124-3
10. Papert S, Harel I (1991) Constructionism. Ables Publishing Corporation, New Jersey
11. Stoffová V (2016) The importance of didactic computer games in the acquisition of new knowledge. In: futureacademy.org.uk [online]. Access: https://www.futureacademy.org.uk/files/images/upload/70_4323_fulltext.pdf
12. Stoffová V (2018) Computer games as a tool for development of algorithmic thinking. In: The European Proceedings of Social & Behavioural Sciences EpSBS. Corresponding Author: Selection and peer-review under responsibility of the Organizing Committee of the conference, eISSN: 2357-1330 (2018)

13. Stoffová V (2019) Educational computer games in programming teaching and learning. In: New technologies and redesigning learning spaces: eLearning and software for education. Bucuresti: Carol 1 National Defence University, 2019. ISSN 2066-026X, CD-ROM, pp 39–45. WoS. https://doi.org/10.12753/2066-026X-19-004
14. Végh L, Takáč O (2021a) Online games to introducing computer programming to children. In: Gómez Chova L, López Martínez A, Candel Torres I (eds) INTED2021 Proceedings. 15th international technology, education and development conference, 8th–9th Mar 2021, pp 10007–10015. ISBN: 978-84-09-27666-0, ISSN: 2340-1079
15. Végh L, Takáč O (2021b) Mobile coding games to learn the basics of computer programming. In: Gómez Chova L, López Martínez A, Candel Torres I (eds) EDULEARN21 Proceedings. 13th international conference on education and learning technology, pp 7791–7799. ISBN: 978-84-09-31267-2, ISSN: 2340-1117
16. Weintrop D, Wilensky U (2016) Playing by programming: making gameplay a programming activity. https://ccl.northwestern.edu/2016/playingbyprogramming.pdf

Cyber Security and Cyber Physical Systems

Ride-Sharing Service Based on Ethereum

Faiyaz Ahmad, Mohd Tayyab, and Mohd Zeeshan Ansari

Abstract A ride-sharing service is a centralized system in which riders share the rides. Such ride-sharing service provides tracking, fare calculation, and cancelation of rides. Such a centralized system suffers from an issue such as single point of failure, fairness of the system, surge pricing, reliance on a third party, and privacy disclosure. Such systems are even at the risk of Sybil and distributed denial of service (DDoS) attacks. In the past, breach of confidential data has also been an issue such as the UBER data attack. Ethereum-based ride-sharing services permit riders and drivers to use ride-sharing systems without depending upon arbitrators. Ethereum allows the creation of a decentralized ride-sharing service. All ride-sharing functions like fare assessment, route management, and rider and driver details are stored on Ethereum. It makes sure the reliability, confidentiality, and integrity of data are achieved by using public and private keys. Smart contracts are used to deploy the functions in the blockchain. The purpose of this paper is to develop an Ethereum-based decentralized system for ride-sharing where all the activity related to ride-sharing such as user, route, and ride-generated data will be stored on the Ethereum blockchain. This makes the ride-sharing network more dependable and unambiguous. As the data stored on Ethereum-based smart contract is immutable, that makes the stored ride-related information more secure and confirms user privacy.

Keywords Ride-sharing · Decentralized · Blockchain · Fairness · Ethereum

1 Introduction

Ride-sharing is one of the important ways of conveyance nowadays as it helps to reach the destination without the trouble of the public transportation system. It facilitates users to share their routes with other users in the system. It helps in enhancing traffic-related complications which ensures less greenhouse effect and more fuel conservation. Existing ride-sharing systems are centralized which makes them prone

F. Ahmad · M. Tayyab · M. Z. Ansari (✉)
Department of Computer Engineering, Jamia Millia Islamia, New Delhi, India
e-mail: mzansari@jmi.ac.in

Y. Singh et al. (eds.), *Proceedings of International Conference on Recent Innovations in Computing*, Lecture Notes in Electrical Engineering 1001,
https://doi.org/10.1007/978-981-19-9876-8_42

to multiple kinds of attacks and such a system lacks fairness in the calculation of fare price and charges high service fees. To tackle such issues, blockchain-based decentralized application is an effective solution. Blockchain is a distributed ledger where every record written on the ledger has a unique key. Every blockchain record is put in writing and authorized by the trusted party which belongs to the blockchain and stamps that record. These records are then chained together to create a chain of multiple data blocks.

1.1 Blockchain

A blockchain could be thought of as an upgrade over the existing centralized systems whether it be banking systems, healthcare data systems, mortgages, supply chain management, etc. Blockchain may be defined as an ever-increasing chain of record blocks where each block is connected to the other to form a chain and there is no centralized server to manage the data. That is to say, the data is stored in a decentralized manner. After the 2007–2008 Global Financial Crisis, a need to bring trust into the financial market gave a chance for blockchain technology to emerge. It differs from the existing data storage systems in how the data is structured in the case of blockchains. Based on access authorization, blockchains could be categorized into three distinct types:

- Public blockchain
- Private blockchain
- Federated blockchain

A public blockchain may be defined as one which allows anyone to join, manage, and access the data. It can be called truly decentralized. In a private blockchain, only the members who have an invitation to join can become a part of the network. Finally, a hybrid blockchain is one that is a combination of the previously defined two. It exists between the public and private blockchain and allows people from multiple organizations to join and manage after getting an invitation. This type of blockchain might also be referred to as the consortium blockchain or federated blockchain. All the transactions stored in the blocks of a blockchain are hashed using cryptography to make the data immutable, hence secure. All the data is transparent because it is always visible to everybody in the network and even if some data is tampered with, it is easily detectable.

Figure 1 lists some of the significant characteristics of a blockchain. Blockchain is programmable in the form of smart contracts which allow us to execute the transaction on the blockchain network. Immutability is one of the main characteristics of blockchain which does not allow anyone to edit the transaction stored over the blockchain. In this distributed architecture, every transaction is timestamped. Every user has access to this shared distributed ledger. Network participants validate the transaction unanimously based on a proof of work consensus mechanism.

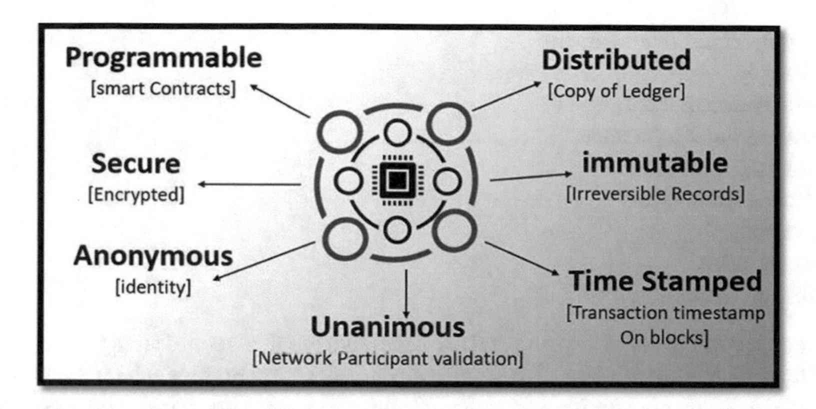

Fig. 1 Blockchain properties

In a blockchain, when a transaction is requested, after its authentication, a block is created. This newly created block is sent to nodes in the blockchain. The nodes are known by the name "miners". These nodes validate each transaction by generating a hash for each block. The node which validates the transaction gets rewards for validating each block. After the validation of a block, the block is added to the existing chain. News of updating or adding a new block to the chain is broadcast across the network to all the miners. This is how the whole transaction gets completed. Every blockchain uses a consensus mechanism to validate blocks. Proof of work is a consensus mechanism used by Bitcoin.

Figure 2 shows the flow of a transaction in the blockchain and its complete working.

Some of the application areas of blockchain are

- Record storage

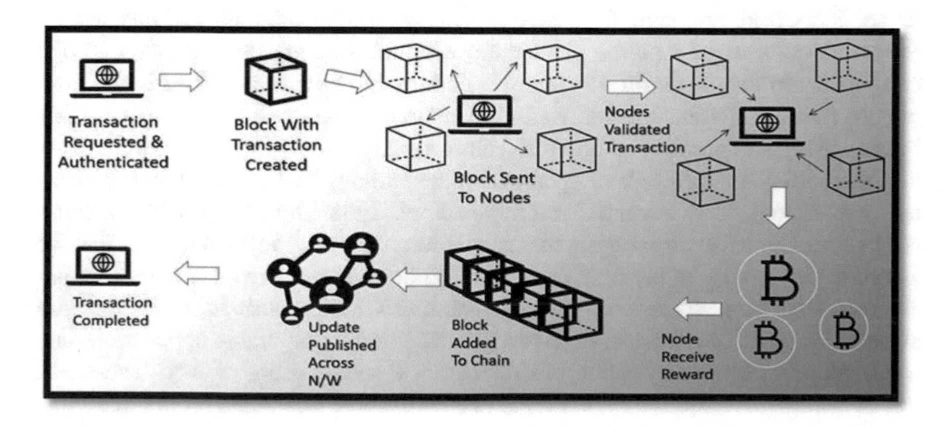

Fig. 2 Blockchain working

- Supply chain management
- Mortgages
- Real estate market
- Employment agreement
- Copyright protection
- Healthcare service
- E-voting
- Smart cities
- Vehicle technology

Record storage could have several use cases like medical record storage, education certificates, and law records. For supply chain management, blockchain technology provides a secure way for the exchange of information between various business entities and record the data. Similarly, mortgages could be managed without any hassle by tokenization of property in a blockchain environment. In a real estate market, the information about ownership of assets could be stored in a blockchain. Lack of trust in the evaluation of a centralized employment system can be aided by employing blockchain technology because of its decentralized nature. The deployment of smart contracts addresses the copyright and platform fee issues. Furthermore, blockchain can be used in other areas for the same reasons that is, transparency and immutability.

Taking into consideration the characteristics and applications of a blockchain, we have proposed an Ethereum-based ride-sharing service based on blockchain technology which stores all the information related to rides from both parties on an Ethereum blockchain to achieve immutability and freedom from a centralized server.

1.2 Ethereum

Ethereum is created for the purpose of an invincible, decentralized, immune to resistance, and world computer. Ethereum is built on similar principles of blockchain. Ethereum consists of a public blockchain which is running on more than 15,000 computers around the world. Ethereum is nothing but a set of protocols in the form of code running on Ethereum software. Ethereum transactions are more than capable in processing and proofing of complex transaction.

Ethereum is a decentralized open-source technology that powers the cryptocurrency ether and many decentralized applications. Ethereum also provides a smart contract functionality. A program that runs automatically by following rules that are stored in the agreement between the driver and the rider will be known as the smart contract. Ethereum was developed by programmer Vitalik Buterin in 2013. Ethereum allows anyone to develop and deploy a permanent and immutable application onto the Ethereum blockchain network. Ethereum 1.0 uses a proof of work consensus mechanism which requires its participant nodes to solve a complex problem to allow them to have a right to add a transaction in Ethereum network.

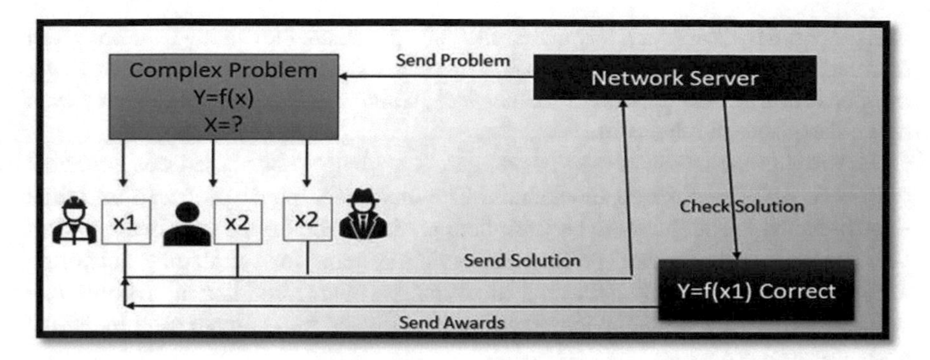

Fig. 3 Proof of work consensus mechanism

Figure 3 shows and explains proof of work mechanism where the first step is to send a complex problem to nodes to solve. Nodes send their solution to the network whichever nodes solve them and get the rewards in return.

The main disadvantage is a huge expenditure and the uselessness of computation. There might be a possibility of a 51 percent attack in which if one entity acquires the 51% computation power in the network, it can manipulate the Ethereum network.

The proof of work consensus mechanism shown in Fig. 3 allows all users to participate in solving the problem which takes 10 min on average to solve it. This mechanism gives all block creators an equal chance of winning. In the proof of work, mechanism creators need to calculate a cryptographic hash from a block of data.

In such services, a smart contract is deployed on the Ethereum network. A smart contract is created between two parties who remain anonymous. A smart contract can be defined as an immutable contract deployed on Ethereum which can receive and execute the transaction. It is used in multiple activities such as record stores and trading activities, the real-estate market, employment agreements, and copyright protection.

2 Literature Review

Mahmoud M. Badr Mohamed Baza works on a ride-sharing system with accurate matching and avoiding a breach of privacy built on blockchain technology in which they proposed the idea of several overlapping grids. Such a system had low communication and computation overheads [1]. Paper [2] proposed a blockchain that enables peer-to-peer rides that ensure the fairness of the ride and maintenance of the inbuilt reputation system. Renat Norderhaug; Shahriar Badsha proposed PEBERS which is a practical Ethereum built on a blockchain for ride-hailing services. Smart contracts build and deploy these functions using such arrangements as generating a ride, automatic payments, canceling a ride, and complete ride methods. In PEBERS, a prior payment from both parties brings trust to the system [3]. Proof of driving approach was proposed by Sowmya Kudva, Shamik Sengupta where the driving coin was given to riders based on the distance traveled by riders. Because of service score, fewer malicious nodes were present in the network, which result in constant latency even on an increasing number of nodes.

E. Vazquez and DL Silva worked on a zero-knowledge proof-based smart contract with a deposit fee built on blockchain. The mentioned paper proposed an initial deposit fee which supposes to be from both parties which brings trust in the system [4]. Paper [5] puts forward privacy-preserving systems for car sharing anchoring zero-knowledge protocols built on a blockchain. This method is quite flexible in a wide variety of use cases having considerable efficacy for example used by BMW and other group projects.

3 Proposed Methodology

The proposed system shown in Fig. 4 consists of multiple phases. In the first phase, the user and ride information such as source and destination location are submitted into the system. The user and rider information are stored in a smart contract, and this data is shared via the application interface. The user first submits the data according to his preferable destination, and this data invokes the smart contract. This data stored in the smart contract is shared with the driver over the mobile application. This mobile application is android based. The driver can either accept or reject the ride. If the driver accepts the provided ride information, then an acceptance message is sent to the user over the user's application interface. The android application then starts the ride generation process in which it initializes the smart contract for the driver and rider. After the invocation of the smart contract, driver details are sent to the rider through the application.

Once the ride is complete, the smart contract processes the ride payment from the user's crypto wallet to the driver's crypto wallet. The driver receives the ride payment notification through the application as a payment confirmation.

The following functions are used to accomplish different objectives in the proposed system shown in Fig. 4

Initialise_userinfo (): This function allows to pass the user and requested ride information to the application.

Initialise_driverinfo (): This function initializes driver information and available vehicle details.

Ride_generation (): This method helps to generate a smart contract based on driver and user details and the path information.

Funds_transfer (): This function transfers the mentioned fund in the smart contract from the user crypto wallet to the driver crypto wallet.

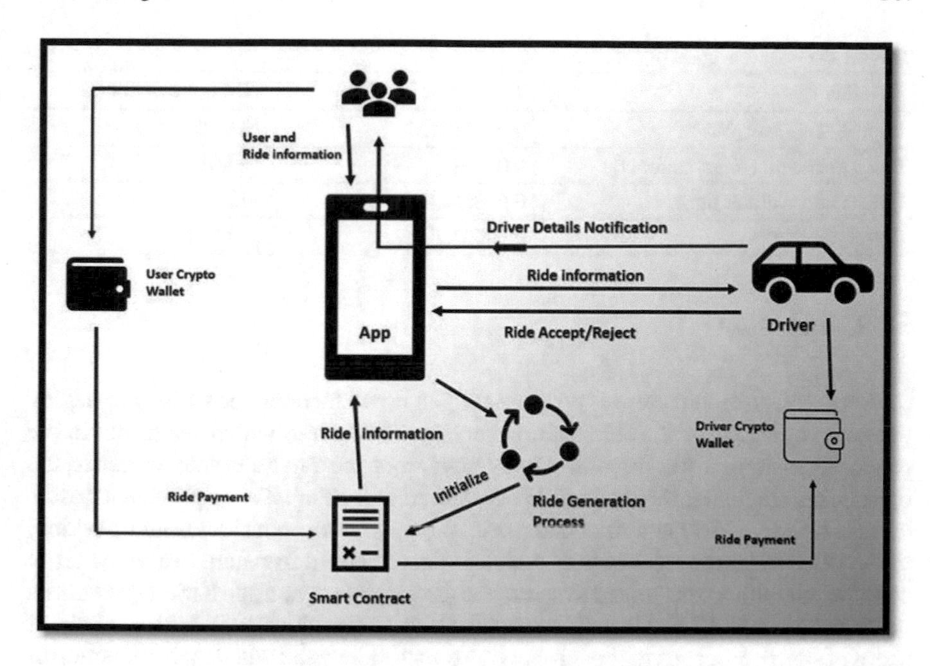

Fig. 4 Proposed ride-sharing architecture

4 Result

This section evaluates the performance of the proposed system which is deployed using a real Ethereum test net. The gas employed in the case of Ethereum is used to determine the associated cost linked to each transaction. Ether is used to pay for gas and is known to be Ethereum currency. There are two types of costs that participate in deploying the smart contract over Ethereum. Transaction cost is the total cost required to send data to Ethereum. The other associated cost is execution cost which is needed to execute the functions in the transaction over the EVM.

The proposed architecture has been implemented on the Ethereum test network using an arrangement having Microsoft Windows-10, the 64-bit OS with 12 GB RAM. The smart contract is developed using remix IDE with the help of Infura API suit 0.1 Ether = 190,000 INR as of 12 Feb 2022. Gwei is a denomination of ether. 1 ether = 1 billion Gwei. Gas is the amount of computation effort for smart contract functions. Table 1 lists the transaction fee for different functionalities of the ride-sharing service because the cost associated with each service is not same.

Table 1 reflects the transaction fee of each function of ride-sharing such as initialization of user info and ride details. Creation and deployment of smart contract also require 0.000437 Eth. Different functions in ride-sharing system cost different transaction fee.

Table 1 Transaction cost for each function

Function	Transaction fee [Eth]	INR/transaction
Initialization[user_info]	0.00007872	14.956
Ride generation (smart contract)	0.000437	83.03
Initialization [driver info]	0.00006953	13.21
Funds_transfer	0.0000386	7.334

5 Conclusion

In this paper, an architecture is proposed which uses Ethereum technology to deploy the smart contract for the ride-sharing service. The feature which stands out in the proposed system is the transfer of payment from the rider's crypto wallet to the driver's crypto wallet through smart contract initialization. The deployment cost is less for both the driver and rider and a prior payment from both sides brings trust into the system. Ethereum technology which is decentralized in nature proves better in comparison with a centralized system. For the future scope, ride-sharing system can be deployed onto ETH 2.0 which provides high transaction speed with the help of proof of stake consensus mechanism. High transaction speed will help in reduction of network backlogs. Building a future proof architecture will improve the ride-sharing system significantly if it will allow to edit and debug the smart contract.

References

1. Baza M, Mahmoud M, Srivastava G, Alasmary W, Younis M (2020) A light blockchain-powered privacy-preserving organization scheme for ride sharing services. In: 2020 IEEE 91st vehicular technology conference (VTC2020-Spring), IEEE, pp 1–6
2. Valaštín V, Kostal K, Bencel R, Kotuliak I (2019) Blockchain based car-sharing platform. https://doi.org/10.1109/ELMAR.2019.8918650
3. Kudva S, Badsha S, Sengupta S, Khalil I, Zomaya A (2021) Towards secure and practical consensus for blockchain based VANET. Inf Sci 545:170–187
4. Joseph R, Sah R, Date A, Rane P, Chugh A (2021) blockwheels-a peer to peer ridesharing network. In: 2021 5th international conference on intelligent computing and control systems (ICICCS), IEEE, pp 166–171
5. https://blockchainlab.com/pdf/Ethereum_white_papera_next_generation_smart_contract_and_decentralized_application_platform-vitalik-buterin.pdf
6. Vazquez E, Landa-Silva D (2021) Towards blockchain-based ride-sharing systems. In: ICORES, pp 446–452
7. Kumar R, Kedia RK, Balodia S, Suvvari SD. Decentralised ride sharing system
8. Rupa C, Midhunchakkaravarthy D, Hasan MK, Alhumyani H, Saeed RA (2021) Industry 5.0: ethereum blockchain technology based DApp smart contract. Math Biosci Eng 18(5):7010–7027
9. Ruch C, Lu C, Sieber L, Frazzoli E (2020) Quantifying the efficiency of ride sharing. In: IEEE transactions on intelligent transportation systems
10. Bodkhe U, Tanwar S, Parekh K, Khanpara P, Tyagi S, Kumar N, Alazab M (2020) Blockchain for industry 4.0: a comprehensive review. IEEE Access 8:79764–79800

11. Pal P, Ruj S (2019) Blockv: a blockchain enabled peer-peer ride sharing service. In: 2019 IEEE international conference on blockchain (blockchain), IEEE, pp 463–468
12. Gudymenko I, Khalid A, Siddiqui H, Idrees M, Clauß S, Luckow A, Miehle D et al (2020) Privacy-preserving blockchain-based systems for car sharing leveraging zero-knowledge protocols. In: 2020 IEEE international conference on decentralized applications and infrastructures (DAPPS), IEEE, pp 114–119
13. Valaštín V, Košťál K, Bencel R, Kotuliak I (2019) Blockchain based car-sharing platform. In: 2019 international symposium ELMAR, IEEE, pp 5–8
14. Baza M, Lasla N, Mahmoud MM, Srivastava G, Abdallah M (2019) B-ride: ride sharing with privacy-preservation, trust and fair payment atop public blockchain. IEEE Trans Netw Sci Eng 8(2):1214–1229
15. Hossan MS, Khatun ML, Rahman S, Reno S, Ahmed M (2021) Securing ride-sharing service using IPFS and hyperledger based on private blockchain. In: 2021 24th International Conference on Computer and Information Technology (ICCIT), IEEE, pp 1–6
16. Shivers RM (2019) Toward a secure and decentralized blockchain-based ride-hailing platform for autonomous vehicles (Doctoral dissertation, Tennessee Technological University)
17. Andoni M, Robu V, Flynn D, Abram S, Geach D, Jenkins D, Peacock A et al (2019) Blockchain technology in the energy sector: a systematic review of challenges and opportunities. Renew Sustain Energy Rev 100:143–174
18. McBee MP, Wilcox C (2020) Blockchain technology: principles and applications in medical imaging. J Digit Imaging 33(3):726–734
19. Raj A, Maji K, Shetty SD (2021) Ethereum for Internet of Things security. Multimedia Tools Appl 80(12):18901–18915
20. Chen H, Pendleton M, Njilla L, Xu S (2020) A survey on ethereum systems security: vulnerabilities, attacks, and defenses. ACM Comput Surv (CSUR) 53(3):1–43
21. Ferretti S, D'Angelo G (2020) On the ethereum blockchain structure: a complex networks theory perspective. Concurrency Comput: Pract Experience 32(12):e5493
22. Dhulavvagol PM, Bhajantri VH, Totad SG (2020) Blockchain ethereum clients performance analysis considering E-voting application. Proc Comput Sci 167:2506–2515
23. Oganda FP, Lutfiani N, Aini Q, Rahardja U, Faturahman A (2020) Blockchain education smart courses of massive online open course using business model canvas. In: 2020 2nd International Conference on Cybernetics and Intelligent System (ICORIS), IEEE, pp 1–6
24. Alladi T, Chamola V, Sahu N, Guizani M (2020) Applications of blockchain in unmanned aerial vehicles: a review. Vehicular Communications 23:100249

Performance Analysis of User Authentication Schemes in Wireless Sensor Networks

Ravi Kumar and Samayveer Singh

Abstract Wireless sensor networks (WSNs) have a broad range of applications in most of the sectors which require a high level of security. These networks provide sensor data to the external users on a real-time basis; thus, it is a good idea to provide them immediate access. User authentication is an essential requirement to access the services of sensor areas to prevent unauthorized access. The network and data security are key concerns in WSNs for secure and authentic data communication. Thus, an analysis of a secure and efficient data transmission scheme is considered. In this paper, an analysis of existing user authentication schemes is discussed and provides comparative studies of user authentication schemes for network security. An explicitly simulated one of the user authentication schemes by the AVISPA tool is discussed, and its demonstration is also given which shows the resistant to the cryptographic attacks. The lower communication costs are one of the main parameters, i.e., considered to compared with the previous relevant techniques. Moreover, future directions are also discussed for the existing works.

Keywords Authentication · AVISPA · Data communication · Cryptographic attacks · Sensor nodes

1 Introduction

Wireless sensor networks (WSNs) are networks of geographically scattered and specialized sensors that monitor and record environmental variables and transmit the gathered data to a central point [1]. The architecture of the WSNs is defined in Fig. 1. The WSNs may monitor environmental factors such as temperature, sound, pollution levels, and humidity. There are many applications of WSNs such as smart home systems, military missions, urgent disaster management, automobile, and agricultural production. The main advantages of WSNs are easy monitoring and flexibility because they can be accessed through the centralized monitor, reduced cost: avoid a

R. Kumar (✉) · S. Singh
Department of Computer Science & Engineering, Dr. B R Ambedkar National Institute of Technology Jalandhar, Punjab, India
e-mail: ravik.cs.19@gmail.com

Y. Singh et al. (eds.), *Proceedings of International Conference on Recent Innovations in Computing*, Lecture Notes in Electrical Engineering 1001,
https://doi.org/10.1007/978-981-19-9876-8_43

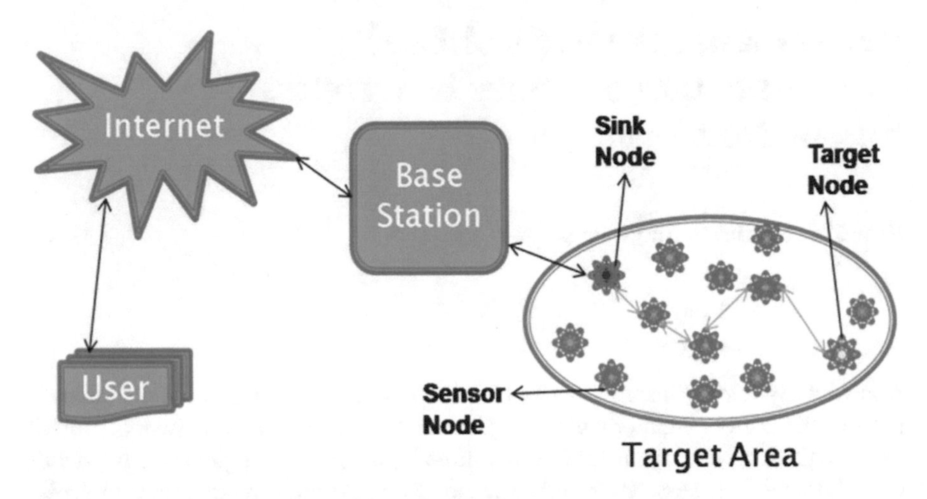

Fig. 1 Architecture of the wireless sensor network [4]

lot of wiring, are extendable, and can add new devices at any time, respectively. Some disadvantages of WSNs such as memory, security, disturbance, and energy: WSNs have limited memory and computational power, are easily hacked, get distracted by various elements, and have low energy, respectively. Authentication is the process of a user proving their identity in order to gain access to a system. For example, WSNs wish to access a certain application along with a specific resource.

After the deployment of sensor nodes and base station along with the ensure the connectivity of the end-user in the WSNs, firstly, user sends an authentication request with his credential information to the base station for the login/registration process which needs to access the sensor information. Now, the base station verifies the authentication request of the user, and after verification, the base station passes the authentication request to the target sensor node. The authentication request will be verified by the sensor node, then the sensor node will generate the session key for the base station as well as the user. After getting the session key, the process of sending the information or collecting the data has started. Nowadays, everyone needs authentication since it protects credential information from being accessed by unauthorized users [2]. If user authentication is not secure, there is a risk of credential information being stolen or leaked [2, 3]. Without authentication, no organization can be secure. As a result, authentication is a critical component of message and credential security. Cybercriminals are always improving their methods to steal secret information. As a result, security teams are having to cope with a variety of authentication problems [3]. There are a few different methods of authentication that are widely used as follow-up authentication.

Authentication using a password: The use of a password is a widely frequent authentication strategy. The passwords can be numeric, characters, special characters, or a combination of these. According to a poll, just 54% of users can use various passwords for separate accounts since remembering all of the passwords for each account is challenging, and users prefer to use easy passwords to complex passwords. The passwords can be obtained using a fishing attack [5, 6].

Authentication using multifactor: Multifactor authentication (MFA) is a security solution that uses several methods to independently validate user authentication [7]. Multifactor authentication is more secure than other standard methods of authentication. The MFA operates on the idea of something users know (such a user password or a user name), and something users have like a mobile phone or biometric information, etc. The authentication using a certificate can be a sort of authentication that relies on digital certificates rather than passwords or biometric information. It is more secure and reliable than password-based authentication, and it is also reusable for user authentication [8]. To sign in to the server, the user first presents his certificate, following which the server validates the user's digital signature and certificate authority. Now, the server utilizes cryptography to determine whether or not the user's private key is connected with the certificate. It is the same as having a driving license or a passport.

Authentication using biometrics: This type of authentication is only required to employ biological information or features for the security procedure. To improve security and authentication, a biometric feature can be integrated into multifactor authentication [8, 9]. Biometric authentication is now being employed in a variety of organizations, including commercial businesses, institutions, military bases, and government agencies. There are certain drawbacks to biometric authentication, such as how users can alter their biometric password if their biometric password is stolen. It is because of certain times devices are unable to distinguish between users and their closed relative faces, they produce incorrect results with the closer or related face image. It also produces incorrect results when viewed from various angles.

Facial recognition: In facial recognition, the device matches the user's face with his database which is a collection of lots of faces, and after that, it verified user's face is an authorized person or not [10]. Sometimes, face recognition gives the wrong result when it compares with a different point of view, or sometimes, a person looks the same then also gives the wrong result.

Scanner of Fingerprints: In the process of fingerprint scanning, the gadget compares the user's fingerprint to his database which has a variety of fingerprints [11, 12]. The fingerprint scanners are now commonly utilized in businesses, colleges, and universities. It is because of erased circles on your finger or greasy fingertips, fingerprint scanners can have trouble scanning fingerprints cleanly.

Speaker Recognition: This gadget compares the user's voice to his database, which contains a variety of pre-recorded voices [13]. When it comes to detecting voices, natural language processing (NLP) is usually used.

Scanners of eyes: We can use IRIS recognition or a retina scanner to scan the eyes. In the iris, a strong light is shined on the eye in order to try to discover a unique pattern in the eye, but in the retina scanner, an AI device scans the eyes with its scanner in order to identify a unique pattern in the database [14, 15].

Authentication using token-based: In token-based authentication, first user sends his credential information to the server and after that server sends a unique random number or string in form of a token to the user [16, 17]. So, for accessing secure information, users use only tokens instead of entering their credential information (like a password) again and again. Basically, API provides tokens to the user and managed them to check whether the user is authorized or not. For secure data transfer, a variety of authentication procedures are used. There are some basic authentication techniques given below-

Message encryption: The cipher text will operate as an authenticator in message encryption. As a result, the sender must be authenticated by the receiver, and the recipient must be authenticated by the sender, all of which will be accomplished only through the use of encrypted text. The plain text will be changed to cipher text using the encryption process, and cipher text will be converted to the plain text using the decryption process as illustrated in Fig. 2. In the encryption process, encrypt key is used for converting the plain text to cipher text. Similarly, in the decryption process, the decrypt key is used for converting the cipher text to plain text. Both the processes are reversible from each other.

Message authentication code (MAC): Basically, MAC is used as an authentication scheme. We can apply these authentication functions to the plain text together with the key and resulting in a fixed-length code known as MAC. The MAC is also called message digest and acts as an authenticator. As a result, the sender or recipient must use this fixed-length code to authenticate (MAC) as shown in Fig. 3. We can apply MAC function with a secret key on a plain message which produces a fixed-length MAC value. The reversibility from the MAC code to the original message is not possible.

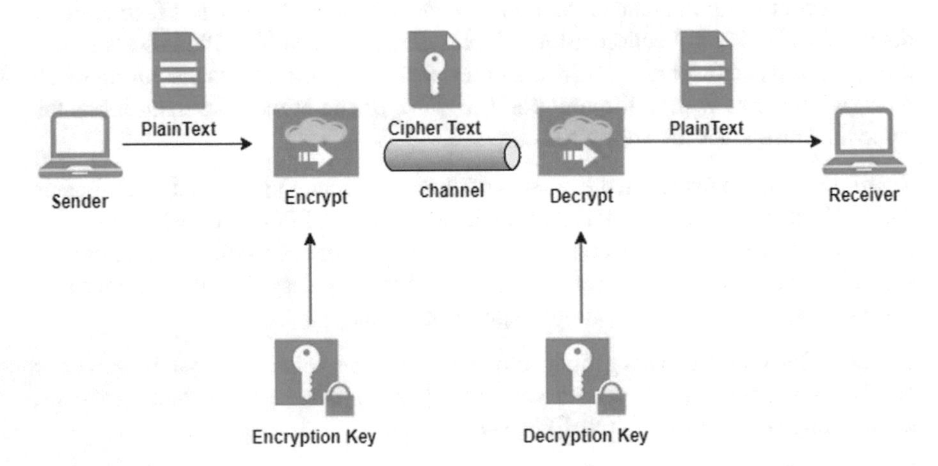

Fig. 2 Message encryption and decryption

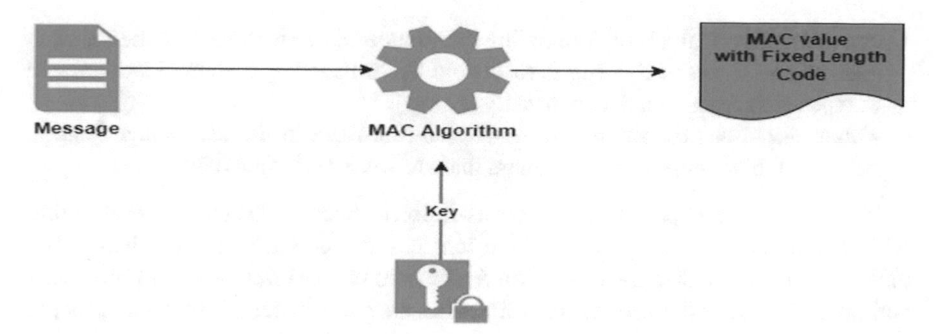

Fig. 3 MAC function process after applying on message

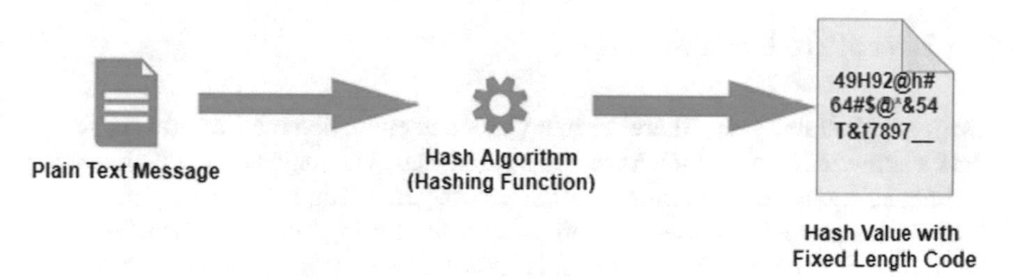

Fig. 4 Apply hash function on the message

Hash functions: In this technique, the hash function can be applied on the plain text which is independent of the key as compared with MAC and also produce fixed-length code which is also called hash code (h). A fixed-length code is used as an authenticator bode in this case. Thus, whether the sender wishes to authenticate the receiver or the recipient wants to authenticate the sender, this authentication code must be provided as shown in Fig. 4; hash function will be applied to plain message which produces a fixed-length hash code (h). This hash code is unidirectional which makes more secure communication.

2 Contribution

In this paper, the existing user authentication work is discussed and also performs their comparative analyzes by considering various matrices. Accordingly, an analysis is discussed of the comparative review for user authentication. The main contribution of the paper is given as follows:

- Analysis of the existing user authentication scheme for developing an improved version of the user authentication scheme than the existing scheme is discussed.

- A comparative analysis of the existing techniques related to the user authentication scheme and network security is discussed by considering communicational and computational cost quality metrics.
- Future directions for the research are also discussed in the user authentication schemes which helps the researchers that are working in this field.

The rest of the paper is organized as follows: Sect. 2 discusses the literature review related to the user authentication schemes. In Sects. 3 and 4, challenges and tools and testing are discussed. Section 5 discusses the performance analysis of user authentication algorithms, and future directions are given in Sect. 6. Finally, the paper is concluded in Sect. 7.

3 Literature Review

During the last few years, there are many user authentication mechanisms for secure data sharing in WSNs which have been introduced. The majority of authentication techniques focus on the communication parties generating a session key. Das et al. [18] discuss a two-factor user authentication scheme using a hash function. However, the limitations of this scheme are denial of service, node compromise attacks, and the problem of updating passwords is possible. Chang et al. [19] discuss an ECC and key agreement-based scheme that removes the drawbacks of Das et al. [18] by privileged insider attacks, offline password guessing attacks, forgery and impersonation attacks, and smart card loss attacks, and so on affect [18]. Singh et al. [4] discuss MAC function and symmetric key cryptosystem which is used one-time-password (OTP) instead of login password, but communicational overhead can be reduced and improve better security. Banerjee et al. [20] can withstand replay attacks, smart card breach attacks, stolen smart card attacks, and stolen verifier attacks; however, it cannot withstand offline password guessing assaults, impersonation attacks, and session key secrecy is not complete. Qi-Xie et al. [21] discussed a scheme that is based on ECC and fuzzy extractor that resist offline password guessing attacks and impersonation attacks and does not achieve session key secrecy and perfect forward secrecy, but limited security attacks considered. The papers [22–24] discuss various methods based on authentication schemes in WSNs.

Shuai et al. [25] discuss a method which is based on 3FAS that shows anonymity and forward secrecy and resist various attacks but suffer from vulnerable to stolen verifier attack, de-synchronization attack, not perfect forward secrecy, denial of service attack vulnerable. Qi-xie et al. [26] consider 3FAS and ECC and fuzzy extractor algorithm which improved computational cost and security and also resist stolen verification attack and perfect forward secrecy while limited security attacks considered. Singh et al. [27] introduce a method based on 3FAS using the Rabin cryptosystem in WSNs which is used 3FAS instead of RSA and ECC. A comparative literature survey related to the user authentications schemes is shown in Table 1 as follows.

Table 1 Literature review based on the user authentications schemes

Year	Objective	Methodology algorithm	Performance and evaluation metrics	Limitation/research gap
2009 [18]	Two-factor authentication is used to authenticate users	Hash function	Computational cost and security	Denial of service is possible, node compromise attack, the problem with update password
2015 [19]	Design a secure and flexible authentication scheme for AWSN	Used ECC and key agreement	Security attacks and computational cost	Offline password guessing attacks, impersonation attacks, and other types of attacks are common
2019 [4]	Secure authentication scheme by using MAC function	MAC function, BAN logic, symmetric key cryptosystem	Communicational and computational overhead, security attacks	Communication overhead is high, Tests on limited security attacks
2019 [20]	In WSN, an improved and secure user authentication scheme based on biometric using smart cards	BAN logic	It is resistant to replay attacks, stolen smart card attacks, and other similar attacks	It is vulnerable to impersonation and password guessing attacks, and the session key's security is not perfect
2021 [21]	For WSNs in smart cities, a safe and privacy- preserving authentication system has been developed	Elliptic curve cryptosystem (ECC), fuzzy extractor	Computational cost, improve security attack	Limited security attacks considered
2021 [25]	A lightweight 3FA scheme with privacy preservation for customized healthcare applications,	Cryptographic primitive as wireless medical sensor networks (WMSNs), pseudonym identity technique, one-way hash chain technique	Computational cost, improve security (smart card loss attack, replay attack)	De-synchronization attack, DoS attack vulnerable
2021 [26]	A 3FA scheme for wireless sensor networks in the Internet of things that is secure and privacy-preserving	ECC, fuzzy extractor algorithm	Computational cost, improve security	Limited security attacks considered

(continued)

Table 1 (continued)

Year	Objective	Methodology algorithm	Performance and evaluation metrics	Limitation/research gap
2021 [27]	In WSNs, 3FAS use the Rabin cryptosystem	Rabin cryptosystem	Resist security attacks and reduce costs than the previous approaches	Complex to implement in large area networks

4 Open Challenges

According to an evaluation of existing user authentication documents, there are many challenges which are given as follows:

- Authenticating the sensors should be in a broadcast fashion to boost security, and it needs to lower the computational and communication costs.
- The user authentications and node authentication schemes need to focus on better security and data communication.
- The process of managing and securing a wide area network is difficult because assaults on gateway nodes are easier in bigger WSNs.
- It is occasionally inefficient to authenticate mobile sensor nodes due to the sensor node's limited energy and computing power, or because the mobility protocol does not always work well with mobile sensor nodes.

5 Tools and Testing

In this section, most of the research works have used the AVISPA tool for verified security attacks. The AVISPA tool is widely used for automated analysis of security-related schemes, and its workflow and analysis are shown in Figs. 6 and 7 [28]. AVISPA uses the modular arithmetic language, i.e., based on the control flow patterns, along with the various complex security properties for specifying the security protocols. It is freely available over the Internet and has a Web-based interface. AVISPA is also considered the various cryptographic primitives and algebraic properties for proving the prepared flow of the exchanging the information which is secure or not. If the process of information is secure, then it passes the process and shows the message which is secure. The HLPSL stands for high-level protocol specification language as depicted in Fig. 5 which uses the HLPSL2IF translator. Then, the HLPSL is translated into intermediate format (IF). This might be fed into backends like on the fly model checker (OFMC), (CL-based attack searcher (CL-AtSe), SAT-based model checker (SATMC), and tree automata-based protocol analyzer (TA4SP) by combining these backends and executing them, and AVISPA now produces a specified output format. Thus, a simulation analysis of the research paper [27] is verified using the AVISPA tool as shown in Fig. 6 which gives the secure result and attacked free by considering the computing costs, communication costs, and running time factors.

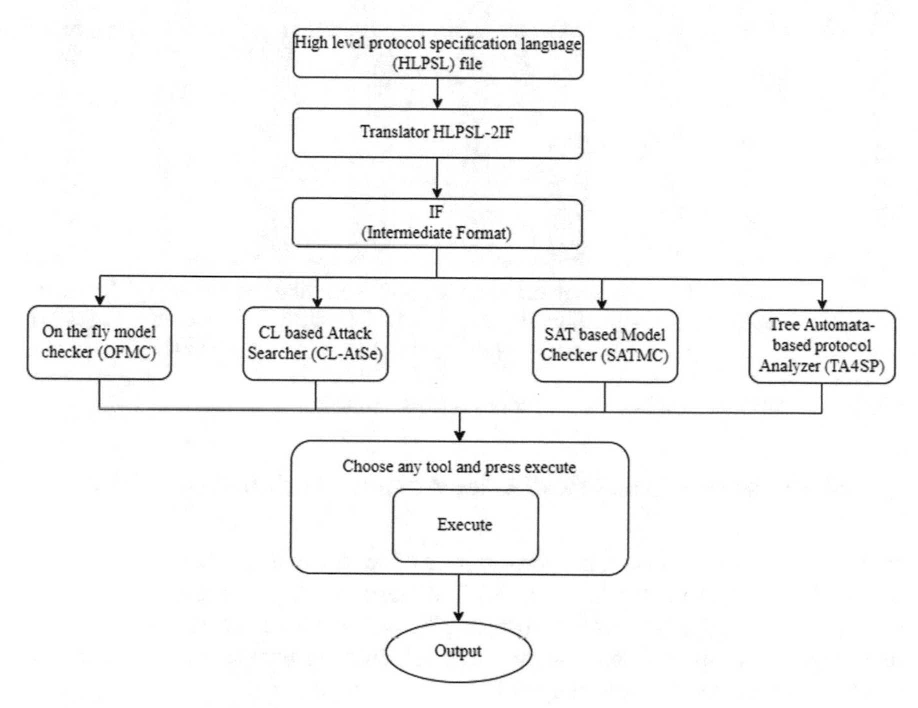

Fig. 5 Architecture of AVISPA

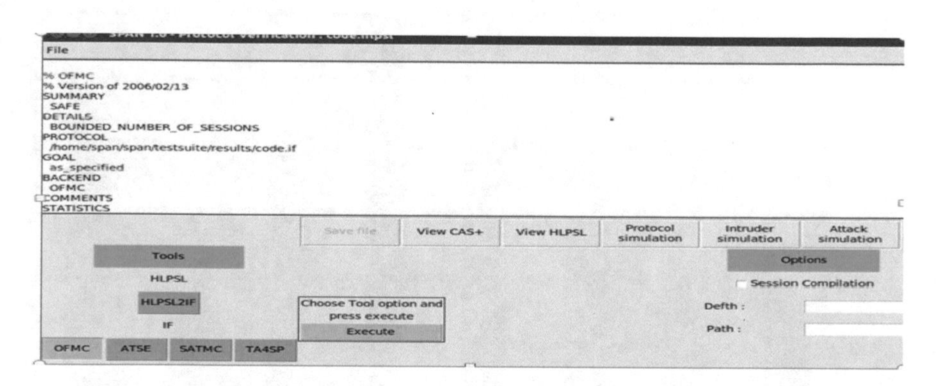

Fig. 6 Tested and verified result [27]

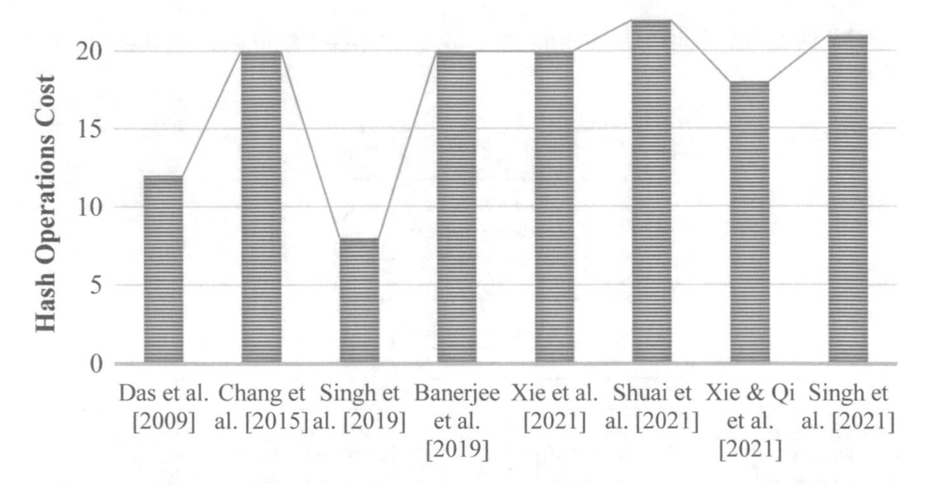

Fig. 7 Computational overhead for the various existing methods

6 Performance Analysis of User Authentication Algorithms

In this section, we compared the performance of some existing schemes such as Das et al. [18], Chang et al. [19], Singh et al. [4], Banerjee et al. [20], Xie et al. [21], Shuai et al. [25], Xie et al. [26], and Singh et al. [27]. We examined various methods based on computing costs, communication costs, and running time factors. Figures 7, 8, and 9 depict the comparison analysis for all the matrices.

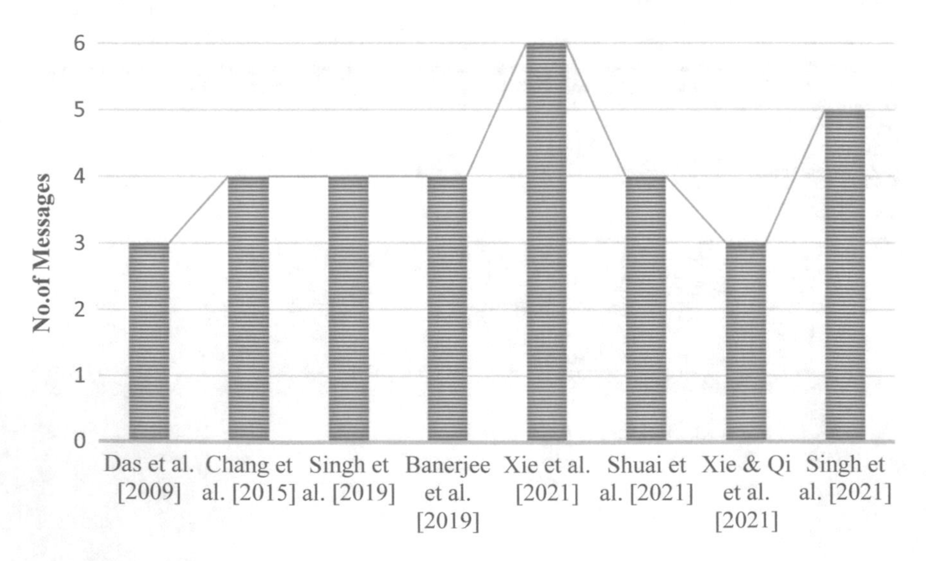

Fig. 8 Communication overhead for the various existing methods

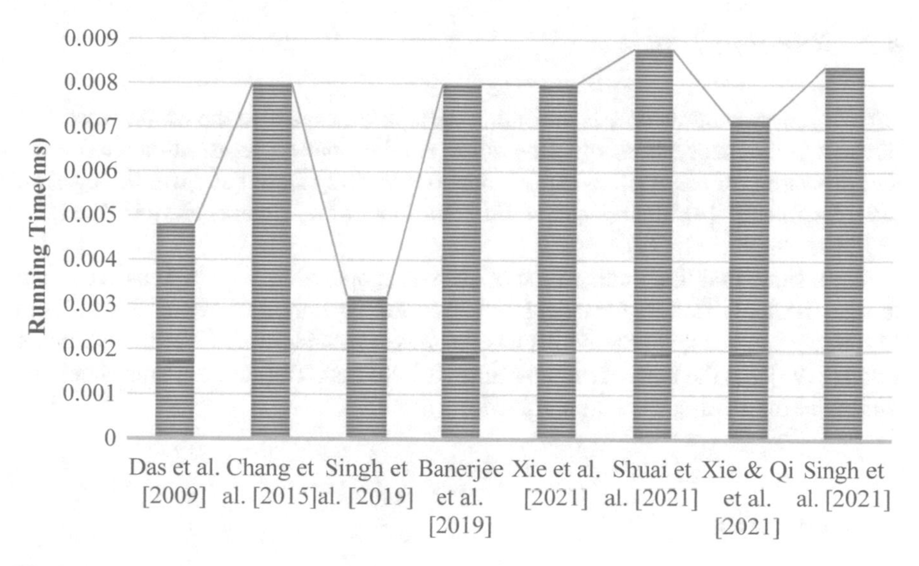

Fig. 9 Running time for the various existing methods

6.1 Computational Overhead

In the computational overhead, various processes are considered, namely encryption, decryption, key generation, key distribution, registration phase, etc. Here in the user algorithm, Singh et al. [4] and Xie et al. [21] have lower computation overhead while Shuai et al. [25] have higher computation overhead, and remaining Chang et al. [19], Banerjee et al. [20], Xie et al. [26], and Singh et al. [27] have medium computation overhead as shown in Fig. 7. The total cost of the computation is directly dependent on the complexity of the encryption and decryption process along with the registration phase and the key generation and key distribution.

6.2 Communication Overhead

It is basically unnecessary information that is used to build communication among sensor nodes. So, the scheme by Das et al. [18] and Xie et al. [26] has lower communication overhead, while scheme [21] has higher communication overhead. The schemes by Chang et al. [19], Singh et al. [4], Banerjee et al. [20], Shuai et al. [25], and Singh et al. [27] have average communication overhead as shown in Fig. 8. The communication overhead should be minimum for an effective communication system.

6.3 Running Time

In this section, running time is indicating the total time taken in the whole process of the encryption, decryption, key generation, key distribution, registration phase, etc., and measured for the existing various methods, namely Das et al. [18], Chang et al. [19], Singh et al. [4], Banerjee et al. [20], Xie et al. [21], Shuai et al. [25], Xie et al. [26], and Singh et al. [27].

We assume that the running time of each hash operation is 0.0004 ms according to data available in [29] for calculating the running time of these schemes. From Fig. 9, we can see that scheme [4] has the lowest running time (0.0032 ms), while scheme [25] has the highest running time (0.0088 ms). The running time should be minimum for an effective communication system.

7 Conclusion

There is a major concern to secure user authenticity with limited energy of sensor nodes in WSNs. As a result, an attacker might compromise a user's authenticity, mislead them, and have an impact on other user nodes. S o, we utilize a variety of user authentication mechanisms to ensure that the user is legitimate. In terms of computational cost, communication overhead, and running time, this study compares the performance of several user authentication schemes. According to a comparative performance study, every strategy tries to reduce overhead but suffers from other issues such as security and complexity. Some schemes do not include all forms of performance and security measurements. As a result, in future, we should offer schemes that are less difficult, high-performing, and secure.

References

1. Ullo SL, Sinha GR (2020) Advances in smart environment monitoring systems using IoT and sensors. Sensors 20(11):3113
2. Al Ameen M, Liu J, Kwak K (2012) Security and privacy issues in wireless sensor networks for healthcare applications. J Med Syst 36(1):93–101
3. Dammak M et al (2019) Token-based lightweight authentication to secure IoT networks. In: 2019 16th IEEE annual Consumer Communications Networking Conference (CCNC), IEEE
4. Singh D et al (2019) SMAC-AS: MAC based secure authentication scheme for wireless sensor network. Wireless Personal Commun 107(2):1289–1308
5. Gupta S, Singhal A, Kapoor A (2016) A literature survey on social engineering attacks: Phishing attack. In: 2016 international conference on computing, communication and automation (ICCCA), IEEE
6. Ahemd MM, Shah MA, Wahid A (2017) IoT security: a layered approach for attacks defenses. In: 2017 international conference on Communication Technologies (ComTech), IEEE
7. Ometov A et al (2018) Multi-factor authentication: a survey. Cryptography 2(1):1

8. Holohan E, Schukat M (2010) Authentication using virtual certificate authorities: a new security paradigm for wireless sensor networks. In: 2010 ninth IEEE international symposium on network computing and applications, IEEE
9. Das AK (2017) A secure and effective biometric-based user authentication scheme for wireless sensor networks using smart card and fuzzy extractor. Int J Commun Syst 30(1):e2933
10. Dahia G, Jesus L, Segundo MP (2020) Continuous authentication using biometrics: an advanced review. Wiley Interdisc Rev: Data Mining Knowl Discovery 10(4):e1365
11. Park YY, Choi Y, Lee K (2014) A study on the design and implementation of facial recognition application system. Int J Bio-Sci Bio-Technol 6(2):1–10
12. Yang C-C, Wang R-C, Liu W-T (2005) Secure authentication scheme for session initiation protocol. Comput Security 24(5):381–386
13. Knox DA, Kunz T (2015) Wireless fingerprints inside a wireless sensor network. ACM Trans Sensor Netw (TOSN) 11(2):1–30
14. Sahidullah M, Kinnunen T (2016) Local spectral variability features for speaker verification. Digital Signal Process 50:1–11
15. Imran M, Said AM, Hasbullah H (2010) A survey of simulators, emulators and testbeds for wireless sensor networks. In: 2010 international symposium on information technology, vol 2, IEEE
16. El-said SA, Hassanien AE (2013) Artificial eye vision using wireless sensor networks. In: Wireless Sensor Networks: Theory and Applications. CRC Press, Taylor and Francis Group, Boca Raton
17. Tanuja R et al (2015) Token based privacy preserving access control in wireless sensor networks. In: 2015 International Conference on Advanced Computing and Communications (ADCOM), IEEE
18. Das ML (2009) Two-factor user authentication in wireless sensor networks. IEEE Trans Wireless Commun 8(3):1086–1090
19. Chang C-C, Le H-D (2015) A provably secure, efficient, and flexible authentication scheme for ad hoc wireless sensor networks. IEEE Trans Wireless Commun 15(1):357–366
20. Banerjee S et al (2019) An enhanced and secure biometric based user authentication scheme in wireless sensor networks using smart cards. Wireless Personal Commun 107(1):243–270
21. Xie Q et al (2021) A secure and privacy-preserving authentication protocol for wireless sensor networks in smart city. EURASIP J Wireless Commun Netw 1:1–17
22. Singh D, Kumar B, Singh S, Chand S (2020) Evaluating authentication schemes for real-time data in wireless sensor network. Wireless Personal Commun 114(1):629–655
23. Singh D, Kumar B, Singh S, Chand S, Singh PK (2021) RCBE-AS: rabin cryptosystem–based efficient authentication scheme for wireless sensor networks. In: Personal and ubiquitous computing, pp 1–22
24. Singh D, Kumar B, Singh S, Chand S (2020) A secure IoT-based mutual authentication for healthcare applications in wireless sensor networks using ECC. Int J Healthcare Inf Syst Informatics (IJHISI) 16(2):21–48
25. Shuai M et al (2021) A lightweight three-factor anonymous authentication scheme with privacy protection for personalized healthcare applications. J Organizational End User Comput (JOEUC) 33(3):1–18
26. Xie Q, Ding Z, Hu B (2021) A secure and privacy-preserving three-factor anonymous authentication scheme for wireless sensor networks in Internet of Things. In: Security and communication networks 2021
27. Singh D et al (2021) RCBE-AS: rabin cryptosystem–based efficient authentication scheme for wireless sensor networks. In: Personal and ubiquitous computing, pp 1–22
28. The AVISPA project. http://www.avispaproject.org
29. Xun Y et al (2019) Automobile driver fingerprinting: a new machine learning based authentication scheme. IEEE Trans Industrial Informatics 16(2):1417–1426

A Secure Authenticated Key Agreement Protocol Using Polynomials

Manoj Kumar Mishra, Varun Shukla, Atul Chaturvedi, Pronaya Bhattacharya, and Sudeep Tanwar

Abstract Data communication is now an inseparable part of our day-to-day life. There are various aspects of data security, and authentication and key agreement are dominating among them in recent security work. Thus, key establishment and agreement is an important problem of study. Intruders always try various decoding methods to crack the security. Motivated from the same, it is now essential to present new authentication and key agreement protocols in order to surprise intruders. In this paper, a couple of innovative key agreement protocols are presented using the platform of non-commutative ring which is relatively a new platform and hence very lucrative for the development of new protocols. The correctness and working example are also given in order to verify the feasibility of the proposed protocols.

Keywords Authentication · Data communication · Key agreement · Non-commutative ring

1 Introduction

In data communication, key distribution is an important problem of study. In a similar direction, key establishment (KE) takes care of the calculation and sharing of the required key for the session [1–4]. Secure KE is classified into two categories, namely the key transport schemes (KTSs) and key agreement schemes (KASs). In KTS-based protocols, only one participating entity calculates the key, while in KAS-based protocols all the entities contribute to the calculation of the key. In this work, our

M. K. Mishra · V. Shukla · A. Chaturvedi
Pranveer Singh Institute of Technology, Kanpur, UP 209305, India

P. Bhattacharya (✉) · S. Tanwar
Institute of Technology, Nirma University, Ahmedabad, Gujarat 382481, India
e-mail: pbhattacharya@kol.amity.edu

S. Tanwar
e-mail: sudeep.tanwar@nirmauni.ac.in

P. Bhattacharya
Amity School of Engineering and Technology, Amity University, Kolkata, WB, India

Y. Singh et al. (eds.), *Proceedings of International Conference on Recent Innovations in Computing*, Lecture Notes in Electrical Engineering 1001,
https://doi.org/10.1007/978-981-19-9876-8_44

prime focus is on KAS in which the key can be shared over open channels which are subjected to attacks by an adversary [5]. Let g be the group generator and p be a large prime number as a requirement. In bidirectional communication, the transmitter and receiver select α and β (random selection of an element in this group) as their secret values. The transmitter will calculate $A = g^{\alpha} \pmod p$ and the receiver will calculate and exchange their values with each other. Now transmitter does $B^{\alpha} \pmod p = (g^{\beta})^{\alpha}$ $mod\, p = g^{\alpha\beta} \pmod p$ and similarly, the receiver does $A^{\beta} \pmod p = (g^{\alpha})^{\beta}\, mod\, p = g^{\alpha\beta}$ $\pmod p$ which is the shared private key for the session. It can be extended too many entities but Diffie-Hellman (DH)-based protocols are not safe enough specifically from quantum computers [6, 7]. Asymmetric cryptography, which is also called public key cryptography (PKC) saw different variants of PKC algorithms. Elliptic curve cryptography (ECC) [8, 9] has attracted the research community because of the minimal requirement of computation power and overheads of computation.

In 1994, Peter [10] proposed a quantum-based scheme to solve the discrete logarithm exponent problem and also produced insights on the factorization problem. In 1996, Kitaev [11] presented a distinct case on the discrete logarithm problem (DLP) which is known as hidden subgroup problem (HSP). Since most of the PKCs utilize a commutative (or abelian) group, it is not beneficial for security. It is obvious that if all the security methods belong to one group, then it can be a big hurdle for information security goals. So, the directions were clear that cryptographic methods other than the abelian group were required and that was the starting of non-commutative cryptography [12]. After that, non-commutative methods were presented for various cryptographic objectives like encryption-decryption, authentication and key agreement and the process replicates commutative ones.

The elliptic curve problem over the HSP or DLP, in general, is known as ECC-DLP [13]. HSP over non-commutative groups performed well for quantum algorithm, and it was realized that HSP over non-commutative groups are very hard for cryptanalysts [14]. The initial progress of non-commutative cryptography was based on braid related cryptography for the development of methods. Further, other structure matrix groups/ring was also developed. The algebraic structure of the group or ring is the backbone method utilizing non-commutative cryptography with related characteristics. In contrast, non-commutative cryptography is dependent on * operation on G, where G is a non-commutative group. The (G, *) consists of a group, ring or semi-ring, where for any two elements x and y, $x * y \neq y * x$. So, it is obvious that non-commutative cryptography is very fruitful for data communication security.

2 State of the Art

In this section, we present the different state-of-the-art schemes. For example, in 1985, Wagner et al. [15] proposed a word problem that relates to semi-group elements in public keys. Birget et al. [16] proposed a scheme which is based on finite groups and is directed toward solutions to computationally hard problems. In 1999, Anshel et al. [17] proposed braid group-based methods, with a specific focus on the proposal of the hardness of algebraic structures. So, it was obvious that braid groups could replace existing PKCs. In 2000, Ko et al. [18] discussed another PKC system using braid groups. The conjugacy search problem (CSP) delivered the essential security needs subject to the point that effective canonical lengths and braid index are chosen carefully. After that, many successions took place like Dehornoy et al. in 2004 [19], in 2003 and 2006 Anshel et al. [20, 21] and Cha et al. in 2001 [22]. In between the above said advancements, various attacks were proposed but in 2002, Ko et al. [23] and in 2003, Cheon et al. [24] and in 2000, Hughes et al. [25] have discussed related initial buoyancy. Many researchers have even discussed the end of braid group cryptography because of security flaws and in 2006, Bohli et al. [26] and in 2004, Dehornoy [27] are named a few.

In the study of the finite non-abelian groups, Paeng et al. [28] proposed a public key cryptosystem that utilizes the discrete log problem over the automorphic property. Magliveras et al. [29] worked on public cryptography to design effective one-way trapdoors on finite algebraic systems. Vasco et al. [30] suggested similarity in factorization and many cryptographic operations on homomorphic cryptosystems for non-abelian groups. Kanso et al. [31] proposed an innovative way to develop PKCs with the help of one-way functions for the finite group. Grigoriev and Ponomarenko [32, 33] increased the hardness of problems on integer matrices for a finite random group of elements.

In 2004, Eick et al. [34] have come up with a serious discussion on the polycyclic group as polycyclic groups are complex in their self-cyclic groups. These characteristics are very complicated. The group $G = G_1 \rhd G_2 \rhd \ldots \rhd G_{n+1} = \{1\}$ suggests the enhancement of subgroups of a group progressively. In this case, the group G is known as polycyclic series with a related cyclic property that is G_i/G_{i+1} recurring $i = 1, 2, \ldots, n$. In 2005, Shpilrain et al. [35] have suggested that Thompson's group could be a wise selection for the development of PKCs because decomposition problem (DP) is intractable with CSP over R. In 2005, Mahalanobis [36] talked about DH key transformation method and a finite non-abelian nilpotent group of class 2. It is a normal series to reach quotient H_i/H_{i+1} remains in the center of G/H_{i+1} as a central succession with shortest nilpotency degree. In 2010, Wang et al. [37] initiated a new idea about one-way left self-distributed (LD) system for all elements $x, y, z \in A$, where $x * (y * z) = (x * y)\ (x * z)$. To find x from given $x * y$ and y, the system behaves like way trapdoor subject to intractability.

The relation between CSP and LD says that if CSP over G is hard, then the derivatives of the LD system can be tackled as a way trapdoor. Cao et al. [38], 2007, have published a way to handle polynomials over non-commutative rings/semirings for the development of various cryptographic protocols which is called $\mathbb{Z}$-the modular method. In 2008, Kubo [39] discussed the dihedral group with some initial order and based on 3D revolutions. The milestone was set by Reddy et al. [40], in 2008, when they used $\mathbb{Z}$-the modular method to develop signature schemes over non-commutative groups. In 2010, Moldovyan et al. [41] discussed the realization of the protocol with security enhancements over non-commutative groups. In 2014, Myasnikov et al. [42] attacked on matrix configuration authentication scheme with the help of the hardness of CSP over non-commutative monoids. In the same year, Svozil [43] suggested the metaphorically hidden variable on non-contextual indecisiveness keeping quantum systems in the picture. In 2016, Du et al. [44] have proposed a number-theoretic transform to develop a class of scalable polynomial multiplier architecture. In 2017, Kumar et al. [45] proposed a non-commutative cryptography scheme based on an extra-special group. The basis of an extra-special group is a hidden subgroup or subfield problem. The order of this group is very important. So, the probability of a brute force attack or length-based attack launched by any intruder is NIL.

3 The Proposed Scheme

3.1 Mathematical Preliminaries

Consider a ring S with $(S, +, \circ)$ and $(S, \bullet, 1)$ as its additive abelian group and multiplication non-abelian semi-group, respectively. For $w \in Z_{>0}$ and an element u from S, define $wu = u + u + \cdots u$ (w times) and for $w \in Z_{<0}$, we can define $wu = (-w)(-u) = (-u) + (-u) + \cdots + (-u)$ ($-w$ times), for $w = 0, wu = 0$. Also, it has some properties as given below:

(i) $au^m \cdot bu^n = (ab)^{m+n} = bu^n \cdot au^m$ for all $a, b, m, n \in Z$ and all $u \in S$.

(ii) Since multiplication in S is non-commutative, therefore $au \cdot bs \neq bs \cdot au$ when $u \neq s$.

We have a polynomial $g(y) = b_0 + b_1 y + \cdots b_m y^n \in Z_{>0}[y]$, where $b_0, b_1, \ldots, b_n$ are the positive integers and also a ring element e in S, $g(e) = b_0 + b_1 e + \cdots b_n e^n \in S$. Now, we have $p(e) \cdot q(e) = q(e) \cdot p(e)$ for all $p\,(e)$ and $q\,(e)$ in S and $p(e) \cdot q(t) \neq q(t) \cdot p(e)$ for $e \neq t$.

Hard problem in non-commutative rings: For a non-commutative ring S and an element $a \in S$, define a set $P_a \subseteq S$ by, $p_a = \{f(a) : f(x) \in Z_{>0}[x]\}$. Suppose there are three elements a, x and y in S and two integers n_1, n_2 in Z, select $z \in P_a$ such that $y = z^{n_1} x z^{n_2}$. It clearly shows that polynomial symmetrical decomposition (PSD) is intractable.

3.2 Problem Formulation

Consider S is a non-commutative ring with two operations + and $\circ$. For any element $\alpha \in S$, let us pick a set $M_\alpha \in S$ such that $M_\alpha = \{g(\alpha) | g(x) \in Z_{>0}[x]\}$. Initially, we discuss DH-based key agreement protocol (KAP) and the proposed protocol will be discussed after that. The entity E_1 selects $g_1(x) \in z_{>0}[x]$ such that $g_1(\alpha) \neq 0$ calculates $\beta(E_1) = g_1(\alpha)^r \delta\, g_1(\alpha)^s$ and transmits $\beta(U_1)$ to an entity E_2. After getting $\beta(U_1)$ from E_1, entity E_2 chooses $g_2(x) \in z_{>0}[x]$ in such a way so that $g_2(\alpha) \neq 0$, compute $\beta(E_2) = g_2(\alpha)^r \delta\, g_2(\alpha)^s$ and sends $\beta(U_2)$ to an entity $E_1 U_1$. U_1 calculates the shared session key: $k(E_1) = g_1(\alpha)^m \beta(E_2) g_1(\alpha)^s$ and the entity E_2 also calculates the required session key:$k(E_2) = g_2(\alpha)^r \beta(E_1) g_2(\alpha)^s$.

3.3 The Proposed KAP Protocol

An innovative KAP is presented here, and the number of users is restricted to four, while it can be generalized for n users. Let four users or entities E_1, E_2, E_3, E_4 want to share a secret key:

Step 1: The entity E_1 randomly picks a polynomial $g_1(x) \in Z_{>0}[x]$ such that $g_1(\alpha) \neq 0$, calculates $K_1 = g_1(\alpha)^r \delta\, g_1(\alpha)^s$ and sends K_1 to the next entity E_2.

Step 2: After getting K_1 from E_1, the entity E_2 randomly picks a polynomial $g_2(x)$ from $Z_{>0}[x]$.

Such that $g_2(\alpha) \neq 0$, calculates $K_2 = g_2(\alpha)^r \delta\, g_2(\alpha)^s$ and $K_{21} = g_2(\alpha)^r K_1 g_2(\alpha)^s$, entity E_2 transmits (K_1, K_2, K_{21}) to the next entity E_3.

Step 3: After receiving the information $(K_1, K_2, K_{21})E_2$, the entity E_3 randomly picks a polynomial $g_3(x) \in Z_{>0}[x]$ such that $g_3(\alpha) \neq 0$ and calculates $K_{31} = g_3(\alpha)^r K_1 g_3(\alpha)^s$, $K_{32} = g_3(\alpha)^r K_2 g_3(\alpha)^s$, $K_{321} = g_3(\alpha)^r K_{12} g_3(\alpha)^s$. Entity E_3 transmits $(K_{31}, K_{32}, K_{21}, K_{321})$ to an entity E_4.

Step 4: Upon getting the information sent by E_3, the entity E_4 randomly selects a polynomial $g_4(x) \in Z_{>0}[x]$ such that $g_4(\alpha) \neq 0$ and calculates

$$K_{431} = g_4(\alpha)^r K_{31} g_4(\alpha)^s$$
$$K_{432} = g_4(\alpha)^r K_{32} g_4(\alpha)^s$$
$$K_{421} = g_4(\alpha)^r K_{21} g_4(\alpha)^s$$

After this calculation, E_4 sends K_{431} to E_2, K_{432} to E_1, K_{421} to E_3.

Step 5: All entities E_1, E_2, E_3, E_4 compute the shared secret keys $K(E_1)$,$K(E_2)$, $K(E_3)$, $K(E_4)$, were

$$K(E_1) = g_1(\alpha)^r K_{432} g_1(\alpha)^s$$
$$K(E_2) = g_2(\alpha)^r K_{431} g_2(\alpha)^s$$
$$K(E_3) = g_3(\alpha)^r K_{421} g_3(\alpha)^s$$
$$K(E_4) = g_4(\alpha)^r K_{321} g_4(\alpha)^s$$

3.4 Flow Diagram and Working Example

An illustration of the above-discussed protocol is required here. Figure 1 given indicates that the presented protocol works smoothly for four participating entities E_1, E_2, E_3, E_4.

Let us consider $\alpha = \begin{bmatrix} 3 & 1 \\ 5 & 2 \end{bmatrix}$ and $\delta = \begin{bmatrix} 3 & 7 \\ 9 & 4 \end{bmatrix}$ $r = 19, s = 23, N = 19 * 23 = 437$.

Step 1: E_1 picks a random polynomial, $g_1(x) = 7x^{413} + 5x^{314} + 9 \in Z_N[x]$ and calculates

$$g_1(\alpha) = \begin{bmatrix} 85 & 292 \\ 149 & 230 \end{bmatrix}$$

E_1 also calculates $K_1 = g_1(\alpha)^r \delta g_1(\alpha)^s = \begin{bmatrix} 122 & 382 \\ 378 & 62 \end{bmatrix}$. Then, he sends it to the second entity E_2.

Step 2: After getting K_1 from E_1, the entity E_2 picks a polynomial, $g_2(x) = 3x^{319} + 7x^{429} + 5 \in Z_N[x]$ and calculates

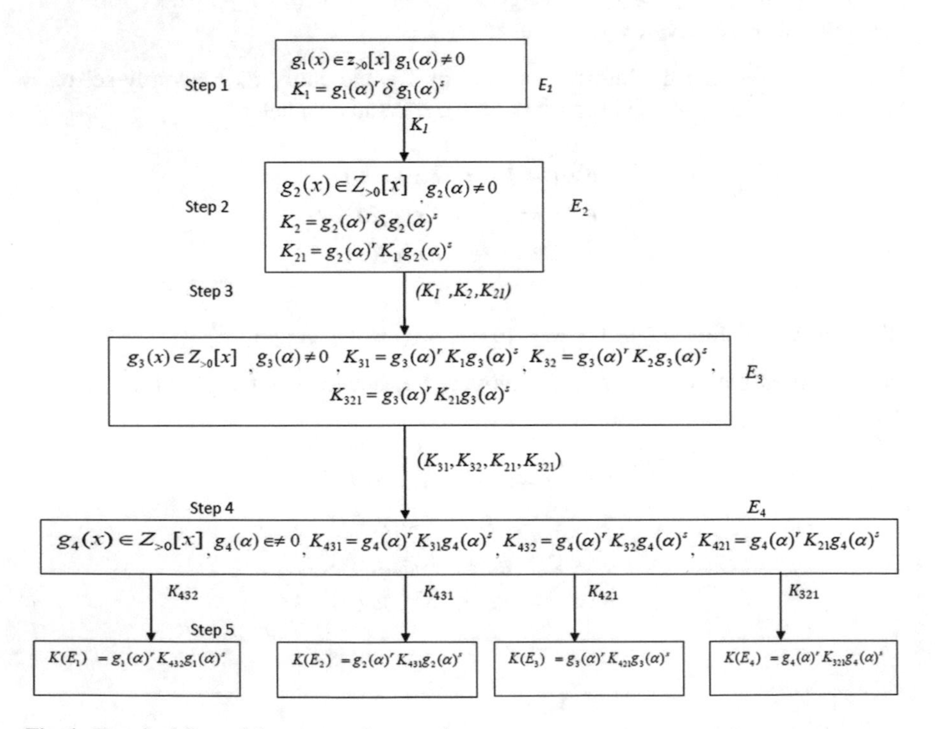

Fig. 1 Required flow of the proposed protocol

$$g_2(\alpha) = \left(3\begin{bmatrix} 3 & 1 \\ 5 & 2 \end{bmatrix}^{319} + 7\begin{bmatrix} 3 & 1 \\ 5 & 2 \end{bmatrix}^{429} + 5I\right) \bmod 437$$

$$= \begin{bmatrix} 186 & 57 \\ 285 & 129 \end{bmatrix}$$

$$K_2 = g_2(\alpha)^r \delta g_2(\alpha)^s$$

$$K_2 = \begin{bmatrix} 148 & 92 \\ 311 & 172 \end{bmatrix}$$

$$K_{21} = g_2(\alpha)^r K_1 g_2(\alpha)^s$$

$$K_{21} = \begin{bmatrix} 268 & 181 \\ 408 & 329 \end{bmatrix}$$

He sends $K_1 = \begin{bmatrix} 122 & 382 \\ 378 & 62 \end{bmatrix}$, $K_2 = \begin{bmatrix} 148 & 92 \\ 311 & 172 \end{bmatrix}$, $K_{21} = \begin{bmatrix} 268 & 181 \\ 408 & 329 \end{bmatrix}$ it to the entity E_3.

Step 3: After getting K_1, K_2 and K_{21} from an entity E_2, the entity E_3 randomly selects a polynomial $g_3(x) = 4x^{138} + 9x^{317} + 7 \in Z_N[x]$ and calculates $g_3(\alpha)$, K_{31}, K_{32} and, K_{321} given as

$$g_3(\alpha) = \left(4\begin{bmatrix} 3 & 1 \\ 5 & 2 \end{bmatrix}^{138} + 9\begin{bmatrix} 3 & 1 \\ 5 & 2 \end{bmatrix}^{317} + 7I\right) \bmod 437$$

$$g_3(\alpha) = \begin{bmatrix} 172 & 350 \\ 2 & 259 \end{bmatrix}$$

$$k_{31} = g_3(\alpha)^r K_1 g_3(\alpha)^s = \begin{bmatrix} 363 & 279 \\ 133 & 395 \end{bmatrix}$$

$$K_{32} = g_3(\alpha)^r K_2 g_3(\alpha)^s = \begin{bmatrix} 365 & 62 \\ 405 & 157 \end{bmatrix}$$

$$K_{321} = g_3(\alpha)^r K_{12} g_3(\alpha)^s = \begin{bmatrix} 200 & 331 \\ 266 & 398 \end{bmatrix}$$

Step 4: After getting K_{31}, K_{32}, K_{21} and K_{312} from entity E_3 E_4 randomly selects a polynomial $g_4(x) = 5x^{149} + 7x^{173} + 8 \in Z_N[x]$ and calculates $g_4(\alpha)$, K_{431}, K_{432} and K_{421} as follows:

$$g_4(\alpha) = \left(5\begin{bmatrix} 3 & 1 \\ 5 & 2 \end{bmatrix}^{149} + 7\begin{bmatrix} 3 & 1 \\ 5 & 2 \end{bmatrix}^{173} + 8I\right) \bmod 437 = \begin{bmatrix} 208 & 216 \\ 206 & 429 \end{bmatrix}$$

$$K_{431} = g_4(\alpha)^r K_{31} g_4(\alpha)^s = \begin{bmatrix} 128 & 184 \\ 310 & 403 \end{bmatrix}$$

$$K_{432} = g_4(\alpha)^r K_{32} g_4(\alpha)^s = \begin{bmatrix} 145 & 313 \\ 378 & 16 \end{bmatrix}$$

$$K_{421} = g_4(\alpha)^r K_{21} g_4(\alpha)^s = \begin{bmatrix} 294 & 187 \\ 156 & 73 \end{bmatrix}$$

Step 5: Now, entity E_4 transmits K_{432} to E_1, K_{431} to E_2, K_{421} to E_3 and K_{321} is utilized E_4 to produce a common key.

$$K(E_1) = g_1(\alpha)^r K_{432} g_1(\alpha)^s$$

$$K(E_1) = \begin{bmatrix} 165 & 122 \\ 277 & 58 \end{bmatrix}$$

$$K(E_2) = g_2(\alpha)^r K_{431} g_2(\alpha)^s$$

$$K(E_2) = \begin{bmatrix} 165 & 122 \\ 277 & 58 \end{bmatrix}$$

$$K(E_3) = g_3(\alpha)^r K_{421} g_3(\alpha)^s$$

$$K(E_3) = \begin{bmatrix} 165 & 122 \\ 277 & 58 \end{bmatrix}$$

$$K(E_4) = g_4(\alpha)^r K_{321} g_4(\alpha)^s$$

$$K(E_4) = \begin{bmatrix} 165 & 122 \\ 277 & 58 \end{bmatrix}$$

So the shared secret key $K(E_1) = K(E_2) = K(E_3) = K(E_4) = \begin{bmatrix} 165 & 122 \\ 277 & 58 \end{bmatrix}$.

3.5 Generalization of the Above Protocol

It is important to mention here that in real-life scenarios, the number of users is variable, so it is essential to generalize the presented protocol for k users.

Step i: (i = 1, 2, 3,…,k − 1), user U_i chooses a random polynomial $f_i(x)$ from $Z_{>0}[x]$ such that $f_i(\alpha) \neq 0$, computes $\left\{X_j = f_i(\alpha)^m \ldots \overline{f_j(\alpha)^m} \ldots f_1(\alpha)^m \beta f_1(\alpha)^n \ldots \overline{f_j(\alpha)^n} \ldots f_i(\alpha)^n, j = 1, 2, \ldots, i\right\}$, were $\overline{f_j(\alpha)^m}$ and $\overline{f_j(\alpha)^n}$ means that $f_j(\alpha)^m$ and $f_j(\alpha)^n$ are not appearing. User U_i also computes $Y_i = f_i(\alpha)^m f_{i-1}(\alpha)^{m \cdot} \ldots f_1(\alpha)^m \beta f_1(\alpha)^n \ldots f_{i-1}(\alpha)^n f_i(\alpha)^n$, user U_i sends the set $\sigma = \{X_1, X_2, \ldots, X_n, Y_i\}$ to U_{i+1} [i = 1,2,…,k − 1].

Step k: After receiving the set σU_{k-1}, the user U_k chooses a random polynomial $f_k(x) Z_{>0}[x]$ such that $f_k(\alpha)$ compute for i = 1,2,…,k − 1.

$\sigma_i = f_k(\alpha)^m \ldots \overline{f_i(\alpha)^m} \ldots f_1(\alpha)^m \beta f_1(\alpha)^n \ldots \overline{f_i(\alpha)^n} \ldots f_k(\alpha)^n$ and sends σ_i to U_i for $i = 1, 2, \ldots, k-1$. Then, each user U_i computes the sharing key, $Key(U_i) = f_i(\alpha)^m (f_k(\alpha)^m \ldots \overline{f_i(\alpha)^m} \ldots f_1(\alpha)^m \beta f_1(\alpha)^n \ldots \overline{f_i(\alpha)^n} \ldots f_k(\alpha)^n) f_i(\alpha)^n = f_k(\alpha)^m f_{k-1}(\alpha)^m \ldots f_1(\alpha)^m \beta f_1(\alpha)^n \ldots f_{k-1}(\alpha)^n f_k(\alpha)^n$. The user U_k also computes the shared key $Key(U_k) = f_k(\alpha)^m \ldots \overline{f_i(\alpha)^m} \ldots f_1(\alpha)^m \beta f_1(\alpha)^n \ldots \overline{f_i(\alpha)^n} \ldots f_k(\alpha)^n$

4 Conclusion and Future Scope

In this paper, a couple of key agreement protocols are presented using NCR. NCR is relatively a new platform, and it can be explored to surprise intruders. The correctness and working example are also given to verify the feasibility. The presented protocols are quite innovative because most of the earlier given protocols were based on number theory or elliptic curve and needless to say that all of them are vulnerable to a quantum computer attack. NCR is an emerging platform for the development of key agreements and other security protocols so a lot of research work is yet to be done in this regard. It makes the future scope of the proposed protocol very bright because other protocols and various implementation approaches can be incorporated to increase the overall throughput.

References

1. Diffie, Hellman ME (1976) New directions in cryptography. IEEE Trans Inf Theor 22(6):644–654
2. Menezes AJ, Oorschot PCV, Vanstone SA (2001) Handbook of applied cryptography, 5th edn. CRC Press Inc, USA
3. Stallings W (2005) Cryptography and network security, principles and practices, 4th edition. Prentice Hall
4. Hwang JY, Eom S, Chang KY, Lee PJ, Nyang D (2012) Anonymity-based authenticated key agreement with full binding property. In: International workshop on information security applications (WISA) (Part of the lecture notes in computer science book series (LNCS), pp 177–1917
5. Krawczyk H (2005) HMQV: a high-performance secure Diffie-Hellman protocol. In: Advances in cryptology-CRYPTO 2005, LNCS, vol 3621, pp 546–566. http://eprint.iacr.org/2005/176
6. Ciou YF, Leu FY, Huang YL, Yim K (2011) A handover security mechanism employing the Diffie-Hellman key exchange approach for the IEEE 802.16e wireless networks. In: Mobile information systems. Hindawi, vol 7, pp 241–269
7. Ryu EK, Yon EJ, Yoo KY (2004) An efficient ID-based authenticated key agreement protocol from pairings. In: International conference on research in networking (Part of the lecture notes in computer science book series (LNCS), vol 3042, pp 1458–1463
8. Koblitz N (1987) Elliptic curve cryptosystems. Math Comput 48(177):203–209
9. Miller VS (1985) Use of elliptic curves in cryptography. In: Proceedings of the ACM advances in cryptology (CRYPTO '85), pp 417–426
10. Peter WS (1994) Algorithms for quantum computation: discrete logarithms and factorings. In: Proceedings of the 35th annual symposium on foundations of computer science, pp 124–134

11. Kitaev A (1996) Quantum measurements and the Abelian stabilizer problem. In: Electronic colloquium on computational complexity, vol 3. http://eccc.hpi-web.de/eccc-reports/1996/TR96-003/index.html
12. Lee E (2004) Braid groups in cryptology. ICICE Trans Fundamentals 87(5):986–992
13. Proos J, Zalka C (2003) Shor's discrete logarithm quantum algorithm for elliptic curves. Quantum Inf Comput 3(4):317–344
14. Rotteler M (2006) Quantum algorithm: a survey of some recent results. Inf Forensic Entwistle 21:3–20
15. Wagner NR, Magyarik MR (1985) A public-key cryptosystem based on the word problem. In: Blakley GR, Chaum D (eds) Advances in cryptology—proceedings of CRYPTO 84, vol 196 of Lecture Notes in Computer Science, pp 19–36
16. Birget JC, Magliveras SS, Sramka M (2006) On public key cryptosystems based on combinatorial group theory. Tatra Mt Math Publ 33:137–148
17. Anshel I, Anshel M, Goldfeld D (1999) An algebraic method for public-key cryptography. Math Res Lett 6(3):287–291
18. Ko KH, Lee SJ, Cheon JH, Han JW, Kang JS, Park C (2000) New public-key cryptosystem using braid groups. In: Bellare M (ed) CRYPTO 2000, vol 1880 of Lecture Notes in Computer Science, pp 166–183
19. Dehornoy P (2004) Braid-based cryptography. Contemp Math 360:5–33
20. Anshel I, Anshel M, Goldfeld D (2003) Non-abelian key agreement protocols. Discret Appl Math 130(1):3–12
21. Anshel I, Anshel M, Goldfeld D (2006) A linear time matrix key agreement protocol over small finite fields. Appl Algebra Eng Commun Comput 17(3):195–203
22. Cha JC, Ko KH, Lee SJ, Han JW, Cheon JH (2001) An efficient implementation of braid groups. In: Boyd C (ed) Advances in cryptology—ASIACRYPT 2001, vol 2248 of Lecture Notes in Computer Science, pp 144–156
23. Ko KH, Choi DH, Cho MS, Lee JW (2002) New signature scheme using conjugacy problem. Cryptology ePrint Archive: Report 2002/168. https://eprint.iacr.org/2002/168
24. Cheon JH, Jun B (2003) A polynomial time algorithm for the braid diffie-hellman conjugacy problem. In: Boneh D (ed) Advances in cryptology—CRYPTO 2003, vol 2729 of Lecture Notes in Computer Science, pp 212–225
25. Hughes J, Tannenbaum A (2000) Length-based attacks for certain group based encryption rewriting systems. Institute for mathematics and its application. http://purl.umn.edU/3443
26. Bohli JM, Glas B, Steinwandt R (2006) Towards provable secure group key agreement building on group theory. Cryptology ePrint Archive: Report 2006/079. https://eprint.iacr.org/2006/079
27. Dehornoy P (2004) Braid-based cryptography. In: Myasnikov AG, Shpilrain V (eds) Group theory, statistics, and cryptography, vol 360 of Contemporary Mathematics, pp 5–33
28. Paeng SH, Ha K-C, Kim JH, Chee S, Park C (2001) New public key cryptosystem using finite non abelian groups. In: Kilian J (ed) Advances in cryptology—CRYPTO 2001, vol 2139 of Lecture Notes in Computer Science, pp 470–485
29. Magliveras SS, Stinson DR, Trung TV (2002) New approaches to designing public key cryptosystems using one way functions and trapdoors in finite groups. J Cryptol 15(4):285–297
30. Vasco MIG, Martinez C, Steinwandt R (2002) Towards a uniform description of several group based cryptographic primitives. Cryptology ePrint Archive: Report 2002/048
31. Kanso A, Ghebleh M (2022) A trapdoor one-way function for verifiable secret sharing. High-Confidence Comput 2(2):100060
32. Grigoriev D, Ponomarenko I (2002) On non-Abelian homomorphic public-key cryptosystems. J Math Sci 126(3):1158–1166
33. Grigoriev D, Ponomarenko I (2003) Homomorphic public-key cryptosystems over groups and rings. https://arxiv.org/abs/cs/0309010v1
34. Eick B, Kahrobaei D (2004) Polycyclic groups: a new platform for cryptology? https://arxiv.org/abs/math/0411077v1
35. Shpilrain V, Ushakov A (2005) Thompson's group and public key cryptography. In: Ioannidis J, Keromytis A, Yung M (eds) Applied cryptography and network security, vol 3531 of Lecture Notes in Computer Science, pp 151–163

36. Mahalanobis A (2005) The Diffie-Hellman key exchange protocol, its generalization and nilpotent groups [Ph.D. dissertation], Florida Atlantic University, Boca Raton, Fla, USA
37. Wang L, Cao Z, Okamoto E, Shao J (2010) New constructions of public-key encryption schemes from conjugacy search problems. In: Information security and cryptology: 6th international conference, Inscrypt 2010, Shanghai, China, 20–24 Oct
38. Cao Z, Dong X, Wang L (2007) New public key cryptosystems using polynomials over noncommutative rings. J Cryptol—IACR 9:1–35
39. Kubo J (2008) The dihedral group as a family group. In: Zimmermann W, Seiler E, Sibold K (eds) Quantum field theory and beyond. World Science Publication, Hackensack, NJ, USA, pp 46–63
40. Reddy PV, Anjaneyul GSGN, Ramakoti Reddy UDV, Padmavathamma M (2008) New digital signature scheme using polynomials over noncommutative groups. Int J Comput Sci Netw Security 8(1):245–250
41. Moldovyan DN, Moldovyan NA (2010) A new hard problem over non-commutative finite groups for cryptographic protocols. In: Computer network security: 5th international conference on mathematical methods, models and architectures for computer network security, MMM-ACNS 2010, St. Petersburg, Russia, 8–10 Sept 2010. Proceedings, vol 6258 of Lecture Notes in Computer Science, pp 183–194
42. Myasnikov AD, Ushakov A (2014) Cryptanalysis of matrix conjugation schemes. J Math Cryptol 8(2):95–114
43. Svozil K (2014) Non-contextual chocolate balls versus value indefinite quantum cryptography. Theoret Comput Sci 560:82–90
44. Du C, Bai G (2016) A family of scalable polynomial multiplier architectures for ring-LWE based cryptosystems. http://eprint.iacr.org/2016/323
45. Kumar G, Saini H (2017) Novel non commutative cryptography scheme using extra special groups. Security Commun Netw 2017:1–21

Randomness Analysis of Secured One-Time Password Generation Scheme for Financial Ecosystems

Manoj Kumar Mishra, Pronaya Bhattacharya, and Sudeep Tanwar

Abstract In financial transactions, digital payments have increased exponentially during the novel coronavirus (COVID-19) pandemic. Different wallets are used in which secured one-time password (OTP) generations are required for login to various banking services. Recent research has suggested an OTP calculation scheme using the Vigenère cipher. The schemes are useful for financial applications, but the generation is not significantly random. As the strength of the OTP depends basically on its generation randomness, so motivated by the same, in this paper, we propose a randomness-based scheme for OTP generation. To validate the security of the generated OTP, we carry out a randomness analysis. We compare our work; they have used only SHA variants to implement the method but we further extend their method by using other algorithms as well.

Keywords Message digest · One-time password · Randomness · Secure hash algorithm · Vigenère cipher

1 Introduction

In the modern scenario, the utility of OTPs is quite high. We use them in a variety of authentication-related scenarios, and they have wide range of applications [5, 10, 11]. Consider a financial transaction between a bank server and a user where the user is supposed to insert unique OTP in order to authenticate [1]. The security of OTPs is very important as any vulnerability makes the entire data communication in danger [6]. In recent works, Kumar et al. [7] have utilized Vigenère cipher for creating the

M. K. Mishra · S. Tanwar
Pranveer Singh Institute of Technology, Kanpur, UP 209305, India
e-mail: sudeep.tanwar@nirmauni.ac.in

Institute of Technology, Nirma University, Ahmedabad, Gujarat 382481, India

P. Bhattacharya (✉)
Amity School of Engineering and Technology, Amity University, Kolkata, WB 700135, India
e-mail: pbhattacharya@kol.amity.edu

Y. Singh et al. (eds.), *Proceedings of International Conference on Recent Innovations in Computing*, Lecture Notes in Electrical Engineering 1001,
https://doi.org/10.1007/978-981-19-9876-8_45

corresponding cipher text. They have selected different plain texts ranging from 5 to 50 characters for illustration, and the encoding time ranges from 0.53 milliseconds (ms) to 4.12 ms, respectively. This cipher text is further utilized as a key for hash-based message authentication codes (HMAC) string generations. They have utilized secured hash algorithms (SHA) variants, namely the SHA1, SHA224, SHA256, SHA384, and SHA512 algorithms.

So the method provides double hybrid security because of the Vigenère key and HMAC output, and at the same time, there is no need to send the OTP via message or SMS. Thus, the proposed method also solves the problem of OTP transportation. However, the authors failed to provide any security analysis which can prove whether the generated OTPs exhibit enough randomness or not. The OTP prediction is very bad for overall data communication security [12]. We further say that the dependence only on SHA variants is not good, and there must be an implementation with a randomness check for other algorithms such as MD2 and MD5.

1.1 Article Layout

The rest of the paper is organized as follows. Section 2 presents the security analysis of Kumar et al. scheme with graphical analysis. Section 3 provides an extension by implementing SHA algorithms, and we generate the corresponding OTPs. We check whether the generated OTPs exhibit enough randomness or not. Finally, Sect. 4 presents the concluding remarks.

2 Security Analysis of Kumar et al. Scheme

Kumar et al. [7] have proposed an innovative method for the generation of OTP, but they did not provide any security analysis. It is essential to check randomness in an OTP generation process. So first of all we present the randomness characteristics involved in the scheme.

Randomness analysis: The analysis of all the outputs shown by Kumar et al. [7] is not possible, but in order to illustrate the procedure, we pick tenth output (serial number 10 in first table) as it is the largest one, i.e., number of characters are 50 with corresponding encoding time of 4.12 ms. Here, it is required to show the corresponding table for the ease of readers of this paper, and it is shown below in Table 1. It is very important to check the frequency of occurrence for a given key and plain text for corresponding algorithms. Table 2 presents the pattern of symbols by SHA-1 algorithm, and Table 3 presents for SHA-224 algorithm. Similarly, Table 4 presents for SHA-256, Table 5 for SHA-384, and Table 6 for SHA-512 algorithm respectively.

Highly random: The generated OTPs are highly random, and in order to prove this, we show a conclusive table which is given below in table. It clearly denotes that occurrence of all the digits is variable in all the algorithms. For example, the

Table 1 Security analysis of Kumar et al. scheme [7]

Plaintext	Key	Algorithm	HMAC	OTP
Hello receiver. It is the time to generate OTP for the session	Dfzpgrgcirlytv Npkhotiehrkeetg Hokitzgptmbf itzcpagqe	SHA 1	4308ae8b0a4be0d0ec67-e51dca8c4aaa3d74f8b6	380065834
		SHA 224	acae465642b27120bf7fe8-6cde623c15007d28fe329f-d486bdde92f	662727463
		SHA 256	d87b79d03582d3bd8d-4353be7b2a939ca41a50e-16c98db184875e1d9267-a1e22	793834579
		SHA384	e7551cee492a1197e1-69068f3e765cd24f94c446-ffe8ce59fd8240c126a-69b0e711a7a42f06e13268-6b0c16037be17db	519191963
		SHA 512	85c20f49e4ef20d3414-bc34404cb0ee7f059-d4802552470f40b31188-f1969fb35ed202295aa06-878a07fefc4ee6c8e17-48b180571b339268dc2a	509231344

Table 2 Pattern of symbols with SHA-1 algorithm

S. No.	Algo.	Symbol	Freq	Occurrence percentage
1	SHA1	0	4	20.0000
2		4	4	20.0000
3		8	4	20.0000
4		3	2	10.0000
5		6	2	10.0000
6		7	2	10.0000
7		1	1	5.0000
8		5	1	5.0000
9		2	0	0
10		9	0	0

occurrence of 1 is 5% in SHA-1, but 15.8730% in SHA-384 while the average is 9.4343% only. Similarly, the occurrence of 3 is 11.9048% in SHA-256, but 4.7619% in SHA-384 while the average is only 8.2121%. The pattern of other digits can be understood in a similar fashion. Table 7 presents the patterns for all digits which are responsible for the OTP generation.

Table 3 Pattern of symbols with SHA-224 algorithm

S. No.	Algo.	Symbol	Freq	Occurrence percentage
1	SHA224	2	7	21.2121
2		6	5	15.1515
3		4	4	12.1212
4		0	3	9.0909
5		7	3	9.0909
6		8	3	9.0909
7		1	2	6.0606
8		3	2	6.0606
9		5	2	6.0606
10		9	2	6.0606

Table 4 Pattern of symbols with SHA-256 algorithm

S. No.	Algo.	Symbol	Freq	Occurrence percentage
1	SHA256	8	6	14.2857
2		1	5	11.9048
3		2	5	11.9048
4		3	5	11.9048
5		7	5	11.9048
6		9	5	11.9048
7		5	4	9.5238
8		4	3	7.1429
9		0	2	4.7619
10		6	2	4.7619

Table 5 Pattern of symbols with SHA-384 algorithm

S. No.	Algo.	Symbol	Freq	Occurrence percentage
1	SHA384	1	10	15.8730
2		6	10	15.8730
3		4	7	11.1111
4		7	7	11.1111
5		0	6	9.5238
6		2	6	9.5238
7		9	6	9.5238
8		5	4	6.3492
9		8	4	6.3492
10		3	3	4.7619

Table 6 Pattern of symbols with SHA-512 algorithm

S. No.	Algo.	Symbol	Freq	Occurrence percentage
1	SHA512	0	12	14.2857
2		4	12	14.2857
3		8	10	11.9048
4		2	9	10.7143
5		5	8	9.5238
6		1	7	8.3333
7		3	7	8.3333
8		9	7	8.3333
9		6	6	7.1429
10		7	6	7.1429

Table 7 Showing the pattern of all digits responsible for the generation of the respective OTP

Digit	SHA1	SHA224	SHA256	SHA384	SHA512	Average
0	20.0000	9.0909	4.7619	9.5238	14.2857	11.5324
1	5.0000	6.0606	11.9048	15.8730	8.3333	9.4343
2	0.0000	21.2121	11.9048	9.5238	10.7143	10.671
3	10.0000	6.0606	11.9048	4.7619	8.3333	8.2121
4	20.0000	12.1212	7.1429	11.1111	14.2857	12.9321
5	5.0000	6.0606	9.5238	6.3492	9.5238	7.2914
6	10.0000	15.1515	4.7619	15.8730	7.1429	10.5858
7	10.0000	9.0909	11.9048	11.1111	7.1429	9.8499
8	20.0000	9.0909	14.2857	6.3492	11.9048	12.3261
9	0.0000	6.0606	11.9048	9.5238	8.3333	7.1645

Chances of brute force: The possibility of predicting OTP is subject to the condition that intruder must be able to predict HMAC output accurately. It is next to impossible task because every algorithm produces different output. Suppose intruder picks SHA-512, then the total possible combinations are 16^{128}. We assume that intruder has tremendous computing strength, and he can check 10^{10} million instructions per second (MIPS), then also it would take $\approx 1.34 \times 10^{138}$ seconds which is close to 4.24×10^{130}, years which is computationally very expensive and infeasible.

Collision resistance: Hash function-based methods enjoy hash characteristics like Avalanche property, collision resistance, and one-way trapdoors [3]. So it can be said that two different inputs cannot generate exactly the same output and that means it is impossible to produce same OTPs having different input ingredients.

Avoidance of length-based attacks: Ordinary hash-oriented realizations suffer with a drawback that they are prone to length-based attacks [9]. In HMAC-based implementation, this drawback is removed.

Table 8 Showing generated OTP from MD and RIPEMD variants

Plain text	Key	Algo.	HMAC	OTP
Hello receiver. It is the time to generate OTP for the session	dfzpgrgcirlytvnpkh otiehrkeetghokitz gptmbfitzcpagqe	MD2	bf4c07fbc4794518-18555cb4d61ac5cf	49588541
		MD4	bdde0ad9b9f833902-36fa4a88be9f644	9.83E+08
		MD5	60860fb5df5328a25-92a099329abe35d	65385293
		RIPEMD128	135f3e8e1d91f0ee-de091ac286deb404	3.31E+08
		RIPEMD160	a9051f2a42c3c06ea-b125ba82ac1c3-24365e9296	14362812

Other benefits: It is very important aspect that a user can create variations in OTP by varying input ingredients. For example, variations can be created if the plain text or secret key is altered between participants. The OTP length can also be made variable. So it can be concluded that these variations will increase the level of difficulty for intruders.

3 The Proposed Scheme

Kumar et al. [7] have only used SHA variants. Here, we extend their scheme for further algorithms for the given plain texts and key as shown in Table 8. In their scheme, they have generated the cipher text for various plain texts combinations ranging from 5 to 50 characters. These cipher texts are further used for HMAC generation. We pick the cipher text (or the key) for plain text which is of 50 characters as it is the largest one will be better for analysis point of view. We use MD2, MD4, MD5, RIPEMD128, and RIPEMD160 for HMAC generations. MD2, MD4, MD5, and RIPEMD128 generate 128 bits or 32 bytes hex characters output while RIPEMD160 provides 160 bits or 40 bytes hex character output. We pick first nine even positioned numeric values for the calculation of corresponding OTP, presented in the Table 8.

Randomness analysis: Now, we show security analysis of the extended scheme. Since only numeric values have been used in the OTP generation process so it is very important to check the frequency of occurrence for a given key and plain text for corresponding algorithms.

Highly random: The generated OTPs in the extended version are highly random, and in order to prove this, we show a conclusive table which is given below in Table

Table 9 Pattern of symbols by the MD2 algorithm

S. No.	Algo.	Symbol	Freq	Occurrence percentage
1	MD2	5	5	26.3158
2		4	4	21.0526
3		1	3	15.7895
4		7	2	10.5263
5		8	2	10.5263
6		0	1	5.2632
7		6	1	5.2632
8		9	1	5.2632
9		2	0	0.0000
10		3	0	0.0000

Table 10 Pattern of symbols with MD4 algorithm

S. No.	Algo.	Symbol	Freq	Occurrence percentage
1	MD4	9	4	22.2222
2		3	3	16.6667
3		4	3	16.6667
4		8	3	16.6667
5		0	2	11.1111
6		6	2	11.1111
7		2	1	5.5556
8		1	0	0.0000
9		5	0	0.0000
10		7	0	0.0000

6. It clearly denotes that occurrence of all the digits is variable in all the algorithms. For example, the occurrence of 1 is 15.7895% in MD2 but 22.2222% in RIPEMD128 while the average is 9.91% only. Similarly, the occurrence of 8 is 9.0909% in MD5 but 3.8462% in RIPEMD160 while the average is 10.2482%. The pattern of other digits can be understood in a similar fashion. Table 9 presents the percentage occurrence of symbol patterns generated from MD2 algorithm, Table 10 shows for the MD4 variants, Table 11 for the MD6 variants, Table 12 for the RIPEMD128 variants, and Table 13 for the RIPEMD160 algorithm variants.

Variations are possible: In the extended method, we have used various versions of MD and RIPEMD algorithm, so overall more variations are possible now. Algorithms can be changed in different protocol runs, and it makes the prediction very difficult for intruders because an intruder has to predict the corresponding algorithm for a particular OTP.

Table 11 Pattern of symbols by the MD5 algorithm

S. No.	Algo.	Symbol	Freq	Occurrence percentage
1	MD5	2	4	18.1818
2		5	4	18.1818
3		9	4	18.1818
4		0	3	13.6364
5		3	3	13.6364
6		6	2	9.0909
7		8	2	9.0909
8		1	0	0.0000
9		4	0	0.0000
10		7	0	0.0000

Table 12 Pattern of symbols with RIPEMD128 algorithm

S. No.	Algo.	Symbol	Freq	Occurrence percentage
1	RIPEMD128	1	4	22.2222
2		0	3	16.6667
3		3	2	11.1111
4		4	2	11.1111
5		8	2	11.1111
6		9	2	11.1111
7		2	1	5.5556
8		5	1	5.5556
9		6	1	5.5556
10		7	0	0.0000

Table 13 Pattern of symbols by the RIPEMD160 algorithm

S. No.	Algo.	Symbol	Freq	Occurrence percentage
1	RIPEMD160	2	6	23.0769
2		1	3	11.5385
3		3	3	11.5385
4		5	3	11.5385
5		6	3	11.5385
6		9	3	11.5385
7		0	2	7.6923
8		4	2	7.6923
9		8	1	3.8462
10		7	0	0.0000

Table 14 Showing the pattern of all the digits responsible for the generation of the corresponding OTP in the extended scheme

Digit	MD2	MD4	MD5	RIPEMD128	RIPEMD160	Average
0	5.2632	11.1111	16.6364	16.6667	7.6923	11.4739
1	15.7895	0.0000	0.0000	22.2222	11.5385	9.9100
2	0.0000	5.5556	18.1818	5.5556	23.0769	10.4739
3	0.0000	16.6667	13.6364	11.1111	11.5385	10.5905
4	21.0526	16.6667	0.0000	11.1111	7.6923	11.3045
5	26.3158	0.0000	18.1818	5.5556	11.5385	12.3183
6	5.2632	11.1111	9.0909	5.5556	11.5385	8.5118
7	10.5263	0.0000	0.0000	0.0000	0.0000	2.1052
8	10.5263	16.6667	9.0909	11.1111	3.8462	10.2482
9	5.2632	22.2222	18.1818	11.1111	11.5385	13.6633

Customized applications: OTPs are the output of this method, and they are suitable for any particular application. OTPs can be used in electronic health database systems (EHDSs), various wireless communication applications, and banking transactions [2, 9].

Plenty of applications: The resultant OTPs can be utilized in various group communication scenarios, where the transmitted message is very important such as army secrets [4, 8]. We show one such illustration in the figure below where an army headquarter is transmitting messages to three active nodes using different OTPs. It is very obvious from the Fig. 1 that three different links are using three distinct OTPs generated from MD5, RIPEMD128, and RIPEMD160, respectively. Table 14 presents the patterns of all digits responsible from the different schemes, with the average value of the OTP generation.

3.1 *Advantages of the Scheme*

The extended scheme exhibits all the advantages present in Kumar et al. [7] scheme. All the hash-related properties like collision resistance, one-way trapdoor, etc., equally hold here. From the intruder's perspective, it becomes more difficult to predict because he has to check for ten algorithms now which was limited to five algorithms earlier. It enhances the difficulty level for intruders. From the user's perspective, now, one has more options to make the OTP variable, and users have a wider range to select the algorithm of their choice. As long as brute force is concerned, it is still very difficult. Suppose an intruder tries to run the above-said attack on RIPEMD160, then he has to run 16^{40} combinations. Keeping intruder capacity limited to 10^{10} MIPS, it would take 1.46×10^{32} s, which is close to 4.50×10^{24} years, and thus the break remains infeasible.

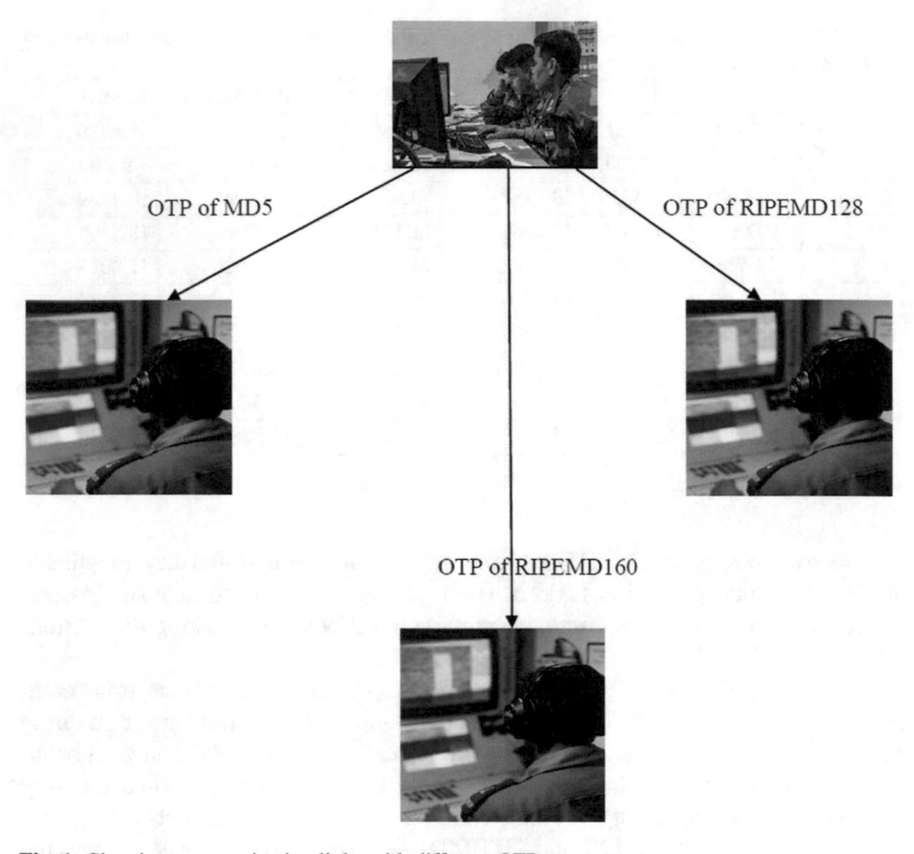

Fig. 1 Showing communication links with different OTPs

4 Conclusions and Future Scope

We have provided the complete security analysis of the Kumar et al. scheme which proves that the generated OTPs are very difficult to predict and easy to generate at the same time. The occurrence of the digits is random for every algorithm, and the brute force is computationally infeasible. We have also provided the extension of the Kumar et al. scheme. It creates more difficulties for intruders. We have shown that the extended scheme enjoys all the benefits of the previous scheme including some value addition, and it provides more options and ease of use from the user's perspective as well.

The future scope of the extended scheme is very rich. One can incorporate other hash algorithms like Whirlpool, Blake, etc. Other poly alphabetic ciphers can also be used. Software tools can also be developed by using this method to perform random numbers or random OTPs.

References

1. Alhothaily A, Alrawais A, Hu C, Li W (2018) One-time-username: a threshold-based authentication system. Proc Comput Sci 129:426–432. https://doi.org/10.1016/j.procs.2018.03.019, https://www.sciencedirect.com/science/article/pii/S1877050918302321, 2017 International conference on identification, information and knowledge in the Internet of Things
2. Alhothaily A, Hu C, Alrawais A, Song T, Cheng X, Chen D (2017) A secure and practical authentication scheme using personal devices. IEEE Access 5:11677–11687. https://doi.org/10.1109/ACCESS.2017.2717862
3. Andreeva E, Preneel B (2008) A three-property-secure hash function. In: International workshop on selected areas in cryptography. Springer, Heidelberg, pp 228–244
4. Dmitrienko A, Liebchen C, Rossow C, Sadeghi AR (2014) On the (in) security of mobile two-factor authentication. In: International conference on financial cryptography and data security. Springer, Heidelberg, pp 365–383
5. Erdem E, Sandıkkaya MT (2019) Otpaas-one time password as a service. IEEE Trans Inf Forensics Security 14(3):743–756. https://doi.org/10.1109/TIFS.2018.2866025
6. Kabra N, Bhattacharya P, Tanwar S, Tyagi S (2020) Mudrachain: blockchain-based framework for automated cheque clearance in financial institutions. Future Gener Comput Syst 102:574–587
7. Kumar M, Tripathi S (2021) A new method for otp generation. In: Healthcare and knowledge management for society 5.0. CRC Press, pp 213–228
8. Sciarretta G, Carbone R, Ranise S, Viganò L (2018) Design, formal specification and analysis of multi-factor authentication solutions with a single sign-on experience. In: International conference on principles of security and trust. Springer, Cham, pp 188–213
9. Shukla V, Chaturvedi A, Srivastava N (2015) Article: a new secure authenticated key agreement scheme for wireless (mobile) communication in an EHR system using cryptography. Commun Appl Electronics 3(3):16–21 (published by Foundation of Computer Science (FCS), NY, USA)
10. Shukla V, Chaturvedi A, Srivastava N (2019) Nanotechnology and cryptographic protocols: issues and possible solutions. Nanomater Energy 8(1):78–83
11. Shukla V, Mishra A, Agarwal S (2021) A new one time password generation method for financial transactions with randomness analysis. In: Favorskaya MN, Mekhilef S, Pandey RK, Singh N (eds) Innovations in electrical and electronic engineering. Springer Singapore, Singapore, pp 713–723
12. Yıldırım M, Mackie I (2019) Encouraging users to improve password security and memorability. Int J Inf Security 18(6):741–759

Social Bot Detection Techniques and Challenges—A Review Approach

N. Menaka and Jasmine Samraj

Abstract As web services and online social networks (OSN) such as Facebook, Twitter, and LinkedIn have grown in popularity, undesirable social bots have emerged as automated social players. In their attempts to emulate the activities of normal accounts, social bot accounts (Sybil's) have become more complex and misleading. In this situation, the disruption of fake news providers and bot accounts that distribute misinformation and sensitive content over the network has sparked. As a result, the research community must focus on developing systems that can detect social bots using neural learning to automatically determine the reliability of social media accounts. The focus of this study is to represent a review of graph-based, crowd sourcing-based, and anomaly based bots detection techniques. The paper also discusses a systematic detailed literature review of the bots identification techniques, the scope challenges, opportunities, and the main objectives clearly. To obtain a clear vision of the survey, the articles from the recent researchers are used. Further, a comparative analysis among the technique is provided based on the evaluation and also initiates future research works and advancement paths to deliver an interesting pasture.

Keywords Anomaly based · Crowd sourcing · Graph based · Social bot account · Online social networks

N. Menaka (✉) · J. Samraj
PG & Research Department of Computer Science, Quaid-E-Millath Government College for Women (Autonomous), Chennai, Tamilnadu, India
e-mail: menaka.research2021@gmail.com

J. Samraj
e-mail: dr.jasminesamraj@qmgcw.edu.in

Y. Singh et al. (eds.), *Proceedings of International Conference on Recent Innovations in Computing*, Lecture Notes in Electrical Engineering 1001,
https://doi.org/10.1007/978-981-19-9876-8_46

1 Introduction

Currently, the technologies are developed in a way for a user to adapt and socialize. Online social networks have become a common path for a user to get the exact information immediately according to their choice. Mobile and other web technologies help users to provide access from anywhere and anytime they want. As public communication channels, OSN plays an important part in recent trends. They give its users a place to participate, interact, and share information. As a result, they lead a fantastic community that understands the importance of recruiting customers for advertisements. Because of their popularity and extensive API, OSNs are also a desirable target for social bots. In its transparency reports, Twitter often mentions' malicious automation as a kind of platform manipulation, as well as how many maliciously automated accounts the firm has terminated. The goal of most studies is not to improve bot detection or develop bot detection methods; rather, they are to analyze how many social bots are active in specific discourses and whether they have any effect on discourse dynamics or opinion formation at all. Humans train bots to do specific tasks or activities. Malicious or non-malicious behavior can be programed into the system. Bots provide valuable information and services, such as emoji-based weather forecasts, horoscopes, and sports scores. Figure 1 shows the various types of bots that requires the continuation of service in social media for the development of business efficiency they are

- Nitreo
- Ingramer
- Instamber

Nitreo: It is an Instagram automation bot that assists in gaining more followers by connecting with random individuals on your account. It used to work very well until Instagram altered its algorithm and banned all bots (along with the humans who utilized them). Nitreo helps the user to choose the location and hashtags they wish to target and the bot contact profiles to meet the criterion (fake or inactive profiles included).

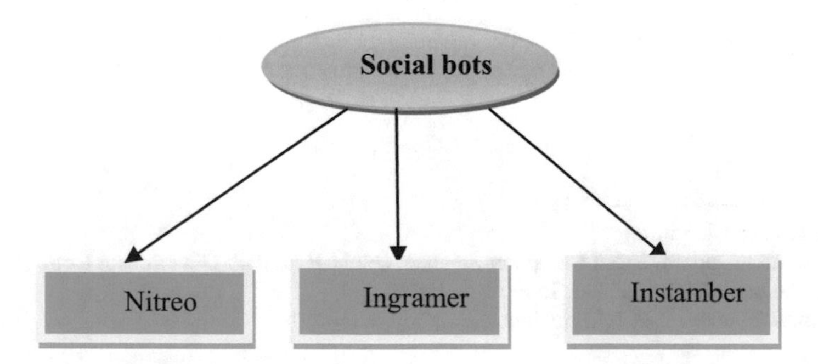

Fig. 1 Types of social bot [1]

Ingramer: Ingramer focuses on metrics like follower count; percentage of people who are engaged; user activity on average; the number of posts each day, amount of uploads; posting time that is most popular; terms used in the top captions; interests of the audience; and the most popular posts.

Instamber: The most effective way to boost social media growth. Instamber creates all-encompassing services to meet your Instagram, TikTok, and Twitter marketing requirements. Instamber is used in cutting costs and increasing productivity by utilizing low-cost tools.

Artificial intelligence (AI) has changed how we go about human's daily lives by developing and evaluating new software and gadgets known as intelligent agents that can perform a variety of tasks. The most crucial motive for bot users is productivity, while other factors such as amusement, social factors, and novelty engagement are also important. Furthermore, bots have become very popular in the business world because they reduce service costs and can handle a large number of customers at the same time [2]. This paper provides a thorough examination of various forms of bots, bot detection approaches, and challenges in bots. It involves in malicious and non-malicious bots activity that could be useful in many applications.

The remaining section of the study is as follows: the section initiates a background. The next Sect. 3 describes the challenges; Sect. 4 describes all the reviewed methodologies of the bots. Section 5 details the review results and evaluation. The conclusion is deliberated in the last section.

2 Background

The use of fake news providers and bot accounts to propagate misinformation and sensitive content over the network has sparked applied research into using artificial intelligence to automatically assess the reliability of social media accounts (AI) [3]. Karatas et al. [1] stated that Sybil accounts are targeted by structure-based detection algorithms. These accounts are used to break into OSN, steal personal information, and spread misinformation and malware. Blaise et al. [4] stated that BotFP is a lightweight bot detection technique that creates signatures based on host behavior in a network. It is found that about 15% of all active accounts on Twitter (i.e., 48 million) are bot accounts [5]. Filtering is largely reactive: a sender is included to a blacklist database only after a new threat has been identified and verified. On Twitter, similar spam-fighting tactics have been proposed, such as blacklisting known dangerous URL content and quarantining known bots [6]. Different components of detection strategies were compared, including many features, the dataset's size, and the data-crawling procedure.

2.1 Social Media Bot Detection Process

Bot developers today are employing latest technology and tools to create complex bots that can evade bot detection systems. Real-time behavioral and pattern analysis is required for effective identification of bot activity in the ever-changing threat landscape. Real-time behavioral assessment and pattern analytics serves as the basis of regular user activity, behaviors, trends, and characteristics. Each visitor's deviant activity is identified, handled, or blocked.

2.2 Static Rule-Based Approach

This approach is applied to detect bot activity on the site; web traffic data are manually checked. Bots are blocked by analyzing parameters including traffic trends, performance of the server, geo-location, bounce rates, and language sources. Bots are now disseminated from well-known private IP addresses. They may also easily imitate human-like signatures, making it nearly hard to tell the difference between their request and that of a genuine user. As a result, this strategy is also ineffective.

3 Challenges

The bot is a piece of malware created by cybercriminals in order to obtain the necessary data. A harmful bot is created with the intent of stealing sensitive information or infecting the host with a virus. Cybercriminals can utilize the data for a variety of purposes, including DDOS violation and spamming.

The following tasks are performed by malicious bots:

- Credential stuffing attacks
- Scraping bots
- DDoS attack
- Inventory denial attacks

3.1 Credential Stuffing Attacks

Credential stuffing is a type of cyber-attack in which intruders break into a server by using a list of acquired user credentials. Figure 2 explains the credential stuffing attacks [7], where fraudsters steal or purchase web account credentials—such as usernames and email addresses along with their passwords, from the web to gain access to user accounts in many applications using automated queries on a big scale, which are frequently carried out by bots. These are automated script attacks, in which

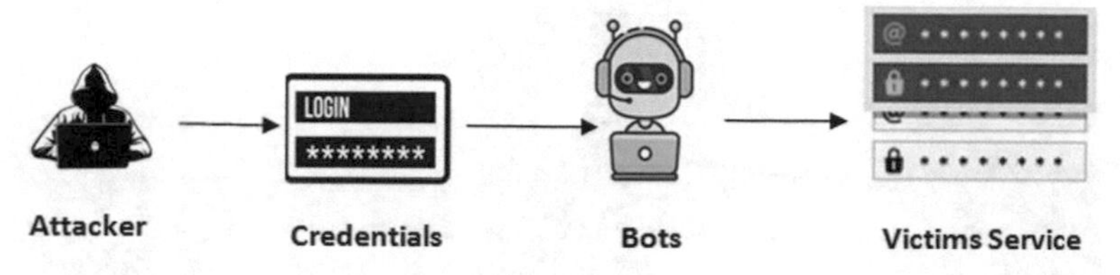

Fig. 2 Credential stuffing attacks [7]

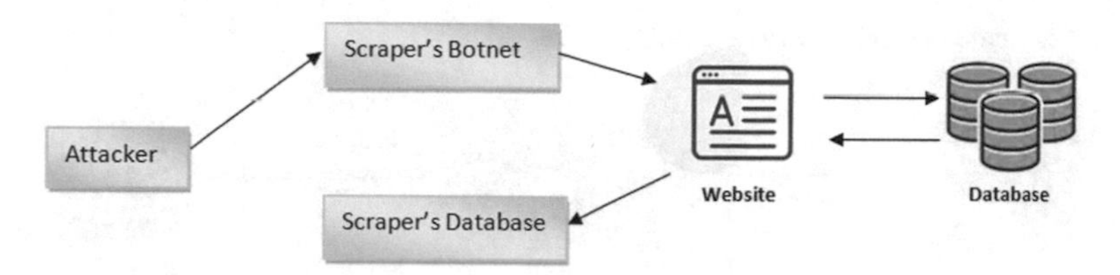

Fig. 3 Scraping bots [2]

a fraudster uses the coding of a website to fill out various application or sign-up forms at once using stolen credentials.

3.2 *Scraping Bots*

Web scraping is the technique of extracting data and information from a website using bots. A number of digital enterprises that depend on information gathering use web scraping. Search engine bots scanning a site, assessing its data, and rating it are examples of valid use cases. Figure 3 states the action of scraping bot [2].

3.3 *DDOS Attack*

DDOS cyber-attacks are a significant threat to the network because a large amount of users send requests to a central computer, causing the server to be unable to deliver adequate services to the customers due to high resource usage; Fig. 4 describes how a common DDOS attack takes place [8]. Systems and other device that have been attacked with spyware and can be manipulated directly by an intruder form a network. Isolated devices are known as bots. A multi-vector DDoS attack employs many attack vectors to overload a target in a variety of ways, perhaps diverting mitigation actions away from any one approach.

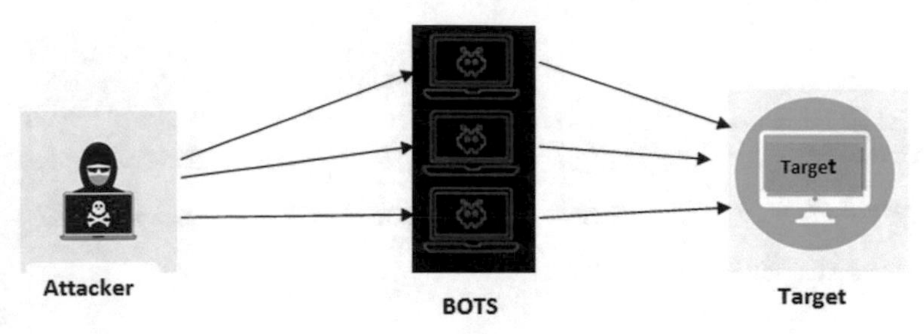

Fig. 4 Distributed denial of service attack (DDOS) [8]

Fig. 5 Inventory denial of service attack [9]

3.4 *Inventory Denial Attacks*

A denial of inventory attack occurs when an automation program (bots) enters an e-commerce goods or service to the online shop repeatedly without ever completing the transaction. This misleads the e-commerce site thinking the goods or service is out of supply, making it unavailable to genuine customers. Shopping bots are specialized automated programs or bots that undertake denial of inventory assaults. The below mentioned Fig. 5 states the action of inventory denial attacks [9]. They can pose a serious threat to an e-commerce site, resulting in not only direct drop in revenue, but also long-term and even irreversible damage to the web's reputation.

4 Systematic Review Methodologies

This section discusses about the systematic methodologies of existing methods.

4.1 *Graph-Based Bot Detection Technique*

Graphs are genuine models of communication networks, and a graph-based method is logical [10]. A graph-based bot detection system is a powerful and effective tool for detecting a wide range of Botnet with diverse behavioral characteristics [11]. Sybil's has a minimal number of links to legal (honest) users, and the graph is weakly interconnected. The Sybil networks use their extensive connections to create a false sense of trustworthiness among OSN members. It is critical to understand how Sybil profiles propagate over the Internet in order to detect them, specifically for this form of detection. The relational graphs which Twitter enable people to create are a good example of graph-based capabilities. The interactions among users are represented by these graphs. The triangle symbolizes the adjacency between nodes in the user network's triangle count; a high number of triangles indicate that the user is genuine. This technique classifies the groups as malware and benign using statistical means and client evaluation.

4.2 Crowd Sourcing-Based Bot Detection Technique

General intelligence is tested against extensive social bots with AI capabilities in crowd sourcing-based detection systems. Crowd generated detection systems face a dilemma when it comes to defending customer privacy [1]. Crowd sourcing is a term used to describe the technique of delegating tasks to an anonymous set of people. In the instance of crowd sourcing bot identification, users are provided data from accounts on OSNs, such as photographs, walls, and profile information, and prompted to categories the accounts as no bot or human based on this data. Other groups, such as cyborgs, might, of course, be included in the labeling in addition to bots and humans. Work is outsourced to an undefined set of people in the process of crowd sourcing. As successful projects like Wikipedia have demonstrated, the web considerably simplifies the effort of assembling virtual groups of labor. Crowd sourcing is effective for any activity that can be broken down into small, easy jobs, and it provides significant benefits for tasks that are difficult to complete using automation algorithms or systems.

4.3 Machine Learning-Based Bot Detection Technique

To counteract its efficacy, a machine learning-based method can be used. A classification system binary detection and can warn of a potentially malicious executable that could put a system at risk earlier [12]. Machine learning technology enhances the experience for users of computer programs while also allowing them to learn. It streamlines the time-consuming documentation process for input of data. The identification of spam is simple in the research and makes it easier to make accurate predictions. Malicious behavior detection is also carried out, using an offline supervised learning and an online detection phase. The majority of bot identification techniques for Twitter accounts use supervised techniques. Because certain bot accounts contain little if any information on Botnet activities, they can elude identification. Because the majority of research focuses on offline identification, machine learning activities, and tasks should be built to be accessible and scalable in terms of dealing with a constant supply of large Twitter data.

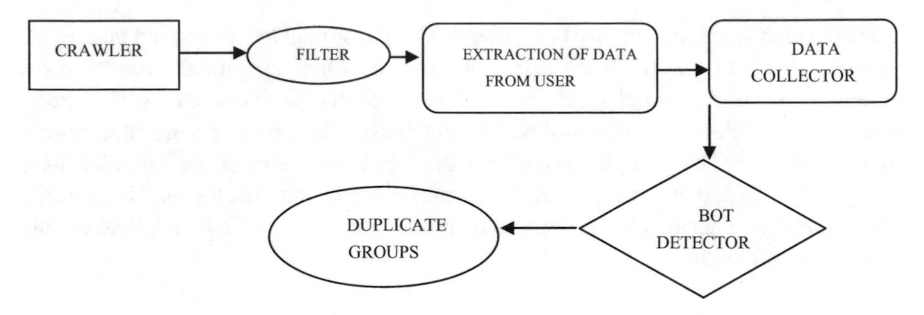

Fig. 6 General architecture of bot detection [1, 10, 12, 13]

4.4 Anomaly Based Bot Detection Technique

This method detects unidentified Botnet, these are extensively employed for Botnet detection. They draw attention to the problem of locating cases in a dataset that do not behave normally. They model a decision engines after creating a baseline of usual routine for the protected system. Any divergence or statistical deviation from the norm can be detected by this decision engine, and it can be flagged as a threat. Anomaly based IDS approaches have been demonstrated to detect Botnet based on high traffic volumes, the amount of Internet traffic anomalies, high network latency (traffic on odd ports), and other unexpected system behavior, according to research. The purpose of anomaly detection is to locate items that are distinct from the majority of other objects. Outliers are anomalous items that are located away from those other pieces of data in a scatter graph. The main drawbacks of anomaly detection are the high rate of false alarms and the limited amount of training data available. Figure 6 stated below describes the general working of bot detector [1, 10, 12, 13].

5 Result Analysis

Result analysis is explained in Table 1 and Fig. 7, which describes the comparison of all the existing techniques. The result is derived by comparative analysis of bot detection techniques such as graph-based technique [10], crowd sourcing [1], anomaly based [13], and machine learning-based techniques [12]. The techniques are proved to be efficient and provide accurate results in bot detection.

Table 1 Accuracy of existing techniques [1, 10, 12, 13]

Existing techniques	Accuracy in percentage (%)
Graph-based bot detection	72
Crowd sourcing bot detection	75
Machine learning-based bot detection	82
Anomaly based bot detection	80

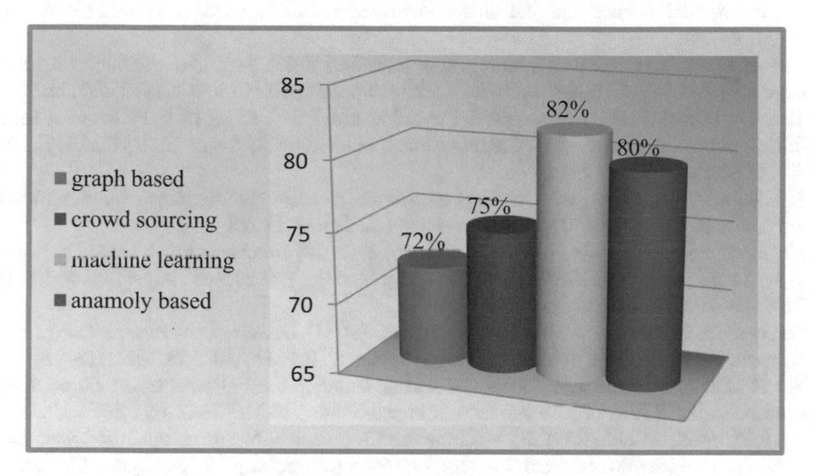

Fig. 7 Comparative analysis of existing techniques in bot detection [1, 10, 12, 13]

6 Conclusion

This survey paper provides a detailed literature review for detecting bots. The deliberated review emphasizes various technologies handle previously for enhancement of the bot detection according in social media. Here, the paper has a detailed discussion about the challenges faced. The paper further analyzes and discusses other researches and emits proper comparative analysis of the best bot detection technique. Moreover, evaluating the research results led us to obtain a clear vision of what actually bot detection technique does and what are the problems faced by them and the techniques to overcome the problems are also well-defined.

References

1. Karatas A, Serap S (2019) A review on social bot detection techniques and research directions. http://www.apple.com.ios/siri/
2. Adamopoulou E, Mossiades L (2020) Chatbots: history, technology, and applications. In: Machine learning with applications, vol 2, ISSN 2666-8270. https://doi.org/10.1016/j.mlwa.2020.100006

3. Martin-Gutierrez D, Hernandez Penaloza G, Hernandez AB, Lozano–Diez A, Alvarez F (2021) A deep learning approach for robust detection of bots in twitter using transformers. In: IEEE Access, vol 9, pp 54591–54601. https://doi.org/10.1109/ACCESS.2021.3068659
4. Blaise A, Bouet M, Conan V, Secci S (2020) Botnet finger printing: a frequency distribution scheme for lightweight bot detection. IEEE Trans Netw Service Manag. https://doi.org/10.1109/TNSM.2020.2996502
5. Farber M, Qurdina A, Ahmedi L (2019) Identification twitter bots using a convolutional neural network, Notebook for PAN at CLEF 2019
6. Van Der Walt E, Eloff J (2018) Using machine learning to detect fake identities: bots vs. humans. In: IEEE access digital object identifier, vol 6, https://doi.org/10.11109/ACCESS.2018.2796018
7. Nathan M (2020) Credential stuffing: new tools and stolen data drive continued attacks. In: Computer fraud security, vol 2020, no 12. https://doi.org/10.1016/S13613723(20)30130-5
8. Tuan TA, Long HV, Son LH, Kumar R, Priyadharshini I, Son NTK (2019) Performance evaluation of Botnet DDos attack detection using machine learning. Springer–Verlag GmbH Germany, Part of Springer Nature 2019
9. API Security Threat Report. Bots and automated attacks explode. https://www.cequence.ai/wp-content/uploads/2022/03/Cequence-Threat-API-Security.pdf
10. Daya AA, Salahuddin MA, Limam N, Boutaba R (2020) Botchase: graph based bot detection using machine learning. IEEE Trans Netw Serv Manage 17(1):15–29. https://doi.org/10.1109/TNSM.2020.2972405
11. Stephens B, Shaghaghi A, Doss R, Kanhere S (2021) Detecting Internet of Things bots: a comparative study. IEEE Access, p 1. https://doi.org/10.1109/ACCESS.2021.3130714
12. Shi P, Zhang Z, Choo KR (2019) Detecting malicious social bots based on clickstream sequences. IEEE Access 7:28855–28862. https://doi.org/10.1109/ACCESS.2019.2901864
13. Shi P, Zhang Z, Alsubhi K (2021) Machine learning based botnet detection in software-defined network: a systematic review. Symmetry 13:866. https://doi.org/10.3390/sym13050866

Digital India

Bidirectional Machine Translation for Punjabi-English, Punjabi-Hindi, and Hindi-English Language Pairs

Kamal Deep Garg, Vandana Mohindru Sood, Sushil Kumar Narang and Rahul Bhandari

Abstract Machine translation (MT) aims to remove linguistic barriers and enables communication by allowing languages to be automatically translated. The availability of a substantial parallel corpus determines the quality of translations produced by corpus-based MT systems. This paper aims to develop a corpus-based bidirectional statistical machine translation (SMT) system for Punjabi-English, Punjabi-Hindi, and Hindi-English language pairs. To create a parallel corpus for English, Hindi, and Punjabi, the IIT Bombay Hindi-English parallel corpus is used. This paper discusses preprocessing steps to create the Hindi, Punjabi, and English corpus. This corpus is used to develop MT models. The accuracy of the MT system is carried out using an automated tool: Bilingual Evaluation Understudy (BLEU). The BLEU score claimed is 17.79 and 19.78 for Punjabi to English bidirectional MT system, 33.86 and 34.46.46 for Punjabi to Hindi bidirectional MT system, 23.68 and 23.78 for Hindi to English bidirectional MT system.

K. D. Garg (✉) · V. M. Sood · S. K. Narang · R. Bhandari
Chitkara University Institute of Engineering and Technology,
Chitkara University, Punjab, India
e-mail: kamaldeep.garg@chitkara.edu.in

V. M. Sood
e-mail: vandana.sood@chitkara.edu.in

S. K. Narang
e-mail: sushilk.narang@chitkara.edu.in

R. Bhandari
e-mail: rahul.bhandari@chitkara.edu.in

Y. Singh et al. (eds.), *Proceedings of International Conference on Recent Innovations in Computing*, Lecture Notes in Electrical Engineering 1001,
https://doi.org/10.1007/978-981-19-9876-8_47

Keywords Machine translation · SMT · Corpus-based · Parallel corpus · BLEU

1 Introduction

Machine translation (MT) is a system that analyzes source text, applies rules and computation to the source text, and converts source text to target text without human contact [1], e.g., Punjabi text is input to the MT system and translated to Hindi text. MT is one of the hard problems in natural language processing (NLP) [2]. The rule-based and the corpus-based are two common techniques for MT [3, 4]. Rule-based MT uses linguistic rules, morphological analyzer, parser, etc., to translate input text to target text [5]. The latter approach uses monolingual and parallel corpus to translate text. Example-based machine yranslation (EBMT) and statistical machine translation (SMT) are based on a corpus-based technique. SMT system is better than a rule-based and EBMT system as it does not require human interpenetration.

In this paper, three bidirectional phrase-based SMT system has been developed by using the parallel corpus. The name of the systems is Punjabi to English bidirectional MT system, Punjabi to Hindi bidirectional MT system, and Hindi to English bidirectional MT system. The researchers have developed various MT systems using the corpus-based approach. Some of the systems are listed here.

Tran et al. [6] developed Chinese to Vietnamese SMT using the phrase-based statistical approach. Chinese dependency relation and Chinese-Vietnamese word alignment are used to improve the accuracy of the baseline SMT model. Azath et al. [7] developed an English to Tigrigna SMT system using Moses's decoder. Abidin et al. [8] developed the Indonesian to Lampung SMT system. The monolingual corpus is used to improve the accuracy of the Indonesian to Lampung MT system. Thu et al. [9] developed a phrase-based, hierarchical phrase-based SMT system, and hybrid MT for English to Myanmar. The author claims the higher accuracy of hybrid MT system as compared to phrase-based and hierarchical-based MT system. Biadgligne et al. [10] developed English to Amharic SMT system. They had augmented the parallel corpus to improve the BELU score of the system. Nagy et al. [11] developed syntax-based data augmentation for English–Hungarian bidirectional MT system.

This is how the paper is structured. The dataset that will be used to train and test the system is discussed in Sect. 2. Section 3 is about the design and architecture of the system. Section 4 discusses the experimental setup and evaluation. The conclusion is discussed in Sect. 5.

2 Development of Dataset

To develop a SMT system, the main requirement is an accurate parallel corpus. A. Kunchukuttan et al. at IIT Bombay [12] had developed a Hindi-English parallel corpus of 1,492,827 sentences. The corpus contains Hindi-English parallel text from various online sources such as Mahashabdkosh and Wiki Headlines. The Hindi and English files from their website link were downloaded. The corpus was manually checked. Corpus had a lot of errors. Without removing these errors, this corpus could not be used in the MT task. Some types of errors in the corpus are listed here.

Error 1: It contained many sentences of a single word only.
Error 2: Same lines were repeated in the corpus.
Error 3: Many lines start with %s and end with %s.
Error 4: Many lines contain ... in the end.

Table 1 gives some content of their Hindi-English parallel corpus that shows errors in the IIT Bombay Hindi-English corpus. Without removing the noisy and incomplete sentences from the corpus, this cannot be used to develop the MT system. The accuracy of the SMT system is dependent on the quality and quantity of parallel corpus used for training and testing the system [13].

2.1 *Preprocessing of Hindi-English Parallel Corpus*

Various preprocessing steps had done to create a clean corpus.

Step 1: English file is inspected for empty lines as well as special characters: '%', ')', '<', '_', '/'in the sentence. To correct these errors, Python code has been written to remove all those sentences in parallel that contain empty lines or lines with special characters. We had left with a Hindi-English parallel corpus of 1,354,333 sentences.

Step 2: Hindi file is inspected for empty lines as well as special characters: '%', ')', '<', '_', '/'in the sentence. By using Python code, all those lines in parallel that contain empty lines or lines with special characters had been removed. We had left with a Hindi-English parallel corpus of 1,251,938 sentences.

Table 1 Errors in Hindi-English parallel corpus

Hindi text	English text	Error description
मान	Val _ ue	_ is in English text
पसंद (_ P)...	_ Preferences..	(_P) and ... in Hindi text
नया % s खाता	New % s account	%s is noise in Hindi and English text
:अर्ड्स्हिप् प्रोविसिओन्	Hardship provision	Noisy sentence

Step 3: English character in the Hindi file was searched and by writing Python code, removed all lines in parallel where the Hindi sentence contains any English character. We had left with a Hindi-English parallel corpus of 1,093,091 sentences. It is also shown in Fig. 1.

Step 4: A parallel corpus of 1,093,091 sentences in Hindi and English is stored in small files, with each file containing a maximum of 50,000 sentences. The Hi2Pu system was used to translate Hindi text to Punjabi [14]. The translation from Hindi to Punjabi took a long time. We got a Punjabi-Hindi-English parallel corpus of 634,468 sentences after translation.

Step 5: Punjabi text is again checked to see it contains English text or not. Then, using Python code, remove all lines in parallel where any English character appears in a Punjabi sentence. We ended up with a parallel corpus of 627,691 sentences in Hindi, English, and Punjabi.

2.2 *Corpus Statistics*

A Punjabi-Hindi-English parallel corpus of 627,691 sentences was developed from the IIT Bombay Hindi-English Parallel Corpus. The statistics of the parallel corpus are given in Table 2.

3 Design and Architecture

The SMT framework assumes that any sentence in the target language is a potential translation of the input text in the source language. The SMT system is composed of three parts (a) language model, (b) translation model, and (c) decoder.

3.1 *Language Model*

It is used to calculate the sentence's probability using the n-gram model [15]. A sentence's probability is calculated by decomposing it into the product of conditional probability. As illustrated in the equation below, the chain rule is applied to the sentence. The probability of a sentence P(S) is divided into the probability of each individual word.

$$\mathrm{P(S)} = \mathrm{P}(\mathrm{word}_1, \mathrm{word}_2, \mathrm{word}_3, \ldots \mathrm{word}_n) \tag{1}$$

$$\mathrm{P(S)} = \mathrm{P}(\mathrm{word}_1)\mathrm{P}(\mathrm{word}_2|\mathrm{word}_1)\ldots \mathrm{P}(\mathrm{word}_\mathrm{n}|\mathrm{word}_1\mathrm{word}_2\mathrm{word}_{n-1}) \tag{2}$$

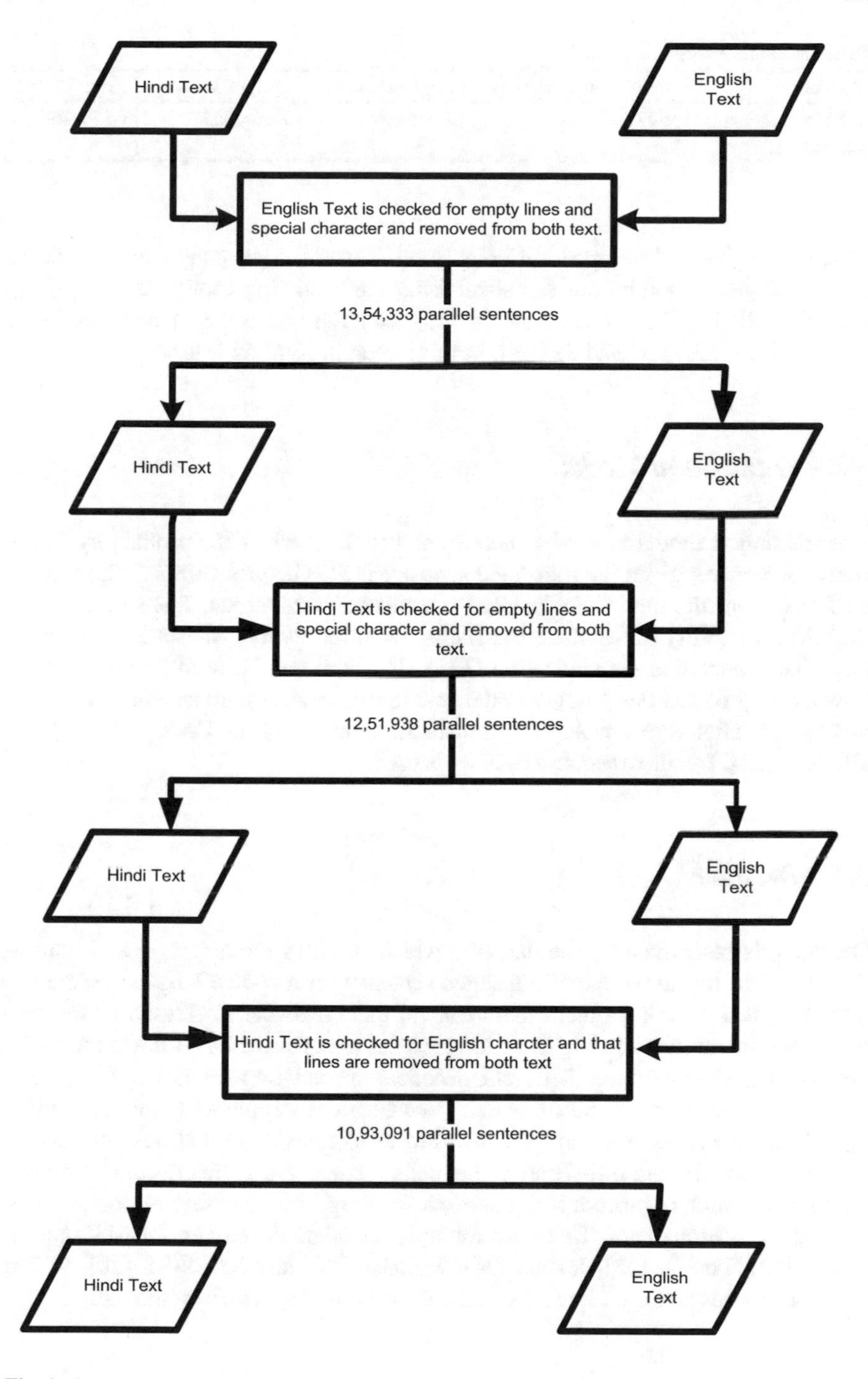

Fig. 1 Preprocessing steps on parallel corpus

Table 2 Parallel corpus statistics

Source	Sentences (parallel)	English tokens	Punjabi tokens	Hindi tokens
IIT Bombay parallel corpus	627,691	7,353,554	7,095,525	7,829,466

Researchers have developed different toolkits for the language model. Some commonly used toolkits are statistical language modeling toolkit (SLMT) [16], SRILM [17], IRSTLM [18], KenLM [19], etc. KenLM is faster and has lower memory than IRSTLM and SRILM. In this research, KenLM is used.

3.2 *Translation Model*

The translation model is used to determine P(f|e), which is the probability of the source sentence f given the translated sentence [15]. The probability P(f|e) can not be found from the number of parallel sentences in the corpus. The solution is to find the probability of sentence translation from the word's translation probabilities. The expectation–maximization (EM) algorithm can be used to extract word translation probabilities from a sentence-aligned parallel corpus. There are various tools available to develop the translation model such as CARMEL [20] and GIZA++ [21]. In this research, GIZA++ is used.

3.3 *Decoder*

Decoding is performed by the decoder, which employs the language and translation model to locate the target translated sentence for a source language sentence. Decoding is a search problem that seeks to maximize the likelihood of the language and translation model. The decoder must look for the best translation in the possible translation space. Different decoders are existing for the SMT system. The greedy decoder and beam search decoder are a couple of them. The initial hypothesis of greedy decoders is a word-to-word translation that has been recursively optimized using hill-climbing heuristics. The beam search decoder employs a heuristic search algorithm that uncovers the graph by widening the most hopeful node in a limited set. There are multiple decoders developed for MT such as Moses [22], Pharoh [23], Joshua [24], MARIE [25], and RAMSES [26]. In this research, Moses is used. Figure 2 depicts the proposed system's architecture.

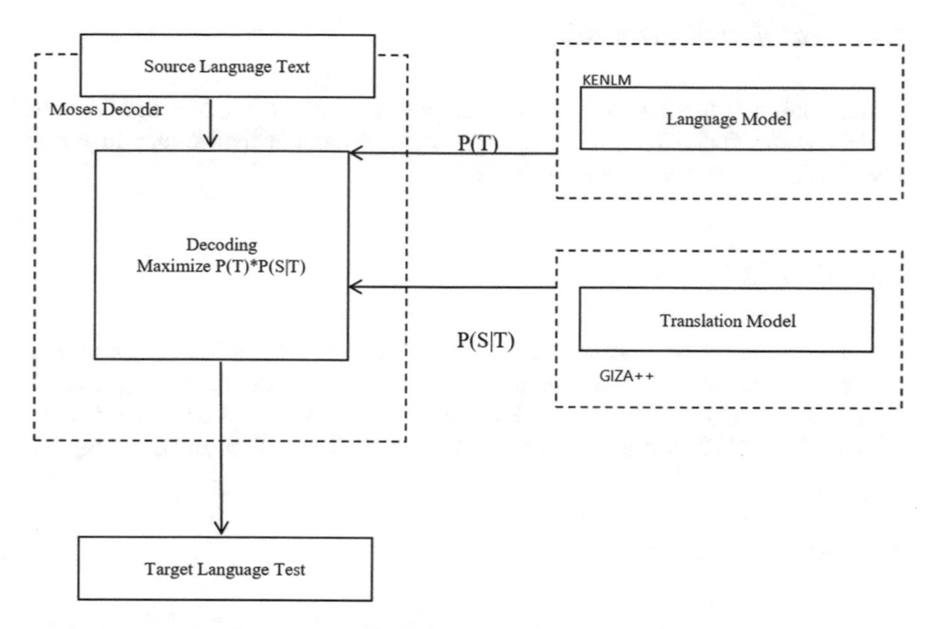

Fig. 2 Architecture of phrase-based SMT system

4 Setup and Analysis of Experiments

In this section, the system is trained using Moses and tested by using an automated evaluation tool known as the BLEU score [27].

4.1 Preprocessing of Dataset

The first step in the development of an MT system is corpus preprocessing. The dataset goes through several steps.

- Tokenization of corpus
- Cleaning of the long sentences from the corpus
- Lowercasing the English corpus.

English, Hindi, and Punjabi text files are tokenized at the word level. After tokenization, a long sentence (having more than 40 words) was removed from all three text files in parallel. English text is also lowercase to increase the accuracy of the MT system.

4.2 Training of Proposed System

Moses' toolkit is being used to train, tune, and test various models. Six different models are developed by using Moses. The whole dataset is divided into three sets. The division of the dataset is given in Table 3.

4.3 Results

The testing set has been separated into three small sets based on the number of words in a sentence. Smaller sentences contain a maximum of 5 words, medium sentences contain 6–14 words, and large sentences contain 15 words or more in a sentence. The BLEU score of all models is shown in Tables 4, 5, and 6 and Fig. 3.

4.4 Analysis of Output

The output obtained from different MT systems is analyzed to check the accuracy of the proposed system. Here, English text is translated to Punjabi and Hindi by using the proposed system. The translation of the word "kings" is missing in the English to Hindi MT system. The output given by the English to Punjabi MT system is accurate.

Input (Eng): He did not know how the world is simplified for kings
Output (Pun): ਉਹ ਨਹੀ ਜਾਣਦਾ ਸੀ ਕਿ ਸੰਸਾਰ ਕਿਵੇਂ ਰਾਜਿਆਂ ਲਈ ਸਰਲ ਹੈ Uha nahīṁ jāṇadā sī ki sasāra kivēṁ rāji'āṁ la'ī sarala hai
Reference (Pun): ਉਹ ਨਹੀ ਜਾਣਦਾ ਸੀ ਕਿ ਇਹ ਰਾਜ ਬਾਦਸ਼ਾਹਾਂ ਲਈ ਬਹੁਤ ਅਸਾਨ ਬਣ ਜਾਂਦਾ ਹੈ Uha nahīṁ jāṇadā sī ki iha rāja bādaśāhāṁ la'ī bahuta asāna baṇa jāndā hai

Table 3 Dataset division

Dataset	Percentage of sentences (%)
Training set	90
Tuning set	5
Testing set	5

Table 4 BLEU score of Punjabi-English bidirectional MT system

	SMT (Pun to Eng)	SMT (Eng to Pun)
Smaller sentences (1–5 words)	26.55	31.42
Medium sentences (6–14 words)	20.07	22.25
Large sentences (15 words or more)	16.02	18.05
For all sentences	17.79	19.78

Table 5 BLEU score of Punjabi-Hindi bidirectional MT system

	SMT (Pun to Hin)	SMT (Hin to Pun)
Smaller sentences (1–5 words)	45.72	50.33
Medium sentences (6–14 words)	35.42	35.62
Large sentences (15 words or more)	32.32	33.13
For all sentences	33.86	34.46

Table 6 BLEU score of Hindi-English bidirectional MT system

	SMT (Hin to Eng)	SMT (Eng to Hin)
Smaller sentences (1–5 words)	52.48	32.83
Medium sentences (6–14 words)	25.65	23.77
Large sentences (15 words or more)	21.77	22.78
For all sentences	23.68	23.78

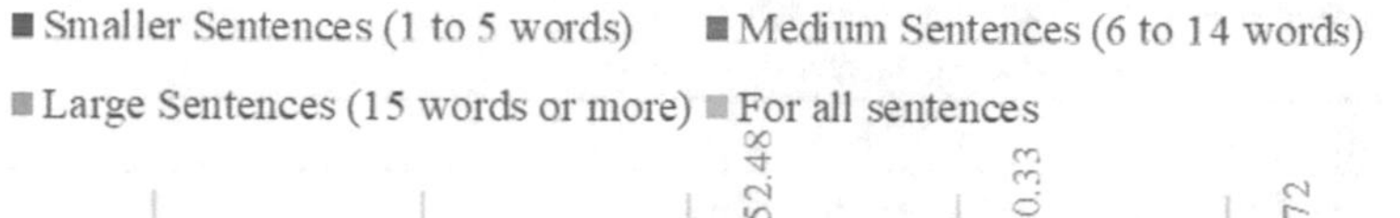

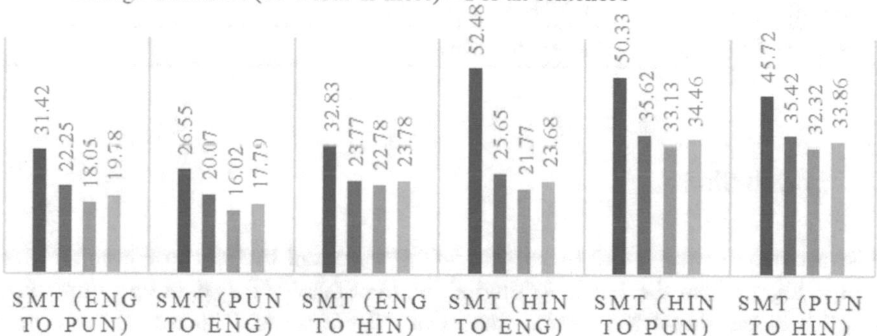

Fig. 3 BLEU score of different models

Output (Hin): वह नहीं जानता है कि किस प्रकार के लिए सरल कर दिया है । vah nahin jaanata hai ki kis prakaar ke lie saral kar diya hai
Reference (Hin): वह यह नहीं जानता था कि राजाओं के लिए दुनिया बहुत आसान हो जाती है । vah yah nahin jaanata tha ki raajaon ke lie duniya bahut aasaan ho jaatee hai

Now, in the next example, the Punjabi text is translated to English and Hindi by using the proposed system. The word ਉੱਚ (uca) is translated to higher by Punjabi to English MT system and it is accurate. But it has skipped the translation of the word ਬਹੁਤ (bahuta). Hindi translation is correct for the given Punjabi text.

Input (Pun): ਇਹ ਬਹੁਤ ਉੱਚ ਪੱਧਰ ਦੀ ਸ਼ੁਰੂਆਤ ਸੀ Iha bahuta uca padhara dī śurū'āta sī
Output (Eng): this was the beginning of a higher level
Reference (Eng): it was the beginning of a much higher degree
Output (Hin): यह बहुत ही उच्च स्तरीय शुरुआत थी yah bahut hee uchch stareey shuruaat thee
Reference (Hin): यह एक बहुत ही उच्च स्तरीय शुरुआत थी yah ek bahut hee uchch stareey shuruaat thee

In the last example, Hindi text is translated to Punjabi and English by using the proposed system. The word जंग (jang) is translated to battle by Hindi to English MT system and it is accurate. But reference contains a synonym of जंग (jang) which is war. Punjabi translation is correct for the given Hindi text.

Input (Hin): जंग ने देश को नष्ट कर डाला । jang ne desh ko nasht kar daala
Output (Pun): ਲੜਾਈ ਨੇ ਦੇਸ਼ ਨੂੰ ਤਬਾਹ ਕਰ ਦਿੱਤਾ Laṛā'ī nē dēśa nū tabāha kara ditā
Reference (Pun): ਜੰਗ ਨੇ ਦੇਸ਼ ਨੂੰ ਤਬਾਹ ਕਰ ਦਿੱਤਾ Jaga nē dēśa nū tabāha kara ditā
Output (Eng): The battle destroyed the country
Reference (Eng): The war destroyed the country

5 Conclusion

The accuracy of the Hindi to Punjabi MT system and Punjabi to Hindi MT system is higher than the Hindi to English bidirectional MT system and Punjabi to English bidirectional MT system based on BLEU score. The reason for higher accuracy for Hindi to Punjabi bidirectional system is due to the word order of both languages being the same subject-verb-object (SVO), whereas English is subject-verb-order (SOV). Due to this, there is a need of reordering in Hindi to English bidirectional MT system and Punjabi to English bidirectional MT system. Further work can be extended to increase the accuracy of the Hindi to English bidirectional MT system and Punjabi to English bidirectional MT system.

References

1. Dowling M, Lynn T, Poncelas A, Way A (2018) SMT versus NMT: preliminary comparisons for Irish. In: Proceedings of AMTA 2018 Workshop Technology MT Low Resource Language (LoResMT 2018), pp 12–20 [Online]. Available: https://ec.europa.eu/cefdigital/

wiki/display/CEFDIGITAL/Machine+Translation%0A. https://www.aclweb.org/anthology/W18-2202
2. Pathak A, Pakray P, Bentham J (2018) English–Mizo machine translation using neural and statistical approaches. Neural Comput Appl 31(11):7615–7631. https://doi.org/10.1007/s00521-018-3601-3
3. Garje GV, Bansode A, Gandhi S, Kulkarni A (2016) Marathi to English sentence translator for simple assertive and interrogative sentences. Int J Comput Appl 138(5):42–45. https://doi.org/10.5120/ijca2016908837
4. Khan NJ, Anwar W, Durrani N (2017) Machine translation approaches and survey for Indian languages, vol 18, no 1, pp 47–78 [Online]. Available: http://arxiv.org/abs/1701.04290
5. Sghaier MA, Zrigui M (2020) Rule-based machine translation from Tunisian dialect to modern rule-based machine translation from tunisian dialect to modern standard Arabic. Proc Comput Sci 176:310–319. https://doi.org/10.1016/j.procs.2020.08.033
6. Tran HA, Huang H, Tran P, Shi S, Nguyen H (2019) Preordering for Chinese-Vietnamese statistical machine translation. IEICE Trans Inf Syst E102-D(2):375–382. https://doi.org/10.1587/transinf.2018EDP7211
7. Azath M, Kiros T (2020) Statistical machine translator for english to tigrigna translation. Int J Sci Technol Res 9(1):2095–2099
8. Abidin Z, Permata, Ahmad I, Rusliyawati (2021) Effect of mono corpus quantity on statistical machine translation Indonesian-Lampung dialect of nyo. J Phys Conf Ser 1751,(1). https://doi.org/10.1088/1742-6596/1751/1/012036
9. Thu YK et al (2021) Hybrid statistical machine translation for English-Myanmar: UTYCC submission to WAT-2021. In: Proceedings of the 8th workshop on Asian Translation, Bangkok, Thailand, pp 83–89. https://doi.org/10.18653/v1/2021.wat-1.7
10. Biadgligne Y, Smaïli K (2022) Offline Corpus augmentation for English-Amharic machine translation. In: The 5th international conference on information and computer technologies, Mar 2022, New York, United States. hal-03547539, pp 1–14
11. Nagy A, Nanys P, Konrád BF, Bial B, Ács J (2022) Syntax-based data augmentation for Hungarian-English machine translation [Online]. Available: http://arxiv.org/abs/2201.06876
12. Kunchukuttan A, Mehta P, Bhattacharyya P (2017) The IIT Bombay English-Hindi Parallel Corpus, pp 2–5 [Online]. Available: http://arxiv.org/abs/1710.02855
13. Mohamed E, Sadat F (2015) Hybrid Arabic-French machine translation using syntactic re-ordering and morphological pre-processing. Comput Speech Lang 32(1):135–144. https://doi.org/10.1016/j.csl.2014.10.007
14. Goyal V, Lehal GS (2010) Web based Hindi to Punjabi machine translation system Vishal. J Emerg Technol WEB Intell 2(2):148–151. https://doi.org/10.1007/978-3-642-19403-0_40
15. Pal M, Mather PM (2001) Decision tree based classification of remotely sensed data. In: 22nd Asian conference on remote sensing, Singapore, vol 7, no 2, pp 1–10. https://doi.org/10.1192/s0368315x00238942
16. Rosenfeld R (1997) Statistical language modeling toolkit. http://www.speech.cs.cmu.edu/SLM/toolkit.html. Accessed 28 May 2020
17. SRI Speech Technology and Research Laboratory (1999) SRILM—The SRI language modeling toolkit. http://www.speech.sri.com/projects/srilm/. Accessed 28 May 2020
18. GNU Library or Lesser General Public License version 2.0 (LGPLv2). IRSTLM. https://hlt-mt.fbk.eu/technologies/irstlm. Accessed 28 May 2020
19. Heafield K (2011) KenLM language model toolkit. https://kheafield.com/code/kenlm/. Accessed 28 May 2020
20. Graehl J (2020) Carmel. https://www.isi.edu/licensed-sw/carmel/. Accessed 28 May 2020
21. Och FJ (2020) GIZA++. http://www.statmt.org/moses/giza/GIZA++.html. Accessed 25 May 2020
22. Hoang H, Koehn P (2008) Design of the Moses decoder for statistical machine translation. In: Software engineering, testing, and quality assurance for natural language processing, Columbus, Ohio, USA, June 2008, pp 58–65 [Online]. Available: papers3://publication/uuid/27F998CB-B861-4CDA-97A0-383F057E3565

23. PHARAOH: a beam search decoder. https://www.isi.edu/publications/licensed-sw/pharaoh/. Accessed 26 Jun 2020
24. Li Z et al (2009) Joshua: an open source toolkit for parsing-based machine translation. In: ACL-IJCNLP 2009—Joint conference on 47th annual meeting association computer linguistics. Proceedings of conference 4th international joint conference on national language processing AFNLP, pp 25–28
25. Crego JM, Mariño JB (2007) Extending MARIE: an N-gram-based SMT decoder Josep. In: Proceedings of the ACL 2007 demo and poster sessions, pp 213–216. https://doi.org/10.3115/1557769.1557831
26. Patry A, Gotti F, Langlais P (2006) Mood at work: Ramses versus Pharaoh Alexandre. In: Proceedings of the workshop on statistical machine translation, New York, pp 126–129. https://doi.org/10.3115/1654650.1654668
27. Banik D, Ekbal A, Bhattacharyya P, Bhattacharyya S (2019) Assembling translations from multi-engine machine translation outputs. Appl Soft Comput J 78:230–239. https://doi.org/10.1016/j.asoc.2019.02.031

A Survey of Machine Learning for Assessing and Estimating Student Performance

Avneet Kaur and Munish Bhatia

Abstract Educational data mining (EDM) contributes cutting-edge methodologies, strategies, and applications to the advancement of the education system, hence playing a crucial part in its development. Utilising machine learning and data mining approaches to explore and utilise educational data, the current advancement gives essential tools for comprehending the student learning environment. Academic institutions in the twenty-first century operate in a highly competitive and complicated environment. Among the prevalent issues faced by universities are performance analysis, the provision of a high-quality education, systems for evaluating the performance of students, and the planning of future activities. Student intervention programmes must be created in these universities in order to address the academic difficulties encountered by students. From 2009 through 2021, the relevant EDM literature relative to predicting student attrition and students at risk is examined in this review. According to the review's results, several machine learning (ML) methodologies are used to discover and address the fundamental challenges of forecasting students at risk and student withdrawal rate. Furthermore, the bulk of studies make use of data from student college/university database and online learning portals. It was determined that ML techniques play crucial roles in forecasting students at risk and withdrawal rates, hence boosting student performance.

Keywords Machine learning · Prediction · Student performance · Deep learning · Education data mining (EDM)

1 Introduction

Recent advancements in the education industry have been substantially influenced by EDM. The diversity of research has uncovered and implemented new chances

A. Kaur
Department of Computer Science and Engineering, Lovely Professional University, Punjab, India

M. Bhatia (✉)
Department of Computer Applications, NIT Kurukshetra, Kurukshetra, India
e-mail: munishbhatia90@gmail.com

Y. Singh et al. (eds.), *Proceedings of International Conference on Recent Innovations in Computing*, Lecture Notes in Electrical Engineering 1001,
https://doi.org/10.1007/978-981-19-9876-8_48

and possibilities for technologically improved learning systems depending on the demands of students. Modern approaches and application strategies employed by the EDM play a major role in developing the learning environment. For instance, the EDM is essential for comprehending the student learning environment since it evaluates both the educational environment and ML algorithms. According to [1, 2], the EDM field is concerned with studying, researching, and implementing data mining (DM) techniques. For its success, the field of DM employs multidisciplinary methodologies. Figure 1 depicts the DM cycle. It provides a complete approach for deriving intellectually useful insights from raw data. Analyses of ML and statistical approaches are performed on educational data to identify relevant patterns that enhance students' understanding and academic institutions in general.

Modern educational institutions function in a very competitive and intricate setting. Consequently, analysing performance, offering a high-quality education, devising ways for evaluating the performance of students, and recognising future requirements are issues encountered by the majority of universities in the present day. Modern educational institutions conduct student intervention strategies in a highly competitive and complicated setting. Universities offer intervention strategies for students in order to overcome academic difficulties. The prediction of student performance at entrance level and during following periods enables institutions to build and adapt intervention strategies in a way that benefits both management and educators. The obtained data is processed and analysed using various ML techniques to enhance the learning platform's usability and construct interactive features. According to Bengio [3] of the University of Monreal, "research utilising machine learning (ML) is a subset of artificial intelligence (AI) that seeks to supply computers with knowledge through data, observations, and intimate interaction with the real world. The accumulated information enables the computer to appropriately generalise to new situations." ML is a subfield of artificial intelligence in which ML systems learn from data, identify patterns, and make inferences. The resuscitation of the machine from a simple pattern recognition algorithm to deep learning (DL) approaches is a result of increasing data quantities, lower storage costs, and more powerful computer systems. Models based on ML can automatically and rapidly assess larger and more complex datasets with precise findings and avoid unforeseen dangers. Using the obtained records, ML methods are beneficial for forecasting children at risk and their likelihood of dropping out of school. This strategy is more effective than the typical on-campus method of evaluating and predicting students' academic performance using data such as quizzes, attendance, examinations, and grades. The EDM research community processes and analyses session logs and student databases in order to forecast student achievement using a ML algorithm. This review examines the application of various DM and ML approaches to:

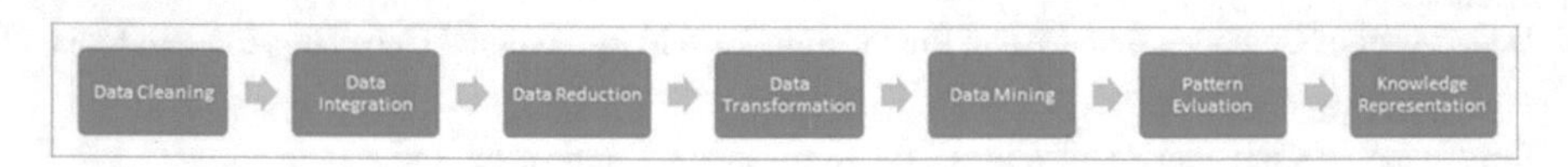

Fig. 1 Stages of DM methodology

1. Estimate the performance of at-risk students in academic settings.
2. Evaluate and estimate students' withdrawal from ongoing classes.
3. Analyse student achievement on the basis of dynamic and static data.

There have been earlier attempts [4, 5] to assess the literature on academic performance; however, the vast majority are broad literature studies aimed at predicting the performance of specific students. Our objective was to compile and assess the finest DM and ML approaches. In addition, we planned to conduct a systematic review of the literature, since the clarity of the approach and research methodology would minimise the repeatability of the review.

2 Methodology

In order to conduct a comprehensive systematic literature evaluation in accordance with the aims of this study, we utilised five research databases to locate the source data and to search for the pertinent publications. Table 1 lists the databases consulted during the whole study procedure. These repositories were explored in depth using a variety of queries linked to ML approaches in order to forecast students at risk and their withdrawal rates from 2009 to 2021. The predetermined searches produced a large number of research articles, which were manually screened to keep just the most pertinent publications for this study. The current piece of work is an outcome of qualitative analysis of 92 current papers on educational themes of interest. Numerous journal articles and conference proceedings have been published on this topic. As a result, the primary purpose of this research is to provide an in-depth analysis of the many strategies and algorithms presented and applied to this issue. The papers lacking significant contributions or quality were filtered out journal articles without an ISI Journal Citation Report impact factor or that were not peer-reviewed and were eliminated. As a result, 31% of the papers evaluated are journal articles; 69% of these have a JCR impact factor, while the remainder are peer-reviewed journals listed in other databases. We used the following descriptors in our search processes for these databases: "artificial neural network," "algorithms analytics students," "estimating algorithm pupil," "machine learning prediction," "predicting pupils' performance," "recommender systems prediction pupil," "students' intervention" or "withdrawal estimation" or "student risks" or "student performance management" or "student categorisation," and "pupil analytics prediction performance," among others.

3 Methods

The use of algorithms viz. CF, RS, ANN, and ML to anticipate student behaviour considers wide range of data, such as socioeconomic variables and task grading. The Hellenic Open University study, in which different machine-supervised learning

Table 1 Database sources

S. No	Databases	URL	Number of articles acquired
1	Scopus	https://www.scopus.com/	40
2	Springer Link	https://link.springer.com/	28
3	Association for Computing Machinery	https://dl.acm.org/	44
4	Research Gate	https://www.researchgate.net/	91
5	IEEE Explore Digital Library	https://ieeexplore.ieee.org/	84

methods were utilised on a specific dataset, provided a suitable beginning point. The Naves Bayes (NB) algorithm was found to be most precise in estimating both performance of students and the probability of dropping out [6]. Nonetheless, because each case study is unique in its own way, many strategies might be chosen as the most effective method for predicting students' behaviour. We divided the approaches into four categories: supervised ML, unsupervised ML, CF, and ANN. An extra grouping addressing with various DM approaches has been included in order to cover those publications that addressed similar aims.

3.1 Machine Learning

ML comprises of step-wise methodology that enables computers to become proficient without utilising man-made programmes [7]. The authors are primarily interested in predictive analysis, where ML enables us to develop complex models. Apparently, such kind of models may be of considerable use to users by giving pertinent data to aid in decision-making. There are two types of ML algorithms: supervised and unsupervised.

Supervised Machine Learning

Supervised machine learning (SML) attempts to create algorithms that can infer from externally provided examples so as to provide general hypotheses that can be used to forecast future occurrences [8]. More precisely the objective of SML is to develop a thorough model of the class label distribution w.r.t. predictive qualities. Rule induction is a time-efficient SML technique for prediction that obtained a 93% accuracy rate for forecasting new nursing student withdrawal utilising 3896 records on 512 students [9].

When utilising classification approaches, caution should be exercised if the datasets are uneven, since this might result in deceptive predicted accuracy. Several enhancements were recommended in [10] for predicting withdrawal, including the exploration of a diverse variety of learning techniques, the selection of qualities, the evaluation of the usefulness of theory, and the examination of aspects that differentiate withdrawal and non-withdrawal students. The study examined the classification methods radial basis networks (RBN), ADTrees, BN, One-R, NB, and C4.5. A potential approach for forecasting the probability of withdrawal rate during the early phases of online courses was described in [11], where higher rate of withdrawal is a significant issue in such kind of courses at the college level. The current approach involves the parallel application of three ML algorithms (K-nearest neighbour (KNN), RBN, and SVM), each of which utilises 25 characteristics per student. Considering students' traits, [12] used a series of ML algorithms (ANN, DT, and BN) to develop prediction models that took into consideration the students' personal characteristics and academic achievement in addition to input variables. The prediction's success was assessed on the basis of indicators like, accuracy rate, the recovery rate, the total accuracy rate, and a specific metric. Additionally, when the cognitive traits of pupils are included, the prediction accuracy significantly improves when applying DT [13].

Several ML algorithms were examined in [6] to predict new student performance, with NB demonstrating the finest behaviour in an online interface. SVM was the most accurate predictor of academic success among the four approaches tested in [14]. Additionally, Bayesian belief networks (BNNs) were utilised to forecast students' early success (grade point average) [15]. Additionally, LR and SVM were used. However, the accuracy of prediction systems may be enhanced by doing extensive research and incorporating various algorithmic properties. As a consequence, preprocessing approaches have been used with classification techniques (SVM, DT, and NB) to enhance prediction accuracy [16].

Unsupervised Machine Learning

The term "unsupervised learning" (UML) is also used to refer to the process of class discovery. Primary distinction between UML and SML is that UML lacks a training dataset. As a result, cross validation appears to have no evident relevance [17]. Another critical distinction is that, while the majority of clustering algorithms are stated in context to optimum criteria, there is no assurance that the optimal solution was produced. By automatically learning many layers of representation, an approach based on a UML sparse auto-encoder produced a classification model to foretell students' performance [18]. Students' performance may be evaluated using classification and clustering methods such as K-means and hierarchical clustering [19]. In this vein, recursive clustering was used to classify students in a programming course according to their performance [20].

3.2 Recommender System

Users' inclination for collection of items (books, apps, websites, travel locations, and e-learning resources) are collected by recommender systems. When it comes to student performance, information can be gathered directly (via the collection of user scores) or indirectly (by the monitoring of user behaviour, such as visits to instructional materials, document downloads, and so on) [21]. RS makes predictions and suggestions based on a variety of sources of information. They attempt to strike a balance between precision, innovation, and dissemination in the suggestions.

Collaborative Filtering

Despite the importance of filtering techniques in recommendation, they are typically used in conjunction with other filtering approaches including knowledge-based, content-based, or social filtering [21]. CF generates predictions in almost the same manner as individuals do based on prior experiences and knowledge. Several research indicated a variety of concerns with students' performance using CF techniques. Thus, commonalities in students' knowledge were discovered in [22, 23], where it was representative of grades gathered from prior courses. CF displayed comparable efficacy to ML, in the above instance. CF was used to produce personalised estimates of student grades in necessary courses [24]. A standard CF methodology was compared to an article recommendation methodology based on a student's grade in order to provide tailored articles in an online forum [25]. Predictive grade models for CF dependent on neighbourhood and MF, as well as categorization approaches based on popularity, might be developed using student groups defined by academic attributes and course-related enrolment patterns [26]. The majority of these studies on predicting student performance make use of big data matrices. Consequently, prediction accuracy was low when CF was utilised at minor academic institutions [27].

We may discover several examples in which CF motivates the development of innovative strategies and technologies aimed at optimising performance in certain contexts. PSFK is an unique student performance prediction model that blends user-based CF with Bayesian knowledge tracing (BKT) user modelling technique [28]. A technique known as hints-model is used to forecast pupils' performance [29]. Based on CF, a technology called grade prediction advisor (pGPA) forecasts grades in forthcoming courses [30]. Two variations of the low range matrix factorization (LRMF) issue as a prediction task have been addressed using the expectation–maximisation procedure: weighted standard LRMF and non-negative weighted LRMF [31]. A CF technique (matrix decomposition) allows students to construct personalised study plans and orientations by predicting grades for previously unseen amalgamation of academic courses [32]. A CF technique forecasts unknown performance values by examining a database of students' performance on certain activities [33]. To increase prediction accuracy, the best parameters for the above mentioned tool (regularisation factor and learning rate) were chosen using several metaheuristics. A prototype of RS for e-learning enhances new students' accomplishments. It makes use of CF and knowledge-based methodologies to leverage previous students' experience and outcomes in order to propose resources and affairs to assist newly admitted students [34].

3.3 *Artificial Neural Network*

A neural network is made up of a collection of densely connected components known as processing elements. The network structure as well as operation are modelled after the actual central nervous system, the main component of brain. Each element is modelled in accordance with neuron, the biological component that accepts a balanced collection of outputs and inputs the corresponding value [35]. As demonstrated by the following instances, ANNs have been used to a range of prediction approaches, the majority of which incorporate student evaluation data. Throughout the course, a feedforward ANN was trained to estimate exam results using partial results. An artificial neural network trained on the cumulative grade point average (CGPA) speculated eighth-semester scholastic achievement, source [36]. Two ANN models (generalised regression neural network and multilayer perceptron) were contrasted to see which was most successful in forecasting students' academic achievement [37]. Finally, in the avenue of clinical education, the predictive capability of ANNs was compared to multivariate LR model [38].

4 Objectives

We classified the various objectives into three broad categories: student withdrawal, student performance, and student risk. The weight assigned to each of these aims in the literature indicates their significance or potential for investigation. In this regard, student performance accounts for the maximum share of prediction attempts (72%), followed by student withdrawal (19%). The aims of enhancing pupil's skills and recommending tasks and assets were in low-demand (5% and 4%, respectively).

4.1 *Student withdrawal*

Accurate assessment of student withdrawal in the early stages aids in the elimination of the underlying problem through the development and implementation of swift and consistent intervention measures. This section addresses withdrawal detection employing ML approaches in depth through a review of relevant research based on datasets, characteristics employed in ML algorithms, and study results. Quadri and Kalyankar [39] conducted an early assessment that employed decision trees and logistic regression to uncover characteristics for attrition detection. The authors employed the students' activity logs database in these research, where a decision tree (DT) was used to identify withdrawal variables and logistic regression was used to estimate attrition. Walia et al. [40] utilised classification methods such as NB, DT, RF, and ZeroR to predict the academic achievement of students. They discovered that the school, the student's attitude, gender, and the amount of time spent reviewing impact the final grade. They conducted a large number of trials using the Weka tool and said that the accuracy of their self-generated dataset exceeded 83%.

Numerous research examining the withdrawal rate from nursing programmes have attempted to identify reasons instead of speculating the chance of students renouncing the programme. A popular technique for attempting this sort of prediction is rule induction, which may be accomplished using IBM SPSS Answer Tree (AT) software [9]. According to the authors [10], family background, family financial status, middle school grade, and test results are all powerful predictors of school withdrawals. Unbalanced class data was identified as a common issue for prediction [41]. Additionally, classification approaches that utilise imbalanced datasets might achieve deceptively high prediction accuracy. The authors addressed this issue by comparing several data balancing strategies in order to increase accuracy. All of these strategies increased prediction accuracy, but SVM paired with the SMOTE data balancing strategy performed the best. Universities and colleges are now attempting to identify students who are at verge of quitting by analysing data obtained from university systems [42]. The data collected in this study is used to verify the Moodle Engagement Analytics Plugin for learning analytics. withdrawal rates have become significant issue for e-learning. Authors suggested a strategy for analysing a set of variables associated with students' behaviours across time using a combination of different classifiers [11]. Other writers [12] used personal qualities and academic achievement of students as input variables. They created prediction models utilising artificial neural networks (ANNs), Bayesian networks (BNs), and decision trees (DTs). In a similar vein, another study [43] found the most significant predictors of school withdrawal risk as those that demonstrated student dedication and regularity in their usage of digital resources. Over 59% of withdrawals happened throughout the first 2 years. One research builds and analyses a survival analysis (SA) paradigm for initial detection and intervention of kids at verge of dropping out of school [44].

Early prediction of student attrition can give management and educators with an opportunity for early intervention. Sara et al. [45] utilised a huge dataset having 72,549 instances with 17 attribute values per instance. The programme Weka was used to implement the RF, CART, SVM, and NB algorithms. Accuracy and AUC were utilised to evaluate the performance of classifiers, with RF achieving high values for both metrics. Work by Kostopoulos [46] pioneered the use of semi-supervised ML algorithms for predicting student withdrawal. The semi-supervised learning approaches were then implemented using the KEEL software tool, and their results were compared. Each of the 244 examples in the sample had 12 characteristics. The investigation's findings revealed that the semi-supervised learning algorithm outperforms in comparison with others.

Predictions of probable student withdrawal must be made as early as feasible in order to determine appropriate corrective procedures. Most researches utilised the characteristics and academic achievement of the student to determine withdrawal characteristics. Using both dynamic and static datasets, early prediction of possible student withdrawal was performed. The most often used withdrawal prediction methods were DT, SVM, CART, KNN, and NB (Table 2).

Table 2 ML strategies and EDM techniques for estimating student withdrawal rate

References	Strategy	Attributes	Algorithms
[5, 40, 47]	Withdrawal factors	Assessment of temporal models	Combination of RNN and LSTM
[45, 46]	Early estimation of student attrition rate	Information about transcript and pre-college details	KNN, DT, ID3, NB, CART, and ICRM2 with SVM
[4, 48]	Retention rate	Freshman students	ANN and DT
[3, 5]	Student performance and curriculum	Student performance disparities in the classroom	KNN and SMOTE

4.2 Student Performance

There are two categories of student performance data used to forecast student performance: (a) static data and (b) dynamic data. The dynamic student outcomes data, according to [49], comprise records of student success and failure gathered as they interact with the learning system. Because the properties of the dataset change over time, engagement logs between students and the e-learning platform are an illustration of dynamic data. Static student outcomes data, on the other hand, is acquired only once and does not change over time. An example of this includes enrolment and demographic information for students. The following sections describe how static and dynamic data are being used in EDM.

Carlos et al. [50] describe a classification-based model for predicting students' success that incorporates a mechanism for collecting student learning and behavioural data from training activities. The SVM algorithm was employed as a classification tool to categorise pupils into three performance-based categories: high, medium, and poor performance levels. Predicting students' performance is a critical and difficult topic for educational institutions. This issue is particularly relevant in university-level e-learning contexts. Numerous ways are available in the literature for this aim. The demographic characteristics of students and their performance on certain activities may be utilised to provide a robust training set for a machine-supervised learning algorithm [6]. To do this, four mathematical models were evaluated for their capacity to estimate students' achievement in a basics course, a high-impact course, and a high-enrolment course in engineering dynamics [14]. Thus, an assessment of statistics from primary school test results in Tamil Nadu indicated an association between ethnicity, regional environment, and student performance [7]. Bydžovská [22, 23] forecasts students' performance based on particular first semester. The purpose of this study was to portray pupils' knowledge as a collection of grades from previously completed courses and to detect patterns of similarity across students in order to anticipate their performance. The research was done at small institutions or classrooms with a particular group of students utilising large sparse matrices representing students, assignments, and grades [27]. The study's findings suggested that prediction accuracy was not as great as projected, necessitating the collection of more data

Table 3 Utilisation of static and dynamic data for predicting educational outcomes

References	Strategy	Attributes	Algorithms
[16, 49, 51]	Static	Enrolment and demographic data	NB, KNN, DT, CART, SVM, DL, and ICRM2
[42, 43, 46–48]	Dynamic	Quiz, student reading, and data related to student performance	SMOTE, BM, BN, SVM, NB, DT, RF, and ADTree
[30, 34, 57]	Static and dynamic	Transcript and pre-college information	NB, KNN, DT, ICRM2 with SVM, DL, CART, and LR

from students or homework. Accuracy is critical because it may aid in the development of educational interventions targeted at enhancing the teaching–learning process's outcomes while conserving government resources and educators' time and effort [51]. Additionally, by preprocessing classification methods, the accuracy of predicting the performance was increased [16].

Identification of at-risk students and early prediction of student performance are critical for predicting the likelihood of student withdrawal and selecting the most effective remedial interventions. SVM, KNN, DT, NB, ID3, RF, and ICRM2 were the most often used algorithms for early identification utilising static and dynamic data. Table 3 depicts the different approaches for estimating the performance of student at early stages.

4.3 Estimating Student's Performance and Knowledge at Risk

Student success prediction provides a number of benefits for improving student retention rates, enrolment management, alumni management, more focused marketing, and entire educational institution performance. School-based intervention programmes assist learners who are on the brink of failing to graduate. The accuracy and timely identification and prioritisation of students in need is critical to the success of such services. This section includes a review of published evidence reporting at-risk student performance using ML approaches from 2009 to 2021.

Using ML approaches, Kovacic [52] investigated the early prediction of student achievement. For successful prediction, the review looked at socio-demographic variables such as education, job, gender, rank, impairment, and so on, as well as course characteristics such as course programme, course block, and so on. Kotsiaritis et al. [53] created a strategy for predicting student performance called the combinational incremental ensemble of classifiers. Three classifiers are merged in the proposed approach, with each classifier calculating the prediction output. The total final forecast is chosen using a voting mechanism. This approach is useful for continually created data, and each classifier estimates the result when a new sample comes. The current trend in educational technology is to analyse the data supplied by pupils

[54]. This technique aims to increase the educational process's efficacy by identifying trends in students' performance. In this vein, proposes an algorithmic technique for detecting students' learning styles on order to deliver customizable courses in Moodle [55]. It is based on the students' answers to the learning style as well as an examination of their Moodle behaviour. In this context, it is critical to know which student traits are linked to test scores and which school features are linked to the student's added value [56]. ML applications, for instance, have been demonstrated for acquiring knowledge about pupils' computer science learning, constructing suitable warning models, and distinguishing behavioural signs from learning analytical reports [57].

5 Discussion

Figure 2 demonstrates that the research communities of Germany and the UK concentrated more on the field than those of other nations. This study provides an introduction of the ML approach used in educational DM, with an emphasis on two crucial aspects:

1. Accurate identification of at-risk pupils and
2. Accurate assessment of withdrawal rate.

Following a comprehensive assessment of scholarly articles published between 2009 and 2021 (Fig. 3), the following findings are drawn:

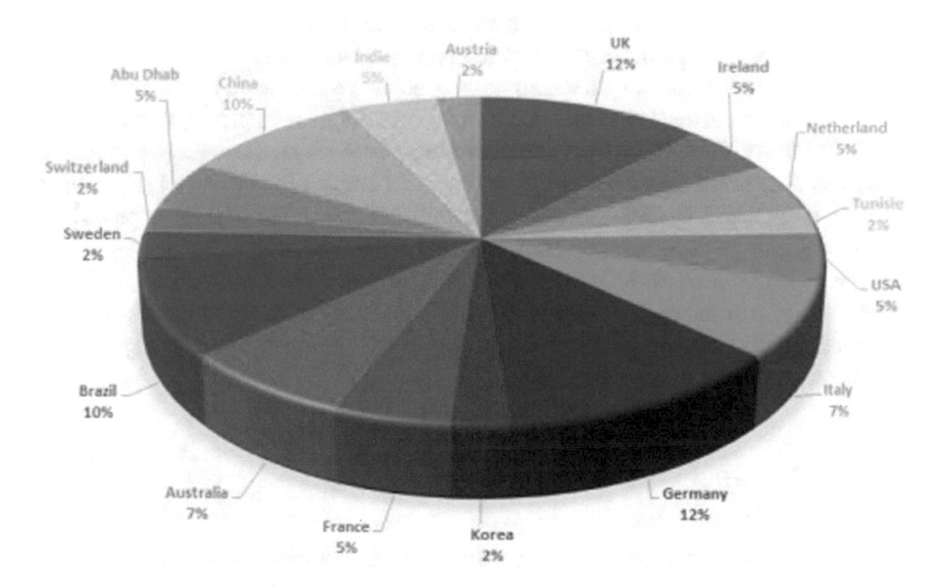

Fig. 2 %age share of country wise publications

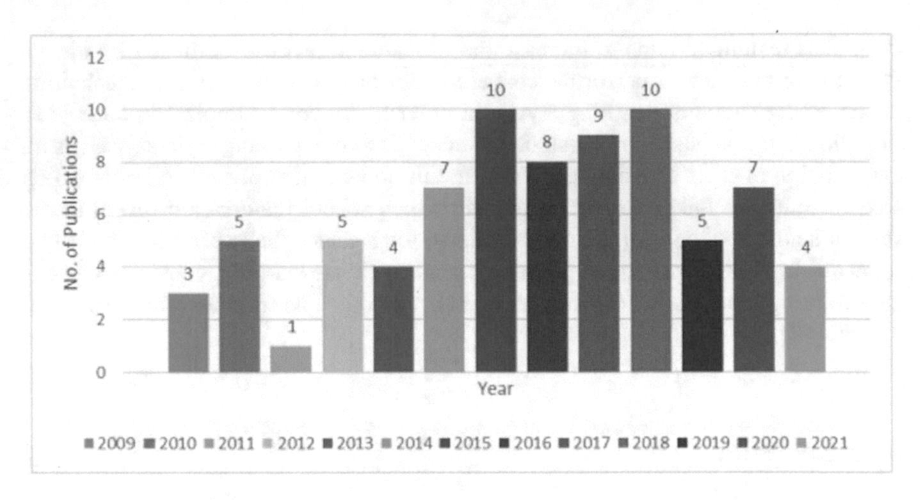

Fig. 3 Yearly publications

1. In the majority of investigations, limited data was utilised to train ML systems. However, it is a truth that ML algorithms require vast amounts of data to work effectively.
2. The majority of research papers approached the issue as a categorization challenge. Whereas relatively few research focused on clustering algorithms that identified the classes of students in a dataset, clustering techniques were the subject of very few investigations.
3. In addition, the aforementioned problems are classified using a binary system, while a number of other classifications would be included to enable the management design more effective intervention strategies.
4. The temporal nature of characteristics used to predict at-risk and withdrawal pupils has not been examined to its fullest potential. Due to their dynamic nature, the values of these traits fluctuate over time.
5. It was also noted that the majority of research employed classic ML algorithms such as KNN, DT, SVM, and NB, but only a small number have explored the possibilities of deep learning methods.
6. The existing literature does not take into account the changing character of student performance. The performance of the students is a dynamic process that constantly improves or declines. It remains to be investigated how well predictors operate on real-time dynamic data.

6 Conclusion

As a result of recent improvements in data collecting technologies and system performance indicators, educational systems are now researched with a great deal less work and with a greater degree of efficiency. For analysing and monitoring huge data, cutting-edge DM and ML approaches have been presented, giving rise to the brand-new discipline of big data analytics. Overall, this evaluation achieved its goal of improving student performance by forecasting students' at-risk status and likelihood of dropping out, underlining the significance of combining both static and dynamic data. This serves as the foundation for future developments in EDM utilising ML and DM techniques. We have studied several research articles during this review that attempt to predict student behaviour in an academic setting. The examination of these publications enables us to make certain conclusions.

We observed a strong trend towards predicting student success at the university level, since almost 72% of the studies included in our study are geared towards this goal. This may prompt us to consider additional research efforts to address deficiencies in other areas. Thus, we believe it would be worthwhile to encourage working lines for applying these predictions at the school level, since this would aid in identifying pupils with low performance at an early stage. The investigation of withdrawal rates among students during their early stages is quite intriguing, since there are still chances to do research on useful predicting tools to allow preventative methods. According to the data gathered for this review, the most often utilised approach for predicting students' behaviour is supervised learning, which produces accurate and dependable findings. The SVM method, in particular, was the most often utilised by the authors and delivered the most accurate predictions. Along with SVM, DT, NB, and RF were all well-studied algorithmic ideas that had positive results. The next effective method in this sector has been recommender systems, namely collaborative filtering algorithms. It should be noted, however, that effectiveness has been measured in terms of providing resources and activities rather than predicting student behaviour. While neural networks are a lesser-used approach, they provide a high degree of precision in forecasting students' performance. We underline that unsupervised learning is an unappealing approach for researchers due to its low predictive accuracy in the scenarios analysed. This fact, however, can serve as an incentive for study, as it allows for further development of these approaches in order to acquire more dependable and accurate findings. The current survey can be beneficial for gaining a broad understanding of the potential for applying ML to forecast students' performance and related issues. Nonetheless, many academics will likely tackle this topic in the next years using alternative and novel ML technologies, as this subject has garnered a lot of interest recently.

Future study will concentrate more on creating an effective ensemble method to practically implement the ML-based predictive approach, as well as searching for dynamic ways or methods to estimate students' performance as well as provide fully automated remedial actions to assist students as soon as possible. Finally, we highlight the possible future paths for research utilising ML approaches to predict student performance. We intend to incorporate some of the outstanding existing works and put more emphasis on the dynamic character of student performance. As a consequence, teachers might obtain additional tips for designing appropriate interventions for students and achieving precise education goals.

References

1. Romero C, Ventura S (2013) Data mining in education. Wiley Interdisc Rev: Data Mining Knowl Discovery 3(1):12–27
2. Han J, Pei J, Kamber M (2011) Data mining: concepts and techniques. Elsevier
3. Bengio Y, Lecun Y, Hinton G (2021) Deep learning for AI. Commun ACM 64(7):58–65
4. Hellas A, Ihantola P, Petersen A, Ajanovski VV, Gutica M, Hynninen T, Knutas A, Leinonen J, Messom C, Liao SN (2018) Predicting academic performance: a systematic literature review. In: Proceedings companion of the 23rd annual ACM conference on innovation and technology in computer science education, pp 175–199
5. Alyahyan E, Düştegör D (2020) Predicting academic success in higher education: literature review and best practices. Int J Educ Technol Higher Educ 17(10:1–21
6. Kotsiantis S, Pierrakeas C, Pintelas P (2004) Predicting students' performance in distance learning using machine learning techniques. Appl Artif Intell 18(5):411–426
7. Navamani JMA, Kannammal A (2015) Predicting performance of schools by applying data mining techniques on public examination results. Res J Appl Sci Eng Technol 9(4):262–271
8. Learning SM (2007) A review of classification techniques, sb kotsiantis. Informatica 31:249–268
9. Moseley LG, Mead DM (2008) Predicting who will drop out of nursing courses: a machine learning exercise. Nurse Educ Today 28(4):469–475
10. Nandeshwar A, Menzies T, Nelson A (2011) Learning patterns of university student retention. Expert Syst Appl 38(12):14984–14996
11. Dewan MAA, Lin F, Wen D et al (2015) Predicting dropout-prone students in e-learning education system. In: 2015 IEEE 12th international conference on Ubiquitous Intelligence and Computing and 2015 IEEE 12th international conference on Autonomic and Trusted Computing and 2015 IEEE 15th international conference on Scalable Computing and Communications and Its Associated Workshops (UIC-ATC-ScalCom), IEEE, pp 1735–1740
12. Tan M, Shao P (2015) Prediction of student dropout in e-learning program through the use of machine learning method. Int J Emerging Technol Learn 10(1)
13. Sultana S, Khan S, Abbas MA (2017) Predicting performance of electrical engineering students using cognitive and non-cognitive features for identification of potential dropouts. Int J Electr Eng Educ 54(2):105–118
14. Huang S, Fang N (2013) Predicting student academic performance in an engineering dynamics course: a comparison of four types of predictive mathematical models. Comput Educ 61:133–145
15. Slim A, Heileman GL, Kozlick J, Abdallah CT (2014) Predicting student success based on prior performance. In: 2014 IEEE symposium on Computational Intelligence and Data Mining (CIDM), IEEE, pp 410–415
16. Chaudhury P, Mishra S, Tripathy HK, Kishore B (2016) Enhancing the capabilities of student result prediction system. In: Proceedings of the second international conference on information and communication technology for competitive strategies, pp 1–6
17. Gentleman R, Carey VJ (2008) Unsupervised machine learning. In: Bioconductor case studies. Springer, Heidelberg, pp 137–157
18. Guo B, Zhang R, Xu G, Shi C, Yang L (2015) Predicting students performance in educational data mining. Int Symp Educ Technol (ISET) 2015:125–128
19. Rana S, Garg R (2017) Prediction of students performance of an institute using classification via clustering and classification via regression. In: Proceedings of international conference on communication and networks. Springer, Heidelberg, pp 333–343
20. Anand V, Rahiman SA, George EB, Huda A (2018) Recursive clustering technique for students' performance evaluation in programming courses. In: 2018 Majan International Conference (MIC), IEEE, pp 1–5
21. Bobadilla J, Ortega F, Hernando A, Gutiérrez A (2013) Recommender systems survey. Knowl-Based Syst 46:109–132

22. Bydžovská H (2015) Student performance prediction using collaborative filtering methods. In: International conference on artificial intelligence in education. Springer, Heidelberg, pp 550–553
23. Bydžovská H (2015) Are collaborative filtering methods suitable for student performance prediction? In: Portuguese conference on artificial intelligence. Springer, Heidelberg, pp 425–430
24. Park Y (2018) Predicting personalized student performance in computing-related majors via collaborative filtering. In: Proceedings of the 19th annual SIG conference on information technology education, pp 151–151
25. Liou C-H (2016) Personalized article recommendation based on student's rating mechanism in an online discussion forum. In: 2016 49th Hawaii International Conference on System Sciences (HICSS), IEEE, pp 60–65
26. Elbadrawy A, Karypis G (2016) Domain-aware grade prediction and top-n course recommendation. In: Proceedings of the 10th ACM conference on recommender systems, pp 183–190
27. Pero S, Horváth T (2015) Comparison of collaborative-filtering techniques for small-scale student performance prediction task. In: Innovations and advances in computing, informatics, systems sciences, networking and engineering. Springer, Heidelberg, pp 111–116
28. Song Y, Jin Y, Zheng X, Han H, Zhong Y, Zhao X (2015) Psfk: A student performance prediction scheme for first encounter knowledge in its. In: International conference on knowledge science, engineering and management. Springer, Heidelberg, pp 639–650
29. Xu K, Liu R, Sun Y, Zou K, Huang Y, Zhang X (2017) Improve the prediction of student performance with hint's assistance based on an efficient non-negative factorization. IEICE Trans Inf Syst 100(4):768–775
30. Sheehan M, Park Y (2012) pgpa: a personalized grade prediction tool to aid student success. In: Proceedings of the sixth ACM conference on recommender systems, pp 309–310
31. Lorenzen S, Pham N, Alstrup S (2017) On predicting student performance using low-rank matrix factorization techniques
32. Houbraken M, Sun C, Smirnov E, Driessens K (2017) Discovering hidden course requirements and student competences from grade data. In: Adjunct publication of the 25th conference on user modeling, adaptation and personalization, pp 147–152
33. Gómez-Pulido JA, Cortés-Toro E, Durán-Domínguez A, Crawford B, Soto R (2018) Novel and classic metaheuristics for tunning a recommender system for predicting student performance in online campus. In: International conference on intelligent data engineering and automated learning. Springer, Heidelberg, pp 125–133
34. Chavarriaga O, Florian-Gaviria B, Solarte O (2014) A recommender system for students based on social knowledge and assessment data of competences. In: European conference on technology enhanced learning. Springer, Heidelberg, pp 56–69
35. Adewale AM, Bamidele AO, Lateef UO (2018) Predictive modelling and analysis of academic performance of secondary school students: artificial neural network approach. Int J Sci Technol Educ Res 9(1):1–8
36. Arsad PM, Buniyamin N et al (2013) A neural network students' performance prediction model (nnsppm). In: 2013 IEEE International Conference on Smart Instrumentation, Measurement and Applications (ICSIMA), IEEE, pp 1–5
37. Iyanda AR, Ninan OD, Ajayi AO, Anyabolu OG (2018) Predicting student academic performance in computer science courses: a comparison of neural network models. Int J Modern Educ Comput Sci 10(6)
38. Dharmasaroja P, Kingkaew N (2016) Application of artificial neural networks for prediction of learning performances. In: 2016 12th International Conference on Natural Computation, Fuzzy Systems and Knowledge Discovery (ICNC-FSKD), IEEE, pp 745–751
39. Quadri MM, Kalyankar N (2010) Drop out feature of student data for academic performance using decision tree techniques. Global J Comput Sci Technol
40. Walia N, Kumar M, Nayar N, Mehta G (2020) Student's academic performance prediction in academic using data mining techniques. In: Proceedings of the International Conference on Innovative Computing & Communications (ICICC)

41. Thammasiri D, Delen D, Meesad P, Kasap N (2014) A critical assessment of imbalanced class distribution problem: the case of predicting freshmen student attrition. Expert Syst Appl 41(2):321–330
42. Liu D, Richards D, Froissard C, Atif A (2015) Validating the effectiveness of the moodle engagement analytics plugin to predict student academic performance In: 21st Americas Conference on Information Systems, AMCIS 2015. Americas Conference on Information Systems, pp 1–10
43. Saqr M, Fors U, Tedre M (2017) How learning analytics can early predict under-achieving students in a blended medical education course. Med Teach 39(7):757–767
44. Chen Y, Johri A, Rangwala H (2018) Running out of stem: a comparative study across stem majors of college students at-risk of dropping out early. In: Proceedings of the 8th international conference on learning analytics and knowledge, pp 270–279
45. Sara N-B, Halland R, Igel C, Alstrup S (2015) High-school dropout prediction using machine learning: a danish large-scale study. In: ESANN 2015 proceedings, European symposium on artificial neural networks, computational intelligence, pp 319–24
46. Kostopoulos G, Kotsiantis S, Pintelas P (2015) Estimating student dropout in distance higher education using semisupervised techniques. In: Proceedings of the 19th Panhellenic conference on informatics, pp 38–43
47. Howard E, Meehan M, Parnell A (2018) Contrasting prediction methods for early warning systems at undergraduate level. Internet Higher Educ 37:66–75
48. Gray CC, Perkins D (2019) Utilizing early engagement and machine learning to predict student outcomes. Comput Educ 131:22–32
49. Thaker K, Huang Y, Brusilovsky P, Daqing H (2018) Dynamic knowledge modeling with heterogeneous activities for adaptive textbooks. In: The 11th international conference on educational data mining, pp 592–595
50. Ahadi A, Lister R, Haapala H, Vihavainen A (2015) Exploring machine learning methods to automatically identify students in need of assistance. In: Proceedings of the eleventh annual international conference on international computing education research, pp 121–130
51. Adán-Coello JM, Tobar CM (2016) Using collaborative filtering algorithms for predicting student performance. In: International conference on electronic government and the information systems perspective. Springer, Heidelberg, pp 206–218
52. Kovacic Z (2010) Early prediction of student success: mining students' enrolment data
53. Kotsiantis S, Patriarcheas K, Xenos M (2010) A combinational incremental ensemble of classifiers as a technique for predicting students' performance in distance education. Knowl-Based Syst 23(6):529–535
54. Villegas-Ch W, Lujan-Mora S, Buenaño-Fernandez D, Roman-Canizares M (2017) Analysis of web-based learning systems by data mining. In: 2017 IEEE second Ecuador Technical Chapters Meeting (ETCM), IEEE, pp 1–5
55. Karagiannis I, Satratzemi M (2018) An adaptive mechanism for moodle based on automatic detection of learning styles. Educ Inf Technol 23(3):1331–1357
56. Masci C, Johnes G, Agasisti T (2018) Student and school performance across countries: a machine learning approach. Eur J Oper Res 269(3):1072–1085
57. Johnson WG (2018) Data mining and machine learning in education with focus in undergraduate CS student success. In: Proceedings of the 2018 ACM conference on international computing education research, pp 270–271

Recent Innovations of Computing in Education: Emerging Collaborative Blended Learning Models in India

Poonam Pandita, Shivali Verma, Sachin Kumar, Ritu Bakshi, and Aman

Abstract Technology has displayed a powerful hold in every facet of life. It has also transformed the mundane learning process into a joyful learning process. Firstly, the teaching–learning process is entirely dynamic; likewise, the methods of teaching should be renovated from time to time. Secondly, our classrooms have a heterogeneous population of students, so it is utterly impossible to teach them only through traditional ways of learning. A profuse number of teaching methods are required for teaching the students. Blended learning is one of the most significant and efficacious strategies for integrating technology into education. Blended learning involves person-to-person learning with online learning. It has multifarious advantages for the education system. The models of blended learning have changed from time to time. Earlier there was a pure offline system of teaching but with the technological advancements, educators and teachers introduced some online interventions in the teaching–learning process. Those interventions were primarily asynchronous. But the recent innovations of computing in education have put forward the new collaborative blended model of teaching–learning. The study reported here is undertaken to explore the emerging collaborative blended model of learning. Further, this paper will also reflect the role of co-teaching in the collaborative blended model of learning.

Keywords Blended learning · Collaborative model · Co-teaching · Joyful learning · Technological advancements

1 Introduction

The advancements in educational technology pushed the concept of blended learning forward, making it constitutive part of the classrooms. It is a boon for quality education. In the current scenario where individual differences and diversities of the learners are taken into consideration, the emphasis on blended learning is increasing. "One size" cannot fit the heterogeneous group of individuals. Now, the teachers have

P. Pandita · S. Verma · S. Kumar · R. Bakshi (✉) · Aman
Department of Educational Studies, Central University of Jammu, Jammu and Kashmir, India
e-mail: ritubakshi.hp@gmail.com

Y. Singh et al. (eds.), *Proceedings of International Conference on Recent Innovations in Computing*, Lecture Notes in Electrical Engineering 1001,
https://doi.org/10.1007/978-981-19-9876-8_49

to be innovative and creative so that they can use plenty of teaching methods in the classroom. By adopting the blended learning, students can learn through the resources that can meet their individual needs and can hold their interest. It also helps teachers to learn from other professionals and to improve their teaching practices. By incorporating multifarious teaching styles, models and different ways of learning, blended learning helps teachers to reinvent their teaching [1]. It is believed that this kind of learning strategy is beneficial for both students and teachers as it also provides solutions to deal with the large classroom [2]. "When the two are thoughtfully integrated, the educational possibilities are multiplied [3].

Teachers should know how to blend online/digital methods of learning with traditional classroom techniques in an appropriate way. The balanced approach of blending the offline and online methods is essential. There are various factors for the emergence of blended learning models. Firstly, when the demand for learner-centered models has increased, the scope of blended learning models has also increased. Secondly, the present generation demands flexibility in learning and feels comfortable with online learning, and this factor is also enhancing the concept of blended learning [2].

In order to foster a better understanding of the blended learning strategy, this study will review:

- What is Blended learning?
- What blended learning is not?
- Blended learning models.
- Emerging collaborative blended learning model.
- Co-teaching through blended learning model.

2 Studies Related to Blended Learning

- Vanicharaoenchai and Tosulkaew compared students' performance in two different teaching methods. It also compared the academic achievement of students who studied through the traditional method and blended learning method. This study was quasi-experimental in nature. The result showed that the experimental group students studied through blended learning method had higher achievement scores when compared to control group students who studied through the traditional method [4].
- Shashi Rekha Muni Reddy [5] in her article, blended learning in India: Are Teachers in India Ready to Go Blended? found that the younger teachers between 21 and 39 years do have the required computer and Internet skills compared to that the older teachers of 40–59 years, and it is imperative for India to adopt blended learning in its school education.
- Nair [4] conducted a study on the effect of the blended learning strategy on achievement in biology and the social and environmental attitude of students at the secondary level. In this study, the researcher analyzes the effect of blended learning strategy on achievement in biology of secondary school students. The

result of this study showed that the majority of the teachers responded that blended learning strategy develops the ability to appreciate the scientific phenomenon through the various learning experience.

- Shand and Farrelly [6] in their study, The Art of Blending: Benefits and challenges of a blended course for pre-service teachers explore the design and delivery of blended social studies teaching methods courses according to principles and core attributes of blended course design. In their study, five overarching themes were identified from the analysis, which includes the benefits of flexibility and pace, access and modeling, peer relationships and community, clear communication and feedback and the challenges of time management and self-discipline.

3 Blended Learning

Blended learning is a contemplative integration of classroom face-to-face learning methods with online methods. Blended learning needs two or more things that can be mixed. Garrison and Kanuka [1] define blended learning as the thoughtful integration of classroom face-to-face learning experiences with online learning experiences" (p. 96). Oliver and Trigwell [7] defined the three types of blended learning:

- Blending e-learning with person-to-person learning,
- Blending online learning with person-to-person,
- Blending different teaching methods and pedagogies.

People have different conceptualizations about this type of learning. Some views reflect that it is simply an integration of online learning with offline learning, whereas some others believe that it is also about other blends like mixing instructional methods and pedagogical approaches. In simple terms, we can say that in blended learning, offline and online methods of teaching/learning complement each other and it is incorporating the best practices of online and offline learning. Blended learning F2F and online are shown in Fig. 1.

3.1 *Different Conceptualizations of Blended Learning*

There is ambiguity in defining the term blended learning. Different persons have different definitions and concepts about blended learning. It is significant to understand how blended learning is being conceptualized. If we comprehend the different concepts of blended learning, only then we will be able to understand what blended learning is not. The various conceptualizations of blended learning are: inclusive conceptualization, quality conceptualization, quantity conceptualization and digital conceptualization. Inclusive conceptualization reflects that any integration of digital learning into real classroom situations is blended learning, quality conceptualization reflects that the amalgamation of online methods with person-to-person learning

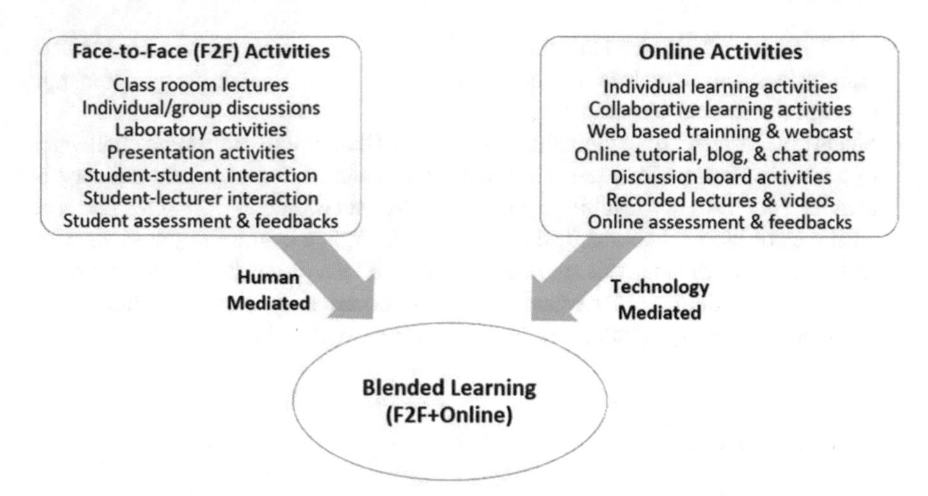

Fig. 1 Blended learning F2F and online [8]

methods is blended learning. Further quantity conceptualization reflects that a considerable part of the course should be presented through a person-to-person mode of learning and other considerable parts of the course should be presented through an online mode of learning. Digital conceptualization reflects the use of computing technologies in the classrooms for making the education system digital [9].

3.2 Benefits of Blended Learning (Mixed Mode Instructions)

The desirable blend of both offline and online methods can enhance classroom practices. It can provide a versatile learning context to the learners. Students can have increased access to the content. The pedagogical practices can also be improved and systematized through a blended learning strategy. It can ensure an active, interactive, collaborative and student-centered classroom environment.

- **What Blended Learning is not?**

There are some conceptual difficulties regarding the term blended learning. To clear these conceptual difficulties, it is important to understand what mixed mode instruction is not. Mixed mode instruction is not completely an online learning. It must have a blend of two methods. And only having technological devices in the classroom does not reflect the blended learning environment. But there should be proper utilization of technological tools, only then it can be termed blended learning. The other most important fact is that the teachers who use this learning approach in the classrooms must ensure the active engagement of the learners in the classroom otherwise blended learning approach will not fully serve its purpose. Personalization also cannot be termed blended learning. It should not be confused with blended learning.

- **Need to adopt the blended learning approach**

The pupil-teacher ratio in Indian classrooms is relatively high. According to NEP 2020, it should be 30:1. Thus, in Indian classrooms, it is a challenge for the teacher to catch the attention of all the learners. The teachers need to adopt teaching methods that can capture the attention of a large no. of students. Blended learning is also required to cater to the needs of heterogeneous learners. To overcome the mundane learning process, it is significant to blend a variety of methods. The traditional classrooms do not hold the capacity to enrich the holistic experience of the learners, so it is required to use this learning approach in the classroom.

- **Blended Learning Models**

Horn and Staker [10] have given the 04 types of blended learning models. These are:

i. Rotation model
ii. Flex model
iii. La Carte model
iv. Enriched virtual model.

The rotation model involves manifold teaching approaches where students are involved in activities and those activities should include a minimum of one online component. The rotation model is divided into 04 kinds:

i. Individual rotation
ii. Station rotation
iii. Lab rotation
iv. Flipped classroom.

The individual rotation model proffers activities that are based on individual needs. In station rotation model, learners perform different activities on a fixed schedule. For example, first students get involved in online learning, then receive face-to-face instruction, and at last, he/she will involve in group projects.

Lab rotation model means physical rotation of space/learners. For example, they will go from classroom to computer lab. And flipped classroom model involves going through the online content at home and then discussing the same in a face-to-face classroom.

Flex model involves delivering the majority of the content online but in traditional classrooms. In Le Carte model, learners have to complete a few online courses that supplement traditional teaching. And at last, enriched virtual blended model means receiving person-to-person instruction once a week and completing the remaining work at home.

3.3 *Emerging Collaborative Blended Learning Model*

The new emerging blended models of learning are providing better opportunities for collaboration that were previously missing. Due to the new technological advancements, collaborative approaches have emerged in blended learning models. The most prominent collaborative blended learning model is the co-teaching model.

India is having a diverse cultural and language background and one model of teaching does not fit in all conditions. So the authors propose a new model based on collaborative teaching where collaborative management of teaching takes place. This model as shown in Fig. 2 proposes to bring the geographically distant teachers and learners to connect meaningfully to make a learning web.

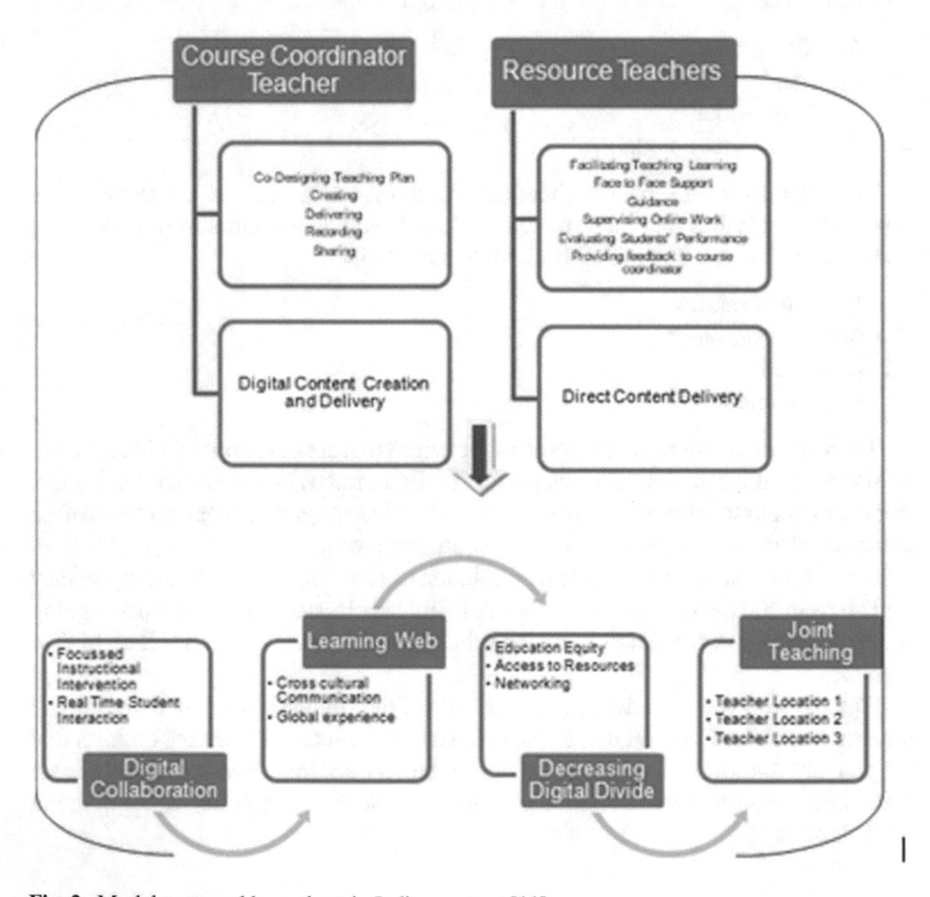

Fig. 2 Model proposed by authors in Indian context [11]

3.4 Co-teaching Model

It is a new model of the blended learning approach. This model connects geographically distant learners and instructors together and it also provides the opportunity to communicate with diverse learners. The teachers can also solve the various problems of students through this model. The teachers can collaborate with other teachers for enhancing the learning of students. Through the co-teaching model, the diverse needs of learners can be addressed. This model will also enhance cross-cultural communication and teamwork. The teachers can also learn various methods and styles of teaching through this model.

4 Reflections

The attempt of this investigation is to understand the emergence of different blended learning models and their importance. The chief focus of this investigation is to understand the various conceptualizations of blended learning. This is important to clear the concepts regarding blended learning because when we comprehend the various meanings of blended learning models, only then we can use them appropriately. The paper also discussed the need of adopting blended learning models. The shortcomings of traditional person-to-person teaching reflect the need for blended learning models. The heterogeneity of classrooms cannot be addressed through only traditional methods of teaching; it requires a profuse no. of teaching methods and a right blend of both offline and online teaching methods. Its emerging collaborative designs are favorable enriching students' experience especially the collaborative co-teaching model is a boon to institutions and learners. The different teachers can address the different needs of learners through this collaborative co-teaching model. The recent innovations in computing have had a profound effect on the teaching–learning process. Through these innovations, new collaborative models of the blended learning approach have emerged. The geographically distant learners can have a huge benefit from the collaborative co-teaching model of blended learning. The teachers need to be trained for using the collaborative blended learning designs. They should know how to appropriately blend two different methods. In nutshell, we can say that this type learning is an innovative trend in the present scenario that is offering opportunities for collaborative learning experiences.

References

1. Garrison DR, Kanuka H (2004) Blended learning: uncovering its transformative potential in higher education. Internet High Educ 7(2):95–105. https://doi.org/10.1016/j.iheduc.2004.02.001

2. Nel L, Wilkinson A (2006) Enhancing collaborative learning in a blended learning environment: applying a process planning model. Syst Pract Action Res 19(6):553–576. https://doi.org/10.1007/s11213-006-9043-3
3. F. Report, F. O. R. The, and A. Year (2004) Blended learning pilot project, pp 1–29
4. Vanicharoenchai V, Toskulkaew T (2010) Effects of blended learning, using online data searches and action learning, upon academic achievement and searching skills of nursing students. Online 28(2):2010–2010
5. Yang HH, MacLeod J, Zhu S (2016) Collaborative teaching approaches: extending current blended learning models. In: International conference on blended learning. Springer, Cham
6. Shand K, Farrelly SG (2017) Using blended teaching to teach blended learning: lessons learned from pre-service teachers in an instructional methods course. J Online Learn Res 3(1):5–30
7. Oliver M, Trigwell K (2005) Can 'Blended Learning' be redeemed? e-Learning 2(1):17. https://doi.org/10.2304/elea.2005.2.1.2
8. Anthony B et al (2022) Blended learning adoption and implementation in higher education: a theoretical and systematic review, vol 27, no 2. Springer, Netherlands. https://doi.org/10.1007/s10758-020-09477-z
9. Hrastinski S (2019) What do we mean by blended learning? TechTrends 63(5):564–569. https://doi.org/10.1007/s11528-019-00375-5
10. Horn MB, Staker H, Christensen CM (2014) Blended: using disruptive innovation to improve schools, p 304
11. Xu M, Liu X, Ye C (2020) A collaborative model of blended learning for the cultivation of qualified pre-service English teachers. In: Proceedings—2020 international symposium on educational technology, ISET 2020, pp 219–223. https://doi.org/10.1109/ISET49818.2020.00055

Online Learning in Teaching Initial Math Education

Milan Pokorný

Abstract The paper summarizes experience with online learning in teaching Initial Math Education at Trnava University, Faculty of Education. In the winter term of the academic year 2021/2022, as a result of COVID-19, face-to-face lessons were limited and teachers had to rely on synchronous and asynchronous forms of online learning, regardless of their previous experience with proper utilization of modern technologies in education. Thus, the subject Initial Math Education was taught by a combination of pre-recorded video lessons, the e-learning course, and face-to-face lessons with a small number of students in a classroom and majority of students connected via Teams. An analysis of the students' results in a final test revealed that this form of teaching was efficient and helped to reduce a possible negative impact of COVID-19 on a students' knowledge level. Moreover, the analysis of students' answers in a questionnaire revealed further benefits of this method of teaching, including an increase of the sense of final evaluation objectivity, as well as an increase of students' satisfaction with their own results.

Keywords Online learning · Blended learning · Video lessons · ICT in education · Teaching mathematics

1 Introduction

It is widely known that modern technologies have significantly influenced teaching methods at universities for more than forty years. Many university teachers try to benefit from proper integration of e-learning, blended learning, mobile learning, or online learning into teaching process.

Blended learning is usually defined as a combination of e-learning and face-to-face teaching in a classroom. Fisher [6] considers blended learning to be a selection of an optimum mix of instructional delivery strategies. This modern method enables a learner to achieve desired learning outcomes. Thorne [20] considers blended learning

M. Pokorný (✉)
Trnava University, Trnava, Slovak Republic
e-mail: mpokorny@truni.sk

Y. Singh et al. (eds.), *Proceedings of International Conference on Recent Innovations in Computing*, Lecture Notes in Electrical Engineering 1001,
https://doi.org/10.1007/978-981-19-9876-8_50

to be the most logical and natural evolution of the learning agenda. There are many studies that confirm suitability of blended learning and its efficiency in teaching process at universities all over the world. Graham and Dziuban [7] give several reasons why to adopt blended learning, for example, an improvement of learning effectiveness, cost effectiveness, and increased access and convenience.

Since we deal with teaching mathematics, we mention the results of blended learning utilization research in mathematics teaching. Lin, Tseng, and Chiang [11] focus on the blended learning pedagogy influence on student learning achievement, as well as on their attitudes toward mathematics. The analysis of their quasi-experiment proved a positive effect of blended learning on learning outcomes, as well as on attitudes toward mathematics. The similar results were reached by Malatinská et al. [12]. Borba et al. [2] focus on e-learning, blended learning, as well as on mobile learning in teaching mathematics and give recommendations for future research. The findings of Cheung and Slavin [4] show a small positive effect of educational technology. The results of Moreno-Guerrero et al. [14] show that e-learning has a positive influence not only on results and grades but also on motivation, autonomy, and participation. The suitability of blended learning was also proved in [13, 15, 16, 21].

By Chiu et al. [5], COVID-19 triggers unprecedented challenges for education. They predict that its impact on education will be resurging and long lasting. Before 2020, despite positive experience with blended learning, majority of subjects at our university was taught only by face-to-face instruction. Thus, a sudden spread of COVID-19 had a great impact on teaching methods. In a really short time, teachers had to rely on online learning, regardless on their previous experience with integration of modern technologies into teaching. We fully agree with Štrbo [18], who states that pandemic situation tested a readiness of Slovak teaching system for online learning. Moreover, Stoffová and Horváth [17] remind that online teaching projected itself also on methods of assessments and exams. The teachers have to adopt methods how to prevent frauds and cheating.

At our faculty, many teachers tried to overcome the reduced number of face-to-face lessons using synchronous online lessons, as well as pre-recorded video lessons. A suitability of these methods has been confirmed by several studies. For example, Insorio [10] proved that teacher-made video lessons help students to understand mathematics lessons through watching conveniently and repeatedly. Bullo [3] reveals that video lessons helped students better understand and comprehend the lessons even without the help of a teacher. A research of Tan and Pearce [19] shows that use of videos is an effective way of supporting learning. By Ichinose and Clinkenbeard [9], students in a flipped class had higher levels of achievement than students in a traditional course.

However, it is clear that teachers have to know how to properly integrate synchronous and asynchronous forms of online learning into teaching. Naturally, majority of teachers need help to be able to properly use these methods. For example, the study of Alsawaie and Alghazo [1] shows that mathematics teachers improved their ability to analyze teaching after the analysis of ten video lessons. Thus, the teachers should be given some good practice examples.

2 Teaching Initial Math Education

At our faculty, Initial Math Education is a compulsory subject for our students of a bachelor study program Pre-School Elementary Pedagogy. In the future, these students will be teachers in kindergartens or will continue in a master study of Primary Education Teaching to become teachers at primary schools. After completing the subject:

1. The students gain an overview of the goals and content of teaching mathematics in pre-primary education.
2. The students are able to characterize the stages of the cognitive process in mathematics and use obtained knowledge in the development of initial mathematical ideas of children in pre-primary education.
3. The students gain knowledge of logic, set theory, and binary relations and are able to apply them in the development of initial mathematical ideas of children.
4. The students are also able to use knowledge about geometric shapes, toys, and kits in activities developing spatial imagination and geometric knowledge of children.
5. The students are able to choose appropriate activities for the development of combinatorial thinking of children.
6. The students are able to apply knowledge of numbering in the educational process in pre-school children.

Before 2021, the subject was taught by blended learning, a combination of the e-learning course with vast majority of face-to-face lessons with a teacher. However, sudden spread of COVID-19 made us to change this teaching method, since a number of students at face-to-face lessons were significantly limited. Thus, we were forced to rely on synchronous and asynchronous forms of online learning.

Based on our experience with integration of modern technologies into teaching as well as on experience of other researchers, we decided to keep high academic standards of the teaching process by a combination of an e-learning course, pre-recorded video lessons, and face-to-face lessons with a small number of students in a classroom and a majority of students connected via Teams.

The pre-recorded video lessons replaced lectures. However, it is generally known that mathematics should be learned by an active work of learners, not by a passive transmission of information. Thus, we strengthened an active student engagement in a learning process by the face-to-face lessons, which replaced seminars. Of course, these face-to-face lessons were also recorded.

However, there are two groups of students of Pre-School Elementary Pedagogy at our faculty. A group of part-time students was taught by a combination of the e-learning course, pre-recorded video lessons, recorded face-to-face lessons for full-time students, and a consultation in Teams at the end of the term, where students could discuss with a teacher and ask questions.

3 Analysis of the Results

In the winter term of the academic year 2021/2022, the subject Initial Math Education was completed by 109 full-time students and 84 part-time students. At the end of the term, the students had to take a final test, which we consider as a measure of a level of their knowledge. The maximum number of points in the final test was 100, the minimal score to pass the test was 50. The results of the final test are given in Table 1 and are depicted in Fig. 1.

Firstly, we wanted to know whether the restriction of face-to-face teaching caused by COVID-19 had a negative impact on an ability of our students to pass the final test. As we can see in Fig. 1, all 193 students were able to pass the final test. Thus, we can conclude that thanks to the combination of e-learning and online learning, our students were able to pass the final test.

Now, we compare the results of full-time and part-time students. The average score of full-time students is 82.34, median 84, and standard deviation 10.56. The average score of part-time students is only 71.27, median is only 70, and standard deviation 13.78. It is clear that the score of full-time students is much higher. To test the significance of this difference, we use a statistical hypothesis testing. Firstly, using Shapiro–Wilk normality test, we reject the null hypothesis about normality of the results of full-time students, as well as part-time students (for full-time students, the value of the test statistic W is 0.9708 and the critical value for 10% probability of type I error is 0.98, for part-time students 0.9535 and 0.9749). The reason for the rejection is visible in Fig. 1, especially in a group of part-time students. We can notice that there are two groups of part-time students. One of them tries to learn as much as possible, the second one tries only to fulfill minimal requirements. Since we rejected the null hypothesis about normality, we use a nonparametric Mann–Whitney U test. The value of the test statistic Z is 5.48, while the 99.9% critical value accepted range is $[-3.29{:}3.29]$. Thus, we reject the null hypothesis in favor of the alternative 'The results of full-time students in the final test are significantly higher than the results of part-time students'. To be able to explain this difference, we compare the results of both groups in other subjects (we compare their average study mean). For full-time

Table 1 Results of a final test

	Full-time students	Part-time students
Number of students	109	84
Number of students who passed the final test	109	84
Average score	82.34	71.27
Median	84	70
Standard deviation	10.56	13.78
W (Shapiro–Wilk normality test)	0.9708	0.9535
Critical value (Shapiro–Wilk normality test)	0.98	0.9749
Null hypothesis about normality	Rejected	Rejected

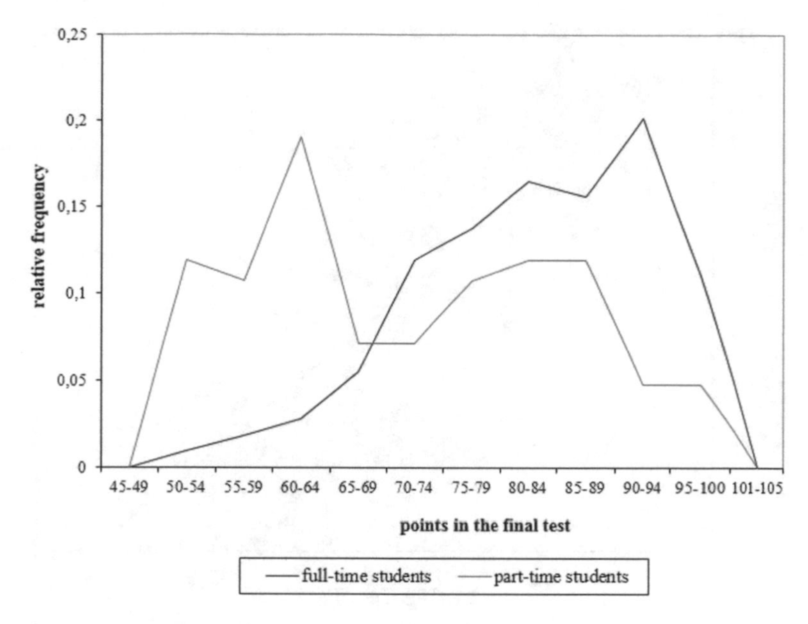

Fig. 1 Results of a final test

students, the average study mean is 1.74, while for part-time students, it is 1.91. Using Mann–Whitney U test, we reject the null hypothesis in favor of the alternative 'The average study mean of full-time students is significantly lower (better) than the average study mean of part-time students' (the value of the test statistic Z is 3.73). Thus, we can conclude that part-time students are worse than full-time students in our subject, because they are generally weaker (also in other subjects).

Secondly, we compare a correlation between average study mean of our students and their score in the final test. The results are depicted in Fig. 2. The value of the correlation coefficient is −0.705. The 95% confidence interval for a correlation coefficient is [−0.770, −0.625]. From the above mention, it follows that there is quite strong positive correlation between the results of the final test and an average study mean of our students. In other words, the better is average study mean, the better is the result of the final test.

To be able to explore other benefits of our method of teaching, we found out opinions of our students in a questionnaire. The questionnaire had fourteen questions. In each of them, the students chose one of the followings: strongly agree (1), agree (2), neutral (3), disagree (4), strongly disagree (5). The frequencies of the answers are given in Table 2 and depicted in Fig. 3.

The questionnaire was anonymous and optional. Thus, it was filled by 61 out of 109 full-time students and 47 out of 84 part-time students. The return rate is 56%.

In questions Q1 and Q2, we asked whether the students consider the final assessment from our subject (Q1) and from other subjects (Q2) to be objective. We can observe a significant difference in favor of our subject, with an average score 1.20

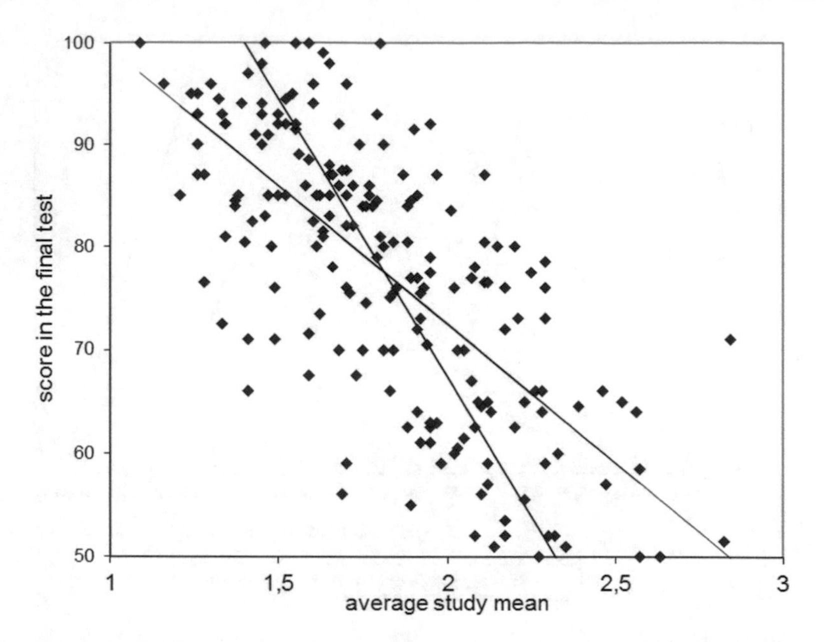

Fig. 2 Dependence of the average study mean and the score of the final test

Table 2 Relative frequency of answers in the questionnaire

Question	Strongly agree	Agree	Neutral	Disagree	Strongly disagree
Q1	0.86	0.10	0.03	0	0.01
Q2	0.49	0.24	0.19	0.06	0.02
Q3	0.81	0.13	0.06	0	0
Q4	0.38	0.36	0.20	0.05	0.02
Q5	0.05	0.05	0.28	0.18	0.45
Q6	0.85	0.13	0.01	0.01	0
Q7	0.70	0.21	0.06	0.01	0.03
Q8	0.88	0.06	0.06	0.01	0
Q9	0.62	0.21	0.08	0.05	0.04
Q10	0.79	0.18	0.04	0	0
Q11	0.87	0.12	0.01	0	0
Q12	0.44	0.19	0.25	0.08	0.03
Q13	0.01	0.04	0.04	0.19	0.72
Q14	0.86	0.09	0.03	0.01	0.01

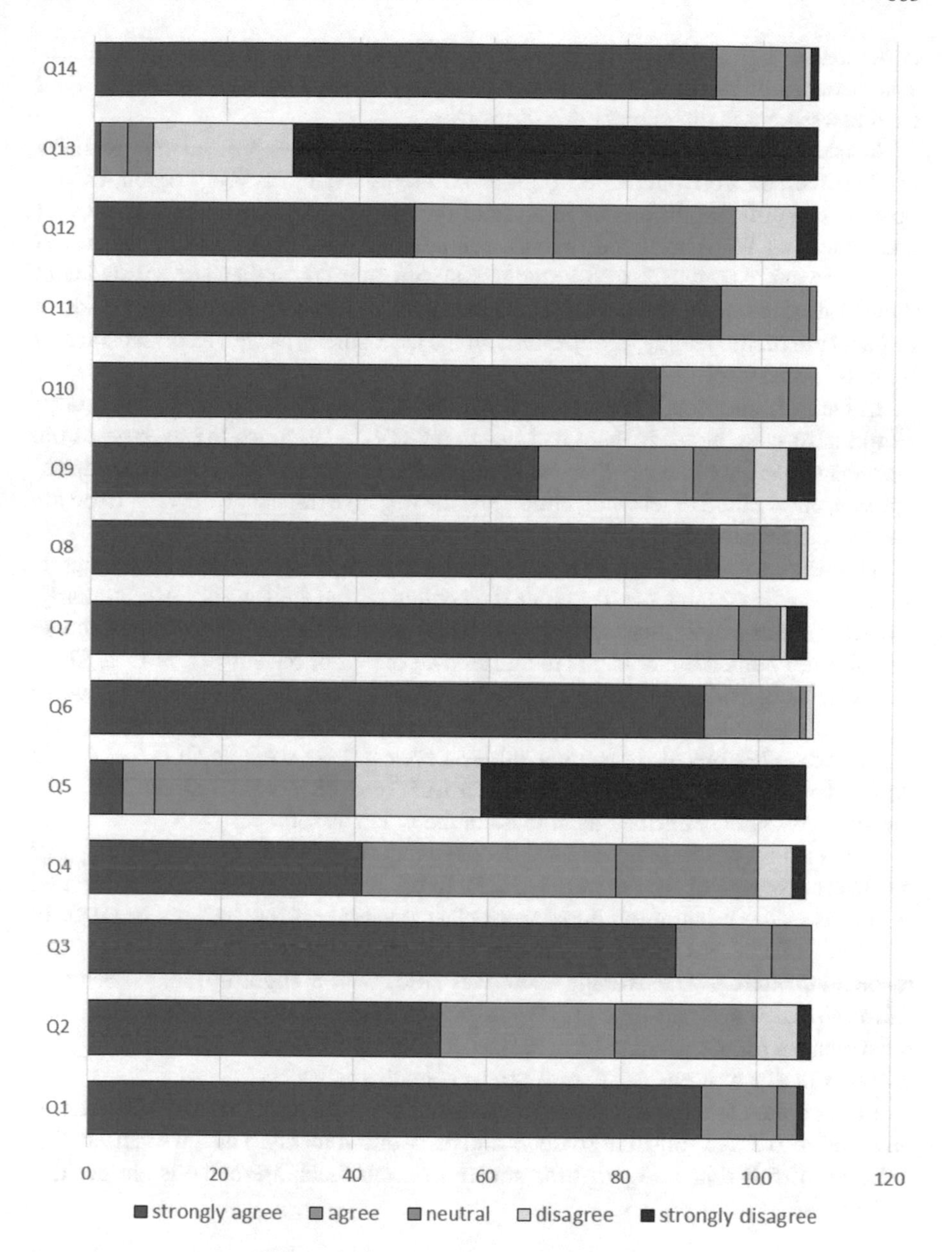

Fig. 3 Answers in a questionnaire

compared to 1.87. This difference was also proved by Pearson's chi-square test of association. Let us notice that about 86% of our students strongly agree that their final assessment in our subject was objective.

In questions Q3 and Q4, we asked whether the students are satisfied with the final assessment from our subject (Q3) and from other subjects (Q4). Again, we can observe a significant difference in favor of our subject, with an average score 1.24 compared to 1.96. Also this difference can be proved by Pearson's chi-square test of association. About 81% of our students strongly agree that they are satisfied with their final assessment. Thus, we can conclude that our teaching method has a positive impact both on the feeling of objectivity of the assessment, as well as on satisfaction with the assessment.

In the fifth question, we asked whether the students think that their assessment would have been better if there had been no COVID-19. Since the average of the answers is 3.93, we can say that our students disagree with that. Thus, in students' opinion, our method of teaching eliminated the negative impact of COVID-19 on the level of their knowledge.

The average score in the sixth question, which asked whether video lessons significantly helped our students to master the content of the subject, is 1.18. Similarly, the average score in the seventh question, which asked whether the e-learning course significantly helped our students to master the content of the subject, is 1.46. Thus, our students consider both the pre-recorded video lessons and the e-learning course to be really useful. Further, in the following two questions (Would you like to have similar video lessons also for other subjects even if there were no COVID? Would you prefer video lessons to a face-to-face instruction even if there were no COVID?), the students again confirmed usefulness of the video lessons.

In the tenth question, the students declared that they exactly knew what they had to master (average of the answers is 1.25). In the following question, they declared that texts in the e-learning course were clear (average of the answers is 1.14). In Q12, 44% of students strongly agree that they prefer online/blended learning to face-to-face instruction. The average score was 2.06, with a slight difference between full-time (2.25) and part-time (1.81) students. In Q13, the students refused that they had technical problems. Finally, in the last question, 86% of students strongly agreed that they prefer written tests to oral exams at mathematics.

Let us notice that in majority of questions, there were only very little differences between answers of full-time students and part-time students. Thus, we can say that opinions of full-time and part-time students on our teaching methods are positive and similar.

4 Conclusion

Before the spread of COVID-19, the subject Initial Math Education was taught by blended learning. We considered this modern method of teaching, which benefits from strengths of both face-to-face instruction and e-learning, to be an ideal way how to teach the subject Initial Math Education. However, a limitation of the number of students at face-to-face lessons made us to change this teaching method to a combination of pre-recorded video lessons, the e-learning course, and synchronous online lessons via Teams with a small number of students in a classroom.

Naturally, the change of teaching methods brings new challenges to both teachers and students. They need to adapt teaching and learning practices to an online learning environment. Although majority of university teachers and students are advanced in operating modern technologies, only some of them were accustomed to use technologies in education. We agree with Chiu et al. [5] that there is an urgent need to expand pedagogical expertise of our teachers and learning repertoire of our students. We also agree with Heng and Sol [8], who state that educational institutions need to offer training programs about online learning to their teachers and students.

Our previous experience with integration of modern technologies into teaching proved to be an advantage over other teachers, who used predominantly face-to-face instruction. These teachers were suddenly forced to use online learning without previous experience, which could have caused different types of problems. The similar situation is with our students. Majority of them studied also the subject Combinatorics and Data Processing, which was also taught by blended learning, so they were used to integration of modern technologies into learning. Thus, the limitation of face-to-face lessons did not cause such serious problems in our subject as in other subjects. Therefore, we suggest that teachers use modern technologies, blended learning, and online learning even when there are no restrictions due to COVID-19.

Another challenge for teachers and students was related to administration of a final test. In their research, Žilková and Žilková [23] warn that students perceived e-testing to be more stressful. Before 2019, at our faculty, it was unimaginable that students would write a test online. The pandemic situation forced teachers and students to take a final test online. It was a completely new situation for both teachers and students. In a very short time, teachers have to learn how to prepare an online test, how to limit opportunities for cheating, and how to test application of knowledge instead of only memorizing facts. In our case, we used a combination of Moodle and Teams. During the final test, students were connected to a meeting with a teacher with cameras and microphones on. Moreover, they took a test in Moodle. Questions in the test were divided into twenty categories and the system randomly generated one question from each category. Thus, each student solved twenty problems and wrote down his/her solution directly into Moodle. Naturally, each student had a different test. We are aware that the students were not accustomed to this new situation, which could have a negative impact on their results. The possibility of this impact could be a matter of further study.

The analysis of the results of the final test, as well as the analysis of the answers in the questionnaire, proves that online learning of Initial Math Education is a proper teaching method, which can eliminate the negatives connected with COVID-19. Although the results of the full-time students were better, it is not proved that it was thanks to the face-to-face lessons, since these students were better also in other subjects. Thus, further research has to be made. Similar results were published also by Žilková and Kondeková [22], who state that an online form of teaching was equivalent to an in-person form.

Another positives of online learning of Initial Math Education were proved by the analysis of the questionnaire. In our subject, the students felt a greater degree of objectivity of the final evaluation, as well as a greater degree of satisfaction with the final evaluation. The students were satisfied also with the e-learning course and pre-recorded video lessons. Moreover, majority of them would prefer online learning even if there were no COVID-19.

As we have already mentioned, in our face-to-face lessons, there was a small number of students in a classroom and majority of students connected via Teams. Thus, all students created one group. However, before COVID, a face-to-face exercises were held with groups of approximately 25 students. The impact of a class size to a students' achievement could be a matter of future research.

In the paper, our experience with online learning of Initial Math Education is presented. We are aware that the results were obtained only on one subject, thus, it is not possible to generalize them to other subjects. To obtain more general results, further research has to be realized.

Acknowledgements The paper was supported by the KEGA 001UMB-4/2020 project entitled Implementation of Blended Learning into Preparation of Future Mathematics Teachers and Future Computer Science Teachers and by the KEGA 004TTU-4/2021 project entitled Teaching Mathematics and Computer Science Using Interactive Components.

References

1. Alsawaie ON, Alghazo IM (2010) The effect of video-based approach on prospective teachers' ability to analyze mathematics teaching. J Math Teacher Educ 13(3):223–241
2. Borba MC, Askar P, Engelbrecht J, Gadanidis G, Llinares S, Aguilar MS (2016) Blended learning, e-learning and mobile learning in mathematics education. ZDM Math Educ 48(5):589–610
3. Bullo M (2021) Integration of video lessons to grade-9 science learners amidst COVID-19 pandemic. Int J Res 10(9):67–75
4. Cheung AC, Slavin RE (2013) The effectiveness of educational technology applications for enhancing mathematics achievement in K-12 classrooms: a meta-analysis. Educ Res Rev 9:88–113
5. Chiu TK, Lin TJ, Lonka K (2021) Motivating online learning: the challenges of COVID-19 and beyond. Asia Pac Educ Res 30(3):187–190
6. Fisher S (2003) Into the mix: the right blend for better learning. Training Dev Austr 30(3):11–13
7. Graham ChR, Dziuban Ch (2008) Blended learning environments. In: Handbook of research on educational communications and technology, pp 269–276
8. Heng K, Sol K (2021) Online learning during COVID-19: key challenges and suggestions to enhance effectiveness. Cambodian J Educ Res 1(1):3–16
9. Ichinose Ch, Clinkenbeard J (2016) Flipping college algebra: effects on student engagement and achievement. Learn Assistance Rev 21(1):115–129
10. Insorio AO, Macandog DM (2022) Video lessons via YouTube channel as mathematics interventions in modular distance learning. In: Contemporary mathematics and science education, vol 3, issue 1

11. Lin YW, Tseng ChL, Chiang PJ (2017) The effect of blended learning in mathematics course. EURASIA J Math Sci Technol Educ 13(3):741–770
12. Malatinská S, Pokorný M, Híc P (2015) Efficiency of blended learning in teaching mathematics at primary school. Inf Commun Educ Appl Adv Educ Res 85:6–11
13. Mišút M, Pokorný M (2015) Does ICT improve the efficiency of learning? Proc Soc Behav Sci 177:306–311
14. Moreno-Guerrero AJ et al (2020) E-learning in the teaching of mathematics: an educational experience in adult high school. Mathematics 8(5):840
15. Pokorný M (2019) Blended learning can improve the results of students in combinatorics and data processing. In: 2019 International Symposium on Educational Technology (ISET), pp 207–210
16. Pokorný M (2021) Video lessons and E-learning can overcome ban of face-to-face lessons in teaching mathematics. In: 2021 International Symposium on Educational Technology (ISET), pp 44–47
17. Stoffová V, Horváth R (2021) How to prevent frauds and cheating at programming exams. ICERI, pp 5388–5394
18. Štrbo M (2020) AI based smart teaching process during the Covid-19 pandemic. In: 3rd International Conference on Intelligent Sustainable Systems (ICISS), pp 402–406
19. Tan E, Pearce N (2011) Open education videos in the classroom: exploring the opportunities and barriers to the use of YouTube in teaching introductory sociology. In: Research in learning technology, vol 19
20. Thorne K (2003) Blended learning: how to integrate online and traditional. Kogan Page, London
21. Voštinár P (2017) GeoGebra applets for graph theory. In: EDULEARN17 conference, Barcelona, pp 10142–10148
22. Žilková K, Kondeková A (2021) The impact of online education on the student's success in the course "Teaching Geometry in Primary Education". In: E-learning in the time of COVID-19 13, pp 114–124
23. Žilková K, Žilková V (2021) Perception of elementary geometry e-test from the perspective of future preservice primary teachers: a qualitative study. In: International symposium elementary mathematics teaching. Charles University, Prague, pp 424–434

Educational Robotics in Teaching Programming in Primary School

Veronika Stoffová and Martin Zboran

Abstract Learning at Slovak primary schools in the last school years takes place in a combined form due to the COVID-19 pandemic. With the distance form of teaching, it was possible to reduce the teaching hours and omit some parts of the study plan. Parts of the subjects often omitted were tied to practical lessons in special facilities, possibly to the use of specific equipment or tools, such as laboratory exercises and experiments, building and construction of robots, programming and physical education. Teachers opted the path of least resistance and often preferred to skip such topics. According to the state educational program in elementary school, it is necessary to develop algorithmic and programming thinking. For this purpose, thematic units focused on algorithmic problem solving and programming are used. The article brings the experience of the authors from the programming of robots in elementary school, which can be equally successfully implemented face-to-face as well as remotely using real or virtual robots. To revive learning and increase its effectiveness, the authors used modern teaching aid and digital educational technologies, which have also proven themselves in the distance form of education. We used virtual and augmented reality, worked in remote laboratories and used implementation of laboratory experiments using visualized simulation models and environments and emulation techniques. The authors report on their experience of teaching programming in primary schools using programmable toys, robots and microcontrollers. Simple tools (buttons marked with symbols) and an interactive environment that offers block or icon programming are used to program the movement and control the activities of such objects—robot toys and own constructed robots. Compiling a functional program does not require high analytical and abstract thinking. The program can be easily and interactively assembled from the offered building elements. This method makes programming more fun for elementary school students and ensures their rapid progress and the joy of success.

V. Stoffová (✉)
Faculty of Education of Trnava University in Trnava, Trnava, Slovakia
e-mail: veronika.stoffa@gmail.com; veronika.stoffova@truni.sk

M. Zboran
Faculty of Mathematics, Physics and Informatics, Comenius University in Bratislava, Bratislava, Slovakia

Y. Singh et al. (eds.), *Proceedings of International Conference on Recent Innovations in Computing*, Lecture Notes in Electrical Engineering 1001,
https://doi.org/10.1007/978-981-19-9876-8_51

Keywords Algorithms · Programming · Teaching and learning programming · Robotics · Programming robots · Programmable robot toys · Microcontrollers

1 Introduction

Programming is considered a creative activity. To master programming, memorized knowledge is not enough, but understanding and recognizing contexts and relationships and their creative application are necessary. A person acquires education mainly in school. It should not only consist a lot of knowledge, but also attitudes, and it should definitely provide the development of such competencies as creativity, teamwork, and the ability to use information and communication technologies. It is also important how the student acquires new knowledge, how to fix it, how he/she can apply it in solving problems, how he/she develops his/her skills using them and how he/she takes his/her skills to a higher level. If knowledge is conveyed directly, the student is not forced to participate in the education process actively and is mostly passive. But if we create space for the learners to actively search for new knowledge, which they apply in solving problems and use in practical activities, they are actively involved in the learning process [2, 3, 7, 11, 14]. If the learner is successful, he/she will improve his self-confidence and interest in the given issue. It is also evident that the assumption and hope that active, viable and creative students will have the right orientation when choosing a field of study for education at the secondary and university level will increase and in the future will find suitable employment in the labor market more quickly. There is a great demand and constant shortage of ICT experts and especially good programmers in the labor market [9].

Employers often point out that absolvent of schools are not prepared for practice, especially in the area of using information and communication technologies [5]. All types of schools provide education in this area, but it is necessary to adapt the curriculum more to the requirements of practice.

New technologies are bringing automation and robotics to many industries. Robotics should become a compulsory part of the curriculum. However, it is too late to encounter robotics until college. The basics of robotics, construction and programming the robots can be taught even in primary school. Teaching robotics should be done in an interesting form, so the students can learn in a playful and fun way. In this way, we can easily motivate and get students to study technical and natural sciences fields, in which there is still a shortage of experts. A prerequisite for success is the need for schools to be equipped with educational technologies, fast Internet and access to both real and virtual robotic kits (also usable in distance learning), as well as having suitable programming environments and tools available.

2 The Role of Educational Robotics in Teaching Programming in Elementary School

Children already in preschool age during compulsory preschool training meet with programmable toys. The use of robotic building blocks in elementary schools as part of the teaching of the subject of informatics combines play with the acquisition of skills necessary for the twenty-first century. Educational robotics thus becomes a suitable tool for preparing for work with robots in the future profession and everyday life. When working with robotic kits, the student learns simple programming to control the robot but also develops his/her creativity, technical skills, fine motor skills and critical thinking.

3 From Toy Programming to Robot Programming

In toy and speciality stores, you can find a whole range of programmable toys for different age groups. We will present only three of them, which are often distributed as programmable toys with various accessories for elementary school students. This combination of fun robots is suitable for different ages, from preschoolers to students who are in the second grade of elementary school. No programming skills and experience are required when using them. Pupils learn symbolic notation of the execution of commands, a kind of coding in a playful way while developing logical, algorithmic and programming thinking [16]. They make it possible to create not only simple, undemanding sequences of commands to be executed but also longer and more complex programs into which sensor control procedures are integrated. For example, drawing shapes, following a line, avoiding obstacles and so on [4]. The more complex ones have a wide range of inputs and outputs for a program that, once created and run—is executed immediately.

3.1 Bee-Bot

Bee-Bot is a simple interactive programmable toy with which pupils can playfully acquire new knowledge in various fields. It is suitable for identifying letters and numbers, practicing orientation in a plane, etc. By creating sequences of "button" commands to control the "bee," children develop logical thinking, spatial orientation and the ability to plan in a playful way. Therefore, the programmable bee plays an important role in programming propaedeutic.

Instructions, represented by buttons on the bee's back, can be used to control the robotic bee (see Fig. 1). Bee-Bot has an easy-to-use user interface. A sequence of 40 instructions can be stored in its memory. The buttons represent the following commands: move forward/backward, turn 90° left/right and wait a given time unit

Fig. 1 Two views on a programmable bee

(approx. 1 s). Commands entered using the control keys can be executed using the Go key, used to start the execution of the program in memory. To create a new algorithm, the Clear button can be used to clear commands from memory. It is possible to create/design game boards/areas based on a square grid (15 × 15) to move the bee. In an area of squares, the bee can move around and carry out the programmer/player's instructions.

Students can create their own apps to move the bee, developing their creativity and imagination. Success in creating programs to control the bee has a strong motivational effect on solving other interesting tasks, which at the same time helps to further develop and consolidate the foundations of logical, algorithmic and programming thinking. The advantage of the Bee-Bot programmable toy is that it also comes with an emulator, which is an invaluable aid for distance learning and for schools that are not equipped with programmable toys. The Bee-Bot emulator can be considered an adequate substitute for a real toy [3, 6].

The execution of a simple program is based on entering sequences of basic instructions using button commands (See Figs. 2 and 3).

3.2 Pro-Bot

Pro-Bot in the shape of a car (Fig. 4) is a more advanced and more perfect version of the robotic Bee-Bot, which can draw lines on the mat using a pen attached to the built-in holder and also has an LCD. In addition to being able to move forward and backwards at programmable distances and turn right and left at programmable angles, it allows you to create more complex and complicated programs for controlling the toy car—a programmable robotic toy [3].

Toy car—Pro-Bot has programmable touch sensors built into the bumpers, a light sensor and a sound sensor. The Pro-Bot can be connected to a computer via USB, with the ability to create programs on the computer, using the separately supplied *Probotix* software. A program created in the *Probotix* environment can contain procedures and

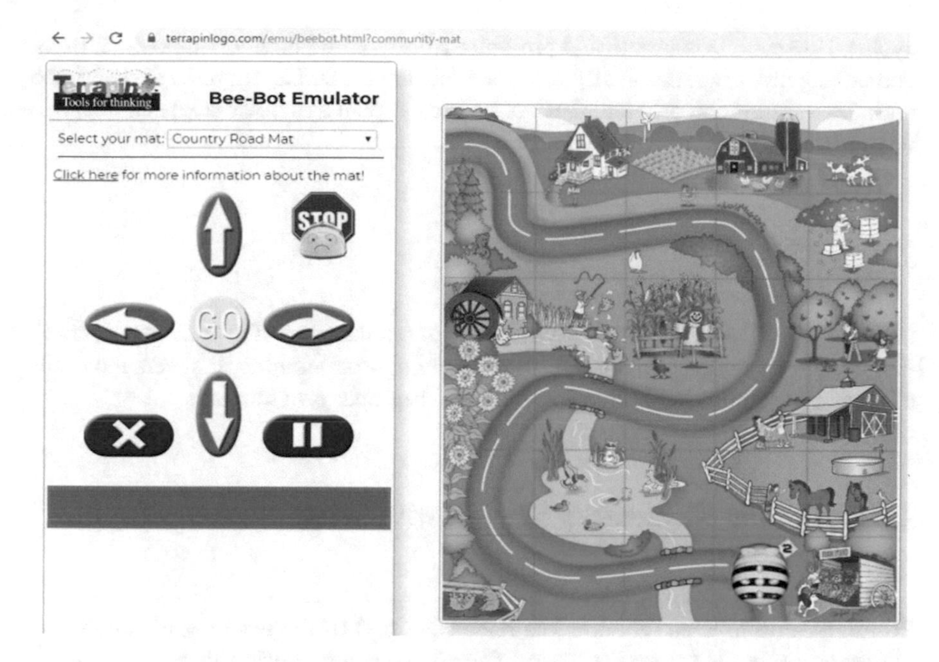

Fig. 2 Bee-Bot emulator (Starting position of the bee)

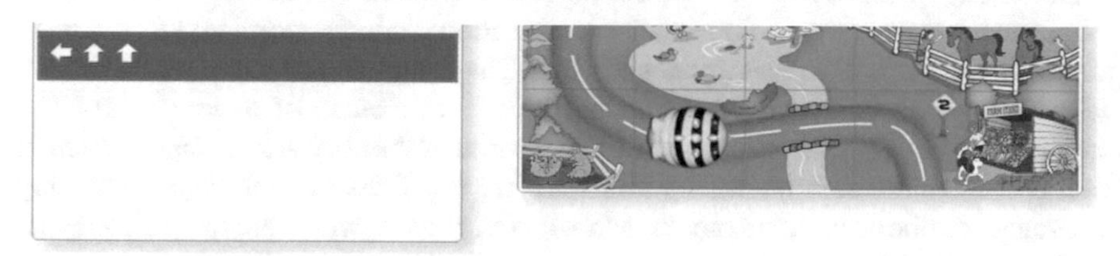

Fig. 3 Bee-Bot emulator (Position of the bee after executing a short sequence of commands on the left side of the picture)

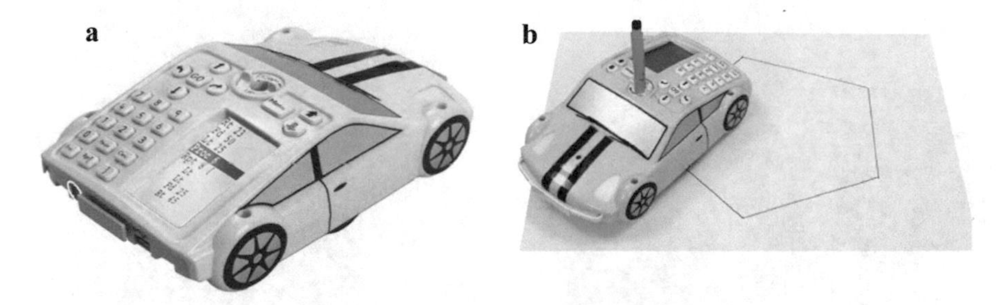

Fig. 4 **a** Pro-Bot programmable toy car, **b** Pro-Bot in action

cycles. However, it is usable and programmable even without a computer. Writing and editing the program can also be done using the LCD. The supplied package also includes a simulation program that can be used to plan a route or create an image to be plotted [3].

3.3 InO-Bot

InO-Bot (Fig. 5) is a more advanced robotic toy in the form of a "transport device." It is designed for the special purposes of teaching programming. It covers the needs of a basic programming course for children—beginner programmers.

4 Equipping Elementary Schools with Programmable Toys and Robotic Kits

Programmable toys and robotic kits have been part of the teaching of robotics in elementary schools for several years. The companies producing these kits monitor technological progress and develop new, more modern kits that meet the requirements of achieving the required competencies in education. Schools can choose from a wide range of programmable toys and robotic kits. The equipment of schools with robotic kits in Slovakia has changed significantly over the course of several years. The kits are used by several schools and their assortment has bee also changed. From a comparison of our surveys in 2017 and 2022, we found that schools were innovating the range of robotic building blocks. More types are used, which are more affordable, easier to assemble and program and more intuitive to operate (Chart 1).

Fig. 5 **a** InO-Bot—a programmable Bluetooth floor robot, **b** InO-Bot in action

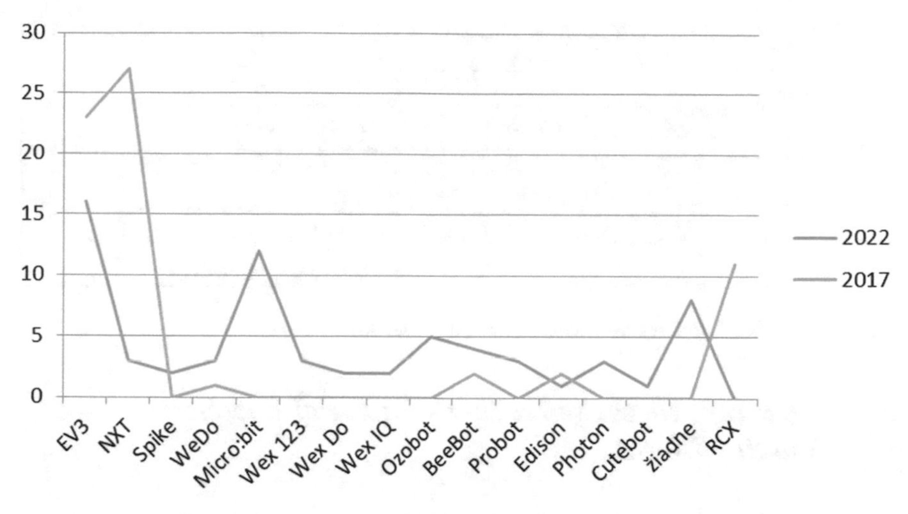

Chart 1 Comparison of robotic construction machines use in 2017 and in 2022

5 Evaluation of Selected Questions of the Questionnaire Survey

In 2022, we conducted an online survey on the teaching of programming in the second grade of elementary school and the use of robotic construction machines. We evaluated only some questions from the questionnaire. We were primarily interested in the state of equipment of elementary schools with programmable robots and robotic kits, what influences the teacher's choice of the robotic building blocks. We were also interested which program language or programming tool the teachers use in schools to program the control of the robots. We needed to get this information in order to prepare a new teacher's study program for accreditation. We wanted to include in the study program subjects that cover the issue of educational robotics, especially the construction and programming of robots. Our goal is that future computer science teachers already during their studies at the university have to be well prepared for practice in primary and secondary schools.

5.1 Robotic Kits in Schools

Wide range of robotic kits used in schools we can see in Chart 1. Computer science teachers work with robotic kits not only during real-time teaching but also during distance study using simulators. Despite the wide offer, schools usually do not have a sufficient number of robot construction kits, no every student have the tool and they need to work in groups, which makes it difficult to acquire the prescribed skills (See Chart 1).

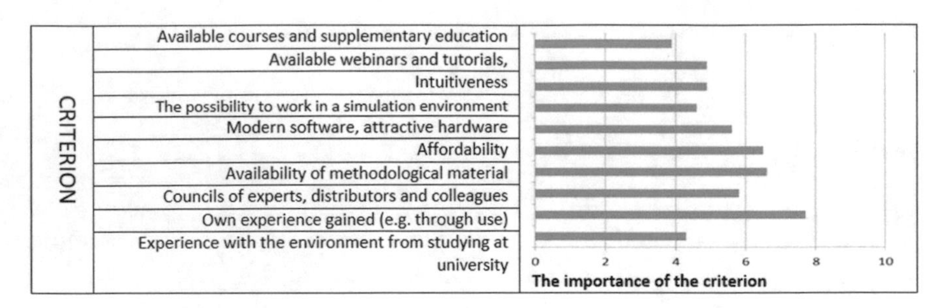

Chart 2 Teachers' preferences when choosing a robotic kit

5.2 What Influences the Teacher's Choice of Robotic Building Blocks

In our online questionnaire, we investigated teachers' preferences for choosing a robotic kit. In the question about the criteria for choosing a kit without any limitation by external factors, most teachers mentioned their own experience with the use of a robotic kit (it would be appropriate to hold workshops aimed at introducing and working with robotic kits for informatics teachers). The next most preferred answer was the availability of methodical material. The price of the kit was also significant. Chart 2 shows a summary of all responses. We took the opinions of practice teachers and their requirements into account when compiling the new study program and when developing study and methodical materials for teachers as well as pupils and students [6, 12].

5.3 In Which Language Do the Teachers Program the Control of the Robot

To program the robot, the teacher can choose from several programming environments. Each robotic kit has its own programming language and environment. Some robot kits are also compatible with other languages and environments. The possibility to work with several languages and environments is offered here. The research shows that the most used programming environments are from the LEGO company, some also use the Scratch environment. The second most used environment is MakeCode created for Micro:bit kits. All these environments are based on visual programming using programming blocks. By stacking these blocks, a program is created using the "drag and drop" method. For students, this method is more intuitive, and they do not make mistakes as often as when programming using text commands with exact syntax [1].

Arduino is the most widely used microcontroller (development board) in Slovak secondary technical and vocational schools [8, 10]. Its programming is more

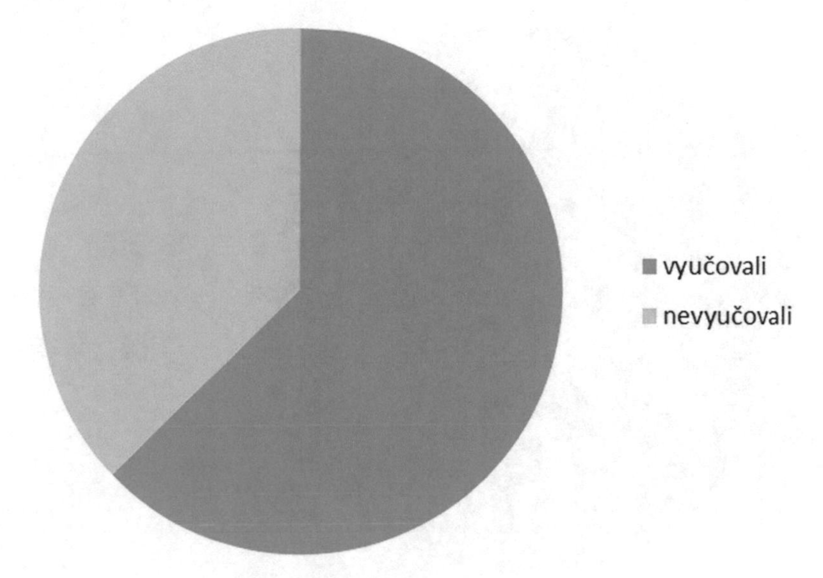

Chart 3 Teaching programming during the pandemic

demanding, so it is only rarely used in primary schools. In our research, we did not observe the occurrence of the Arduino microcontroller at the primary school.

5.4 How Programming Was Taught During the Pandemic

At the time of the pandemic, teaching was mostly done by distance learning, so we were interested in whether programming was taught in computer science classes in schools. About 62.7% of respondents said that they also taught programming (dark green color in Chart 3).

When asked about the method of teaching programming, 62.5% of respondents said online through screen sharing, 21.9% offline through assignments and video tutorials and 15.6% said other methods ("iný spôsob"), such as commented tutorials (Chart 4).

Teaching programming remotely requires basic digital literacy and the necessary hardware and software equipment on the part of both the teacher and the student. Both teachers and students tend to overestimate their own digital literacy. It often turns out that intuition is not enough to effectively use many digital technologies, but serious theoretical knowledge and practical skills and experience are needed [13, 17].

Both teachers and students had the biggest problems installing the software on their computer, controlling its operation and using its functions effectively. The most common communication tool for online teaching at primary schools in Slovakia is MS Teams and for offline Edupage.

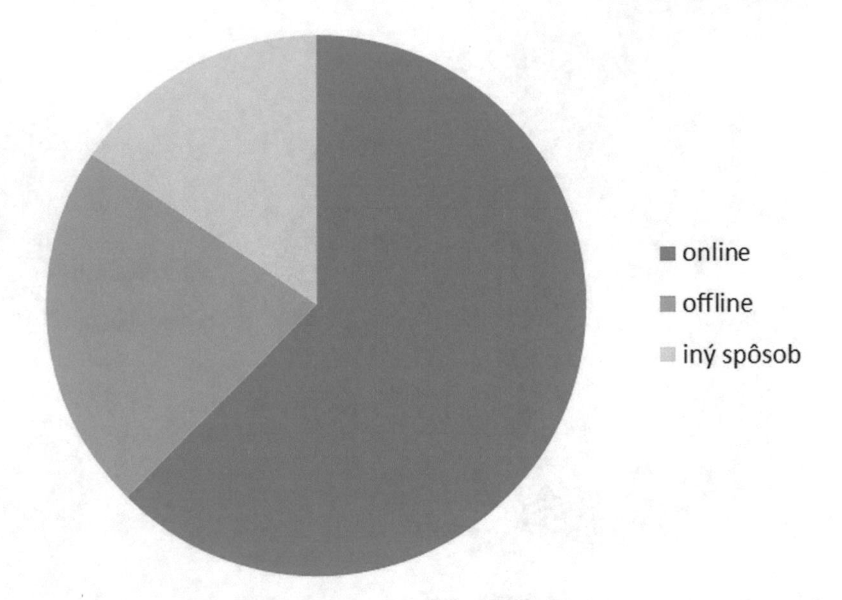

Chart 4 Method of teaching programming during the pandemic

6 An Example of a Control Program for a Simple Robot

Due to the financial simplicity and the fact that the micro:bit kit appears more and more often in elementary schools during the busy period, we chose this robotic kit as an example for programming. Several sensors are also included in this microcontroller, which increases its popularity among users. Among the sensors included are a pressure sensor, a light intensity sensor, an acceleration and tilt sensor, a temperature sensor and a magnetic field sensor. For this task, we will use the acceleration and tilt sensor to program a digital cube that, when shaken, will print a random number 1–6 on the LED screen as a dice. The number is deleted after 1 s (Figs. 6, 7 and 8).

From the menu of INPUT blocks, we selected the shake block, from the basic blocks show number, in order to ensure that the same number is not always displayed, we use the Mathematics block—randomly select 0–6. To achieve better observability of the displayed number, we selected pause and delete from the basic block screen, thus ensuring the deletion of the displayed number on the LED panel after the set time has elapsed (Figs. 7 and 8).

When assembling the code from a block, the blocks fit into each other, so their assembly is intuitive and easier than in the case of programming in the Python language (Fig. 8), where the student also has to watch out for syntax errors. Figure 9 is a view of the interactive environment while creating a program for the Micro:bit.

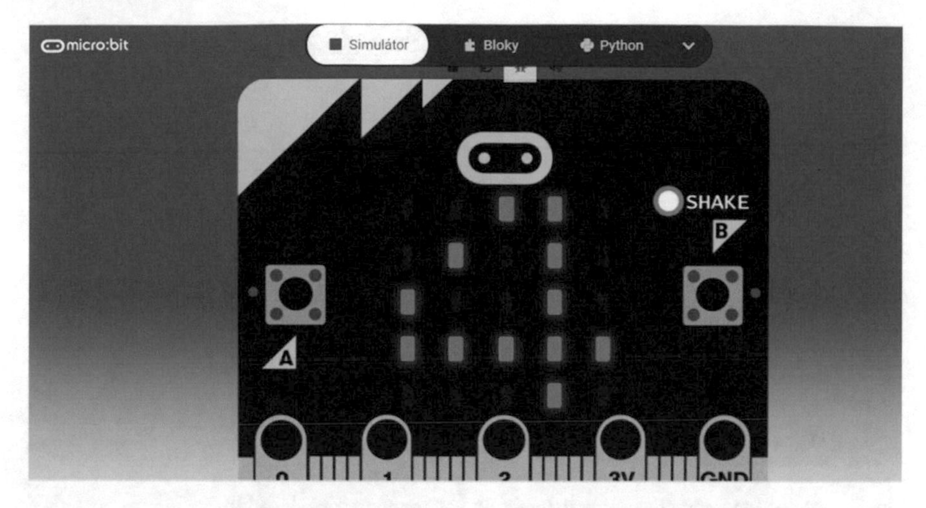

Fig. 6 Digital dice program

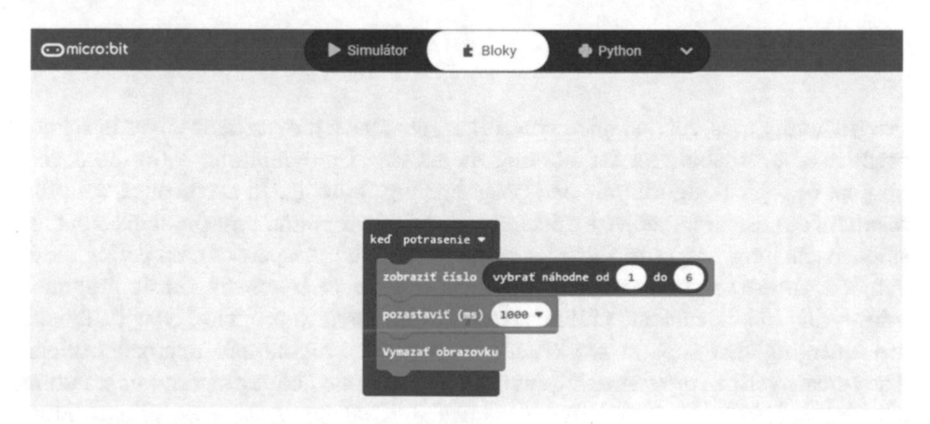

Fig. 7 Programming with blocks

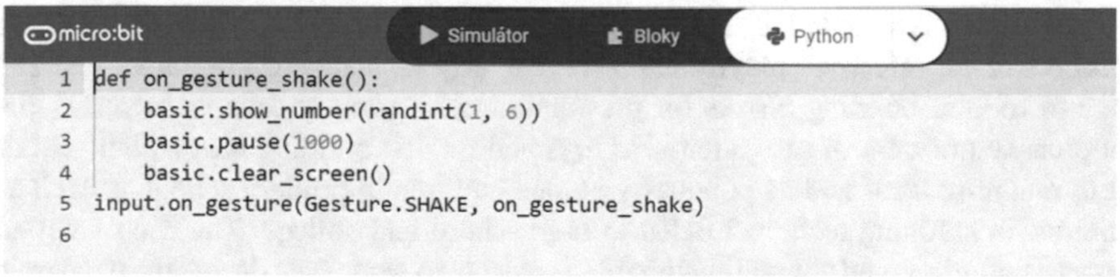

micro:bit Simulátor Bloky Python

```
1 def on_gesture_shake():
2     basic.show_number(randint(1, 6))
3     basic.pause(1000)
4     basic.clear_screen()
5 input.on_gesture(Gesture.SHAKE, on_gesture_shake)
6
```

Fig. 8 "Playing cube" program in the Python programming language

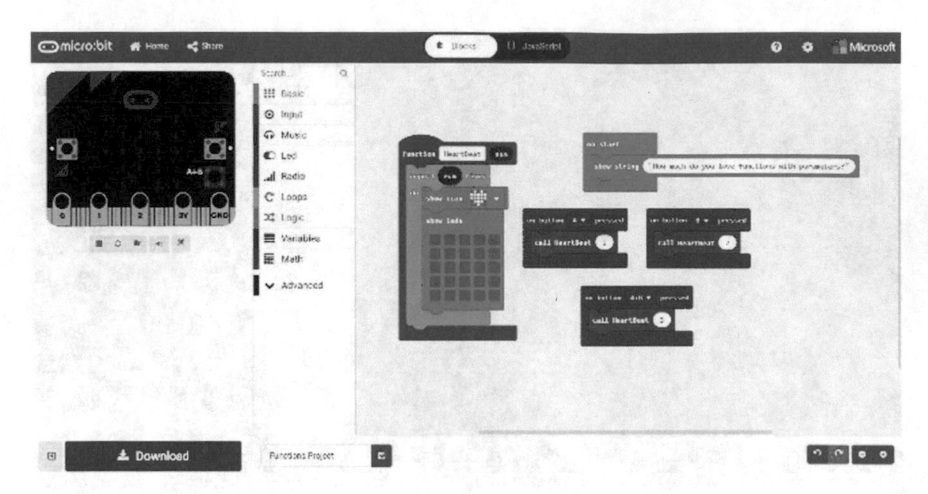

Fig. 9 Look at the interactive program creation environment for Micro:bit

7 Conclusions

Programming toys, robots, microcontrollers, etc., have proven themselves in school practice as a suitable tool for learning the basics of programming—for the development of logical algorithmic and programming thinking. In accordance with the results of our research, subjects oriented to practical programming appear in computer science teaching programs (also for primary schools). As part of completing these subjects, future teachers will gain practical experience in creating basic programming skills in students and children, teaching children to program. New textbooks are emerging that support work with robotic toys, robots and microcontrollers. They create various programming environments for interactive programming. Future teachers and teachers in practice receive methodical support from universities to introduce these tools into the teaching of informatics. As a part of the lifelong education and retraining of teachers in practice, various courses and tutorials are organized to help them overcome the first obstacles on the way to modernizing the teaching process to an effective, playful and efficient way of teaching programming. The use of robotic building blocks for programming in primary schools increases the popularity not only of programmable toys and robotic building blocks, but also of programming itself and its popularity among children in primary school, as well as interest in studying technical fields in high school and college. The use of virtual tools, emulations and simulations enabled students to work even during the pandemic in the form of distance online learning of programming [12].

Acknowledgements The paper was supported by the national project, KEGA 013TTU-4/2021 "Interactive animation and simulation models for deep learning."

References

1. Csóka M, Czakóová K (2021) Innovations in education through the application of raspberry pi devices and modern teaching strategies. In: INTED 2021 Proceedings of the 15th international technology, education and development conference. IATED Academy, Valencia
2. Czakóová K (2021) Game-based programming in primary school informatics. In. INTED 2021 Proceedings of the 15th international technology, education and development conference. IATED Academy, Valencia
3. Czakóová K, Stoffová V (2020) Hravá forma rozvíjania algoritmického myslenia na základnej škole = A playful form of developing algorithmic thinking in primary school. In: Didinfo 2020: Liberec: Technická univerzita v Liberci, pp 104–111. ISBN 978-80-7494-532-8. ISSN 2454-051X (online)
4. Czakóová K, Udvaros J (2021) Applications and games for the development of algorithmic thinking in favor of experiential learning. In. EDULEARN21: Proceedings of the 13th international conference on education and new learning technologies. IATED Academy, Valencia, pp 6873–6879. https://doi.org/10.21125/edulearn.2021.1389. ISBN 978-84-09-31267-2. ISSN 2340-1117
5. Hašková A, Zatkalík D, Zatkalík M (2021) Employers requirements for graduates of vocational education and training in study branches transport and automotive service and repair. In: 24th international conference on Interactive Collaborative Learning (ICL 2021); 50th IGIP international conference on Engineering Pedagogy, pp 1642–1652
6. Hyksová H (2021) Programování robotů na základní škole. Robot programming at elementary school In: DIDINFO 2021. Univerzita Mateja Bela, Banská Bystrica, pp 81–85. ISBN 978-80-557-1823-1. ISSN 2454-051X
7. Koreňová L, Lavicza Z, Veress-Bágyi I (2020) Chapter 5: Augmented reality applications in early childhood education. In: Augmented reality in educational settings. Brill|Sense, Leiden, The Netherlands. https://doi.org/10.1163/9789004408845_005
8. Kuna P, Palaj M (2021) Programovanie vývojovej dosky Arduino: Zbierka úloh pre stredné školy. FP Constantine the Philosopher University, Nitra, p 113. ISBN 978-80-558-1827-6
9. Kuna P, Danko F, Hašková A (2021) Riešenie aspektov diverzity PLC systémov v príprave programátorov. In: The impact of Industry 4.0 on job creation: Proceedings of scientific papers from the international scientific conference, s 229–238. FSEV TnUAD, Trenčín. ISBN 978-80-8075-XXX-X
10. Pšenáková I, Minárik M (2020) Use of the microcontroller as a teaching device. 2020 Budapest. ISBN 978-963-489-244-1
11. Stoffová V (2019) Educational computer games in programming teaching and learning. In: New technologies and redesigning learning spaces: eLearning and software for education. Carol 1 National Defence University, Bucuresti, pp 39–45. ISSN 2066-026X, CD-ROM. https://doi.org/10.12753/2066-026X-19-004
12. Stoffová V, Zboran M (2021) Teaching constructuon and programming of robots in a distance form. In: Proceedings of the 15th international technology, education and development conference. IATED Academy, Valencia, pp 4911–4918. ISBN: 978-84-09-27666-0. ISSN: 2340-1079
13. Štrbo M (2020) Self-evaluation of knowledge from basic programming. In: Proceeding of the 3rd international conference on inventive computation technologies, pp 340–343. ISBN 978-1-5386-5384-5 (online). Access: http://ukftp.truni.sk/epc/16065.pdf
14. Végh L, Takáč O (2021) Online games to introducing computer programming to children. In: Gómez Chova L, López Martínez A, Candel Torres I (eds) INTED2021 Proceedings. 15th international technology, education and development conference, pp 10007–10015. ISBN: 978-84-09-27666-0, ISSN: 2340-1079
15. Végh L, Takáč O (2021) Mobile coding games to learn the basics of computer programming. In: Gómez Chova L, López Martínez A, Candel Torres I (eds) EDULEARN21 Proceedings. 13th international conference on education and learning technology, pp 7791–7799. ISBN 978-84-09-31267-2, ISSN: 2340-1117

16. Udvaros J, Czakóová K (2021) Developing of computational thinking using microcontrollers and simulations. In. EDULEARN21: Proceedings of the 13th international conference on education and new learning technologies. IATED Academy, Valencia, pp 7945–7951. https://doi.org/10.21125/edulearn.2021.1619. ISBN 978-84-09-31267-2. ISSN 2340-1117
17. Záhorec J, Hašková A, Munk M (2021) Self-reflection of digital literacy of primary and secondary school teachers: case study of Slovakia. Eur J Contemporary Educ 10(2):496–508 (Academic Publishing House Researcher). ISSN 2305-6746. https://doi.org/10.13187/ejced.2021.2.496

Author Index

Y. Singh et al. (eds.), *Proceedings of International Conference on Recent Innovations in Computing*, Lecture Notes in Electrical Engineering 1001,
https://doi.org/10.1007/978-981-19-9876-8

Printed in the United States
by Baker & Taylor Publisher Services